(Part B)

LOW TEMPERATURE PHYSICS
LT 9

PART A
PAGES 1-620

Superfluidity

Quantization and Vortices in Superconductors and Liquid Helium

Quantum Liquids

Helium

Superconductivity

PART B
PAGES 621-1255

Low Temperature Transitions

Fourth Fritz London Award

Fermi Surfaces

Magnetism

Dilute Alloys

Solids and Metals at Low Temperatures

Liquid and Gaseous Hydrogen

Techniques

(Part B)

LOW TEMPERATURE PHYSICS LT9

Proceedings of the IXth International Conference
on Low Temperature Physics
Columbus, Ohio, August 31 - September 4, 1964

Edited by

J. G. Daunt, D. O. Edwards, F. J. Milford, M. Yaqub

Springer Science+Business Media, LLC
1965

ISBN 978-1-4899-6217-1   ISBN 978-1-4899-6443-4 (eBook)
DOI 10.1007/978-1-4899-6443-4

Library of Congress Catalog Card Number 65-14085

©*1965* Springer Science+Business Media New York
Originally published by **Plenum Press** in 1965.
Softcover reprint of the hardcover 1st edition 1965
All rights reserved

No part of this publication may be reproduced in any form without written permission from the publisher

CONTENTS OF PART B

List of Contributors xvii

6. Low Temperature Transitions

Low Temperature Transitions, E. W. MONTROLL [not available for publication]
Phase Transitions, R. BROUT 623
λ-Point Transitions, C. DOMB 637
An Investigation of the Nature of the Specific Heat in the Neighborhood of the Critical Point of ^{4}He, M. R. MOLDOVER AND W. A. LITTLE 653
Ultrasonic Propagation Near the Critical Point in Helium, C. E. CHASE AND R. C. WILLIAMSON 657

7. Fourth Fritz London Award

Presentation Address, L. D. ROBERTS 663
Fritz London Award Address, D. SHOENBERG 665

8. Fermi Surfaces

8.1. General

The Role of the Fermi Surface in Low Temperature Physics, A. F. KIP 679
Present Experimental Knowledge of Fermi Surfaces in Metals, D. SHOENBERG 680
Calculation of Band Structures and Fermi Surfaces, V. HEINE 698
Magnetic Breakdown in Magnesium and Zinc, R. W. STARK..................... 712
Interaction of Helicon Waves and Sound Waves in Potassium, C. C. GRIMES... 723

8.2. Magnetoresistance, Cyclotron Resonance, de Haas–van Alphen Effect

Longitudinal Magnetoresistance of Copper, R. L. POWELL....................... 732
Negative Magnetoresistance in Tellurium, D. C. PEARCE, R. M. WASILIK, AND DONALD A. NEEPER 736
Magnetoresistance of Bismuth, A. M. FORREST AND A. C. HOLLIS-HALLETT... 740
New Oscillatory Magnetoresistance Effect in Gallium, J. A. MUNARIN AND J. A. MARCUS 743
Magnetoresistance in Antiferromagnetic Chromium, A. J. ARKO AND J. A. MARCUS 748
The Fermi Surface of Aluminum and Dilute Aluminum–Zinc Alloys, J. P. G. SHEPHERD, C. O. LARSON, D. ROBERTS, AND W. L. GORDON 752
Observation of Cyclotron Resonance in Magnesium, T. G. ECK AND M. P. SHAW 759
Cyclotron Resonance Investigation of the Fermi Surface of Cadmium, M. P. SHAW AND T. G. ECK... 761
Extremal Dimensions of Cyclotron Orbits in Tungsten, W. M. WALSH, JR., C. C. GRIMES, G. ADAMS, AND L. W. RUPP, JR. 765

Doppler-Shifted Cyclotron Resonance with Helicon Waves, M. T. TAYLOR... 770
de Haas–van Alphen Effect in Pyrolytic and Single Crystal Graphite, S. J. WILLIAMSON, S. FONER, AND M. S. DRESSELHAUS 771
de Haas–van Alphen Effect in β'-CuZn, J. P. JAN, W. B. PEARSON, AND M. SPRINGFORD 776
The de Haas–van Alphen Effect in Chromium, B. R. WATTS..................... 779
High-Field Galvanomagnetic Properties of Rhenium, E. FAWCETT AND W. A. REED 782

8.3. Acoustic Absorption, Transport Phenomena, Size Effects, Theory

Landau Level Oscillations in the Magnetoacoustic Absorption in Zinc, H. V. BOHM AND L. MACKINNON 786
Magnetoacoustic Effect in Tungsten and Molybdenum near 1 Gcps, C. K. JONES AND J. A. RAYNE 790
Magnetothermal Oscillations in Semimetals, B. D. MCCOMBE AND G. SEIDEL... 794
Magnetothermal Oscillations in Beryllium, J. LEPAGE, M. GARBER, AND F. J. BLATT 799
Electron Transport Phenomena and Oscillatory Behavior in the Lattice Conductivity of an Antimony Single Crystal in a Magnetic Field at Low Temperature, C. G. GRENIER, J. R. LONG, J. M. REYNOLDS, AND N. H. ZEBOUNI 802
Oscillatory Magnetomorphic Behavior in the Electron Transport Effects of a Cadmium Single Crystal at Low Temperatures, J. M. REYNOLDS, K. R. EFFERSON, C. G. GRENIER, AND N. H. ZEBOUNI................................. 808
The Thermal Hall Effect in Copper, S. G. LIPSON 814
The Surface Impedance of Metals in a Weak Magnetic Field, J. F. KOCH AND A. F. KIP... 818
A RF Size Effect in Gallium, D. M. SPARLIN AND D. S. SCHREIBER 823
The Electronic Mean Free Path and the Areas of the Fermi Surface in Aluminum and Indium, G. BRÄNDLI, P. COTTI, E. M. FRYER, AND J. L. OLSEN 827
The Effect of Tension on the Fermi Surfaces of the Noble Metals, D. SHOENBERG AND B. R. WATTS 831
Fermi Surface of Sodium and Lithium by Positron Annihilation, J. J. DONAGHY, A. T. STEWART, D. M. ROCKMORE, AND J. H. KUSMISS 835
The Lorenz Number of Single Crystals of Lead and Indium in Transverse Magnetic Fields, L. J. CHALLIS, J. D. N. CHEEKE, AND P. WYDER 839
Effect of Collisions on the Conductivity Tensor of a Quantum Plasma in a Uniform Magnetic Field, S. TOSIMA, J. J. QUINN, AND M. A. LAMPERT 844
Electromagnetic Excitation Modes in Pure Metals in High Magnetic Fields, Including Their Contribution to the Specific Heat, M. A. LAMPERT, J. J. QUINN, AND S. TOSIMA 848
Collision Broadening of Giant Quantum Oscillations, J. J. QUINN 851

8.4. Magnetic Breakdown

Magnetic Breakdown in the Long DHVA Periods in Zinc, J. R. LAWSON AND W. L. GORDON 854

Effect of Alloying on Magnetic Breakdown in Zinc, R. J. HIGGINS, J. A. MARCUS, AND D. H. WHITMORE 859
Possible New Open Orbits in the Fermi Surface of Zinc, J. E. SCHIRBER......... 863
Giant Orbits in Zinc, A. C. THORSEN, A. S. JOSEPH, AND L. E. VALBY 867

9. Magnetism

Complex Antiferromagnetism, J. VAN DEN HANDEL AND H. C. MEIJER......... 873
Antiferromagnetic Transitions in Some Thiourea Coordinated Compounds, R. AU, J. A. COWEN, R. D. SPENCE, AND H. VAN TILL 877
Thermal Conductivity of Crystals with Magnetic Linear Chains at Very Low Temperatures, A. R. MIEDEMA, J. N. HAASBROEK, AND F. W. GORTER...... 880
Thermal Behavior of the Antiferromagnet $CoCl_2 \cdot 6H_2O$ at its Néel Point, J. SKALYO, A. F. COHEN, AND S. A. FRIEDBERG............................. 884
Magnetic Susceptibility of Single Crystal $FeCl_2$ Parallel and Perpendicular to the Trigonal Axis, J. W. STOUT, C. L. BRANDT, AND CHARLES TRAPP 887
Magnetic and Caloric Properties of K_3MoCl_6, P. A. VAN DALEN, H. M. GIJSMAN, N. LOVE, AND H. FORSTAT 888
Hyperfine Structure Coupling in Ferric Ammonium Sulfate as a Function of Magnetic Field and Temperature, F. E. OBENSHAIN, L. D. ROBERTS, C. F. COLEMAN, D. W. FORESTER, AND J. O. THOMSON 892
Specific Heat of Four Rare Earth Compounds Between 0.4° and 5°K, D. G. ONN, R. GONANO, AND H. MEYER 897
The Specific Heat of Ytterbium, Terbium, and Dysprosium Metals Between 3° and 25°K, O. V. LOUNASMA 901
Specific Heat of Scandium Between 0.15° and 3°K, P. LYNAM, R. G. SCURLOCK, AND E. M. WRAY 905
Magneto-Optical Behavior of Ferromagnetic Metals, S. DONIACH AND D. H. MARTIN.. 908
Magnetic Susceptibility of Mn^{++} in CdS and Effects of Antiferromagnetic Exchange, M. KREITMAN, F. J. MILFORD, AND J. G. DAUNT 909
Magnetoresistance Measurements on Cadmium Sulfide with Manganese Impurity, M. M. KREITMAN 914
Antiferromagnetic Structure of Cobalt Fluosilicate Below 1°K, A. OHTSUBO AND E. KANDA.. 917
Suppression of Nuclear Dynamic Polarization by RF Radiation, G. E. G. HARDEMAN AND G. GERRITSEN 921
Internal Local Fields on Antimony and Yttrium Nuclei in Iron. B. N. SAMOILOV, V. N. AGUREEV, V. D. GOROBCHENKO, V. V. SKLYAREVSKII, AND O. A. CHILASHVILI 925
Paramagnetism of Hemoglobin and Myoglobin Derivatives at Low Temperatures, M. KOTANI 929

10. Dilute Alloys
10.1. General

Localized States in Dilute Alloys, E. DANIEL................................... 933
Anomalies in Dilute Metallic Solutions of Transition Metals, G. J. VAN DEN BERG 955
A Correlation of the Mössbauer Isomer Shift and the Residual Electrical Resistivity for ^{197}Au Alloys, L. D. ROBERTS, R. L. BECKER, F. E. OBENSHAIN, AND J. O. THOMSON 985

10.2. Magnetic Scattering, Localized Moments, Hyperfine Splitting, Other Magnetic Phenomena

On the Question of Magnetic Ordering in Dilute Alloys, A. M. DE GRAAF AND R. LUZZI 994
Magnetic Scattering in Dilute Metals, F. T. HEDGCOCK AND D. MATHUR ... 995
Inelastic Impurity Scattering of Electrons in Gold Alloys at Low Temperatures, D. H. DAMON AND P. G. KLEMENS 996
Anomalous Electron Scattering in Dilute Magnetic Alloys, S. H. LIU 1001
Anomalous Properties of Dilute Magnetic Alloys, J. KONDO 1004
Ferromagnetism of Dilute Solutions of Cobalt in Palladium, R. D. DUNLAP, J. G. DASH, P. M. HIGGS, D. G. HOWARD, AND J. D. SIEGWARTH 1007
Localized Moments of Very Dilute Iron Impurities in Metals from Mössbauer Effect Studies over the Range 0.5° to 300°K, R. D. TAYLOR, T. A. KITCHENS, AND W. A. STEYERT..................... 1012
Polarization of Silver Nuclei in Ferromagnetic Metals, G. A. WESTENBARGER AND D. A. SHIRLEY..................... 1016
Mössbauer Studies of a Dilute Iron in Palladium Alloy in External Magnetic Fields, R. SEGNAN, P. P. CRAIG, AND R. C. PERISHO 1019
Contribution of Magnetization (down to 0.05°K) and Specific Heat Measurements to the Study of Segregation in a Gold–Iron Alloy, O. BETHOUX, Y. ISHIKAWA, J. SOULETIE, R. TOURNIER, AND L. WEIL 1023
Magnetization Discontinuities Below 1°K in Alloys, B. DREYFUS, Y. ISHIKAWA, R. TOURNIER, AND L. WEIL 1026
Heat Capacity Below 1°K: Observation of the Linear Term and the hfs Contribution in some Dilute Alloys, F. J. DU CHATENIER AND A. R. MIEDEMA 1029
Hyperfine Fields in Dilute Transition Metal Alloys, J. A. CAMERON, I. A. CAMPBELL, J. P. COMPTON, M. F. GRANT, R. W. HILL, AND R. A. G. LINES 1033
Specific Heat of $V_{10}Fe_{90}$ Alloy Between 0.4° and 7°K, D. F. BREWER, D. R. HOWE, AND B. G. TURRELL 1037

10.3. Electrical Resistivity, Thermoelectric Power, Electronic Specific Heat

Magnetic Scattering in Dilute Metals, F. T. HEDGCOCK AND D. P. MATHUR 1039
The Electrical Resistance of Some Dilute Gold Alloys, H. J. M. VAN RONGEN, B. KNOOK, AND G. J. VAN DEN BERG 1041
The Resistance of Dilute Alloys of Magnesium and Gadolinium or Neodymium, A. N. GERRITSEN AND S. B. DAS 1042
Resistivity and Magnetoresistance of Dilute Alloys of Iron in $Mo_{0.8}Nb_{0.2}$, M. P. SARACHIK 1044
Resistivity and Thermoelectric Power of Silver–Palladium Alloys, P. R. F. SIMON 1045
The Thermoelectric Power of Nickel and Dilute Nickel–Copper Alloys, D. GREIG AND J. P. HARRISON 1050
Specific Heat Measurements on a Series of Gadolinium–Praseodymium Alloys, B. DREYFUS, J. C. MICHEL, AND A. DE COMBARIEU 1054
Low Temperature Specific Heat of Solid Solutions of the Third Transition Series, E. BUCHER, F. HEINIGER, AND J. MULLER 1059

Precision Measurement of Electronic Specific Heat in Dilute Alloys, S. Shinozaki and A. Arrott 1066
Electronic Specific Heat of Dilute Noble Metal Alloys, M. L. Glasser 1067
Specific Heat of a Dilute Palladium–Cobalt Alloy and of Pure Palladium, B. M. Boerstoel, F. J. du Chatenier, and G. J. van den Berg 1071
Heat Capacity of Ordered and Disordered CuPt Below 4.2°K, J. Rayne and B. Roessler 1074

11. Solids and Metals at Low Temperatures
11.1. Rare Gas Solids, Solid Hydrogen, Solid Methane

Thermodynamics of the Inert Gas Solids Using Three-Parameter Interatomic Potentials, J. W. Leech and J. A. Reissland 1081
The Crystal Structures of Argon and its Alloys, C. S. Barrett and Lothar Meyer 1085
Spectroscopic Observation of the 1-Phonon Spectrum of Solid Argon, G. O. Jones and J. M. Woodfine 1089
Variation with Temperature of the Velocity of Transverse Elastic Waves in Solid Argon, G. O. Jones and A. R. Sparkes 1090
Variation with Temperature of the Refractive Index and Lorentz–Lorenz Function of Solid Argon, A. J. Eatwell and G. O. Jones 1091
The Specific Heat of Solid Neon, C.-H. Fagerstroem and A. C. Hollis Hallett 1092
Nuclear Magnetic Resonance in Solid Hydrogen and Deuterium Under High Pressures, S. A. Dickson and H. Meyer 1095
A Search for Isotopic Phase Separation in Hydrogen–Deuterium Mixtures, E. M. de Castro, D. Husa, J. R. Gaines, and J. G. Daunt 1099
Liquid–Solid Phase Equilibria in the Hydrogen–Deuterium System, D. White and J. R. Gaines 1104
Phase-Separation in Solid Hydrogen–Deuterium and ^{3}He–^{4}He Mixtures, R. A. Coldwell-Horsfall 1110
Phase Diagram of Solid Methane, J. S. Rosenshein and W. M. Whitney 1114
Proton Magnetic-Resonance Absorption in Solid Methane, R. P. Wolf and W. M. Whitney 1118
Ultrasonic Propagation in Solid Methane, A. A. Thiele, W. M. Whitney, and C. E. Chase 1122

11.2. Specific Heat, Phonon Scattering, Ionic Solids

Surface Contribution to the Specific Heat of Crystals at Low Temperatures, A. A. Marudin and M. Ashkin 1126
Phonon Spectroscopy Down to 0.3°K, W. D. Seward 1130
The Effect of Spin-Phonon Interactions on the Thermal Conductivity of Chromium-Doped MgO, L. J. Challis and D. J. Williams 1135
Observations of a Phonon Bottleneck in CuCs$_2$(SO$_4$)$_2$ · 6H$_2$O at Temperatures Near 0.1°K, A. R. Miedema and K. W. Mess 1140
The Effect of Electron-Phonon Scattering on the Thermal Conductivity of InSb, L. J. Challis, J. D. N. Cheeke, and D. J. Williams 1145
Grüneisen γ versus Temperature from Elastic Coefficients, K. Brugger 1151
The Scattering of Phonons by Small-Angle Grain Boundaries, K. A. McCarthy 1155

Photodielectric Effects in Thallous Halides at Low Temperatures, I. LEFKOWITZ, R. P. LOWNDES, D. H. MARTIN, AND A. D. YOFFE 1158
The Static Dielectric Constant of Alkali Halides at Low Temperatures, M. C. ROBINSON AND A. C. HOLLIS HALLETT 1162
The Thermal Expansion and Related Properties of Rubidium Bromide at Low Temperatures, B. W. JAMES AND B. YATES 1165
Far Infrared Impurity Modes in Potassium Iodide, A. J. SIEVERS 1170
Line Width Transitions in the Deuteron Magnetic Resonance of Polycrystalline ND_4Cl and ND_4Br, V. HOVI AND P. PYYKKÖ 1175
X-Ray Investigation of the Modifications II and III of NH_4Br at Temperatures Between 22° and −125°C, V. HOVI, K. HEISKANEN, AND M. VARTEVA... 1179
X-Ray Investigation of the Transition I–II of NH_4I at Temperatures Between 22° and −163°C, V. HOVI AND M. VARTEVA 1184
Heat Pulses in Alkali Halides at Low Temperatures, R. J. VON GUTFELD AND A. H. NETHERCOT, JR. 1189

11.3. Metals at Low Temperatures

Temperature Dependence of the Electron Mean Free Path in Tin at Liquid-Helium Temperatures, V. F. GANTMAKHER AND YU. V. SHARVIN 1193
Heat Conductivity of Pure Metals Below 1°K, G. DAVEY, K. MENDELSSOHN, AND J. K. N. SHARMA 1196
Failure of Matthiessen's Rule in Plutonium, E. KING, J. A. LEE, K. MENDELSSOHN, AND D. A. WIGLEY 1200
Pressure Dependence of Electrical Conductivity at Low Temperatures, W. S. GOREE AND T. A. SCOTT 1205
Phase Transformation in Sodium and its Effect on the Electronic Specific Heat, D. L. BHATTACHARYA AND E. A. STERN 1210
Effect of Spin-Orbit Coupling on the Hyperfine-Contact Contribution to the Knight Shift, J. C. APPEL 1215

12. Liquid and Gaseous Hydrogen

Summary of Recent Determinations of the Ultrasonic Velocity in Fluid Parahydrogen, B. A. YOUNGLOVE 1223
Measurements of the Viscosity of Parahydrogen, D. E. DILLER 1227
The Dielectric Polarizability of Fluid Parahydrogen, J. W. STEWART 1230
The Temperature Dependence of the Relaxation Time for Rotational Transitions in H_2, HD, and D_2, H. F. P. KNAAP, C. G. SLUIJTER, AND J. J. M. BEENAKKER 1233
The Difference in Polarizability Between Orthohydrogen and Parahydrogen in Their Ground State, H. F. P. KNAAP, L. J. F. HERMANS, AND J. J. M. BEENAKKER 1237

13. Techniques

Dielectric Dissipation Measurement Below 7.2°K, W. H. HARTWIG AND D. GRISSOM 1243
Use of Low Temperature Techniques to Measure Gravitational Forces on Charged Particles, F. C. WITTEBORN, L. V. KNIGHT, AND W. M. FAIRBANK 1248
A Realization of a London–Clarke–Mendoza Type Refrigerator, P. DAS, R. DE BRUYN OUBOTER, AND K. W. TACONIS 1253

CONTENTS OF PART A

1. Superfluidity

Introduction to the Symposium on Superfluids, JOHN BARDEEN	3
Weak Superconductivity and Josephson Tunneling, P. W. ANDERSON	8
Hydrodynamics and Kinetic Theory of Superfluids, PAUL C. MARTIN	9
Vortex Lines in Superconductors, P. G. DE GENNES	19

2. Quantization and Vortices in Superconductors and Liquid Helium

Quantized Magnetic Flux in Superconductors, WILLIAM M. FAIRBANK	33
Quantized Vortices in Superconductors, R. D. PARKS	34
Investigation of Quantized Vortex Rings in Superfluid Helium, F. REIF	46

3. Quantum Liquids

Correlations and Excitations in Quantum Liquids, DAVID PINES	61
Superfluidity of ^{3}He, V. P. PESHKOV	79
Strong Coupling Cell Theory of Liquid Helium, LYLE B. BORST	84
Energy Fluctuations and the Nature of the Rotons in He II, O. K. RICE	88
Inequalities for the Ground State of Liquid ^{4}He, O. PENROSE	91
Ground-State Energy of a One-Dimensional Bose System, T. R. KOEHLER	92
On the Line Width of Cold Neutrons Scattered by He II, S. FRANCHETTI	97
A Third Sum Rule for the Density Correlation Function, R. PUFF	101
Virial Expansion for Bose and Fermi Fluids, T. MORITA and T. TANAKA	104
Many-Body Effects at Metallic Densities, T. M. RICE	108
On the Quasi-Particle Energy Spectrum in Liquid ^{3}He, P. M. RICHARDS	113
The Attenuation of Sound in Liquid ^{3}He at Low Temperatures, C. E. GOUGH AND W. F. VINEN	118
The Acoustic Impedance of Liquid ^{3}He, G. A. BROOKER	121
The Acoustic Impedance of Liquid ^{3}He Under Pressure, B. E. KEEN AND J. WILKS	125
The Viscosity of Liquid ^{3}He Above 0.04°K, D. S. BETTS AND J. WILKS	129
The Heat Capacity of Liquid ^{3}He, B. M. ABRAHAM, M. DURIEUX, C. J. N. VAN DEN MEIJDENBERG, AND D. W. OSBORNE	133
Nuclear Susceptibility of Liquid ^{3}He, B. T. BEAL AND J. HATTON	137
Nuclear Magnetic Susceptibility in ^{3}He Vapor and Liquid, J. E. OPFER, K. LUSZCZYNSKI, AND R. E. NORBERG	143
Diffusion of ^{3}He in Liquid ^{3}He–^{4}He Mixtures, P. HORVITZ AND H. E. RORSCHACH	147

4. Helium
4.1. Second Sound, Rotation, Vortices, Films

Attenuation of Second Sound in Rotating Liquid Helium at the Passage of the Point of Phase Transition, E. L. ANDRONIKASHVILI, R. A. BABLIDZE, AND J. S. TSAKADZE 155
Relaxation of Onsager–Feynman's Vortices at Heating of Rotating He II Above the Point of the Phase Transition, E. L. ANDRONIKASHVILI, G. V. GUJABIDZE, AND J. S. TSAKADZE 159
Megacycle Frequency Second Sound, H. A. NOTARYS AND J. R. PELLAM ... 164
Wavemode Modification in Liquid Helium with Clamped Normal Fluid, G. L. POLLACK AND J. R. PELLAM 166
Second Sound in Square and Cubical Cavities, P. J. BENDT 170
Mutual Friction Measurements in Liquid He II, R. H. BRUCE 174
A Study of the Laminar and Turbulent Flow of He II in Two Closed Oscillating Geometries, S. KNIGHT AND J. D. REPPY 179
Displacement–Time Curves of an Oscillating Pile of Disks in He II, A. R. CONSTABLE AND C. B. BENSON 184
Angular Momentum of He II in a Rotating Cylinder, G. B. HESS AND W. M. FAIRBANK 188
Rotational Anomaly in Liquid He II, J. R. PELLAM 191
A Search for Vortex Lines in Rotating Superfluid Helium by Positron Annihilation, L. O. ROELLIG, T. KELLY, H. H. MADDEN, AND J. MCNUTT ... 195
An Attempt to Observe the Creation of a Single Quantized Vortex Ring, W. J. TRELA AND W. M. FAIRBANK 200
The λ Transition in Adsorbed Helium Films at Temperatures Below 1°K, F. D. MANCHESTER 202
The Isothermal Helium Film Below 1°K, C. F. MATE, R. HARRIS-LOWE, AND J. G. DAUNT 206
Enhanced Transfer Rates and Thick He II Films, C. C. MATHESON AND J. TILLEY 210
Oscillations of the He II Film, F. I. GLICK AND J. H. WERNTZ, JR. 214

4.2. Solid Helium Including Isotopic Mixtures

The Exchange Interaction and Related Relaxation Process in Solid ^{3}He, M. G. RICHARDS, J. HATTON, AND R. P. GIFFARD 219
Spin-Lattice Relaxation in Solid ^{3}He, E. D. ADAMS, R. A. SCRIBNER, M. F. PANCYZK, A. SAMEC, AND E. A. GARBATY 226
Properties of the bcc–hcp Phase Transition of ^{3}He and ^{4}He, A. F. SCHUCH AND W. C. OVERTON, JR. 229
Solidification of ^{3}He–^{4}He Mixtures, C. LE PAIR, K. W. TACONIS, R. DE BRUYN OUBOTER, E. DE JONG, AND J. PIT 234
The Melting Curve Slope of ^{4}He and ^{3}He–^{4}He Mixtures, G. O. ZIMMERMAN 240
The Specific Heat of ^{3}He–^{4}He Solid Mixtures, G. O. ZIMMERMAN 244
On the Phase Diagram of ^{3}He–^{4}He Mixtures, P. M. TEDROW AND D. M. LEE... 248
Evidence for a Quadruple Point in ^{3}He–^{4}He Mixtures, F. P. LIPSCHULTZ, P. M. TEDROW, AND D. M. LEE 254
Specific Heat of ^{4}He Under Pressure Below 1°K, J. WIEBES AND H. C. KRAMERS 258

The Thermal Conductivity of bcc ^{4}He and of Isotopic Mixtures, R. BERMAN AND S. J. ROGERS 262
Thermal Conductivity of Isotopic Mixtures of Solid Helium, B. BERTMAN, C. W. WHITE, AND HENRY A. FAIRBANK.. 266
Isotopic Solutions of Solid Helium, P. G. KLEMENS, R. DE BRUYN OUBOTER, AND C. LE PAIR 270
^{4}He Melting Curve Minimum, S. G. SYDORIAK AND R. L. MILLS 273
Calculations of the Ground-State Energies of Crystals of the Light Elements, L. H. NOSANOW 277

4.3. Superfluid Flow, Critical Velocities, Turbulence

Momentum of Liquid Helium and Critical Velocity Calculations, YU. G. MAMALADZE.. 281
Turbulent Effects in Liquid Helium Flow, T. M. WIARDA, G. VAN DER HEYDEN, AND H. C. KRAMERS 284
The Flow of He II at Moderate Reynolds Numbers, J. T. TOUGH, W. D. MCCORMICK, AND J. G. DASH 287
Thermohydrodynamic Flow Similarity of He II in Wide Channels at Supercritical Velocities, T. H. K. FREDERKING AND J. D. SCHWEIKLE............ 291
Helium Superflow Through Saran Charcoal, M. H. EDWARDS, A. S. MCKIRDY, AND W. C. WOODBURY 295
Vorticity in He II in Narrow Channels, P. A. DZIWORNOOH, E. S. R. GOPAL, K. MENDELSSOHN, AND S. M. A. TIRMIZI 299
The Superflow of He II in Very Narrow Channels, J. F. ALLEN AND D. J. WATMOUGH 304
Isothermal Flow of Liquid He II in Narrow Channels, W. E. KELLER AND E. F. HAMMEL 307
Critical Velocities in the Isothermal Flow of Superfluid Helium, G. CARERI, M. CERDONIO, AND F. DUPRE 311
Heat Transport by He II Through Small Capillaries, H. H. MADDEN, R. H. HAMMERLE, AND H. V. BOHM 312
Heat Transport by He II at Low Temperatures, P. L. J. CORNELISSEN AND H. C. KRAMERS 316
Superfluid Vorticity and Critical Velocities in Heat Currents in He II, D. J. GRIFFITHS, D. V. OSBORNE, AND J. F. ALLEN................................... 320
Superfluid Flow in Liquid Helium, W. M. VAN ALPHEN, W. VERMEER, K. W. TACONIS, AND R. DE BRUYN OUBOTER.. 323
Stability of Superfluid Flow Near the λ-Point, JAMES CLOW, JAMES C. WEAVER, DAVID DEPATIE, AND JOHN D. REPPY 328
Gravitational Flow of Superfluid Helium at 0.45°K, J. N. KIDDER AND H. A. BLACKSTEAD 331

4.4. Ions in Liquid Helium

Hot Ions in Liquid He II, G. CARERI, S. CUNSOLO, P. MAZZOLDI, AND M. SANTINI 335
Ions in He II in a Magnetic Field, LOTHAR MEYER 338
Change of the Mobility of Positive Ions in Rotating He II, I. MODENA, A. SAVOIA, AND F. SCARAMUZZI... 342

Ions in Rotating He II: Some Exploratory Experiments, D. J. Tanner, B. E. Springett, and R. J. Donnelly ... 346
Dependence of Ionic Mobilities in Liquid ^{3}He on Temperature and Density, P. de Magistris, I. Modena, and F. Scaramuzzi ... 349
Ion Mobility Discontinuities in Liquid Helium, J. A. Cope and P. W. F. Gribbon ... 353
Localized Excitations in Condensed Neon, Argon, Krypton, and Xenon, J. Jortner, L. Meyer, S. A. Rice, and E. G. Wilson ... 356
Anomalous Scintillation Effects in He II, F. E. Moss and F. L. Hereford ... 360

4.5. Adsorbed ^{3}He

Nuclear Magnetic Susceptibility of Adsorbed ^{3}He, G. Careri, M. Santini, and G. Signorelli ... 364
Search for a Magnetically Ordered State of Adsorbed ^{3}He, H. Weinstock and C. E. Long ... 366
The Specific Heat of Adsorbed Monolayers of ^{3}He, D. L. Goodstein, W. D. McCormick, and J. G. Dash ... 368
Surface Effects in ^{3}He at Low Temperatures, D. F. Brewer, A. J. Symonds, and A. L. Thomson ... 370

5. Superconductivity

5.1. Ultrasonics, Microwaves, Tunneling

Ultrasonic Absorption in Superconducting Lanthanum, T. Olsen, L. T. Claiborne, and N. G. Einspruch ... 375
On the Problem of Ultrasonic Attenuation in Superconductors Containing Magnetic Impurities, L. P. Kadanoff and I. I. Falko ... 378
Microwave Transmission and Reflection Coefficients of Thin Superconducting Films, N. M. Rugheimer, C. V. Briscoe, A. Lehoczky, and R. E. Glover, III ... 381
Magnetic Field Dependence of the Microwave Surface Impedance of Superconducting Aluminum, R. Glosser and D. H. Douglass, Jr. ... 385
Millimeter Wave Absorption in Superconducting Silver-Doped Aluminum Single Crystals, M. A. Biondi, M. P. Garfunkel, and W. A. Thompson ... 387
Effect of a Static Magnetic Field on Millimeter Wave Absorption in Superconducting Aluminum, W. V. Budzinski and M. P. Garfunkel ... 391
RF Losses in Superconducting Lead and Niobium, J. M. Pierce, H. A. Schwettman, W. M. Fairbank, and P. B. Wilson ... 396
The Superconducting Energy Gap of Lead Above H_c, Y. Goldstein ... 400
Effect of Magnetic Fields on Energy Gap of Superconductors, K. Maki ... 405
Anomalous Densities of States in Normal Tantalum and Niobium, A. F. G. Wyatt ... 411
Tunnel Junction RF Modes Driven by the AC Josephson Current, R. E. Eck, D. J. Scalapino, and B. N. Taylor ... 415
Persistent Currents and Flux Quantization in Superconducting Rings Interrupted by Josephson Junctions, A. M. Goldman ... 421
Tunneling Between a Type II Superconductor and a Very Thin Superconducting Aluminum Film in a Magnetic Field, W. J. Tomasch ... 424

Method for Obtaining Second Derivatives in Electron Tunneling, J. W. T. DABBS 428

Flux Quantization and Josephson Tunneling, T. I. SMITH AND H. E. RORSCHACH, JR. 432

Tunneling Density of States for a Superconductor Carrying a Current, P. FULDE 438

DC Pair-Tunneling Between Two Superconductors at Finite Voltages, C. B. SATTERTHWAITE, M. G. CRAFORD, R. N. PEACOCK, AND R. P. RIES 443

5.2. Quantized Flux, Rotating Superconductors, Superconducting Transition

DeBroglie Wavelength of a Current-Carrying Superconductor, R. C. JAKLEVIC, J. J. LAMBE, A. H. SILVER, AND J. E. MERCEREAU 446

Quantized Flux in Superconducting Cylinders, A. L. KWIRAM AND B. S. DEAVER, JR. 451

Measurement of the Absolute Value of the Penetration Depth Using Flux Quantization, R. MESERVEY 455

Superconducting Transition of a Rotating Superconductor—The Hollow Cylinder, A. F. HILDEBRANDT AND M. M. SAFFREN 459

The Magnetic Field Generated by a Rotating Superconductor, C. A. KING, JR., J. B. HENDRICKS, AND H. E. RORSCHACH, JR. 466

Measurement of the London Moment, M. BOL AND W. M. FAIRBANK 471

Experiments Concerning the Effect of Surface Charging on the Superconducting Transition Temperature, W. RÜHL 475

The Superconducting Transition Width in Pure Lead, J. E. NEIGHBOR, J. F. COCHRAN, AND C. A. SHIFFMAN 479

Anomalies in the Superconducting Transition of hcp Titanium Alloys, E. BUCHER, F. HEINIGER, AND J. MULLER 482

Positron Annihilation in V_3Si, T. W. MIHALISIN AND R. D. PARKS 487

Nuclear Spin Relaxation in Superconducting Gallium, R. H. HAMMOND AND G. M. KELLY 492

The Temperature Dependence of the Ginzburg–Landau Coefficient in Type I Superconductors, A. PASKIN, M. STRONGIN, P. P. CRAIG, AND D. G. SCHWEITZER 496

Field Penetration in Superconductors with Surface Layers, D. G. SCHWEITZER, P. P. CRAIG, M. STRONGIN, AND A. PASKIN 503

Effect of Impurities on the Critical Temperature of Superconductors, D. M. GINSBERG 508

Neutron Diffraction by Vortex Lines in Superconducting Niobium, D. CRIBIER, B. FARNOUX, B. JACROT, L. MADHAV RAO, B. VIVET, AND M. ANTONINI 509

5.3. Thermal Phenomena

Magnetocaloric Effects in a High-Field Superconductor, R. R. HAKE AND L. J. BARNES 513

The Efficiency of Superconducting Thermal Switches, V. P. PESHKOV AND A. YA. PARSHIN 517

Specific Heat Evidence for Gapless Superconductivity, D. K. FINNEMORE AND F. H. SPEDDING 521

Electronic Thermal Conductivity and Other Properties of Superconductors Containing Paramagnetic Impurities, A. GRIFFIN AND V. AMBEGAOKAR   524

5.4. Type II Superconductors, Thin Films

Flux Penetration and Mobility in Type I and II Superconductors, W. DeSorbo ... 530

Oscillatory Longitudinal Paramagnetic Effect in a Semireversible Type II Superconductor, M. A. R. LeBlanc ... 531

Magnetothermal Effects in Type II Superconductors, N. H. Zebouni, A. Venkataram, G. N. Rao, C. G. Grenier, and J. M. Reynolds ... 535

Experiments on Type II Superconductors, R. A. French, J. Lowell, and K. Mendelssohn ... 540

Crystallographic Properties of Lattices of Vortex Lines in a Type II Superconductor, J. Matricon ... 544

On the Mixed State of a Type II Superconductor, P. M. Marcus ... 550

Density of States of Second-Kind Superconducting Films Near Their Upper Critical Fields, J. P. Burger and E. Guyon ... 556

Surface Superconductivity and Supercooling in Weak and Strong Coupling Type I Superconductors, M. Cardona and B. Rosenblum ... 560

Distinctive Properties of Quantized Vortices in Superconducting Films, J. Pearl ... 566

Dynamic Vortex Effects in Thin Superconductors, J. M. Mochel and R. D. Parks ... 571

Critical Fields of Superconducting Films, F. Odeh and W. Liniger ... 575

Superconductivity in the Systems Nb–Sn, Nb–Al, V–Si, and V–Ga, W. Kunz and E. Saur ... 581

The Influence of Magnetic Fields on the Transition of High-Field Superconductors with A-15 Structure, G. Meyer and H. Wizgall ... 584

Superconducting Behavior of Indium–Lead Alloys, S. Gygax, J. L. Olsen, and R. H. Kropschot ... 587

Magnetic Field Penetration Into Nb_3Sn Disks, K. G. Petzinger and J. J. Hanak ... 591

The Motion of Magnetic Flux in a Type II Superconductor, P. H. Borcherds, C. E. Gough, W. F. Vinen, and A. C. Warren ... 596

5.5. Superconducting Materials

Superconductivity in GeTe and in the Monoxides of Titanium and Niobium, J. K. Hulm, C. K. Jones, R. Mazelsky, R. C. Miller, R. A. Hein, and J. W. Gibson ... 600

Superconductivity in the SnTe System, T. A. Hein, J. W. Gibson, R. S. Allgaier, B. B. Houston, Jr., R. Mazelsky, and R. C. Miller ... 604

Theory of the Thermodynamic Behavior of the Strong-Coupling Superconductors Lead and Mercury, J. C. Swihart, D. J. Scalapino, and Y. Wada ... 607

Niobium Oxy-Nitrides with High Transition Temperatures, G. K. Gaulé, J. T. Breslin, R. L. Ross, J. R. Pastore, and J. R. Shappirio ... 612

Experimental Relations for Electron Interactions in Transition Metals Determined from Superconductivity Data, E. Bucher, F. Heiniger, J. Muller, and J. L. Olsen ... 616

LIST OF CONTRIBUTORS
PART A PAGES 1–620, PART B PAGES 621–1255

Abraham, B. M., 133
Adams, E. D., 226
Adams, G., 765
Agureev, V. N., 925
Allen, J. F., 304, 320
Allgaier, R. S., 604
Ambegaokar, V., 524
Anderson, P. W., 8
Andronikashvili, E. L., 155, 159
Antonini, M., 509
Appel, J. C., 1215
Arko, A. J., 748
Arrott, A., 1066
Ashkin, M., 1126
Au, R., 877

Bablidze, R. A., 155
Bardeen, J., 3
Barnes, L. J., 513
Barrett, C. S., 1085
Beal, B. T., 137
Becker, R. L., 985
Beenakker, J. J. M., 1233, 1237
Bendt, P. J., 170
Benson, C. B., 184
Berman, R., 262
Bertman, B., 266
Bethoux, O., 1023
Betts, D. S., 129
Bhattacharya, D. L., 1210
Biondi, M. S., 387
Blackstead, H. A., 331
Blatt, F. J., 799
Boerstoel, B. M., 1071
Bohm, H. V., 312, 786
Bol, M., 471
Borcherds, P. H., 596
Borst, L. B., 84
Brändli, G., 827
Brandt, C. L., 887
Breslin, J. T., 612
Brewer, D. F., 370, 1037
Briscoe, C. V., 381
Brooker, G. A., 121
Brout, R., 623
Bruce, R. H., 174
Brugger, K., 1151
Bucher, E., 482, 616, 1059
Budzinski, W. V., 391
Burger, J. P., 556

Cameron, J. A., 1033
Campbell, I. A., 1033
Cardona, M., 560
Careri, G., 311, 335, 364
Cerdonio, M., 311
Challis, L. J., 839, 1135, 1145

Chase, C. E., 657, 1122
Cheeke, J. D. N., 839, 1145
Chilashvili, O. A., 925
Claiborne, L. T., 375
Clow, J., 328
Cochran, J. F., 479
Cohen, A. F., 884
Coldwell-Horsfall, R. A., 1110
Coleman, C. F., 892
Compton, J. P., 1033
Constable, A. R., 184
Cope, J. A., 353
Cornelissen, P. L. J., 316
Cotti, P., 827
Cowen, J. A., 877
Craford, M. G., 443
Craig, P. P., 496, 503, 1019
Cribier, D., 509
Cunsolo, S., 335

Dabbs, J. W. T., 428
Damon, D. H., 996
Daniel, E., 933
Das, P., 1253
Das, S. B., 1042
Dash, J. G., 287, 368, 1007
Daunt, J. G., 206, 909, 1099
Davey, G., 1196
Deaver, B. S., Jr., 451
de Bruyn Ouboter, R., 234, 270, 323, 1253
de Castro, E. M., 1099
de Combarieu, A., 1054
de Gennes, P. G., 19
de Graaf, A. M., 994
de Jong, E., 234
de Magistris, P., 349
Depatie, D., 328
DeSorbo, W., 530
Dickson, S. A., 1095
Diller, D. E., 1227
Domb, C., 637
Donaghy, J. J., 835
Doniach, S., 908
Donnelly, R. J., 346
Douglass, D. H., Jr., 385
Dresselhaus, M. S., 771
Dreyfus, B., 1026, 1054
du Chatenier, F. J., 1029, 1071
Dunlap, R. D., 1007
Dupre, F., 311
Durieux, M., 133
Dziwornooh, P. A., 299

Eatwell, A. J., 1091
Eck, R. E., 415
Eck, T. G., 759, 761

Edwards, M. H., 295
Efferson, K. R., 808
Einspruch, N. G., 375

Fagerstroem, C.-H., 1092
Fairbank, H. A., 266
Fairbank, W. M., 33, 188, 200, 396, 471, 1248
Falko, I. I., 378
Farnoux, B., 509
Fawcett, E., 782
Finnemore, D. K., 521
Foner, S., 771
Forester, D. W., 892
Forrest, A. M., 740
Forstat, H., 888
Franchetti, S., 97
Frederking, T. H. K., 291
French, R. A., 540
Friedberg, S. A., 884
Fryer, E. M., 827
Fulde, P., 438

Gaines, J. R., 1099, 1104
Gantmakher, V. F., 1193
Garbaty, E. A., 226
Garber, M., 799
Garfunkel, M. P., 387, 391
Gaulé, G. K., 612
Gerritsen, A. N., 1042
Gerritsen, G., 921
Gibson, J. W., 600, 604
Giffard, R. P., 219
Gijsman, H. M., 888
Ginsberg, D. M., 508
Glasser, M. L., 1067
Glick, F. I., 214
Glosser, R., 385
Glover, R. E., III, 381
Goldman, A. M., 421
Goldstein, Y., 400
Gonano, R., 897
Goodstein, D. L., 368
Gopal, E. S. R., 299
Gordon, W. L., 752, 854
Goree, W. S., 1205
Gorobchenko, V. D., 925
Gorter, F. W., 880
Gough, C. E., 118, 596
Grant, M. F., 1033
Greig, D., 1050
Grenier, C. G., 535, 802, 808
Gribbon, P. W. F., 353
Griffin, A., 524
Griffiths, D. J., 320
Grimes, C. C., 723, 765
Grissom, D., 1243

Gujabidze, G. V., 159
Guyon, E., 556
Gygax, S., 587

Haasbroek, J. N., 880
Hake, R. R., 513
Hammel, E. F., 307
Hammerle, R. H., 312
Hammond, R. H., 492
Hanak, J. J., 591
Hardeman, G. E. G., 921
Harris-Lowe, R., 206
Harrison, J. P., 1050
Hartwig, W. H., 1243
Hatton, J., 137, 219
Hedgcock, F. T., 995, 1039
Hein, R. A., 600, 604
Heine, V., 698
Heiniger, F., 482, 616, 1059
Heiskanen, K., 1179
Hendricks, J. B., 466
Hereford, F. L., 360
Hermans, L. J. F., 1237
Hess, G. B., 188
Higgins, R. J., 859
Higgs, P. M., 1007
Hildebrandt, A. F., 459
Hill, R. W., 1033
Hollis Hallett, A. C., 740, 1092, 1162
Horvitz, P., 147
Houston, B. B., 604
Hovi, V., 1175, 1179, 1184
Howard, D. G., 1007
Howe, D. R., 1037
Hulm, J. K., 600
Husa, D., 1099

Ishikawa, Y., 1023, 1026

Jacrot, B., 509
Jaklevic, R. C., 446
James, B. W., 1165
Jan, J. P., 776
Jones, C. K., 600, 790
Jones, G. O., 1089, 1090, 1091
Jortner, J., 356
Joseph, A. S., 867

Kadanoff, L. P., 378
Kanda, E., 917
Keen, B. E., 125
Keller, W. E., 307
Kelly, G. M., 492
Kelly, T., 195
Kidder, J. N., 331
King, C. A., Jr., 466
King, E., 1200
Kip, A. F., 679, 818
Kitchens, T. A., 1012
Klemens, P. G., 270, 996

Knaap, H. F. P., 1233, 1237
Knight, L. V., 1248
Knight, S., 179
Knook, B., 1041
Koch, J. F., 818
Koehler, T. R., 92
Kondo, J., 1004
Kotani, M., 929
Kramers, H. C., 258, 284, 316
Kreitman, M. M., 909, 914
Kropschot, R. H., 587
Kunz, W., 581
Kusmiss, J. H., 835
Kwiram, A. L., 451

Lambe, J. J., 446
Lampert, M. A., 844, 848
Larson, C. O., 752
Lawson, J. R., 854
LeBlanc, M. A. R., 531
Lee, D. M., 248, 254
Lee, J. A., 1200
Leech, J. W., 1081
Lefkowitz, I., 1158
Lehoczky, A., 381
LePage, J., 799
le Pair, C., 234, 270
Lines, R. A. G., 1033
Liniger, W., 575
Lipschultz, F. P., 254
Lipson, S. G., 814
Little, W. A., 653
Liu, S. H., 1001
Long, C. E., 366
Long, J. R., 802
Lounasma, O. V., 901
Love, N., 888
Lowell, J., 540
Lowndes, R. P., 1158
Luszczynski, K., 143
Luzzi, R., 994
Lynam, P., 905

Mackinnon, L., 786
Madden, H. H., 195, 312
Madhav Rao, L., 509
Maki, K., 405
Mamaladze, Yu. G., 281
Manchester, F. D., 202
Maradudin, A. A., 1126
Marcus, J. A., 743, 748, 859
Marcus, P. M., 550
Martin, D. H., 908, 1158
Martin, P. C., 9
Mate, C. F., 206
Matheson, C. C., 210
Mathur, D. P., 995, 1039
Matricon, J., 544
Mazelsky, R., 600, 604
Mazzoldi, P., 335
McCarthy, K. A., 1155

McCombe, B. D., 794
McCormick, W. D., 287, 368
McKirdy, A. S., 295
McNutt, J., 195
Meijer, H. C., 873
Mendelssohn, K., 299, 540, 1196, 1200
Mercereau, J. E., 446
Meservey, R., 455
Mess, K. W., 1140
Meyer, G., 584
Meyer, H., 897, 1095
Meyer, L., 338, 356, 1085
Michel, J. C., 1054
Miedema, A. R., 880, 1029, 1140
Mihalisin, T. W., 487
Milford, F. J., 909
Miller, R. C., 600, 604
Mills, R. L., 273
Mochel, J. M., 571
Modena, I., 342, 349
Moldover, M. R., 653
Morita, T., 104
Moss, F. E., 360
Muller, J., 482, 616, 1059
Munarin, J. A., 743

Neeper, D. A., 736
Neighbor, J. E., 479
Nethercot, A. H., Jr., 1189
Norberg, R. E., 143
Nosanow, L. H., 277
Notarys, H. A., 164

Obenshain, F. E., 892, 985
Odeh, F., 575
Ohtsubo, A., 917
Olsen, J. L., 587, 616, 827
Olsen, T., 375
Onn, D. G., 897
Opfer, J. E., 143
Osborne, D. V., 320
Osborne, D. W., 133
Overton, W. C., Jr., 229

Pancyzk, M. F., 226
Parks, R. D., 34, 487, 571
Parshin, A. Ya., 517
Paskin, A., 496, 503
Pastore, J. R., 612
Peacock, R. C., 443
Pearce, D. C., 736
Pearl, J., 566
Pearson, W. B., 776
Pellam, J. R., 164, 166, 191
Penrose, O., 91
Perisho, R. C., 1019
Peshkov, V. P., 79, 517
Petzinger, K. G., 591

List of Contributors

Pierce, J. M., 396
Pines, D., 61
Pit, J., 234
Pollack, G. L., 166
Powell, R. L., 732
Puff, R., 101
Pyykkö, P., 1175

Quinn, J. J., 844, 848, 851

Rao, G. N., 535
Rayne, J. A., 790, 1074
Reed, W. A., 782
Reif, F., 46
Reissland, J. A., 1081
Reppy, J. D., 179, 328
Reynolds, J. M., 535, 802, 808
Rice, O. K., 88
Rice, S. A., 356
Rice, T. M., 108
Richards, M. G., 219
Richards, P. M., 113
Ries, R. P., 443
Roberts, D., 752
Roberts, L. D., 663, 892, 985
Robinson, M. C., 1162
Rockmore, D. M., 835
Roellig, L. O., 195
Roessler, B., 1074
Rogers, S. J., 262
Rorschach, H. E., Jr., 147, 432, 466
Rosenblum, B., 560
Rosenshein, J. S., 1114
Ross, R. L., 612
Rugheimer, N. M., 381
Ruhl, W., 475
Rupp, L. W., Jr., 765

Saffren, M. M., 459
Samec, A., 226
Samoilov, B. N., 925
Santini, M., 335, 364
Sarachik, M. P., 1044
Satterthwaite, C. B., 443
Saur, E., 581
Savoia, A., 342
Scalapino, D. J., 415, 607
Scaramuzzi, F., 342, 349
Schirber, J. E., 863
Schreiber, D. S., 823
Schuch, A. F., 229
Schweitzer, D. G., 496, 503
Schwettman, H. A., 396
Schweikle, J. D., 291
Scott, T. A., 1205
Scribner, R. A., 226
Scurlock, R. G., 905
Segnan, R., 1019
Seidel, G., 794

Seward, W. D., 1130
Shappirio, J. R., 612
Sharma, J. K. N., 1196
Sharvin, Yu. V., 1193
Shaw, M. P., 759, 761
Shepherd, J. P. G., 752
Shiffman, C. A., 479
Shinozaki, S., 1066
Shirley, D. A., 1016
Shoenberg, D., 665, 680, 831
Siegwarth, J. D., 1007
Sievers, A. J., 1170
Signorelli, G., 364
Silver, A. H., 446
Simon, P. R. F., 1045
Skalyo, J., 884
Sklyarevskii, V. V., 925
Sluijter, C. G., 1233
Smith, T. I., 432
Souletie, J., 1023
Sparkes, A. R., 1090
Sparlin, D. M., 823
Spedding, F. H., 521
Spence, R. D., 877
Springett, B. E., 346
Springford, M., 776
Stark, R. W., 712
Stern, E. A., 1210
Stewart, A. T., 835
Stewart, J. W., 1230
Steyert, W. A., 1012
Stout, J. W., 887
Strongin, M., 496, 503
Swihart, J. C., 607
Sydoriak, S. G., 273
Symonds, A. J., 370

Taconis, K. W., 234, 323, 1253
Tanaka, T., 104
Tanner, D. J., 346
Taylor, B. N., 415
Taylor, M. T., 770
Taylor, R. D., 1012
Tedrow, P. M., 248, 254
Thiele, A. A., 1122
Thompson, W. A., 387
Thomson, A. L., 370
Thomson, J. O., 892, 985
Thorsen, A. C., 867
Tilley, J., 210
Tirmizi, S. M. A., 299
Tomasch, W. J., 424
Tosima, S., 844, 848
Tough, J. T., 287
Tournier, R., 1023, 1026
Trapp, C., 887
Trela, W. J., 200
Tsakadze, J. S., 155, 159
Turrell, B. G., 1037

Valby, L. E., 867
van Alphen, W. M., 323
van Dalen, P. A., 888
van den Berg, G. J., 955, 1041, 1071
van den Handel, J., 873
van den Meijdenberg, C. J. N., 133
van der Heyden, G., 284
van Rongen, H. J. M., 1041
van Till, H., 877
Varteva, M., 1179, 1184
Venkataram, A., 535
Vermeer, W., 323
Vinen, W. F., 118, 596
Vivet, B., 509
von Gutfeld, R. J., 1189

Wada, Y., 607
Walsh, W. M., Jr., 765
Warren, A. C., 596
Wasilik, R. M., 736
Watmough, D. J., 304
Watts, B. R., 779, 831
Weaver, J. C., 328
Weil, L., 1023, 1026
Weinstock, H., 366
Werntz, J. H., Jr., 214
Westenbarger, G. A., 1016
White, C. W., 266
White, D., 1104
Whitmore, D. H., 859
Whitney, W. M., 1114, 1118, 1122
Wiarda, T. M., 284
Wiebes, J., 258
Wigley, D. A., 1200
Wilks, J., 125, 129
Williams, D. J., 1135, 1145
Williamson, R. C., 657
Williamson, S. J., 771
Wilson, E. G., 356
Wilson, P. B., 396
Witteborn, F. C., 1248
Wizgall, H., 584
Wolf, R. P., 1118
Woodbury, W. C., 295
Woodfine, J. M., 1089
Wyatt, A. F. G., 411
Wray, E. M., 905
Wyder, P., 839

Yates, B., 1165
Yoffe, A. D., 1158
Younglove, B. A., 1223

Zebouni, N. H., 535, 802, 808
Zimmerman, G. O., 240, 244

6
LOW-TEMPERATURE TRANSITIONS

PHASE TRANSITIONS

R. Brout

Faculté des Sciences Université Libre de Bruxelles
Brussels, Belgium

This paper is concerned with the theory of those phase transitions which can be satisfactorily dealt with in terms of simple self-consistent equations. The transition itself can be thought of as arising from nontrivial (usually symmetry breaking) solutions of self-consistent equations. Phenomena falling into this category and reviewed below are ferromagnetism and antiferromagnetism, gas–liquid condensation, freezing, and superconductivity. Presumably many other phenomena can be similarly accounted for—ferroelectricity, superlattice formation, unmixing of solutions, etc.—but time does not permit discussion of all of these. Crystal phase transitions are probably of a more delicate variety and most likely cannot be handled in the rather crude approximations which seem so successful in the first class. Finally, a most unfortunate omission is the λ-point transition of liquid helium. This transition is associated with the Bose–Einstein condensation of the ideal gas, modified of course, by the interatomic interactions. It is not a transition of the type which falls naturally into the class which we have chosen to discuss.

Ferromagnetism

As an introduction to the subject, let us consider the classic Weiss theory of ferromagnetism. Consider a set of spins fixed at lattice sites, interacting through a spin–spin exchange force and with an external field. Two types of models, called Heisenberg and Ising models, respectively, are usually studied,

$$H = -\tfrac{1}{2}\sum v_{ij}\mathbf{S}_i \cdot \mathbf{S}_j - \mathscr{H}\sum \mathbf{S}_i \quad \text{Heisenberg} \quad (1a)$$

$$H = -\tfrac{1}{2}\sum v_{ij}\mu_i\mu_j - \mathscr{H}\sum \mu_i \quad \text{Ising} \quad (1b)$$

Here v_{ij} is the exchange force between spins i and j. $\mathbf{S}_i$ is the spin operator of the spin at site i in the Heisenberg model. In this model, the spins are isotropically coupled. $\mathscr{H}$ is the external field measured in units of energy. μ_i, in the Ising model, is taken to be the z-component of spin $\tfrac{1}{2}$ and hence takes on values ± 1. The coupling is then completely anisotropic. Since all operators μ_i commute, the Ising model is classical.

The Ising model is useful in magnetic theory only insofar as it serves as a mathematical guinea pig. On the other hand, it is also very useful in providing a simple model of order–disorder superlattice transitions as well as, somewhat surprisingly, a model which yields many hints on the theory of liquid–gas condensation. The Heisenberg model is less general in providing lessons over a wide domain of physics, but often provides an adequate description of realistic physical situations: antiferromagnetism and ferrimagnetism in insulating crystals, as well as ferromagnetism in the rare earths. The various applications come about according to the chosen form of v_{ij}.

The physical content of the Heisenberg and Ising models is quite similar for temperatures in the region of the Curie point and above. At low temperatures, however, the models differ drastically due to the spin wave excitations of the Heisenberg model. We shall touch upon this point somewhat later.

The idea of the Weiss theory is as follows: We may view the Hamiltonian (1a) in terms of an exchange field $\mathscr{H}_i$ acting on spin i such that

$$\mathscr{H}_i = \sum v_{ij} \mathbf{S}_j + \mathscr{H}_{\text{ext}} = \mathscr{H}_{\text{exch}} + \mathscr{H}_{\text{ext}} \tag{2}$$

The energy is then

$$-\sum (\tfrac{1}{2}\mathbf{S}_i \cdot \mathscr{H}_i^{\text{exch}} + \mathbf{S}_i \cdot \mathscr{H}^{\text{ext}}) \tag{3}$$

The factor of a half comes from not counting the same pair of interactions twice.

Let us suppose that v_{ij} is positive and nonvanishing over a certain range of spins. We shall take Z spins to be in the range. (A common model is to take v_{ij} to be nonvanishing for near neighbors only, in which case Z is the number of sites in the first coordination shell.) The ground state in this example is totally ferromagnetic, so we look for the onset of a ferromagnetic phase as the temperature is lowered. Such a phase is characterized by $\langle \mathbf{S}_i \rangle \neq 0$, where $\langle \mathbf{S}_i \rangle$ is independent of the position i, and for definiteness we will take to point in the z-direction. (We take $\mathscr{H}_{\text{ext}}$ to be in the z-direction.)

The Weiss idea is to approximate $\mathscr{H}_{\text{exch}}$ on any site by $\langle \mathscr{H}_{\text{exch}} \rangle$ according to

$$\langle \mathscr{H}_{\text{exch}} \rangle = -\sum v_{ij} \langle \mathbf{S}_j \rangle = -(\sum v_{ij}) \mathbf{R} \equiv -v(0) \mathbf{R} \tag{4}$$

The average value of $\mathbf{S}_i$ is then given by

$$\langle \mathbf{S}_i \rangle = \mathbf{R} = \frac{\text{trace} \exp\{-\beta[\mathbf{S}_i \cdot (\langle \mathscr{H}_{\text{exch}} \rangle + \mathscr{H}_{\text{ext}})]\} \mathbf{S}_i}{\text{trace} \exp\{-\beta[\mathbf{S}_i \cdot (\langle \mathscr{H}_{\text{exch}} \rangle + \mathscr{H}_{\text{ext}})]\}} \tag{5}$$

or

$$R = B_s\{\beta[v(0)R + \mathscr{H}_{\text{ext}}]\} \tag{6}$$

Here B_s is the Brillouin function for spin S. For simplicity, we will deal from now on with spin $\tfrac{1}{2}$ and for convenience will redefine v_{ij} so that the Heisenberg interaction is $-\tfrac{1}{2} \Sigma v_{ij} \boldsymbol{\sigma}_i \cdot \boldsymbol{\sigma}_j$, where σ_i are the Pauli matrices. Then (6) becomes

$$R = \tanh\{\beta[v(0)R + \mathscr{H}_{\text{ext}}]\} \tag{7}$$

A schematic plot of the solution $R(\mathscr{H}_{\text{ext}})$ is given for $T > T_c$ and $T < T_c$ in Fig. 1. The dotted sections are in fact thermodynamically incorrect; however the sections for which $dR/d\mathscr{H}_{\text{ext}} > 0$ may be observed in a metastable phase. The correct equilibrium curve is the solid section.

In Fig. 2, we plot the spontaneous magnetization vs. T at $\mathscr{H}_{\text{ext}} = 0$. It vanishes at T_c where

$$kT_c = v(0) \tag{8}$$

We find, therefore, a ferromagnetic phase for $T < T_c$. As $T \to T_c$, R vanishes like $(T_c - T)^{\tfrac{1}{2}}$, and for $T \ll T_c$, R approaches unity like $(1 - R) \sim e^{-2\beta v(0)}$. This latter result is to be expected since at very low temperatures almost all the spins are aligned; to turn a single one then costs a gap energy of $2v(0)$. This result is in fact rigorous for the Ising model, but is incorrect for the Heisenberg model. We will discuss this point subsequently in connection with spin waves. The above theory is

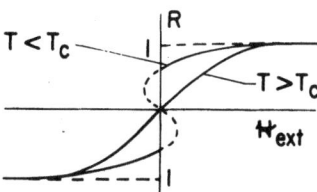

Fig. 1. Schematic plot of R versus H_{ext}.

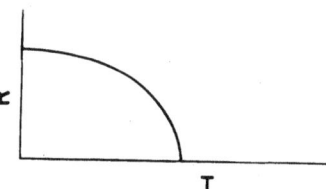

Fig. 2. Spontaneous magnetization as a function of temperature.

of pedagogical value only, since there are no pure Heisenberg ferromagnets around (i.e., with v_{ij} everywhere positive). To compare the idea with experiment, we must then apply it to more realistic situations. Our plan is to run down the list of transitions, showing how fact and molecular field theory agree and disagree when comparison is possible. We will then present a critique of the theory which will lead us into the subject of critical fluctuations.

Antiferromagnetism

Let us take a model of two sublattices A, B, such that $v_{ij} < 0$ for i, j on different sublattices, and $v_{ij} > 0$ for ij on the same sublattice. This seems typical of antiferromagnets such as MnF_2. In this case, we define two order parameters R_A and R_B and two exchange fields

$$\mathcal{H}^A_{exch} = \alpha R_B - \gamma_A R_A$$
$$\mathcal{H}^B_{exch} = -\gamma_B R_B + \alpha R_A$$

$$\alpha \equiv +\sum_{j_B} v_{i_A j_B} \qquad \gamma_A \equiv -\sum_{j_A} v_{i_A j_A} \qquad \gamma_B \equiv -\sum_{j_B} v_{i_B j_B} \tag{9}$$

The Weiss equations are

$$R_A = B_{s_A}(\mathcal{H}^A_{exch})$$
$$R_B = B_{s_B}(\mathcal{H}^B_{exch}) \tag{10}$$

If the spins on the sublattices are the same, then $R_A = -R_B$. For spin $\tfrac{1}{2}$, we then have

$$R_A = \tanh \beta[\alpha R_A - \gamma R_B] = \tanh [\beta(\alpha + \gamma)R_A]$$
$$R_B = \tanh [\alpha R_B - \gamma R_A] = \tanh [\beta(\alpha + \gamma)R_B] \tag{11}$$

We then have a Curie point at $kT_c = \alpha + \gamma$. A similar analysis is obviously possible for ferrimagnets where in this case $R_A \neq -R_B$, a situation which leads to coupled equations.

In Fig. 3, we compare the molecular field theory prediction with the observed sublattice magnetization of MnF_2.[1] It is seen that the fit is excellent up to about

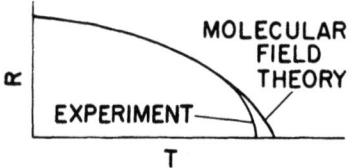

Fig. 3. Sublattice magnetization of MnF_2 as a function of temperature.

5% of the Curie point; the deviations are due to singular effects in the critical region which we will discuss subsequently.

Liquid–Gas Condensation

Here the molecular field idea is due to van der Waals. The instantaneous value of the potential energy is

$$V = \tfrac{1}{2} \sum_{i \neq j} v(r_i - r_j) \tag{12}$$

It is convenient to consider the potential energy density and write V as

$$= \tfrac{1}{2} \int v(r - r')\rho(r)\rho(r') \, dr \tag{13}$$

where $\rho(r) = \Sigma_i \delta(r - r'_j)$. (The term $i = j$ in the double sum will not contribute due to hard core effects.) The molecular field idea comes from the interpretation of (13) in terms of the potential acting at r due to all other particles. This potential is

$$(r) = \int v(r - rV)\rho(r') \tag{14}$$

Straightforward application of our previous development would then indicate that it is a suitable approximation to replace $V(r)$ by $\langle V(r) \rangle$, where

$$\langle V(r) \rangle = \int v(r - r') \langle \rho(r') \rangle \, dr' \tag{15}$$

Since a fluid is homogeneous, we then have $\langle \rho(r') \rangle = \rho$, the ambient density, and $\langle V \rangle$ = constant.

This procedure would make sense if $v(r)$ was everywhere well-defined. However, the intermolecular potential appears as in Fig. 4. For $r < c$, we will approximate the potential by a simple hard core. The modification in (15) due to the hard core is a simple thing to come by and some analysis shows that the correct molecular field is given by[2]

$$\langle V \rangle = \int v(r - r') g_{HC}(r - r') \langle \rho(r') \rangle = \text{constant} \tag{16}$$

$g_{HC}(r)$ is the hard core radial distribution function. It vanishes for $r < c$ and has some local packing structure immediately outside c. A typical $g_{HC}(r)$ is sketched in Fig. 5.

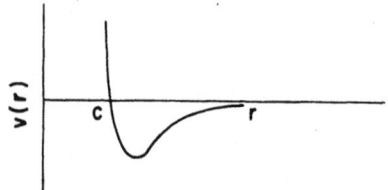

Fig. 4. Intermolecular potential characteristic of rare gas atoms.

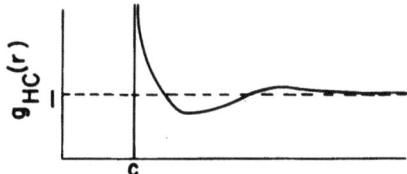

Fig. 5. Hard core distribution function.

From (16), we then find one ingredient contributing to the balance of internal and external pressure in a gas in equilibrium. We write

$$P_{\text{ext}} = P_{\text{thermal}} + P_{\text{attractive}} \tag{17}$$

P_{thermal} is the pressure due to the thermal motion of a system of hard spheres and $P_{\text{attractive}}$ is the internal pressure due to the particles sitting in a density dependent potential. We get $P_{\text{attractive}}$ most easily by direct differentiation of the energy

$$P_{\text{attractive}} = -\frac{\partial \langle E \rangle}{\partial \Omega} = +\rho^2 \frac{\partial}{\partial \rho}\left[\tfrac{1}{2}\rho \int g_{\text{HC}}(r)v(r)\,dr\right]$$

$$= -a\rho^2 + [\text{term in } \partial g_{\text{HC}}/\partial \rho] \tag{18}$$

The van der Waals "a" is given by

$$a = -\tfrac{1}{2}\int v(r)g_{\text{HC}}(r)\,dr^3 > 0 \tag{19}$$

The term in $\partial g_{\text{HC}}/\partial \rho$ is negligible when the range of the attractive force is long compared to c. Actually this is not the case, and this term must in general be calculated as well. When we say van der Waals theory, we include this term as well.

P_{thermal} in (17) was approximated by van der Waals by $[kT/(v-b)]$. This is a poor approximation as revealed by direct calculation of the cluster expansion for hard spheres. The cluster expansion converges well at the critical density.

We may now compare theory and experiment. The parameters of the potential are obtained from high temperature gas data. The potential used in the calculation is of the form

$$v(r) = 4\varepsilon\left[\left(\frac{c}{r}\right)^{12} - \left(\frac{c}{r}\right)^{6}\right] \qquad r > c \tag{20}$$

$$v(r) = \infty \qquad r < c$$

where ε is the depth at the minimum which is at $r \cong (1.12)c$. P_{thermal} and $g_{\text{HC}}(r)$ were calculated from the hard core cluster expansion. The results for the critical parameters in reduced units are given in Table I.[2] In the last column, we have tabulated for comparison the results obtained if the usual van der Waals equation is used. It is seen that van der Waals' idea is correct, and with a little modification locates

Table I

	Experimental	Coopersmith and Brout[2]	van der Waals
kT_c/ε	1.28	1.42	0.30
V_c/c^3	3.09	3.04	8.88
$\beta_c P_c V_c$	0.42	0.29	0.38

the critical region very well. More recently, Dr. H. Frisch, in a private communication to the author, has confirmed this agreement between theory and experiment. In fact, Frisch can fit all three critical parameters with excellent precision. He used a three-parameter potential and the Percus–Yevick theory for the hard sphere gas. Since the P–Y and cluster theories are in excellent agreement at the critical density, the hard core problem is handled in essentially the same way. The difference is in the shape of the attractive tail. The results of Frisch's calculation indicate that at the present stage of our uncertain knowledge of intermolecular potentials, we cannot hope to improve upon the calculation of the critical parameters.

As far as the shape of the coexistence curve is concerned, the van der Waals equation is in error to about the same extent as the magnetization curve of Fig. 3. The van der Waals theory predicts $(V^{liq} - V^{gas}) \sim |T - T_c|^{\frac{1}{2}}$. For $(|T - T_c|/T_c) \lesssim 5\%$, experiment shows up deviations and present evidence indicates that in the critical region $(V^{liq} - V^{gas}) \sim |T - T_c|^{0.33}$. This peculiar behavior is surely related to the similar observation in the magnetic case and is attributable to critical fluctuations which are not accounted for in the van der Waals theory.

Melting and Freezing[3]

In a solid, the molecular field of equation (16) is a periodic function since $\langle \rho(\mathbf{r}) \rangle$ is periodic of the form

$$\langle \rho(\mathbf{r}) \rangle = \sum_{\mathbf{G}} \rho_{\mathbf{G}} e^{i \mathbf{G} \cdot \mathbf{r}} \qquad (21)$$

Here $\mathbf{G}$ are the points of the reciprocal lattice. The quantities $\rho_{\mathbf{G}}$ for $\mathbf{G} \neq 0$ then constitute a set of order parameters. The self-consistent equation is then

$$\langle \rho(\mathbf{r}) \rangle = e^{-\beta V(\mathbf{r})} \Big/ \int_{\text{unit cell}} e^{-\beta V(\mathbf{r}')} d^3 r' \qquad (22)$$

In the well-formed solid, it is convenient to use the representation

$$\langle \rho(\mathbf{r}) \rangle = \sum_i \varphi(\mathbf{r} - \mathbf{R}_i) \qquad (23)$$

where $\mathbf{R}_i$ are the lattice sites. We assume that in the ith cell $\langle \rho(\mathbf{r}) \rangle$ is approximated by only one term in (23). Furthermore, the condition of self-consistency dictates that, when the hard core edge overlaps into neighboring cells, the function $\varphi(\mathbf{r} - \mathbf{R}_i)$ vanishes. Thus, (22) becomes

$$\varphi(\mathbf{r}) = C \exp\left[-\beta \sum_i \int v(\mathbf{r} - \mathbf{r}') \varphi(\mathbf{r}' - \mathbf{R}_i)\right] \qquad (24)$$

where C normalizes φ to one particle per cell. The boundary condition is then imposed that $\varphi(\mathbf{r}) = 0$ when the hard core edge overlaps the cell edge. For a close-packed lattice, the sum over neighbors in a given coordination shell is well approximated by an integration.

At low temperatures, (24) becomes the Einstein oscillator theory, modified in the manner prescribed by Grueneisen to allow for expansion. At higher temperatures, numerical solution is necessary. The solutions are presented in the graphs of Fig. 6 at varying temperatures. Once $kT/\varepsilon \geqslant 0.6$, the approximation goes over to a free volume theory (i.e., $\varphi(r) = $ constant in the permitted domain). Since this is in the region of the observed melting temperature, we may conjecture that melting occurs when $\varphi(r)$ becomes approximately constant. Under these conditions, the solid is equivalent to a cell model liquid. Melting then occurs because of the increase

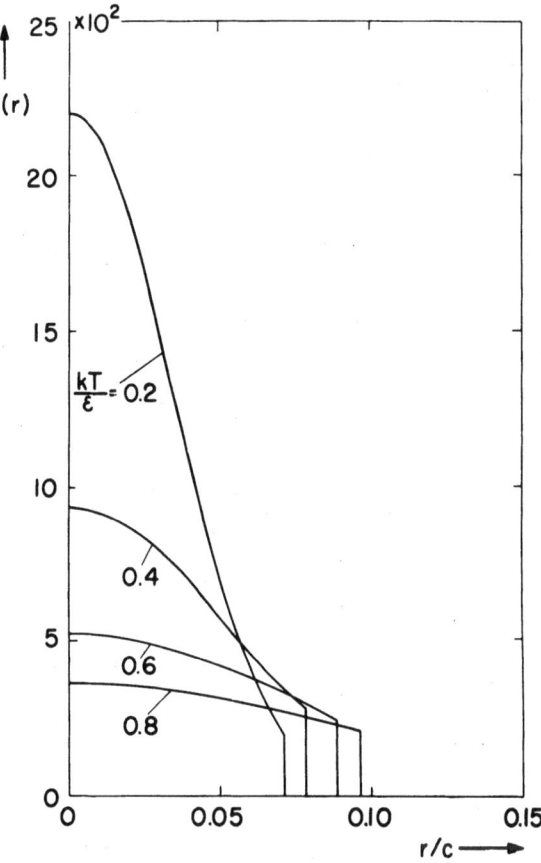

Fig. 6. Cell distribution functions in a crystal.

of entropy in going over to the usual fluid phase. At the time this paper was written, the melting point was being calculated by direct comparison of solid free energy and liquid free energy. The two expressions being used were

$$G_{\text{liquid}} = F_{\text{HC}} + N\rho/2 \int v(\mathbf{r} - \mathbf{r}') g_{\text{HC}}(\mathbf{r} - \mathbf{r}') \, d^3r' + pV \tag{25}$$

$$G_{\text{solid}} = \frac{N}{2} \sum_{R_i \neq 0} \int_{\text{unit cell}} d^3r \, d^3r' \, \varphi(\mathbf{r}) \varphi(\mathbf{r}' - \mathbf{R}_i) v(\mathbf{r} - \mathbf{r}') + N \int_{\text{unit cell}} \varphi(\mathbf{r}) \ln \varphi(\mathbf{r}) \, d^3r + pV \tag{26}$$

We hope to report on the outcome of the calculation at the conference.* Confidence in (26) is based on the graph presented in Fig. 7 for the interparticle distance as a function of temperature at low pressures. This is found by minimizing (26) with respect to density at $p = 0$. Excellent agreement is obtained in the melting region, whereas at low temperatures quantum effects occur which are not included in the theory.

Confidence in (25) is based on the success of the calculation of the gas–liquid condensation.

* See note added in proof.

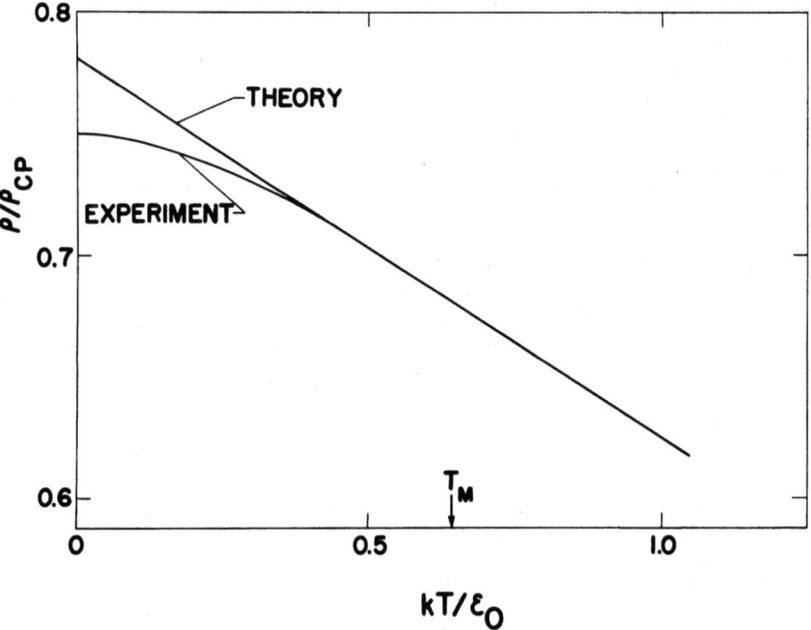

Fig. 7. Volume as a function of temperature in the solid state.

Bloch–Stoner Theory of Ferromagnetism

We present the self-consistent field theory of band ferromagnetism not only for its intrinsic interest as a band model of magnetism but in the hope that it may serve as an introduction to the ideas contained in the BCS theory of superconductivity.

In Hartree–Fock approximation the energy is

$$E = \sum E(k)\langle n(k,\sigma)\rangle$$
$$+ \tfrac{1}{2}v(0)\sum \langle n(k,\sigma)\rangle\langle n(k',\sigma')\rangle - \tfrac{1}{2}\sum v(k-k')\delta\sigma\,\sigma'\langle n(k,\sigma)\rangle\langle n(k',\sigma')\rangle \qquad (27)$$

where $\langle n(k,\sigma)\rangle$ is the average number of electrons in a Bloch state of wave number **k** and spin σ, and $E(k)$ is the Bloch energy. The Hartree–Fock field is then

$$V_{k,\sigma} = v(0)\sum \langle n(k',\sigma')\rangle - \delta\sigma\,\sigma' \sum v(k-k')\langle n(k',\sigma')\rangle \qquad (28)$$

The second term is the exchange field. The self-consistent energies are then

$$\varepsilon_{k,\sigma} = E_k + V_{k,\sigma} \qquad (29)$$

and the self-consistent equations are

$$\langle n_{k,\sigma}\rangle = \frac{1}{e^{\beta(\varepsilon_{k,\sigma}-\mu)}+1} \qquad (30)$$

where μ is fixed by $\Sigma\langle n_{k,\sigma}\rangle = N$.

Stoner has shown that this theory can explain some features of the band ferromagnetism in the transition metals. However, there are still many doubtful points both in theory and experiment. Nevertheless, it is most likely that this idea contains the necessary germ of truth which will serve as a nucleus for theoretical development.

Equation (30) shows the usual critical behavior, yielding a critical temperature at the temperature for which

$$1 = V \sum (dn_k/d\varepsilon_k) \tag{31}$$

where V is some average of the exchange potential $v(k - k')$.

BCS Theory of Superconductivity[5]

The essential idea of Bardeen, Cooper, and Schrieffer is that the superconducting ground state is a condensation of electrons into a coherent set of bound state pairs. The cooperative process responsible for pairing lies in the "reduced" Hamiltonian. This is a special set of terms in the total interaction Hamiltonian just as is the Hartree–Fock set in the band theory of ferromagnetism. It is given by

$$\sum v(k, k') b_{k'}^+ b_k \tag{32}$$

Here, b_k is the annihilation operator $a_{k\sigma} a_{-k, -\sigma}$. The term in question then yields a bound state for $v(k, k')$ attractive. In particular for $v(k, k') = $ constant within a skin about the Fermi surface, we get S-states. We will assume this approximation in what follows. (The full theory contains dynamical effects due to phonon exchange which are not contained in the simple approach presented here.)

The novel feature of the BCS theory is the invention of the set of order parameters $\langle b_k \rangle$. These quantities will be automatically zero in a gauge invariant theory, just as is the magnetization $\langle \mathbf{R} \rangle$ in the Heisenberg model in the absence of external field. Nevertheless, they are interesting quantities to introduce. Due to Anderson[6] and others, it is thoroughly understood how to maintain the necessary gauge symmetry after it is deliberately broken. Time does not not allow discussion of this point here.

Using the BCS idea of nonvanishing $\langle b_k \rangle$, we then find a molecular BCS field on the pair k whose value is*

$$\mathscr{H}_k = \sum_{k'} v(k, k') \langle b_{k'}^+ \rangle \equiv -V \sum_{k' \varepsilon \text{ skin}} \langle b_{k'}^+ \rangle \tag{33}$$

The total Hamiltonian of the pair is

$$H_k = (E_k - \mu)(n_k + n_{-k}) + b_k \mathscr{H}_k^+ + b_k^+ \mathscr{H}_k \tag{34}$$

We may then use straightforward statistical mechanics to calculate $\langle b_k \rangle$

$$\langle b_k \rangle = \frac{\text{trace } b_k e^{-\beta E_k}}{\text{trace } e^{-\beta E_k}} \tag{35}$$

The trace is over the four states whose occupation numbers of $\{n_k, n_{-k}\}$ are $\{1, 1\}$, $\{1, 0\}$, $\{0, 1\}$, and $\{0, 0\}$. Calculation of (35) then gives

$$\langle b_k \rangle = \frac{\mathscr{H}_k}{\sqrt{(E_k - \mu)^2 + |\mathscr{H}_k|^2}} \tanh \left[\frac{\beta \sqrt{(E_k - \mu)^2 + |\mathscr{H}_k|^2}}{2} \right] \tag{36}$$

* It should be pointed out that the fundamental approximation in Bloch–Stoner and BCS theories is the restriction to only certain small parts of the Hamiltonian. Once this restriction is made, the molecular field approximation is correct to $0(N)$ in the free energy.

In the approximation (33), $\mathcal{H}_k$ is independent of k. Let us call it ε_0; then (36) becomes

$$1 = V \sum_k \frac{\tanh[\beta\sqrt{(E_k - \mu)^2 + \varepsilon_0^2}/2]}{\sqrt{(E_k - \mu)^2 + \varepsilon_0^2}} \tag{37}$$

The critical temperature found by setting $\varepsilon_0 = 0$ is given by $kT_c \simeq 1.7\varepsilon_0(T = 0)$. This is in good agreement for most superconductors and in general, the BCS theory, even in its crude form, has met with success wherever put to the test.

Local Order

The essential approximation of the molecular field theories is the neglect of local order around the critical point. For example, in the Heisenberg model of ferromagnetism, we assume that the field on $i = \sum_j v_{ij}S_j$ is independent of the configuration of the spin i itself. This is false since if i is up it will tend to polarize its neighbors also up. This effect will become increasingly smaller as the range of the force becomes larger, since the effect of the single spin i will then become diluted.

A simple first correction to Weiss theory is easily developed in all models presented. We will run through the argument for the Ising model and present the results for the other cases—all for $T > T_c$. We will then discuss what happens for $T < T_c$ in the various cases. Finally, we compare with experiment and moment calculations.

In the Ising model, we consider a problem with, say, spin 1 known to be up at the origin. The correlation function $\langle \mu_1\mu_2 \rangle$ is then equal to $\langle \mu_2 \rangle$ in this ensemble. The field on 2 is the field due to 1 and the other spins. Thus

$$\mathcal{H}_2 = v_{12} + \sum_{j \neq 1} v_{2k}\mu_k \tag{38}$$

The value of $\langle \mu_2 \rangle$ (i.e., the correlation $\langle \mu_1\mu_2 \rangle$) is then

$$\langle \mu_1\mu_2 \rangle = \tanh(\beta\mathcal{H}_2) \simeq \beta\mathcal{H}_2 = \beta v_{12} + \sum_{k \neq 1} \beta v_{2k}\langle \mu_k\mu_1 \rangle \tag{39}$$

In (39), we have used a molecular field "local" approximation in that we have neglected the effect of the configuration of spin 2 in the evaluation of $\langle \mu_1\mu_k \rangle$. This is the essential approximation. The linearization of the tanh in (39) is valid at and above the critical point.

Equation (39) is solved by Fourier transforms. We define

$$\mu_q = \frac{1}{\sqrt{N}} \sum_i \mu_i e^{i\mathbf{q} \cdot \mathbf{R}_i} \quad (\mathbf{q} \text{ in first Brillouin zone}) \tag{40}$$

Then (39) solves to

$$\langle |\mu_q|^2 \rangle = \frac{1}{1 - \beta v(\mathbf{q})} \tag{41}$$

The energy is found from

$$E = \tfrac{1}{2}\sum v_{ij}\mu_i\mu_j = \tfrac{1}{2}\sum v(\mathbf{q})\langle |\mu_q|^2 \rangle \tag{42}$$

We find $\beta_c E = 0(1/Z)$, where Z is the number of atoms in the range of force. Thus at the critical point, we are left with $\sim(1/Z)$ of the total interaction energy instead of the Weiss value which vanishes. We also find a specific heat which diverges like

$(1/\sqrt{T-T_c})$. The susceptibility can be found from the fluctuation formula

$$kT\chi = \lim_{q \to 0} \langle |\mu_q|^2 \rangle = \frac{1}{1 - \beta v(0)} \tag{43}$$

This coincides with the Weiss value of $(dR/d\mathcal{H})_{\mathcal{H}=0}$ for $T > T_c$, and predicts $\chi^{-1} \sim T - T_c$). Finally, we remark that the spin scattering form factor is $\langle |\mu_q|^2 \rangle$, so that we get critical opalescence.

All of these results were derived about fifty years ago on macroscopic grounds by Ornstein and Zernike.[7] At the end of the talk we will present a brief critique.

We now tabulate the analogous expressions in all the models we have handled in this talk. In all cases of critical phenomena the specific heat diverges like $(T - T_c)^{-\frac{1}{2}}$, and the response function (susceptibility or compressibility) like $(T - T_c)^{-1}$.

Heisenberg Model

$$\langle |S_q|^2 \rangle = \frac{3}{1 - \beta v(q)} \tag{44}$$

Condensation

$$\langle |\rho_q|^2 \rangle = \frac{\langle |\rho_q|^2 \rangle_{HC}}{1 + \beta v(q) \langle |\rho_q|^2 \rangle_{HC}} \tag{45}$$

Here, $\rho_q = (1/\sqrt{N}) \Sigma_i e^{i\mathbf{q} \cdot \mathbf{r}_i}$ is the Fourier transform of the instantaneous density $\Sigma_i \delta(\mathbf{r} - \mathbf{r}_i)$, $\langle |\rho_q|^2 \rangle_{HC}$ is its mean square value in the hard core medium, and $v(q)$ is the Fourier transform of the attractive part of the potential and is negative. Equation (45) gives critical opalescence; it also gives a compressibility from

$$\rho kT\kappa = \lim_{q \to 0} \langle |\rho_q|^2 \rangle \tag{46}$$

which coincides with that of the van der Waals equation. Hence, the fluctuation rises to infinity at the critical point and then goes back $\langle |\rho_q|^2 \rangle_{HC}$ on either side. (In the range of coexistence this is guaranteed by the Maxwell construction.)

Freezing. Equation (45) is also valid for this case. For $q = 2\pi/c$, the fluctuation is enhanced and we will see a rise in specific heat. However, this is an intrinsic first-order transition and the point of instability is never realized.

Bloch Ferromagnetism. We define

$$\sigma_q^+ = \sum_k a_{k+q\uparrow}^+ a_{k\downarrow} \tag{47}$$

The fluctuation for small q is then (for $v(k - k') = V = $ constant)

$$\langle \sigma_q^+ \sigma_{-q} \rangle = \frac{\chi_q^0}{1 - V\chi_q^0} \tag{48}$$

where χ_q^0 is the susceptibility of the ideal Fermi gas given by

$$\chi_q^0 = \frac{1}{N} \sum_k \frac{\langle n(k) \rangle - \langle n(k+q) \rangle}{E(k+q) - E(k)} \tag{49}$$

At the critical point of the Bloch–Stoner model, this fluctuation again blows up in a manner strictly analogous to the Heisenberg model.

BCS Theory. We define

$$B_q = \sum a_{-k+q\downarrow} a_{k\uparrow} \tag{50}$$

Then the fluctuation in question is

$$\langle |B_q|^2 \rangle = \frac{\eta_q^0}{1 - V\eta_q^0} \tag{51}$$

where

$$\eta_q^0 = \frac{1}{N} \sum_{(k \varepsilon \text{ skin})} \frac{\langle n(k) \rangle + \langle n(k+q) \rangle + 1}{E(k) + E(k+q)} \tag{52}$$

The behavior is again analogous to the previous results.

For $T < T_c$, all the above results for fluctuations are drastically modified due to the presence of long-range order. The modification is classed in one of two categories, depending upon whether the condensed phase breaks a group of symmetry operations which is continuous or discrete. Examples of the latter case are the Ising model or the liquid–gas condensation. For the Ising model, the modification is simply to change (41) to[8]

$$\langle |\mu_q|^2 \rangle = \frac{1 - R^2}{1 - \beta v(q)(1 - R^2)} \tag{53}$$

Equation (53) is always positive and finite. Once outside the critical region, it shows that the fluctuation drops rapidly to zero, thereby validating the Weiss approximation on the low-temperature side of the critical region.

For the liquid–gas condensation, (45) is to be used as it stands. It should be correct everywhere but within a few percent of the critical region. $\langle |\rho_q|^2 \rangle$ is always positive and blows up at $q = 0$ at the limits of liquid–gas metastability (minimum and maximum of the van der Waals loop).

In the case of continuous symmetry, the mode of suppressing fluctuations is more interesting and less drastic. We establish a set of boson eigenmodes (spin waves in ferromagnets, phonons in crystals, and Anderson modes in superconductors). The theory of how this is done in the critical region is not well established, whereas it is well known at low temperatures. An attempt to bridge the two regions is given by the random phase approximation of Englert.[9] We will not pursue this matter here. The important point is that the fluctuations, though suppressed, are done so much more mildly than in the discrete case, since a continuum of excitations arises rather than the gap which characterizes the discrete case. Thus, whereas the specific heat of the Ising model at low temperatures falls off exponentially ($\sim e^{-2\beta v(0)}$), that of the Heisenberg model falls off as a power law $\sim [kT/v(0)]^{\frac{3}{2}}$.

We now discuss agreement with experiment and moment calculations of the ideas presented above. Domb, Sykes, and Fisher (DSF),* in an important series of calculations, have evaluated, by power series techniques, various critical properties of the three-dimensional Ising model with near neighbor interactions. These calculations will be reviewed by Professor Domb at this conference. We present a brief rundown of the results together with experiment in the critical region. We set these calculations into the context of the molecular field theories discussed above.

* A recent review is that by M. Fisher.[10]

First, for $T > T_c$, DSF have established that the correlation function for large R is given by

$$\langle \mu(0)\mu(R) \rangle = e^{-KR}/R^n \tag{54}$$

$$n = 1.06$$

$$K \sim (T - T_c)^{0.65} \quad (T \text{ very near to } T_c) \tag{55}$$

This is in contrast to the Fourier transform of (41) which for large R yields

$$e^{-K'R}/R$$

$$K' \sim (T - T_c)^{\frac{1}{2}} \tag{56}$$

It is seen that the approximate asymptotic distribution resembles the true one very closely. Experiment on light scattering from density fluctuations cannot yet distinguish between the two behaviors; however, careful experiment should yield the DSF behavior in the critical region. Once five percent outside of the critical region, the behavior characterized by (56) should take over. Experiments on spin systems by neutrons will also be useful in this field.

The specific heat calculated by Domb and Sykes on the Ising model is predicted to be either logarithmically divergent or to diverge as a weak power law $[(T - T_c)^{-0.2}]$. The square root behavior $[(T - T_c)^{-0.5}]$ again takes over outside the critical region. This behavior has in fact been found by Bagatskii et al.[11] for the case of the condensation of argon, indicating that critical behavior seems to be model independent.

The susceptibility in the critical region is predicted to proceed as $(T - T_c)^{-\frac{5}{4}}$, in contrast to $(T - T_c)^{-1}$ of the Weiss theory. A careful experiment by Habgood and Schneider[12] measures the compressibility of xenon in the critical region. This is plotted in Fig. 8. We see the confirmation of the $(\frac{5}{4})$ exponent as well as the bend to the Weiss exponent of unity once we are a few percent outside the critical region. Again the Ising model fits, indicating model independence of the critical behavior.

Finally, we mention again that the DSF calculation predicts a magnetization drop-off like $(T_c - T)^{0.32}$ in the critical region. This has been confirmed on MnF_2 (Fig. 3) and also by the observed shape of the coexistence curve in condensation.

For near neighbors and $Z = 12$, the Weiss value of the critical point is calculated to be 18% too high. Since the molecular field plot of MnF_2 yields a Curie–Weiss point which is only 5% too high, it is probable that second neighbors give an important contribution to the molecular field for this substance.

We may summarize the present situation as follows: The elementary molecular field calculation seems to locate the critical region to within about ten percent. Outside of 10% of the critical region (and very often only 4 or 5%), the Ornstein–Zernike fluctuation type theory describes adequately the short-range order. Within the critical region, the existent experiments show that the DSF Ising model calculations give a correct description of the analytic character of the critical anomalies.

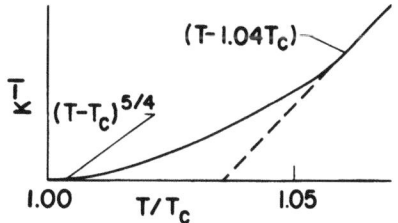

Fig. 8

This indicates that critical anomalous behavior is probably model independent, at least for a wide class of systems.

Further theoretical work is necessary, then, in establishing the conjectured model independence and finally in developing a true analytic solution for the Ising model. Further experiment on light scattering and neutron scattering is essential as well as thermodynamic data in various critical systems. Finally, in quantum systems both theory and experiment are necessary to explore the quenching of fluctuations in the critical region through the collective modes.

Note Added in Proof

In order to get agreement with experiment, we have found it necessary to let the hard sphere overlap the cell edge and take explicit hard-core excluded volume factors into account. Present indications are that a simple theory can be obtained which does account for the experimental data.

References

1. P. Heller and G. B. Benedek, *Phys. Rev. Letters* **8**, 428, 1962.
2. M. Coopersmith and R. Brout, *Phys. Rev.* **130**, 2539, 1963.
3. R. Brout, *Physica* **29**, 1041, 1963; R. Brout, S. Nettel, and H. Thomas, *Phys. Rev. Letters* (submitted).
4. E. Stoner, *Proc. Roy. Soc. (London) Ser. A* **165**, 372, 1938.
5. J. Bardeen, L. Cooper, and J. R. Shrieffer, *Phys. Rev.* **108**, 1175, 1957.
6. P. W. Anderson, *Phys. Rev.* **112**, 1900, 1958.
7. L. Ornstein and F. Zernike, *Proc. Acad. Sci. Amsterdam* **17**, 793, 1914.
8. R. Brout, *Phys. Rev.* **118**, 1009, 1960.
9. F. Englert, *Phys. Rev. Letters* **5**, 102, 1960.
10. M. Fisher (to be published).
11. M. I. Bagatskii, A. V. Voronel, and B. G. Gusak, *Zh. Eksperim. i Teor. Fiz.* **43**, 728, 1962. (*Soviet Phys. JETP* (English Transl.) **16**, 517, 1963.)
12. A. W. Habgood and W. G. Schneider, *Can. J. Chem.* **32**, 98, 1954.

λ-POINT TRANSITIONS*

C. Domb

Wheatstone Laboratory, King's College
London, England

Introduction

I propose to take a broad-minded interpretation of the term "λ-point transition" and to assume that it includes all transitions involving no discontinuity in entropy but only in its derivatives. When such transitions were first observed experimentally, Ehrenfest[1] introduced his well-known classification scheme in which second order transitions were associated with a discontinuity in the second derivative of the Gibbs free energy (and hence first derivative of entropy), third order transitions were associated with a discontinuity in the third derivative of free energy, and so on. Gradually, as more experimental and theoretical results were assembled, it became clear that this classification was too narrow, and that a large variety of transition points are theoretically possible. Figure 1 (taken from Pippard's *Classical Thermodynamics*) illustrates a generalization of Ehrenfest's scheme which is needed to take account of these additional transitions. For example, the standard Ehrenfest transition is represented by Class 2; for Class 2a $\partial c_p/\partial T$ becomes infinite on the low temperature side; in 2b it becomes infinite as the transition is approached

* Research supported (in part) by the United States Army through its European Research Office.

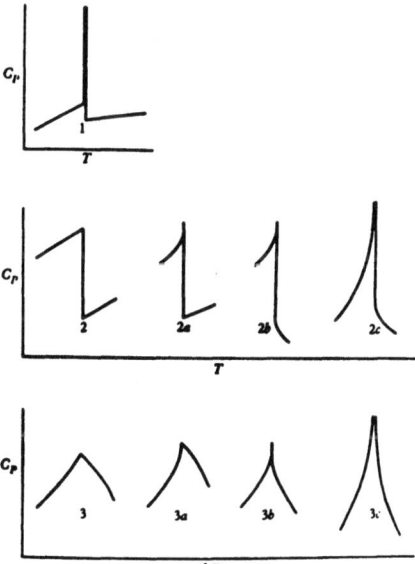

Fig. 1. Extended classification scheme for λ-point transitions (after Pippard).

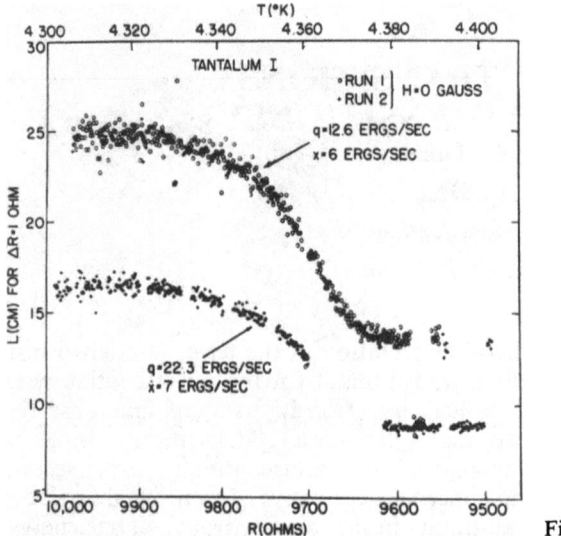

Fig. 2

from both sides; in Class 2c the discontinuity itself becomes infinite. Similar variants of third order transitions are illustrated.

The experimental identification of the precise shape of a specific heat curve associated with a λ-point transition is difficult since very precise measurements must be made in the region close to the transition. However, there is strong evidence that the superconducting transition corresponds to a standard Ehrenfest second order transition[2,3] (Fig. 2); some ferroelectric transitions also seem to be of this kind (Fig. 3). The careful investigations of Buckingham, Kellers, and Fairbank[3] (Fig. 4) seem to show clearly that the transition in liquid helium is proportional to $\ln |T - T_c|$ on both sides of the transition temperature so that this transition belongs to Class 3c. A similar result was obtained by Robinson and Friedberg for

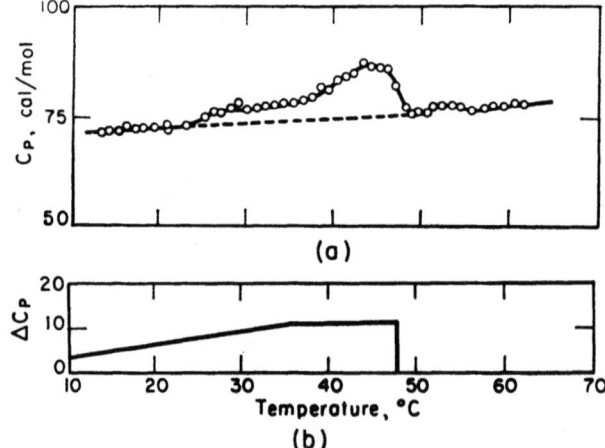

Fig. 3. Specific heat of triglycine sulfate as a function of temperature. (a) Experimental. (b) Theoretical (according to Hoshino).

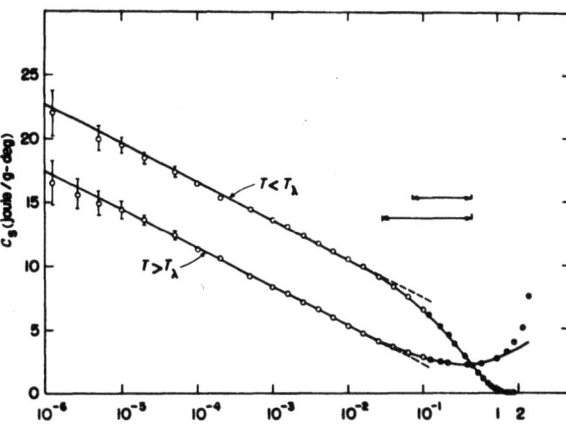

Fig. 4. The specific heat of liquid heat of helium versus $\log|T - T_\lambda|$. $\bigcirc$ represents data of Kellers, Fairbank, and Buckingham. $\oplus$ represents, above 1.5°K, data of Hill and Lounasmaa and Lounasmaa and Kojo. $\oplus$ represents, below 1.5°K, data of Kramers, Wasscher, and Gorter. Solid line represents empirical equations (1) and (2). For aid in visualization, the straight lines of equation (1) have been extended as dotted lines. ($\leftrightarrow$) indicate equivalent temperature range of the measurements on $NiCl_2 \cdot 6H_2O$. Upper arrow refers to $T < T_N$, lower arrow to $T > T_N$.

the antiferromagnetic transitions in $NiCl_2 \cdot 6H_2O$ and $CoCl_2 \cdot 6H_2O$. Other identifications, e.g., of ferromagnetic Curie points as standard Ehrenfest third order transitions, and of order–disorder transitions in alloys and transitions of ammonium salts with Class 2c, are based largely on old experimental data and need re-examination.

On the theoretical side, statistical mechanical calculations in terms of specific models present great difficulties and several false lines have been pursued. However, in the past few years more reliable information has been forthcoming and a coherent pattern is beginning to emerge. The purpose of the present paper is to describe theoretical features which lead to a clearer understanding of the nature of λ point transitions, and to summarize the results of theoretical calculations which have been undertaken for various specific models.

Basic Theoretical Concepts

As mentioned above, Ehrenfest described a second order transition as a point of second order contact between the Gibbs functions of two phases; this is illustrated in Fig. 5. From this he derived the well-known analogues of the Clausius–Clapeyron equation relating to the p–T curve of the transition to the discontinuity in specific heat and to changes in compressibility and thermal expansion. But his discussion was subject to the serious criticism that the line corresponding to phase 1 (in Fig. 5) would always correspond to lower free energy and would therefore be stable; hence thermodynamic reasoning would lead us to expect that no transition to phase 2 would occur.

The answer to this criticism became clear only after a microscopic description had been suggested for the mechanism of a λ-point transition. In attempting a statistical theory of order–disorder transitions in alloys, Bragg and Williams[5] introduced the very important concept of "long-range order," which they represented by a parameter S. When $S = 0$, lattice sites a great distance apart are uncorrelated in occupancy; when $S \neq 0$ they are correlated. Bragg and Williams calculated the variation of S with temperature and showed that it decreased steadily from unity at $T = 0$ to zero at $T = T_c$, the transition temperature.

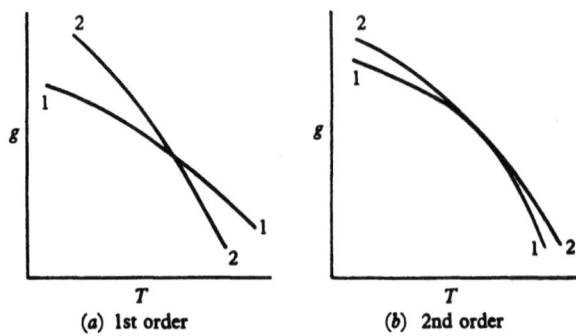

Fig. 5. Suggested variation of the Gibbs function in first and second order transitions.

Although their statistical treatment was primitive for a short-range force model and has subsequently been substantially improved, their basic description of the background to an order–disorder transition has remained valid. Long-range order disappears at the Curie temperature, and there is no conceivable way in which the low-temperature phase can be continued.

An analogous property is associated with every λ-point transition; for example, Gorter suggested the use of a liquid and gas enclosed in a constant volume as an illustration of a standard second order transition, and in this case the transition temperature is associated with the disappearance of the liquid phase; there is again no possibility of continuing the low-temperature phase beyond the transition.

If we wipe out the portion of phase 1 above T_c in Fig. 5, the thermodynamic validity of Ehrenfest's discussion is re-established; there is every reason to think that his relations will be satisfied by any standard second order transition. It is natural to enquire at the same time whether phase 2 can be continued into the low-temperature region; we shall defer discussion of this point until we have a better grasp of the behavior of this phase at T_c for specific models.

It was pointed out by Bethe[6] that the Bragg–Williams theory was deficient in that it failed to take account of short-range correlation in the occupancy of lattice sites. Bethe introduced a second parameter σ to account for short-range order which was unity at $T = 0$ but did not become zero until T reached infinity. The nonzero value of σ above T_c gave rise to a small residual specific heat but was absent in the Bragg–Williams treatment. Both the Bragg–Williams and Bethe accounts led to a standard second order transition at the Curie point.

A major theoretical landmark was the solution by Onsager[7] in 1944 of the two-dimensional Ising model in zero field for the SQ lattice. The Ising model can be used as a representation of ferromagnetism, antiferromagnetism, solid solutions, and order–disorder transitions in alloys, and hence Onsager's solution could be compared with other approximate treatments of these problems. The mathematical apparatus used in deriving the solution was intricate, and there has been a steady series of simplified treatments subsequently; solutions for a variety of other two-dimensional lattices have also become available, and this whole aspect has been adequately reviewed elsewhere.[8] The aim here is to summarize the physical consequences of these solutions and to emphasize the insight they have provided into the nature of the transition. The properties to which I shall refer are common to all two-dimensional lattices.

The specific heat C is logarithmically infinite on both sides of the Curie temperature:

$$C \simeq A \ln |T - T_c| \qquad (1)$$

It has a large "tail," the entropy change above T_c being appreciably greater than that below T_c. The long-range order goes steeply to zero at T_c:

$$S \simeq B(1 - T/T_c)^{\frac{1}{8}} \qquad (2)$$

The closed-form approximations are all incorrect in their detailed description of critical behavior. This is because they fail to deal adequately with the "propagation of order" in the lattice. The degree of correlation between any two lattice sites distance $\mathbf{R}$ apart can usefully be measured by a parameter $\sigma(\mathbf{R})$ analogous to the Bethe short-range order parameter (at temperatures less than T_c, $\sigma(R)$ tends to S as $R \to \infty$). Closed-form approximations represent these correlations as functions of a finite number of parameters; the exact solution shows that an infinite number of parameters is required for adequate representation.

At the Curie temperature T_c, Onsager showed that $\sigma(\mathbf{R})$ fell off asymptotically with distance as $R^{-\frac{1}{4}}$, and for $T > T_c$ the average distance to which order extended was proportional to $1/(T - T_c)$. Fisher made use of the Onsager solution to suggest the general relation:

$$\sigma(\mathbf{R}) \simeq R^{-\frac{1}{4}} \exp - [R(1 - T_c/T)] \qquad (3)$$

It is convenient to define more precisely a "range of order" L by means of the relation:[10]

$$L^2 = \frac{\sum_{\mathbf{R}} R^2 \sigma(\mathbf{R})}{\sum_{\mathbf{R}} \sigma(\mathbf{R})} \qquad (4)$$

and for two dimensional models:

$$L \simeq \frac{F}{1 - T_c/T} \qquad (5)$$

The asymptotic behavior of $\sigma(\mathbf{R})$ is related to the singularity in magnetic susceptibility at T_c which becomes infinite for a ferromagnet:

$$\chi_0 \simeq C(1 - T_c/T)^{-\frac{7}{4}} \qquad (6)$$

and has an infinite slope for an antiferromagnet:

$$\chi_0 \simeq D(1 - T_c/T) \ln (1 - T_c/T) \qquad (7)$$

The exact solution of the Ising model for two-dimensional lattices differs substantially in its physical properties from most experimentally observed λ-point transitions in solids. It therefore became important to determine whether the properties of the solution should be attributed to the form of the Ising interaction, or to the two-dimensional nature of the solution. For this purpose it was necessary to assess the characteristics of the solution for three-dimensional lattices.

Nearest Neighbor Force

The exact methods which led to the solution of the Ising model for two-dimensional lattices are inapplicable in three dimensions. Several different approaches have been used to explore the properties of the three-dimensional model; the method which has so far yielded the most reliable information is that of exact series expansions at high and low temperatures of the thermodynamic properties of interest. A substantial number of terms is needed before such information can be derived, and special techniques have been devised to enable the calculations to be pushed forward. To quote a specific example,[11] the series for the susceptibility of the fcc lattice is given (except for a multiplying factor) by

$$1 + 12w + 132w^2 + 1404w^3 + 14{,}652w^4 + 151{,}116w^5$$
$$+ 1{,}546{,}332w^6 + 15{,}734{,}460w^7 + 159{,}425{,}580w^8 + \ldots (T > T_c) \quad (8)$$

where $w = \tanh(J/kT)$.

The critical behavior of the susceptibility is determined by the asymptotic behavior of the coefficients a_n in this series. It is reasonable to try a relation of the form:

$$\frac{a_n}{a_{n-1}} \simeq \left(\frac{1}{w_c}\right)\left(1 - \frac{g}{n}\right) \quad (9)$$

where w_c corresponds to the Curie temperature and $(g + 1)$ is the susceptibility index $[\chi_0 \sim A(1 - T_c/T)^{-(g_n)}]$.

Figure 6 shows that higher order coefficients are tending well to a relation of type (9). It is particularly instructive to pursue independent calculations for a variety of three-dimensional lattices and then to compare the results obtained. From this the conclusion can reasonably be drawn that the susceptibility index is $\tfrac{5}{4}$ for all such lattices.

For an antiferromagnet in a loosely packed lattice, a much more careful analysis is needed, since the dominant ferromagnetic singularity must first be subtracted. However, such an analysis undertaken by Fisher and Sykes[12] has suggested that formula (7) remains a good approximation in three dimensions.

When the terms of the series corresponding to a thermodynamic property are not consistent in sign, estimation of critical properties is more difficult, but the introduction by Baker[13] of the Padé approximant represented a major advance.

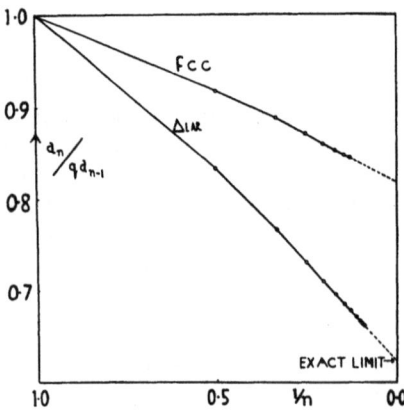

Fig. 6. Ising model. Successive ratios in the susceptibility expansions of the triangular and fcc lattices as functions of $1/n$.

In this way, the conclusion has been drawn that for three-dimensional lattices the long range order goes to zero as:[14]

$$S \simeq B'\left(1 - \frac{T}{T_c}\right)^{\frac{5}{16}} \tag{10}$$

and this conclusion was substantiated by a study of the diamond lattice,[15] for which all terms of the corresponding series are positive.

Table I summarizes the final estimates (based on high-temperature series expansions) of critical parameters of the Ising model of spin $\frac{1}{2}$ for a variety of three-dimensional lattices. The exact results for two-dimensional lattices are included for comparison purposes.

In order to assess the difference between the specific heat curves for the various lattices, it is convenient to take T_c as the unit of temperature, and consider the specific heat C as a function of $\tau(= T/T_c)$. Then:

$$S_c = \int_0^1 \frac{C}{\tau} d\tau$$

$$S_\infty - S_c = \int_1^\infty \frac{C}{\tau} d\tau \tag{11}$$

so that a tabulation of S_c and $S_\infty - S_c$ for various lattices permits comparison of the magnitude of the specific heat curves below and above the Curie temperature. The sum of the two terms S_∞ is the same for all lattices and is equal to $k \ln 2 (= 0.693\,k)$. The quantities in equation (11) are also particularly useful for comparison with experimental results, and hence for testing the validity of a given model, since they do not depend on the magnitude of the interaction energy J. It is similarly useful to tabulate:

$$\frac{E_c - E_0}{kT_c} = \frac{1}{k}\int_0^1 C\,d\tau \qquad -\frac{E_c}{kT_c} = \frac{1}{k}\int_1^\infty C\,d\tau \tag{12}$$

which directly represent the areas under the specific heat curve below and above the Curie temperature. The sum of these two terms, $-E_0/kT_c$, is no longer constant, but decreases from ∞ to a limiting value of $\frac{1}{2}$ as the coordination number of the lattice q tends to infinity.

Table I. Critical Values for Ising Model ($s = \frac{1}{2}$)

Lattice structure	q	$\dfrac{kT_c}{qJ}$	$\dfrac{S_c}{k}$	$\dfrac{S_\infty - S_c}{k}$	$\dfrac{E_c - E_0}{kT_c}$	$\dfrac{-E_c}{kT_c}$
Honeycomb	3	0.506	0.265	0.428	0.227	0.761
Simple quadratic	4	0.567	0.306	0.387	0.258	0.623
Triangular	6	0.607	0.330	0.363	0.275	0.549
Diamond	4	0.676	0.511	0.182	0.418	0.322
Simple cubic	6	0.752	0.560	0.133	0.447	0.218
Body-centered cubic	8	0.794	0.586	0.107	0.460	0.169
Face-centered cubic	12	0.816	0.591	0.102	0.463	0.150

It will be seen that there is a major difference between the results for two and three-dimensional lattices. The "tail" of the specific heat curve is much smaller for the latter; we can see this more clearly in Fig. 7, which is an estimate of the specific heat for the fcc lattice based on series expansions.

The detailed behavior of the specific heat C near T_c has not yet been properly established, since the convergence of the series expansions is slow and many additional terms are needed. It seems clear that C will tend to infinity on both sides of T_c; its low-temperature behavior is probably logarithmic,[15,16] and its high-temperature side is sharper than for two-dimensional lattices,[17] and may be of the form $(1 - T_c/T)^{-1/n}$, with $n \geqslant 5$.

Dr. Miedma and I recently investigated the experimentally determined properties of a number of low temperature magnetic transitions to see if they fitted reasonably with the predictions of simple models. For comparison with the three-dimensional Ising model, we looked at cobalt tutton salts, all of which exhibit strong magnetic anisotropy and have spin $\frac{1}{2}$. For the susceptibility index, we found an estimated 1.27 ± 0.05 for the $CoNH_4$ salt, and 1.22 ± 0.03 for the CoK salt. The entropies corresponding to loss of short-range order above the Curie temperature $(S_\infty - S_c)/k$ for the CoK, $CoNH_4$, CoRb, and CoCs salts were 0.105, 0.13,

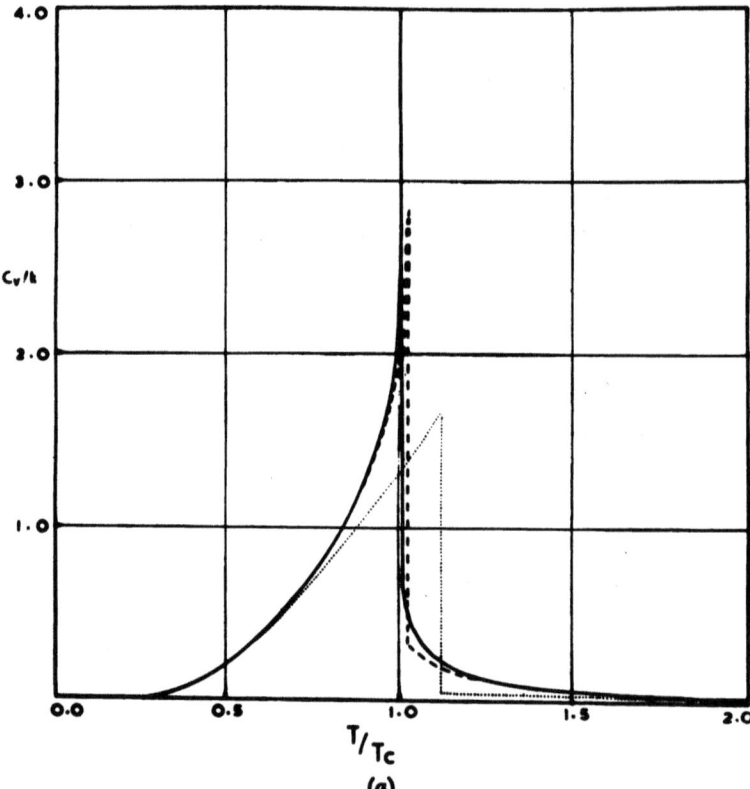

Fig. 7a. Comparison of specific heat curves given by various approximations. Face-centered cubic lattice. ——— based on series expansions. ----- Kikuchi approximation. ...... first order Bethe approximation.

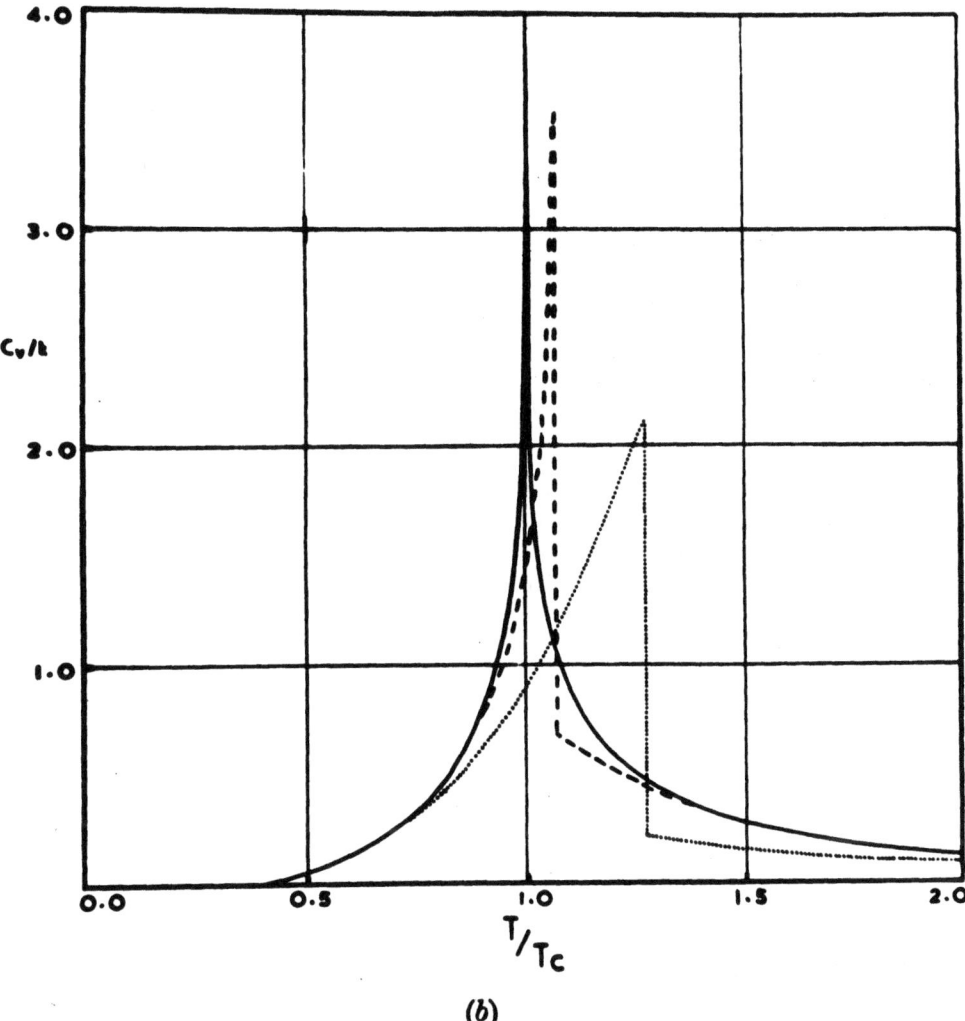

Fig. 7b. Comparison of specific heat curves given by various approximations. Simple quadratic lattice (after Onsager). ——— exact. - - - - - Kramers and Wannier variation (≡ Kikuchi). ...... Bethe.

0.09, and 0.14, respectively. This certainly represents reasonable agreement with the predictions of the model.

The short-range order for temperatures above T_c has recently been investigated by Fisher and Burford,[10] who found that the range of order L is given by:

$$L \simeq F'(1 - T_c/T)^{-\nu} \qquad (\nu \simeq 0.645) \qquad (13)$$

The more detailed statistical picture represented by the function $\sigma(\mathbf{R})$ is:

$$\sigma(\mathbf{R}) \simeq \frac{E'}{R^{1+\eta}} \exp - [R(1 - T_c/T)^\nu] \qquad (\eta \simeq \tfrac{1}{16}) \qquad (14)$$

The most immediate generalization of this model is the extension to spin s. We consider the general Hamiltonian:

$$\mathcal{H} = -\frac{J}{s^2}\sum_{i,j} S_{\zeta i}S_{\zeta j} - \frac{mH}{s}\sum_i S_{\zeta i} \qquad (15)$$

the summation being taken over all nearest neighbor pairs i, j in the lattice. The constant factors in equation (15) have been chosen so that as s varies the maximum interaction between neighboring spins remains equal to J and the maximum magnetic moment remains equal to m; this is useful for comparing the thermodynamic and magnetic properties of the model for different values of s, and particularly for considering the limiting case $s \to \infty$.

High temperature expansions can be derived without too much additional difficulty, and the critical behavior of magnetic susceptibility remains unchanged.[19] The total entropy change in the transition is now increased from $k\ln 2$ to $k\ln(2s+1)$; however, nearly all of this increase in entropy takes place in the region below the Curie temperature. Even when $s \to \infty$, the increase in entropy above the Curie temperature is not more than 30%. This is shown clearly in Table II, where critical parameters are presented for different values of s for the fcc lattice.

The variation of specific heat above T_c with change of s can best be assessed from the following expansions of C for the fcc lattice:

$$\begin{aligned}
C/k &= 0.06257t'^2(1 + 0.8170t' + 0.6779t'^2 \\
&\quad + 0.5680t'^3 + 0.4819t'^4 + 0.4191t'^5 \\
&\quad + 0.3727t'^6 + 0.3365t'^7 + \ldots) \qquad (s = \tfrac{1}{2}) \\
&= 0.05756t'^2(1 + 0.7838t' + 0.7882t'^2 + 0.7052t'^3 \\
&\quad + 0.6268t'^4 + 0.5625t'^5 + 0.5091t'^6 \ldots) \qquad (s = 1) \quad (t' = T_c/T) \\
&= 0.05456t'^2(1 + 0.7629t' + 0.8414t'^2 + 0.7660t'^3 \\
&\quad + 0.7055t'^4 + 0.6487t'^5 + 0.5985t'^6 + \ldots) \qquad (s = \infty)
\end{aligned} \qquad (16)$$

It will be seen that as s increases, there is an increase in sharpness of C as $T \to T_{c+}$, although at sufficiently high temperatures the magnitude of C decreases with increasing s.

It is perhaps not surprising that the high temperature thermodynamic properties are insensitive to changes in the form of interaction. These properties can be expressed as averages over the interaction, and are not likely to vary greatly as the form of the interaction changes. By contrast, the behavior at very low temperatures is determined by the excitation spectrum, which is extremely sensitive to the form of interaction. However, when the Curie temperature is approached

Table II. Critical Values for Ising Model with General s (fcc Lattice)

s	$\dfrac{3skT_c}{qJ(s+1)}$	$\dfrac{S_c}{k}$	$\dfrac{S_\infty - S_c}{k}$	$\dfrac{E_c - E_0}{kT_c}$	$\dfrac{-E_c}{kT_c}$
$\tfrac{1}{2}$	0.816	0.591	0.102	0.463	0.150
1	0.851	0.983	0.116	0.721	0.160
2	0.864	1.486	0.123	0.990	0.167
∞	0.874	∞	0.131	1.541	0.175

even from the low temperature side, it seems possible that such sensitive features will have been "ironed out," and hence even on the low temperature side the critical behavior may well remain insensitive to the form of interaction. Detailed calculations of the low temperature behavior of the Ising model for general s would be useful in this connection.

The generalization of the above results to the Heisenberg model is more difficult; the Hamiltonian is now:

$$\mathcal{H} = -\frac{J}{s^2}\sum_{i,j}\mathbf{s}_i \cdot \mathbf{s}_j - \frac{mH}{s}\sum_i s_{\zeta i} \tag{17}$$

and the spin operators do not commute. Hence, the standard method of deriving series expansions at high temperatures involves taking account of all possible permutations of such operators, which increase very rapidly in number. A new method has recently[20] been introduced which avoids this difficulty and offers promise of extending the series expansions which have so far been derived.

The most comprehensive data on high temperature series expansions are those of Rushbrooke and Wood;[21] even though fewer terms are available than for the Ising model and the series behave less smoothly, the broader features of the model can still be assessed on the high temperature side. The susceptibility index changes a little from $\frac{5}{4}$ to $\frac{4}{3}$ so that we now have:[19,22]

$$\chi_0 \simeq C''(1 - T_c/T)^{-\frac{4}{3}} \tag{18}$$

For an antiferromagnet, general arguments have been advanced by Fisher[23] to show that the pattern of behavior is similar to that in the Ising model.

The "tail" of the specific heat curve is nearly three times as large as for the Ising model, as can be seen from Table III, where critical parameters are presented. For comparison with equations (16) we now have:

$$\begin{aligned}
C/k &= 0.2712 t'^2 (1 + 0.7364 t' + 0.07533 t'^2 - 0.08629 t'^3 \\
&\quad + 0.08511 t'^4 + 0.18792 t'^5 + \ldots) \quad (s = \tfrac{1}{2}) \\
&= 0.2240 t'^2 (1 + 0.8088 t' + 0.5133 t'^2 + 0.3616 t'^3 \\
&\quad + 0.2810 t'^4 + 0.2325 t'^5 + \ldots) \quad (s = 1) \quad (t' = T_c/T) \\
&= 0.1964 t'^2 (1 + 0.8357 t' + 0.7006 t'^2 + 0.5836 t'^3 \\
&\quad + 0.4866 t'^4 + 0.4130 t'^5 + \ldots) \quad (s \to \infty)
\end{aligned} \tag{19}$$

and the general pattern of behavior for changing s is similar to that for the Ising

Table III. Critical Values for Heisenberg Model with General s (fcc Lattice)

s	$\dfrac{3skT_c}{qJ(s+1)}$	$\dfrac{S_c}{k}$	$\dfrac{S_\infty - S_c}{k}$	$\dfrac{E_c - E_0}{kT_c}$	$\dfrac{-E_c}{kT_c}$
$\tfrac{1}{2}$	0.679	0.473	0.220	0.297	0.439
1	0.747	0.810	0.289	0.555	0.449
2	0.774	1.305	0.304	0.833	0.459
∞	0.798	∞	0.322	1.406	0.474

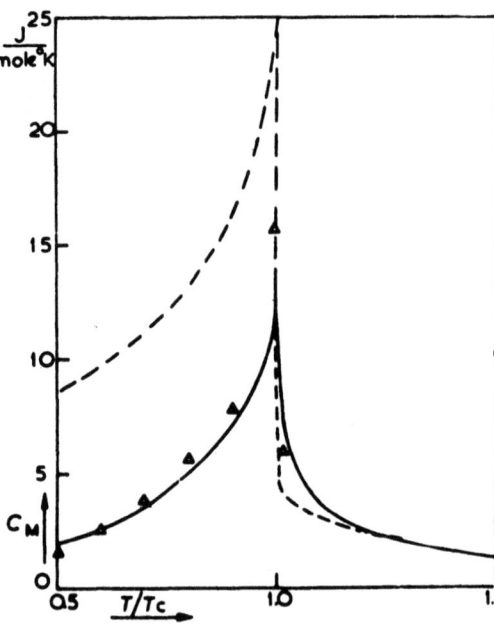

Fig. 8. Specific heat anomalies of three ferromagnets. The curves for nickel ($s = \frac{1}{2}$), CuK$_2$Cl$_4 \cdot$2H$_2$O ($s = \frac{1}{2}$), and GdCl$_3$ ($s = \frac{7}{2}$) are plotted vs. T/T_c. △ nickel, corrected for 60% spins. ——— CuK$_2$Cl$_4 \cdot$2H$_2$O, Cu(NH$_4$)$_2$Cl$_4 \cdot$2H$_2$O. ---- GdCl$_3$.

model. For most values of s it again seems as if the specific heat becomes infinite as $T \to T_{c+}$, although the behavior for $s = \frac{1}{2}$ merits closer investigation.

The low temperature side of the Curie point is much more difficult to deal with because it involves taking account of the statistics of interacting spin waves.

There are a number of salts whose properties seem to correspond to those of the Heisenberg model. The most promising examined so far[18] are CuK$_2$Cl$_4 \cdot$2H$_2$O ($s = \frac{1}{2}$), Cu(NH$_4$)$_2$Cl$_4 \cdot$2H$_2$O ($s = \frac{1}{2}$), and GdCl$_3$ ($s = \frac{7}{2}$). Estimates of the susceptibility indices of these salts are 1.36, 1.37, and 1.33, respectively, and these values are quite close to $\frac{4}{3}$. The critical values of $(S_\infty - S_c)/k$ are 0.22, 0.22, and 0.29; these are in fair agreement with the values given in Table III and demonstrate how the change from $s = \frac{1}{2}$ to $s = \frac{7}{2}$ produces only a small change in entropy in the region $T \geqslant T_c$. This is illustrated in Fig. 8, where the specific heats of two of these salts are plotted on the same scale. However, there is a difference between the theoretical and experimental results quoted for GdCl$_3$, since the experimenters suggest[24] that the high temperature value remains finite. This would seem to merit a closer experimental investigation in the neighborhood of T_c.

We may sum up the results of this section by suggesting that when only nearest neighbor interactions are taken into account the physical properties are not very sensitive to the form of interaction, and the specific heat seems to be infinite at the Curie temperature for all the models investigated.

Above the Curie temperature, the magnetic susceptibility exhibits a marked curvature as $T \to T_{c+}$ for a ferromagnet; it passes through a maximum above the Néel temperature and has an infinite slope at the Néel temperature for an antiferromagnet.

Longer-Range Forces

The precision of our knowledge of the effect of longer-range forces has increased appreciably during the past few years. The idea had often been expressed intuitively that if the number of neighbors participating equally in the interaction tended to infinity, the mean field or Bragg–Williams approximation would result exactly. This has now been established rigorously[25] by taking an interaction of the form $\gamma J e^{-\gamma R}$ and allowing γ to tend to zero; the result is also valid for a one-dimensional model.

A solution of Bethe type including a small residual specific heat above T_c results if the force is long range in one direction and short range in the other.[26] It is significant to observe this physical interpretation of the Bethe result, which had previously been regarded as a first approximation to the solution for a nearest neighbor model, or the exact solution for an unphysical "pseudolattice" containing no closed circuits.[27]

It is interesting to see how the mean field solution is approached as the range of force increases. For this purpose, we have found it convenient to introduce an "equivalent neighbor (r-shell)" model in which all interactions are equal up to the r-th neighbor shell, and zero outside. The series expansion methods described in the previous section can be applied to this model,[28] and we have been able to take account of the second and third neighbor shells for a number of two- and three-dimensional lattices. We find that in a given dimension the solution rapidly becomes independent of the detailed lattice structure, and depends only on the number of interacting neighbors q. As might be expected, the residual entropy for $T > T_c$ tends to zero as $q \to \infty$, detailed asymptotic formulas for various models being as follows:

$$\frac{S_\infty - S_c}{k} \simeq \frac{3.0}{q} \quad \text{(two-dimensional Ising)}$$

$$\frac{1.25}{q} \quad \text{(three-dimensional Ising)} \quad (20)$$

$$\frac{3.7}{q} \quad \text{(three-dimensional Heisenberg)}$$

But for any equivalent neighbor model of finite order, the critical behavior (e.g., susceptibility index, specific heat singularity) is identical with that for a nearest neighbor model. It is only in the limit when the range of interaction becomes infinite that the inverse susceptibility becomes linear at the Curie point and the specific heat becomes finite and discontinuous. This limit is therefore approached in a singular manner, as illustrated in Fig. 9.

In the light of the above discussion it is instructive to consider the effect of a $1/R^n$ force for varying n; one of my research students, G. S. Joyce, has recently applied to this problem a method suggested by Hiley and myself[29] (based on a development originally introduced by Yvon[30]). He has been able to show, for example, that a force of $1/R^3$ in two dimensions completely changes the susceptibility index, which drops from the value of $\frac{7}{4}$ for a nearest neighbor model to about 1.13. He is in the process of investigating the critical behavior of the specific heat.

Of more direct physical application is the problem of dipolar forces, which are particularly difficult to deal with because they give rise to shape-dependent effects. Hiley and Joyce[31] have shown that the series for the inverse susceptibility

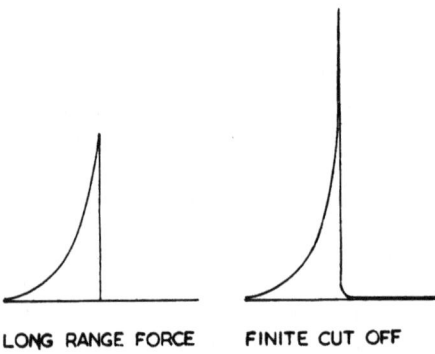

Fig. 9. Specific heat of long-range force model illustrating the difference between a finite cut-off and an infinite range.

LONG RANGE FORCE   FINITE CUT OFF

at high temperatures can be cast in such a form that only the zero order term is shape dependent and all higher order terms are shape independent. The corresponding specific heat expansion is completely shape independent. We thus have a picture in which the specific heat is represented by a continuous function, and the point corresponding to the transition is determined by the shape-dependent singularity in the susceptibility. This would seem to be consistent with the properties of a standard second order transition.

Conclusions

I shall first return to the question of whether any supercooling of the high-temperature portion is possible in a λ-point transition. For the models discussed, the transition point corresponds to the range of order becoming infinite, and there is no possibility of a continuous extension of the high temperature "phase" below T_c. It is possible for "frozen-in" order to be maintained below T_c, but this represents a non-equilibrium state analogous to a glass rather than to a supercooled liquid. If thermal equilibrium is maintained, long-range order must take over when the short-range order reaches infinity.

Pippard[32] has suggested that the absence of a supercooled phase is characteristic of all λ-point transitions, that for a first order transition the spectrum of fluctuations from equilibrium is such that it is possible to define fairly unambiguously two phases with different entropies, and that surface effects cause the supercooled phase to be locally metastable; such conditions are absent in a λ-point transition. It is clear that supercooling is possible for the Gorter model of a liquid and vapor enclosed in a constant volume, which gives a standard Ehrenfest second order transition; but Pippard argues that this is the one aspect of the model which is not truly representative.

A more detailed investigation of the superconducting transition would be of interest in this connection to find if any property of the normal phase becomes singular at the transition point. Supercooling has been experimentally possible only in the presence of a magnetic field when the transition becomes first order. However, the concept of a superconducting and normal "phase" below the transition temperature has been useful, and one must differentiate between a phase which does not exist and one which is difficult to realize in practice (such as the superheated phase of a solid above its melting point[33]). For a particular limiting model of superconductivity, Thouless[34] has shown that no metastable state is possible below T_c.

Since the discussion began with the thermodynamic treatment of Ehrenfest, it is of interest to examine the more comprehensive thermodynamic theory of Landau[35] in the light of recent theoretical developments. Landau first made the important observation that a λ-point transition is associated with a change in symmetry; a parameter η can be chosen which would be zero in the more symmetric phase and nonzero in the less symmetric phase (for the models we have considered, η can be identified with long-range order). He then proceeded to expand the free energy as a function of p, T, as follows:

$$G = G_0 + \alpha\eta + A\eta^2 + B\eta^3 + C\eta^4 + \ldots \qquad (21)$$

where the coefficients are functions of p, T. From considerations of symmetry he deduced that α must be identically zero. Thermodynamic reasoning then led to the conclusion $A > 0$ for the symmetric phase (usually the high temperature phase) and $A < 0$ for the unsymmetric phase. At temperature T_c, $A(p, T)$ vanishes; hence B must also vanish and $C > 0$. The theory leads to a critical dependence of η of the form $(T_c - T)^{\frac{1}{2}}$, and to a simple discontinuity in specific heat.

Landau and Lifshitz observe that for the Onsager solution this treatment is clearly incorrect, but have suggested that it might be valid for a three-dimensional model. In fact, the detailed evidence which we have collected for the three-dimensional short-range force model shows that it is equally invalid here; it may possibly be valid for a standard second order Ehrenfest transition.[36] We may note, incidentally, that this theory also leads to the conclusion that no supercooled phase is possible, since the value $\eta = 0$ below T_c corresponds to a maximum of G.

From the thermodynamic properties of short-range force models, it is clear that the transition point corresponds to singular behavior and is not a good starting point for series expansions of the type (21). However, the concept of going beyond the range of equilibrium and looking for the dependence of the thermodynamic functions on the degree of order is very enlightening and has not been sufficiently explored. For the Ising model, it is closely related to investigating the critical behavior in the presence of a magnetic field; this is more difficult than the critical properties so far discussed, but extensive series expansions are available, and it should be possible to make effective use of them to draw some reliable conclusions.

In summing up we may say that for lattice models an infinite specific heat seems to be associated with short-range forces, whereas a discontinuity arises from long-range forces. This conclusion is put forward only tentatively, and more refined theoretical and experimental information in the immediate neighborhood of the critical point is therefore quite desirable. It would be of particular help if the latter could be correlated with knowledge of the forces giving rise to the transition.

References

1. P. Ehrenfest, *Commun. Kamerlingh Onnes Lab.*, University of Leiden, Supplement 75b, 1933; *Proc. Acad. Sci. Amsterdam* **36**, 153, 1933.
2. J. F. Cochran, *Ann. Phys.* **19**, 186, 1962.
3. See M. J. Buckingham and W. F. Fairbank, in: *Progress in Low-Temperature Physics*, Vol. III, North Holland Publishing Co., Amsterdam (1961), p. 109.
4. W. K. Robinson and S. A. Friedberg, *Phys. Rev.* **117**, 402, 1960; *Phys. Rev. Letters* **13**, 133, 1964.
5. W. L. Bragg and E. J. Williams, *Proc. Roy. Soc. (London) Ser. A* **145**, 699, 1934.
6. H. A. Bethe, *Proc. Roy. Soc. (London) Ser. A* **150**, 552, 1935.
7. L. Onsager, *Phys. Rev.* **65**, 117, 1944.
8. C. Domb, *Advan. Phys.* **9**, 149, 1960; M. E. Fisher, *J. Math. Phys.* **4**, 278, 1963; E. W. Montroll, "Lattice Statistics" to appear in *Applied Combinatorial Statistics* by E. F. Beckenbath.

9. M. E. Fisher, *Physica* **25**, 521, 1959.
10. M. E. Fisher and R. J. Burford, Aachen Conference on Statistical Mechanics (1964). (To be published.)
11. C. Domb and M. F. Sykes, *J. Math. Phys.* **2**, 63, 1961.
12. M. E. Fisher and M. F. Sykes, *Physica* **28**, 939.
13. G. A. Baker, Jr., *Phys. Rev.* **124**, 768, 1961.
14. J. W. Essam and M. E. Fisher, *J. Chem. Phys.* **38**, 802, 1963.
15. J. W. Essam and M. F. Sykes, *Physica* **29**, 378, 1963.
16. G. A. Baker, Jr., *Phys. Rev.* **129**, 99, 1963.
17. C. Domb and M. F. Sykes, *Phys. Rev.* **108**, 1415, 1957.
18. C. Domb and A. R. Miedma, "Magnetic Transitions," *Progress in Low-Temperature Physics, Vol. IV* (1964), p. 296.
19. C. Domb and M. F. Sykes, *Phys. Rev.* **128**, 168, 1962.
20. C. Domb and D. W. Wood, *Phys. Letters* **8**, 20, 1964; *Proc. Phys. Soc.* July 1965; G. A. Baker, Jr., and G. S. Rushbrooke, *Phys. Rev.* (in press).
21. G. S. Rushbrooke and P. J. Wood, *Mol. Phys.* **1**, 257, 1958.
22. W. Marshall, J. Gammel, and L. Morgan, *Proc. Roy. Soc. (London) Ser. A* **275**, 257, 1963.
23. M. E. Fisher, *Phil. Mag.* **7**, 1731, 1962.
24. M. J. M. Leask, W. P. Wolf, and A. F. G. Wyatt, *Proceedings of the Eighth International Conference on Low-Temperature Physics*, Butterworths, London (1963), p. 230.
25. M. Kac, *Phys. Fluids* **2**, 8, 1959; A. J. F. Siegert, "On the Ising Model with Long Range Interaction," Technical Report No. 7, Northwestern University; G. A. Baker, Jr., *Phys. Rev.* **122**, 1477, 1961; **126**, 2072, 1962.
26. G. A. Baker, Jr., *Phys. Rev.* **130**, 1406, 1963; M. Kac and E. Helfand, *J. Math. Phys.* **4**, 1078, 1963.
27. C. Domb, *Advan. Phys.* **9**, 284, 1960, Fig. 13(a).
28. N. W. Dalton and C. Domb (to be published).
29. C. Domb and B. J. Hiley, *Proc. Roy. Soc. (London) Ser. A* **268**, 506, 1962.
30. J. Yvon, *Cahiers Phys.*, No. 28, 1945; No. 31, 32, 1948.
31. B. J. Hiley and G. S. Joyce, *Proc. Phys. Soc.* **85**, 493, 1965.
32. B. Pippard, *Classical Thermodynamics*, Cambridge (1957), p. 152–159.
33. P. W. Bridgman, *The Physics of High Pressure*, G. Bell (1949), p. 210.
34. D. J. Thouless, *Phys. Rev.* **117**, 1256, 1960.
35. L. D. Landau and E. M. Lifshitz, *Statistical Physics*, Pergamon Press, Ltd., London (1958), p. 434.
36. An alternative approach using similar basic concepts to Landau has been developed by Tisza, *Ann. Phys.* **13**, 1, 1961.

AN INVESTIGATION OF THE NATURE OF THE SPECIFIC HEAT IN THE NEIGHBORHOOD OF THE CRITICAL POINT OF ^{4}He

M. R. Moldover and W. A. Little*

Physics Department, Stanford University
Stanford, California

Introduction

Recently, Bagatskii, Voronel', and Gusak[1] showed that the specific heat at constant volume of argon exhibited a logarithmic singularity at the critical temperature, T_c, for measurements taken at the critical density. This singular behavior is in sharp contrast to the predictions of the accepted theory of phase transition of Landau and Lifshitz.[2] However, the behavior is precisely that to be expected for the so called "lattice gas" model for the liquid–gas transition. Lee and Yang[3] have shown that for a classical gas of particles moving on a discrete lattice with a repulsive force preventing double occupancy of any site and a near neighbor attraction, the partition function can be mapped precisely onto that of an Ising model in an external magnetic field. The specific heat for this Ising model in zero field exhibits a logarithmic singularity at the Curie point. The critical point of the lattice gas corresponds to the Curie point of the corresponding Ising model. The measurements on argon then indicate that for a real gas the specific heat behaves in a similar manner to that of a lattice gas. We have investigated this point further by studying the specific heat at constant volume, C_v, of ^{4}He at densities close to the critical density. We have done this for two main reasons: First, to see whether the behavior observed for argon, which, because of its large mass, behaves essentially classically, is also observed for helium, in which quantum effects should be important; and, second, to investigate the detailed nature of the singularity in the pressure–density plane, not only on the critical density, but also in its immediate neighborhood. Yang[4] has conjectured that quantum effects would reduce the magnitude of the singular contribution to the specific heat. Our results support this view.

Experimental Procedure

The calorimeter was built of two OFHC copper parts. The lower part contained the helium in 50 slots milled 0.3 cm deep and 0.01 cm wide. The large area of these slots facilitated good thermal contact between the helium and the calorimeter. The lower part was wound with a constantan heater and had a carbon resistor cemented and clamped to it. A tight-fitting cap formed the upper part of the calorimeter and was sealed to the lower part with an indium O-ring. The calorimeter

---

* Work supported in part by the National Science Foundation and the Office of Naval Research.

was supported on nylon threads in an evacuated chamber. Thermal contact could be made to the bath with a mechanical heat switch. A 6-mil-ID stainless steel capillary 5 in. in length led from the calorimeter to a needle valve. The dead volume was about $\frac{1}{2}$% of the total volume of the calorimeter. Helium was drawn from a storage dewar and admitted to the calorimeter via a Toeppler pump which was used to measure the volume of the gas to an accuracy of about 0.2%. The calorimeter was filled with a measured amount of gas by condensing the helium in, while the temperature was held below the λ-point with the heat switch closed. The capillary was then sealed with the needle valve and the heat switch opened. Thus, it was possible to change the quantity of helium in the calorimeter without warming it.

A $\frac{1}{10}$-W ohmite carbon resistor of nominal resistance 560 Ω was used as a secondary thermometer. Its resistance was measured with a 100-cps-AC bridge with a phase sensitive detector. Temperature changes of 1×10^{-6}°K could be observed at 5.2°K. At the conclusion of a run, exchange gas was admitted to the vacuum space, permitting calibration of the resistor against the bath pressure. This was done as the bath pressure was reduced from about 1670 to about 760 mm mercury. After the hydrostatic head corrections had been made, the three-constant Clement–Quinnell[5] resistance temperature relation could be fitted over the temperature range 4.2°–5.15°K with all residuals less than 8×10^{-4}°K, or a four-constant formula could be used with all residuals less than 3×10^{-4}°K.

On four separate runs near the critical density, the "singularity" in the specific heat vs. temperature curve occurred at 5.189 ± 0.001°K (from extrapolating the resistance–temperature relation). Although this value differs from the value assigned to the critical temperature by the T_{58} temperature scale ($T_c = 5.1994$ in T_{58}), it was called the critical temperature for further data analysis.

Results

Figure 1 gives a typical result for measurements of C_v at a density within 0.5% of the critical density for ^{4}He. First, for temperatures below T_c one observes that this exhibits a similar logarithmic behavior to that observed in argon[1] over the region from 0.4° to 5×10^{-4}°K of T_c. It was found impossible to get reliable measurement of C_v at temperatures closer to T_c than this because of the enormously long thermal relaxation times close to T_c. Relaxation times of the order of many minutes were observed in this region and became progressively longer the closer one came to T_c. Second, for temperatures above T_c, the specific heat within 5×10^{-3}°K of T_c appears to follow a similar logarithmic behavior to that below T_c. However, the curve for $T > T_c$ deviates from logarithmic behavior closer to T_c than the curve for $T < T_c$. This too is similar to the result for argon. The specific heat over the whole range above and below T_c may be expressed as

$$\frac{C_v}{R} = -\alpha \log_e |T - T_c| + \beta$$

where $\alpha = 0.57$, $\beta = 3.0$ for $T < T_c$ and -0.8 for $T > T_c$.

Argon and oxygen[6] can be described by similar expressions and in Table I the appropriate values are given for α.

We see that the coefficient, α, for helium is substantially smaller than that for argon and oxygen, in agreement with Yang's conjecture. We also note that the coefficient 0.57 is very close to that observed for C_s at the λ-point for ^{4}He (0.63)

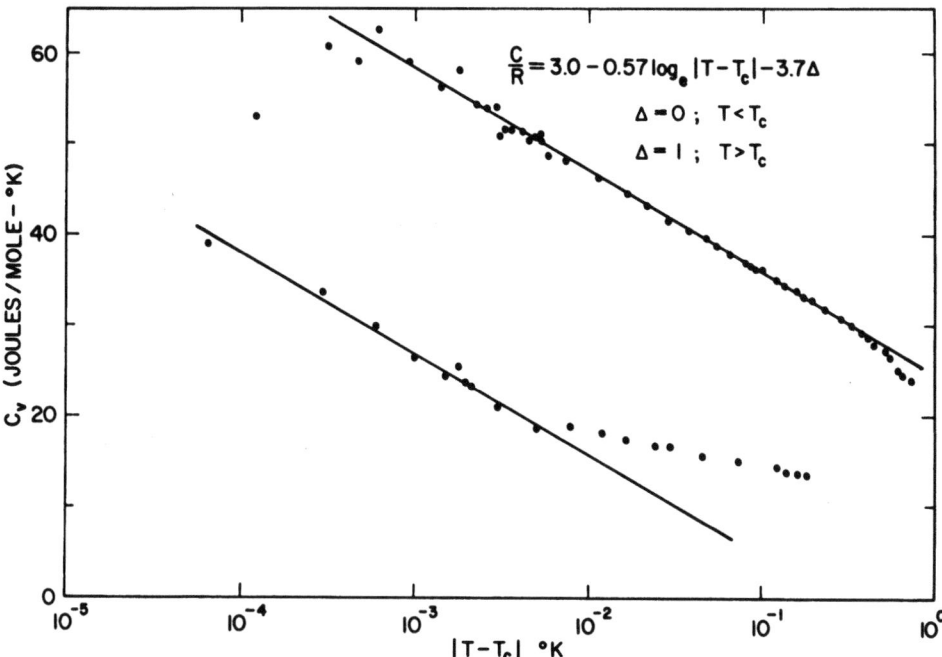

Fig. 1. C_v is plotted vs. $\log|T - T_c|$ for density within 1% of ρ_c. The upper curve is for temperatures below T_c. T_c for this plot is 5.189°K.

by Buckingham and Fairbank[7] and may have particular significance in explaining the quantum transition.

At densities other than the critical density the specific heat vs. temperature curve differs from that described above. As the temperature is raised it increases logarithmically until some temperature below T_c (depending upon the density), where it falls abruptly (within 10^{-4}°K in some instances). The fall is greater at higher densities. The specific heat continues to fall more gradually, at least until 5.4°K at the densities measured. Details of this will be published elsewhere.

In Fig. 2 the temperature at which the precipitous fall in the specific heat occurs is plotted for a number of densities in the immediate neighborhood of the critical density. These points are compared with the coexistence curve extrapolated by Edwards and Woodbury[8] to 5.1994°K, beyond the last two points where they could obtain measurements of the density. If one assumes that the coexistence curve is defined by the precipitous fall in C_v, then from our data it appears to be much flatter than their extrapolation. This lends some support to theoretical models of the phase transition such as the one given by Hemmer, Kac, and Uhlenbeck,[9] who showed on a particular soluble model that exactly such behavior could occur.

Table I

Element	α	β below T_c
Argon	1.8	14
Oxygen	2.4	16
Helium	0.57	3.0

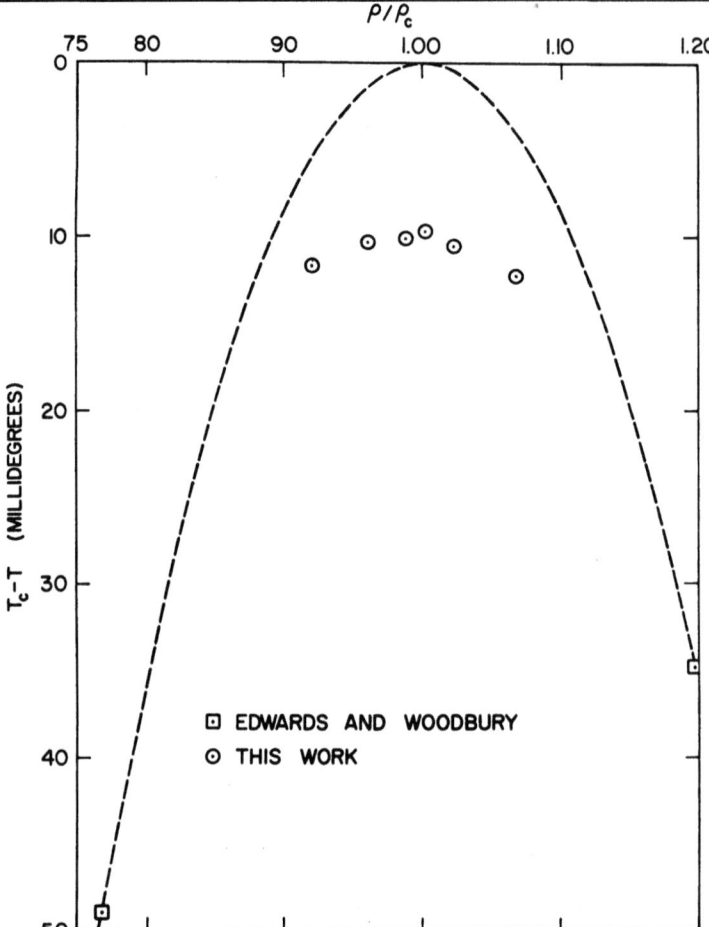

Fig. 2. The temperature at which the break in $C_v(T)$ occurs is plotted against ρ/ρ_c. For this plot $T_c - T = 0$ has been chosen at $T = 5.1994°K$, the value used in the T_{58} temperature scale.

Acknowledgments

We wish to thank Dr. Derek Griffiths, Dr. R. S. Safrata, Dr. C. F. Kellers, and Dr. M. H. Edwards for several useful discussions, and Dr. Derek Griffiths and Donald Moldover for much help in analyzing data. We wish to thank Professor C. N. Yang for his suggestions, interest, and advice.

References

1. M. I. Bagatskii, A. V. Voronel', and V. G. Gusak, *Zh. Eksperim. i Teor. Fiz.* **43**, 728, 1962. (*Soviet Phys. JETP* (English Transl.) **16**, 517, 1963.)
2. L. D. Landau and E. M. Lifshitz, *Statistical Physics*, Addison-Wesley Publishing Co., Inc., Reading, Mass. (1958).
3. T. D. Lee and C. N. Yang, *Phys. Rev.* **87**, 410, 1952.
4. C. N. Yang and C. P. Yang, *Phys. Rev. Letters* **13**, 303, 1964.
5. J. R. Clement and E. H. Quinnell, *Rev. Sci. Instr.* **23**, 213, 1952.
6. A. V. Voronel', Yu R. Chashkin, V. A. Popov, and V. G. Simkin, *Zh. Eksperim. i Teor. Fiz.* **45**, 828, 1963. (*Soviet Phys. JETP* (English Transl.) **18**, 568, 1964.)
7. M. J. Buckingham and W. M. Fairbank, "The Nature of the λ-Transition," *Progress in Low Temperature Physics, Vol III*, North Holland Publishing Co., Amsterdam (1961), p. 86.
8. M. H. Edwards and W. C. Woodbury, *Phys. Rev.* **129**, 1911, 1963.
9. P. C. Hemmer, M. Kac, and G. E. Uhlenbeck, *J. Math. Phys.* **5**, 60, 1964.

ULTRASONIC PROPAGATION NEAR THE CRITICAL POINT IN HELIUM

C. E. Chase

National Magnet Laboratory,* Massachusetts Institute of Technology
Cambridge, Massachusetts

and

R. C. Williamson†

Department of Physics, Massachusetts Institute of Technology
Cambridge, Massachusetts

A logarithmic singularity in the specific heat at constant volume C_v at the critical point has recently been reported to occur in argon[1] and oxygen.[2] Such a result is in direct contradiction to standard theory,[3,4] according to which C_v must remain finite at the critical point. It can readily be shown[5] that these results imply the existence of a corresponding singularity in the adiabatic compressibility κ_s, accompanied by a zero in the sound velocity u and large absorption. Sound velocity measurements might thus serve to confirm the reported behavior of C_v; in addition, they have several other advantages. The available resolution ($\approx 0.01\%$) is much greater than in specific-heat measurements, and each observation is made at a fixed temperature instead of over a temperature interval. If measurements are made along a suitable line in the phase diagram, e.g., an isotherm or an isobar, it is possible to remain in a homogeneous phase at all times, whereas measurements of C_v along an isopycnal are necessarily carried out in the mixed phase below T_c. In the latter circumstances small departures of ρ from the critical value may lead to large contributions from the latent heat close to T_c. We have accordingly undertaken a study of the sound velocity at 1 Mcps in the critical region of helium ($T_c = 5.1994°K$, $P_c = 1718$ mm of mercury). Similar measurements have previously been made in xenon,[6] but with insufficient resolution to detect the presence of a singularity in the compressibility.

Changes in sound velocity were measured by a phase-sensitive method[7] which provided a resolution of about 0.01% with the 0.396-cm path length used in the experiment. The absolute value of the velocity was chosen to agree with earlier measurements along the vapor pressure curve.[8] The ultrasonic chamber was contained in a vacuum-jacketed pressure vessel, the temperature of which was controlled within $\pm 10^{-4}°K$ by an automatic regulator.[9] This vessel was initially filled with liquid helium from the bath through a needle valve; subsequent changes in density were effected by removing gas or adding it from a room-temperature reservoir. The pressure was controlled manually, and was measured with a resolution of ± 0.1 mm of mercury on a Wallace and Tiernan gauge calibrated against

---

* Supported by the U.S. Air Force Office of Scientific Research.
† Supported by the Advanced Research Projects Agency under Contract SD-90.

a mercury manometer. Absolute values are believed to be accurate to within ±1 mm of mercury. All pressures refer to mercury at 0°C, and temperatures are given on the 1958 scale.[10]

The following precautions were taken to promote equilibrium: (1) The pressure vessel was made of copper; (2) the fluid was stirred continuously during the course of the experiments; (3) the velocity was observed during the establishment of equilibrium, and measurements were not made until it was essentially steady (this required from 5 min to about $\frac{1}{2}$ hr close to the critical point); and (4) measurements were made both with increasing and decreasing pressure or temperature as a check of reproducibility. Although at no time did we wait several hours for equilibrium (as was done by Bagatskii et al.[1] and Voronel' et al.[2]), we believe that the above precautions, especially (4), are sufficient to ensure the reliability of the data.

Results of measurements along the isobar $P = 1718 \pm 1$ mm of mercury and the isotherm $T = 5.200 \pm 0.002°K$ are shown in Fig. 1. The observed rapid fall in sound velocity u as the critical point is approached from any direction is in agreement with expectation. Because, as anticipated, the attenuation is high in the critical region, measurements are impossible with the present apparatus over a region of width about $2 \times 10^{-3}°K$ or 2.5 mm of mercury; nevertheless, the drop outside this region is so rapid that the vanishing of u at the critical point does not appear unreasonable.

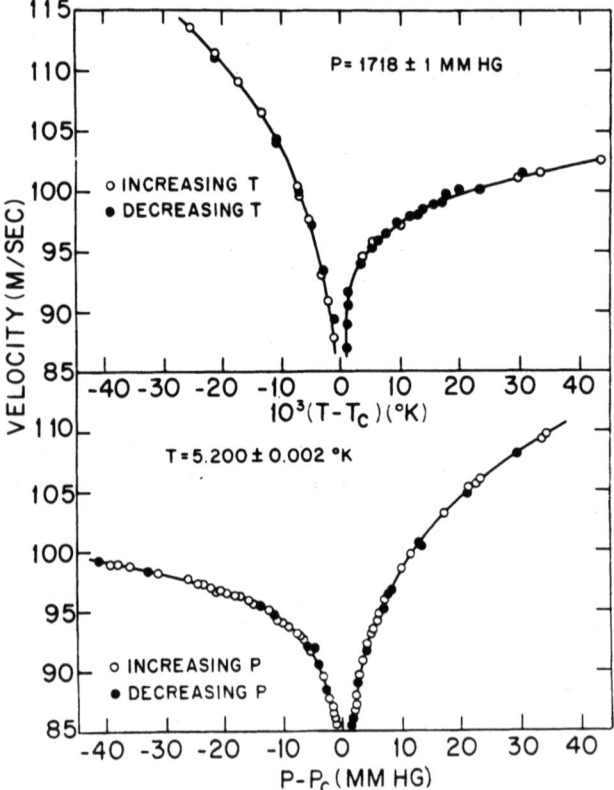

Fig. 1. Sound velocity as a function of temperature (a) and pressure (b) in the critical region.

The behavior of the adiabatic compressibility κ_s is more revealing, but its evaluation from u requires knowledge of the density. We have estimated $\rho(P, T)$ from the expansion given by Edwards and Woodbury,[11]

$$P = -Atv - \tfrac{1}{3}Bv^3 - \tfrac{1}{2}Ctv^2 - Dt^2v + f(t)$$

where $t = T - T_c$, $v = V - V_c$, and A, B, C, and D are given by Edwards and Woodbury.[11] In order to carry out this calculation, we chose $f(t)$ so as to give Edwards and Woodbury's density values along the vapor pressure curve, and for $T > T_c$ we adjusted it (somewhat arbitrarily) so that the maximum in $(\partial V/\partial P)_T$ falls on the extrapolated vapor pressure curve. The errors involved in this procedure are difficult to estimate, but it is unlikely that they can be so large as to change the qualitative nature of the singularity in κ_s.

Figure 2 shows κ_s along the above isobar and isotherm plotted against $\log|T - T_c|$ or $\log|P - P_c|$. The presence of a logarithmic singularity over $1\tfrac{1}{2}$ to 2 decades in the liquidlike phase ($T < T_c$ or $P > P_c$) is clearly indicated. In the gaslike phase the slope is much smaller, but a singularity is still evident. Indeed, along the isotherm the singularity may even be stronger than logarithmic. Although the shape of these curves near the critical point is very sensitive to the precise values chosen for P_c and T_c, no reasonable assignment of these quantities would suggest that κ_s is leveling off to a finite value at the critical point. We therefore conclude that κ_s exhibits just such a singularity at the critical point as would be expected on the basis of the reported behavior of C_v.

We have also measured the velocity along isotherms at 5.221°, 5.296°, 5.393°, and 5.492°K. On the lowest of these, which is about 0.021°K above T_c, u falls to a

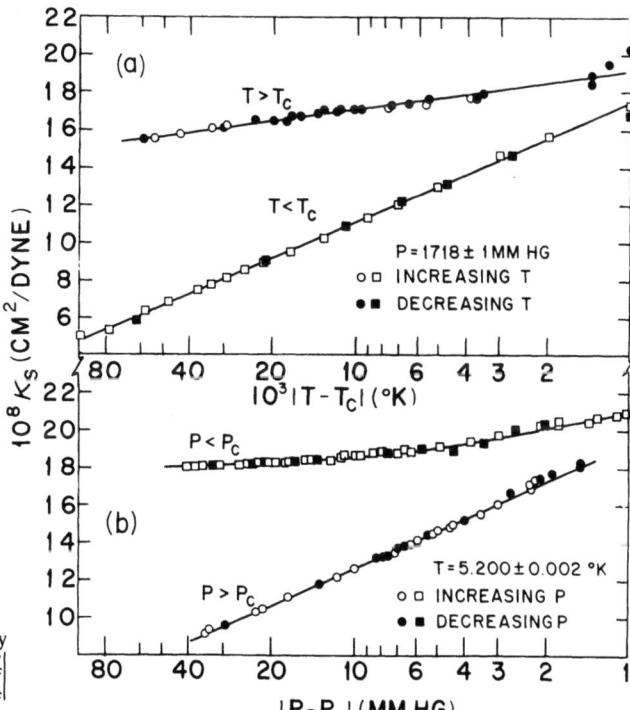

Fig. 2. Adiabatic compressibility (a) as a function of $\log|T - T_c|$ and (b) as a function of $\log|P - P_c|$ in the critical region.

sharp minimum at a pressure of 1744 mm of mercury and the attenuation remains small enough so that the signal can be continuously observed. The other isotherms display clearly rounded minima, which become increasingly broad as the temperature increases, located, respectively, at 1827, 1930, and 2030 mm of mercury. These minima (which imply the presence of corresponding maxima in the compressibility) lie, as expected, very nearly on the extrapolated vapor pressure curve.

Acknowledgment

The authors are grateful to Professor L. Tisza for suggesting this experiment and for enlightening discussions of the consequences of the C_v singularity.

References

1. M. I. Bagatskii, A. V. Voronel', and V. G. Gusak, *Zh. Eksperim. i Teor. Fiz.* **43**, 728, 1962. (*Soviet Phys. JETP (English Transl.)* **16**, 517, 1963.)
2. A. V. Voronel', Yu. R. Chashkin, V. A. Popov, and V. G. Simkin, *Zh. Eksperim. i Teor. Fiz.* **45**, 828, 1963. (*Soviet Phys. JETP (English Transl.)* **18**, 568, 1964.)
3. L. D. Landau and E. M. Lifshitz, *Statistical Physics*, Pergamon Press, Ltd., London (1958), p. 259ff.
4. L. Tisza, *Ann. Phys.* **13**, 1, 1961.
5. L. Tisza, private communication.
6. A. G. Chynoweth and W. G. Schneider, *J. Chem. Phys.* **20**, 1777, 1952.
7. C. E. Chase, *Phys. Fluids* **1**, 193, 1958.
8. A. van Itterbeek and G. Forrez, *Physica* **20**, 133, 1954.
9. C. Blake and C. E. Chase, *Rev. Sci. Instr.* **34**, 984, 1963.
10. H. van Dijk, M. Durieux, J. R. Clement, and J. K. Logan, *Physica* **24**, S129, 1958.
11. M. H. Edwards and W. C. Woodbury, *Phys. Rev.* **129**, 1911, 1963.

7
FOURTH FRITZ LONDON AWARD

PRESENTATION ADDRESS FOR THE FOURTH FRITZ LONDON AWARD

L. D. Roberts
Oak Ridge National Laboratory
Oak Ridge, Tennessee

It is my very pleasant duty this afternoon to make the presentation of the Fourth Fritz London Award to Dr. David Shoenberg, for outstanding accomplishment in the field of low temperature physics. In doing this I am acting on behalf of the Fourth Fritz London Award Committee. The members of this committee are Dean Henry Boorse of Barnard College, also Professor of Physics, Columbia University; Dr. D. W. Osborne of the Argonne National Laboratory; Dr. Paul Marcus of the International Business Machines Corp.; Dr. James Nicol of Cryonetics Corporation; with myself as chairman. Furthermore, I am acting on the behalf of this Ninth International Conference on Low Temperature Physics, Batelle Memorial Institute, and the Ohio State University. I am also acting on behalf of the Arthur D. Little Foundation which, through its outstanding generosity and public spirit, has sponsored the three previous awards and is also the sponsor of this Fourth Fritz London Award.

Professor London was known, and is remembered by his friends and associates, not only as a most distinguished scientist but also as an individual of human warmth and great sincerity. His death was felt keenly by his many friends and this affection for him led to the establishment of the award being presented this afternoon. The three previous awards have been presented to Dr. Nicholas Kurti, Professor Lev Davidoff Landau, and to Professor John Bardeen. I know that all of those persons whom I represent are pleased that this award today is being made to Dr. Shoenberg. It is a special pleasure to me to have the privilege of making this presentation to a friend whom I have known for so many years.

Many, or I feel most of you, know Dr. Shoenberg well as a warm friend and as an outstanding scientist. Nevertheless, I would like to speak at least briefly about his many accomplishments. Dr. Shoenberg received his Ph.D. degree at Cambridge in 1935. His first publication in 1935 was on the relationship of magnetostriction to crystal symmetry, and in succeeding years Dr. Shoenberg has made numerous ingenious and perceptive contributions to the experimental investigation of magnetic phenomena and of superconductivity. Although he has numerous joint publications, a very large proportion of his work has been done individually. I do not intend to read to you the titles of all of Dr. Shoenberg's publications, but I would like to note that his second publication in 1936 was on the magnetic properties of bismuth. With this paper he began his pioneer work on the de Haas–van Alphen effect which has continued for almost three decades. This outstanding sustained experimental work has led recently to the observation of the de Haas–van Alphen effect in the noble metals and has contributed greatly to the present understanding of the Fermi surface. His work has been instrumental in making the de Haas–van Alphen effect a most powerful tool for the investigation of the Fermi surface of metals. In addition to the above work, he has contributed steadily to the knowledge of the superconducting state especially through his pioneering studies of the penetration depth and through his book entitled *Superconductivity*, which has appeared in two editions. He has always chosen measurements which are most important and fundamental even though these experiments were often exceedingly difficult.

Dr. Shoenberg is a fellow of Gonville and Caius College, Cambridge, and Reader in Physics, Cambridge University. In recognition of his outstanding accomplishments in physics, he is a Fellow of the Royal Society.

The Fritz London Award consists of $1000 and a certificate, both of which have been most generously and graciously provided by the Arthur D. Little Foundation. The certificate is signed by Louis D. Roberts, Chairman, Fritz London Award Committee; J. F. Allen, Secretary of Commission for Very Low Temperature Physics, International Union of Pure and Applied Physics; John G. Daunt, Chairman, Organizing Committee of Ninth International Conference on Low Temperature Physics; Novice G. Fawcett, President, Ohio State University; B. D. Thomas, President, Battelle Memorial Institute; and Howard O. McMahan, President, Arthur D. Little, Inc.

I now make the presentation of $1000 and the certificate to Dr. Shoenberg.

Dr. Shoenberg has consented to address this conference. His subject now is "The de Haas–van Alphen Effect."

THE DE HAAS–VAN ALPHEN EFFECT
(London Award Address)

D. Shoenberg

Royal Society Mond Laboratory
Cambridge, England

The subject of my talk is not very closely linked with any of Fritz London's interests, but there is nevertheless a slight link provided by flux quantization, which London so brilliantly predicted for superconductors. As Onsager pointed out, the quantization condition for metallic electrons moving in a magnetic field can be very neatly expressed as a flux quantization condition. The flux through the electron orbit has in fact to be quantized in units of hc/e, the same units (apart from a rather significant factor 2) as are relevant for the flux through a superconducting loop. From this quantization condition it follows in a fairly straightforward manner that the magnetic properties of a metal should oscillate as the field H is varied (Fig. 1) and that the oscillations should be periodic in $1/H$, with a period

$$\Delta(1/H) = 4\pi^2 e/hA$$

where A is an extremal area of cross section of the Fermi surface normal to the field.

This simple relation, due to Onsager and to I. M. Lifshitz, is of course the basis of the real importance of the de Haas–van Alphen effect. For any given direction of H, the area A is given absolutely by measurement of the de Haas–van Alphen period, and in favorable circumstances a knowledge of all its extremal areas of cross section determines the shape and size of the Fermi surface.

For me, however, an important incentive in studying the de Haas–van Alphen effect has been its interest as a curious physical phenomenon in its own right and a rather childish delight in the pretty patterns it offers. Indeed, I still get a thrill to see the pen of an X–Y recorder drawing out a series of wriggles or to watch the oscillations flash out on the screen of an oscilloscope, and to feel that in some sense I am witnessing almost directly the peculiar antics of the electrons themselves. The fact that such an aesthetically satisfying effect has also proved a powerful tool in unraveling the electronic structure of metals, I always regard as rather a stroke of luck.

What I should like to talk about today is not the determination of Fermi surfaces—much has indeed been achieved and this will be reviewed tomorrow—but rather some of the problems offered by the actual detection and measurement of the oscillations and some of the interesting side effects that turn up.

The theory shows that the field and temperature dependence of the free energy, and hence of its derivatives such as magnetic moment or torque (if the Fermi surface is nonspherical), is mainly controlled by a factor, which in simplified form is

$$\exp[-2\pi^2 k(T + x)/\beta H] \sin(2\pi F/H)$$

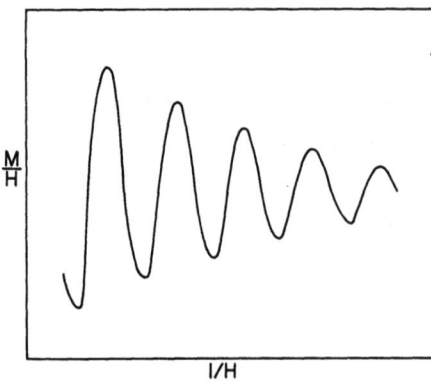

Fig. 1. The de Haas–van Alphen effect (schematic).

Here, F, the de Haas–van Alphen "frequency," is the reciprocal of the period $\Delta(1/H)$ and β is $e\hbar/mc$, where m is the cyclotron mass (for a free electron β is just 2 Bohr magnetons); x is a parameter proportional to the broadening of the Landau levels due to the finite relaxation time of the electron and other causes.

The exponential factor immediately puts a severe restriction on the practical possibility of observing the oscillations. If we put in numerical values, assuming that the cyclotron mass is just the free electron mass m_0 (as is roughly true for major portions of the Fermi surface), and make T as low as is easily possible, say 1°K, we find that for the field of an ordinary electromagnet, say $H = 10^4$ G, the exponential factor (even if we put $x = 0$, i.e., ignore level broadening) is $\exp(-15)$, which proves to be cripplingly small. Even if T were made much lower, the situation would not be greatly improved because usually x is of the order of 1°K.

The reason that oscillations have nevertheless been observed at $T = 1$°K and even higher temperatures, with fields of order 10^4 G, is that many polyvalent metals have small pieces of Fermi surfaces with abnormally low cyclotron masses, typically only $0.2 m_0$ or even less. In this situation the exponential factor is of order $\exp(-3)$ and is no longer small enough to kill the effect. To observe oscillations from a "major" piece of Fermi surface with a normal cyclotron mass ($\sim m_0$), it is essential to go to somewhat higher magnetic fields, usually well above 3×10^4 G.

The importance of high magnetic fields is emphasized also by another consideration. This is that the phase $2\pi F/H$ of the oscillations must not vary by much more than 2π over the specimen, or the amplitude will be destroyed by interference. For a major portion of Fermi surface (e.g., for that of copper), a typical value of F is 5×10^8 G, and we see that at once the phase constancy condition makes severe demands both on the uniformity of field and the perfection of the specimen. Thus for $H = 10^4$ G, H and F must not vary by more than 1 part in 5×10^4 over the specimen, but as H is increased, the severity of this requirement diminishes in inverse proportion (e.g., for $H = 5 \times 10^4$ G the requirement is only 1 in 10^4).

It is interesting to note that Landau, in his classical paper[1] on the theory of diamagnetism of metals, actually predicted the oscillatory field variation of susceptibility before he knew of de Haas and van Alphen's experiments, but took it for granted that in practice field inhomogeneity would kill the effect.* Fortunately

---

* "This last case ($\mu H > kT$) would therefore lead to a complicated nonlinear field dependence of magnetic moment, with a strongly periodic character. Because of this periodicity, however, it would hardly be possible to observe this effect experimentally because the homogeneity of available fields would always lead to an averaging."

Landau's assessment of experimental possibilities has proved overpessimistic in two ways. First, he reckoned only on major pieces of Fermi surface, and as I have just said, many metals have small pieces, with F values 100 times smaller than for copper, and second, the necessary field homogeneity, even for high F values, can usually be achieved by using sufficiently small specimens.

Although it is difficult to draw hard and fast lines, we can say roughly that with fields less than 3×10^4 G it is possible to measure only "small" pieces of Fermi surface such as occur in polyvalent metals. Bigger fields are essential to study monovalent metals or larger pieces of Fermi surface in polyvalent metals. A great deal of valuable work on polyvalent metals has indeed been done by studies at low fields, but usually high-field studies have been essential to obtain a full understanding of the Fermi surface, and I shall therefore mention the low-field techniques only rather briefly.

The Faraday method used by de Haas and van Alphen in their original discovery of the effect in bismuth (Fig. 2) is now mainly of historical interest, since it requires an inhomogeneous field to produce a force, and as I have already emphasized, inhomogeneity inevitably limits the values of F that can be observed. In practice, it would be difficult to observe oscillations with F much above 10^6 G, but for bismuth, F is only of order 10^4 to 10^5 G so that this is not a severe limitation. Another method of measuring the magnetic moment directly is that due to Foner, in which the specimen is vibrated in a suitable pickup coil system. So far, this has not been used much for de Haas–van Alphen studies, but probably has considerable potentialities.

The development of the torque method, now the most widely used "low-field" method, came about through a curious conjunction of international circumstances. In 1937, the late K. S. Krishnan came from India on a visit to Cambridge and I was fascinated by the ingenious methods he had devised for studying magnetic anisotropy. At about that time I had an invitation to spend a year in Kapitza's laboratory in Moscow and I was looking for something to do there which would be sufficiently simple to get done in the time. Nothing could be simpler than measuring a torque, for it is only necessary to hang the crystal on a sufficiently stiff bit of wire and watch how much it turns as the field is varied, and much to my delight I was able to see the oscillatory variation in bismuth almost the first time I tried. I was particularly lucky that Landau was on the spot and had just worked out an explicit form of the theory which made a rather detailed interpretation possible. As a matter of fact, I think I can claim that this was the very first determination of a Fermi surface.

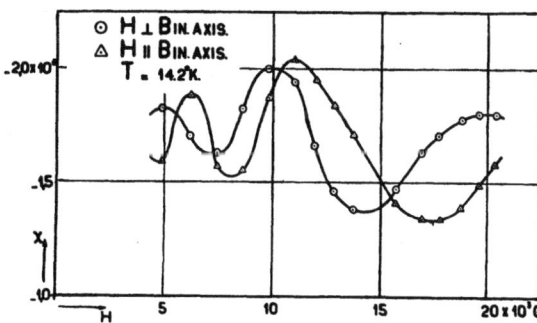

Fig. 2. Oscillations in bismuth (after de Haas and van Alphen).

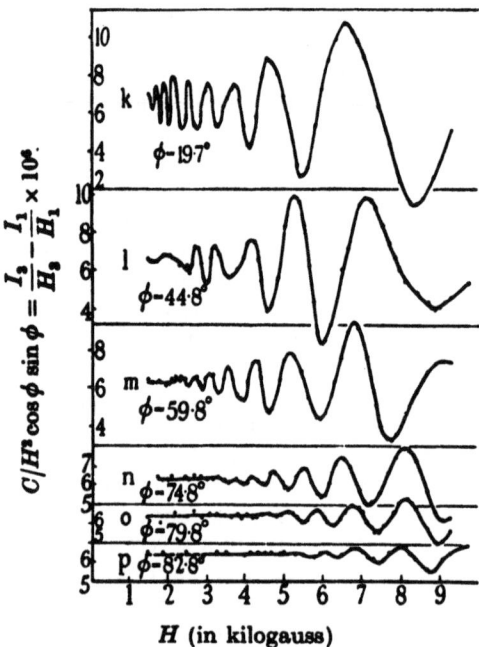

Fig. 3. Oscillations in bismuth by the torque method.

In spite of its "blind spot," that the torque has to vanish whenever F is a maximum or minimum with respect to angle (e.g., at symmetry positions), the torque method has yielded a great deal of useful information. The technique has been vastly refined since the days when I used to take point-by-point measurements of the deflection of a blurry spot of light all day and then burn the midnight oil reducing the data to plot the kind of graphs shown in Fig. 3. Nowadays, the spot of light activates a photocell which feeds back a countertorque and after a great deal of electronics the answer comes out on an X–Y recorder while the observer relaxes (Fig. 4).

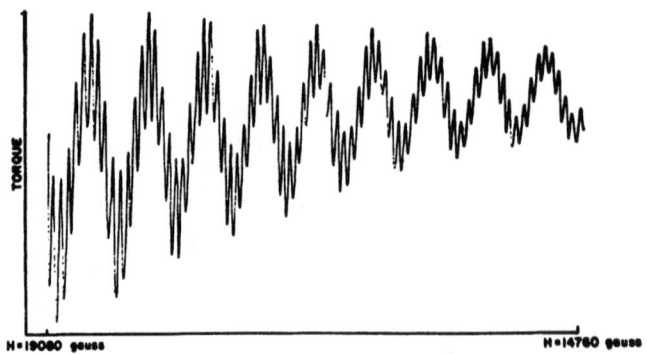

Fig. 4. X–Y recorder trace of oscillations in zinc by the torque method (after Joseph and Gordon).

It was about 15 years ago that I first started brooding about how it might be possible to go to higher fields and look for the really high-frequency oscillations characteristic of large pieces of Fermi surface. Superconducting solenoids were not then in the picture and with water-cooled solenoids powered by motor generators, I anticipated considerable difficulty in stabilizing the field to the necessary 1 in 10^4 or better. It then suddenly struck me that Kapitza's trick of producing high fields only momentarily might provide a simple solution not only to the problem of achieving the field but also of detecting the oscillations. The very fact that the field varies with time, which at first sight seems to make measurement difficult, could be turned to advantage if an inductive method were used. Thus the emf induced in a suitable pickup coil system round the specimen is proportional to dM/dt, i.e., to $(dM/dH)(dH/dt)$, and if M oscillates with H so too will dM/dH. Indeed, the amplitude of oscillation of dM/dH is $2\pi F/H$ times that of M/H, so that the method is particularly sensitive for high values of F.

The impulsive field (of the order 10^5 G and lasting 10 msec or so) is very simply produced by discharging a large condenser through a suitably designed coil. The de Haas–van Alphen oscillations are then shown up by feeding the output of the pickup coil to an oscilloscope, after going through a high-pass filter to remove any residual out-of-balance emf (usually much larger than the de Haas–van Alphen signal, but with much lower frequency). Eventually, the method worked quite successfully though it took nearly 3 years before any oscillations were observed at all (mainly because of an ill-conceived notion that sodium was the best metal to try) and an additional few years before the technique was properly developed to become a useful tool. The oscillations for a copper crystal are shown in Fig. 5, and I might add that the observation of the de Haas–van Alphen effect in the noble metals, which gave a great impetus to the exploitation of the method, came about through another international contact. This was a visit to General Electric at Schenectady, in 1958, when Ethel Fontanella prepared some copper whiskers for me which eventually did the trick. It now appears that whiskers are not really essential, but without them I probably should have been discouraged by my abortive attempts with other crystals.

Measurement of frequency when only one or two widely separated frequencies are present is straightforward enough and an accuracy of 1 or 2% is fairly easily achieved by just counting off oscillations between two known fields. However, it turned out, in the study of the Fermi surfaces of the noble metals, that the F values (i.e., the central areas of cross section) did not vary by more than 2 or 3% as the field direction was varied, and an accuracy of at least 0.1% was needed to yield a detailed description of the surface. This was achieved by beating the de Haas–van

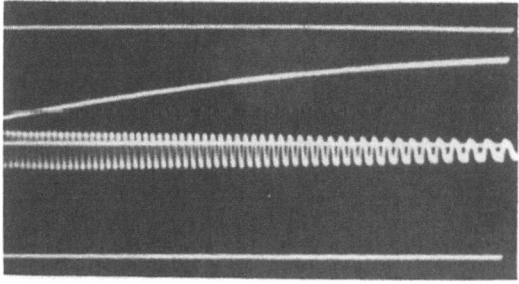

Fig. 5. Oscillations in copper by the impulsive field method.

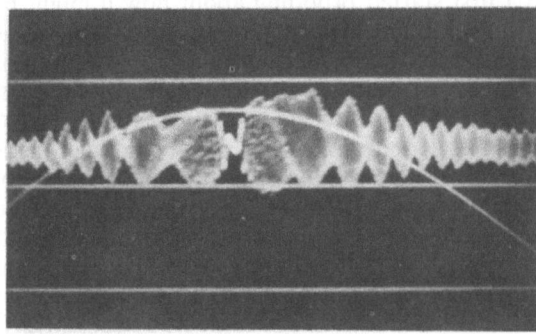

Fig. 6. Beats between two silver specimens with slightly different orientations.

Alphen oscillations of the specimen against those of a reference specimen of only slightly different frequency, say 2% different (Fig. 6). As the specimen is rotated in the field, the reference specimen being kept fixed, changes of F show up as fiftyfold changes of the beat frequency and thus very small changes can be sufficiently accurately followed.

Another trick which has proved invaluable in studying the complicated Fermi surfaces of polyvalent metals, where as many as 10 different frequencies, coming from different pieces of the Fermi surface, may be simultaneously present, is to make the amplifier resonant at some particular time frequency. Since the rate of change of field is variable, the time frequency of the de Haas–van Alphen oscillations varies during the discharge and when it hits off the resonant frequency, f, the amplitude blows up into a "blip" (Fig. 7). If ΔH is the field interval for one oscillation, it is evident that resonance occurs when

$$\Delta H/|dH/dt| = 1/f$$

or

$$1/F = |dH/dt|/fH^2$$

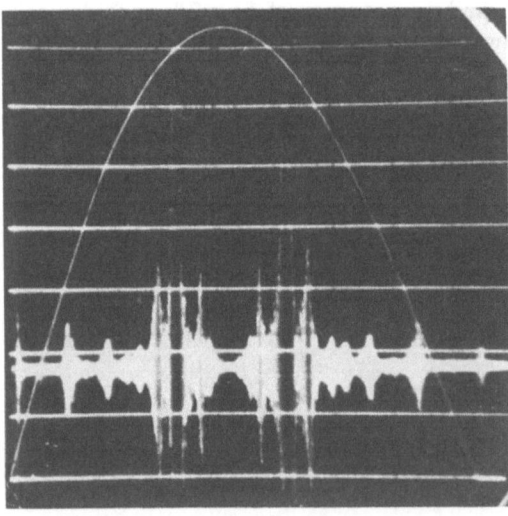

Fig. 7. De Haas–van Alphen "spectrum" in tungsten (after Girvan and Gold).

Thus each F value produces its own pair of blips (one for H rising and one for H falling) and since (dH/dt) varies approximately linearly with time from maximum H, the distance of the blips on the record from the position of maximum H is roughly proportional to $1/F$. In other words, this is something like a roughly linear spectrometer for measuring de Haas–van Alphen periods. Unfortunately, its resolution cannot be made very high (not much better than 5 or 10) because the passage through resonance is too rapid. Another use of the resonant technique is simply to increase the amplification; this is often valuable in showing up a weak signal above the noise level.

One serious complication of the impulsive field method arises from the eddy currents induced in the specimen by the varying field. These can be harmful in two ways: first, by causing heating and so reducing amplitude, and second, by causing inhomogeneity of field through the specimen, since the field in the interior lags behind that at the outside. Both these effects can be largely overcome by making the condenser discharge sufficiently slowly and by using sufficiently thin specimens. However, both these measures reduce sensitivity, since a slow discharge means lower field and lower dH/dt, while a thinner specimen obviously reduces the signal in proportion to the volume. The very thin specimens required (typically 0.1 to 0.5 mm diameter according to whether there is little or much magnetoresistance) are also very difficult to handle, and any damage causes harmful complications which I shall come to in a moment.

With the growing availability of static high-field installations and in particular of superconducting magnets, I began to think again how an inductive method might be used in a static field. Instead of using the dH/dt of an impulsively varying field to produce an emf, one could superimpose a small alternating field $h \cos(\omega t)$ on a high steady field H. The emf in a pickup coil round the specimen should contain a component proportional to $(dM/dH)h\omega \sin(\omega t)$, and so the rectified signal should show de Haas–van Alphen oscillations as H is varied. Unfortunately, however, this is not the whole story: ω has to be high, usually at least of the order of 10^6 sec^{-1}, in order to get sufficient sensitivity; thus the alternating field penetrates only into a skin depth of order 10^{-4} cm.* This not only reduces the sensitivity (which varies with $\omega^{\frac{1}{2}}$ rather than ω) but adds into the signal a component proportional to the square root of the resistivity ρ of the specimen. Since for many metals ρ varies rapidly with H, the rather small oscillations of dM/dH will appear superimposed on a steeply rising background due to the variation of ρ, and will be correspondingly awkward to detect. This was the difficulty encountered by Gunnersen when he tried the idea about 10 years ago.

I was on sabbatical leave at the University of Pittsburgh, in 1962, when another conjunction of two propitious events gave the whole idea a new impetus. In the first place, John Hulm, in a sentimental moment over cocktails, said that Westinghouse would like to present his former laboratory with a superconducting solenoid. The other event was that Phil Stiles, then at the University of Pennsylvania, told me that he would like to spend a year in Cambridge and moreover that he was interested in developing just such an RF method as I had in mind. We discussed various possible ways of overcoming the skin effect difficulty and Phil came up with a very neat solution. Since the variation of M with H is nonlinear, the signal from the pickup coil should contain not only the fundamental frequency ω (accompanied by its unwanted ρ-dependent part), but also a component of frequency 2ω of amplitude

---

* For very large de Haas–van Alphen amplitudes, however, much lower values of ω can be used, and if ρ is high enough, the skin effect is not marked in a thin enough specimen.

proportional to $h^2 d^2 M/dH^2$, and indeed higher harmonics. Thus by tuning the receiver to 2ω rather than ω, we should obtain a signal only if the $M-H$ relation is nonlinear, and the amplitude of the signal would show oscillations if there were a de Haas–van Alphen effect.

In due course both the superconducting solenoid and Phil arrived in Cambridge, and with his drive and skill in electronics the project very quickly got under way. As can be seen (Fig. 8), even at only 5×10^4 G, the "neck" and "belly" oscillations of gold show up very nicely, and incidentally we managed at long last to find the de Haas–van Alphen effect in sodium. There is indeed some chance that this kind of technique may soon put the impulsive field method out of business, except perhaps for very weak oscillations where the higher fields of the impulsive method may still win out.

The skin effect difficulty may be overcome in other ways. For instance, if the field is modulated not only by the RF, but also by a low frequency Ω, then the amplitude of the Ω signal will be effectively proportional to the field derivative of the rectified RF signal. In other words, the picture of oscillations on a steeply rising background will effectively be differentiated and the oscillations will now stand out. This scheme is essentially similar to the idea of an NMR spectrometer, and recently Volski has in fact successfully shown up de Haas–van Alphen oscillations by putting an aluminum crystal into an only slightly adapted version of a standard NMR outfit. Yet another scheme involving modulating at two frequencies and detecting at the sum frequency is being developed by Priestley. Incidentally, an interesting point in all these RF methods is that it is by no means obvious whether the final oscillatory curve is due more to the oscillations of M (de Haas–van Alphen effect) or ρ (Shubnikov–de Haas effect); calculation shows that usually van Alphen wins, but even if Shubnikov should be dominant, it doesn't matter very much since both have the same periodicity.

One novelty of the superconducting solenoid is the possibility of putting it into the persistent mode and effectively making it into a permanent magnet. This makes possible, in a very simple way, new studies of the de Haas–van Alphen effect which with ordinary magnets would require stabilization to much better than 1 part in 10^4. Instead of the oscillations being produced in the usual way by varying H, they can be produced at constant H by varying some other parameter, such as orientation or stress, which affects the phase $2\pi F/H$ through F rather than H. This is the principle of the study that Stiles and I have made of the nonsphericity of the alkali Fermi surfaces. The slight nonsphericity (of the order of 1 in 1000 for potassium) means that F varies slightly as the crystal is rotated in fixed field and because F/H is of the order of 4000, at $H = 5 \times 10^4$ G, one full oscillation is gone through for a change of F of only 1 in 4000 (Fig. 9). Later at this conference, Watts will report some preliminary results obtained in a similar way, which show how the Fermi surfaces of the noble metals change under tension.

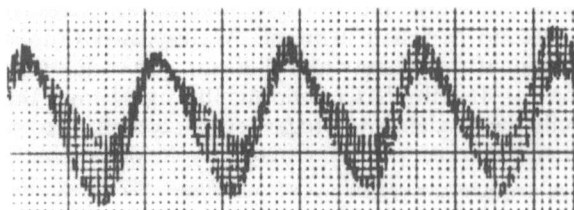

Fig. 8. Neck and belly oscillations in gold by the RF method.

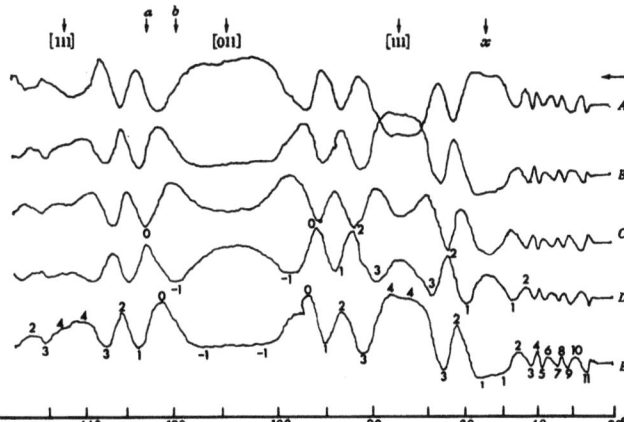

Fig. 9. Oscillations for rotation of a potassium crystal in a fixed field.

I mentioned earlier that the metal crystal has to be perfect if the amplitude is not to suffer. This comes about in two ways: First of all, if the specimen is impure or full of faults, the relaxation time becomes short and the level broadening effect becomes serious. This makes it difficult to study alloys unless they are very dilute, and usually 1% added impurity is enough to kill the amplitude. This is a topic of considerable interest in itself, for by studying how the amplitude depends on perfections and purity, one might learn a great deal about the electron scattering mechanism. For instance, King-Smith finds that the "neck" oscillations in gold are much more sensitive to deformation than to impurity, while for the "belly" oscillations it is the other way round.

The second influence of imperfections is of a more elementary nature. If the specimen is bent or consists of a number of "mosaic" units, these will in general have slightly different F values (because of the varying orientation of the crystal axes with respect to the field) and so will interfere with each other. A simple example is of a specimen consisting of two equal straight pieces at a small angle β to each other, for which it is easily seen that the amplitude will be modulated by a factor $\cos[\pi\beta(\partial F/\partial\theta)/H)]$. For silver, in the orientation of the example of Fig. 10, $\partial F/\partial\theta \sim 0.03F$ and $\beta \sim 2°$, i.e., 0.03, so that the factor is $\cos(10^{-3}\pi F/H)$, which means that beats should occur with 1000 oscillations per beat. These show up

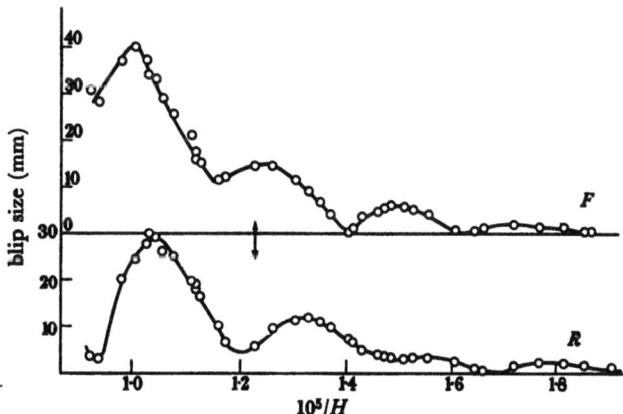

Fig. 10. Slow beats in a bent silver crystal.

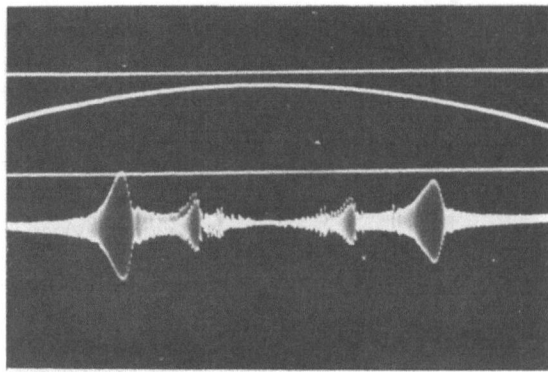

Fig. 11. Abnormal harmonic content in oscillations of a copper crystal.

clearly in the field variation of the size of the resonant blips. It is interesting that calculations of such interference effects in nonuniform specimens closely parallel those of well-known optical problems. Thus the calculation for a specimen bent into the arc of a circle is exactly analogous to that of Fresnel diffraction for a slit, including the transition to the Fraunhofer limit when the field is at a large angle to a symmetry axis.

Finally, I want to mention rather briefly some puzzles which have turned up at various times, and whose explanation has turned out to require refinements of the theory, of some interest in themselves.

First, there is the question of harmonic content of the de Haas–van Alphen oscillations. In some of the high-field measurements on copper (Fig. 11), the harmonic content was much greater than it should have been according to the detailed theory. It turned out that this discrepancy comes from a deficiency in the theory, in that it assumes that the field acting on the electron is H, while really it is B. I feel it is appropriate that I should mention this since I have just been visiting Göttingen as Gauss Professor! At first sight, it might seem that this effect is of no importance, since $4\pi M$, the difference between B and H, is so small (of the order of $10^{-5}H$), but it can easily be shown that it is really $4\pi(dM/dH)$ that matters and that strong harmonics are produced if this approaches unity, as with copper indeed it does.

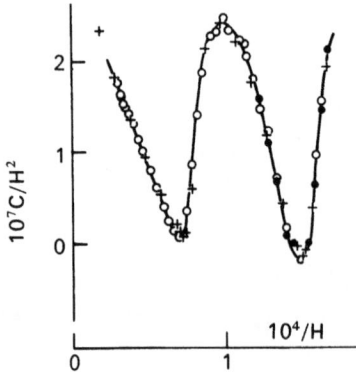

Fig. 12. Oscillations in bismuth at high fields.

Another puzzle came up in my first study of bismuth in Moscow, when it seemed as if the whole oscillatory curve was upside down as compared with theory (Fig. 12), and later, at the N.P.L. in New Delhi, Dhillon and I confirmed that this was really so. The detailed theory takes account of electron spin; this has the effect of splitting the Landau levels into two sets, according to whether the spin is "up" or "down." The result is as if two oscillations of the same frequency but different phases were interfering and introduces a factor $\cos(\pi m/m_s)$, where m_s is a "spin" mass, into the theoretical formula. Ordinarily, m_s is just about the same as a free-electron mass, and since the cyclotron mass m is much smaller for bismuth, it seemed puzzling why the cosine factor should be so close to -1. Later, M. H. Cohen showed that spin-orbit coupling should be particularly strong in bismuth, so that m_s should really be closely equal to m. While on this topic, I should like to show a pretty result obtained by Joseph and Thorsen, by the torque method, for the neck oscillations in copper, where for a particular orientation, $m/m_s = \frac{1}{2}$. One can see (Fig. 13) that the amplitude vanishes just where it should, because of this spin effect.

Finally, there is magnetic breakthrough, discovered by Priestley when he was analyzing the Fermi surface of magnesium. When the field was exactly along the hexad axis, he found a "giant" orbit which for quite some time seemed to make no sense, since it was much too large to be compatible with the scheme which otherwise fitted so well. Later, Cohen and Falicov showed that the cause was simply that at high enough fields the electrons in their orbits prefer to skip the energy gaps which would normally keep them going round a number of smaller orbits, and instead go round the giant orbit (Fig. 14). Magnetic breakthrough proves to be a great complication in the working out of Fermi surfaces, for it means roughly that the structure of the Fermi surface changes as the field is increased and there is a horrid "no-man's land," where it is not very meaningful to speak of a surface at all.

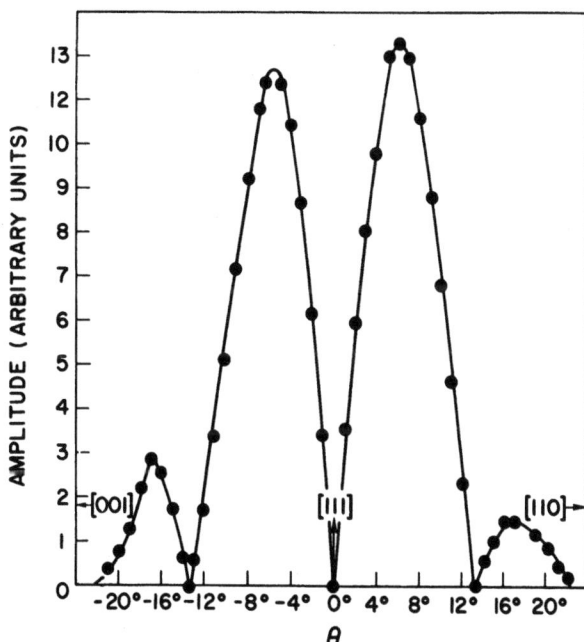

Fig. 13. Vanishing of neck amplitude in copper due to spin effect (after Joseph and Thorsen).

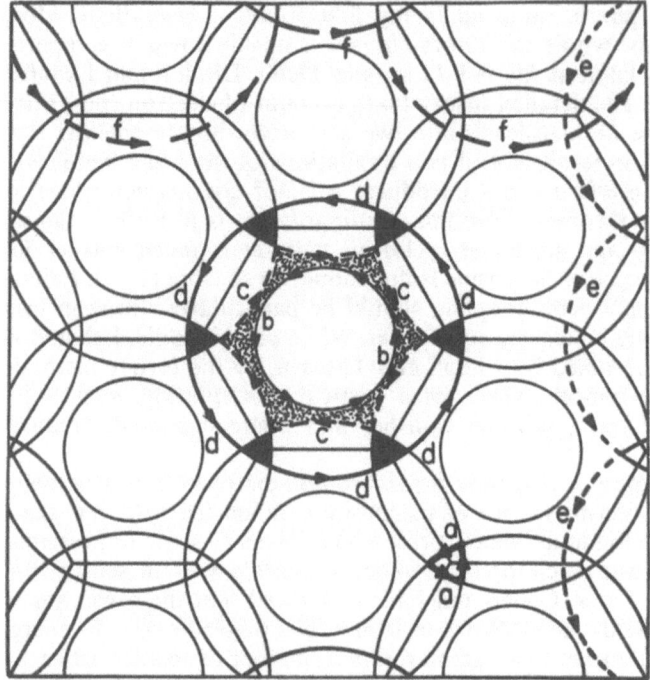

Fig. 14. The giant orbit in magnesium (after Priestley, Falicov, and Weiss).

I should like to end by acknowledging how much I owe to others, in the various experiments with which I have been concerned. It would take too long to mention everyone who has contributed, but I must mention Emil Laurmann (who died in 1954) and Harry Davies, who between them not only designed and built most of the equipment and helped in the measurements, but supplied many ingenious ideas which made a successful outcome possible. I have also benefited greatly from having a succession of bright research students (some of whom I am glad to see here today), not only in the results they have themselves achieved, but in the "feedback" of their developing understanding into my own researches.

Reference

1. L. D. Landau, Z. Phys. **64**, 629, 1930.

8
FERMI SURFACES

8.1. General

THE ROLE OF THE FERMI SURFACE IN LOW TEMPERATURE PHYSICS

Arthur F. Kip

University of California
Berkeley, California

A short discussion is given of the connection between low temperature physics and Fermi surface phenomena. The obvious importance of the Fermi surface in all considerations of transport phenomena is suggested. Advances in the art of Fermi surface investigations have been closely tied to advances in low temperature techniques. The Fermi surface symposium has been designed to summarize usefully the present state of knowledge, both experimental and theoretical, of the Fermi surface, and to describe briefly two additional interesting new experiments.

PRESENT EXPERIMENTAL KNOWLEDGE OF FERMI SURFACES IN METALS

D. Shoenberg

Royal Society Mond Laboratory
Cambridge, England

The study of Fermi surfaces first acquired an air of respectability four years ago at the Cooperstown conference. A number of methods of study were reviewed and a few solid achievements noted, but much of the time was spent over puzzles and future program. Since then, the area of solid achievements has greatly increased, many of the puzzles have been solved and in some ways the future program looks like a mopping-up operation, or at least, mopping up at our present level of precision, but perhaps opening up quite new areas of enquiry at a higher level.

First, I must say a brief word about the methods and give a rough and ready assessment of their relative merits.

1. The de Haas–van Alphen effect and allied effects (i.e., Shubnikov–de Haas effect and indeed oscillations with field of every conceivable property of the metal) have a characteristic frequency of oscillation as the field varies. This frequency measures A, the extremal cross-sectional area of the Fermi surface. The temperature and field dependence of amplitude also give (though more tediously), dA/dE and information about electron scattering.

2. The magnetoacoustic effect, the oscillatory field dependence of sound attenuation, measures a caliper dimension of the Fermi surface.

3. The galvanomagnetic effects indicate whether or not open orbits exist in particular directions and thus tell about the topology of the Fermi surface. Under favorable conditions certain linear dimensions can be estimated and the Hall effect gives us information about the volume of the Fermi surface, or the difference of the volumes of certain slabs.

4. Cyclotron resonance, the oscillatory field dependence of the high-frequency impedance, measures dA/dE. If the form of the Fermi surface is already known, knowledge of dA/dE for all directions enables us to determine the electron velocity at all points of the Fermi surface, at least in principle.

5. If cyclotron resonance is measured in a thin plate of very pure metal (i.e., of very long mean free path) the oscillations suddenly die out when the orbit size just fits into the thickness. This size effect, studied by Khaikin, gives a caliper dimension of the Fermi surface.

6. Another size effect, studied by Gantmacher and Sharvin, is that anomalies occur at radio frequencies much lower than those used in cyclotron resonance in the surface impedance of a thin plate (again very pure) whenever the orbit size or a multiple of it just fits into the thickness. This, of course, again gives the caliper dimension.

7. The anomalous skin effect (measurement of the surface impedance when

the skin depth is much smaller than the mean free path) gives an integral over an appropriate belt of the Fermi surface of the radius of curvature.

8. Positron annihilation, in an oversimplified interpretation, gives the areas of all sections of the Fermi surface (i.e., not merely the extremal ones).

9. In the Kohn effect, kinks in the ω–q relation for the lattice vibrations are expected to appear over a surface in a q-space corresponding to q-values which measure "diameters" of the Fermi surface.

10. For completeness, I should point out that there are a number of properties which measure an integral over the whole Fermi surface, and thus cannot determine the shape of the surface, but can serve as useful checks of a determination by other methods. For instance, electronic specific heat and spin paramagnetism measure the density of states, which is an integral over this surface of the reciprocal velocity. Again, measurement of the anomalous skin effect of a polycrystalline sample measures approximately the total surface area of the Fermi surface.

At first sight, it would appear that the most powerful methods should be those which measure linear dimensions rather than areas, since an area is already an integral over the surface and inevitably irons out the detailed form of the surface to some extent. Of the various linear methods, the size effects (5) and (6) demand such high skill in specimen preparation and such high-purity material that it is doubtful if they will find very wide application for some time yet, so I shall consider only the magnetoacoustic effect. Compared with the de Haas–van Alphen effect, it has some drawbacks in principle quite apart from the fact that it is technically somewhat more complicated to study. First, the magnetoacoustic oscillations are in principle of much lower phase (typically 1000 times lower) than those of the de Haas–van Alphen effect. This means that accurate frequency measurement is more difficult and moreover that it is much less precisely the extreme linear dimension which is measured than it is the extreme area in the de Haas–van Alphen effect. In other words, the relation of the observed quantity to the geometry of the Fermi surface is less definite. Another contrast is that in a complicated Fermi surface, the number of possible caliper dimensions which might produce characteristic magnetoacoustic oscillations tends to be rather more than the number of extreme areas which might produce de Haas–van Alphen oscillations. Paradoxically, more information beyond a certain point turns out to be a nuisance because with increasing possibilities of identifying observed frequencies with geometrical features, mistaken identification becomes more and more probable.

The galvanomagnetic effects can hardly be said in themselves to determine Fermi surfaces, but they can be exceedingly useful in checking proposed models and in clarifying complicated situations. The anomalous skin effect cannot really determine a Fermi surface unequivocally, since it measures rather too complicated a geometrical property, but for a comparatively simple situation (a Fermi surface with only a single sheet) it can give some strong hints, and with the inspiration of a Pippard even produce nearly the right answer. As I shall explain later, it can, however, be very valuable in suggesting slight adjustments of a surface determined by other methods. Positron annihilation and the Kohn effect have the merit that they do not require low temperatures, but because of interpretational difficulties in the former and the feebleness of the kinks in the latter they cannot yet be regarded as more than interesting phenomena in their own right.

To summarize, though perhaps not entirely without bias, the de Haas–van Alphen effect is probably the most useful tool for determining Fermi surfaces, with the magnetoacoustic effect a fairly close second and the galvanomagnetic and anomalous skin effects as auxiliaries. Once the Fermi surface is worked out, cyclotron resonance enables the velocity distribution to be found out, and hence the density of states at the Fermi surface, which can be checked against the experimental value from the electronic specific heat. I should perhaps end this review of methods with a word of caution. All I have said has been based on the assumption that the electrons in the metal behave as if they were equivalent to an equal number of noninteracting quasi-particles, and the theoreticians seem happy that this is justified at the Fermi surface itself, so that, for instance, the simple interpretation of the de Haas–van Alphen effect is probably valid. It is quite another matter, however, when we dive into the murky waters of the Fermi sea and consider neighboring surfaces of constant energy. Thus, the possibility I have mentioned of deducing velocities and the density of states from cyclotron resonance data may be considerably complicated by many body effects, and we should not be too surprised if sufficiently complete and accurate measurements show up discrepancies. This, however, is a matter for "future program," since there is very little sufficiently precise evidence at present.

The present state of the art is indicated by the chart (Fig. 1) in which the various effects studied (not necessarily thoroughly) for each metal are indicated;* a full circle indicates (again perhaps not without bias) a fairly complete determination of the Fermi surface and a wavy circle a partial or less certain determination. It is characteristic of the rapid rate of progress that had this chart been drawn up for the Cooperstown meeting in 1960, only about 3 of the present 12 or so full

* A fairly detailed bibliography will be found at the end of the article.

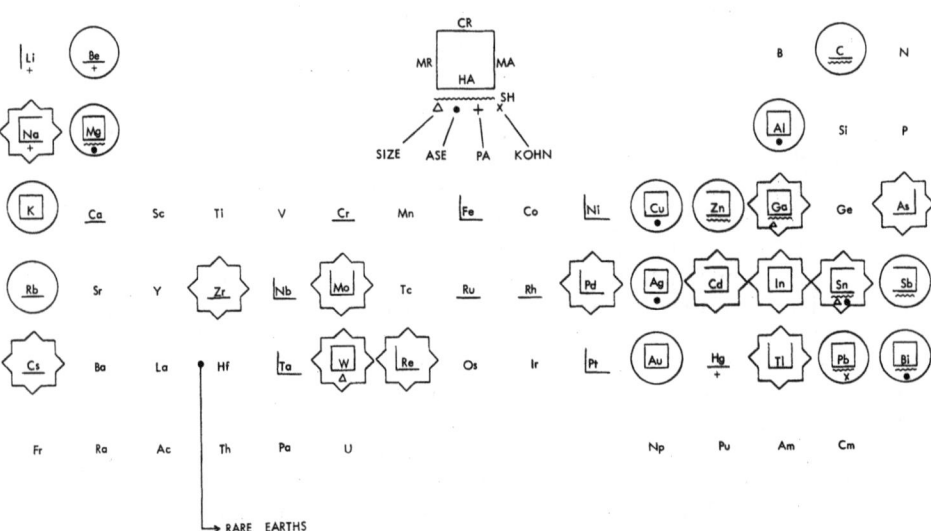

Fig. 1. Chart showing present state of the art. The code at the top indicates which methods have been used for each metal (HA = de Haas–van Alphen effect, MA = magnetoacoustic effect, MR = magnetoresistance, SH = Shubnikov–de Haas and similar effects, ASE = anomalous skin effect, PA = positron annihilation). A full circle indicates a fairly detailed Fermi surface determination; a wavy circle indicates a partial determination.

circles could have been put in, and even since May of this year, when I first drafted the chart, quite a few promotions have been necessary.

In reviewing the progress recorded on the chart, I should like to reverse the historical order and put the simplest Fermi surfaces, those of the monovalent metals, first, and then come only later to the much more complicated surfaces of the polyvalent metals, even though, for the reasons I have explained, it was the latter which were the first to be studied. The very simplest of all are the alkali metals since their Fermi surfaces are nearly spherical. Using the RF method which I have described, Stiles and I could show up the slight departures from a sphere by rotating a crystal in a fixed field and following the de Haas–van Alphen oscillations (Fig. 2) due to the consequent slight changes of frequency (proportional to cross-sectional area). The variations in area with orientation for potassium are shown in Fig. 3 and by suitable analysis in cubic harmonics, we could deduce the form of the Fermi surface (Fig. 4). As one can see, the departures from a sphere are only of the order of 1 in 1000, which I have been told by no less an authority than Kapitza is somewhat less than the tolerance allowed in the manufacture of billiard balls. Note, by the way, that the Fermi surface is pulled out from the sphere along the $\langle 110 \rangle$ direction, where the sphere comes closest to the Brillouin zone boundaries.

The variations are several times larger in rubidium, while they are larger still (of the order of 2 or 3%) in cesium (studied particularly by Okumura and Templeton using the impulsive field method). Because of the martensitic transformation, no oscillations have yet been observed in lithium and only feeble ones in sodium, so that the rotation technique has not yet given fully reliable results. However, there are fairly strong indications that the departures from a sphere are considerably less for sodium than for potassium.

The observed departures of the alkali Fermi surfaces from sphericity are several times smaller and rather different in character from those predicted by Ham's band structure calculations, while Heine and Abarenkov's calculation with a different potential agrees qualitatively for potassium and rubidium, but predicts that sodium should be less spherical than potassium. Thus it will be of particular interest to persevere with sodium, since it may provide a useful clue to the manner in which the band structure calculation needs to be improved.

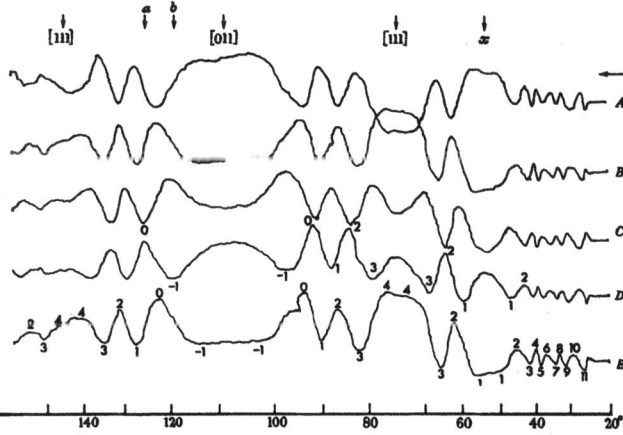

Fig. 2. De Haas–van Alphen oscillations produced by rotation of a potassium crystal in a fixed field.

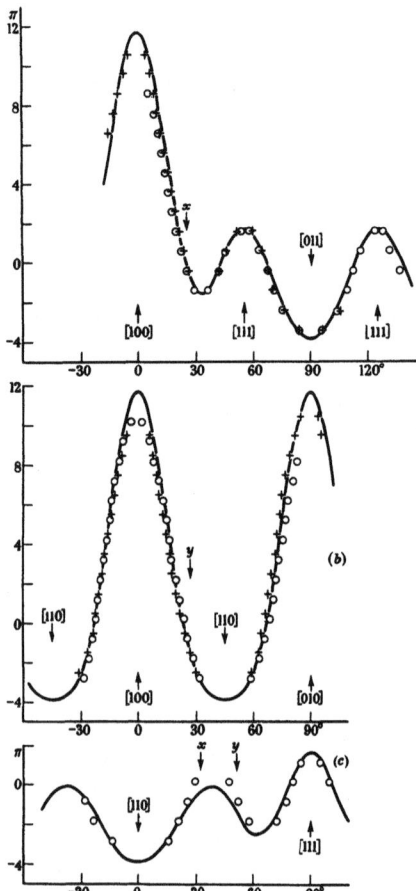

Fig. 3. Variation of frequency with orientation for potassium. Each unit of π represents a change of 1 part in 7400.

The de Haas–van Alphen effect also gives the absolute radius of the sphere of the same volume as the Fermi surface and although this agrees to within a percent or so with the value expected for 1 electron per atom, there seem to be just significant discrepancies, which may yet point to some many-body effect (though more probably they are of experimental origin). Grimes and Kip have studied cyclotron resonance in sodium and potassium and have shown that the cyclotron mass is isotropic to 1 or 2%. This is to be expected in view of the sphericity of the Fermi surface, but the observed mass, although in fair agreement with the specific heat mass (i.e., to within a few percent), is appreciably higher than indicated by band structure calculations. A discrepancy in this sense turns up for nearly every metal studied, and recent calculations by Ashcroft and Wilkins indicate that it is mainly a consequence of the electron–phonon interaction. Incidentally, since the Fermi surfaces of the alkalis are so near spherical, the electron velocity is almost isotropic too and the measurement of the cyclotron mass immediately determines it.

The noble metals, copper, silver, and gold, show more features of interest in the geometry of their Fermi surfaces and have been the most intensively studied (especially copper). Following Pippard's anomalous skin effect study of copper

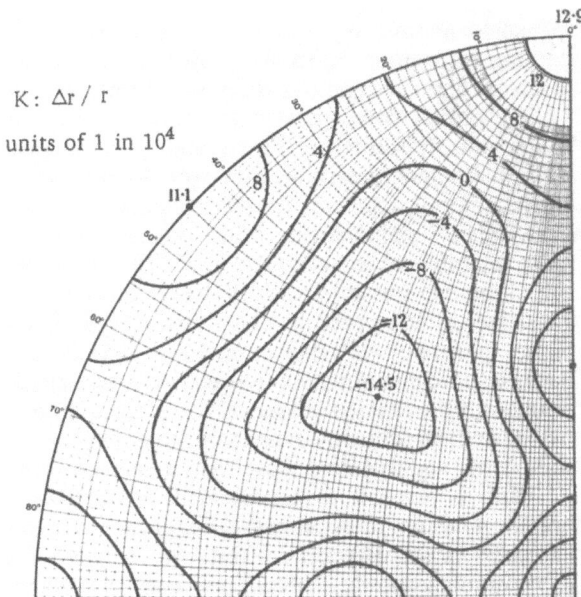

Fig. 4. Stereogram of $\Delta r/r$ for the Fermi surface of potassium.

from which he was able to infer that the Fermi surface was probably sufficiently distorted to make substantial contact on the hexagon faces of the Brillouin zone (Fig. 5), a definitive determination of the surfaces of all three metals has been made by studying the de Haas–van Alphen effect with the impulsive field method. This study fully confirmed Pippard's suggestion by showing up the small minimum areas at the "necks" and other features such as the "dog's bone" and "four-cornered rosette," which are automatic consequences of the implied connectivity of a surface with necks in the $\langle 111 \rangle$ directions. By using the beat method, we could

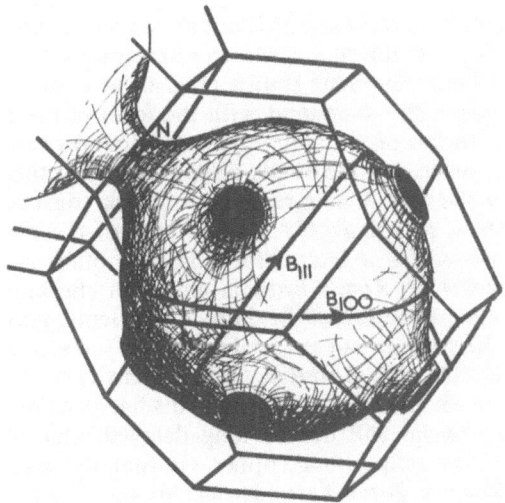

Fig. 5. Schematic model of Fermi surface for the noble metals.

measure the small variations in area of the major sections with direction accurately enough to permit an explicit representation of the surfaces of all three metals in the form of Fourier expansions with 4 to 6 terms. To within 1 or 2%, the volumes of the surfaces are again as they should be for 1 electron per atom.

To within the experimental accuracy, the magnetoacoustic effect, studied by Morse and by Bohm and Easterling, gives almost exactly the same Fermi surface, while the magnetoresistance data of Alekseevski and Gaidukov and of Coleman's group not only confirm the topology but are consistent with the observed neck sizes. When I spoke earlier of the anomalous skin effect, I mentioned that it is not really possible to go uniquely from the measurements to the Fermi surface, but it is of course possible to do the reverse and in fact the Fermi surfaces of copper and silver as computed by Roaf from the de Haas–van Alphen effect led to qualitative agreement with the anomalous skin effect results of Pippard and Morton. Because the anomalous skin effect involves curvature, it is very sensitive to small changes of the detailed shape of the surface. Roaf found it possible to get quantitative agreement with the anomalous skin effect data by adjustments of his formulas, sufficiently slight not to upset the agreement with the de Haas–van Alphen data.

This is a result of considerable importance, for it shows that to within the present precision of experimental data, one and the same Fermi surface will account not only for the magnetic effects but also for such a very different phenomenon as the anomalous skin effect. This confirms that in spite of many-body effects, the independent particle picture is a consistent and meaningful one. I think, however, that we should not be complacent and that it would be of considerable interest to push the experimental accuracy up by a factor of ten and see if the same happy picture still prevails. I suspect that it may very well not, and that something useful may be learned about many-body effects from the discrepancies.

Kip and his students have studied cyclotron resonance in copper and silver, and though at first it was most satisfying to see such beautiful correspondence of the experimental curves with the theoretical ones predicted by Azbel and Kaner, doubts and difficulties began to creep in as the measurements were refined and extended. It is only quite recently that it became clear that extraordinary precautions are required in regard to the flatness of the sample surface and its parallelism to the field, not only to get an accurate cyclotron mass value, but to avoid getting quite false indications for some directions of the field. The physics behind these tilt effects is quite an interesting story in itself, but there is no time to discuss it here. The final results (Fig. 6) in contrast to the earlier ones, seem to be very reasonably consistent with the form of the Fermi surface, inasmuch as the various branches of the mass-orientation curves can all be associated with particular types of orbit, and the orbits disappear at just the angles they should when the necks get in the way. It now remains to check consistency at a somewhat more sophisticated level, by using only a small part of the mass data (essentially only 6 masses are needed for a 6-term formula) to obtain values of the differentials of the 6 parameters in Roaf's formula and then checking that these differentials are consistent with all the rest of the data. If sufficiently good consistency can be obtained, then the differentials can be used to derive the electron velocity at each point of the Fermi surface and, of course, also the density of states, which can then be checked against the electronic specific heat. This has been on the program for some time and I hope fulfilment will not be long delayed. One final point about the noble metals—or rather only about copper—is that the band structure calculations of Segall and Burdick give a Fermi surface in surprisingly good agreement with experiment.

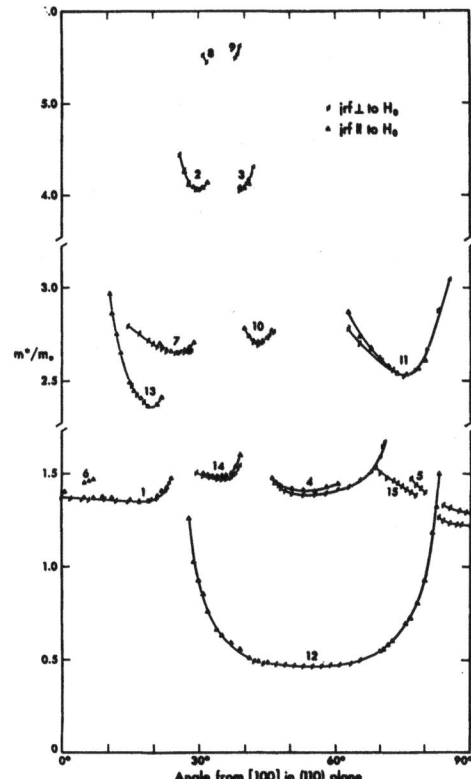

Fig. 6. Orientation variation of cyclotron mass for copper (after Koch, Stradling, and Kip).

It is impossible in a few minutes to do justice to all the work that has been done on the much more complicated polyvalent metals, but I will try to give a general picture and point out a few of the highlights. I will not say much about the semimetals of the bismuth group, beyond mentioning that their Fermi surfaces are very small and to a fair approximation ellipsoidal; it seems that a theoretical explanation of their peculiar electronic structure is now beginning to emerge. One interesting feature of bismuth, in particular, is that it offers the possibility of experimenting at and beyond the quantum limit in quite modest magnetic fields, because the Fermi energy is so small. I will say even less about the transition elements which are only just beginning to be systematically explored, though in some ways they are perhaps of the greatest interest from a theoretical point of view.

This leaves the metals which (until recently) Morrel Cohen has described as "simple" and of which, as he pointed out, there are fortunately quite a few. Here, it turns out that a surprisingly simple generalization gives quite a good guide to the nature of the Fermi surface. Heine first noticed, in a band structure calculation for aluminum, that the Fermi surface of aluminum was rather close to that of the free-electron model, and Gold found that the free-electron model gave a surprisingly good basis for interpreting the rather complicated patterns of angular variation of de Haas–van Alphen frequencies in lead which he was studying by the impulsive field method. A little later, Harrison suggested that this model might be a good guide to the Fermi surfaces of many other metals and worked out the predictions of the model for various crystal structures and valences.

This free-electron model in extended k-space is just a sphere of the right radius to hold the relevant number of valence electrons (e.g., 3 for aluminum, 4 for lead, etc.). The energy gaps where the sphere crosses zone planes are ignored, but the various parts of the spherical surface are assigned to different bands according to where they come. The bits in each band are then remapped into the reduced zone scheme to produce a number of odd-looking surfaces (Fig. 7) which have been given even odder sounding names like "monster," "cigar," "lens," etc., according to the taste and imagination of the author. The observed de Haas–van Alphen or magneto-acoustic frequencies can be fitted surprisingly well to the exact free-electron model or to a recognizable distortion of it in quite a few of the metals so far investigated. Sometimes when the departures from the free-electron model are not too great, as in aluminum and lead, it is possible to do even better and compute the form of the surface in detail, in terms of only very few parameters, chosen to fit the experimental data.

As an example, I should like to outline very briefly Ashcroft's calculation for aluminum. He assumed that the departure from free-electron behavior can be ascribed to a weak effective potential of which only two Fourier components V_1 and V_2 are significant. For any pair of assumed values of V_1 and V_2, he could then carry out a perturbation calculation and compute the detailed form of the

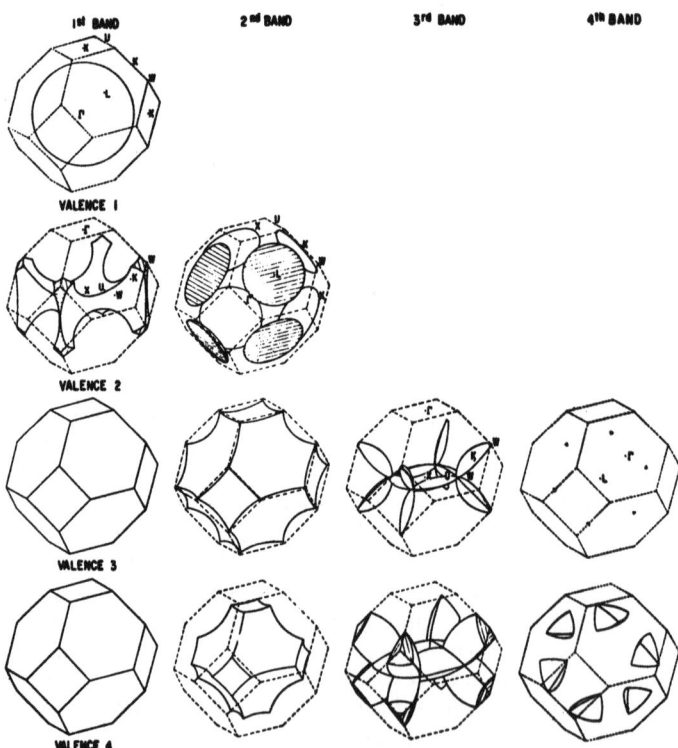

Fig. 7. Free electron Fermi surfaces for fcc metals (after Harrison).

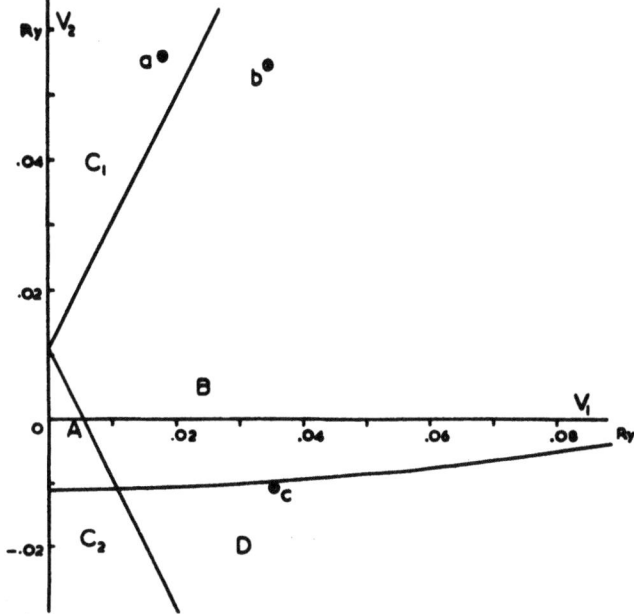

Fig. 8. V_1, V_2 plot for aluminum (after Ashcroft).

corresponding Fermi surface in the regions of interest. It turns out that the connectivity of the monster, which is very thin near its junction points (round the corners W of the zone), is quite sensitive to the choice of V's and the $V_1 V_2$ plane can be divided into regions which have characteristic features (Fig. 8). Thus in region A, the monster has the same connectivity as in the strictly free-electron case ($V_1 = V_2 = 0$); in C_1, the monster breaks up into rings of four (Fig. 9); in D, the

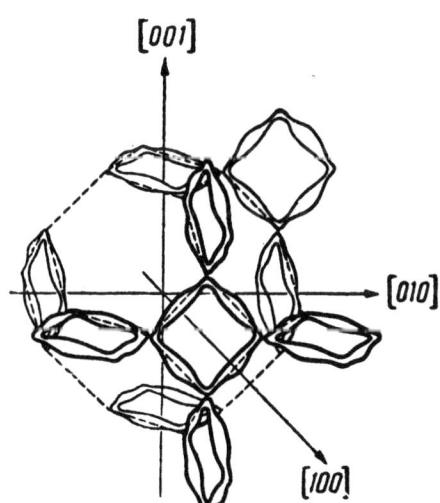

Fig. 9. "Rings of four" for aluminum (after Volski).

monster is completely broken up into "sausages" while tunnels form between the second-zone hole surfaces, and so on.

Now, as I have just said, the regions around the places where the monster can break up or where tunnels may form are exceedingly thin for plausibly small values of V_1 and V_2, and the correspondingly small areas of cross section give rise to very low de Haas–van Alphen frequencies. Such frequencies were observed in the earliest de Haas–van Alphen experiments and have been studied by Gunnersen and more recently by Gordon and Larson and by Volski. They provide a very delicate index to the exact form of the Fermi surface monster around W, and essentially what Ashcroft was able to do was to show that only one kind of connectivity (the rings of four) could account for the general form of the angular variation of the low frequencies and that one particular combination of V_1 and V_2 (the point marked a in C_1, Fig. 8) was needed to give a quantitative fit. What is really remarkable is that the fit is not only perfect for the low frequencies (Fig. 10), but also for the medium frequencies (which come from orbits round the tentacles of the monster) and for the high frequencies observed in impulsive field experiments. It should perhaps be emphasized that these higher frequencies are much less sensitive to the choice of V_1 and V_2, since they are very nearly free-electronlike, but nevertheless they are appreciably affected and the agreement with experiment is improved by the particular choice made. This is a rather beautiful example of making the tail wag the dog, inasmuch as the whole Fermi surface is fixed by forcing a fit with its most insignificant details. A somewhat similar approach has also been used with considerable success by J. R. Anderson and Gold to calculate the Fermi surface of lead, though the situation there is more complicated because spin–orbit coupling is important.

My last example is the group of divalent hexagonal metals, beryllium, magnesium, zinc, and cadmium, whose Fermi surfaces are gradually getting sorted out in more and more detail by de Haas–van Alphen, magnetoacoustic, and galvanomagnetic studies. The situation is complicated by the fact that the whole picture is rather different as one takes into account spin–orbit coupling and uses a "single" zone picture (Fig. 11) or ignores it, in which case the energy gaps over the hexagon planes disappear and a "double" zone picture (Fig. 12) is appropriate with the first and second bands combined and the third and fourth zones also combined. At

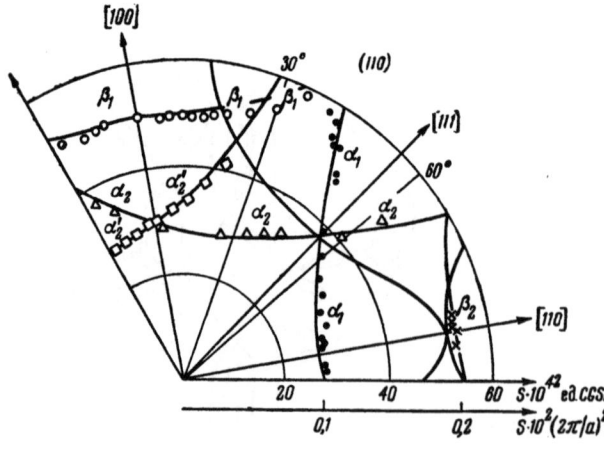

Fig. 10. Low-frequency de Haas–van Alphen periods compared with Ashcroft's theory (after Volski).

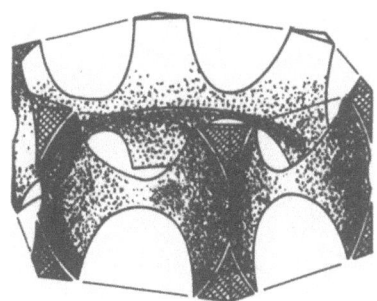
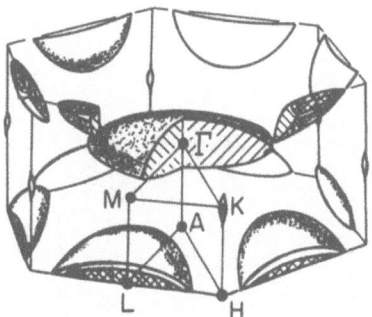

Fig. 11. "Single zone" free-electron Fermi surface for divalent hexagonal metals; the first and second zones are above and the third and fourth zones below (after Stark).

sufficiently low magnetic fields, the single-zone picture is appropriate, but the gaps are small and as the field is increased, they are broken through and eventually the double-zone picture applies. In the sorting out of how this magnetic breakthrough changes the effective connectivity, galvanomagnetic studies have proved invaluable, but there is no time to go into this and Stark will be telling us about his beautiful experiments himself.

If we assume the field is high enough for the double-zone picture to be relevant, we see that the main features are:

1. A hole monster, not open in the hexad direction.
2. Two cigars or needles parallel to the hexad.
3. A lens with its axis along the hexad.
4. Three V-shaped surfaces.

These give rise to a rich variety of extremal areas—as many as 10 for some field directions—and it is quite an intricate business to sort out the experimental data and reach a plausible interpretation. The cigars come out surprisingly close in length and diameter to the free-electron predictions; the most astonishing case is zinc, where the cigar comes from the overlap of a sphere of radius 1.005 times the hexagon radius, so that it is really a fine needle, and yet the experimentally found dimensions are quite close to those of the free-electron needle (0.7 times the mean radius and 0.8 times the length). The lens agrees fairly well for cadmium but gets smaller relative to the free-electron model as we go to zinc and magnesium, and vanishes altogether for beryllium. I will not go into the complicated geometry of

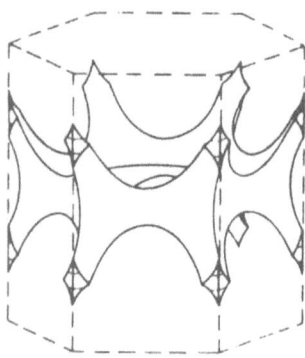

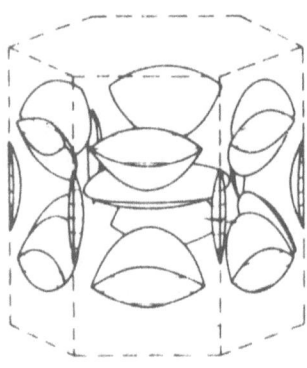

Fig. 12. "Double zone" free-electron Fermi surface for divalent hexagonal metals; the first and second bands are above and the third and fourth bands below (after Harrison).

the V-shaped pieces and the monster arms beyond the general remark that a good deal of plausible interpretation of the de Haas–van Alphen data (the low-field torque data are particularly relevant here) has been achieved, though quite a few puzzles remain.

Of the four metals, beryllium proves to be the simplest because its Fermi surface is sufficiently distorted from the free-electron model to eliminate the lens and the V-shaped pieces and to leave only the rather aristocratic coronet and cigars. Because of this relative simplicity, the de Haas–van Alphen data are somewhat more manageable and Watts has been able to work out the shapes and the detailed dimensions of the surfaces by a purely empirical approach, bearing in mind only the general requirements of crystal symmetry. By somewhat of a coincidence, two band structure calculations (by Loucks and Cutler and by Terrell) appeared just when Watts had completed his interpretation and it is gratifying that the computed surfaces agree almost perfectly with his experimentally determined one.

What remains to be done? First, there is obviously enough to do for several generations of research students in the way of mopping-up operations to complete the studies of partly determined Fermi surfaces, to resolve ambiguities, to explore more thoroughly the bewildering variety of complicated orbits that can occur by magnetic breakthrough, and so on. Second, as I have already said, it would be well worthwhile repeating some of the earlier studies with the improved accuracy now becoming more readily available, in order to check more delicately how far the independent particle model can be carried. Then there is the whole field of the transition metals which is only just beginning to be explored; as the techniques of metal purification and crystal preparation improve, this field should become more readily accessible and should yield handsome dividends. Finally, there is the whole question of the electron scattering mechanism of which I have said nothing in this talk, but which is of course for many phenomena as vital a part of the concept of electronic structure as is the Fermi surface itself. Here, studies of the de Haas–van Alphen amplitude and of magnetoresistance (particularly longitudinal magnetoresistance as Pippard has recently emphasized) may be valuable and (apart from reconnaissances by King–Smith and by R. L. Powell) the territory is virtually unexplored.

Bibliography

In general only the latest or most complete investigations are quoted; these usually contain detailed references to earlier work. The references to "work in progress" are based on private information and so will probably prove incomplete; it is hoped however that they may serve as a useful guide to current activity. In view of the recent publication of Fawcett's review on galvanomagnetic effects, no references to papers in this field are given. The section on band structure calculations, etc., is limited on the whole to papers in which the discussion is closely linked with experimental results; a more detailed bibliography of band structure calculations is given by Heine in his contribution to this conference.

De Haas–van Alphen effect

Na 1   K 1   Rb 1   Cs 1, 2   Be 3   Mg 4, 5   Ca 6
Zr 7   Nb 8, 9   Ta 8   Cr 10   Mo 11, 12, 13, 14
W 11, 12, 13, 14   Re 15   Fe 16, 17   Ni 17, 18   Ru 19
Rh 19   Pd 20   Pt 19, 21   Cu 22, 23   Ag 22, 24
Au 22, 24   Zn 25, 26, 27, 28, 29, 30   Cd 31, 32, 33   Hg 34
Al 35, 36, 37, 38   Ga 34, 39, 40   In 34   Tl 34, 41
C 34, 42, 43   Sn 34, 44   Pb 45, 46   As 47, 48, 49
Sb 34, 49   Bi 50, 51, 52, 53, 54

Shubnikov–de Haas and similar effects

Mg 55   Zn 55, 56, 57   Ga 58, 59   C 60, 61   Sn 62
Pb 62   Sb 63, 64, 65, 66, 67   Bi 68, 69

Magnetoacoustic effect

K 70   Mg 71   Mo 72   W 72, 73   Cu 74, 75   Ag 74
Au 74   Zn 76   Cd 76, 77   Al 77a   Ga 59, 78   In 79
Tl 71, 80   Pb 79, 81   As 59, 71   Sb 82, 83   Bi 84

Galvanomagnetic effects   85

Cyclotron resonance
Na 86   K 86   Mg 87   W 88   Cu 89   Ag 90   Au 91
Zn 92   Cd 93   Al 94, 95, 96   Ga 97   In 98   Sn 99
Pb 100, 101, 102   Sb 103   Bi 104, 105, 106

Size effects
W 107   Ga 108   Sn 109, 110, 111

Anomalous skin effect
Mg 112   Cu 113   Ag 114   Al 112   Sn 115   Bi 116

Positron annihilation
Li 117   Na 117   Be 118   Hg 119

Kohn effect   120
Pb 121, 122

Band structure calculations, etc.
General   123, 124
Alkalis   125, 126   Be 127, 128   Mg 129
Transition metals   130   Cu 131, 132, 133   Zn 134   Al 135, 136
C 61   As 137   Bi 138, 139, 140, 141

References

1. D. Shoenberg and P. J. Stiles, *Proc. Roy. Soc. (London) Ser. A* **281**, 62, 1964.
2. K. Okumura and I. M. Templeton, *Phil. Mag.* **89**, 889, 1963.
3. B. R. Watts, *Proc. Roy. Soc. (London) Ser. A* (in press).
4. M. G. Priestley, *Proc. Roy. Soc. (London) Ser. A* **276**, 258, 1963.
5. M. G. Priestley, L. M. Falicov, and G. Weisz, *Phys. Rev.* **131**, 617, 1963.
6. J. H. Condon and J. A. Marcus, *Phys. Rev.* **134**, A446, 1964.
7. A. C. Thorsen and A. S. Joseph, *Phys. Rev.* **131**, 2078, 1963.
8. A. C. Thorsen and T. G. Berlincourt, *Phys. Rev. Letters* **7**, 244, 1961.
9. D. D. McDonald (Jet Propulsion Laboratory), work in progress.
10. B. R. Watts, *Phys. Letters* **3**, 284, 1963.
11. D. Shoenberg, *Progress in Low Temperature Physics, Vol. 2,* (C. J. Gorter, editor), North Holland Publishing Co., Amsterdam, 1957, p. 226.
12. D. M. Sparlin and J. A. Marcus, *Bull. Am. Phys. Soc.* **9**, 250, 1964.
13. G. B. Brandt and J. A. Rayne, *Phys. Rev.* **132**, 1945, 1963.
14. R. F. Girvan and A. V. Gold (Iowa State University), work in progress.
15. A. S. Joseph and A. C. Thorsen, *Phys. Rev.* **133**, A1546, 1964.
16. J. R. Anderson and A. V. Gold, *Phys. Rev. Letters* **10**, 277, 1963.
17. A. V. Gold, Magnetism Conference, Nottingham, England (1964).
18. A. S. Joseph and A. C. Thorsen, *Phys. Rev. Letters* **11**, 554, 1963.
19. P. T. Coleridge (R. S. Mond Laboratory, Cambridge), work in progress.
20. M. G. Priestley and J. Vuillemin (University of Chicago), work in progress.
21. B. R. Watts, quoted in D. Shoenberg, *Proc. Phys. Soc.* **79**, 1, 1962.
22. D. Shoenberg, *Phil. Trans. Roy. Soc. London, Ser. A* **255**, 85, 1962.
23. A. S. Joseph and A. C. Thorsen, *Phys. Rev.* **134**, A979, 1964.
24. A. S. Joseph and A. C. Thorsen, *Phys. Rev. Letters* **13**, 9, 1964.
25. I. M. Dmitrenko, B. I. Verkin, and B. G. Lazarev, *Zh. Eksperim. i Teor. Fiz.* **35**, 328, 1958. (*Soviet Phys. JETP* (English Transl.) **8**, 229, 1959.)
26. A. S. Joseph and W. L. Gordon, *Phys. Rev.* **126**, 489, 1962.

27. J. R. Lawson and W. L. Gordon, this volume, p. 854.
28. M. Mondino and M. G. Priestley (University of Chicago), work in progress.
29. R. J. Higgins and J. A. Marcus, this volume, p. 859.
30. A. C. Thorsen, this volume, p. 867.
31. A. S. Joseph, W. L. Gordon, J. R. Reitz, and T. G. Eck, *Phys. Rev. Letters* **7**, 334, 1961.
32. A. D. C. Grassie, *Phil. Mag.* **9**, 847, 1964.
33. J. G. Anderson and W. F. Love, *Phys. Rev.* (to be published).
34. D. Shoenberg, *Phil. Trans. Roy. Soc. London, Ser. A* **245**, 1, 1952.
35. E. M. Gunnersen, *Phil. Trans. Roy. Soc. London, Ser. A* **249**, 299, 1957.
36. M. G. Priestley, *Phil. Mag.* **7**, 1205, 1962.
37. E. P. Volski, *Zh. Eksperim. i Teor. Fiz.* **46**, 123, 1964. (*Soviet Phys. JETP* (English Transl.) **19**, 89, 1964.)
38. J. P. G. Shepherd, C. O. Larson, D. Roberts, and W. L. Gordon, this volume, p. 752.
39. J. H. Condon, *Bull. Am. Phys. Soc.* **9**, 239, 1964.
40. A. Goldstein and S. Foner (National Magnet Laboratory), work in progress.
41. M. G. Priestley (University of Chicago), work in progress.
42. D. E. Soule, *IBM J. Res. Develop.* **8**, 268, 1964.
43. S. J. Williamson, S. Foner, and M. S. Dresselhaus, this volume, p. 771.
44. A. V. Gold and M. G. Priestley, *Phil. Mag.* **5**, 1089, 1960.
45. A. V. Gold, *Phil. Trans. Roy. Soc. London, Ser. A* **251**, 85, 1958.
46. J. R. Anderson and A. V. Gold (Iowa State University), work in progress.
47. T. G. Berlincourt, *Phys. Rev.* **99**, 1716, 1955.
48. A. S. Joseph and L. S. Lerner (North American Aviation and Hughes), work in progress.
49. M. G. Priestley and J. R. Windmiller (University of Chicago), work in progress.
50. D. Shoenberg, *Proc. Roy. Soc. (London) Ser. A* **170**, 341, 1939.
51. J. S. Dhillon and D. Shoenberg, *Phil. Trans. Roy. Soc. London, Ser. A* **248**, 1, 1955.
52. W. C. Overton and T. G. Berlincourt, *Phys. Rev.* **99**, 1165, 1955.
53. N. V. Brandt and M. V. Razumenko, *Zh. Eksperim. i Teor. Fiz.* **39**, 276, 1960.
54. N. V. Brandt, T. F. Dolgolenko, and N. N. Stupochenko, *Zh. Eksperim. i Teor. Fiz.* **45**, 1319, 1963. (*Soviet Phys. JETP* (English Transl.) **18**, 908, 1964.)
55. R. W. Stark (University of Chicago), work in progress.
56. C. J. Bergeron, C. G. Grenier, and J. M. Reynolds, *Phys. Rev.* **119**, 925, 1960.
57. C. G. Grenier, J. M. Reynolds, and N. H. Zebouni, *Phys. Rev.* **129**, 1088, 1963.
58. J. Yahia and J. A. Marcus, *Phys. Rev.* **113**, 137, 1959.
59. Y. Shapira, *Phys. Rev. Letters* **13**, 162, 1964.
60. D. E. Soule, *Phys. Rev.* **112**, 708, 1958.
61. D. E. Soule, J. W. McClure, and L. B. Smith, *Phys. Rev.* **134**, A453, 1964.
62. A. D. C. Grassie, Ph.D. Thesis, Cambridge University (1963).
63. J. B. Ketterson and Y. Eckstein, *Phys. Rev.* **132**, 1885, 1963.
64. J. B. Ketterson, *Phys. Rev.* **129**, 18, 1963.
65. G. N. Rao, N. H. Zebouni, C. G. Grenier, and J. M. Reynolds, *Phys. Rev.* **133**, A141, 1964.
66. L. S. Lerner and P. C. Eastman, *Can. J. Phys.* **41**, 1523, 1963.
67. M. S. Dresselhaus and J. G. Mavroides, *Phys. Rev. Letters* (in press).
68. L. S. Lerner, *Phys. Rev.* **127**, 1480, 1962.
69. J. E. Kunzler and F. S. L. Hsu, in: *The Fermi Surface*, John Wiley & Sons, Inc., New York (1960), p. 88.
70. H. W. Foster, Ph.D. Thesis, Catholic University of America, Washington D.C. (1964).
71. Y. Eckstein and J. B. Ketterson (Argonne National Laboratory), work in progress.
72. C. K. Jones and J. A. Rayne, this volume, p. 790.
73. J. A. Rayne and H. Sell, *Phys. Rev. Letters* **8**, 199, 1962.
74. H. V. Bohm and V. J. Easterling, *Phys. Rev.* **128**, 1021, 1962.
75. E. V. Mielczarek, D. L. Shelley, R. Meister, and P. H. E. Meijer, Navy Report No. 2 Nonr—2249 (03) (1962).
76. D. F. Gibbons and L. M. Falicov, *Phil. Mag.* **8**, 177, 1963.
77. M. R. Daniel and L. Mackinnon, *Phil. Mag.* **8**, 537, 1963.
77a. G. N. Kamm and H. V. Bohm, *Phys. Rev.* **131**, 111, 1963.
78. B. W. Roberts, *Phys. Rev. Letters* **6**, 453, 1961, and work in progress.
79. J. A. Rayne, *Phys. Rev.* **129**, 652, 1962.
80. J. A. Rayne, *Phys. Letters* **2**, 128, 1962.
81. A. R. Mackintosh, *Proc. Roy. Soc. (London) Ser. A* **271**, 88, 1963.
82. O. Beckman, L. Eriksson, and S. Hörnefeldt, *Solid State Comm.* **2**, 7, 1964.

83. Y. Eckstein, *Phys. Rev.* **129**, 12, 1963.
84. D. Reneker, *Phys. Rev.* **115**, 303, 1959.
85. E. Fawcett, *Advan. Phys.* **13**, 139, 1964.
86. C. C. Grimes and A. F. Kip, *Phys. Rev.* **132**, 1991, 1963.
87. T. G. Eck and M. P. Shaw, this volume, p. 759.
88. E. Fawcett, *Phys. Rev. Letters* **8**, 476, 1962.
89. J. F. Koch, R. A. Stradling, and A. F. Kip, *Phys. Rev.* **133**, A240, 1964.
90. D. G. Howard, Ph.D. Thesis, University of California, Berkeley (1964).
91. D. N. Langenberg and S. M. Marcus, *Phys. Rev.* (to be published).
92. J. K. Galt and F. R. Merritt, in: *The Fermi Surface*, John Wiley & Sons, Inc., New York (1960), p. 159.
93. M. P. Shaw and T. G. Eck, this volume, p. 761.
94. T. W. Moore and F. Spong, *Phys. Rev.* **125**, 846, 1962.
95. C. C. Grimes, A. F. Kip, F. Spong, R. A. Stradling, and P. Pincus, *Phys. Rev. Letters* **11**, 455, 1963.
96. E. Fawcett, in: *The Fermi Surface*, John Wiley & Sons, Inc., New York (1960), p. 166.
97. T. W. Moore (G.E. Research Laboratories), work in progress.
98. J. G. Castle, B. S. Chandrasekhar, and J. A. Rayne, *Phys. Rev. Letters* **6**, 409, 1961.
99. M. S. Khaikin, *Zh. Eksperim. i Teor. Fiz.* **42**, 27, 1962. (*Soviet Phys. JETP* (*English Transl.*) **15**, 18, 1962.)
100. M. S. Khaikin and R. T. Mina, *Zh. Eksperim. i Teor. Fiz.* **42**, 35, 1962. (*Soviet Phys. JETP* (*English Transl.*) **15**, 24, 1962.)
101. M. S. Khaikin, *Zh. Eksperim. i Teor. Fiz.* **45**, 1304, 1963. (*Soviet Phys. JETP* (*English Transl.*) **18**, 896, 1964.)
102. R. C. Young, *Phil. Mag.* **7**, 2065, 1962.
103. W. R. Datars and R. N. Dexter, *Phys. Rev.* **124**, 75, 1961.
104. J. E. Aubrey, *J. Phys. Chem. Solids* **19**, 321, 1961.
105. M. S. Khaikin, R. T. Mina, and V. S. Edelman, *Zh. Eksperim. i Teor. Fiz.* **43**, 2063, 1962.
106. M. S. Khaikin, *et al.*, (Inst. Phys. Problems, Moscow), work in progress.
107. W. M. Walsh, Jr., C. C. Grimes, G. Adams, and L. W. Rupp, this volume, p. 765.
108. D. M. Sparlin and D. S. Schreiber, this volume, p. 823.
109. M. S. Khaikin, *Zh. Eksperim. i Teor. Fiz.* **41**, 1773, 1961. (*Soviet Phys. JETP* (*English Transl.*) **14**, 1260, 1962.)
110. M. S. Khaikin, *Zh. Eksperim. i Teor. Fiz.* **43**, 59, 1962. (*Soviet Phys. JETP* (*English Transl.*) **16**, 42, 1963.)
111. V. F. Gantmacher, *Zh. Eksperim. i Teor. Fiz.* **43**, 345, 1962 (*Soviet Phys. JETP* (*English Transl.*) **16**, 247, 1962); this volume, p. 1193.
112. E. Fawcett, *J. Phys. Chem. Solids* **4**, 320, 1961.
113. A. B. Pippard, *Phil. Trans. Roy. Soc. London, Ser. A* **250**, 325, 1957.
114. V. M. Morton, Ph.D. Thesis, Cambridge University (1960).
115. E. Fawcett, *Proc. Roy. Soc.* (*London*) *Ser. A* **232**, 519, 1955.
116. G. E. Smith, *Phys. Rev.* **115**, 1561, 1959.
117. J. J. Donaghy, A. T. Stewart, J. H. Kusmiss, and D. M. Rockmore (University of North Carolina), work in progress.
118. A. T. Stewart, J. B. Shand, J. J. Donaghy, and J. H. Kusmiss, *Phys. Rev.* **128**, 118, 1962.
119. D. R. Gustafson, A. R. Mackintosh, and D. J. Zaffarino, *Phys. Rev.* **130**, 1455, 1963.
120. W. Kohn, *Phys. Rev. Letters* **2**, 393, 1959.
121. B. N. Brockhouse, K. R. Rao, and A. D. B. Woods, *Phys. Rev. Letters* **3**, 93, 1961.
122. A. Paskin and R. J. Weiss, *Phys. Rev. Letters* **9**, 199, 1962.
123. V. Heine, this volume, p. 698.
124. W. A. Harrison, *Phys. Rev.* **118**, 1190, 1960.
125. F. S. Ham, *Phys. Rev.* **128**, 2524, 1962.
126. V. Heine and A. Abarenkov, *Phil. Mag.* **9**, 451, 1964.
127. T. L. Loucks and P. H. Cutler, *Phys. Rev.* **133**, A819, 1964.
128. J. H. Terrell, *Phys. Letters* **8**, 149, 1964.
129. L. M. Falicov, *Phil. Trans. Roy. Soc. London, Ser. A* **255**, 55, 1962.
130. W. M. Lomer, *Proc. Phys. Soc.* (*London*) *Ser. A* **80**, 489, 1962.
131. D. J. Roaf, *Phil. Trans. Roy. Soc. London, Ser. A* **255**, 135, 1962.
132. B. Segall, *Phys. Rev.* **125**, 109, 1962.
133. G. A. Burdick, *Phys. Rev. Letters* **7**, 156, 1961.
134. W. A. Harrison, *Phys. Rev.* **126**, 497, 1962.
135. V. Heine, *Proc. Roy. Soc.* (*London*) *Ser. A* **240**, 340, 354, 261, 1957.

136. N. W. Ashcroft, *Phil. Mag.* **8**, 2055, 1963.
137. L. M. Falicov and S. Golin, *Phys. Rev.* (to be published).
138. A. L. Jain and S. H. Koenig, *Phys. Rev.* **127**, 442, 1962.
139. E. Behrens, *Z. Physik* **161**, 279, 1961.
140. A. A. Abrikosov and L. A. Falkovski, *Zh. Eksperim. i Teor. Fiz.* **43**, 1089, 1962. *JETP (English Transl.)*
141. M. H. Cohen, L. M. Falicov, and S. Golin (in press).

CALCULATION OF BAND STRUCTURES AND FERMI SURFACES

V. Heine

Cavendish Laboratory
Cambridge, England

In recent years, several quite detailed band structure calculations have been performed for both ordinary and transition metals. Some of the results will be reviewed in the following section. These calculations accurately solve the Schroedinger equation in the periodic structure, any weakness lying in the slight arbitrariness of the potential which is put into the equation.

At the same time, a new approach has been developed for nontransition metals with two important features (see *Pseudism* and *Fermiology* sections): First, it entails a much more careful procedure for setting up the potential, making it self-consistent and including to a good approximation exchange and correlation effects. Second, the theory is expressed in terms of the Fourier components $v(q)$ of the pseudopotential, which gives not only the band structure but all other properties depending on the electron–ion interaction, such as the phonon spectrum, the resistance of the molten metal, and the superconducting transition temperature. It is therefore possible to use information about $v(q)$ gained from one set of experimental data to help interpret some other property.

Brute Force

The calculation of electron energy bands in solids always resolves itself into two more or less distinct parts: first, how to solve the Schroedinger equation in a periodic structure and, second how to determine what potential to solve it for. It may seem illogical to state them in that order, but it is in that order that progress has been made. It still remains true that band structures are computed numerically correct to 0.001 Ry (1 Ry = 13.6 eV) or even 0.0001 Ry, but the results are not physically significant to better than 0.01 Ry or more often 0.1 Ry or even worse, depending on circumstances, because of uncertainties about the potential used. This can be seen by comparing calculations with different potentials for the alkali metals (Ham[19] and Heine and Abarenkov[24]), for silicon (Kleinman and Phillips[26]), and a transition metal (Mattheiss[31]) as typical cases. On the other hand, as most authors have preferred to stress, the calculations can be relied on with regard to the general outlines of the band structure, and in that sense are certainly extremely useful. For instance, in transition metals the shape of the d-bands does not appear to depend sensitively on the potential, but the relative positions of the s- and d-bands do and so have to be adjusted empirically.

The difficulty of solving the Schroedinger equation in a solid stems from the two radically different situations that have to be simultaneously accommodated. Inside the core of an atom, the potential is practically spherically symmetric and the wave function ψ_k has the rapid oscillations and angular dependence of an atomic wave function. In the region between the atomic cores, the potential reflects the structure of the crystal; the wave function is slowly varying in the manner of a plane wave, joining up correctly from cell to cell. Such a separation into two regions lies behind the success of the Kohn–Rostocker (KR or variational or Green's function) method and the augmented plane wave (APW) method (Callaway[16]). Both of these methods have been perfected to a high degree and thoroughly tested during the last few years (compare, e.g., Segall[42] and Burdick[12]). In each case, a "muffin tin" potential is employed. A set of touching spheres is drawn around the atoms, inside of which is performed an atomic-like integration of the wave equation. Between the spheres the potential is set equal to a constant, and the wave function expanded in plane waves. A massive secular equation then weds the two halves together. Taking the potential as a constant between the spheres is not so drastic an approximation as it may seem if the constant is correctly chosen as the mean value of the real potential in that region. In any case, the variation of the potential about that mean can be taken into account if desired, though less easily in the Kohn–Rostocker method than with augmented plane waves.

The orthogonalized plane wave (OPW) method remains considerably simpler than either of the two methods discussed above. If care is taken about the core functions used in the method, the convergence of the secular equation is reasonably good, but not so excellent as in the Kohn–Rostocker or augmented plane wave methods: for instance, in silicon the energy of the $\Gamma_{25'}$ level at the top of the valence band dropped by 0.009 Ry in going from a secular equation of order 6×6, including 65 plane waves, to one with 469 plane waves (Kleinman and Phillips[27]). The s-levels converge more rapidly than this, and since all the p-levels converge rather similarly, it is possible to make reasonable extrapolations and reduce the computational errors below the uncertainties stemming from the potential.

With respect to other methods, Schlosser and Marcus[30] have pioneered a variant of the augmented plane wave method and found it to give good results when tested on a potential for which the band structure was already accurately known. Altmann[1–3] has resurrected the cellular method, which has been out of favor for some years with other workers because of the difficulties with convergence and boundary conditions (see Reitz[37] and Callaway[16]). Apparently, these difficulties have still not been completely overcome. Among seven energy levels discussed in detail (Altmann[1]), the energy decreased by up to 0.008 Ry in enlarging the expansion from six to eight terms and in two cases actually went *up* by 0.016 and 0.007 Ry, respectively. Differences between results with different boundary conditions were up to 0.05 Ry between a 9-point and a 16-point fitting and 0.03 Ry between two different 16-point fittings.

Apart from the work to be described in the next section, this completes the list of reasonably accurate methods of calculating energy band structures that have been found useful in recent years. The results with metallic elements, and a few others, are listed in Table I. In addition, there are the calculations performed by Mattheiss[32] on a range of seven metallic V_3X alloys ($X = $ Ga, Ge, As, etc.) and on a number of semiconducting compounds, which are not relevant to Fermi surface studies. The potentials are generally determined from that of a free atom or ion in some suitable configuration. (In this connection, recent unpublished calculations on heavy atoms by Herman, using the Hartree–Fock–Slater approximation, and by Mayers, with the Hartree–Fock method, are most useful.) Thus, it can be seen that workers have skated with varying speed over many problems of exchange, correlation,

Table I. Recent Energy Band Calculations of Elements

Element	Method	Reference
Alkali metals	OPW	Callaway[13,15]
	KR + quantum defect	Ham[19]
	special	Schlosser and Marcus[39]
	model potential	Heine and Abarenkov[24]
Be	OPW	Loucks,[29] Loucks and Cutler[30]
Mg	OPW	Falicov[18]
Zn	pseudopotential	Harrison[21]
Ca and others	pseudopotential	Harrison[22]
Al	KR	Segall[41]
Ga	APW	Wood[51]
Tl	OPW	Soven[46]
C	OPW	Kleinman and Phillips[26]
Si	OPW	Kleinman and Phillips[27]
Sn	OPW	Bessani and Liu[9]
	pseudopotential	Miasek[33]
Cu	KR	Segall[42]
	APW	Burdick[12]
Ag, Au	KR	Segall[40]
Fe	APW	Wood[50]
Ni	KR	Yamashita et al.[52]
Ti	cellular	Altmann and Cohan[3]
Ar, Ti, V, Cr, Fe, Co, Ni, Cu, Zn	APW	Mattheiss[31]
Fe transition series	cellular	Altmann[2]
Zr	cellular	Altmann[1]
W	APW	Mattheiss[32]

For earlier calculations, including those on compounds, see Herman[25] and Callaway.[14,16]

self-consistency, and angular dependence of the potential. With this limitation about the appropriateness of the potential, it appears that calculating a band structure becomes purely a question of time and computer time. Some order of magnitude figures for these are as follows: To develop from scratch an APW program, two years. To learn to use an existing program and adapt it, six months. A recent calculation on body-centered cubic tungsten for 1024 points in the Brillouin zone (55 distinct points in $\frac{1}{48}$ part of the zone) took two weeks' work and $1\frac{1}{2}$ hours of 7094 computer time at $650 per hour (Mattheiss, private communication). The time for the KR method would work out similarly, the most important factor being whether the structure constants have been previously calculated for the required crystal structure (Ham and Segall[20] and Ham[19]). An OPW calculation can be done completely *ab initio* by a graduate student using a modest computer in two years. By using the model potential described in the next section, this time can be reduced further. Moreover, as we shall see, it is not difficult to calculate in greater detail the shape of the Fermi surface for nontransition metals from the energies at nearby symmetry points in the zone. Thus, a band-structure calculation is becoming an everyday aid in research to help understand approximately what is going on in a solid, though the computer time and professional theoretician's guidance required may not yet be within reach of every laboratory. For instance, it is difficult to believe that much sense would have been made out of some recent data on the Fermi surface of tungsten without a theoretical band structure as guide (Walsh et al.[49]). Incidentally, these remarks apply not only to elements but also to insulating and magnetic compounds, as has been shown by recent work on the V_3X compounds and $SrTiO_3$ with the APW method (Mattheiss[32]).

Pseudism

Even more important than the progress with brute force band-structure calculations, there has been a veritable flux of cross-currents in ideas about the electronic structure of materials. At the center stands the effective potential of an atom in the material, as seen by an incoming electron which is pictured more or less as a plane wave. Now it is far from obvious that such a concept has any meaning—that it is possible to set up the potential in the ideal solid and in the molten metal, around a dislocation or in a phonon wave, in terms of a basic atomic potential. Nor is it clear that the electrons can be reduced to something like plane waves. However, brushing aside for the moment all the qualifications and limitations we shall have to make, the fact remains that it *is* possible, at least in nontransition metals to which we shall restrict ourselves in the following. We shall want to look into how this comes about, into its mathematical formulation, and how to calculate with it. It has been termed by Ziman,[53] "the method of neutral pseudoatoms." Its significance goes far beyond calculating band structures or Fermi surfaces. The interaction of an electron with an atom is the basic component determining the electronic structure and properties of the material. It allows one to formulate in a unified way all these properties, from the band structure of the solid to the thermoelectric power of the molten metal, from the energy of a defect to the electron–phonon enhancement of the effective mass. Here, the experimental study of Fermi surfaces plays a crucial role. Since the band structure $E(\mathbf{k})$ is a one–electron property and the periodicity of the lattice picks out reciprocal lattice vectors, the shape of the Fermi surface gives some of the cleanest experimental information about the potential. This information can then be applied to understanding the other properties which depend on the potential in a more scrambled way. As already noted, there have been several cross-currents of ideas which can be permuted in multiple ways. For the sake of coherence, therefore, we shall not review all the contributions that have been made, but try to bring out the salient points, primarily using for illustration the work that has been done in Cambridge.

We start by considering the fact that diamond is in some ways a closer approximation to a free electron gas than, for instance, copper and lithium are. It does not need to be stressed that all (nontransition) metals can be treated as perturbations of a free electron gas. As several abstracts for this conference bear witness, it is now the universal custom to analyze experiments on Fermi surfaces in terms of a modest distortion of a free electron sphere. By a "modest" distortion we mean one which can produce considerable changes in the connectivity and shape of the Fermi surface, without distorting the overall band shapes beyond recognition. It may therefore be instructive to look at a substance like diamond which may legitimately be regarded as rather an extreme case. One look at the energy bands and the corresponding free-electron bands shows their similarity (Kleinman and Phillips[28] and Herman[25]). The Fourier coefficients $V(\mathbf{K})$, which in the sense of the nearly free electron approximation produce the band gaps, are of the order of 0.5 Ry compared with a free-electron Fermi energy E_F of 2.2 Ry. We thus have $V(\mathbf{K})/E_F$ in diamond of 0.2, compared with 0.3 for lithium and copper. The band gaps in diamond can honestly be described as "modest" in our sense. Moreover, as already pointed out by Mott and Jones,[34] the (220) reciprocal lattice planes are the first ones with the maximum structure factor of 2 and form a Jones zone of nearly spherical shape containing exactly four electrons per atom. The Fermi sphere distorts itself somewhat to fit into the zone and the resultant lowering of the energy makes a substantial contribution to the cohesive energy of diamond, which from one point of view can be regarded at least partially as an account of why diamond takes on the covalent tetrahedral structure (Anderson,[4] p. 41). Of course, the free-electron model of diamond must not be taken too literally. The high degeneracy of the plane wave states near the Fermi energy and the large (1, 1, 1) Fourier component of V produce a strong mixing of the plane wave states, resulting in a picture of bond formation and a quite nonuniform charge density (Kleinman and Phillips[28]). The point is that, for all the hybridizing and the potential too strong to be treated adequately by perturbation theory, the whole system can be described systematically and quantitatively on the basis of a plane-wave analysis. Recently, Bennemann[10] has applied this approach to calculate, among other properties, the energies of formation and migration of vacancies and interstitials in diamond, silicon, and germanium.

Now, as already discussed in the *Brute Force* section, when we talk about representing the conduction electrons in terms of plane waves, this only refers to the part of the wave function between the ion cores. In a good metal this may be $\frac{7}{8}$ of the total volume, since the ionic radius is typically half the atomic one. Figure 1 shows a Bloch function ψ_k in sodium, with the atomic-like oscillations inside the core. However, we can make an extrapolation of the plane wave part into the core, producing the smooth pseudo wave function φ_k also shown in the figure. It is to the φ functions that the plane wave analysis must be applied, and it is to the bogus φ functions that the atoms present an equally bogus weak pseudopotential. Making the pseudopotential weak goes automatically with eliminating the atomic–like oscillations in ψ_k. The real potential energy V for an electron inside an atom is very great and also negative due to the nuclear attraction. The electron consequently has great kinetic energy there, equal to $E - V(\mathbf{r})$, which by the de Broglie relation is reflected in the short wavelength oscillations of ψ_k inside the core (Fig. 1). Conversely, the smooth pseudo wave function φ_k must relate to a pseudopotential which is too weak to produce any nodes.

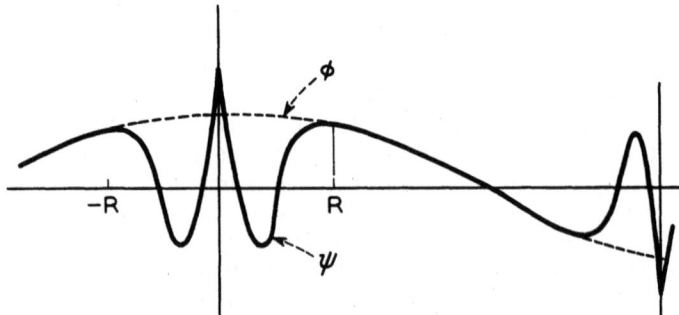

Fig. 1. True wave function ψ and pseudo wave function φ in sodium. R marks a radius slightly outside the ion core.

There are several different ways of looking at the change from the ψ to the φ. It is, for instance, possible to make a mathematical transformation of the Schroedinger equation:

$$\left(-\frac{\hbar^2}{2m}\nabla^2 + V\right)\psi = E\psi \tag{1}$$

to an equation for φ which has exactly the same eigenvalue E:

$$\left(-\frac{\hbar^2}{2m}\nabla^2 + V_{ps}\right)\varphi = E\varphi \tag{2}$$

Here, V_{ps} is the pseudopotential (Phillips and Kleinman,[36] Cohen and Heine,[17] and Austin et al.[7]). There are actually many different forms for V_{ps}, all producing slightly different φ functions but still the same energy. One form for V_{ps} is defined by:

$$V_{ps}\varphi = V\varphi - \sum_c (\psi_c, V\varphi)\psi_c \tag{3}$$

where the parentheses denote the matrix element of ψ_c with $V\varphi$. The summation is over all the atomic states ψ_c in the ion core (with suitable precautions; see Heine[23]). If φ refers to an s-like state, φ is approximately constant over the core (Fig. 1), and equation (3) becomes

$$V_{ps} \approx V - \sum_c (\psi_c, V)\psi_c \tag{4}$$

Here, the second term represents an expansion of V in terms of the functions ψ_c. If they formed an infinite complete set of functions, the expansion would be identically equal to V, and V_{ps} would vanish. The same argument can be applied to equation (3), but it is easier to visualize in terms of just V. Now the ψ_c functions are only a finite set of atomic functions, so that equations (3) and (4) do not vanish completely; but the ψ_c functions are quite a good set for expansions inside the core radius, and so we have:

$$\begin{aligned} V_{ps} &\approx 0 \quad \text{inside the core} \\ &\approx V \quad \text{outside the core} \end{aligned} \tag{5}$$

An actual calculated V_{ps} is given in Fig. 2, showing how the center is chopped out of V, thus making an effectively weak V_{ps}. Incidentally, the sharp spike remaining at the origin is of no moment.

Quite another way of looking at the pseudopotential is in terms of phase shifts. As is well known (Schiff,[38] p. 105), the scattering of plane waves from a short-range potential can be described entirely in terms of δ_l, irrespective of the details of the potential. Moreover, the scattering is determined completely by $\exp(2i\delta_l)$, so that phase shifts differing by integer multiples of π produce the same effect. In Fig. 1, the real potential is strong enough to produce a phase shift of about 2π, corresponding to the two nodes in ψ_k at small r. If we write for $l = 0$:

$$\delta_0 = 2\pi + \Delta_0 \tag{6}$$

then Δ_0 for sodium would be quite small. We now subtract the 2π, and define the pseudopotential as a weak potential producing the small $l = 0$ phase shift of Δ_0 with no nodes in the wave function; the procedure is the same for higher l.

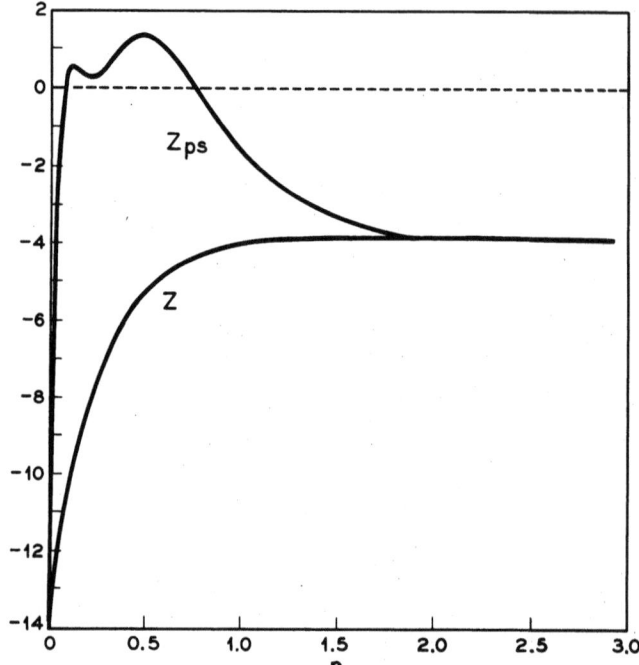

Fig. 2. True potential V and pseudopotential V_{ps} for Si^{4+}, given by $V = 2Z(r)/r$ and $V_{ps} = 2Z_{ps}(r)/r$ in rydbergs. r is in atomic units.

A similar approach is that of the quantum defect method for calculating band structures. If we integrate the Schroedinger equation out from the center of the atom, the boundary condition that ψ is finite at $r = 0$ means there is only one solution (for each l) at given E. Thus, if we suppose we have accurately determined in a band structure calculation some Bloch state ψ_k and its energy E, then it must have the prescribed form inside the atom. In particular, it must have the right radial derivative $\psi'/\psi (\psi' \equiv \partial \psi/\partial r)$ at some radius R a bit outside the core. Thus, if we knew the required ψ'/ψ at R, the band structure problem would reduce to solving for ψ in the region between the spheres of radius R, while making it satisfy the correct boundary condition ψ'/ψ at $r = R$. There is no need for any reference to the potential or wave function inside the spheres, and in the quantum defect method the required ψ'/ψ at R is deduced from the observed spectroscopic term values of the free atom without ever calculating a potential. We can also deduce that an atomic pseudopotential V_{ps} giving a pseudo wave function φ will determine exactly the same band structure for the solid as the real potential V, so long as V_{ps} and V are identical for $r > R$ and produce identical radial derivatives at R, as shown for ψ and φ in Fig. 1. Inside the spheres, ψ and φ may differ, and in particular V_{ps} may be chosen such that φ is smooth without nodes there.

The preceding discussion may be summarized in three points: First, the use of a pseudopotential is not a qualitative or empirical swindle, but a respectable mathematical transformation from the real Schroedinger equation with V and ψ to one with V_{ps} and φ. Moreover, there is considerable arbitrariness left among the allowed pseudopotentials. Second, by definition the pseudo wave functions φ have had eliminated from them all atomic-like nodes inside the ion core. Consequently, the φ functions in a solid can be expanded in terms of a small number of plane waves. Along with this, V_{ps} is weak. If the cancellation in equations (3) and (4) is good, it may be sufficiently weak to be treated well by a simple perturbation theory, but that is a point of detail and not always so. Finally, the pseudopotential gives only energies and scattering amplitudes correctly, but not of course the wave function inside the core or anything like the conduction electron charge density depending on it. The ψ has to be constructed from the φ, and is given to a very good approximation by:

$$\psi \approx \varphi - \sum_c (\psi_c, \varphi)\psi_c \qquad (7)$$

All the 1s, 2s, 3s, etc., wave functions of one atom have their inner maxima and minima at about the same radii, and so the core states ψ_c can be used in this way to make an excellent approximation to ψ. The coefficients are conveniently chosen to make ψ orthogonal to the ψ_c, as it has to be.

The form of the pseudopotential which has been developed in Cambridge illustrates most of the above ideas. Actually, we call it the Model Potential V_M for historical reasons. At this stage, we are concerned with the bare ion, e.g., Al^{+++}: The potential from the conduction electrons comes later. In the spirit of equation (5), we choose a radius R_M somewhat larger than the ion core radius, and set:

$$V_M(r) = -A \qquad \text{for} \quad r < R_M \qquad (8a)$$
$$= -Z/r \qquad \text{for} \quad r > R_M \qquad (8b)$$

as shown in Fig. 3 (Heine and Abarenkov[24]). The constant A is determined by adding one extra electron and fitting A to the observed spectroscopic term values of the free atom or ion Al^{++} in our example. This is quite simple. Outside R_M, the

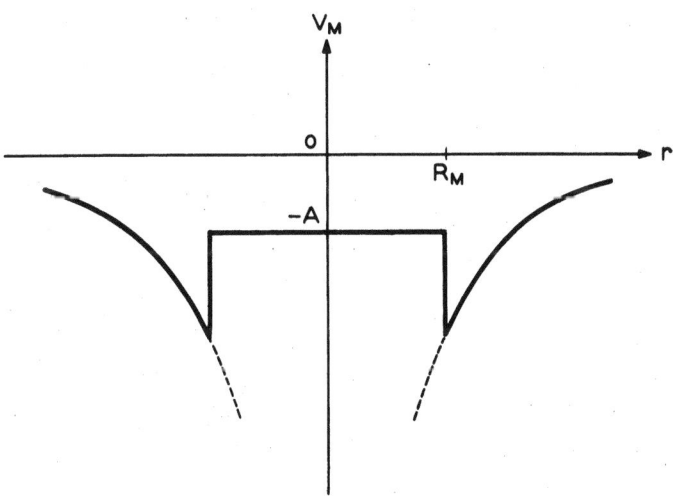

Fig. 3. The model potential.

potential is a pure coulomb one, for which wave functions are tabulated. The table gives the radial derivative ψ'/ψ at R_M. The wave function at $r < R_M$ can be written explicitly in sines and cosines and A adjusted to make the derivatives match at R_M. Actually, A will depend on l and slightly on energy. A more correct statement of equation (8a) is therefore:

$$V_M = - \sum_l A_l(E) P_l \qquad \text{for} \qquad r < R_M \qquad (8a')$$

where P_l is a projection operator which picks out the lth component of the wave function. Since we have used the experimental term values, we have a pseudopotential which represents the ion core *exactly*, including all exchange and correlation effects. The formalism has also been extended to include spin-orbit coupling.

Fermiology

We are now in a position to set up the potential in the metal. We start by representing the conduction electrons as a uniform negative jelly into which we place the bare positive ions at arbitrary positions R_j. The potential in the system can be Fourier analyzed in the form:

$$(\text{const}) + \Sigma' C(\mathbf{q}) \exp(i\mathbf{q} \cdot \mathbf{r}) \qquad (9)$$

where Σ' excludes the term $\mathbf{q} = 0$. Since the electron density is uniform, it does not contribute to the $\mathbf{q} \neq 0$ components of the charge density and hence by Poisson's equation not that of the potential either. In fact we have:

$$C(\mathbf{q}) = S(\mathbf{q}) V_{\text{ion}}(q) \qquad (10)$$

where:

$$S(\mathbf{q}) = \frac{1}{N} \sum_j \exp(-i\mathbf{q} \cdot \mathbf{R}_j) \qquad (11)$$

is the structure factor, and:

$$V_{\text{ion}}(q) = (1/\Omega) \int V_{\text{ps}}(\mathbf{r}) \exp(-i\mathbf{q} \cdot \mathbf{r}) \, dv \qquad (12)$$

depends only on the pseudopotential of the bare ion. N is the total number of atoms and Ω the volume per atom.

We can now calculate what the conduction electrons do. In principle, we should calculate all the occupied Bloch states and sum their charge density, but we can obtain much more simply a result which is an excellent approximation for all applications discussed here and which in other cases like diamond is the starting point for a higher order calculation (Bennemann[10]). Since we are using a pseudopotential, the conduction electrons can be treated as a free electron gas. If the system is perturbed by an external potential:

$$C(\mathbf{q}) \exp(i\mathbf{q} \cdot \mathbf{r}) \qquad (13)$$

the electrons will rearrange themselves so as to screen it, the resultant total self-consistent potential becoming:

$$\frac{C(\mathbf{q})}{\varepsilon(q)} \exp(i\mathbf{q} \cdot \mathbf{r}) \qquad (14)$$

where $\varepsilon(q)$ is some screening factor or "dielectric constant." The $\varepsilon(q)$ can be calculated simply by perturbation theory in the Hartree approximation (Bardeen[8]). Sham has shown how to include screened exchange so that the result represents what an incoming electron sees, including the exchange with the heaped-up electron gas (Sham and Ziman[44] and Sham[43]). Also, Animalu[5] has generalized the calculation to take account of the nonlocal l-dependence of the pseudopotential. Another correction comes from the fact that the oscillations of ψ in the core result in some charge deficit there; and there are others (Heine and Abarenkov[24])!

Now, in all the applications considered here, $C(q)$ is small. In the solid at the reciprocal lattice vectors $V_{ion}(q)$ is small, and in the liquid metal at small q where $V_{ion}(q)$ is large there $S(q)$ is small. Thus the linearity approximation in formula (14) is justified, and so is the superposition approximation of screening each $C(q)$ in expression (9) individually. We therefore obtain the self-consistent screened pseudopotential:

$$U(r) = (\text{const}) + \sum_{q}{}' \frac{S(\mathbf{q})V_{ion}(q)}{\varepsilon(q)} \exp(i\mathbf{q} \cdot \mathbf{r}) \tag{15}$$

We now define:

$$v(q) = \frac{V_{ion}(q)}{\varepsilon(q)}$$

so that the final potential becomes:

$$U = \sum_{q}{}' S(\mathbf{q})v(q) \exp(i\mathbf{q} \cdot \mathbf{r}) \tag{16}$$

We term this the "screened ion" method of setting up the potential. In the solid, of course, $S(\mathbf{q})$ is nonzero only for reciprocal lattice vectors $\mathbf{g}_n$. The potential is:

$$U = \sum_{n}{}' v(\mathbf{g}_n) \exp(i\mathbf{g}_n \cdot \mathbf{r}) \tag{17}$$

There are two points to make at this juncture. The first is to transform equation (16) back into real space:

$$U = \sum_{j} v(\mathbf{r} - \mathbf{R}_j)$$
$$v(\mathbf{r}) = (1/N)\Sigma_q v(q) \exp(i\mathbf{q} \cdot \mathbf{r}) \tag{18}$$

We have therefore proved that in our approximation U can be written as a superposition of neutral pseudopotentials $v(\mathbf{r})$, each representing an ion screened by the conduction electrons. Actually, the pseudopotentials are nonlocal operators and not just functions of r as written here, so that one has to be a bit careful about all these Fourier transforms, but that is a matter of detail for the professionals. In practice, it means that the $v(q)$ relevant to the distortion of the Fermi surface is not exactly equal to half the band gap, and neither is quite equal to the potential determining the frequency of a phonon of wave number q. The second point is that it is possible to calculate the charge density and kinetic energy of the electron gas at the same time as the self-consistent potential in expression (14) is calculated, and hence to write down an expression for the total energy of the whole system in terms of the positions $\mathbf{R}_j$.

Table II. Fourier Components of the Pseudopotential for Aluminum (in Rydbergs)

	$\bar{v}(111)$	$\bar{v}(200)$
From experiment (Ashcroft[6])	0.0179	0.0562
Calculated (Animalu[5])	0.012	0.045

We turn at last to the actual applications. The first is of course to use equation (17) as a conceptual scheme in order to analyze band structures and Fermi surfaces in terms of a perturbed free-electron model. This is already so well known that we will only stop to note two outstanding examples: the work of Brust[11] in correlating all the relevant data on silicon and germanium, and Ashcroft's[6] fitting of the Fermi surface of aluminum. By using the fact that the band structure must fit reasonably closely to a nearly free-electron model, Ashcroft was able to resolve ambiguities about the data and arrive at a unique and consistent interpretation of all the observations. The values deduced for the Fourier components $v(111)$ and $v(200)$ are shown in Table II. To be precise, since he used a 4×4 secular equation near the Fermi energy, his values do not correspond exactly to $v(\mathbf{g}_n)$ but to:

$$\bar{v}(\mathbf{g}_n) = v(\mathbf{g}_n) + \sum_{m}{}'' \frac{\langle \mathbf{k}_F + \mathbf{g}_n | U | \mathbf{k}_F + \mathbf{g}_m \rangle \langle \mathbf{k}_F + \mathbf{g}_m | U | \mathbf{k}_F \rangle}{E - E(\mathbf{k}_F + \mathbf{g}_m)} \quad (19)$$

where the second term represents the effect of higher components and the summation excludes the four reciprocal lattice vectors included in the original 4×4 equation.

The second application of the pseudopotential is to calculate band structures *ab initio*. As a test of the screened ion model potential, Table I includes values of $\bar{v}(\mathbf{g})$ calculated by Animalu.[5] The agreement is really quite good. Typical electron energies in aluminum are of the order of 1 Ry, the $\bar{v}$ values being small because of much cancellation; it is doubtful whether an accuracy better than a few thousandths of a rydberg can be hoped for without an elaborate analysis of many minor effects. Such an error corresponds approximately to twice the thickness of the line in Fig. 4

Table III. Extension of the Fermi Surface in the (110) Direction Beyond the Free Electron Sphere (%)

	Li	Na	K	Rb	Cs
Calculated from screened ion model potential*	2.7–8	0.10–0.45	0–0.18	0.25–0.75	1.0–2.0
Calculated from Wigner–Seitz–Ham potential†	2.3	0.00	0.7	1.8	8.0
From experiment		~3‡	0?§	0.11§	0.6¶	3¶‖
					0.95§	

* Heine and Abarenkov.[24]
† Ham.[19]
‡ Stewart.[47]
§ Schoenberg and Stiles.[45]
¶ Okumura and Templeton.[35]
‖ Templeton and Okumura.[48]

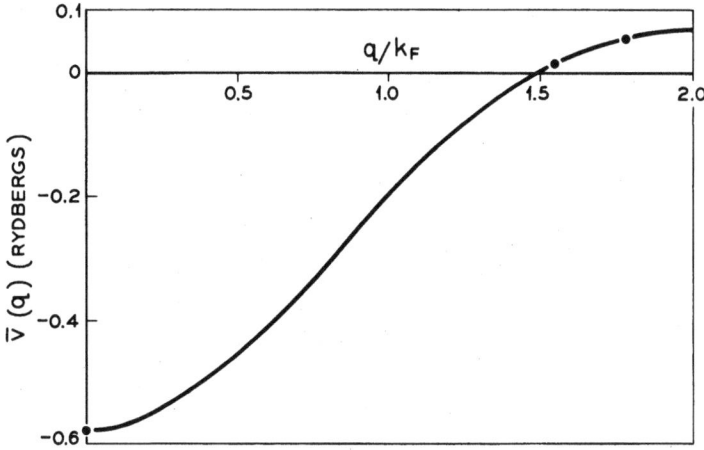

Fig. 4. Screened pseudopotential in aluminum metal.

showing $v(q)$. The agreement is therefore a quantitative validation of the whole structure we have built up. Table III shows some calculated results for the distortion of the Fermi surfaces of the alkali metals, together with experimental results and the calculations by Ham.[19] Here, the screened-ion–model-potential calculations were not done quite so carefully as in the work on aluminum above, but the somewhat better agreement with experiment compared to Ham's calculation probably indicates the superiority of the screened ion method over the Wigner–Seitz potential in representing the correlation and exchange hole. The next task will be to make calculations for more complicated polyvalent metals where the theoretical results may aid materially in interpreting the data. With the model potential of equation (8), one can deal with a heavy atom such as bismuth just as easily as with a simple one like sodium, so long as the atomic term values are available. In any case, the quantitative validation of the screened ion model potential by the Fermi surface data on aluminum and the alkalis is an important step toward its wider use for other properties.

As the third application, we turn to the study of other properties of aluminum. It is necessary to construct $\bar{v}(q)$ for the whole range of q, at least up to $2 k_F$. The Fermi surface analysis gives values at $g(111)$ and $g(200)$, and at $q = 0$ it is $-n/n(E_F)$, where n is the electron density and $n(E_F)$ the density of states at the Fermi level, including exchange and correlation corrections but excluding the electron–phonon contribution to the effective mass. The rest of the curve can be interpolated by taking the calculated one and moving it slightly to pass through the required points (Fig. 4). Incidentally, the difference between $\bar{v}$ and v is like the difference between the real scattering matrix and the first Born approximation in scattering theory, and it is $\bar{v}$ we want. We also note that Ashcroft's matrix elements are those relevant to scattering on the Fermi surface, i.e., $\langle \mathbf{k} + \mathbf{g}|v(E)|\mathbf{k}\rangle$ with $|\mathbf{k}| = |\mathbf{k} + \mathbf{g}| = k_F$ and $E = E_F$: this is just what is required for the following. It is now possible to calculate the resistance of liquid aluminum using the theory of Ziman,[55] and Bradley et al.[55]: the answer comes out within a few percent of the experimental figure (Ashcroft, prive communication). This is important because it constitutes the

Table IV. Electron–Phonon Enhancement of the Density of States at E_F, Expressed as an Effective Mass

	m^*
Calculated	1.47
Experiment (specific heat)	1.46
Experiment (cyclotron resonance)	1.49

first empirical check on Ziman's theory, and dispels the doubts expressed by Bradley *et al.* At that time there appeared to be a serious and systematic discrepancy when rough estimates were made: the values of $|\bar{v}|^2$ deduced by them probably appear low because these represent a mean of $|\bar{v}(q)|^2$ over a region which includes the point where $\bar{v}(q)$ passes through zero (Fig. 4). Ashcroft and Wilkins (private communication) are now using the experimentally fitted $\bar{v}(q)$ to calculate the superconducting transition temperature and the electron–phonon enhancement of the effective mass at low temperature for aluminum. Their results for the latter are shown in Table IV.

Acknowledgments

It is a pleasure to acknowledge the generous hospitality of Bell Telephone Laboratories, where this paper was written, and in particular the helpful conversations with Dr. L. Mattheiss and Professor J. C. Phillips.

References

1. S. L. Altmann, *Proc. Roy. Soc. (London) Ser. A* **244**, 141, 153, 1958.
2. S. L. Altmann, unpublished communication.
3. S. L. Altmann and N. V. Cohan, *Proc. Phys. Soc. (London)* **71**, 383, 1958.
4. P. W. Anderson, *Concepts in Solids*, W. A. Benjamin, Inc., New York (1963).
5. A. E. O. Animalu, *Phil. Mag.* **11**, 379, 1965.
6. N. W. Ashcroft, *Phil Mag.* **8**, 2055, 1963.
7. B. J. Austin, V. Heine, and L. J. Sham, *Phys. Rev.* **127**, 276, 1962.
8. J. Bardeen, *Phys. Rev.* **52**, 688, 1937.
9. F. Bassani, and L. Liu, *Phys. Rev.* **132**, 2047, 1963.
10. K. H. Bennemann, *Phys. Rev.* **133**, A1045, 1964; *Phys. Rev.* **137**, A1497, 1965.
11. D. Brust, *Phys. Rev.* **134**, A1337, 1964.
12. G. A. Burdick, *Phys. Rev.* **129**, 138, 1963.
13. J. Callaway, *Phys. Rev.* **112**, 322, 1958.
14. J. Callaway, in: *Solid State Physics, Vol. 7*, Academic Press, Inc., New York (1958), p. 99.
15. J. Callaway, *Phys. Rev.* **124**, 1824, 1961.
16. J. Callaway, *Energy Band Theory*, Academic Press, Inc., New York (1964).
17. M. H. Cohen and V. Heine, *Phys. Rev.* **122**, 1821, 1961.
18. L. M. Falicov, *Phil. Trans. Roy. Soc. London, Ser. A* **255**, 55, 1962.
19. F. S. Ham, *Phys. Rev.* **128**, 82 and 2524, 1962.
20. F. S. Ham and B. Segall, *Phys. Rev.* **124**, 1786, 1961.
21. W. A. Harrison, *Phys. Rev.* **126**, 497, 1962.
22. W. A. Harrison *Phys. Rev.* **131**, 2433, 1963.
23. V. Heine, *Proc. Roy. Soc. (London) Ser. A* **240**, 354, 1957.
24. V. Heine and I. Abarenkov, *Phil. Mag.* **9**, 451, 1964.
25. F. Herman, *Rev. Mod. Phys.* **30**, 102, 1958.
26. L. Kleinman and J. C. Phillips, *Phys. Rev.* **116**, 880, 1959.
27. L. Kleinman and J. C. Phillips, *Phys. Rev.* **118**, 1153, 1960.
28. L. Kleinman and J. C. Phillips, *Phys. Rev.* **125**, 819, 1962.

29. T. L. Loucks, *Phys. Rev.* **134**, A1618, 1964.
30. T. L. Loucks and P. H. Cutler, *Phys. Rev.* **133**, A819, 1964.
31. L. F. Mattheiss, *Phys. Rev.* **133**, A1399, 1964; **134**, A970, 1964.
32. L. F. Mattheiss (to be published).
33. M. Miasek, *Phys. Rev.* **130**, 11, 1963.
34. N. F. Mott and H. Jones, *The Theory of the Properties of Metals and Alloys*, Clarendon Press, Oxford (1936).
35. K. Okumura and I. M. Templeton, *Phil. Mag.* **7**, 1239, 1963; **8**, 889, 1963.
36. J. C. Phillips and L. Kleinman, *Phys. Rev.* **116**, 287, 1959.
37. J. R. Reitz, in: *Solid State Physics, Vol. 1*. Academic Press, Inc., New York (1955), p. 1.
38. L. I. Schiff, *Quantum Mechanics*, second edition, McGraw-Hill, Book Co., New York (1955).
39. H. Schlosser and P. M. Marcus, *Phys. Rev.* **131**, 2529, 1963.
40. B. Segall, *Bull. Am. Phys. Soc.* **6**, 145, 1961; and private communication.
41. B. Segall, *Phys. Rev.* **124**, 1797, 1961.
42. B. Segall, *Phys. Rev.* **125**, 109, 1962.
43. L. J. Sham, thesis, University of Cambridge (available on microfilm from Micromethods Ltd., East Ardsley, Wakefield, Yorkshire, England).
44. L. J. Sham and J. M. Ziman, *Solid State Physics, Vol. 15*, Academic Press, Inc., New York (1963), p. 221.
45. D. S. Shoenberg and P. J. Stiles, *Proc. Roy. Soc. (London) Ser A*, **281**, 62, 1964.
46. P. Soven, *Phys. Rev.* **137**, A1717, 1965.
47. A. T. Stewart, this volume, p. 835.
48. I. M. Templeton and K. Okumura, *Bull. Am. Phys. Soc.* **9**, 239, 1964.
49. W. M. Walsh, C. C. Grimes, G. Adams, and L. W. Rupp, this volume, p. 765.
50. J. H. Wood, *Phys. Rev.* **126**, 517, 1962.
51. J. H. Wood, *Bull. Am. Phys. Soc.* **8**, 221, 1963.
52. J. Yamashita, M. Fukuchi, and S. Wakoh, *J. Phys. Soc. Japan* **18**, 999, 1963.
53. J. M. Ziman, *Advan. Phys.* **13**, 89, 1964.
54. J. M. Ziman, *Principles of the Theory of Solids*, Cambridge University Press, Cambridge (1964).
55. J. M. Ziman, *Phil. Mag.* **6**, 1013, 1960; see also C. C. Bradley, T. E. Faber, E. G. Wilson, and J. M. Ziman, *Phil. Mag.* **7**, 865, 1962.

MAGNETIC BREAKDOWN IN MAGNESIUM AND ZINC*

R. W. Stark[†]

Department of Physics and Institute for the Study of Metals
University of Chicago, Chicago, Illinois

Introduction

The concept of magnetic breakdown (MB) was introduced by Cohen and Falicov[1] to explain the presence of the "giant orbit" which Priestley[2-3] observed in his pulse field de Haas–van Alphen (DHVA) measurements in magnesium. Since then, the effects of MB have been observed in several metals such as zinc,[4-8] aluminum,[9,10] thallium,[11-13] beryllium,[14] rhenium,[15-16] tin,[17] and iron.[18] Thus, the effect is quite common and must be taken into consideration when experiments are performed to study the electronic band structure of metals.

In this paper, we will be concerned with the effects of MB on the galvanomagnetic properties of magnesium[19] and zinc.[4] The electronic band structures of these two hexagonal close-packed metals are very similar in many respects, if one takes into account the difference in their c/a ratios. Both have small band gaps which are affected by MB in fields as small as 1 kG. This MB causes essential changes in the gross behavior of the galvanomagnetic properties and, in addition, reveals a wealth of information about the Landau level structure in the bands involved in the MB process.

Theory

In general, the tranverse magnetoresistance, $\rho_T(H)$, of a metal single crystal will follow one of two characteristic patterns of behavior in the limit of high magnetic fields. Either $\rho_T(H)$ will saturate, i.e., $\rho_T(H) = C$ for large H or $\rho_T(H)$ will increase without bound as H^2. These properties have been uniquely related to the type of electronic trajectories and to the relative number of electrons and holes that exist for a given direction of H by Lifshitz et al.[20,21] The following cases are of interest:

1. All of the electronic trajectories are closed for a given direction of H.
   Here we have two cases, depending on the number of electrons, n_1, and the number of holes, n_2.
   a. For $n_1 \neq n_2$, $\rho_T(H)$ saturates for large H.
   b. For $n_1 = n_2$, $\rho_T(H) = CH^2$.
2. A layer of open trajectories with a single average direction exist for a given direction of H.

---

* Supported in part by the Army Research Office (Durham), the Advanced Research Projects Agency, and the Alfred P. Sloan Foundation.
† Alfred P. Sloan Research Fellow.

In this case, the transverse magnetoresistance is expressed by the equation:

$$\rho_T(H) = A + BH^2 \cos^2 \alpha \tag{1}$$

where A and B are constants, and α is the angle between the current direction, J, and the open orbit direction in reciprocal space.

When there are regions where two bands of the Fermi surface of a metal are separated by relatively small band gaps ($\sim 10^{-2}$ eV), the dynamics of the electrons' motion in the presence of a high magnetic field ($\sim 10^4$ G) is such that there is a high probability for the electrons to ignore such gaps and to follow trajectories which combine segments of the two bands. Thus, the essential effect of MB is to change the character of the orbits on the Fermi surface. Groups of open orbits can be changed into segments of a new group of closed orbits; groups of closed orbits can be changed into new groups of open orbits, etc. In addition, the number of electrons and holes are defined in terms of the electron trajectories, so that another effect of MB is that it can change the relative numbers of holes and electrons. Thus, the effects of MB on $\rho_T(H)$ can cause a transition from an initial H^2 behavior to a final saturation behavior, from an initial saturation behavior to a final H^2 behavior, etc.

The probability for MB, that is, the probability that an electron will ignore a Bragg diffraction condition and make a transition between two different orbits has been shown by Blount[22] to be

$$P = \exp[-(H_0/H)] \tag{2}$$

$$H_0 = \frac{kmcE_g^2}{\hbar e E_F} \tag{3}$$

where E_g is the energy gap, E_F the Fermi energy, and k is a constant of order unity.

The Giant Orbit

A giant orbit whose cross-sectional area is larger than the cross-sectional area of the Brillouin zone has been observed in both magnesium[1,2,3,19] and zinc[4,6] when the magnetic field direction is parallel to the hexad axis. This orbit is, in effect, the orbit of an electron which ignores all Bragg diffraction conditions present at various points along its path. Hence, the orbit arises as a result of MB. The gross effects of this MB on $\rho_T(H)$ for magnesium and zinc can be readily understood if we consider the model shown in Fig. 1.

Figure 1 shows how the free-electron orbit segments are coupled in the low and high field limits in a hexagonal grid of crystallographic planes. The situation shown is very similar to that which exists in magnesium and zinc. Orbit 1 corresponds to an orbit around the outside of the second band hole surface (the "monster"), and orbit 2 corresponds to an orbit about portions of the third band electron surfaces (the "cigar" in magnesium, the "needle" in zinc). Bragg diffraction conditions are satisfied at the points labeled B. In the low-field limit (orbits 1 and 2), the electron suffers Bragg diffraction at each point B and ends up with the possibility of forming one of two separate closed orbits. On orbit 1, it travels with a net counterclockwise motion. On orbit 2, it travels with a net clockwise motion. In the high-field limit (orbit 3), MB causes the electron to ignore the Bragg diffraction

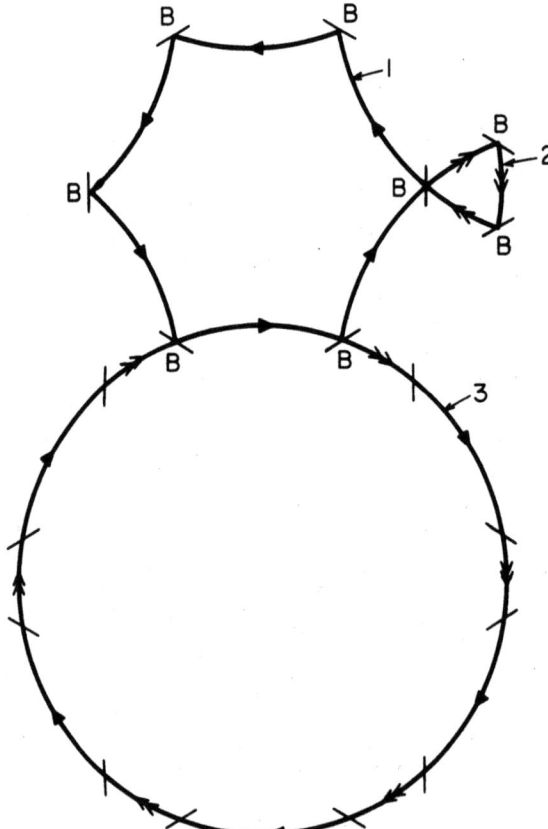

Fig. 1. Possible real space orbits for an electron in a hexagonal crystallographic grid. The points, B, are those points for which Bragg diffraction conditions are satisfied. The lines represent crystallographic planes. Orbits 1 and 2, only, exist when $P = 0$. Orbit 3, only, exists when $P = 1$.

conditions and to form a free-electronlike orbit (the giant orbit) on which it travels with a net clockwise motion.

The important point to note here is that, although all of these orbits are closed, the electron on orbit 1 in the low-field limit travels around its path in the opposite direction to the high-field limit free electron. The net effect, then, is that it acts as if it had a positive charge instead of a negative charge. Because of this, we define this orbit to be a hole orbit.

Both magnesium and zinc are even-valent metals with two atoms and four electrons per unit cell. In the low-field limit, when H is parallel to the hexad axis, all of the orbits on the Fermi surface are closed and $n_1 = n_2$. Thus, $\rho_T(H)$ exhibits class Ib behavior and increases as H^2. When H is increased to a few kilogauss, MB resulting in a giant orbit, such as discussed above, becomes prevalent. The net effect of this MB is that it recouples the segments of closed-hole orbits and closed-electron orbits to form a new closed-electron orbit. This changes the relative numbers of electrons and holes, so that $n_1 \neq n_2$ in the high-field limit. Thus, $\rho_T(H)$ will exhibit case Ia behavior and saturate in high magnetic fields.

Figure 2 shows a plot of $\rho_T(H)$ of magnesium for J in the $(10\bar{1}0)$ direction and H parallel to (0001). When $H \lesssim 1$ kG, $\rho_T(H)$ increases as H^2. When $H \simeq 1$ kG, MB begins to become significant and $\rho_T(H)$ begins to depart from its low-field behavior. When $H = 1950$ G, $\rho_T(H)$ has reached its maximum value, and any further increase

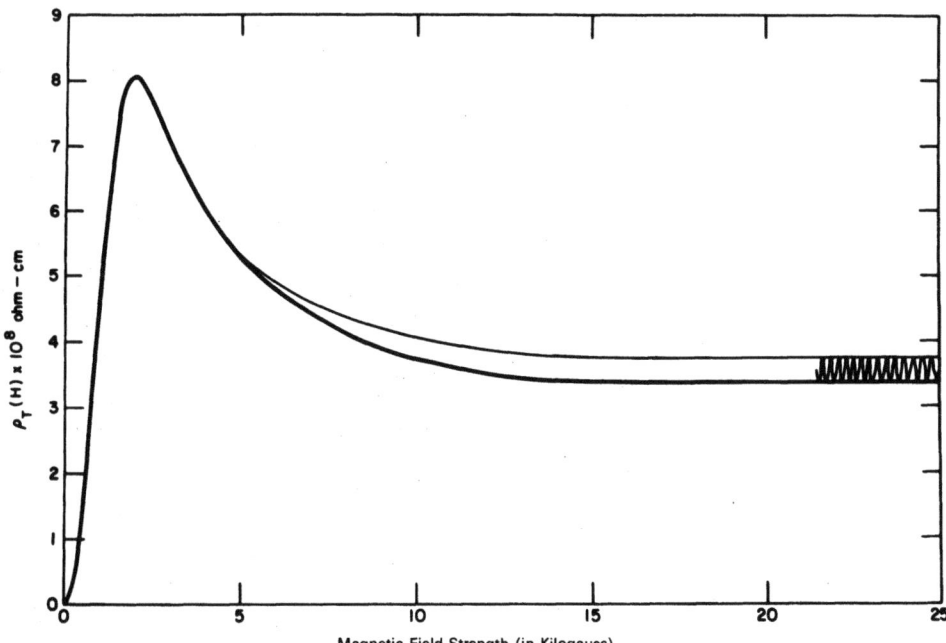

Fig. 2. $\rho_T(H)$ for a magnesium specimen with J parallel to $(10\bar{1}0)$ and H parallel to (0001). The residual resistance ratio of this specimen was about 3×10^6.

in H causes $\rho_T(H)$ to decrease until MB becomes essentially complete, and it reaches its high-field saturation value at $H \simeq 15\,\text{kG}$.

Quantum oscillations which are periodic in H^{-1} with a period of 4.38×10^{-7} G^{-1} are superimposed on this background behavior. At 1.2°K, these oscillations occur within the envelope shown in Fig. 2. As the temperature is increased, the amplitude of these oscillations decreases in a fashion such that the lower curve of the envelope remains fixed and the upper curve approaches it. The amplitude of oscillation at 4.2°K is only about 10% of that at 1.2°K. These oscillations arise from the quantized density of states of the cigar portion of the Fermi surface (orbit 2 in Fig. 1) and are seen here due to the fact that this quantization perturbs the MB process, since P depends on the relative density of states at the Fermi surface in the two bands between which the transition occurs.

One of the most important features of Fig. 2 is the fact that when MB is nearly complete, $\rho_T(H)$ approaches a saturation value of $3.4 \times 10^{-8}\,\Omega\,\text{cm}$. This value is typical of that observed in several other specimens of different purity for similar orientation of J and H. The residual resistance ratios of these specimens ranged from 10^4 to 3×10^6, so that it appears that this saturation value for $\rho_T(H)$ is nearly independent of the electronic relaxation time τ. This is certainly a major departure from the behavior one observes in a metal when MB is not present. In such a case, the saturation value of $\rho_T(H)$ is inversely proportional to τ.

Falicov and Sievert[23] showed that this particular effect is a result of MB and can be considered to result from an additional electronic scattering mechanism associated with incoherent Bragg diffraction after MB is complete. They used a modification of Chambers' path integral method to compute the semiclassical transverse conductivity and resistivity tensors for the model shown in Fig. 1 and found that the saturation value of $\rho_T(H)$ satisfied the relation

$$\rho_T(H)_{sat} \alpha (\tau^{-1} + c\omega_0) \qquad (4)$$

where c is a constant of order unity, $\omega_0 = eH_0/mc$, and H_0 is defined in (2) and (3). Thus, when $\omega_0\tau \gg 1$, $\rho_T(H)_{sat}$ is essentially independent of τ and is determined by the effects of MB.

Figure 3 shows $\Delta\rho_T(H)/\rho_T(0)$ of zinc for J in the $(11\bar{2}0)$ direction and H parallel to (0001). The two curves shown were taken at two different temperatures. The lower curve was obtained at 4.2°K, while the upper curve was obtained at 1.6°K. The difference between the curves results from the fact that the specimen's resistivity was limited by electron–phonon interactions at 4.2°K. The residual resistance ratio at 4.2°K was 3.2×10^4, whereas the ratio at 1.6°K had risen to 5×10^4. These curves have the same basic features as the curve shown in Fig. 2 and arise for the same reasons. However, in contrast with Fig. 2, the dominant feature of these curves is the large amplitude of the oscillatory component whose period is $6.42 \times 10^{-5} \, \text{G}^{-1}$. These oscillations have been discussed in detail in ref. 4. The contrast with Fig. 2 arises as a result of the fact that the effective mass, m_c^*, of the electrons on the cigar in magnesium is about $0.1m_0$, while the effective mass, m_N^*, of the electrons on the needle in zinc is $m_N^* = 0.007m_0$. Thus, the Landau levels of the needle will be separated in energy by a factor of about 15 times the separation of the Landau levels of the cigar and will cause a much greater perturbation on the MB. The secondary structure shown in the oscillations in Fig. 3 results from spin splitting of the original Landau levels of the needle and are associated with an effective g^* factor of $g^* = 90$.[4,24] The energy equation for the position of the

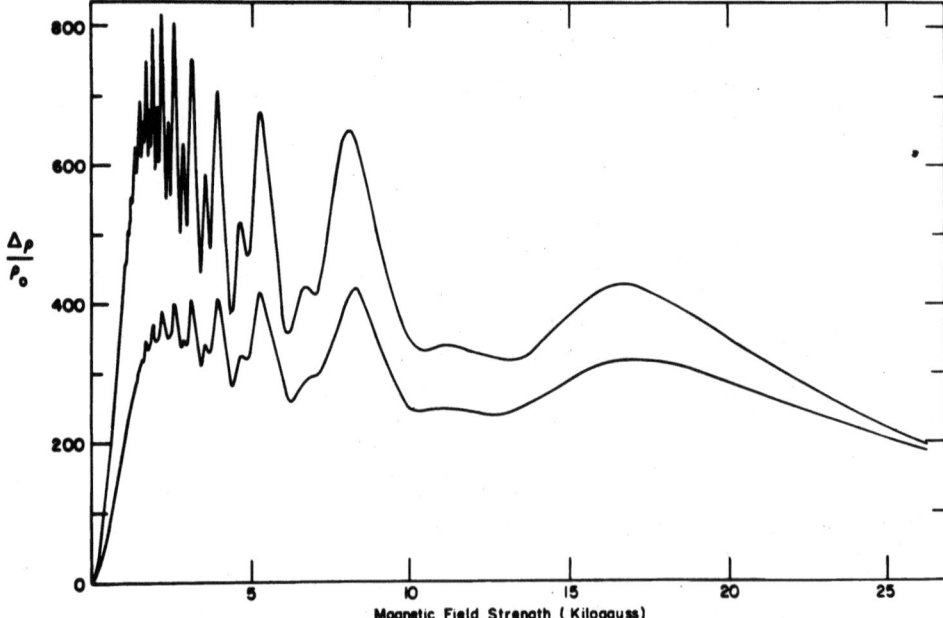

Fig. 3. $\Delta\rho_T(H)/\rho_T(0)$ for a zinc specimen with J parallel to $(11\bar{2}0)$ and H parallel to (0001). Both curves were taken from the same specimen. The lower curve was obtained at 4.2°K and the upper curve was obtained at 1.6°K. The average electronic relaxation time at 1.6°K was 1.56 times greater than at 4.2°K.

Landau levels of the needle for this value of g^* is

$$E = (n + \tfrac{1}{3} \pm \tfrac{1}{6})\hbar\omega_N \tag{5}$$

$$\omega_N = \frac{eH}{m_N^* c} \tag{6}$$

Falicov and Sievert[25] have obtained theoretical curves which are very similar to those shown in Fig. 3. They introduced the oscillatory behavior into their semiclassical theory by assuming that the electron behaves in a semiclassical manner everywhere, except when on the needle, where one must keep track of the phase of its wave function in the manner discussed by Pippard.[26] They used energy levels corresponding to (5) and found a fine structure in the oscillations similar to that shown in Fig. 3. Figure 4 shows the result of their calculations. They have obtained similar curves which agree equally well with the experimental data for magnesium.

It is thus becoming apparent that many of the MB effects on the galvanomagnetic properties can be treated semiclassically. One of the most important results of Falicov and Sievert's theory is the concept of a magnetic relaxation time $\tau_m \alpha \omega_0^{-1}$. In the case discussed above, this relaxation effect causes $\rho_T(H)_{\text{sat}}$ to be determined by ω_0, independent of τ when $\omega_0 \tau \gg 1$. In other cases where MB recouples orbit segments to form new orbits (for instance, by changing one group

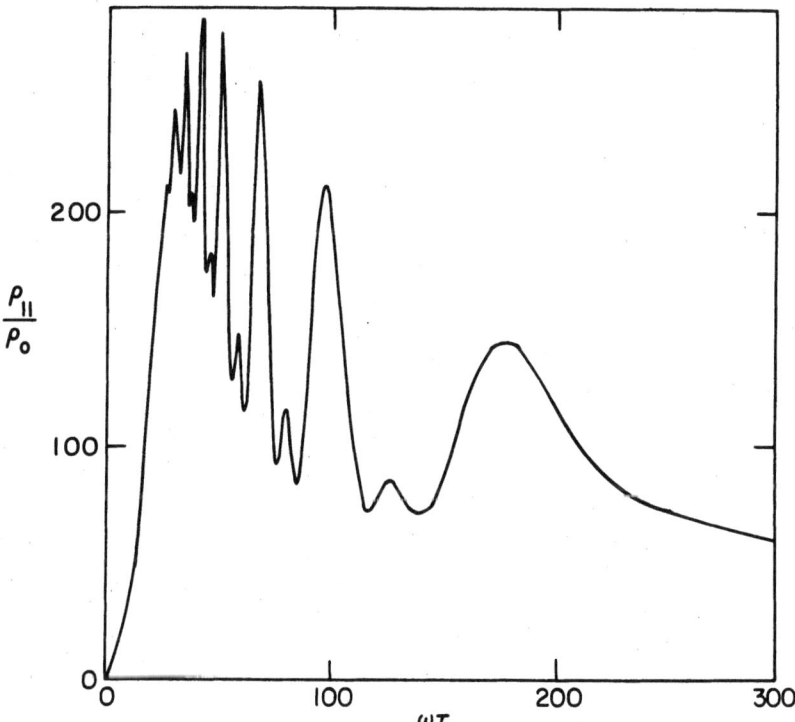

Fig. 4. $\rho_T(H)/\rho_T(0)$ as calculated by Falicov and Sievert. They used energy levels corresponding to equation (5) to introduce the effects of an effective g^* factor of $g^* = 90$. This theoretical curve corresponds to the case shown in Fig. 3.

of closed orbits into a new group of closed orbits in such a fashion that $n_1 = n_2$ in both the low and high field limit), $\rho_T(H)$ is dominated by MB effects if $\omega_0\tau \gg 1$. Because of this and the fact that the MB probability, P, depends on the density of states in the two bands involved, $\rho_T(H)$, in the limit of large τ, will show oscillatory behavior which directly reflects the level structure of the bands in the region where MB occurs. We will consider this oscillatory behavior next.

Oscillatory Magnetic Breakdown

We have used a field modulation technique to separate the oscillatory portion of $\rho_T(H)$ from the basic background behavior which is proportional to H^2 for arbitrary directions of H in magnesium. Thus far, we have been able to accurately sort about 20 different periods for H in the $(10\bar{1}0)$ plane of magnesium. All of these periods are observed because of their influence on the process of MB. Comparing the data taken from $\rho_T(H)$ with DHVA data obtained with a field modulation technique, we found that they were completely different. The dominant DHVA periods which arise from extremal cross sections of the Fermi surface in regions where no MB occurs have not been seen in $\rho_T(H)$, except for one period which arises from the lens-shaped piece of the Fermi surface centered about Γ in the third zone. The band gaps around the lens are large and almost certainly are not affected by MB. However, the lens contains nearly one-half of all the charge carriers in magnesium. Thus, as the Landau levels of the lens pass through its surface, the redistribution of electrons over the rest of the Fermi surface will cause major fluctuations in the Fermi energy, and these fluctuations will affect P in the same manner as the variation in density of states at the Fermi surface discussed earlier.

With the exception of the lens, all of the periods which we have observed arise in the region where MB is prevalent, and of these only five periods arise from orbits which exist on the Fermi surface in the low-field limit. All of the rest result from orbits which arise when MB recouples the segments of the orbits which exist in the low-field limit. This effect has been discussed in detail by Pippard.[26]

Since this recoupling of orbit segments creates new orbits whose areas are uniquely related to the areas of the low-field limit orbits, for the remainder of this discussion we will convert all measured periods to cross-sectional areas by the relation

$$A = \frac{2\pi e}{\hbar c P} = \frac{9.55 \times 10^{-9}}{P} \quad (7)$$

where A is area in Å^{-2} when the period P is given in G^{-1}.

Figure 5a shows three of the areas which we have observed for H in the $(10\bar{1}0)$ plane of magnesium. The area A_c arises from an orbit about the extremal cross section of the cigar. The area A_E arises from an orbit (labeled ε by Priestley[2]) which goes around the junction of two horizontal and two vertical tentacles of the monster. The area A_{MB} is caused by an orbit which arises when MB recouples the orbit segments of A_c and A_E. All of these orbits are shown in Fig. 5b. The solid curve which fits the area A_{MB} fairly well was obtained from the expression:

$$A'_{MB} = A_E - A_c \quad (8)$$

The minus sign results from the fact that A_E arises on a hole surface, while A_c arises on an electron surface.

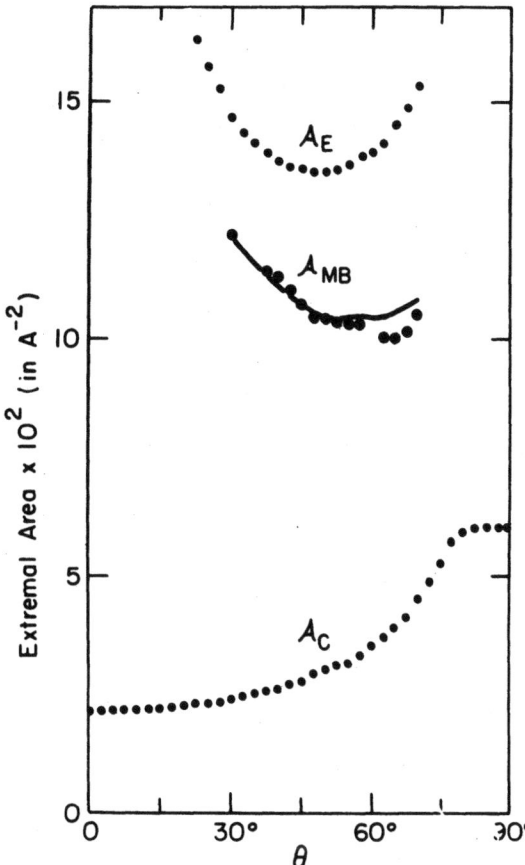

Fig. 5a. Three of the extremal areas which were observed in the oscillatory portion of $\rho_T(H)$ for a magnesium specimen with J in the $(10\bar{1}0)$ direction and H in the $(10\bar{1}0)$ plane. $\theta = 0°$ corresponds to H parallel to (0001) and $\theta = 90°$ corresponds to H parallel to $(11\bar{2}0)$. The area A_{MB} results when MB couples the areas A_c and A_E by an orbit which encloses both. The solid curve is the area which one would expect for the coupled orbit.

When H was parallel to the $(11\bar{2}0)$ direction, we found eight different areas. The largest of these, $A_{lens} = 0.262\,\text{Å}^{-2}$, is due to the lens. The smallest, $A_{cap} = 1.34 \times 10^{-3}\,\text{Å}^{-2}$, is associated with the cap and is seen in fields as small as 1 kG. Two more of these areas arise from fundamental pieces of the Fermi surface in the region where MB occurs. One, whose area $A_{cigar} = 6.04 \times 10^{-2}\,\text{Å}^{-2}$, comes from the extremal orbit around the cigar, while the other with an area $A_{monster} = 1.97 \times 10^{-2}\,\text{Å}^{-2}$ comes from the extremal orbit about the junction of the monster tentacles with the AHL plane of the Brillouin zone. The other four areas result from a complex recoupling of these orbits by MB. These areas and their associated orbits are shown for the free-electron model in Fig. 6.

The first coupled orbit whose area $A_{MB-1} = 1.05 \times 10^{-2}\,\text{Å}^{-2}$ results from MB of the spin-orbit induced gap across the AHL plane of the Brillouin zone between the monster and the cap. This MB begins in fields of a few hundred gauss and destroys the topological openness of the monster parallel to the hexad axis. The area of this coupled orbit is given by

$$A'_{MB-1} = \tfrac{1}{2}A_{monster} + \tfrac{1}{2}A_{cap}$$
$$A'_{MB-1} = 1.05 \times 10^{-2}\,\text{Å}^{-2} \tag{9}$$

which agrees very well with the experimental value.

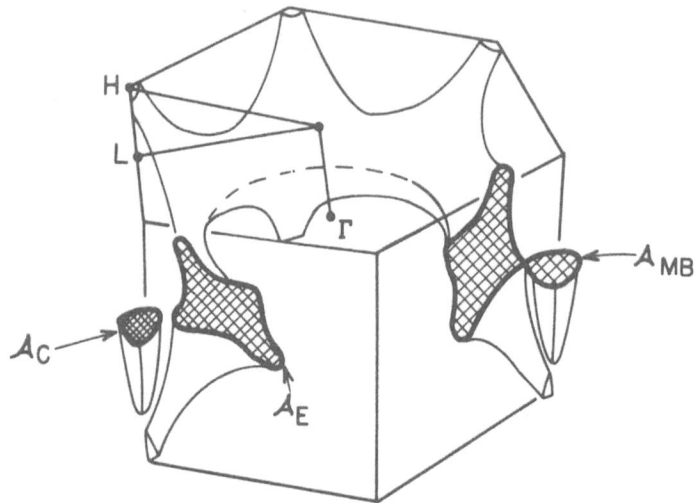

Fig. 5b. Portions of the monster in the second zone and cigars in the third zone for magnesium. The surface is cut by a plane to show the extremal areas A_c and A_E on the left and the combined extremal area A_{MB} on the right. The arrows indicate the direction in which the orbits are traversed. Some of the symmetry points of the hexagonal Brillouin zone are shown.

The remaining three-coupled orbits result from MB between the cigar and the monster. In this case, the extremal orbit about the cigar and the extremal orbit about the monster tentacle junction do not occur in the same plane, although their planes are only separated by about 5×10^{-2} Å^{-1}. The cross sections of both of these pieces of the surface change quite rapidly as one goes from the extremal sections, so that we cannot hope to obtain the expected values for the periods of these coupled orbits by algebraic sums of the areas A_{cigar} and $A_{monster}$.

If we let the direction of H define the Z-axis, then a coupled orbit which goes periodically around a nonextremal section of the monster junction m times, and a nonextremal section of the cigar n times will have an extremal cross section if

$$\frac{d}{dZ}[m\,A_{monster}(Z) - n\,A_{cigar}(Z)] = 0 \tag{10}$$

where m and n are integers. In this case, the extremal cross-sectional areas of the coupled orbits occur in planes fairly close to the plane containing the extremal area of the cigar. The area assignments which are shown in Fig. 6 are the only self-consistent assignments which we could find.

The other ten areas which we have observed for H in the $(10\bar{1}0)$ plane correspond to different orbits which link various nonextremal areas of the monster and cigar in such a fashion that the areas of the coupled orbits are extremal. These orbits will be discussed in detail elsewhere.

Conclusions

When MB occurs in a metal, its effects can be expected to dominate the galvanomagnetic properties of the metal, if it is sufficiently pure so that the electronic

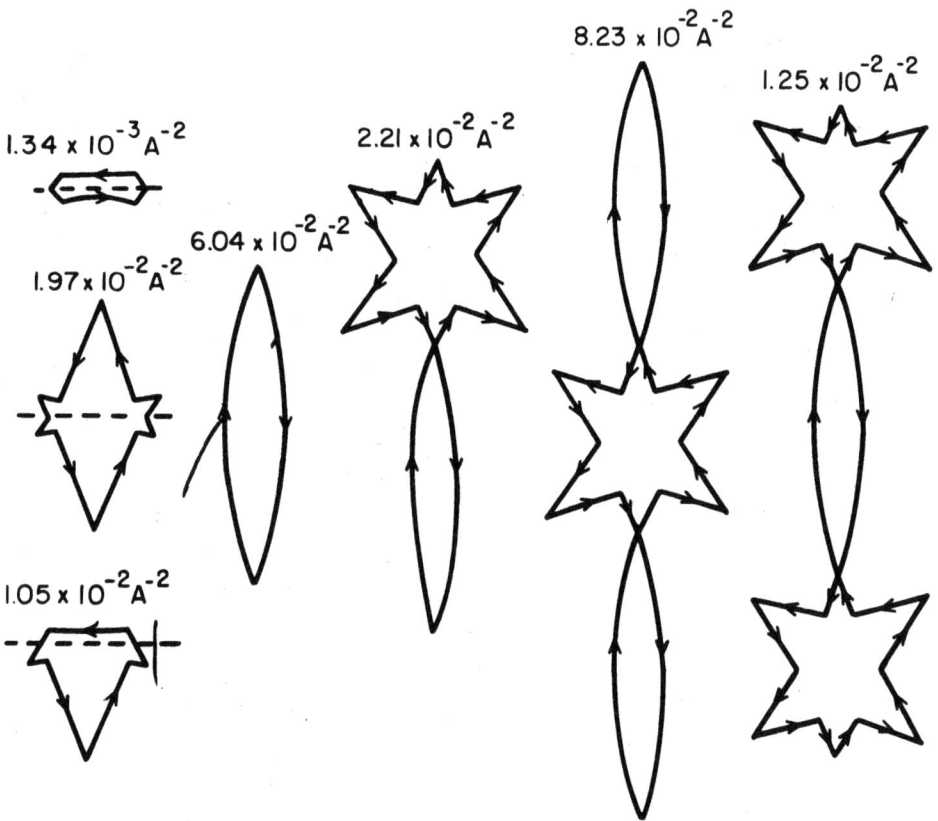

Fig. 6. Various experimental areas and their associated orbits on the free-electron model for H parallel to $(11\bar{2}0)$ in magnesium.

relaxation time, τ, is of the same order of magnitude as the effective MB relaxation time, τ_m. For metals with energy gaps of about 10^{-2} eV, τ_m will be about 10^{-8} to 10^{-9} sec. The gross effects of MB are observable in specimens of much less purity if MB takes the galvanomagnetic properties of the metal between two of the limiting cases discussed earlier. An example would be the case in which $\rho_T(H)$ is changed from an initial H^2 to a final saturation behavior by MB.

The energy-level spectrum in the region of the Fermi surface where MB is occurring can be more conveniently explored by observing oscillatory MB effects in the galvanomagnetic properties than by observing the DHVA effect, since MB effects will dominate the galvanomagnetic properties but will likely be obscured in the DHVA effect by those periods which arise from regions of the Fermi surface where MB is not occurring.

Acknowledgments

I am most grateful to Professors L. M. Falicov and M. G. Priestley for several enlightening conversations.

References

1. M. H. Cohen and L. M. Falicov, *Phys. Rev. Letters* **7**, 231, 1961.
2. M. G. Priestley, *Proc. Roy Soc. (London) Ser. A*, **276**, 258, 1963.
3. M. G. Priestley, L. M. Falicov, and G. Weisz, *Phys. Rev.* **131**, 617, 1963.
4. R. W. Stark, *Phys. Rev. Letters* **9**, 482, 1962; *Phys. Rev.* **135**, A1698, 1964.
5. W. A. Reed and G. F. Brennert, *Phys. Rev.* **130**, 565, 1963.
6. A. C. Thorsen, A. S. Joseph, and L. E. Valby, this volume, p. 867.
7. J. R. Lawson and W. L. Gordon, this volume, p. 854.
8. R. J. Higgins and J. A. Marcus, this volume, p. 859.
9. R. J. Balcombe, *Proc. Roy. Soc. (London) Ser. A* **275**, 113, 1963.
10. E. S. Borovik, V. G. Volotskaya, and N. Ya. Fogel, *Soviet Phys. JETP (English Transl.)* **18**, 34, 1964.
11. A. R. Mackintosh, L. E. Spanel, and R. C. Young, *Phys. Rev. Letters* **10**, 434, 1963.
12. M. G. Priestley (to be published).
13. P. Soven, *Phys. Rev.* **137**, A1717, 1965.
14. N. E. Alekseevskii and V. S. Egorov, *Soviet Phys. JETP (English Transl.)* **18**, 268, 1964.
15. A. S. Joseph and A. C. Thorsen, *Phys. Rev.* **133**, A1546, 1964.
16. E. Fawcett and W. A. Reed, this volume, p. 782.
17. M. G. Priestley, private communication.
18. R. A. Reed and E. Fawcett, *Phys. Rev.* **136**, A422, 1964.
19. R. W. Stark, T. G. Eck, and W. A. Gordon, *Phys. Rev.* **133**, A443, 1964.
20. I. M. Lifshitz, M. I. Azbel, and M. I. Kaganov, *Soviet Phys. JETP (English Transl.)* **4**, 41, 1957.
21. I. M. Lifshitz and V. G. Peschanskii, *Soviet Phys. JETP (English Transl.)* **8**, 875, 1959; **11**, 137, 1960.
22. E. I. Blount, *Phys. Rev.* **126**, 1636, 1962.
23. L. M. Falicov and P. R. Sievert, *Phys. Rev. Letters* **12**, 558, 1964.
24. A. Bennett and L. M. Falicov, *Phys. Rev.* **136A**, 998, 1964.
25. L. M. Falicov and P. R. Sievert, private communication; and article to be published.
26. A. B. Pippard, *Proc. Roy. Soc. (London) Ser. A* **270**, 1, 1962; *Phil. Trans. Roy. Soc. (London)* **256**, 317, 1964.

INTERACTION OF HELICON WAVES AND SOUND WAVES IN POTASSIUM

C. C. Grimes

Bell Telephone Laboratories
Murray Hill, New Jersey

Recently there has been a rapid growth of interest in the study of electromagnetic waves in solids. It has been scarcely three years since the initial observation of helicon waves by Bowers, Legendy, and Rose[1] and the observation of Alfven waves by Buchsbaum and Galt.[2] Yet, in a recent review article,[3] Bowers lists more than 30 articles on these waves. Despite this spurt of activity, the nature of electromagnetic waves in solids has remained rather obscure to many of the workers in other areas of physics. In this paper I will endeavor to convey a physical feeling for the nature of the helicon wave, which is the most easily studied of the electromagnetic waves in solids. A simple experimental technique for the study of helicon waves in metals will be described, and then the application of the experimental technique to the observation of the helicon wave–sound wave interaction will be discussed.

Let us consider what happens when an electromagnetic wave is incident on a conductor. We would like to know what fraction of the incident radiation is transmitted into the conductor and how the wavelength in the conductor varies with frequency. To answer such questions we need to know the dielectric constant that characterizes the conducting medium. Let us consider a simple model for a conductor: A semi-infinite free-electron gas neutralized by a uniform positive background of charge. We assume that radiation is incident along the z-axis, which is normal to the surface of the medium, and also assume a magnetic field $\mathbf{H}_0$ applied along the z-axis. The dispersion relation, which can be expressed in terms of the dielectric constant, is found by requiring that there exist simultaneous solutions of the electrodynamic wave equation obtained from Maxwell's equations,

$$\nabla^2 E = \frac{\varepsilon_0}{c^2} \ddot{E} + \frac{4\pi}{c^2} \frac{\partial j}{\partial t} \tag{1}$$

and a constitutive equation which is obtained from a solution of the Boltzmann transport equation*

$$\mathbf{j} = G\sigma_0 \mathbf{E} \qquad G^{\pm} = \tfrac{3}{4} \int_0^\pi \frac{\sin^3\theta \, d\theta}{1 - i\tau(\omega \pm \omega_c - kv_F \cos\theta)} \tag{2}$$

Here, σ_0 is the DC conductivity ($\sigma_0 = ne^2\tau/m$), ω and ω_c are the signal and cyclotron angular frequencies, respectively, $k = 2\pi/\lambda$ is the wave number, and v_F is the Fermi velocity. If it is assumed that the field quantities vary as $e^{i(kz - \omega t)}$, the determinant of

---
* For a derivation of the frequency and wave number dependent conductivity-tensor, see M. Cohen et al.[4]

coefficients yields the dispersion relation

$$\varepsilon^\pm = \left(\frac{ck^\pm}{\omega}\right)^2 = \varepsilon_0 - \frac{4\pi i\sigma_0 G^\pm}{\omega} \qquad (3)$$

where the superscripts denote left hand (+) or right hand (−) circular polarization. To illustrate some of the features of this dispersion relation, we make the long-wavelength approximation that $k = 0$. The integral in equation (2) is then trivial and we can express ε in the form

$$\varepsilon^\pm = \varepsilon_0 - \frac{\omega_p^2}{\omega(\omega \pm \omega_c + i\nu)} \qquad (4)$$

where $\omega_p^2 = 4\pi n e^2/m$ is the plasma frequency squared and $\nu = 1/\tau$. In Fig. 1, we have sketched ε vs. ω for $H_0 = 0$ and two different values of ν. At room temperature, $\nu \sim 10^{13}$/sec for metals, and ε is predominantly imaginary for microwave and

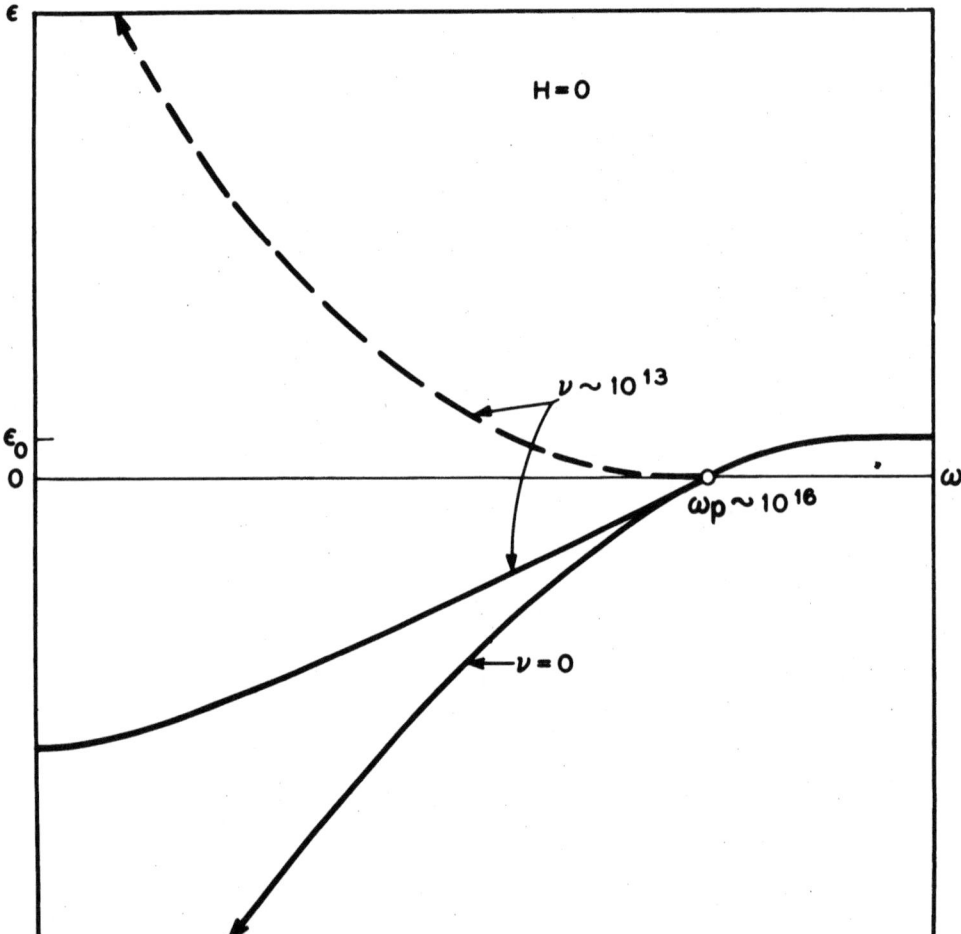

Fig. 1. Sketch of the dielectric constant, ε, vs. frequency for an electron gas in the $k = 0$ approximation. The solid curves represent the real part of ε and the dashed curve the imaginary part. The collision frequency is denoted by ν.

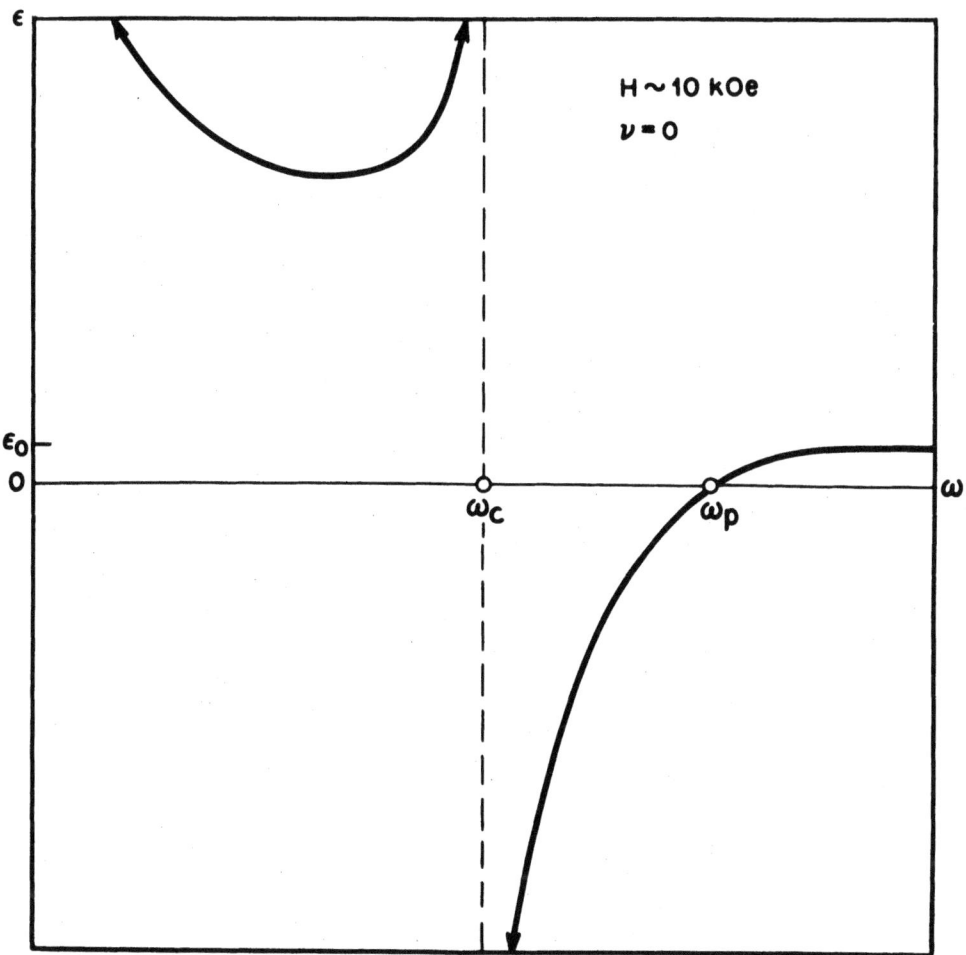

Fig. 2. Sketch of the dielectric constant, ε^-, of a collisionless electron gas in a magnetic field in the $k = 0$ approximation. The helicon wave branch occurs in the frequency interval $0 < \omega < \omega_c$.

lower frequencies. Consequently, our conductor is largely reflecting and does not support a propagating wave; this is the familiar ordinary skin effect regime. If the collision frequency goes to zero, the dielectric constant becomes real, but negative, for $\omega < \omega_p$, and the medium is totally reflecting. In both cases, the dielectric constant becomes real and positive for $\omega > \omega_p$, so that the medium becomes transparent to ultraviolet, X-ray, and γ-radiation. When a strong magnetic field is applied to the electron gas, the dielectric constant for right-hand circularly polarized radiation displays a resonance at the electron cyclotron frequency as illustrated in Fig. 2. The dielectric constant ε^-, is then real and positive for frequencies less than the cyclotron frequency. Consequently, the electron gas will now support a propagating wave for $\omega < \omega_c$, and it is this wave that Aigrain[5] has termed the helicon wave. From equation (4) we have $\varepsilon^- \approx \omega_p^2/\omega\omega_c$ for $\omega \ll \omega_c \gg \nu$. For metals $\omega_p^2 \sim 10^{31}/\text{sec}^2$, and $\omega_c \sim 10^{11}/\text{sec}$ when $H_0 = 10^4$ Oe so that ε^- lies in the range from 10^{18} to 10^{10} for audio and radio frequencies. This huge dielectric constant indicates that the phase velocity of a helicon wave is much smaller than the velocity of light, and the phase velocity can be readily varied over four orders of magnitude by varying the signal frequency and the magnitude of the applied magnetic field.

The physical nature of the helicon wave can be illustrated by considering the motion of a magnetic flux line immersed in an electron gas. Recalling that a flux line moves as though it is under a tension $H_0^2/4\pi$ (in Gaussian units), we can write the equation of motion for a perturbed flux line as

$$\left(\frac{H_0^2}{4\pi}\right)\frac{\partial^2 S}{\partial z^2} \approx -\frac{ne}{c}\dot{S} \times H_0 \qquad (5)$$

where S is the transverse displacement of the flux line from its equilibrium position. The term on the right side arises from the Lorentz force acting on the electrons trapped on the flux line. In this equation we are neglecting collisions and we neglect the inertial term arising from the mass of the electrons (the latter approximation is equivalent to assuming $\omega \ll \omega_c$). Now, if we assume that S varies as $e^{i(kz-\omega t)}$, we obtain the dispersion relation $(ck^\pm/\omega)^2 = \pm\omega_p^2/\omega\omega_c$, which is identical to equation (4) in the low-frequency limit. Hence, we can visualize a helicon wave as consisting of the transverse motion of a flux line which is acted upon by the Lorentz force and the tension in the flux line.

The fact that the helicon phase velocity can be varied by simply varying H_0 allows one to employ a simple experimental technique to study the helicon waves. A block diagram of our experimental apparatus is shown in Fig. 3. The specimen is in the shape of a square slab about 2×2 cm by 0.1 to 3.0 mm thick. The steady magnetic field, H_0, is applied normal to the surface of the specimen. The RF current from an oscillator passes through a transmitter coil and excites a helicon wave at one surface of the specimen. The helicon wave propagates along the magnetic field, through the specimen, and induces a voltage in a carefully shielded receiver coil at the opposite side of the specimen. The signal voltage and a reference voltage are summed and applied to the input terminals of the receiver. The AVC voltage developed by the receiver is proportional to the amplitude of the input signal, and it is the AVC voltage that is plotted along the y-axis of the X–Y recorder, while H_0 is plotted along the x-axis. The transmitted signal and the reference signal alternately interfere constructively and destructively as H_0 is varied and the number of wavelengths contained in the specimen changes. The interference of the two signals

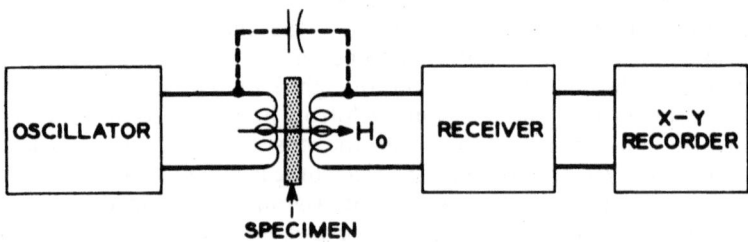

Fig. 3. Block diagram of the apparatus used to study helicon wave propagation.

results in the beat pattern illustrated in curve (a) of Fig. 5. This curve is characteristic of a "pure" helicon wave, and was obtained with a sodium specimen 0.48 mm thick at a frequency of 50 Mcps. The absence of helicon wave transmission below about 45 kOe is due to nonlocal or cyclotron damping effects[6] arising from the $\mathbf{k} \cdot \mathbf{v}_F$ term contained in $G^{\pm}$ in equation (3). The monotonic increase in amplitude of the transmitted helicon wave with increasing $\mathbf{H}_0$ for $\mathbf{H}_0 > 45$ kOe is due to three effects:

1. The decrease in the collision damping as $\omega_c \tau$ increases.
2. The decrease of the number of wavelengths contained in the specimen.
3. The decrease of damping due to cyclotron damping.*

We now ask what will happen if we perform a similar experiment under conditions such that the phase velocity of the helicon wave becomes comparable to the velocity of a transverse sound wave propagating in the same direction. To answer this question we need to consider the dispersion relation for interacting helicon waves and sound waves. The dispersion relation is obtained by requiring that simultaneous solutions of the electrodynamic wave equation, the constitutive equation, and the sound wave equation exist. The dispersion relation for an electron gas immersed in a uniform, positively charged elastic continuum was derived first by Akramov[8] and subsequently treated in more detail by a number of authors.[9] The interaction between the helicon wave and the elastic wave arises from the self-consistent electric and magnetic fields accompanying the waves, and through direct exchange of momentum due to scattering of the electrons by phonons and impurities. The dispersion relation for the interacting waves can be written

$$\left(\frac{V_H^2}{U^2} \pm 1\right)\left(\frac{V_S^2}{U^2} - 1 \pm \frac{\Omega_c}{\omega}\right) = \frac{\Omega_c}{\omega} \tag{6}$$

when $\omega \ll \omega_c \gg kv_F$ and $v = 0$. Here, $U = \omega/k$, V_H is the phase velocity of a pure helicon wave $V_H = (c^2 \omega \omega_c/\omega_p^2)^{\frac{1}{2}}$, V_S is the phase velocity of a pure sound wave $V_S = (C_{ij}/nM)^{\frac{1}{2}}$, where nM is the mass density of the continuum, and Ω_c is the "ion" cyclotron frequency, eH/Mc. When the parameter Ω_c/ω is negligibly small, the two factors on the left side of equation (6) yield independent solutions corresponding to pure helicon waves and pure sound waves, and the dispersion curves cross at $V_H = V_S$. When Ω_c/ω is not negligible, the solutions are coupled and the dispersion relations for the two waves no longer cross. Figure 4 is a plot of the dispersion relation for the right circularly polarized helicon and sound waves. (The left circularly polarized electromagnetic wave does not propagate, and the left circularly polarized sound wave doesn't interact with the helicon wave.) In Fig. 4, we see that the lower branch of the dispersion relation changes from a helicon wave at small $\mathbf{H}_0$ to a wave which resembles a sound wave at large $\mathbf{H}_0$. Conversely, the upper branch changes from a sound wave at $\mathbf{H}_0 = 0$ to a helicon wave at large $\mathbf{H}_0$. In the crossover region where $V_H \approx V_S$, the two excitations have mixed helicon and sound wave character.

In an experiment designed to observe the coupling of the helicon and sound waves, the coupling parameter, Ω_c/ω, should be maximized to produce a readily observable effect. At the crossover point, $V_H = V_S$, the coupling parameter can be written $(\Omega_c/\omega) = \mathbf{H}_0^2/4\pi C_{ij}$. Thus the coupling is maximized when the largest attainable $\mathbf{H}_0$ is employed and a specimen having the smallest possible stiffness

* A complete discussion of the properties of helicon waves is presented in the review articles by S. J. Buchsbaum.[7]

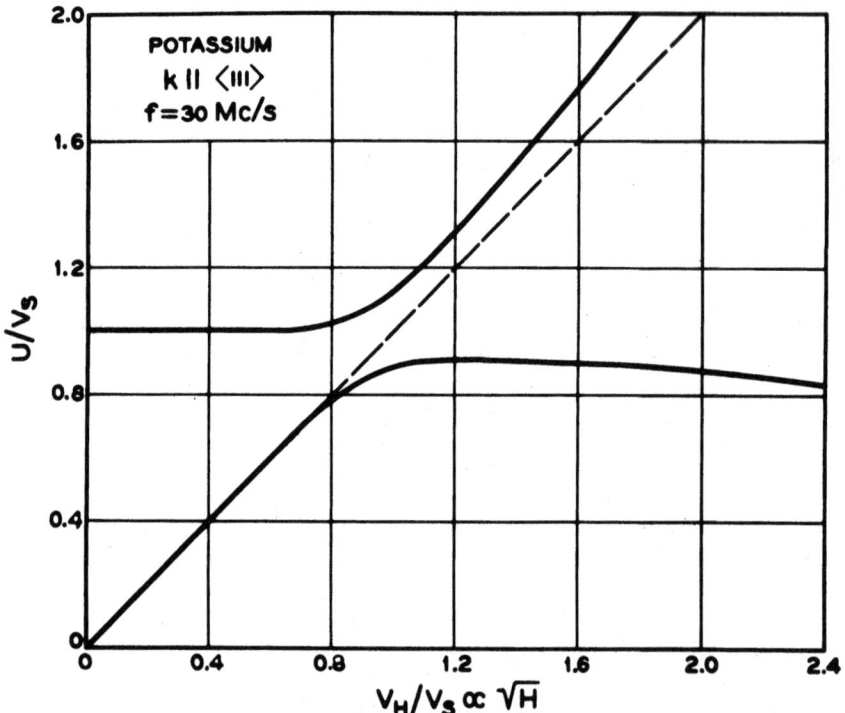

Fig. 4. Theoretical plot of the dispersion relation for interacting helicon and sound waves propagating in the [111] direction in potassium. The quantity U/V_S plotted along the ordinate is the phase velocity of an excitation normalized to the velocity of sound, while V_H/V_S plotted along the abscissa is proportional to the square root of the magnetic field intensity. The lower branch of the dispersion relation changes from a helicon wave to a sound wave with increasing V_H/V_S, while the upper branch does the opposite.

modulus is used. Of the high-purity, uncompensated metals currently available, potassium has the smallest stiffness modulus so that it was selected for the experiments. Further consideration of the elastic properties of potassium reveals that C_{ij} is minimal for the slow shear wave which propagates in the [110] direction with particle motion in the [1$\bar{1}$0] direction. Hence, slab-shaped single-crystal specimens having (110) surfaces were employed in the experiments.

Three representative experimental curves are shown in Fig. 5. Curve (a) was obtained in a polycrystalline sodium slab and has already been discussed. Since, in sodium, there is no interaction with the sound wave over the range of H_0 used, curve (a) is characteristic of a pure helicon wave and is included for contrast with curves (b) and (c). Note that in curve (a) there is a monotonic increase in separation of successive maxima as H_0 is increased which results from the fact that $V_H \propto H_0^{\frac{1}{2}}$. Curves (b) and (c) of Fig. 5 were obtained with wave propagation in the [110] direction in potassium at frequencies of 30 and 20 Mcps, respectively, and at a temperature of 4.2°K. At the higher frequency [curve (b)], the coupling between the helicon and sound waves is relatively weak and manifests itself by an increasing separation of the successive maxima as the magnetic field is decreased at low values, that is, the decrease of the phase velocity of the wave with decreasing magnetic

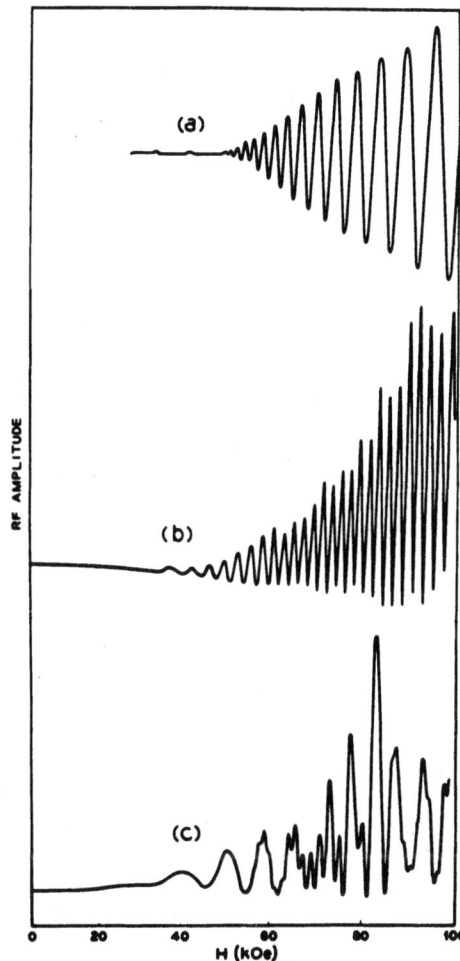

Fig. 5. Experimental traces of RF amplitude across the receiver coil as a function of H_0. Curve (a), obtained in sodium, represents "pure" helicon wave propagation and is to be contrasted with traces (b) and (c), which represent helicon waves interacting with sound waves in single-crystal potassium.

field is less than that dictated by the 'pure' helicon wave formula. Only the upper branch of the dispersion relation of Fig. 4 is discernible here. In curve (c), at 20 Mcps, the coupling is sufficiently strong for both waves to appear, that is, the beat pattern in Fig. 5c can be resolved into two series of peaks corresponding to two coupled branches of the dispersion relation. The series of peaks that are widely spaced at lower fields and become more closely spaced at higher fields belong to a branch of the dispersion relation that resembles a sound wave at low fields and a helicon wave at high fields. The second series of peaks becomes more widely spaced with increasing H_0 and belongs to a branch of the dispersion relation which is changing from a helicon wave to a sound wave.

Data points deduced from the experimental curves in Figs. 5b and 5c are compared with theoretical curves in Fig. 6. The theoretical curves were calculated from the dispersion relation derived by Grimes and Buchsbaum[10] which is applicable to the anisotropic elastic medium appropriate for wave propagation along a twofold axis. The dispersion relation for the anisotropic medium is similar in form to equation (6), but it is considerably more complicated so that we will not write it

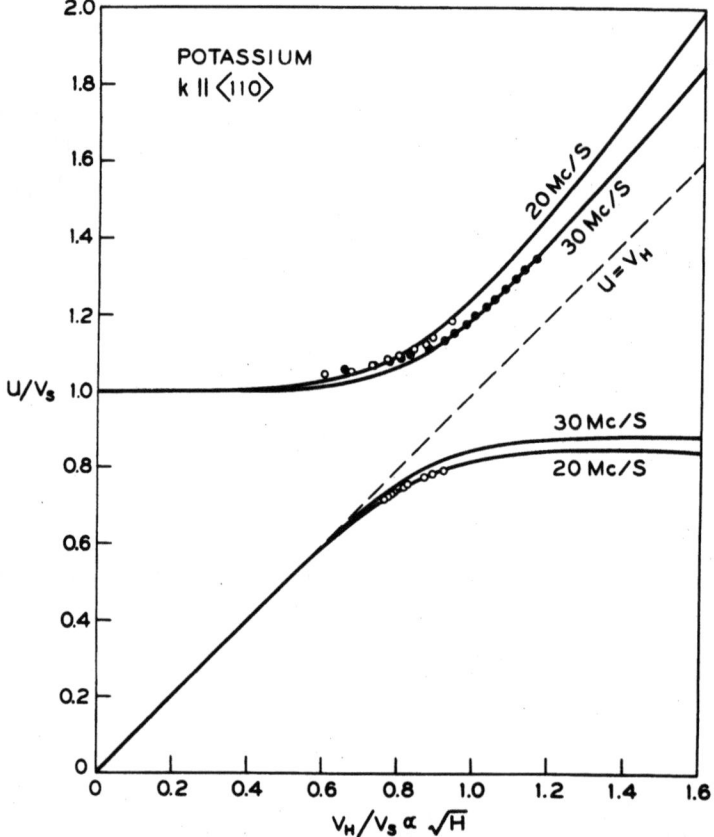

Fig. 6. Comparison of the calculated dispersion curves with data points deduced from the experimental traces in Figs. 5b and 5c.

down here. In calculating the theoretical curves we have used the elastic constants for potassium as listed by Mason[11] and the lattice constant measured by Barrett.[12]

The experimental data points in Fig. 6 contain an adjustable parameter in the following sense: The magnitude of the wave number, k corresponding to a given peak is not accurately known. However, in a series of peaks in the beat pattern two successive peaks differ in wave number by $2\pi/d$, where d is the thickness of the specimen. Since k decreases with increasing H_0, we arbitrarily assign a k-value to one, and only one, peak in each series of peaks. In Fig. 6, we have arbitrarily assigned a k-value to that peak in each series of peaks which appears at the highest magnetic field. In each case the k-value was chosen to make the data point lie near the corresponding theoretical curve. The adjustable parameter enters essentially as a translation of each series of data points along the ordinate in Fig. 6. Consequently, the data points are to be compared only with the slope and curvature of the theoretical curves.

In summary, the helicon–sound wave interaction was observed as anticipated, and there is good agreement between experiment and a simple phenomenological theory.

In conclusion, let me remind you that the helicon wave–sound wave interaction is just one of the many effects that can be studied using helicon waves. Although a number of helicon wave phenomena have been observed and are currently being studied, there remain several interesting phenomena that have been predicted theoretically, but not yet observed.[7]

References

1. R. Bowers, C. Legendy, and F. Rose, *Phys. Rev. Letters* **7**, 339, 1961.
2. S. J. Buchsbaum and J. K. Galt, *Phys. Fluids* **4**, 1514, 1961.
3. R. Bowers, *Plasma Effects in Solids*, Dunod, Paris (1965), p. 19.
4. M. H. Cohen, M. J. Harrison, and W. A. Harrison, *Phys. Rev.* **117**, 937, 1960.
5. P. Aigrain, *Proceedings of the International Conference on Semiconductor Physics*, Prague, 1960, Czechoslovakian Academy of Sciences, Prague (1961).
6. C. C. Grimes, *Plasma Effects in Solids*, Dunod, Paris (1965), p. 87.
7. S. J. Buchsbaum, *Plasma Effects in Solids*, Dunod, Paris (1965), p. 3.
8. G. Akramov, *Fiz. Tverd. Tela* **5**, 1310, 1963, (*Soviet Phys.–Solid State* (English Transl.) **5**, 955, 1963.)
9. V. G. Skobov and E. A. Kaner, *Zh. Eksperim. i Teor. Fiz.* **46**, 273, 1964. (*Soviet Phys. JETP* (English Transl.) **19**, 189, 1964); D. N. Langenberg and J. Bok, *Phys. Rev. Letters* **11**, 549, 1963; J. J. Quinn and S. Rodriquez, *Phys. Rev. Letters* **11**, 552, 1963; and *Phys. Rev.* **133**, A1589, 1964.
10. C. C. Grimes and S. J. Buchsbaum, *Phys. Rev. Letters* **12**, 357, 1964.
11. W. P. Mason, *Physical Acoustics and the Properties of Solids*, D. Van Nostrand Company, Inc., Princeton, New Jersey (1948).
12. C. S. Barrett, *Acta Cryst.* **9**, 671, 1956.

8.2. Magnetoresistance, Cyclotron Resonance, de Haas–van Alphen Effect

LONGITUDINAL MAGNETORESISTANCE OF COPPER*

Robert L. Powell†

Cryogenics Division, National Bureau of Standards
Boulder, Colorado

As pointed out by Pippard[1] in his les Houches lecture, a complete explanation of galvanomagnetic properties depends on a quantitative analysis of the experiments and a detailed knowledge of the shape of the Fermi surface and of the electronic velocities and mean free paths on that surface. Because the theoretical guidelines are not at all clear (in fact the basic assumption that a mean free path is uniquely definable is in doubt), the metals to be studied in detail must have simple, single-sheet, well-known Fermi surfaces, and must be available with sufficient purity to allow magnetic field saturation. These criteria practically eliminate all metals but copper and silver. At the suggestion of Professor Pippard, research was begun on the longitudinal magnetoresistance of copper in an attempt to obtain detailed information on the variation of electronic mean free path over the Fermi surface.

Single crystals of copper in the form of cylindrical slugs about 5 cm long and 2 cm in diameter were prepared by induction melting in graphite-lined quartz crucibles. The basic stock was AS & R 99.999% pure copper. High purities (with resistance ratios, as determined by eddy-current decay, from 500 to 4500) were only obtained after a cycle of controlled oxidation was introduced into the melting routine. From three to six samples, 2 cm long and 1 mm wide with a square cross section, were obtained by spark-trepanning from each slug. The samples were planed by further spark machining to their final shape. With auxiliary equipment developed by Stephen Lipson and the author, we were able to prepare crystals aligned to a given crystallographic direction within 0.1° or better. Galvanomagnetic measurements were made on samples with orientations of 100, 111, 110, 321, and 521 and with residual resistance ratios of 630 to 1520.

All of the measurements were carried out at the normal boiling point of helium, 4.2°K. The magnet used for all but one set of measurements on a [111] crystal was a 3-in. bore Helmholtz pair, mounted horizontally (transverse to the cryostat), that provided fields up to 30 kG. For the one exception we used a 3-in. bore solenoid, longitudinal to the cryostat, providing fields up to 50 kG. Both magnets were designed and developed at the Cavendish laboratory and installed during the course of the preliminary experiments. Sample currents were measured with a calibrated milliammeter; potentials, with a simple Lindeck potentiometer network that had a photocell galvanometer for a null detector.

Results on three [100] crystals of different purity are given in Fig. 1. All three crystals were measured up to 30 kG. However, for crystal R-3, only the curve up to

---
* The experimental part of the research was performed at the University of Cambridge, England; the analysis and interpretation at the National Bureau of Standards.
† A National Science Foundation Fellow, 1962–1963, while at the University of Cambridge.

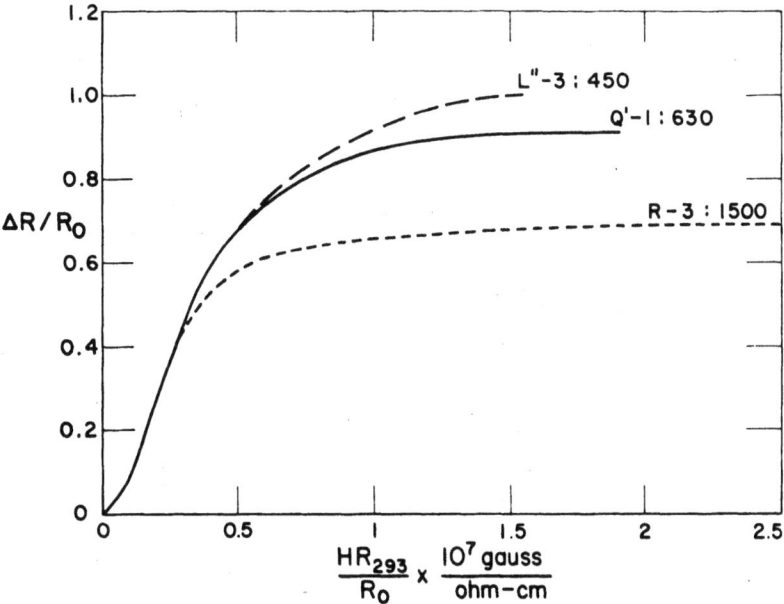

Fig. 1. The magnetoresistance of three [100] copper single crystals. The latter-number code identifies the samples; the number after the colon indicates the residual resistance ratio.

18 kG is shown because of saturation at a lower field. The strong dependence of the saturation resistance ratio on purity was unexpected. Even the lowest value, 0.70, is considerably higher than the previous estimate by Pippard, 0.1, based on a simple model. For all three, the behavior is quadratic for very low fields.

Results for three crystals of the same purity but different orientations, [100], [110], and [111], are given in Fig. 2. The residual resistance ratios were all approximately 1500. Crystals R-2 and R-3 were measured up to 30 kG, as indicated; crystal R-1 was measured in the longitudinal magnet for fields up to 50 kG, but no indication of saturation was observed even at that high field. Curve R-1 is an average of two sets of readings with the sample orientation reversed. The sample holder, designed for use in a transverse magnet, did not allow precise alignment in the longitudinal magnet used for crystal R-1 only.

For crystals with a residual resistance ratio below 1000, longitudinal magnetoresistance saturation was not observed for fields below 30 kG (except for the orientation [100]). In some sample crystallographic directions, such as [111], a misalignment of as little as $\frac{1}{4}$% in the cryostat holder caused a qualitative change in the shape of resistance vs. field-angle plots near high-symmetry directions. Except for the [111] crystal R-1, reversals of sample, field, or current direction did not affect the measured magnetic resistances.

The easiest theoretical analysis of the results utilizes the equations developed by Pippard.[1] In the quasi-classical, impulsive field approximation, with an assumed existence of a mean free path or relaxation time for every point on the Fermi surface, the ratio of resistance at saturation to the resistance at zero field is given

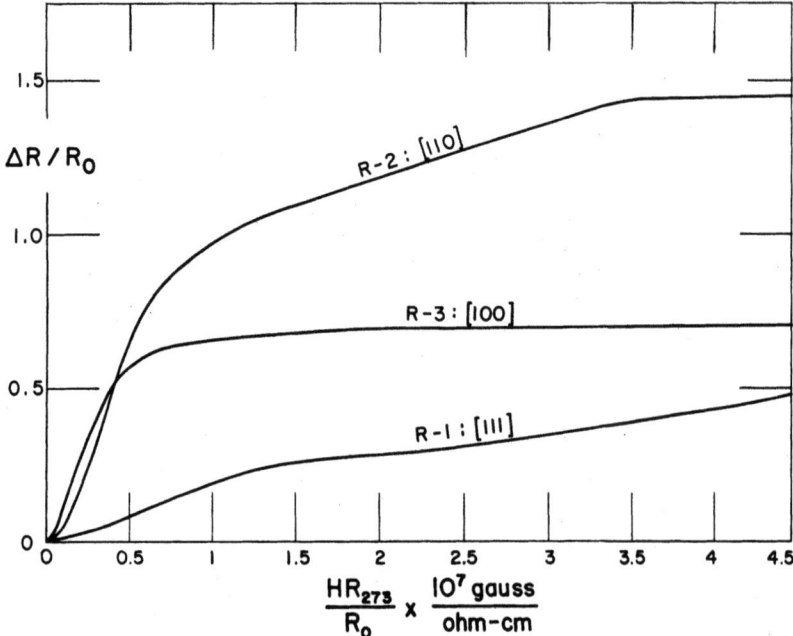

Fig. 2. The magnetoresistance of three copper single crystals of the same purity but different orientations.

by the expression

$$\frac{\rho_{zz}(\infty)}{\rho_{zz}(0)} = \frac{\int \oint l \cot \theta \cos \theta \, ds dk_z}{\int \frac{[\oint \cot \theta \, ds]^2}{\oint \csc \theta \, ds/l} dk_z}$$

where z is the magnetic field direction, l is the mean free path, ds is an incremental distance along a contour on the surface at the height k_z from the center as expressed in cylindrical coordinates, and θ is the angle between the normal to the surface at a given point and the z-axis.

The relatively large values for longitudinal magnetoresistance at saturation can be explained by an estimate of the fraction of the total number of orbits that intersect necks and therefore do not contribute to the high-field conductivity. For the three principal crystallographic directions, this effect is most important for [110] and least for [111] crystals. A simple approximation of a sphere with 18.8° caps in the appropriate directions gives $\Delta R/R$ values of 0.11, 0.29, and 0.54 for the [111], [100], and [110] directions, respectively. Pippard[2] has recently given a discussion of the above effect. The remaining added resistance must be explained by the variation of the mean free path over the surface.

The strong dependence of the saturation magnetoresistance on purity for one given direction can be interpreted as indicating a large variation of the relative mean free path between "neck" and "belly" regions for rather small variations in the type of impurity scattering. A detailed numerical analysis using an IBM 7090 computer is now underway. The R-1, R-2, and R-3 samples will be analyzed for trace impurities, particularly silver and iron. A complete analysis will be published later.

Acknowledgments

We would like to thank Professor Pippard for the original suggestion of the problem and his constant advice and encouragement; Drs. D. Shoenberg, J. M. Ziman, and P. L. Taylor for enlightening discussions; and S. Lipson, D. D. Stewart, and E. Collins for assistance during the cryostat assembly and experimental runs.

References

1. A. B. Pippard, in: *Low Temperature Physics*, C. de Witt, B. Dreyfus, and P. S. de Gennes (eds.), Gordon and Breach, Science Publishers, Inc., New York (1962), pp. 3–148.
2. A. B. Pippard, *Proc. Roy. Soc. (London) Ser. A* **282**, 464, 1964.

NEGATIVE MAGNETORESISTANCE IN TELLURIUM

Douglas C. Pearce, Robert M. Wasilik, and Donald A. Neeper

U.S. Army Electronics Laboratories, HQ, U.S. Army Electronics Command
Fort Monmouth, New Jersey

Introduction

Anomalous positive or negative magnetoresistance at low temperatures has been observed in germanium, InSb, gray tin, InAs, GaAs, and C.[1-7] Some of the more extensive investigations have revealed certain similarities which may be common to the anomalous magnetoresistance in semiconductors: (1) The magnitude of the anomalous magnetoresistance increases with decreasing temperature. (2) For a given semiconductor the anomalous magnetoresistance reaches a maximum for a particular impurity concentration. (3) In some cases the magnetoresistance can be represented as the sum of an ordinary term with quadratic field dependence and an extraordinary term which saturates with increasing field. A negative magnetoresistance was once observed in tellurium,[8] but in subsequent investigations, using zone refined specimens, no such anomaly was detected.[9] This paper is a progress report of some new features of this anomalous negative magnetoresistance in tellurium.

Experimental Methods

The samples, described below, were chemically etched, and probes of either platinum or copper wire were attached with indium solder or by spot welding. Spot welding with 2-mil platinum proved to be the most successful method. DC magnetic fields up to 3500 Oe were supplied by a solenoid. The electric current in the specimen was always parallel to the c-axis. For measurements above 1°K, the specimen under study rested on a cotton cloth inside a helium vapor pressure capsule. For measurements below 1°K, the specimens were glued with GE 7031 to copper wires which were cast into a chrome alum epoxy refrigeration pill. Temperatures were determined with a carbon thermometer which was calibrated against the susceptibility of the salt. The susceptibility, in turn, was calibrated against the helium vapor pressure. The electrical properties were determined as a function of the time while the specimen warmed in constant magnetic field. Each experimental value, for temperatures either below or above 1°K, was an average for two directions of current and two directions of magnetic field.

Observations on Cast Crystalline Tellurium

Specimens were cleaved from cast crystalline tellurium, frozen from the melt. This material was 99.999% pure, supplied by the American Smelting and Refining Co. All of these specimens displayed similar galvanomagnetic behavior. From 0.09° to 4.2°K, the resistivity could be represented by $\rho = A + BT^{-\frac{1}{2}}$, where A and B were

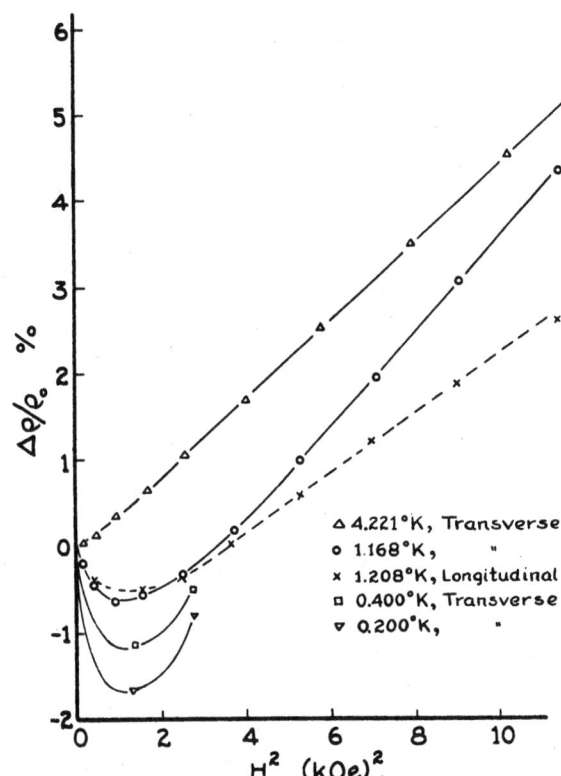

Fig. 1. Magnetoresistance of unannealed cast crystalline tellurium.

constants for a given specimen. The Hall voltage was nearly a linear function of the magnetic field. No Hall measurements were obtained below 1°K, but above 1.2°K the Hall voltage was independent of the temperature. The Hall constant at low magnetic fields indicated a hole density of about $2 \times 10^{14}/cm^3$. Below 4.2°K, both the longitudinal and transverse magnetoresistances were negative for small values of the field, and became positive quadratic functions of the field at higher fields, as shown in Fig. 1. Whenever data could be obtained in the region of quadratic field dependence, it was possible to represent the longitudinal and transverse magnetoresistances by expressions of the form $\Delta\rho/\rho_0 = bH^2 - g(H)f(T)$, where b is a function of temperature, $g(H)$ is a function of magnetic field only, and saturates above 1.5 kOe. $f(T)$ decreases with increasing temperature. The function $g(H)$ is shown in Fig. 2. The extraordinary component of the magnetoresistance could not be represented as a single function of the variable H/T, as would be expected if the extraordinary magnetoresistance were a consequence of a magnetic ordering process. This result does not contradict the recent theory of Toyozawa,[10] as his theory deals explicitly with metallic impurity conduction. At the present time, the conduction process in pure tellurium at low temperatures is uncertain.

Since the galvanomagnetic properties of tellurium are known to be sensitive to crystalline imperfections, additional specimens were annealed for about 150 hr at 370°C. The annealed specimens have not yet been studied below 1°K; however

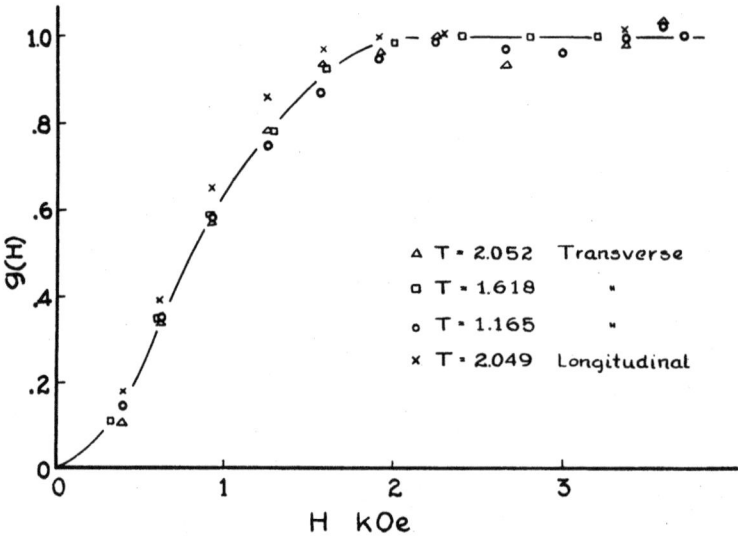

Fig. 2. Field dependence of the negative component of the magnetoresistance.

above 1°K, the resistivity could not be described by $\rho = A + BT^{-\frac{1}{2}}$, although $d\rho/dT$ remained negative. The Hall voltage was not linear with H, and no negative value of the magnetoresistance was detected. Annealing left the ratio (at 3.5 kOe) of transverse to longitudinal magnetoresistance little changed at about two, but increased the magnetoresistance by about eight. Above 1 kOe, the magnetoresistance varied linearly with the field and the Hall angle tended to saturate, in agreement with other measurements on annealed tellurium specimens.[11] This saturation of the Hall angle was more pronounced at lower temperatures. Annealing increased the low-field Hall coefficient of the cast crystalline material by about a factor of two.

Observations of Tellurium Crystals Grown by the Czochralski Method

Measurements have also been made in the temperature range 1.2° to 4.2°K on unannealed and annealed specimens cleaved from a crystal grown by the Czochralski method. This crystal was furnished by Dr. D. E. Brown of the University of Michigan. For an unannealed specimen of this crystal the resistivity and Hall coefficient below 2.1°K exhibited behavior similar to that of the unannealed cast crystalline material, but with a Hall concentration of about 2×10^{13} holes/cm^3. On the other hand, the longitudinal and transverse magnetoresistance of the unannealed material grown by the Czochralski method varied linearly with the magnetic field, similar to that of the annealed cast crystalline material.

After the material grown by the Czochralski method was annealed, the Hall coefficient became temperature and field dependent, but for small fields was little changed by the annealing, while the magnetoresistance increased by a factor of four. In all respects the galvanomagnetic behavior of the annealed specimens was similar to that of the annealed cast crystalline material.

Discussion

These experiments show that the negative magnetoresistance of tellurium is closely associated with crystalline imperfections. This conclusion was further borne out by experiments on unannealed, spectroscopically pure, polycrystalline specimens from the Johnson, Matthey Co. These specimens had temperature-independent Hall concentrations of about 10^{16} holes/cm^3, and displayed a negative magnetoresistance similar to that of the unannealed cast crystalline material. However, the magnitude of the negative component was twice that of the cast crystalline specimens and tended toward saturation above 3.5 kOe. Annealed polycrystalline material has not yet been studied.

No annealed specimen ever displayed a negative magnetoresistance of magnitude greater than 0.01%. The resistivity of unannealed specimens at low temperatures could always be described by $\rho = A + BT^{-\frac{1}{2}}$, while the Hall coefficient was only slightly dependent on temperature and magnetic field. Increasing crystalline perfection caused the Hall coefficient to become temperature and field dependent, the Hall angle to saturate with increasing field, and the magnetoresistance to become a linear function of the magnetic field. An attempt to describe the properties of annealed specimens in terms of a two-band model was unsuccessful.

References

1. H. Roth, W. D. Straub, W. Bernard, and J. E. Mulhern, Jr., *Phys. Rev. Letters* **11**, 328, 1963.
2. K. Sugiyama, *J. Phys. Soc. Japan* **18**, 1555, 1963.
3. W. Sasaki, C. Yamanouchi, and G. M. Hotayama, *Proceedings of the International Conference on Semiconductor Physics*, Prague, 1960, Czechoslovakian Academy of Sciences, Prague (1961), p. 159.
4. E. D. Hinkley and A. W. Ewald, *Phys. Rev.* **134**, A1261, 1964.
5. N. V. Zotova, T. S. Lagunova, D. N. Nasledov, *Fiz. Tverd. Tela* **5**, 3329, 1963. (*Soviet Phys.–Solid State* (English Transl.) **5**, 2439, 1964.)
6. M. Pollack and D. H. Watt, *Phys. Rev.* **129**, 1508, 1963.
7. B. J. C. van der Hoeven, Jr. and P. H. Keesom, *Phys. Rev.* **135**, A631, 1964.
8. R. A. Chentsov, *Zh. Eksperim i Teor. Fiz.* **18**, 374, 1948.
9. S. S. Shalyt, *Zh. Techn. Fiz.* **27**, 189, 1957. (*Soviet Phys.–Tech. Phys.* (English Transl.) **2**, 166, 1959.)
10. Y. Toyozawa, *J. Phys. Soc. Japan* **17**, 986, 1962.
11. R. V. Parfenev, A. M. Pogarskii, I. I. Farbshtein, and S. S. Shalyt, *Fiz. Tverd. Tela* **4**, 3596, 1962. (*Soviet Phys.–Solid State* (English Transl.) **4**, 2630, 1963.)

MAGNETORESISTANCE OF BISMUTH*

A. M. Forrest and A. C. Hollis Hallett

Department of Physics, University of Toronto
Toronto, Ontario, Canada

The resistance of single crystals of bismuth has been measured at 1.2°K along the trigonal axis, while various constant magnetic fields up to about 18 kG were rotated in the plane perpendicular to that axis. The specimens were cut from seeded single crystals, grown from zone-refined bismuth, and the current and voltage leads were soldered to them with low melting-point solder. Specimens with resistance ratios as large as 1000 (in the earth's magnetic field) have been obtained with this process. The resistance measurements were made by conventional methods, using a potentiometer.

The results obtained in a typical run at 17.02 kG are shown in Fig. 1a, where the resistance is plotted against the angle, θ, between the magnetic field and the binary axis. The pattern shown in Fig. 1a repeats itself every 60°, but generally the repetition was not exact due to small misalignment of the specimen's trigonal axis and slight misplacing of the leads. From preliminary experiments it was clear that if a well-resolved pattern of peaks was to be observed, the trigonal axis of the specimen had to be normal to the plane of rotation of the field within 3°; smaller misalignments distorted the central peak at some of the 60° positions and reduced the symmetry of the pattern about the central peak, but did not affect the angular separations of the peaks. The peaks denoted by A and B in Fig. 1a were particularly sensitive to errors in alignment. Furthermore, the resolution of the peaks improved markedly with increasing purity. The results at 17.02 kG were reproduced several times on different occasions and with different specimens.

In Fig. 1b the similar pattern observed with a field of 15.41 kG is also shown. This pattern is very similar to that shown by Kunzler *et al.*† and obtained in a rotational study of magnetothermal oscillations in bismuth at 15.5 kG. The major difference is that the peaks designated by A were not observed by Kunzler *et al.*[1] Otherwise, the relative magnitudes of the peaks observed by Kunzler *et al.* and their angular separation from the central one agree well with the magnetoresistance results shown in Fig. 1b.

As the strength of the magnetic field is increased, the angular separation, ϕ, of the peaks from the central one alters. The dependence of ϕ upon magnetic field strength is shown in Fig. 2 for each of the peaks. The absence of a symbol to indicate a D or E peak at any of the fields investigated does not mean that the peak was not resolved, but simply that they were not studied in that particular run. However, where the A or B peaks are not indicated, they were not found. An extra peak, between the D and E peaks, was observed for one specimen used once at 17.06 kG.

---

* Research supported by the National Research Council of Canada, the Ontario Research Foundation, and the University of Toronto.
† See Kunzler *et al.*,[1] Fig. 16.

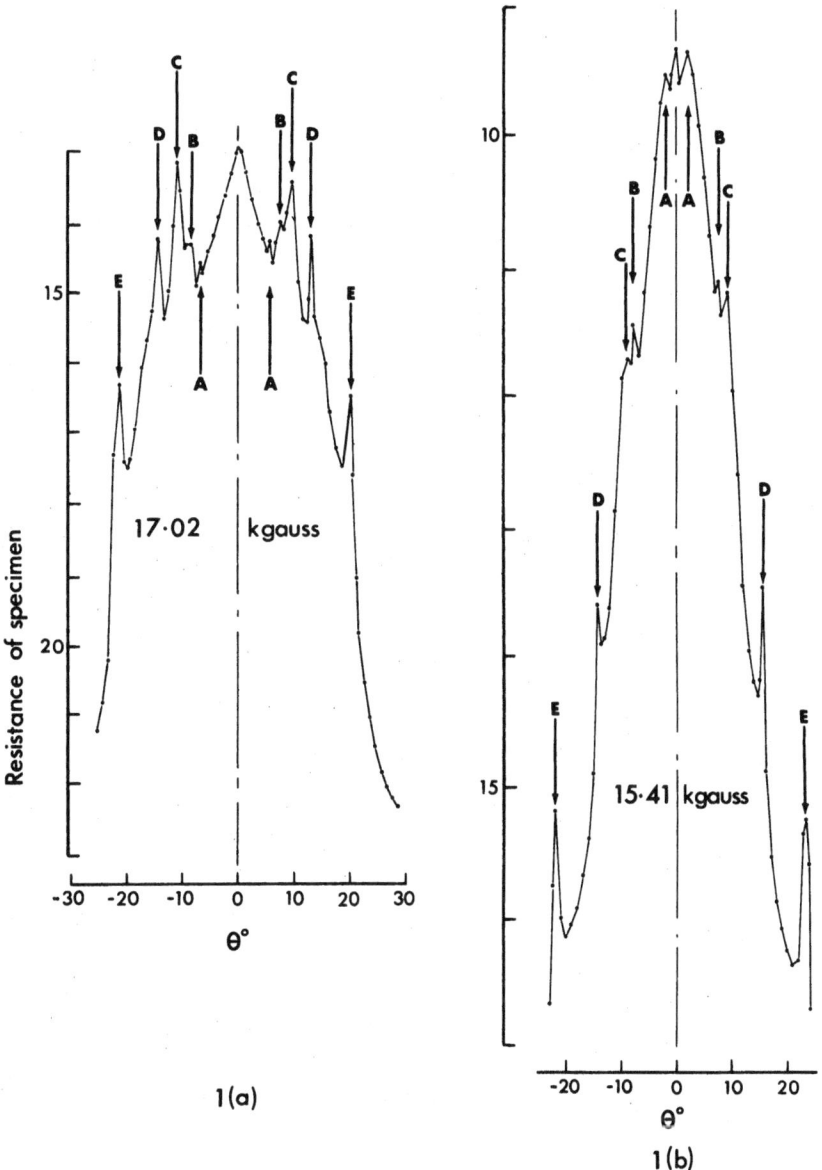

Fig. 1. The variation of the resistance in arbitrary units along the trigonal axis with the angle, θ, between the applied field and the binary axis.

Figure 2 shows that the general tendency of the angular separation of the peaks is first to decrease with increasing field and then to increase again, except for the A peaks, in the field range investigated so far.

The magnitudes or heights of the various peaks did not vary in a simple manner as the magnetic field was changed, and the situation was further complicated by the fact that these heights were very sensitive to the alignment of the specimen.

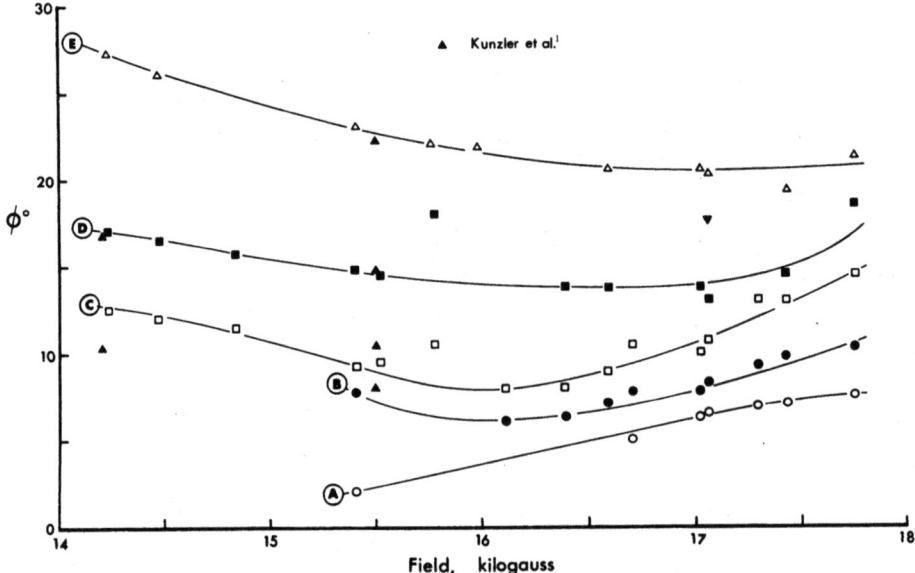

Fig. 2. The variation with magnetic field of the angular separation, ϕ, of the magnetoresistance peaks from the central one. The letters refer to the peaks identified in Fig. 1.

The oscillations in resistance which occur as the field is varied steadily have been observed at fixed orientations in the plane perpendicular to the trigonal axis. Spin-splitting of the Landau levels has been observed with the field parallel to both the binary axis and the bisectrix axis, and from the results, the following spin-splitting parameters have been deduced:

$$\Delta_{\text{bin}} = 0.46 \qquad \Delta_{\text{bisec}} = 0.46$$

These figures are in good agreement with those given by Kunzler et al.,[1] Lerner,[2] and Saito.[3]

References
1. J. E. Kunzler, F. S. L. Hsu, and W. S. Boyle, *Phys. Rev.* **128**, 1084, 1962.
2. L. S. Lerner, *Phys. Rev.* **130**, 605, 1963.
3. Y. Saito, *J. Phys. Soc. Japan* **18**, 1845, 1963.

NEW OSCILLATORY MAGNETORESISTANCE EFFECT IN GALLIUM*

J. A. Munarin and J. A. Marcus

Northwestern University
Evanston, Illinois

Magnetic size effects have increasingly become the subject of experimental and theoretical investigations[1–8] because they provide a means of studying the Fermi surface, supplementing the information obtained from de Haas–van Alphen and other related effects. For size-effect oscillations to occur in the magnetoresistance, the inequality $t_0 > a/|\bar{v}_a| \gg T$ must be satisfied,[1] where t_0 is the volume relaxation time of the conduction electrons, a is the thickness of the crystal, $|\bar{v}_a|$ is the time average of the a component of the electron velocity, and T is the cyclotron period. These conditions are generally difficult to satisfy, requiring very pure, strain-free single crystals, low temperatures, and high magnetic fields. In gallium, the situation is particularly favorable. Cochran and Yaqub[9] have estimated, from a study of the resistivity of a series of square "wires" ranging in size from 0.0115 to 0.102 cm, that the mean free path of conduction electrons in the bulk material is as large as 1 cm at 0°K. The bulk magnetoresistance, however, is so large in gallium (7×10^4 at 15 kG for $I \parallel c$, $H \parallel a$[10]) that the surface scattering term in the resistivity is completely obscured when the magnetoresistance is measured by conventional AC or DC methods. This suggests using a derivative technique[11] which eliminates the monotonic increase characteristic of the bulk magnetoresistance and permits the investigation of the oscillatory part of the resistivity.

This paper reports the observation of a new magnetoresistive size effect in gallium single crystals at liquid-helium temperatures which is periodic in the magnetic field with period inversely proportional to thickness and which, at least for certain field directions, exhibits a growth of amplitude with magnetic field up to 16 kG.

Crystals CB-II and BA-II were prepared by W. A. Reed[10] from 99.999% pure gallium obtained from the Aluminum Company of America; these crystals measure $1 \times 1 \times 20$ mm and the crystallographic axes are aligned with the specimen axes to within 1°. Sample CB-211 was grown from the same batch of raw material with dimensions $0.53 \times 0.50 \times 20$ mm in the a, b, and c directions, respectively. Crystal BA-651 is a rectangular plate $0.15 \times 3 \times 20$ mm grown from 99.9999% pure gallium supplied by Eagle-Picher. The crystals are labeled as follows: The first letter indicates the crystallographic axis of current flow, the second indicates the axis normal to the mirror face.[10] Current and potential leads were attached in a manner described by Yahia and Marcus.[12] Longitudinal probes are spaced 3 mm apart.

AC modulation fields of 1 to 100 G at a frequency of 36 cps are superimposed on the DC magnetic field. The signals are amplified and detected by a lock-in

* Supported by the National Science Foundation and the Advanced Research Projects Agency of the U.S. Department of Defense.

amplifier driven either at the driving frequency, f, of the modulation coils, or at harmonics, nf, of the driving frequency if it is desired to record derivatives of the magnetoresistance higher than the first. The output of the lock-in amplifier is fed into the y-axis of an x–y recorder; the x-axis is driven by a high-linearity Hall probe. The data are simultaneously recorded on paper data processing tape to be used for a digital computer Fourier analysis of the frequencies of oscillation.

Large-amplitude oscillations periodic in the magnetic field appear in the transverse magnetoresistance and Hall effect when H is in the ab plane (CA or CB type crystals) for ϕ less than 45°, where ϕ is the angle between the direction of the magnetic field and the a-axis of the crystal. Four periods have been identified in crystal CB-II in this angular range (see Fig. 2, right-hand section). In addition, a fifth period of approximately 450 G has been observed for ϕ between 69° and 71°. At other angles one observes only de Haas–Shubnikov oscillations periodic in the reciprocal of the magnetic field. The data for CB-II exhibits an apparent linear increase of amplitude with magnetic field; this is especially evident for $\phi = 10°$, where the beat pattern due to the two largest amplitude periods disappears, and for $\phi = 70°$, where the long period is observed.

Measurements made on BA-type crystals (magnetic field in the ac plane) with $\phi < 37°$ also show pronounced oscillations in H. The amplitude of the dominant period falls off quite abruptly at 9°, a factor of 15 in 1° finally vanishing at 37° from the a-axis. A long period is seen in the data for BA-II; this period rises sharply with angle and has not been followed beyond 9° because of the small amplitude of the oscillations. No oscillations periodic in the magnetic field are observed beyond 37°; however, strong de Haas–Shubnikov oscillations are detected in this range (see Fig. 1, upper part, 80°). The amplitude of the magnetoresistance oscillations in BA-II at 7° appears to be falling off for magnetic fields higher than 7 kG. A possible explanation for this behavior of the amplitude is that the surfaces of crystal BA-II were roughened by sandblasting before it was mounted, thereby changing the nature of the scattering of electrons at the surface, or perhaps the parallelness of the opposite faces.

Preliminary measurements have failed to show size-effect oscillations in crystals of type AB, either in the transverse or longitudinal magnetoresistance, although pronounced de Haas–Shubnikov oscillations are again observed.

A plot of period vs. angle for H in the ab and ac plane for crystals CB-II and BA-II is given in Fig. 2, right. The data were reduced visually and spot checked at several angles with a Fourier analysis performed on an IBM 709 computer. The agreement between the hand-reduced data and the results of the computer analysis is better than $\frac{1}{2}\%$. This plot is not intended to be a complete distillation of the data but rather an indication of the angular variation of the periods of oscillation as has been determined at this stage of the investigation.

In Fig. 2, (left), the dominant period for $H \parallel a$ has been plotted vs. the reciprocal of the crystal dimension parallel to a. The four points shown lie within 3% of the best straight line through the points. Within the limits of experimental error, the dominant period scales with inverse thickness, indicating that the oscillations arise in some manner from scattering at the surface of the crystal. However, all calculations that have been made of the oscillating part of the magnetoresistance predict that the amplitude falls off with increasing magnetic field.

The first calculation of the oscillatory size effect in metallic films was made by E. H. Sondheimer[12] for a free-electron model of the Fermi surface; the period of the oscillations with H normal to the film surface was found to be $2(pc/ae)$, where

New Oscillatory Magnetoresistance Effect in Gallium

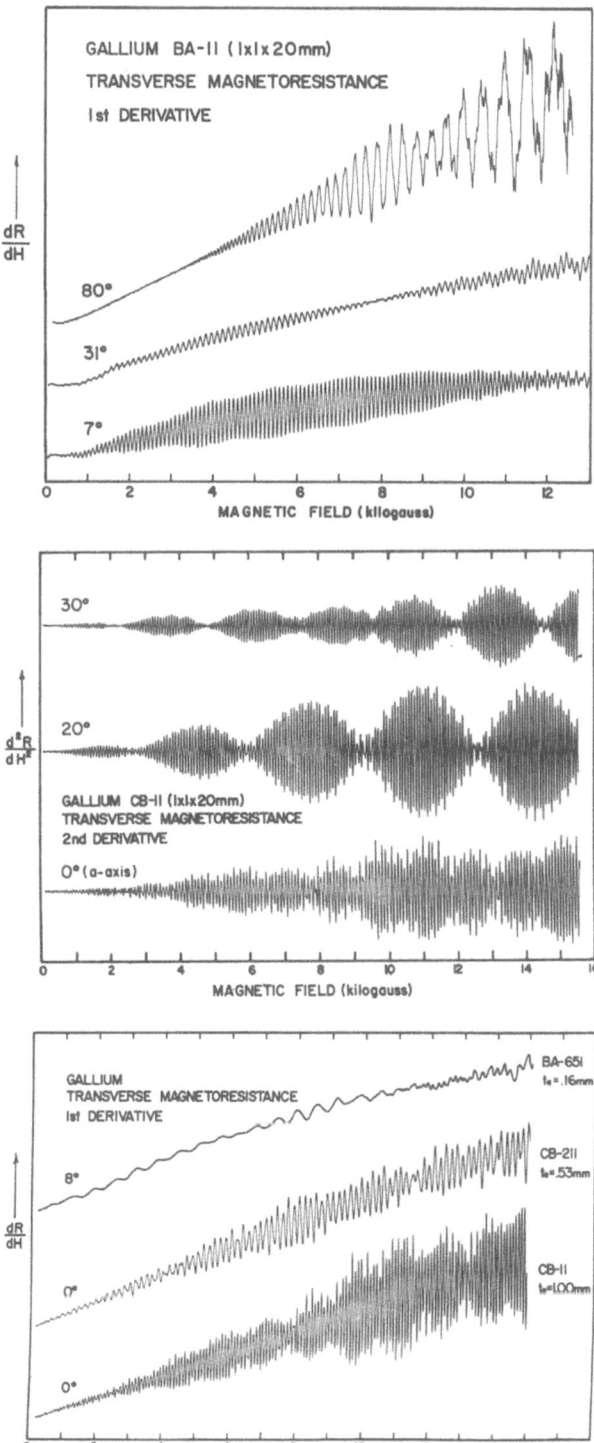

Fig. 1. Transverse magnetoresistance at 1.3°K. All angles are measured from the a-axis. Top: first derivative data for H in the ac plane. Oscillations at 80° are de Haas–Shubnikov effect. Middle: second derivative data for H in the ab plane. Bottom: comparison of data at or near the a-axis for crystals of different dimension in the a-direction.

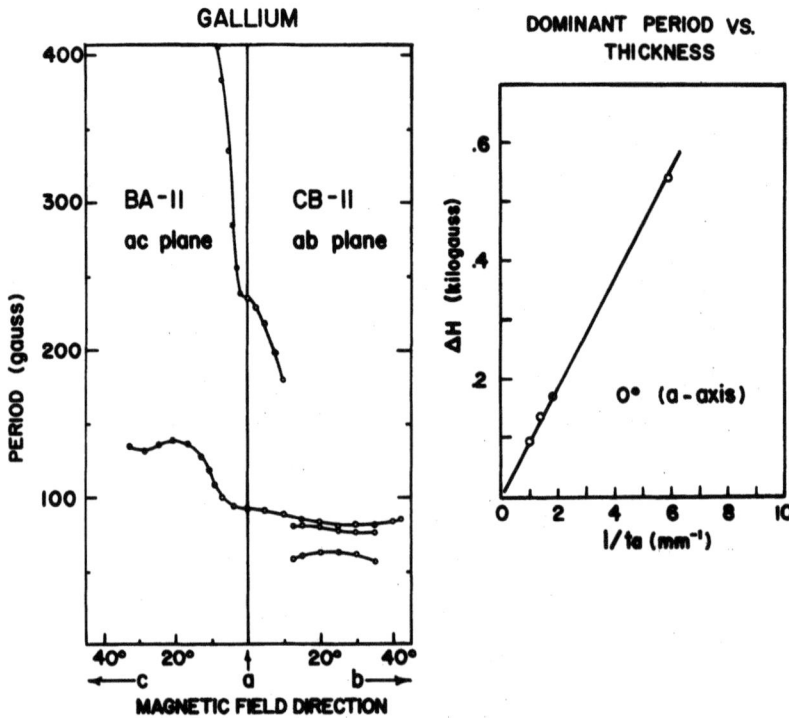

Fig. 2. Left: period vs. magnetic field direction for CB-II and BA-II, 1 × 1 × 20 mm. Right: dominant period vs. thickness for $H \parallel a$. The abscissa is the reciprocal of the crystal dimension parallel to a.

p is the extremal Fermi momentum in the field direction and a is thickness of the film. On this basis, the periods measured for $H \parallel a$ correspond to momenta $p_1 = 0.234 \times 10^{-19}$ gm-cm/sec and $p_2 = 0.598 \times 10^{-19}$ gm-cm/sec for $\Delta H = 92$ and 235 G, respectively. These values were compared with the extremal dimensions of the Fermi surface obtained by J. H. Condon[13] from de Haas–van Alphen data. The smaller momentum value shows possible agreement with the largest sheet of the surface constructed from the de Haas–van Alphen data, but the larger value apparently does not.

Gurevich[1] has calculated the oscillatory dependence on magnetic field of the conductivity for a general Fermi surface. The existence of at least three varieties of oscillations periodic in the magnetic field is indicated. The first type, presumably corresponding to Sondheimer's calculation, falls off asymptotically with field as H^{-4} and has periodicity determined by the Gaussian curvature of the Fermi surface at its elliptic points. These calculations apparently do not explain the linear amplitude growth of the resistivity oscillations reported here, and their validity as an explanation of the gallium results is questionable.

It is hoped that further investigation will reveal the nature of the effect, and lead to an eventual interpretation of both the angular dependence of period and amplitude which appear to contain a great amount of information about the Fermi surface.

Acknowledgment

We are indebted to J. H. Condon for many enlightening discussions concerning the development of the method.

References

1. V. L. Gurevich, *Soviet Phys. JETP (English Transl.)* **8**, 464, 1959.
2. K. Forsvoll and I. Holwech, *Phys. Letters* **3**, 66, 1962.
3. M. Yaqub and J. F. Cochran, *Phys. Rev. Letters* **10**, 390, 1963.
4. V. F. Gantmakher, *Soviet Phys. JETP (English Transl.)* **17**, 549, 1963.
5. B. N. Aleksandrov, *Soviet Phys. JETP (English Transl.)* **16**, 286, 871, 1963.
6. N. H. Zebouni, R. E. Hamburg, and H. J. Mackey, *Phys. Rev. Letters* **11**, 260, 1963.
7. V. F. Gantmakher and E. A. Kaner, *Soviet Phys. JETP (English Transl.)* **18**, 988, 1964.
8. E. A. Kaner, *Soviet Phys. JETP (English Transl.)* **17**, 700, 1963.
9. J. F. Cochran and M. Yaqub, *Phys. Letters* **5**, 307, 1963.
10. W. A. Reed and J. A. Marcus, *Phys. Rev.* **126**, 1298, 1962.
11. L. S. Lerner, *Phys. Rev.* **127**, 1480, 1962.
12. J. Yahia and J. A. Marcus, *Phys. Rev.* **113**, 137, 1959.
13. J. H. Condon, *Bull. Am. Phys. Soc.* **9**, 239, 1964.

MAGNETORESISTANCE IN ANTIFERROMAGNETIC CHROMIUM*

A. J. Arko and J. A. Marcus

*Northwestern University**
Evanston, Illinois

In measurements of the anisotropy of magnetic susceptibility, the de Haas–van Alphen effect and the magnetoresistance of chromium single crystals cooled through the Néel temperature in a magnetic field, Montalvo and Marcus[1] observed that the symmetry of the magnetic structure is reduced from cubic to tetragonal. Subsequently, the effect of field cooling was observed in the pulsed field de Haas–van Alphen effect by Watts,[2] in neutron diffraction by Møller et al.,[3] and in Young's modulus by Munday et al.[4]

It is the purpose of this paper to give a more extensive treatment† of the magnetoresistance of chromium and the changes induced in it by field cooling. These changes in the magnetoresistance are found to be of the order of several hundred percent as compared to changes of only a few percent in magnetic susceptibility.

All measurements of magnetoresistance were made at 4.2°K on a single crystal of chromium with a residual resistance ratio of 650 and the current along the [001] direction. The sample was 10 mm long, with four approximately symmetric though irregular faces around the long axis. X-ray measurements have shown that the faces correspond to $\langle 110 \rangle$ type directions while the edges are $\langle 100 \rangle$ type. The long axis was chosen as the [001] axis. Forty gauge copper potential leads were soldered to the sample, separated by 3.5 mm. The experimental procedure consisted of heating the sample above the Néel temperature (to 50°C) and then cooling it in the presence of a constant magnetic field, H_c, to 4.2°K. The potential drop across the sample was recorded continuously as the y-ordinate on an x–y recorder whose x-axis measured either the magnet angle or the current through the magnet.

With a measuring field, H_m, of 32 kOe in the (001) plane and $H_c = 0$, the transverse magnetoresistance shows cubic symmetry with minima at $\langle 100 \rangle$ and maxima at $\langle 110 \rangle$ type directions (see Fig. 1a). $\Delta\rho/\rho$ is very nearly proportional to H^n, with n varying from about 0.80 at the [100] minimum to 1.6 at the maxima. Subsequent experiments in which stress was applied to the sample have shown that the slight difference between the [100] and [010] directions is probably due to some residual strain in the crystal.

If the sample is cooled through the Néel point in a field, $H_c = 32 \text{ kOe} \| [100]$, a twofold anisotropy is induced in the rotation diagram of magnetoresistance with

---

* Supported by the National Science Foundation and the Advanced Research Projects Agency of the Department of Defense.

† The only other investigation of magnetoresistance in chromium was made by Borovik,[5] who found only a very small anisotropy in magnetoresistance and as a result, concluded that the properties of chromium do not differ much from those of the nontransition metals.

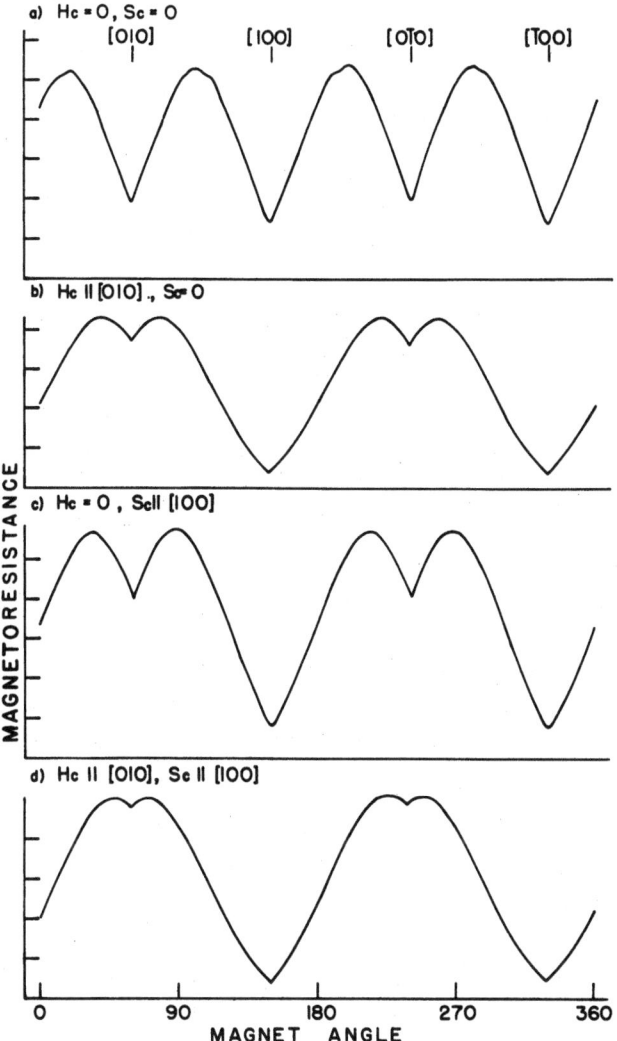

Fig. 1. Rotation diagrams of magnetoresistance for various types of stress and field cooling; all $H_c = 32$ kOe; units are arbitrary but the same for all curves. The slight departure from cubic symmetry in (a) is probably due to strain.

$\Delta\rho/\rho$ no longer tending toward saturation for H_m along [100]. The value of n changes from 0.80 to 1.40 with $\Delta\rho/\rho$ itself increasing by a factor of 2.3 at $H_m = 32$ kOe. With H_m in the [010] direction, however, the magnetoresistance tends more toward saturation with $\Delta\rho/\rho$ being decreased by a factor of 6 at $H_m = 32$ kOe, while n changes little from its value of 0.9. The situation is reversed if H_c is along [010] (Fig. 1b). Except for the slight differences presumed to be due to strain, all results are consistent with an interchange of the [100] and [010] axes. The field-induced anisotropy can be completely erased simply by heating the sample above 312°K.

If H_c is oriented along [110], the rotation diagram maintains its cubic symmetry but $\Delta\rho/\rho$ is decreased in value at all points. On the other hand, with H_c along [001], the entire rotation diagram is increased considerably while still maintaining its cubic symmetry. In either case, $\Delta\rho/\rho$ shows a tendency to saturate along the crystal axes, though in the latter case n increases to 1.0 and 1.1 for [100] and [010] directions of H_m, respectively.

The amount of induced twofold anisotropy is strongly dependent on the strength of the cooling field and on the temperature interval of cooling. Preliminary experiments have shown that a field of 8 kOe will produce little effect, while at 24 kOe it is already tending toward its limiting value. In like manner, failure to cool the sample from above the Néel point for these fields* results in only a small percentage of induced twofold anisotropy, while cooling below 77°K in the presence of H_c adds almost nothing to the effect.

Measurements of longitudinal magnetoresistance were made with a 33 kOe Westinghouse superconducting magnet in a way similar to that described above.

* Watts[2] has found that for sufficiently large fields (80 to 100 kOe), simply placing the sample in the field at any temperature below the Néel point constitutes a treatment similar to that described above.

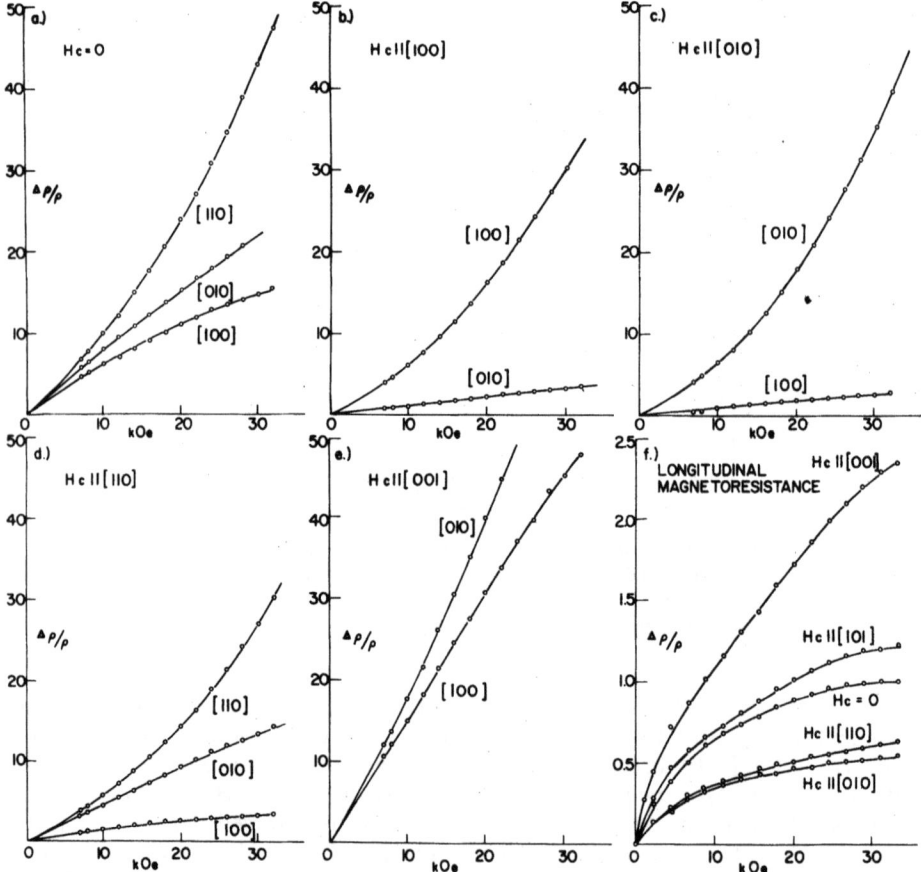

Fig. 2. $\Delta\rho/\rho$ vs. H curves for various directions of H_m and H_c; (a) through (e) are transverse magnetoresistance; all $H_c = 32$ kOe.

Field cooling for various directions of H_c raises or lowers the entire $\Delta\rho/\rho$ vs. H curve, but saturation occurs in all cases (Fig. 2f). Just as with transverse magnetoresistance, H_c along [001] raises the entire curve while H_c along [110] lowers it.

Figure 1c shows the results of preliminary experiments to determine the effects of stress, S_c, applied during cooling. Although the sample was not regularly shaped and it was impossible to determine the actual stress on the sample, the gross features are sufficiently interesting to be published at this time. The experimental procedure consisted of merely placing a weight (3.86 kg) on an edge of the sample, heating to 323°K, and then cooling it to 77°K. The weight was then removed, the sample further cooled to 4.2°K, and measurements again made as before. The effect of S_c applied along $\langle 100 \rangle$ appears to be the same as that of H_c applied 90° away, which would be consistent with positive magnetostriction. Orienting H_c along [010] while at the same time applying S_c along [100] gives the results of Fig. 1d. The final result is not a simple addition of curves b and c but there is a greater tendency toward complete twofold symmetry in the magnetoresistance.

We are tempted to explain the results on the basis of the Lifshitz, Azbel, and Kaganov (LAK) theory of magnetoresistance of a compensated metal having a limited number of open orbits, and assume that the open orbits arise from modifications to the Fermi surface produced by additional Brillouin zone planes associated with magnetic ordering. By changing the distribution of domains among the $\langle 100 \rangle$ directions, H_c then changes this magnetic ordering with an accompanying change in the relative number of open orbits along the $\langle 100 \rangle$ directions. However, due to a lack of true quadratic rise and true saturation* it is not yet certain that the high-field conditions are satisfied and that the LAK theory can be applied. Further research is needed and, in particular, the Hall voltage and the transverse-even voltage remain to be investigated. We are now in the process of making these measurements on samples of regular shape.

Acknowledgment

We are indebted to Dr. W. A. Reed for providing us with the chromium crystals.

References

1. R. A. Montalvo and J. A. Marcus, *Phys. Letters* **8**, 151, 1964.
2. B. R. Watts, *Phys. Letters* **10**, 275, 1964.
3. H. B. Møller, K. Blinkowski, A. R. Mackintosh, and T. Brun, *Solid State Comm.* **2**, 109, 1964.
4. B. C. Munday, A. R. Pepper, and R. Street, *Brit. J. Appl. Phys.* **15**, 611, 1964.
5. E. S. Borovik and V. G. Volotskoya, *Zh. Eksperim. i Teor. Fiz.* **36**, 1650, 1959. (*Soviet Phys. JETP (English Transl.)* **36**, 1175, 1959.)

---

*Though the high-field conditions may be satisfied, it may still be possible to get departure from true quadratic rise and true saturation as a result of an averaging process over the various magnetic domains of different orientations.

THE FERMI SURFACE OF ALUMINUM AND DILUTE ALUMINUM–ZINC ALLOYS*

J. P. G. Shepherd, C. O. Larson, D. Roberts, and W. L. Gordon

Case Institute of Technology
Cleveland, Ohio

We have employed the de Haas–van Alphen effect (DHVA) to study the third zone portion of the Fermi surface of pure aluminum and of dilute alloys of zinc in aluminum.

Measurements were made with a self-balancing torsion balance[1] in magnetic fields up to 23 kG. The magnetic field is swept so that its reciprocal varies in an approximately linear manner in time in order to produce DHVA oscillations with a constant frequency in time. Either broad or narrow band filters are then used to sort out the component frequencies. This technique is very useful in the study of polyvalent metals such as aluminum, since there are as many as 10 different sets of oscillations present with periods ranging from 1×10^{-7} to $40 \times 10^{-7} G^{-1}$ in aluminum.

Now, although the period in $1/H$ of a set of DHVA oscillations is inversely proportional to an extremal cross section of the Fermi surface[2] normal to the magnetic field, it is not generally possible to assign unambiguously these extremal cross sections to definite regions of k-space. Instead, a comparison must be made with a theoretical model of the Fermi surface.

The first DHVA results for aluminum of Gunnersen[3] could be mostly interpreted on a nearly-free electron model of Harrison.[4] In a more detailed nearly-free electron calculation, Ashcroft[5] has shown that although the general features of the surfaces remain the same, the details of the connectivity and the size of the cross sections of the portions of the Fermi surface in the third Brillouin zone depend very much on the choice of the first Fourier components of the effective potential at the [111] and [200] zone faces. By fitting the existing experimental data, Larson and Gordon[6] for the short periods ($3.50 \times 10^{-7} G^{-1}$) and Gunnersen[3] for the long periods (30–$40 \times 10^{-7} G^{-1}$), Ashcroft was able to eliminate 3 of the 4 possible models for the Fermi surface of aluminum.

In this final model, Harrison's[4] "monster" in the third zone is dismembered into rings consisting of four arms having slightly thickened junctions. The planes of the rings are normal to [100] directions. These third zone surfaces account for all of the DHVA periods present at low fields.

---

* Work supported by the Air Force Office of Scientific Research and by the Case Center for the Study of Materials.

Pure Aluminum

The data presented here are for a specimen mounted so that the magnetic field is rotated in a [100] plane. The pure aluminum specimen was grown by the strain anneal method from 6-9 s commercially zone refined aluminum and had a resistance ratio of 5000.

In Fig. 1a we compare our results for the short periods with Ashcroft's prediction for electron orbits around the middle of the arms. As can be seen, agreement is excellent, both for the magnitude and for the variation with angle.

Figure 1b shows the long periods and again the agreement with Ashcroft's model is excellent. The calculated $20 \times 10^{-7} \, G^{-1}$ periods here correspond to orbits around the thickened part of the junction between two arms. The $40 \times 10^{-7} \, G^{-1}$ periods correspond to orbits about the waists close to this junction. The latter period does not have a maximum at a symmetry direction, but at approximately 20° from the [110] direction in a [010] plane; this is because the narrow ends of the arms bend away from the [110] directions. Similar agreement with Ashcroft's model is reported by Volskii,[7] who observed the oscillations in the surface resistance of aluminum.

A careful search for evidence of junctions between the rings, as in Harrison's model, has yielded negative results. This is to be expected from the observed angular dependence of the two-arm joint period. Finally, we note one point which has not yet been resolved. In torque measurements, electron orbits on Fermi surface cross sections with a small angular dependence, such as these two arm junctions show, are expected to have small amplitudes of oscillations while these are very strong except near [100].

The effective masses, obtained from the temperature dependence of the amplitude of the DHVA oscillations, are shown in Table I, together with values

Table I

Period	Pure aluminum			Alloys	
	This work	Ashcroft calculated	Other workers	Al-0.27 at. %Zn	Al-0.50 at. %Zn
Long periods					
$36 \times 10^{-7} G^{-1}$ [100]	0.093	0.0605	0.097[a]	0.092	0.096
$21 \times 10^{-8} G^{-1}$ [100]		0.0660			
[110]	0.12			0.112	0.114
Short periods					
[100]	0.18	0.1390	0.149[a] 0.18[b]	0.19	0.17
[110]	0.130	0.0897	0.146[a] 0.13[b]		

[a]See Gunnersen.[3]
[b]See Moore and Spong.[8]

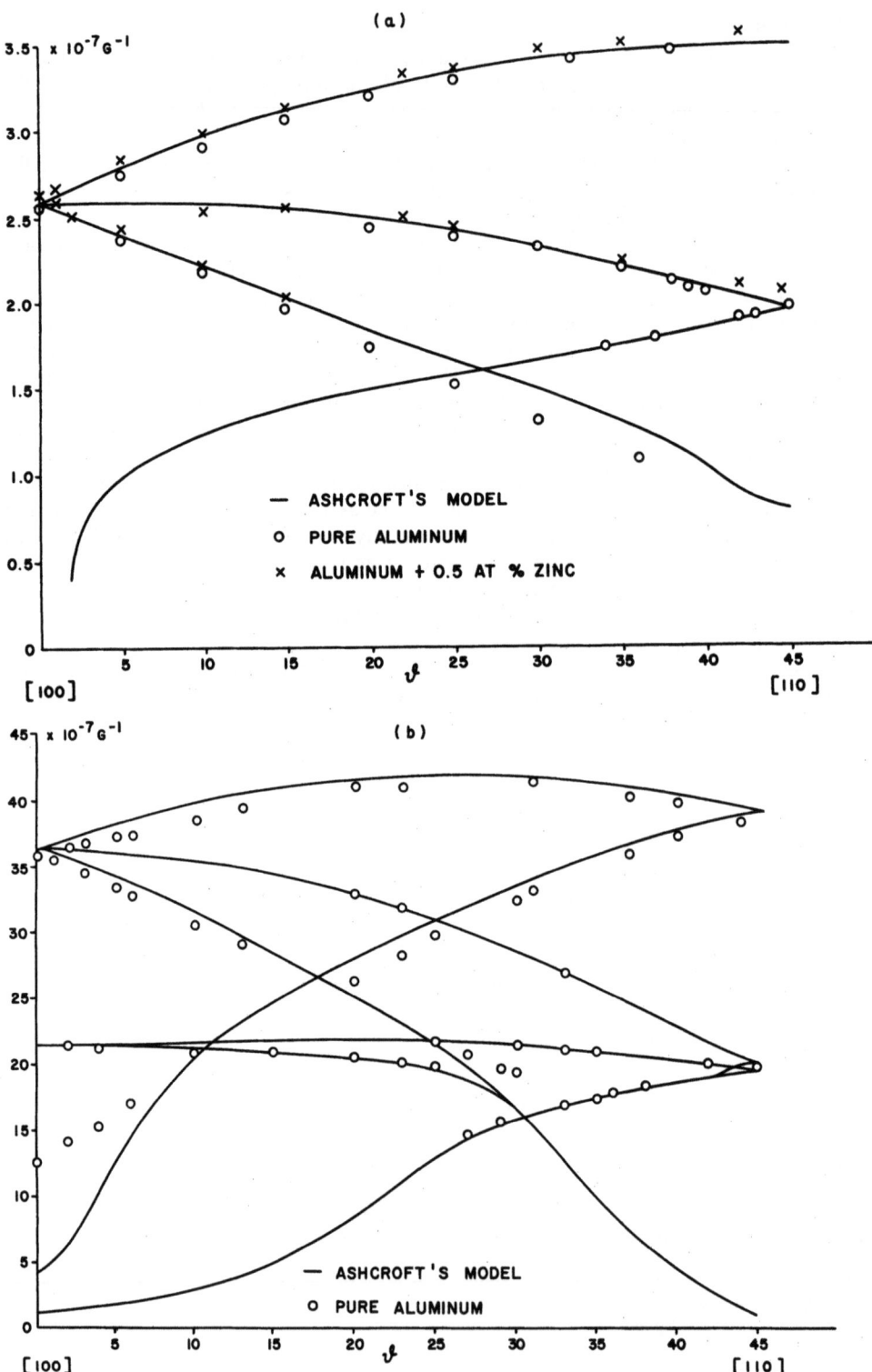

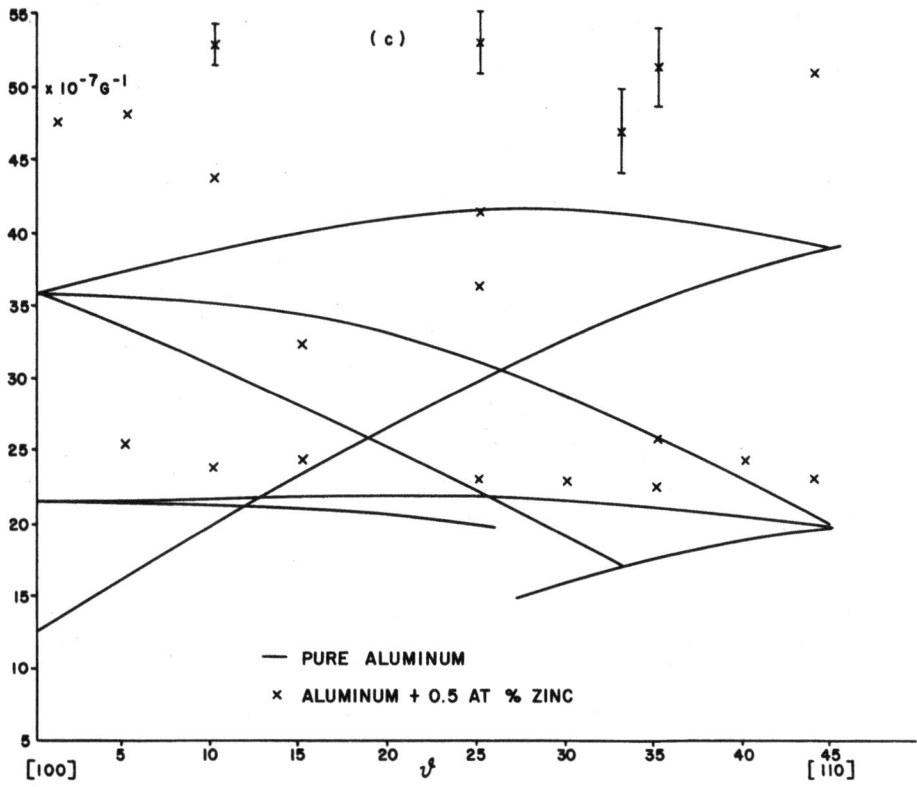

Fig. 1. DHVA periods as a function of magnetic field direction in a [100] plane showing the comparison of Ashcroft's model with pure aluminum and of the Al + 0.5 at.% Zn with pure aluminum. (a) short periods; (b) long periods; (c) long periods.

obtained by other workers.[3,8] The agreement with other experiments is good, but the values of Ashcroft are 20-35% too low.

Aluminum Alloys

The alloy specimens were prepared by adding zinc to molten aluminum in an atmosphere of helium or nitrogen. After thoroughly stirring, the melt was chill cast. Chill casting was adopted to prevent segregation of the zinc. One 0.5-at.% ingot so prepared was checked by surveying longitudinal and transverse sections with a movable aperture (sampling areas of approximately 1 mm square) mounted on an X-ray fluorescent spectrometer. The inhomogeneity in the zinc concentration was less than 2%.

Crystals were then grown by the strain anneal method and specimens spark-cut from them. A test sample was cut from beside the specimen and the surface of this was mechanically worked to give a polycrystalline layer. X-ray fluorescent measurements were then made on these to determine the homogeneity and the percentage composition by comparison with spectroscopic standards. Crystals grown in the

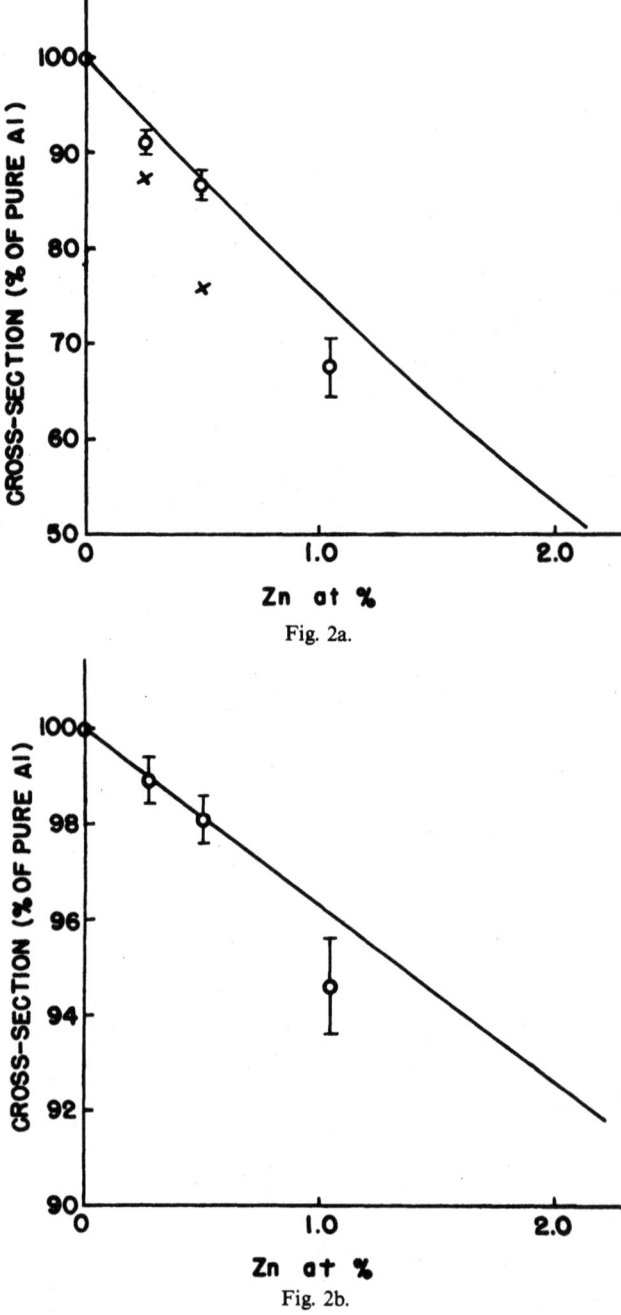

Fig. 2. Frequencies in definite symmetry direction for three alloys expressed as a percentage of the pure aluminum frequency are plotted against the at.% Zn. The calculated percentage variation of the appropriate cross sections are shown by the solid lines. (a): short periods at [110]; (b): long periods at [110]—○, aluminum period $20 \times 10^{-7} G^{-1}$; ×, aluminum period $38.3 \times 10^{-7} G^{-1}$.

above manner showed variations in the zinc content of less then 3% of the mean. One crystal prepared by a rapid molten zone pass did, however, show a small gradient.

The results are again presented for samples mounted so that the field rotated about a [100] axis. In Fig. 1a the short periods measured in an aluminum +0.5 at.% zinc alloy are plotted to be compared with the pure aluminum periods. Due to the large increase in scattering and hence decrease in signal amplitude (resistance ratio was 40), fewer points could be obtained. The general trend, however, is clearly discernible. There is an increase in period of 2% in the [110] direction and 3% in the [100] direction.

In Fig. 1c the long periods are plotted for the same alloy. The change here is much greater being 10% for the 20×10^{-7} G periods and 25% for the longest periods.

The choice of which period in aluminum to compare with an alloy period has been guided by similarities in the rotation diagram and relative amplitudes as compared with the other periods for the same sample.

We have interpreted these data in the following manner: The addition of zinc to aluminum reduces the number of electrons in the conduction band and thus reduces the volume of the Fermi surface on the assumption of a rigid energy band structure. This causes a reduction in the size of the third zone segments.

To put this on a semiquantitative basis, we assume that aluminum supplies 3 electrons per atom while zinc gives 2 to the conduction band, and then calculate the radius of the free electron sphere for various concentrations of zinc. A remapped free-electron model for aluminum is then considered. The area of large cross sections of the third zone arms is assumed to be proportional to the square of the distance of the free-electron sphere from the point K in the Brillouin zone. The cross section of the junction between the arms is similarly assumed proportional to the square of the distance of the surface from the W point of the zone.

In Fig. 2 these calculated cross sections have been plotted, against the at.% of zinc, together with the results for 3 alloys.

In Fig. 2a the value for the short period extrapolated to [110] is interpreted as the orbit around the middle of the third zone arm, and is compared to the calculated variation for this orbit. It should be noted that the period at [100] will not give the same result, since this involves an orbit along the arm and thus sees an averaged change between the waist and the middle of the arm.

In Fig. 2b the period for the orbit around the junction of two arms is compared with the predicted variation of the junction of third zone arms. The agreement is surprisingly good considering the simple model used.

The variation of the longest period corresponding to an orbit around the thinnest parts of the arms is also shown in Fig. 2b. This is a more rapid variation and indicates that the arms may pinch off near 2 at.% zinc. It is interesting to note that the α-phase boundary occurs at 1.7 at.%. The effective masses for the alloy data are shown in Table I and they agree quite well with the results for pure aluminum.

In conclusion, it has been shown that agreement between our values of the DHVA periods for aluminum and those calculated by Ashcroft is very good. Thus, Ashcroft's description of the Fermi surface of aluminum is very probably correct and the third zone arms are connected only in groups of 4.

The alloy data can readily be interpreted by a free-electron perturbation from this surface.

References

1. A. S. Joseph and W. L. Gordon, *Phys. Rev.* **126**, 489, 1962.
2. L. Onsager, *Phil. Mag.* **43**, 1006, 1952.
3. E. M. Gunnersen, *Phil. Trans. Roy. Soc. (London) Ser. A* **249**, 299, 1957.
4. W. A. Harrison, *Phys. Rev.* **118**, 1182, 1960.
5. N. W. Ashcroft, *Phil. Mag.* **8**, 2055, 1963.
6. C. O. Larson and W. L. Gordon, (unpublished).
7. E. P. Volskii, *Soviet Phys. JETP (English Transl.)* **19**, 89, 1964.
8. T. W. Moore and F. W. Spong, *Phys. Rev.* **125**, 846, 1962.

OBSERVATION OF CYCLOTRON RESONANCE IN MAGNESIUM*

T. G. Eck and M. P. Shaw

Case Institute of Technology
Cleveland, Ohio

We have observed Azbel–Kaner cyclotron resonances at 24 Gcps and 2°K in a single crystal of magnesium† with a residual resistance ratio of 8500. The normal to the specimen surface was within 1° of [10$\bar{1}$0]. The resonant cavity and sample mounting technique were the same as those used for the study of cadmium.[1]

Three distinct subharmonic series were seen for most orientations of the magnetic field, H, in the sample surface. Figure 1 shows recorder traces of dR/dH, the field derivative of the surface resistance, as a function of H for H 20° from [11$\bar{2}$0]. The upper trace is for E_{RF}, the microwave electric field, parallel to H, and the lower trace for E_{RF} perpendicular to H. For this orientation $m_{a^*} \cong 2m_{c^*}$ and the resonance peaks of c coincide with the even subharmonics of a.

* Work supported by the Air Force Office of Scientific Research.

† We are indebted to Dr. R. W. Stark for supplying the magnesium crystal from which our specimen was prepared.

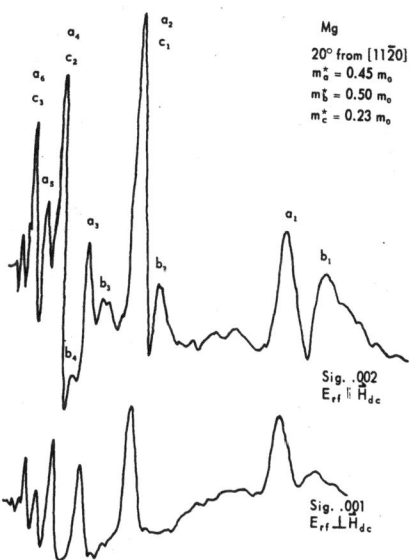

Fig. 1. Recorder traces of dR/dH vs. H for H 20° from [11$\bar{2}$0] in a (10$\bar{1}$0) plane.

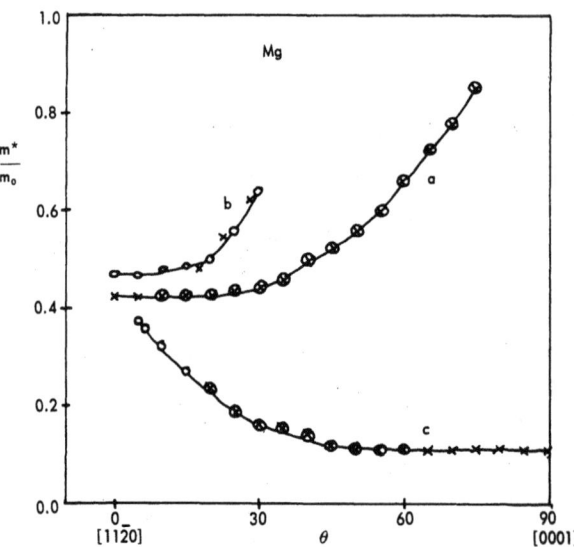

Fig. 2. m^*/m_0 as a function of the orientation of H in a (1010) plane. × is for E_{RF} perpendicular to H and ○ for E_{RF} parallel to H.

Figure 2 is a plot of m^*/m_0 for the three observed mass series as a function of the orientation of H. An × indicates that a resonance series was obtained with this value of m^*/m_0 for E_{RF} perpendicular to H and an ○ for E_{RF} parallel to H.

We associate branch a of Fig. 2 with the extremal cross section of the lens-shaped electron surface in the third band of magnesium. The value of $m^*/m_0 = 0.42 \pm 0.02$ obtained for H parallel to [11$\bar{2}$0] is in good agreement with the value of 0.44 ± 0.02 obtained by Priestley[2] from the temperature dependence of the amplitude of his de Haas–van Alphen oscillations. However, his value of 0.54 ± 0.05 for the lens mass when H is 4° from [0001] is in conflict with our observation of a mass of 0.85 ± 0.04 with H 15° from [0001].

Branch c of Fig. 2 almost certainly arises from the electron "cigars" in the third band. The mass of 0.11 ± 0.01 for H parallel to [0001] is to be compared with Priestley's value of 0.20 ± 0.03 and the value of 0.13 ± 0.04 from the DHVA investigation of Gordon et al.[3]

We attribute branch b to resonances from an orbit around two arms of the second band "monster," the η orbit of Priestley,[2] though our m^*/m_0 of 0.47 ± 0.02 for H parallel to [11$\bar{2}$0] is substantially less than his value of 0.78 ± 0.04.

References

1. M. P. Shaw and T. G. Eck, "Cyclotron Resonance Investigation of the Fermi Surface of Cadmium," this volume, p. 761.
2. M. G. Priestley, Proc. Roy. Soc. (London) Ser. A **276**, 258, 1963.
3. W. L. Gordon, A. S. Joseph, and T. G. Eck, in: The Fermi Surface, W. A. Harrison and M. B. Webb (eds.), John Wiley & Sons, Inc., New York (1960), p. 84.

CYCLOTRON RESONANCE INVESTIGATION OF THE FERMI SURFACE OF CADMIUM*

M. P. Shaw and T. G. Eck

Case Institute of Technology
Cleveland, Ohio

Azbel–Kaner resonances in cadmium have been investigated at 2°K and 24 Gcps, using a standard reflection spectrometer employing derivative detection. Samples with surfaces parallel to the principle crystal planes were mounted with conducting silver paint to the end wall of a cylindrical, side-coupled, rotatable cavity of the type described by Koch et al.[1] Eight different specimens were used in this work, two with surfaces parallel to (11$\bar{2}$0), two parallel to (0001), and four parallel to (10$\bar{1}$0). The residual resistance ratios of the samples ranged from 18,000 to 26,000.

Figure 1 shows the observed values of m^*/m_0 as a function of the orientation of the magnetic field, H, in the three principal crystal planes. Branch a of the mass plot exhibited the strongest resonance signals, with as many as eighteen subharmonics visible for most orientations. For all orientations of H in the basal plane, the mass of branch a remains constant at $0.53m_0$. This branch can be followed to the [0001] axis where m^*/m_0 takes on its maximum value of 1.25. This mass variation plus the variations of signal intensity with RF polarization serve to identify branch a as arising from orbits on the extremal cross section of the lens-shaped electron surface in the third band (ζ of Fig. 2).

Branch k, observed only with E_{RF} parallel to H, persists with a negligible mass variation ($m^*/m_0 = 1.36$) out to approximately 30° from the [0001]. It shows the inverted line shape and "mass doubling" which Grimes et al.[2] have associated with limiting point resonances from spherical regions of the Fermi surface, and almost certainly is due to resonances from the limiting point electrons on the top and bottom of the lens.

The fact that (1) the lens mass is constant in the basal plane, (2) the lens mass varies smoothly in the (11$\bar{2}$0) and (10$\bar{1}$0) plane traversals, and (3) limiting point resonances are observed out to 30° from [0001], leads us to conclude that the cadmium lens has a smooth spherical cap and a nearly free electron-like shape. Grassie's DHVA data, which exhibit a smooth variation of cross-sectional area with crystallographic angle for this section of the Fermi surface, support this conclusion.

We associate branch c with orbit δ or Fig. 2, a hole orbit on the second band "monster." Its mass reaches a minimum of $0.71m_0$ 40° from [10$\bar{1}$0] and is rising rapidly when the branch appears to terminate 19.0° from [10$\bar{1}$0]. This behavior is consistent with the plane of the extremal m^* for this orientation of H becoming tangent to the Fermi surface where two arms of the monster join in a "saddle"

---

* Work supported by the Air Force Office of Scientific Research.

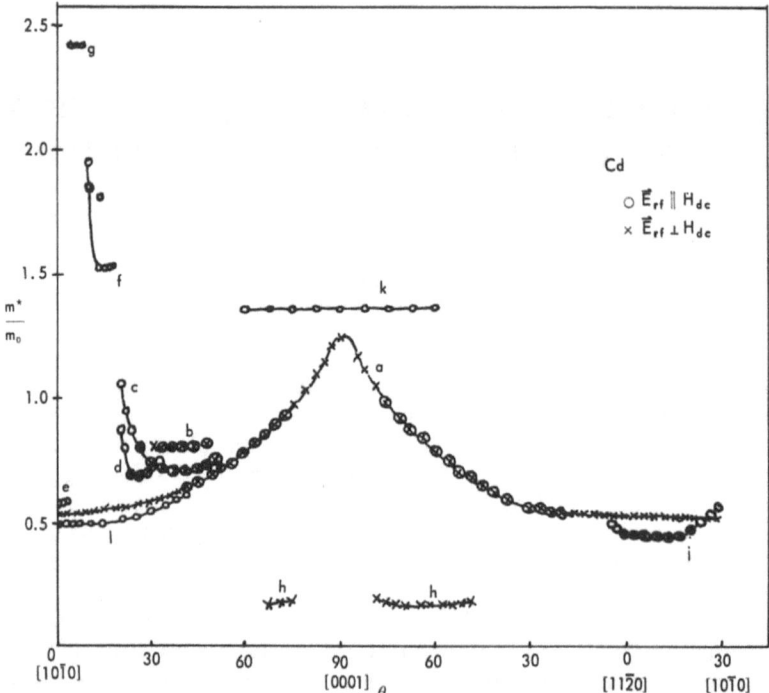

Fig. 1. Cyclotron mass ratio as a function of orientation of H with respect to the principal crystallographic axes. The upper limit to the uncertainty in m^*/m_0 for an individual mass point is $\pm 5\%$.

near the symmetry point H in the zone. For orbits in the vicinity of such a saddle point, m^* approaches infinity.

The mass and angular position of h identifies this branch with the monster arm orbit γ. Branch h is observed with E_{RF} perpendicular to H, is centered approximately 28° from [0001] in a (10$\bar{1}$0) plane, and has a minimum mass value $m^* = 0.16 m_0$. These observations are in excellent agreement with the DHVA results of Grassie[3] and Joseph et al.[4]

In the basal plane traversal with a minimum $m^* = 0.46 m_0$, j is observed and can again be identified with an orbit on the monster. This series probably arises from the "2-arm" orbit denoted by σ in Fig. 2. The branch can only be followed 5° from [11$\bar{2}$0] in the (10$\bar{1}$0) plane and the signal is extremely broad and of very low intensity in this plane. Note that it is observed with both RF polarizations in the basal plane specimen and only with E_{RF} parallel to H in the (10$\bar{1}$0) plane. This is a reasonable result since the carriers associated with this orbit enter into the skin depth with a velocity nearly parallel to H in the (10$\bar{1}$0) plane specimen, whereas in the basal plane specimen they enter the skin depth region with appreciable velocity components both parallel and perpendicular to H. Branch e very likely also arises from the orbit σ.

Branch f, which has only been observed with E_{RF} parallel to H, begins approximately 17° from [10$\bar{1}$0]. Its mass remains essentially constant at $1.53 m_0$ for 7°, then rises rapidly when the branch appears to terminate at 7.5° from [10$\bar{1}$0]. We associate branch f with an extended monster orbit passing through three zones. There is a saddle at the point where branch c terminates. Another such saddle

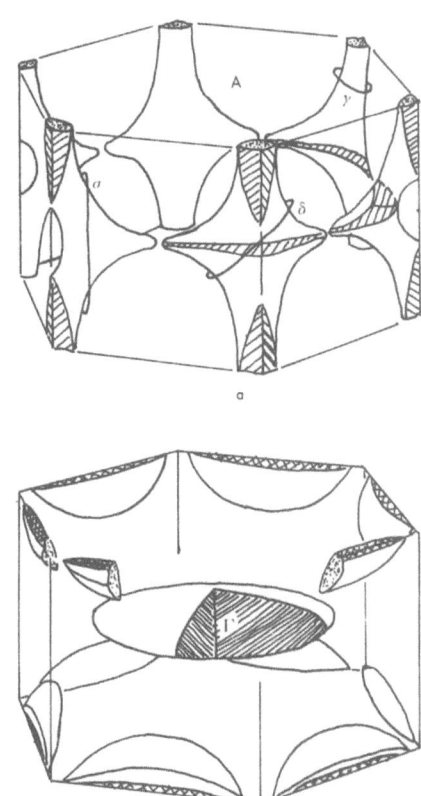

Fig. 2. The Fermi surface for cadmium. (a) First and second bands. (b) Third and fourth bands.

exists at a distance "up" along [0001] equal to a reciprocal lattice translation vector. The orbit attributed to branch f is "pinned" between these two saddle points. Branch g is of the same type as branch f, an extended monster orbit. We attribute g to carriers that pass through five zones. These carriers are also pinned between two saddle points, the saddle pinning f at 7.5° and the next saddle point "up" along [0001]. Branch g is centered approximately 6° from [10$\bar{1}$0] in the (11$\bar{2}$0) plane and has a minimum $m^* = 2.4m_0$. The fact that f and g are observed only with E_{RF} parallel to H supports the present interpretation since the carriers that we associate with these branches enter into the skin depth with a velocity component nearly parallel to H. The present interpretation of these branches is also in agreement with existing evidence that cadmium is "open" in the [0001] direction.

We have thus far identified five orbits that involve carriers (holes) on various parts of the monster and orbits that originate on the lens. No masses have been observed that can be associated with the butterfly–cigar complex ε in Fig. 2, nor have we been able to identify branches b or d of Fig. 1. Branch l, which appears in the (11$\bar{2}$0) plane, has also not been identified. The derivative line shape of the resonance series due to this orbit is neither that of the Azbel–Kaner maxima nor minima of m^* with respect to p_H. The fundamental resonance is always the largest amplitude signal and the series is only observed with E_{RF} parallel to H. It is worth pointing out that we have observed behavior of this type (a branch of the type cadmium-l) in a (11$\bar{2}$0) plane of zinc as well. No such series was observed in the (10$\bar{1}$0) plane of either cadmium, zinc, or magnesium.

References
1. J. H. Koch, R. A. Stradling, and A. F. Kip, *Phys. Rev.* **133**, A240, 1964.
2. C. C. Grimes, A. F. Kip, F. Spong, and R. A. Stradling, *Phys. Rev. Letters* **11**, 455, 1963.
3. A. D. C. Grassie, *Phil. Mag.* **9**, 847, 1964.
4. A. S. Joseph, W. L. Gordon, J. R. Reitz, and T. G. Eck, *Phys. Rev. Letters* **7**, 334, 1961.

EXTREMAL DIMENSIONS OF CYCLOTRON ORBITS IN TUNGSTEN

W. M. Walsh, Jr., C. C. Grimes, G. Adams, and L. W. Rupp, Jr.

Bell Telephone Laboratories
Murray Hill, New Jersey

Two size-effect techniques for determining extremal linear dimensions of cyclotron orbits in metals have recently been demonstrated. Khaikin[1,2] has observed the cutting off of Azbel–Kaner cyclotron resonance series in thin tin samples due to collision of the carriers with the sample surfaces when the size of the resonant cyclotron orbit exceeded the sample thickness at low magnetic field values. Gantmakher[2-6] has detected well-defined anomalies in the radio-frequency surface reactance of similar samples under the same conditions, i.e., when the orbits just "fit" in the sample. The latter type of experiment should find general application since the requirement $\omega_c \tau \gtrsim 1$ is less stringent than that for well-defined, high-order cyclotron resonance subharmonics, i.e., $\omega_c \tau \gtrsim 10$, $\omega \tau \gtrsim 100$.

We have observed the nonresonant size-effect phenomena in tungsten using a somewhat different technique:* A radio-frequency oscillator operating at a fixed frequency in the range 2 to 100 Mcps drives a transmitter coil placed near one face of the thin, planar sample. A receiver coil on the opposite side of the sample picks up a transmitted signal when RF currents flow on that face due to cyclotron orbits exactly spanning the sample thickness. Maxima in the transmission occur for values of the magnetic field H lying in the plane of the sample which cause extremal dimension orbits to span the sample. By extremal dimension is meant those orbits on the Fermi surface whose real-space linear dimension in the direction of the sample normal is maximum or minimum. Anisotropy may be studied by changing the orientation of the field in the plane of the sample. In practice, the vector sum of the transmittal signal and a field-independent leakage signal is amplified and detected in a radio receiver. By modulating H at an audio frequency, the field-dependent part of the receiver output may be synchronously detected to yield the derivative with respect to field of the received envelope. Finally, the derivative is displayed vs. H on an X–Y recorder. A typical run at liquid-helium temperature is illustrated in Fig. 1. The magnetic field was applied parallel to a $\langle 211 \rangle$ axis in a (110) plane sample of thickness $t = 0.235$ mm cut from a tungsten single crystal. The crystal of resistance ratio $\rho_{295°K}/\rho_{4.2°K} \simeq 37,000$ was obtained from Dr. H. Sell of the Westinghouse Lamp Research Division. The sharp peaks designated H_e and H_h are believed to be produced by extremal dimension orbits on the principal electron and hole sheets of the Fermi surface for reasons to be given below. The peaks at $H_h + H_e$, $2H_h$, $2H_h + H_e$, etc., arise from creation of current sheets within the bulk of the sample and consequent excitation of other orbits,[4] the sum of the extremal dimensions just spanning the sample thickness. It is curious that little or

---
* This method was developed by C. C. Grimes in order to study helicon propagation in metals and has been used to observe the size-effect phenomena in thin samples of sodium and potassium.

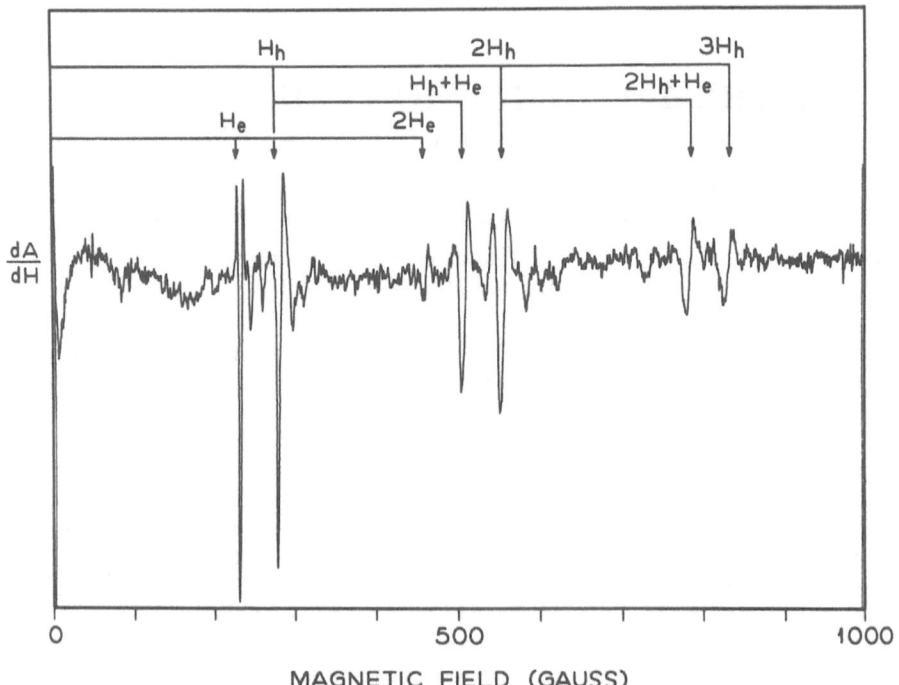

Fig. 1. Recorder plot of the field derivative of the 4 Mcps signal amplitude transmitted through a thin (110) plane tungsten sample vs. the magnetic field. The experiment was performed at 4.2°K with the field parallel to a ⟨211⟩ axis in the plane of the sample. The harmonic pattern is discussed in the text.

no signal is observed at $2H_e$, $2H_e + H_h$, etc. Apparently the holes are capable of re-exciting both holes and electrons, but the electrons do not produce appreciable re-excitation. We are unable to explain this distinction.

The relationship between the measured field positions and the orbit dimensions in crystal momentum space Δk is given by

$$\Delta k = \frac{e}{\hbar c} Ht = 1.519 Ht \text{ Å}^{-1}$$

where H is expressed in gauss and t in millimeters. It is necessary to remember that when H is applied in a given direction it determines k-vectors rotated 90° away from the field in the plane of the sample, i.e., $k_{[111]}$ is measured by directing H along $[2\bar{1}\bar{1}]$ in the $(01\bar{1})$ plane.

The identification of the signals as due to holes and electrons is based on the anisotropy study summarized in Fig. 2 and the present state of knowledge of the Fermi surface of tungsten. The peak designated H_h in Fig. 1 can be observed for all directions of H in the (110) plane and shows the anisotropy expected of a nearly octahedral surface. This is evident from the close fit of the solid curve (a true octahedron) with the H_h data, using the value measured with $H \parallel \langle 211 \rangle$ for normalization. Such a surface is very similar to the principal hole sheet predicted qualitatively by Lomer[7] and recently established more quantitatively by Mattheiss,[8] using the augmented plane wave (APW) method.

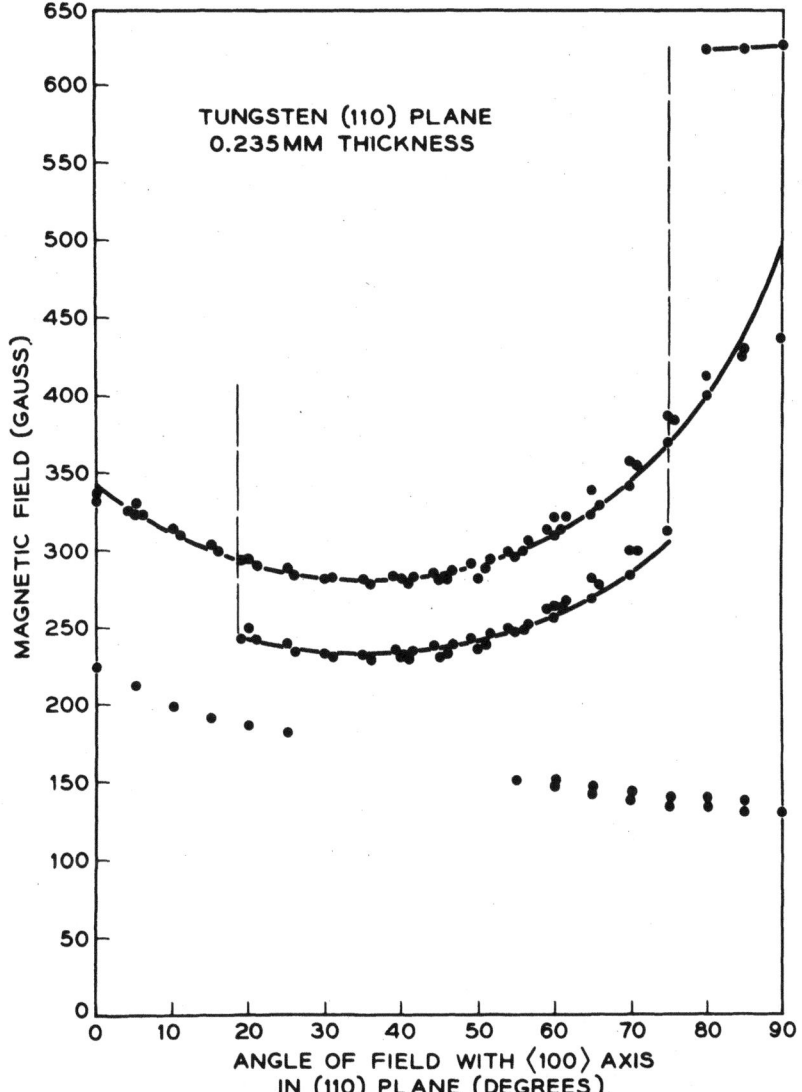

Fig. 2. Anisotropy of the most intense fundamental peak positions in the (110) plane. The results of several distinct runs have been superposed. The peak labeled H_h in Fig. 1 is seen in all orientations and may be well represented by an octahedral surface (solid curve) save near the vertices where some rounding off occurs. The peak H_e of Fig. 1 displays a similar anisotropy over a more restricted range of angles. The discontinuous increase in this orbit's extremal dimension for $H \sim 15°$ from $\langle 110 \rangle$ is characteristic of a jacklike surface as discussed in the text. The unconnected points indicate that several other signals have been observed but not interpreted as yet.

The peak designated H_e in Fig. 1 exhibits an anisotropy very similar to that of H_h, but is observable only over a restricted range of angles. The intensity drops to zero when H comes within 18° of the $\langle 100 \rangle$ axis and within 15° of the $\langle 110 \rangle$ axis. Similar behavior has been observed in cyclotron resonance experiments,[9] where it has been possible to establish the electron character of the orbits. The shape of this electron surface consists of an octahedral "body" (roughly 20% smaller in linear dimensions that the hole octahedron) whose six vertices terminate in ball-like protrusions. The existence of the "balls" was deduced by Sparlin and Marcus[10] from their de Haas–van Alphen measurements and led them to describe the surface as a "jack" in analogy with the child's toy. The re-entrant character of this principal electron surface has also been independently suggested by Lomer[11] in a revision of his qualitative theoretical study and it emerges clearly from Mattheiss' APW calculation.

A transmission peak at 624 G has been observed when H lies within 10° of the $\langle 110 \rangle$ axis (see Fig. 2). It is believed to be produced by large central orbits on the electron jack which pass over two balls on opposite vertices of the body. This peak should therefore determine the maximum dimension of the jack.

Several smaller sheets of Fermi surface as well as extremal but noncentral orbits on the electron jack are expected on the basis of the model and do indeed manifest themselves in a variety of experiments. It is likely that a number of weak transmission signals observed in the present experiment, which are not harmonics or combinations of strong signals, will find interpretation in this company.

A summary of the principal wave vectors deduced from our measurements as well as values calculated by Mattheiss is given in Table I. Comparison has not been made with Rayne's published magnetoacoustic results,[12] as his original interpretation must be modified in view of the re-entrant nature of the electron jack. It is worthwhile to emphasize in this regard that data obtained in the transmission experiment are more easily interpreted than magnetoacoustic oscillations, since individual peaks are observed rather than complex beat patterns.

While the agreement between our data and the APW calculation is fairly satisfactory, there is a discrepancy which may prove significant. It is inherent in the calculation that the electron jack which is situated at the Brillouin zone center Γ should touch the hole octahedron centered at H, the intersection of the $\langle 100 \rangle$ axis with the zone boundary, along the line ΓH. Thus, the sum of the $\langle 100 \rangle$ dimensions of these two sheets should be equal to the $\langle 100 \rangle$ dimension of the Brillouin zone. This sum is 3.78 Å^{-1} with an estimated total error of $\pm 2\%$. The $\langle 100 \rangle$ BZ dimension is 3.974 Å^{-1}. The discrepancy of $\sim 5\%$ implies that an appreciable gap exists between the principal electron and hole sheets. That such a gap should appear due to spin-orbit interaction has been suggested by Herring[13] prior to reduction of our data. This point will be investigated more thoroughly in the near future

Table I. Experimental and Theoretical k-Vectors* of the Principal Electron and Hole Sheets of the Tungsten Fermi Surface

Direction	Electron "jack"		Hole "octahedron"	
	Experimental†	Theoretical‡	Experimental†	Theoretical‡
$\langle 100 \rangle$	1.112	1.16	0.777	0.81
$\langle 111 \rangle$	0.415	0.47	0.501	0.50
$\langle 110 \rangle$		0.51	0.597	0.60

* These are radii expressed in Å^{-1}.
† Results of the experiments reported here should be accurate to within at least $\pm 2\%$.
‡ Results of an APW calculation by Mattheiss.

and may lead to a direct determination of spin-orbit coupling in this transition metal.

Note Added in Proof

Our expectation concerning spin-orbit coupling has been fulfilled.[14-16]

References

1. M. S. Khaikin, *Zh. Eksperim i Teor. Fiz.* **41**, 1773, 1961. (*Soviet Phys. JETP* (*English Transl.*) **14**, 1260, 1962.)
2. M. S. Khaikin, *Zh. Eksperim. i Teor. Fiz.* **43**, 59, 1962. (*Soviet Phys. JETP* (*English Transl.*) **16**, 42, 1963.)
3. V. F. Gantmakher, *Zh. Eksperim. i Teor. Fiz.* **42**, 1416, 1962. (*Soviet Phys. JETP* (*English Transl.*) **15**, 982, 1962.)
4. V. F. Gantmakher, *Zh. Eksperim. i Teor. Fiz.* **43**, 345, 1962. (*Soviet Phys. JETP* (*English Transl.*) **16**, 247, 1962.)
5. V. F. Gantmakher, *Zh. Eksperim. i Teor. Fiz.* **44**, 811, 1963. (*Soviet Phys. JETP* (*English Transl.*) **17**, 549, 1963.)
6. V. F. Gantmakher and E. A. Kaner, *Zh. Eksperim. i. Teor. Fiz.* **45**, 1430, 1963. (*Soviet Phys. JETP* (*English Transl.*) **18**, 988, 1964.)
7. W. M. Lomer, *Proc. Phys. Soc.* (*London*) **80**, 489, 1962.
8. L. F. Mattheiss (unpublished).
9. W. M. Walsh, Jr., *Phys. Rev. Letters* **12**, 161, 1964.
10. D. M. Sparlin and J. A. Marcus, *Bull. Am. Phys. Soc.* **9**, 250, 1964.
11. W. M. Lomer, *Proc. Phys. Soc.* (*London*) **84**, 327, 1964.
12. J. A. Rayne, *Phys. Rev.* **133**, A1104, 1964.
13. C. Herring, private communication.
14. W. M. Walsh, Jr., and C. C. Grimes, *Phys. Rev. Letters* **13**, 523, 1964.
15. L. F. Matheiss and R. E. Watson, *Phys. Rev. Letters* **13**, 526, 1964.
16. T. L. Loucks, *Phys. Rev. Letters* **14**, 693, 1965.

DOPPLER-SHIFTED CYCLOTRON RESONANCE WITH HELICON WAVES*

M. T. Taylor

Cornell University
Ithaca, New York

Cyclotron resonance, with the static magnetic field normal to the surface of the metal, has been performed at frequencies several thousand times smaller than the electron cyclotron frequency, $\omega_c = eB/m$. The low frequency at which resonance occurs (9 Mcps with a field of 23 kG for sodium) arises from the Doppler effect and the large difference between the Fermi velocity ($\sim 10^8$ cm/sec) and the helicon wave velocity ($\sim 10^3$ cm/sec). The condition for resonance is related to the Gaussian curvature at the point on the Fermi surface with the maximum component of velocity along the static magnetic field. The experiment measures the surface impedance of the metal as a function of magnetic field. The sample in the form of a plate (10 × 10 × 1 mm) is placed in a coil at liquid-helium temperatures; the inductance and the Q of the coil are measured using a Twin-T radio-frequency bridge. Measurements were made on polycrystalline sodium, potassium, and indium over the frequency range 1–50 Mcps. It is found that the magnetic field at which resonance occurs is proportional to the cube root of the frequency. The values obtained for the radius of the Fermi surface for sodium and potassium are in very good agreement with the theoretical values computed, assuming a spherical surface. The fractional width of the resonance was found to be inversely proportional to the electron mean free path.

This work has been extended to single crystal aluminum and indium by J. R. Merrill of this laboratory. Preliminary data seem to agree with the known Fermi surfaces of these metals. Merrill also reports that the method can be used to study de Haas–van Alphen oscillations sensitively.

---

* Work supported by the U.S. Atomic Energy Commission and the Advanced Research Projects Agency.

DE HAAS–VAN ALPHEN EFFECT IN PYROLYTIC AND SINGLE-CRYSTAL GRAPHITE*

S. J. Williamson† and S. Foner

National Magnet Laboratory ‡
Massachusetts Institute of Technology
Cambridge, Massachusetts

and

M. S. Dresselhaus §

Lincoln Laboratory ¶
Massachusetts Institute of Technology
Lexington, Massachusetts

Introduction

The de Haas–van Alphen oscillations in the magnetic susceptibility of pyrolytic graphite (PG) have been observed for the first time and a comparison made with those of single-crystal graphite (SCG). These preliminary measurements were performed at temperatures between 1.2° and 4.2°K with fields, H_0, up to 55 kG applied along the c-axis. The usual beat pattern[1] attributed to majority electron and hole carriers was found to be identical for PG and SCG between 1.5 and 20 kG. This equality between the PG and SCG frequencies, f_M, and effective masses, m^*, is significant in establishing the graphite band parameters, because the most complete band parameter determination[2] has been made by combining magneto-reflection data in PG with Shubnikov–de Haas data on SCG.[3] Low-temperature de Haas–van Alphen measurements also yield a comparison of the local order in PG with that in SCG. Furthermore, the oscillations of unusually low frequency discovered by Soule[4] were confirmed by observation in both PG and SCG. The existence of such low f_M oscillations is shown to be consistent with existing band structure calculations.

Experimental Results

Disk-shaped samples of PG with 3 mm diameter and 1 mm thickness were cut by a sharp knife from much larger slabs which had been stress-annealed at 3600°C. The audio frequency field modulation technique[5] was used to monitor the derivative of the magnetization dM/dH as H_0 was varied. In this method, a

---

* A portion of this paper is taken from a thesis to be submitted by one of the authors (S.J.W.) as partial fulfillment for the degree of Sc.D. in the Department of Physics at the Massachusetts Institute of Technology.
† National Science Foundation Predoctoral Fellow.
‡ Supported by the U.S. Air Force Office of Scientific Research.
§ Visiting Scientist, National Magnet Laboratory.
¶ Operated with support from the U.S. Air Force.

small AC field of about 10 G at 100 cps was superimposed on H_0, and the resulting time-varying magnetization was monitored by a phase sensitive detector which processes the signal induced in a pickup coil surrounding the sample. An arrangement of two other coils and electronic circuitry was used to buck out undesired background signals.

Typical curves of dM/dH vs. H_0 for different temperatures are shown in Fig. 1. Enhancement of the oscillations by lowering the temperature is evident. The reproducibility of the measurements on different PG samples is illustrated by the fact

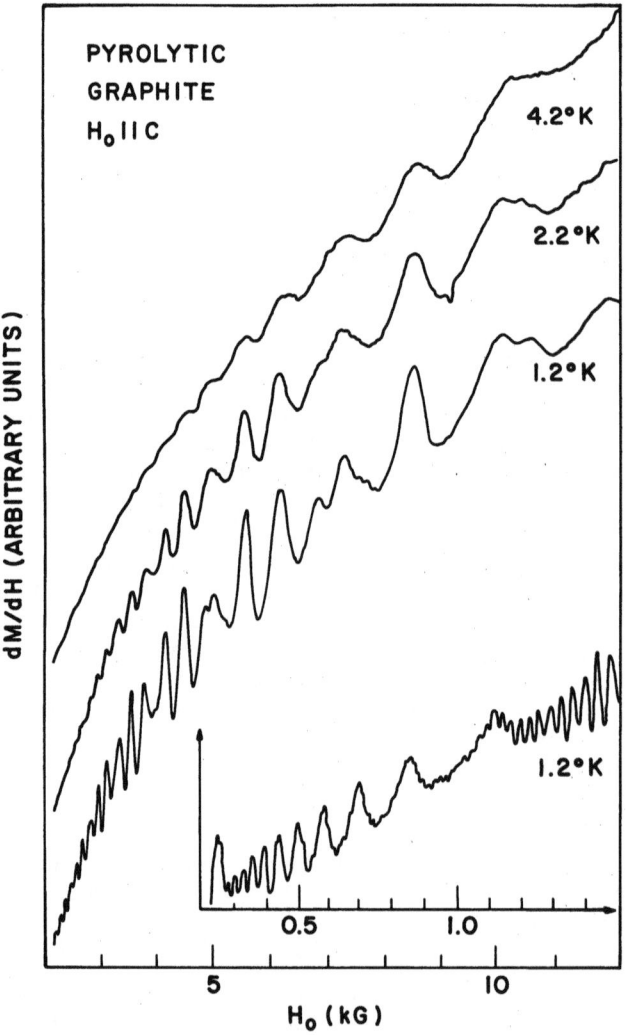

Fig. 1. The de Haas–van Alphen oscillations at various temperatures in pyrolytic graphite. The inset trace, for which the ordinate has been expanded by about a factor of 4, shows the minority carrier oscillations of frequency $f_M = 4.40 \pm 0.05$ kG, which are evident at lower applied fields, and majority carrier oscillations, which appear for $H_0 \gtrsim 1$ kG.

that two samples grown in different molds, but annealed under similar conditions, exhibited identical patterns, the maxima and minima deviating by less than 0.2% for $1.5 < H_0 < 20$ kG. Similar measurements were performed on natural SCG. The majority carrier beat pattern maxima and minima over this same field region at 1.2°K is in agreement with that from PG to within the 2% experimental uncertainty. This larger uncertainty is due to the reduced signal-to-noise ratio resulting from the smaller SCG samples. It is noteworthy that the layer plane Dingle temperature (collision time)[6,7] for PG is nearly that of SCG, since the DHVA amplitude of SCG exceeded that of PG by only $\sim 20\%$.

Despite the remarkable similarity in the field range $1 < H_0 < 20$ kG for which majority carrier frequencies of $f_M = 66.2$ kG for the holes and $f_M = 48.3$ kG for the electrons[3] have been observed, the oscillations in the lower field ranges exhibit differences between PG and SCG. For $0.3 < H_0 < 1$ kG, an oscillation of unusually low frequency $f_M = 7.4 \pm 0.6$ kG has been observed in SCG. This low f_M oscillation has now also been observed on PG; however, it has a lower frequency than SCG. We find that for PG, $f_M = 4.40 \pm 0.05$ kG and $m^* = (4.0 \pm 0.4) \times 10^{-3} \, m_0$ (see inset of Fig. 1).

Our SCG f_M agrees with the low f_M oscillation discovered by Soule,[4] who reports $f_M = 7.4 \pm 1.7$ kG for which $m^* \simeq 2.3 \times 10^{-3} \, m_0$. Soule has observed that the SCG low-frequency oscillation f_M increases by a factor of about 8 as H_0 is tipped into the layer plane and that f_M, for $H_0 \parallel c$, is sensitive to the impurity contents of the sample, increasing by a factor of 2 with the addition of 13 ppm of the acceptor boron.

Interpretation

In the past few years, measurements of the transport properties of PG have revealed that when the sample is heat treated at sufficiently high temperatures, its characteristics approach that of SCG.[8,9] Recently, quantum effects in PG have been observed for the first time in the magnetoreflection by Dresselhaus and Mavroides[2] (hereafter referred to as D.M.) These data were supplemented by Shubnikov–de Haas results on SCG obtained by Soule, McClure and Smith[3] to yield the most complete band parameter determination to date for PG. The present experiment was undertaken to ascertain to what extent the band parameters found for the best available PG pertain to SCG.

The observation of a bulk quantum effect in PG—DHVA oscillations—shows the essentially single-crystal nature of this material. A lower bound to the layer plane local order can be estimated from the lowest field at which oscillations were observed. If $v \sim 0.5 \times 10^8$ cm/sec and $m^* = 0.06 \, m_0$ for the majority hole,[9,10] then for $H_0 = 1$ kG the classical cyclotron orbit diameter is 4×10^4 Å. Furthermore, since the amplitude of DHVA oscillations for PG and SCG are essentially the same, we have additional evidence that the local order is comparable to that in SCG. Orientation studies now in progress are expected to give estimates for the order between layer planes.

The identical nature of majority carrier DHVA oscillations in PG and SCG for $\mathbf{H}_0 \parallel \mathbf{c}$ implies the equality of specific band parameters for these samples. In the Slonczewski–Weiss model,[11] the energy band parameters, γ_0^2/γ_1 and γ_2 (the band overlap parameter), are sensitive to m^* and f_m, respectively, for the electron and hole,[12] and the ratio of electron to hole frequencies is sensitive to the magnitude and sign of γ_4.[2] We therefore conclude that these quantities are the same for PG and SCG. In particular, this justifies the use of Shubnikov–de Haas information on SCG to determine γ_2 in the band calculations for PG.[2]

Furthermore, the existence of a low-frequency minority carrier oscillation for all orientations of $\mathbf{H}_0$ is consistent with the Slonczewski–Weiss model. The magnetic energy levels based on this model[13] show a complicated dependence on the dimensionless wave vector ξ, directed along $\mathbf{c}$.* The features of these energy levels which are pertinent to the de Haas–van Alphen experiment are the occurrence of broad stationary values near point K in the Brillouin zone, for which $\xi = 0$ (corresponding to the majority hole oscillations), and near $\xi = 0.35$ (corresponding to the majority electron oscillations). A very narrow region of stationary values of the magnetic energy levels occurs near point H or $\xi = 0.5$, with which we associate the present low-frequency oscillations.

* See M.S. Dresselhaus and J. G. Mavroides,[2] Fig. 4.

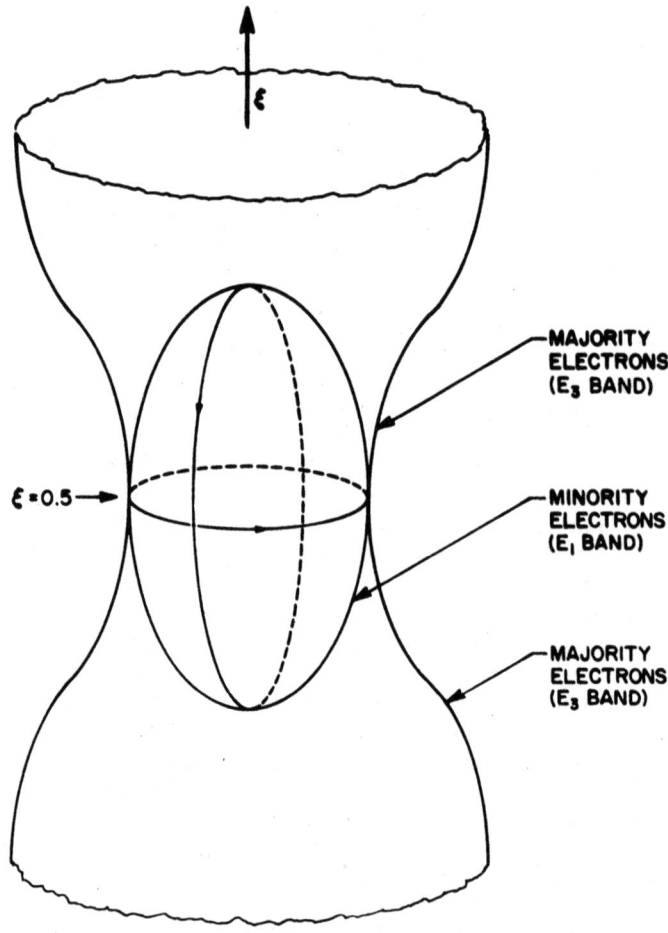

Fig. 2. Sketch of the reassembled graphite Fermi surface near $\xi = 0.5$, the Brillouin zone boundary. Orbits of electrons producing de Haas–van Alphen oscillations are shown for $\mathbf{H}_0$ parallel and perpendicular to the layer plane. This figure is intended to show only the qualitative aspects of the surfaces.

The Fermi surface constructed from the band model[2,12] has a circular cross section at $\xi = 0.5$, where the Fermi surface touches the Brillouin zone boundary. It is possible to reassemble the pieces of Fermi surface on either side of the zone boundary, as shown in Fig. 2, to obtain pieces of Fermi surface with a degenerate extremal cross-sectional area. The low-frequency oscillations are attributed to the small piece of closed Fermi surface. This construction also predicts a second low-frequency oscillation associated with the major electron surface as $\mathbf{H}_0$ is tilted away from $\mathbf{c}$. Orientation experiments are now in progress to observe this additional oscillation, which is expected to exist over only a limited angular range. For the band parameters given by D.M., it is found that for $\mathbf{H}_0 \parallel \mathbf{c}$, $f_m = 6.5\,\text{kG}$, and $m^* = 5.9 \times 10^{-3}\,m_0$, which is in qualitative agreement with experiment.

If the above interpretation is correct, the DHVA effect for $\mathbf{H}_0 \parallel \mathbf{c}$ affords a sensitive and direct determination of Δ, the difference in potential energy between A and B lattice sites. In particular, the degenerate extremal area is $4\pi E_F(E_F - \Delta)/3a_0^2\gamma_0^2$, where E_F and γ_0 are parameters fairly well known from other experiments. Using $f_m = 7.15\,\text{kG}$ for SCG, we calculate that $\Delta = -0.024\,\text{eV}$ and $m^* = 6.3 \times 10^{-3}\,m_0$. For PG, with $f_m = 4.40\,\text{kG}$, we find that $\Delta = -0.0072\,\text{eV}$ and $m^* = 4.6 \times 10^{-3}\,m_0$.

Although these preliminary calculations show agreement with the experimental results for the frequencies, there remain unsolved problems associated with m^* of SCG, and the boron doping experiments of Soule have not been explained. These experiments suggest that the minority carriers are holes, and not electrons as the negative value of Δ would seem to suggest.

Acknowledgments

We are grateful to Dr. R. G. Diefendorf of the General Electric Research Laboratory, who provided the pyrolytic graphite, and Dr. D. E. Soule of the Union Carbide Corp. Research Laboratory, who provided the natural single crystals. Without these generous contributions, these experiments would not have been possible.

References

1. D. Shoenberg, *Phil. Trans. Roy. Soc. (London), Ser. A* **245**, 1, 1952.
2. M. S. Dresselhaus and J. G. Mavroides, Proc. APS Top. Conf. on Semimetals, *IBM J. Res. Develop.* **8**, 262, 1964; and article to be published.
3. D. E. Soule, J. W. McClure, and L. B. Smith, *Phys. Rev.* **134**, A453, 1964.
4. D. E. Soule, Proc. APS Top. Conf. on Semimetals, *IBM J. Res. Develop.* **8**, 268, 1964.
5. A. Goldstein and S. Foner, *Bull. Am. Phys. Soc.* **9**, 239, 1964.
6. R. B. Dingle, *Proc. Roy. Soc. (London) Ser. A* **211**, 500, 517, 1952.
7. S. J. Williamson, S. Foner, and R. A. Smith, *Phys. Rev.* **136**, A1065, 1964.
8. C. A. Klein, W. D. Straub, and R. J. Diefendorf, *Phys. Rev.* **125**, 468, 1962.
9. C. A. Klein, *Rev. Mod. Phys.* **34**, 56, 1962.
10. J. W. McClure, *Phys. Rev.* **112**, 715, 1958.
11. J. C. Slonczewski and P. R. Weiss, *Phys. Rev.* **109**, 272, 1958.
12. J. W. McClure, *Phys. Rev.* **108**, 612, 1957.
13. J. W. McClure, *Phys. Rev.* **119**, 606, 1960; M. Inoue, *J. Phys. Soc. Japan* **17**, 808, 1962.

DE HAAS–VAN ALPHEN EFFECT IN β'–CuZn

J. P. Jan, W. B. Pearson, and M. Springford

National Research Council, Division of Pure Physics
Ottawa, Canada

De Haas–van Alphen oscillations of the low-temperature magnetic susceptibility have been observed on several samples of β'–CuZn at the equiatomic composition.

Samples were made by melting together high-purity copper and zinc in a helium atmosphere and quenching to room temperature; a long annealing just below the solidus then produced grains 1 cm or more in length, from which oriented single-crystal samples were spark-cut and etched to size. These samples exhibit a residual resistance ratio of about 100. They are free from any martensitic transformation. X-ray rotating crystal pictures taken in liquid helium confirm that the ordered CsCl structure is retained at that temperature.

Measurements were made by the pulsed field technique, with a peak of about 150 kG. The signal-to-noise ratio for the higher frequencies was low and the best signals were obtained with samples exhibiting the least mosaic structure, as shown by an X-ray study of crystal perfection.

The de Haas–van Alphen effect was recorded as a function of crystallographic orientation by rotation of [100] crystals in (100) and (110) planes and of [110] crystals in (110) and (111) planes. Some results are given in Figs. 1 and 2, where the frequencies of the de Haas–van Alphen oscillations have been converted into extremal cross sections of the Fermi surface; the unit chosen in wave-vector space is $2\pi/a$, where a is the lattice parameter; the value $2.94_4 \pm 0.01$ Å was measured at 4.2°K. Three distinct sets of frequencies are found, the F_1 oscillations around 3.3×10^7 G, F_2 around 8×10^7 G, and F_3 around 10^8 G. The amplitude of the F_1 oscillations becomes small when the magnetic field is approaching either [110] or [100], and is maximum along [111]. No beats were observed. F_2 oscillations have only been recorded in and close to the [110] direction, and F_3 in and close to the [100] direction, both with a very low signal-to-noise ratio.

In an attempt to explain these results, the empty lattice model was worked out for three electrons per primitive cell in the following two cases: (a) neglecting ordering of the lattice; the first Brillouin zone is then the dodecahedron of the body-centered cubic lattice; (b) taking ordering into account; the first Brillouin zone is a cube, and the dodecahedron bounds the second zone. It was immediately apparent that case (a) is unable to explain our results. In case (b), the model predicts orbits in four bands, but the overlap into the third and fourth bands is so small that it is undoubtedly removed by the lattice potential.

We then propose the following model. The first band is almost full, but for holes in the corners of the cubic Brillouin zone. The second band is slightly more than half-filled by all the remaining electrons, and the Fermi surface contacts all 12 {110} faces of the second Brillouin zone. We attribute the F_1 oscillations to the

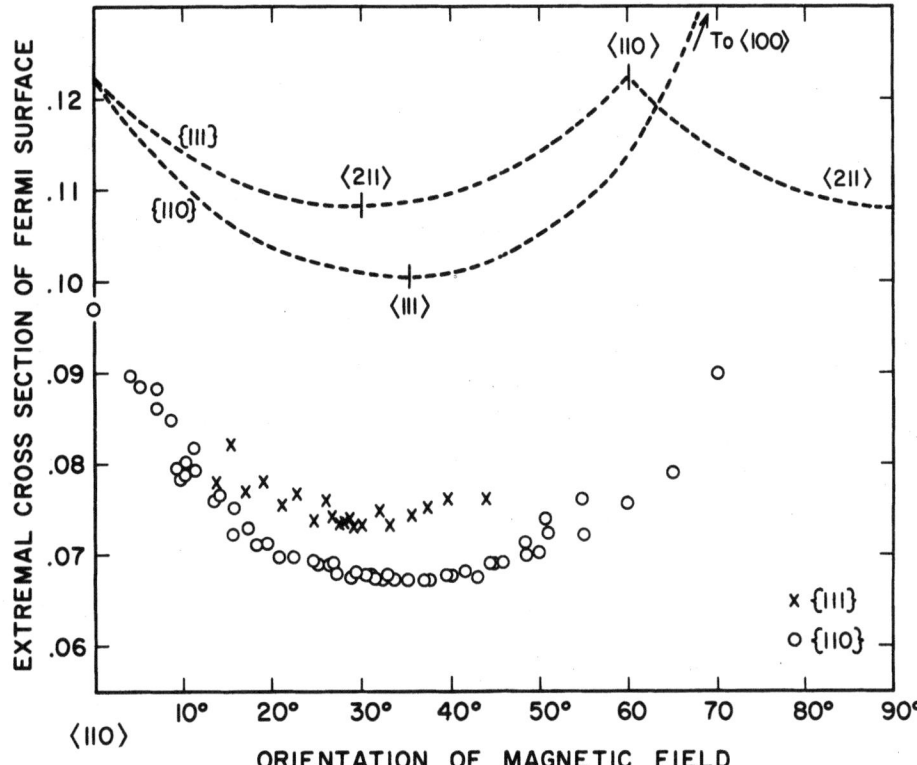

Fig. 1. Extremal cross-sectional areas of Fermi surface *vs.* magnetic field orientation, for scanning in {111} and {110} planes. The dotted lines correspond to the empty lattice model in the first band, and the measured points, to frequency F_1. The face of the first cubic Brillouin zone is taken as unity, and corresponds to a frequency of 4.74×10^8 G.

holes in the first band. Figure 1 shows the predictions of the empty lattice model; the observed values are smaller as expected, and exhibit the correct angular variation. Calculations using two OPW's show that the departure from the free-electron model can be explained by assuming reasonable values of the effective potential, but a more complete OPW calculation is needed in the vicinity of the cube corner to obtain quantitative results.

The second band, in a repeated zone scheme, is made of a multiply-connected set of lenses with which we associate the F_2 and F_3 oscillations. The F_2 oscillations are ascribed to the [110] "necks," and the F_3 oscillations to a hole orbit in agreement with the empty lattice prediction. There is also a large orbit predicted in this second band which has not been observed so far, and work is proceeding to obtain better crystals.

Veal and Rayne[1] have explained their results on the electronic specific heat of β-brass alloys using a copperlike Fermi surface contacting the {110} faces of the dodecahedral zone. They ignored the cubic subzone due to ordering. Viewed in an extended zone scheme, their Fermi surface looks very much like the one we are proposing except for our discontinuities along the {100} planes, and the predicted density of states would be quite similar on both models. The two models, however, differ fundamentally in a repeated zone scheme; topology and connectivity are

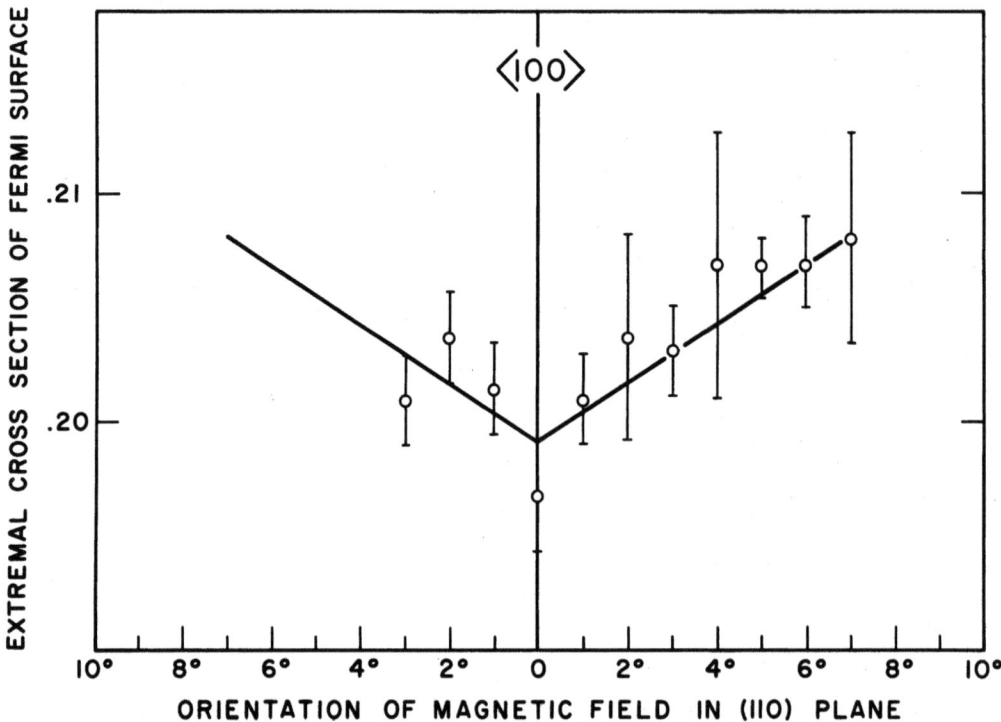

Fig. 2. Extremal cross-sectional areas of Fermi surface in the vicinity of [100], corresponding to frequency F_3.

totally modified by energy gaps along the {100} planes, and our model results in observed de Haas–van Alphen orbits not predicted by Veal and Rayne's model.

A neck orbit in the [110] direction is predicted by Veal and Rayne, with values of 0.14 or 0.19 in our units, corresponding to (110) energy gaps of 2.25 or 4.7 eV, respectively. Our F_2 oscillation has the value 0.17 ± 0.01 in the [110] direction. It is, however, interesting to note that our model also predicts this same (110) orbit in our second band.

Additional evidence for the existence of an important first cubic zone will undoubtedly be found in band structure calculations,[2,3] and optical studies are also of importance.[4,5] A copperlike Fermi surface would lead to a Hall coefficient close to the free-electron value, whereas the room-temperature value is an order of magnitude lower,[6] pointing to a significant hole contribution. We have also computed Fourier components of the effective potential, applying to ordered β-brass the method used by Harrison for elements,[7] and found $V_{100} = 0.42$ eV and $V_{110} = 0.14$ eV.

References
1. B. W. Veal and J. A. Rayne, *Phys. Rev.* **128**, 551, 1962.
2. H. Amar and K. H. Johnson, *Bull. Am. Phys. Soc.* **9**, 250, 1964.
3. H. Amar, private communication.
4. L. Muldawer, *Phys. Rev.* **127**, 1551, 1962.
5. K. H. Johnson and R. J. Esposito, *J. Opt. Soc. Am.* **54**, 474, 1964.
6. V. Frank, *Kgl. Danske Videnskab. Selskab, Mat.-Fys. Medd.* **30**(4), 1955.
7. W. A. Harrison, *Phys. Rev.* **131**, 2433, 1963.

THE DE HAAS–VAN ALPHEN EFFECT IN CHROMIUM

B. R. Watts

Royal Society Mond Laboratory
Cambridge, England

An investigation of the Fermi surface of magnetic chromium has been started by measuring the de Haas–van Alphen effect from single crystals in pulsed magnetic fields. The specimens were previously treated in a way similar to that described by Montalvo and Marcus[1] by cooling through the Néel temperature, T_N (about 40°C), in a magnetic field of 65 kG parallel to one of the ⟨100⟩ directions. The angular variations of the de Haas–van Alphen frequencies have so far been studied in two different planes. One plane contains the directions z (the ⟨100⟩ cooling field direction) and x (one of the other two ⟨100⟩ directions), and the other contains the directions x and y (the two ⟨100⟩ directions perpendicular to the cooling field direction). The results are shown in Fig. 1, where it can be seen that every de Haas–van Alphen frequency curve has tetragonal symmetry with z as the tetragonal axis; this is the same symmetry as that of the magnetic domains. This symmetry has been demonstrated by neutron diffraction studies[2–5] which have shown that below the spin flip temperature, T_F (about 120°K), the magnetic structure in any domain consists of spins lined up parallel and antiparallel to the relevant ⟨100⟩ direction λ along which the magnetic structure has a long-wavelength modulation (wavelength about 25 unit cells). Such domains clearly have tetragonal symmetry with λ as tetragonal axis. It follows that since the de Haas–van Alphen frequency curves have the same symmetry as the magnetic domains, all the observed parts of the Fermi surface are from sheets which are due to energy gaps arising from the magnetic structure. Furthermore, it is possible to distinguish between the two alternative ways in which the arrangement of the magnetic domains could produce overall tetragonal symmetry; these alternatives are (1) a single domain with λ parallel to z and (2) an equal mixture of two kinds of domains with λ parallel to x and y, respectively. The complete tetragonal symmetry of the de Haas–van Alphen oscillations rules out the second alternative, which would require all the frequencies observed at z to be observed at x and y with their amplitudes reduced by a factor of two.

The next stage in deducing the shape of the Fermi surface is to decide how the experimental points should be joined up. The symmetry of the magnetic energy gaps (which are believed to be the important ones) allows only certain kinds of connectivities for the frequency curves, depending on the position of the relevant piece of Fermi surface with respect to the tetragonal axes in **k** space. Figure 2 shows the different possible ways that frequency curves must join up for any tetragonal zone scheme. In the actual zone scheme for chromium it is possible that the position of a particular part of the Fermi surface with respect to the energy gaps could give it extra symmetry beyond that required by the tetragonal symmetry alone. For

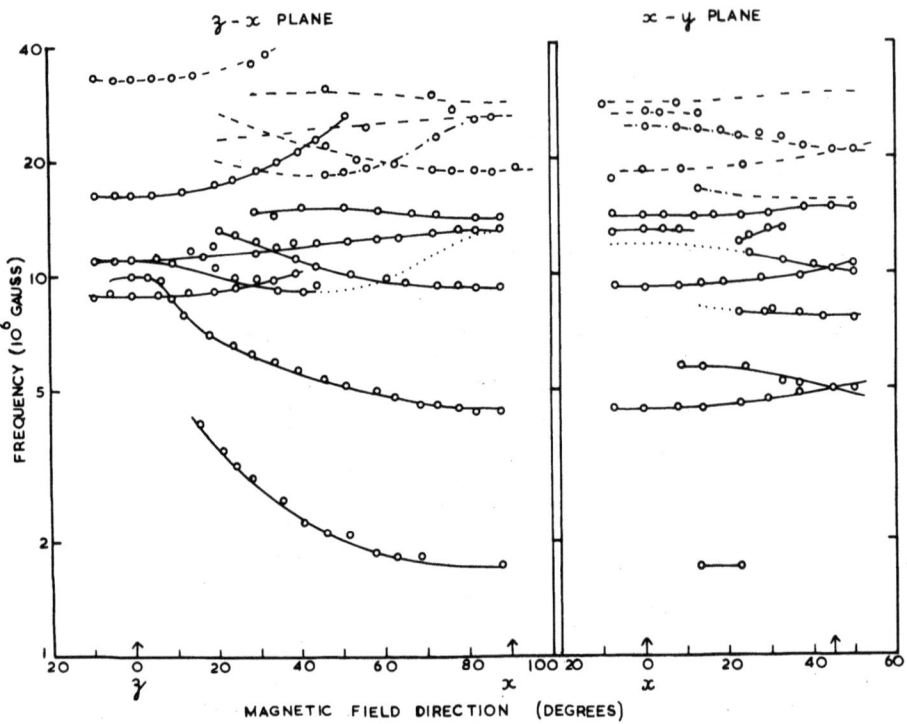

Fig. 1. Angular variation of frequency in the two planes containing z and x, and x and y, respectively. The points, of which only a representative selection are included for the sake of clarity, come from measurements on five differently oriented specimens. The different lines drawn through the points are explained in the text.

example, a piece of Fermi surface on the tetragonal axis need only have fourfold symmetry about that axis unless, for example, it is on a zone face giving it an extra mirror plane normal, say, to the tetragonal axis. In this case, as is usual, this extra symmetry has no effect on the type of connectivity to be expected. In the cases where the extra symmetry could have effect, it could only be to reduce a more complicated class of connectivity to a simpler one. Therefore, it follows that in deciding the curves to draw through the points, we must choose curves which belong to one of the classes shown in Fig. 2.

The solid lines shown in Fig. 1 have been drawn through the points to illustrate what are believed to be the correct ways of joining up the points. The dashed lines are drawn at exactly twice the frequencies of the solid lines, i.e., where the harmonic frequencies should be. It can be seen that all the higher frequencies lie nearly enough on the dashed lines and are therefore assumed to be harmonics. This is an important conclusion because they must now be disregarded so far as giving extremal cross-sectional areas of the Fermi surface. In one or two cases, however, more points exist on these harmonic curves than exist on the fundamental curves. These extra harmonic points are joined by dotted and dashed lines, and the half values (i.e., the deduced fundamental values) are drawn in as dotted curves. If this is the correct

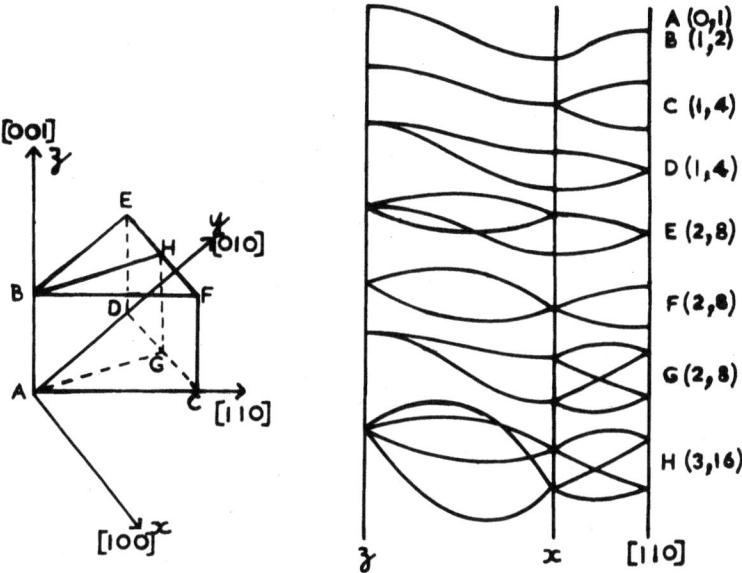

Fig. 2. The classes of connectivity of the frequency curves which would result from pieces of Fermi surface situated at the 8 distinct sites (A, B, C, etc.) in a tetragonal environment. The numbers in the brackets after each site letter give the number of degrees of freedom the site has and the number of such sites in **k** space, e.g., F (2, 8) means that F can lie anywhere in a plane (actually the {110} plane), except for the special high symmetry points of that plane (A, B, and C), and that 8 equivalent points exist owing to the tetragonal symmetry.

interpretation of these curves, it can be seen that the harmonics give some rather useful information about the fundamental curves. The absence of these particular points on the fundamental curves can be explained if the cyclotron mass m round the orbit is about 0.5 times the free electron mass m_0, since there is a term cos $(\pi p m/m_0)$ in the theoretical expression for the de Haas–van Alphen amplitude (p is the number of the harmonic). Clearly a measurement of the cyclotron mass should help to clear up this point. The number of distinct solid or dotted lines in Fig. 1 which either cut the z direction or can be extrapolated to cut it at distinct values gives the number of distinct kinds of orbit round the Fermi surface which are observed. This number appears to be at least 6. Since the frequency curves have not been measured in a sufficient number of planes, no attempt has been made to analyze the data yet in terms of a detailed zone structure, but since it has been possible to connect the data so far in a reasonable way, it seems that in due course it will be possible to infer many of the details of the Fermi surface in antiferromagnetic chromium.

References

1. R. A. Montalvo and J. A. Marcus, *Phys. Letters* **8**, 151, 1964.
2. G. E. Bacon, *Acta Cryst.* **14**, 823, 1961.
3. V. N. Bykov, V. S. Golovkin, N. V. Ageef, V. A. Levdik, and S. I. Vinogradov, *Dokl. Akad. Nauk SSSR* **128**, 1153, 1959.
4. L. M. Corliss, J. M. Hastings and R. J. Weiss, *Phys. Rev. Letters* **3**, 211, 1959.
5. G. Shirane and W. J. Takei, *J. Phys. Soc. Japan* **17** (suppl. B III), 35, 1962.

HIGH-FIELD GALVANOMAGNETIC PROPERTIES OF RHENIUM

E. Fawcett and W. A. Reed

Bell Telephone Laboratories
Murray Hill, New Jersey

The high-field galvanomagnetic properties of a metal can reveal the existence of open cyclotron orbits on an open sheet of its Fermi surface and the occurrence of magnetic breakdown between two sheets. In the medium-field region, where $\overline{\omega_c \tau} \sim 1$ (ω_c being the cyclotron frequency in a closed orbit, which is proportional to the magnetic field B, τ the relaxation time, and the average being taken over all orbits on all sheets), the galvanomagnetic properties also exhibit characteristic effects if the "cyclotron mobility," $\mu = \omega_c \tau / B$, is very different for different sheets.

In Fig. 1 we show the field dependence of the magnetoresistance in the medium-field region of a single crystal of rhenium having a residual resistivity ratio, $RRR = \rho_{295°K}/\rho_{4.2°K}$, of 920 for several transverse field directions.[1] We find that the field dependence of the transverse magnetoresistance of rhenium for any nonsymmetry field-direction approaches a quadratic power law at sufficiently high fields. For this sample the power law becomes quadratic only when $B \sim 100$ kG, but for purer samples with $RRR \sim 25,000$, the power law is quadratic within the limits of experimental error when $B \sim 10$ kG. We conclude that rhenium is compensated, having equal numbers of electrons and holes, as one expects for a hcp metal with two atoms per unit cell.[2] The approach to saturation of the experimental curve, $\theta = 90°$, in Fig. 1 indicates that rhenium has open orbits along the hexagonal axis [0001]. Measurements on other samples confirm the existence of these open orbits, since the magnetoresistance saturates when both $\mathbf{B}$ and the current direction $\mathbf{J}$ are in the basal plane, but is quadratic in B when $\mathbf{B}$ is in the basal plane and $\mathbf{J}$ is along the hexagonal axis.

We have constructed a simple model, which reproduces the behavior in the medium-field region, but is not intended to include all topological features of the Fermi surface of rhenium. To conform with the experimental observations, the model has equal numbers of electrons and holes, and one electron sheet is a cylinder with its axis along the hexagonal axis. The other two sheets, one electron and one hole, are assumed to be spherical with equal isotropic cyclotron mobilities μ^c, whereas the open electron sheet has a cyclotron mobility $\mu^o \cos \theta$ which depends upon the angle θ between $\mathbf{B}$ and the hexagonal axis (see Fig. 1) in accordance with its cylindrical shape. The unusual shape of the experimental curves, $\theta = 0°$, $60°$, and $75°$, in Fig. 1 is reproduced in the model by assuming that there is a large mobility ratio between the closed and open sheets, i.e., $\mu^c/\mu^o = R \gg 1$. Thus the spherical sheets begin to enter the high-field region when $\mu^c B \sim 1$, and the magnetoresistance, which is quadratic in B in the low-field region, begins to saturate as for an uncompensated metal in the high-field region. Only when B is raised by a factor $R/\cos \theta$ does the open electron sheet begin to enter the high-field region, and

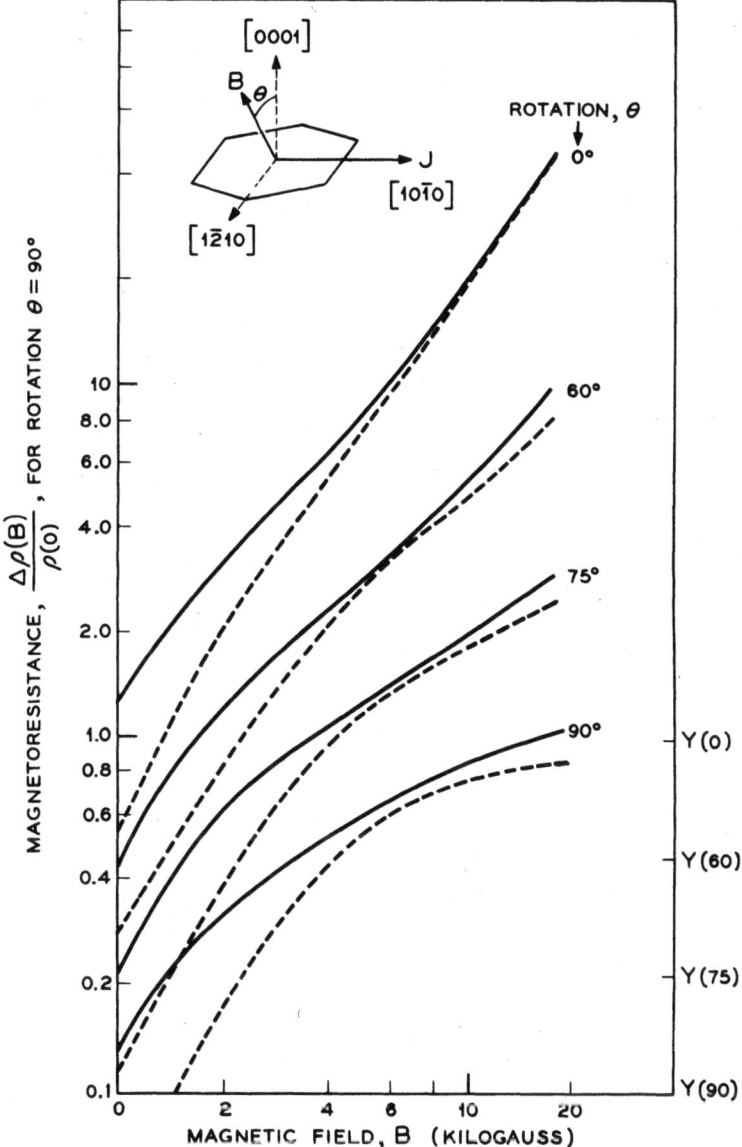

Fig. 1. Magnetoresistance of rhenium. The continuous lines show the field dependence of the transverse magnetoresistance at 4.2°K of a sample of rhenium with **J** along the [10$\bar{1}$0] axis for several values of the rotation angle θ between **B** and the [0001] axis, as shown schematically in the inset. The points $Y(\theta)$ on the right-hand ordinate axis show, for each value of θ, where the magnetoresistance $\Delta\rho(B)/\rho(0)$ has the value 0.1, and the left-hand ordinate scale should be shifted accordingly (except for $\theta = 90°$). The dashed lines show the corresponding field-dependence curves for the model of the Fermi surface described in the text, with the following values of the parameters: $R = 12.8$, $n_e^0 = 0.056$, $K = 0.707$, and $\omega_c\tau = 6.1$ when $B = 18$ kG.

the field-dependence curve, after going through a point of inflection, rises at higher fields toward the quadratic power law characteristic of a compensated metal in the high-field region.

Besides the mobility ratio R, the other disposable parameters in the model are the sign and number n_e^0 per unit cell of carriers in the open sheet, a ratio

$$K = \frac{n_h^c - n_e^c}{n_h^c + n_e^c} = \frac{n_e^0}{n_h^c + n_e^c}$$

where n_h^c and n_e^c are the numbers of carriers of opposite sign in the closed sheets, and the mobility μ^c (or equivalently the value of $\omega_c\tau$ on the closed sheets for a given value of B). The experimentally determined sign of the Hall voltage establishes the electron nature of the open sheet, and the optimum values of the parameters are chosen to give the best fit with the experimental data for samples with **J** along the three symmetry axes. A comparison of single crystals of the same orientation, but with values of RRR ranging from ~ 1000 to $\sim 40,000$ in fields up to 100 kG, shows that the magnetoresistance obeys Kohler's rule, which states that the magnetoresistance is a function of the product $B \times RRR$. Since $\omega_c \sim B$, and $\tau \sim 1/\rho_{4.2°K} \sim RRR$, this product is proportional to the value of $\omega_c\tau = \mu B$ for each sheet, so that the model necessarily gives a magnetoresistance consistent with Kohler's rule.

The measurements in fields up to 100 kG reveal the existence of another set of open orbits,[3] which are not incorporated in the model. In Fig. 2 we show the anisotropy of the transverse magnetoresistance of a sample of the same orientation as the one whose field-dependence curves are illustrated in Fig. 1, but having a

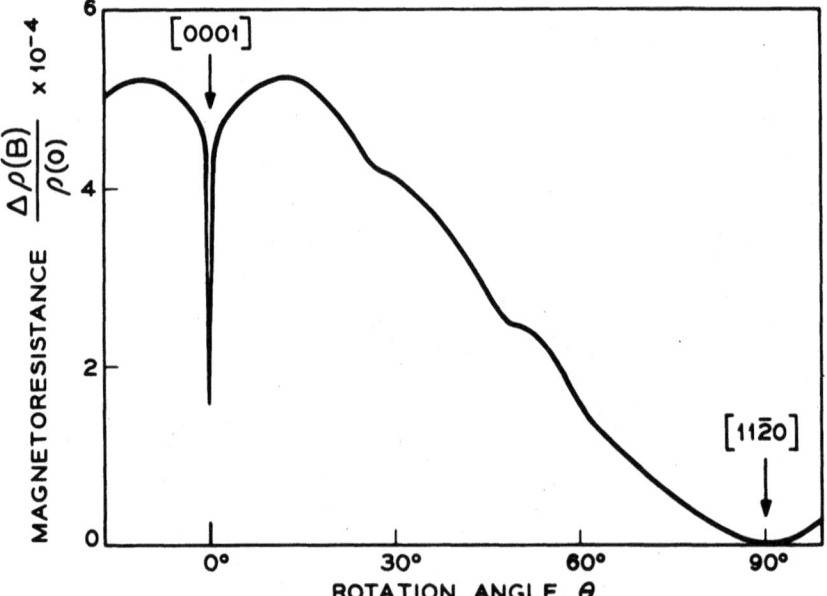

Fig. 2. Magnetoresistance of rhenium. The curve shows the anisotropy of the transverse magnetoresistance at 4.2°K in a field $B = 100$ kG of a sample of rhenium with **J** along the [10$\bar{1}$0] axis and RRR = 24,600. The arrows are labeled to show the direction of **B** at the minima.

value of RRR about 30 times larger. When the sample is rotated about its axis in a transverse field of 100 kG, a narrow minimum, at which the magnetoresistance saturates, appears in the anisotropy curve at $\theta = 0°$, where **B** is along the hexagonal axis. When the sample axis is tilted toward **B** and the sample rotated about its axis, and when samples with **J** along the other symmetry axes are measured in the same way, a similar narrow minimum appears where **B** passes through a $\{10\bar{1}0\}$ plane, so long as the angle α between **J** and the corresponding $\langle 10\bar{1}0 \rangle$ axis is not near zero. The magnetoresistance at the minimum approaches saturation only when $\alpha = 90°$, and the saturation value is roughly proportional to RRR. We conclude that open orbits occur on the Fermi surface of rhenium which have their open directions along the $\langle 10\bar{1}0 \rangle$ axes. The dependence of the saturation value of the magnetoresistance upon the value of RRR, which constitutes a deviation from Kohler's rule, shows that these open orbits result from magnetic breakdown.[4]

A sample with **J** along the hexagonal axis and RRR = 25,800 also exhibits a narrow maximum in its magnetoresistance anisotropy curve where **B** passes through the basal plane, which suggests the occurrence of a second set of open orbits along the hexagonal axis. When **J** is also in the basal plane, the narrow minimum one expects to be associated with these open orbits is masked by the broad minimum associated with the cylindrical electron sheet, since at this minimum ($\theta = 90°$ in Figs. 1 and 2) the magnetoresistance saturates at low fields. The second set of open orbits along the hexagonal [0001] axis and the open orbits along the $\langle 10\bar{1}0 \rangle$ axes probably involves the same sheet of the Fermi surface, since there appears to be a two-dimensional region of aperiodic open orbits centered on the $\langle 11\bar{2}0 \rangle$ axes where the (0001) and $\{10\bar{1}0\}$ lines of periodic open orbits intersect. The anisotropy of the magnetoresistance here is complicated because of the effects of magnetic breakdown.

Acknowledgments

We wish to thank R. R. Soden and E. Buehler for purifying the rhenium and growing oriented single crystals, and G. F. Brennert for mounting the samples and measuring their residual resistivity ratios.

References

1. W. A. Reed and E. Fawcett, *Bull. Am. Phys. Soc.* **7**, 478, 1962.
2. E. Fawcett and W. A. Reed, *Phys. Rev.* **131**, 2463, 1963.
3. N. E. Alekseevskii, V. S. Egorov, and B. N. Kazak, *Zh. Eksperim. i Teor. Fiz.* **44**, 1116, 1963. (*Soviet Phys. JETP* (*English Transl.*) **17**, 752, 1963.)
4. L. M. Falicov and P. R. Seivert, *Phys. Rev. Letters* **12**, 558, 1964.

8.3. Acoustic Absorption, Transport Phenomena Size Effects, Theory

LANDAU LEVEL OSCILLATIONS IN THE MAGNETOACOUSTIC ABSORPTION IN ZINC*

H. V. Bohm and L. Mackinnon[†]

*Department of Physics, Wayne State University
Detroit, Michigan*

The de Haas–van Alphen or Shubnikov–de Haas effect periods were first observed in the electron component of low-temperature ultrasonic absorption by Reneker[1] in bismuth. These observations were made with the magnetic field (H) perpendicular to the propagation direction (q) of longitudinal megacycle ultrasonic waves. Further observations of these periods were made in zinc and cadmium by Gibbons and Falicov,[2] using magnetic fields of about 30 kOe, again with **H** $\perp$ **q**.

Gurevich, Skobov, and Firsov[3] have suggested a mechanism whereby such oscillations could become pronounced for the orientation **H** $\parallel$ **q**. They assume first that, when the electron mean free path is very long compared with the sound wavelength, the only electrons absorbing the sound are those on the Fermi surface traveling with their component of velocity in the direction of the sound wave equal to the velocity of sound. Then, when a magnetic field is applied to a metal in the z-direction, which is also the direction of sound propagation, the energy E of these electrons may be written as

$$E = (n + \varphi)\hbar\omega_c + \frac{\hbar^2 k_z^2}{2m}$$

where n is an integer, φ a phase factor, $\hbar$ is Planck's constant divided by 2π, ω_c (proportional to H) is the cyclotron frequency of the electron motion in the xy-plane, k_z is the z-component of the wave vector describing the relevant electrons, and m is the appropriate electron mass. The second term in this expression is a constant, so that once $\hbar\omega_c > kT$, where k is Boltzmann's constant and T is the absolute temperature, it is possible that there may be no such electrons in the range of energies kT around the Fermi energy E_F, and hence no ultrasound absorption. As the field H changes, E changes and a Landau level may cross the Fermi surface at the appropriate point; then there will be an absorption. Thus, for any one sheet of Fermi surface, the absorption will oscillate from maximum to minimum as Landau levels pass through the surface on varying the magnetic field strength. Such oscillations will have periods (in $1/H$) essentially the same as those of the de Haas–van Alphen effect. They were first observed experimentally by Korolyuk and Pruschak[4] with ultrasound propagation in the [10$\bar{1}$0] and [11$\bar{2}$0] directions in zinc, and have also been reported at high fields in gallium by Shapira and Lax,[5] and by others. It will be noted that most of the work so far published or described has been confined to

---
* Supported by U.S. Air Force Office of Scientific Research Grant No. 62-379.
† National Science Foundation Senior Foreign Scientist Fellow, on leave from the University of Leeds; now at the University of Essex, England.

the orientations $\mathbf{H} \perp \mathbf{q}$ and $\mathbf{H} \parallel \mathbf{q}$. This paper will describe some experimental results obtained for other relative orientations of $\mathbf{H}$ and $\mathbf{q}$.

The ultrasonic apparatus used was that described by Kamm and Bohm,[6] operating at frequencies between 230 and 350 Mcps. The apparatus selected a pulse which had been through the specimen, integrated it, and passed the integrator output onto one pen of a two-pen recorder. The other pen recorded simultaneously the output of a Rawson–Lush rotating coil fluxmeter. The magnetic field around the specimen could be slowly varied to give fields up to about 13.7 kOe.

The specimens used were zinc single crystals, grown from 99.9999% pure zinc by Semi-Elements Inc., and cut by them to the approximate dimensions 1 cm × 1 cm × 3 mm. The ultrasonic path was through the 3 mm thickness. Three directions of ultrasound propagation, [0001], [10$\bar{1}$0], and [11$\bar{2}$0], were studied. Longitudinal ultrasonic pulses from a 10-Mcps fundamental frequency X-cut quartz transducer passed through both the specimen and a 1-cm quartz delay line to a second matched transducer. The electromagnet could be rotated around the dewar vessel to vary the angle between $\mathbf{H}$ and $\mathbf{q}$, and the temperature could be varied between 4.2° and 1.4°K.

The first experiments were done with $\mathbf{q}$ along [0001]—the direction not tried by Korolyuk and Pruschak. $\mathbf{H}$ was varied in direction between [0001] and [11$\bar{2}$0]. At 1.4°K, with magnetic fields above about 8 kOe, the recorded amplitude of the transmitted pulse showed oscillations as the magnetic field was varied. These oscillations were markedly temperature dependent. An example is shown in Fig. 1; at 4.2°K they were not observable at all, but their amplitude increased steadily

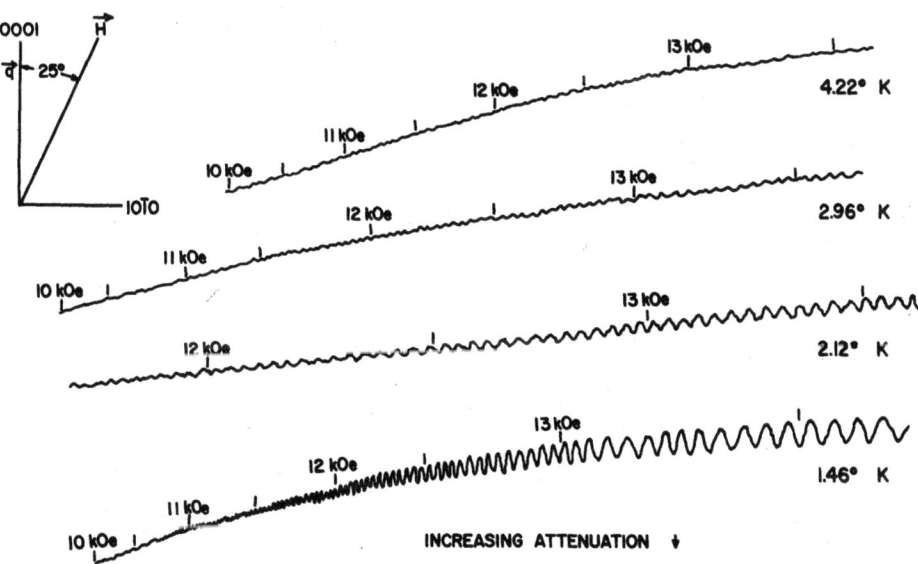

Fig. 1. Temperature dependence of Landau level oscillations at 350 Mcps. The temperature dependence for the orientation referred to in the text is similar. The oscillations shown here have periods of the order of 10^{-6} Oe^{-1}.

as the temperature was lowered. When the angle between **H** and **q** was greater than about 75°, the attenuation in the specimen rose at high fields to too great a value for observations with our apparatus. At lower angles, the overall attenuation was decreasing with increasing field.

Similar measurements were carried out with **q** along [0001] and **H** variable between [0001] and [10$\bar{1}$0], and with **q** along [10$\bar{1}$0] and [11$\bar{2}$0] and **H** variable from these axes to the [0001] axis. Figure 2 shows an example of our results in the [0001] to [11$\bar{2}$0] plane, together with the corresponding de Haas–van Alphen effect periods measured by Joseph and Gordon.[7] It will be seen that for this group there is good agreement to within the experimental accuracy.

It is noteworthy that within our present experimental accuracy, we obtain the same periods for a given direction of **H**, independent of whether **q** is propagated along the [0001] or the [11$\bar{2}$0] axis. However, it can be seen from Fig. 2 that in some field directions the periods were observable only with one or the other **q** propagation direction. Further, shorter periods (of the order of 10^{-7} Oe^{-1}) could only be detected with **q** along [0001] (not plotted in Fig. 2).

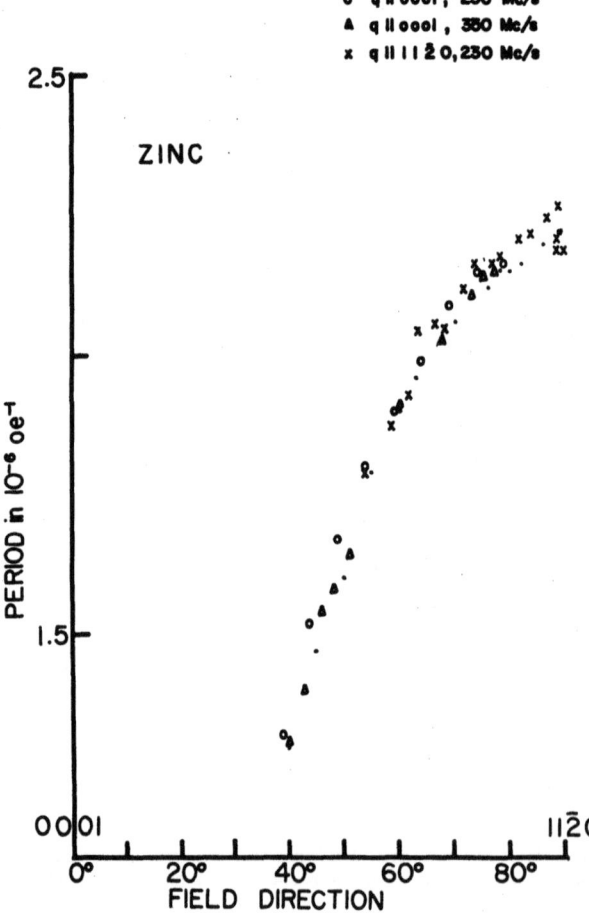

Fig. 2. Comparison of observed ultrasonic periods with one of the periods (P_2) of Joseph and Gordon's de Haas–van Alphen effect periods.

For other **q** and **H** configurations, our internal consistency was about the same as that shown in Fig. 2, but in the [0001] to [10$\bar{1}$0] configurations, our results show periods several percent smaller than those of Joseph and Gordon. To date, we do not know the reasons for these differences. All periods that have been identified so far are assignable to the second Brilliouin zone Fermi surface (the monster); with higher magnetic fields, it should become possible to observe other periods.

Oscillations were also observed, but not as well, at lower frequencies than 230 Mcps. Standard magnetoacoustic geometrical resonances did in fact suggest that the specimen purity was not quite as good as hoped; cutting, of course, may have introduced defects. In passing, it may be noted that with both **q** and **H** along [0001], while the overall attenuation decreased markedly as the field was increased, the decrease did not seem to be as great as that observed in cadmium by Mackinnon and Daniel,[8] nor were the corresponding oscillations as peaked or as distinct. It is not absolutely clear whether this is a consequence of lower specimen purity or of a fundamental difference in the electronic structure of the two metals.

From these experiments, we conclude that the magnetoacoustic technique is intrinsically capable of as precise observations of Landau level oscillation effects as other techniques, and is therefore capable of providing information on certain Fermi surface cross-sectional areas as well as on caliper dimensions.

The marked temperature dependence of the effect clearly indicates the importance of the condition $kT < h\omega_c$. At 1.4°K, for free electrons, $kT = h\omega_c$ at about 11,000 Oe; at 4.2°K, the equivalent field is 33,000 Oe. With our present limited magnetic field strength, we have only seen periods of cyclotron masses less than 1. In any case, although to date we have not attempted to do so, it may be possible to obtain some information about the effective masses of the electrons involved from the temperature dependence noted above.

Note Added in Proof

The periods mentioned above as being several percent smaller than those of Joseph and Gordon have since been found to be in general agreement with the corresponding data of Joseph and Gordon. The statement in the paper was based on our mistaken identification of a Joseph and Gordon graph.

Acknowledgments

We are grateful to Mr. Don Wallace and Mr. George Hovey for assistance in the measurements, to Mr. Joseph Mantel for assistance in the reduction of the experimental data, and to Dr. W. L. Gordon for kindly providing de Haas–van Alphen effect data for comparison purposes. One of us (L.M.) is also grateful to the National Science Foundation for the opportunity to do this work.

References

1. D. H. Reneker, *Phys. Rev.* **115**, 303, 1959.
2. D. F. Gibbons and L. M. Falicov, *Phil. Mag.* **8**, 177, 1963.
3. V. L. Gurevich, V. G. Skobov, and Y. A. Firsov, *Zh. Eksperim. i Teor. Fiz.* **40**, 786, 1961. (*Soviet Phys. JETP (English Transl.)* **13**, 552, 1961.)
4. A. P. Korolyuk and T. A. Pruschak, *Zh. Eksperim. i Teor. Fiz.* **41**, 1689, 1961. (*Soviet Phys. JETP (English Transl.)* **14**, 1201, 1962.)
5. Y. Shapira and B. Lax, *Phys. Rev. Letters* **12**, 166, 1964.
6. G. Kamm and H. V. Bohm, *Rev. Sci. Instr.* **33**, 957, 1962.
7. A. S. Joseph and W. L. Gordon, *Phys. Rev.* **126**, 489, 1962.
8. L. Mackinnon and M. R. Daniel, *Phys. Letters* **1**, 157, 1962.

MAGNETOACOUSTIC EFFECT IN TUNGSTEN AND MOLYBDENUM NEAR 1 GCPS*

C. K. Jones

Westinghouse Research Laboratories
Pittsburgh, Pennsylvania

and

J. A. Rayne

Carnegie Institute of Technology
Pittsburgh, Pennsylvania

The band structure of tungsten and molybdenum has recently been treated theoretically by Lomer[1] and Mattheiss.[2] Figure 1 shows a (100) section of the proposed model in an extended zone scheme. It can be seen that the principal parts of the Fermi surface are electron and hole surfaces centered at Γ and H, respectively. These surfaces may or may not be in osculating contact along $\langle 100 \rangle$, depending on the exact position of the Fermi level relative to the bands. The existence of the lens-shaped electron pockets along the cube axes is also strongly dependent on the location of the Fermi level. Small hole surfaces exist at the points N situated at the centers of the $\{110\}$ faces of the BZ.

Previous magnetoacoustic experiments on tungsten have verified some aspects of the theoretical model. The data are not, however, sufficiently detailed to establish unequivocally the relation between the main electron and hole surfaces and to determine whether the former are necked down as shown in Fig. 1. The present work was undertaken to elucidate these questions and to obtain similar information regarding the Fermi surface of molybdenum.

Experiments were performed at 930 Mcps with specimens of tungsten and molybdenum having resistance ratios $\rho_{273}/\rho_{4.2}$ of 60,000 and 5000, respectively. Measurements were made in transmission, using samples about 3 mm in length bonded to Z-cut quartz delay lines. Both the transmitting and receiving transducers were 30 Mcps X-cut crystals excited at an odd harmonic. The attenuation was obtained automatically as a function of $1/H$, typical data obtained for tungsten being shown in Fig. 2 with $\mathbf{q} \parallel [100]$, $H \parallel [010]$.

As may be seen, the oscillating behavior in this case is quite complex, at least four periods being observed. The extremal dimensions calculated from the formula[4]

$$k_{\text{ext}} = \frac{e}{2\hbar c} \frac{\lambda}{\Delta(1/H)} \qquad (1)$$

where λ is the sound wavelength and $\Delta(1/H)$ the period in $1/H$, are given in Table I. The largest dimension appears over a very narrow angular range of approximately

---

* Supported in part by National Science Foundation grant.

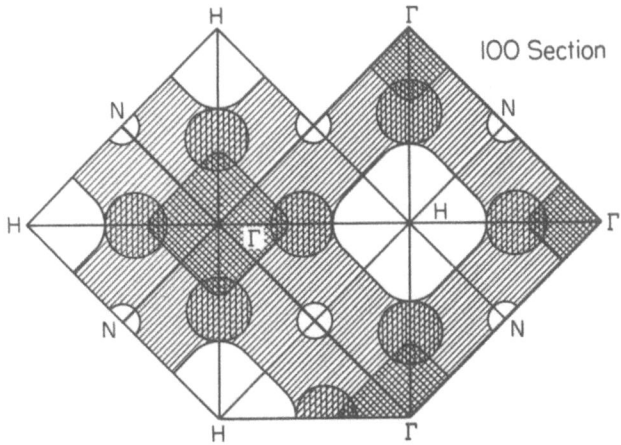

Fig. 1. Extended zone [100] section for the Fermi surface of tungsten, according to Lomer.

14° around $\langle 100 \rangle$ and is presumably associated with the neck of the electron surface at Γ. If the situation is as shown in Fig. 1, we have

$$k_{\text{electron}} + k_{\text{hole}} = \frac{2\pi}{a} \qquad (2)$$

which gives $k_{\text{hole}} = 0.90 \times 10^8$ cm^{-1}. Beats in the high-frequency oscillation give $k_{\text{hole}} = 0.70 \pm 0.10 \times 10^8$ cm^{-1} in approximate agreement with this value. It is possible that the discrepancy is due to the effects of spin-orbit coupling as has been suggested by Walsh et al. The next largest value of k_{ext} for this field direction is

Table I. Extremal Dimensions of Tungsten and Molybdenum Obtained from Magnetoacoustic Data at 930 Mcps

Metal	Direction	Extremal radius, 10^8 cm^{-1}
Tungsten	[100]	1.04 ± 0.03 (0.70 ± 0.10*)
		0.43 ± 0.02
		0.24 ± 0.005
	[110]	0.59 ± 0.05
		0.25 ± 0.005
Molybdenum	[100]	0.31 ± 0.02
		1.20 ± 0.10
	[110]	0.24 ± 0.03
		0.60 ± 0.05

* Obtained from analysis of beat pattern in high-frequency oscillation. This dimension is presumably due to the main hole surface at H.

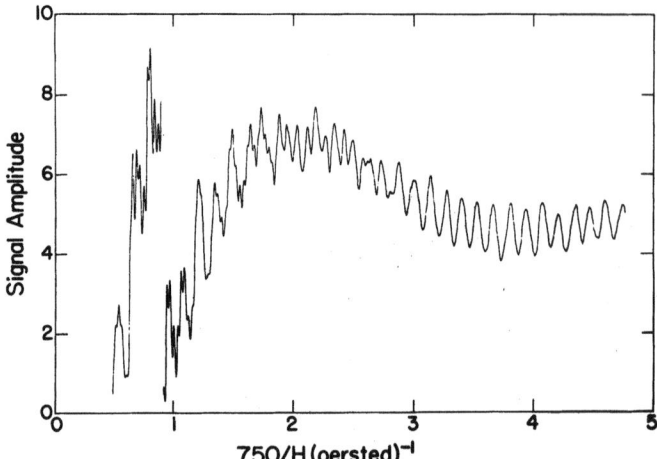

Fig. 2. Representative data obtained for tungsten at 930 Mcps. The sound propagation direction is [100] and the magnetic field direction [010].

approximately 0.45×10^8 cm^{-1}. This latter dimension cannot be associated with the surface at H, since the resulting hole volume would be too small to ensure charge compensation.* It is of course possible that it is associated with some orbit around the throat of the electron surface, but such an assignment is highly conjectural.

The smallest dimension listed in Table I is roughly isotropic with respect to the magnetic field direction and can reasonably be associated with the existence of approximately spherical balls on the extremities of the electron surface at Γ. There is very good agreement between the estimated extremal areas and those obtained from de Haas–van Alphen measurements.[6] The latter also provide a very clear indication of the existence of constrictions along the necks of the main electron surface, after the manner in Fig. 1. The present work gives no indication of the existences of such orbits even with H along the symmetry direction. This result could be due to the fact that magnetic breakdown effects are present and that at low fields the electron pockets lie outside the main surface at Γ. Such an interpretation, while consistent with the observed extremal dimensions along $\langle 100 \rangle$, again meets with the difficulty of satisfying the condition of charge compensation.

The results of the measurements on molybdenum are subject to much greater uncertainty. Here the oscillatory behavior is much less pronounced, owing to the relatively low purity of the specimen and the difficulty of insuring the condition $ql \gg 1$. The magnitude of the extremal dimensions in Table I are in reasonable agreement with those expected from theory and the results of de Haas–van Alphen measurements. Further work is needed before any precise comparison is possible. Experiments on both tungsten and molybdenum are continuing and will be reported at a future time.

Acknowledgment

One of us (JAR) wishes to acknowledge support of the National Science Foundation.

---

* This requirement is imposed by the results of magnetoresistance measurements.[5]

References

1. W. M. Lomer, *Proc. Phys. Soc. (London) Ser. A* **80**, 489, 1962.
2. L. F. Mattheiss, *Phys. Rev.* **134**, A970, 1964.
3. J. A. Rayne, *Phys. Rev.* **133**, A1104, 1964.
4. A. B. Pippard, *Proc. Roy. Soc. (London) Ser. A* **257**, 165, 1960.
5. E. Fawcett and W. A. Reed, *Phys. Rev.* **134**, A723, 1964.
6. D. M. Sparlin and J. A. Marcus, *Bull. Am. Phys. Soc.* **9**, 250, 1964; A. V. Gold, private communication.

MAGNETOTHERMAL OSCILLATIONS IN SEMI-METALS*

B. D. McCombe and G. Seidel

Brown University
Providence, Rhode Island

Experiment

The original magnetothermal experiments of Kunzler et al.[1] measured slowly varying temperature changes of a thermally isolated sample of bismuth at ^{4}He temperature as the magnetic field was adiabatically varied. An oscillatory variation of the temperature was observed as Landau levels passed through the Fermi surface. As an experimental technique for studying Fermi surfaces, this type of measurement is seriously impaired by extraneous heat inputs to the sample, such as those resulting from mechanical vibrations, fluctuations in the bath temperature, and induced eddy currents in the sample itself. While eddy current heating is not a major problem in bismuth because of its relatively high resistivity, it could easily mask magnetothermal phenomena in higher conductivity metals.

In a previous paper we reported preliminary observations of magnetothermal oscillations in bismuth by a sensitive derivative technique.[2] With this method, instead of thermally isolating the sample, it is coupled to a heat reservoir (a ^{3}He bath in this experiment) with a relatively short relaxation time of approximately 1 sec. A small modulating magnetic field at low audio frequency is superimposed upon the linearly swept DC field, and the temperature variation of a carbon resistance thermometer in good thermal contact with the sample is then detected synchronously with the modulating field. The short relaxation time between sample and bath maintains the sample at a constant average temperature. The predominant effect of eddy currents, which produce heating at twice the modulation frequency, is removed from the detected signal. The small eddy current heating at the signal frequency, which results from the product of the linearly swept and AC fields, is 90° out of phase with the desired signal and can be removed by the phase sensitive detector. The effects of thermal fluctuations from other sources are also greatly reduced by the narrow-band detection system.

For such a field modulation technique to work, certain conditions must be satisfied. The thermal relaxation time between sample and bath must be long compared to the inverse of the modulation frequency in order not to damp the signal. At the same time, the thermal relaxation time between sample and thermometer must be much less than the inverse of the modulation frequency; that is, the thermal contact between sample and thermometer must be good in order to ensure that the temperature of the thermometer will follow the sample at the modulation frequency. In addition, the detecting thermometer must be of high sensitivity, low noise, and have a small heat capacity compared with that of the sample. These

---

* This work was supported by the Advanced Research Projects Agency.

experiments were performed with an 0.008-in. slice of a nominal 10Ω, $\frac{1}{10}$ W Allen–Bradley resistor. Thermal contact was achieved by gluing the resistor to the sample with G.E. 7031 varnish. This procedure yielded sample-thermometer relaxation times of less than 0.01 sec; consequently, modulating frequencies up to 100 cps could be employed. Painted carbon-film thermometers were found to be unsuitable because of decreased sensitivity and unexplained noise.

The quantity that is measured in such an experiment is the rate of change of temperature with magnetic field at constant entropy. By the Maxwell relations, this quantity is related to the magnetization by the equation

$$\left(\frac{\partial T}{\partial H}\right)_{S,\vartheta} = -\frac{T}{C_{H,\vartheta}}\left(\frac{\partial M}{\partial T}\right)_{H,\vartheta}$$

where T is the absolute temperature, H the magnetic field intensity, S the entropy, M the magnetization, and C the heat capacity. The angle ϑ measures the direction of the magnetic field with respect to a crystalline axis and is specifically indicated as being held constant for the case of the linearly polarized modulating field applied parallel to the DC field.

In many cases it is advantageous to apply the AC field perpendicular to the DC field for purposes of sorting different periods of the magnetothermal oscillations associated with various extremal areas of the Fermi surface. With perpendicular modulation, the quantity measured is the rate of change of temperature with angle at constant entropy. This quantity is related to the torque $\mathcal{T}$ by the equation

$$\left(\frac{\partial T}{\partial \theta}\right)_{S,H} = -\frac{T}{C_{H,\vartheta}}\left(\frac{\partial \mathcal{T}}{\partial T}\right)_{H,\vartheta}$$

Since the torque is proportional to the derivative of the period with respect to angle, this measurement will emphasize periods having a rapid angular variation and will discriminate against periods that change slowly with angle. Judicious use of both parallel and perpendicular modulation can be of great help in following and measuring different periods throughout a given angular range. Although not yet attempted, it appears feasible to do much of this signal separation electronically before recording by applying both types of modulation simultaneously.

While the temperature variation of a metal with magnetic field appears easily measurable, calculations indicate that the inverse effect, the variation of magnetization with temperature, should with care be easier to detect. The quantity $\partial M/\partial T$ can be measured by varying the temperature of the sample sinusoidally and detecting synchronously the voltage induced in a pickup coil surrounding the sample. The sample is maintained at constant average temperature by coupling it to the bath with a relaxation time comparable to the inverse of the modulation frequency, and the induced voltage in the coil resulting from the variation of the magnetization is at twice the frequency of the current in the heater.

Results

Measurements of the magnetothermal oscillations in antimony are shown in Fig. 1 to illustrate the considerable advantage of employing both parallel and perpendicular modulation. The magnetic field is approximately 2° from a bisectrix axis in the binary–bisectrix plane. The top trace is taken with parallel modulation

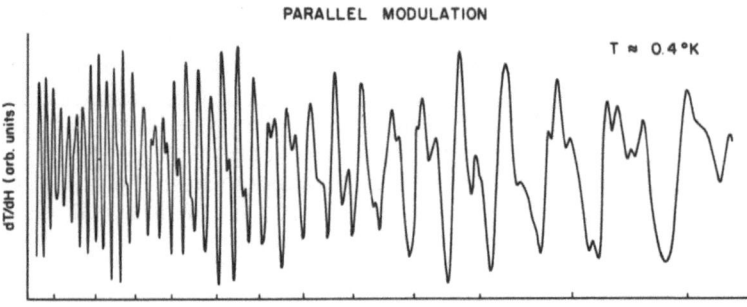

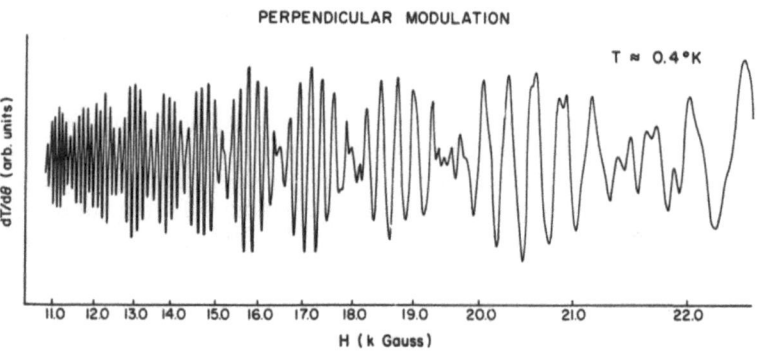

Fig. 1. Magnetothermal oscillations in antimony. Magnetic field in the binary–bisectrix plane.

and the bottom with perpendicular modulation in the same field region. The dominant period is clearly different in the two cases; hence, two distinct periods can be determined with ease. In addition, the beating effect in both traces yields two more periods, and the high field data indicate the presence of still shorter period oscillations. As yet, only preliminary analysis has been made of the angular dependence of the periods. The more obvious features of the data are in agreement with the Shubnikov–de Haas measurements,[3,4] and the tilted ellipsoid model for both electrons and holes.

The angular variation of the period of the magnetothermal oscillations in bismuth is shown in Fig. 2 for the magnetic field in the binary–trigonal plane. Through the use of both parallel and perpendicular modulation, the periods due to two of the electron ellipsoids and the single-hole ellipsoid have been followed throughout the entire plane. No magnetothermal oscillations arising from the third electron ellipsoid were observed. Spin splitting of the light electrons near the binary axis and of the holes in the vicinity of the trigonal axis was evident in the data. The splittings are in good agreement with other measurements.[1,5] At the angles $4.5° \pm 1°$ and $17.5° \pm 1°$ from the binary axis, the first harmonic content of the magnetothermal oscillations of the holes decreased to zero and the second harmonic was observed. At these angles the spin splitting of the holes is a half odd-integral

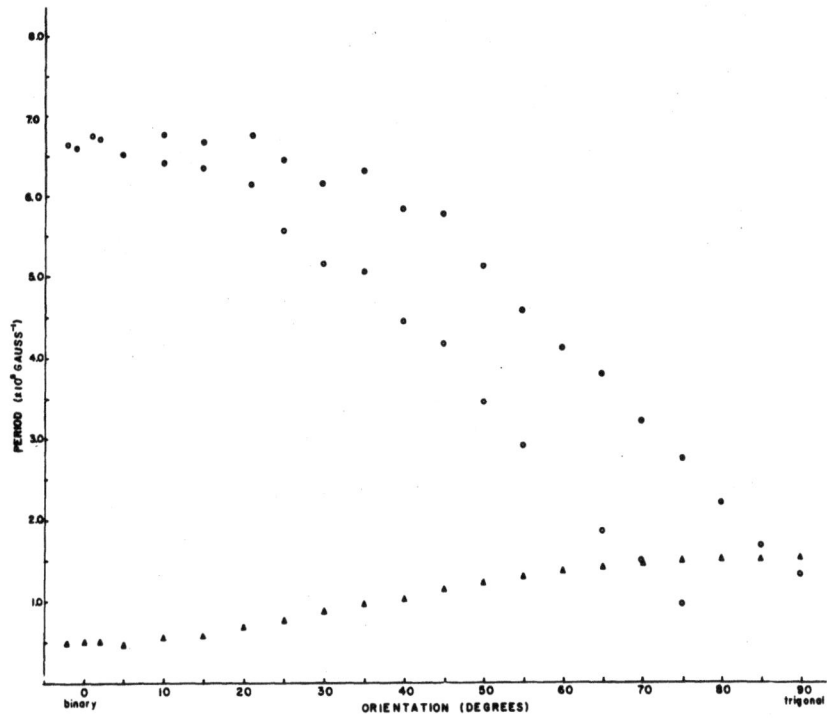

Fig. 2. Angular variation of the periods in bismuth. Magnetic field in the binary–trigonal plane; circles: electron periods; triangles: hole periods.

multiple of the orbital splitting. If the orbital effective mass tensor of Kao[6] and the spin effective mass tensor of Smith et al.[5] are used, the calculated angles for this condition are 4.3° and 17.1°, in agreement with the measurements. A calculation using the orbital effective mass tensor of Smith et al. yields 4.6° and 20.3°.

The magnetothermal oscillations of the holes in the vicinity of the binary axis was observed to be nonperiodic in $1/H$. This phenomenon has been previously reported and explained by Smith et al. as being due to the variation of the Fermi level with magnetic field.

The agreement between the theoretically predicted and observed amplitudes of the magnetothermal oscillations is poor for both the electrons and holes in bismuth. In each case the measured values are an order of magnitude smaller than the calculated values, a fact for which we have no explanation at present. Better agreement is obtained in antimony, where the calculated and measured value of $(\partial T/\partial H)_S$ for the electrons along the bisectrix axis is 10^{-4} °K/G at 10 kG.

Order of magnitude calculations indicate that magnetothermal oscillations should be observable in most metals. In view of this, a preliminary attempt was made, which was unsuccessful, to observe the oscillations due to the neck orbits in copper in a field of 20 kG. The lack of success of this measurement is attributed to the probable deformation of the 0.010-in.-thick sample during preparation. The major problem associated with field modulation techniques in the study of the magnetic behavior of high-conductivity metals is the shielding of the field variation from the bulk of the material by the induced currents. It is for this reason that a thin copper sample was used.

When the AC techniques are used, temperature variations of the order of 10^{-7} °K can currently be detected below 1°K. An improvement of greater than an order of magnitude is expected with refinements in thermometry and the reduction of electrical noise. Thus, with reasonable amplitudes of field modulation, magnetothermal oscillations having $(\partial T/\partial H)_S$ of the order of 10^{-8} °K/G can be expected to be observed. In comparison, the body orbits of copper are calculated to produce magnetothermal oscillations having $(\partial T/\partial H)_S$ of greater than 10^{-4} °K/G at 60 kG. Due to the obvious simplicity and sensitivity of this technique, it should prove to be a very useful tool in the study of the electronic properties of metals.

References

1. J. E. Kunzler, F. S. L. Hsu, and W. S. Boyle, *Phys. Rev.* **128**, 1084, 1962.
2. B. D. McCombe and G. Seidel, *Bull. Am. Phys. Soc.* **9**, 264, 1964.
3. J. Ketterson and Y. Eckstein, *Phys. Rev.* **132**, 1885, 1963.
4. G. Rao, N. W. Zebouni, C. G. Grenier, and J. M. Reynolds, *Phys. Rev.* **133**, A141, 1964.
5. G. E. Smith, G. A. Baraff, and J. M. Rowell, *Phys. Rev.* **135**, A1118, 1964.
6. Y. H. Kao, *Phys. Rev.* **129**, 1122, 1962.

MAGNETOTHERMAL OSCILLATIONS IN BERYLLIUM*

J. LePage,† M. Garber, and F. J. Blatt

Michigan State University
East Lansing, Michigan

We were stimulated by the work of Kunzler et al.[1] on bismuth to attempt to observe magnetothermal oscillations in a metal. Beryllium was chosen for this study because of its high Debye temperature (1150°K) and correspondingly low-lattice specific heat. The effect is associated with the oscillations in a magnetic field of the free energy of the conduction electrons; but since one measures changes in the lattice temperature due to an exchange of energy between the lattice and the electron system, the effect will be enhanced by a low lattice specific heat. Recently, we reported our first observations of magnetothermal oscillations in beryllium.[2] The amplitude of the oscillations was about 3 times that observed in a bismuth crystal under similar experimental conditions, even though the frequency was 8 times greater. Since then we have improved our experimental techniques and have observed frequencies from 10^5 G to 1.5×10^7 G. The measurements were made at 0.9°K in fields up to 21.5 kG. The maximum amplitude of the low-frequency oscillations was about 10^{-2}°K and of the highest frequency about 3×10^{-5}°K. The specimen was an 0.7-g single crystal of beryllium which was generously loaned to us by the Franklin Institute.‡ Although we have not finished our study of the Fermi surface of beryllium, we would like to report here a brief description of the method and the principal results.

The sample was supported in vacuum by a thin-walled graphite tube. The major thermal contact to the ^{4}He bath was through the manganin leads to the carbon resistance thermometer. These leads were wound several times around the beryllium sample and glued with GE-7031 varnish. This technique thermally isolated the carbon thermometer from direct contact with the bath. The thermal time constant of this arrangement was 5 min. The time constant was reduced to 1 min by connecting the sample to the bath with a 3 cm length of #46 copper wire. With the shortest relaxation time, eddy current heating effects were minimized. The magnetic field was swept at rates varying from 2 kG/min to 100 G/min. Neither the variation in sweep rate nor the change in time constant (both of which affect the amplitude of the oscillations) measurably affect the frequency of the oscillations. With the shorter time constant, the amplitudes of the low-frequency oscillations were drastically reduced at the slower sweep rates, thus facilitating the observations of the high-frequency oscillations. The shorter time constant also reduced the magnitude of a

---

* Supported by the National Science Foundation.
† National Science Foundation Predoctoral Fellow.
‡ The crystal was prepared under a purification contract from the Bureau of Naval Weapons from S–R grade beryllium with two zone passes.

field-dependent temperature change. This temperature change consisted of monotonic heating with increasing magnetic field, and cooling with decreasing field. We attribute this effect to conduction electron spin paramagnetism.

In addition to the sample thermometer, a similar thermometer was glued to the 0.9°K container. These two thermometers were selected in another experiment to have the same magnetoresistance and temperature dependence. They formed one pair of arms of an AC bridge, the balance arms being at room temperature. The thermometer power was 2×10^{-10} W. The small unbalance voltages due to the thermal oscillations were amplified, detected, and displayed on a strip chart recorder. Both temperature drifts ($\sim 0.005°$K) and magnetoresistance ($\Delta R/R \sim 5\%$ at 20 kG) had negligible effect on the bridge output. Thermal noise in the present experiment was a few microdegrees and was probably due to fluctuations in the magnetic field sweep rate.

The Fermi surface of beryllium was first described by Watts[3] on the basis of pulsed high-field de Haas–van Alphen measurements: "The first zone is full and the Fermi surface consists of a coronet-shaped hole surface in the second zone and cigar-shaped electron overlaps into the double third and fourth zone." Recently, Loucks and Cutler[4] and Terrell[5] published theoretical Fermi surfaces in substantial agreement with the picture of Watts. We have measured the area of the necks of the coronet of holes through an angular range of 55° with the field in the basal plane. The electron cigars have been observed over an angular range of 30° from the hexagonal axis. We also have a few points for the bellies of the coronet of holes. The electron cigars and the bellies of the coronet give rise to much higher frequencies than the necks. Figure 1 shows cigar oscillations superimposed on neck oscillations. A summary of the frequencies observed to date is shown in Fig. 2. We have also observed what appears to be beating between two frequencies from the cigars, which seems to agree with Watts' observation of waists on the cigars. Additional data need to be taken in order to confirm this. We cannot account for the appearance

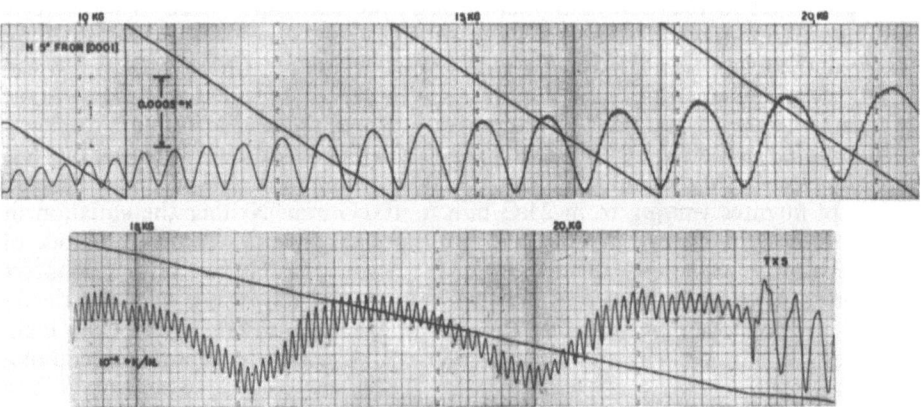

Fig. 1. Magnetothermal oscillations in beryllium at 0.9°K. The upper trace was taken with a sweep rate of 1 kG/min. The lower trace shows the effect on the relative amplitudes of the two frequencies of decreasing the sweep rate to 180 G/min. The noise can be seen in the last trace where the sweep rate was reduced to 50 G/min.

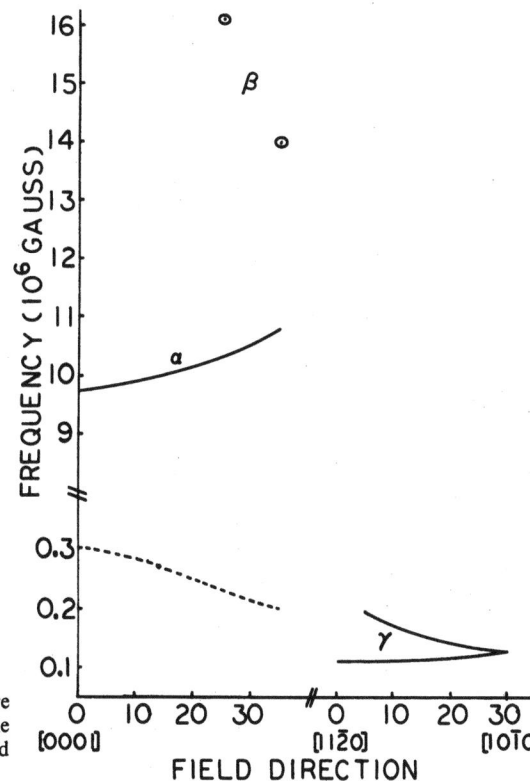

Fig. 2. Frequency vs. rotation. The labels are those used by Watts. The dashed line is the set of frequencies which cannot be identified with Watts' picture of the Fermi surface.

of the low-frequency oscillations with the field along the hexagonal axis.* The orientation of the crystal was rechecked after the experiment and cannot be off by more than 2°. The crystal was etched and also examined with X-rays using Laue backscattering and divergent beam technique. No evidence of twinning could be found.

Acknowledgment

We would like to thank J. Abele for his assistance in calculating frequencies from the data.

References
1. J. E. Kunzler, F. S. L. Hsu, and W. S. Boyle, Phys. Rev. **128**, 1962.
2. J. LePage, M. Garber, and F. J. Blatt, Phys. Letters **11**, 102, 1964.
3. B. R. Watts, Phys. Letters **3**, 284, 1963.
4. T. L. Loucks and P. H. Cutler, Phys. Rev. **133**, A819, 1964.
5. J. H. Terrell, Phys. Letters **8**, 149, 1964.

---

* M. Halloran and W. L. Gordon informed us that they have observed these low-frequency oscillations using magnetothermal oscillations and the de Haas–van Alphen torque method, respectively. They suggest that these oscillations arise from a nonlinearity due to the difference between the two-cigar orbits, for which $\partial M/\partial T$ (or $\partial M/\partial H$) is large enough so that one must use B instead of H in the analysis.

ELECTRON TRANSPORT PHENOMENA AND OSCILLATORY BEHAVIOR IN THE LATTICE CONDUCTIVITY OF AN ANTIMONY SINGLE CRYSTAL IN A MAGNETIC FIELD AT LOW TEMPERATURE*

C. G. Grenier, Jerome R. Long, J. M. Reynolds, and N. H. Zebouni

Louisiana State University
Baton Rouge, Louisiana

With the magnetic field parallel to the trigonal direction, the experimental coefficients of the transport effects have been measured in the basal plane of an antimony crystal, i.e., in a transverse field. The zero-field resistance ratio was $\rho_{300°}/\rho_{1.3°} \approx 9000$.

The fundamental kinetic equations of the transport effects are given as equation (1a)

$$\begin{align}
\text{(a)} \quad & \mathbf{J} = \hat{\sigma}\mathbf{E} - \hat{\varepsilon}''\mathbf{G} & \mathbf{W} = -\hat{\pi}''\mathbf{E} + \hat{\lambda}''\mathbf{G} \\
\text{(b)} \quad & \mathbf{E} = \hat{\rho}\mathbf{J} + \hat{\varepsilon}\mathbf{G} & \mathbf{W} = -\hat{\pi}\mathbf{J} + \hat{\lambda}\mathbf{G} \quad (1) \\
\text{(c)} \quad & \mathbf{E} = \hat{\rho}'\mathbf{J} + \hat{\varepsilon}'\mathbf{W} & \mathbf{G} = \hat{\pi}'\mathbf{J} + \hat{\gamma}\mathbf{W}
\end{align}$$

where $\mathbf{J}$ is the density of electric current, $\mathbf{W}$ is a modified heat current density,† $\mathbf{G} = -\nabla T$, the negative of the absolute temperature gradient, and $\mathbf{E}$ is a modified electric field.† Isothermal and adiabatic experimental transport coefficients are defined by equations (1b) and (1c).

Symmetries and fundamental (Onsager–Kelvin) relations reduce to 6 the number of experimental coefficients necessary for computation of the kinetic coefficients, i.e., those coefficients of (1a) which may be directly compared with transport theories. Simple relations between the kinetic coefficients and experimental quantities exist. For example,

$$\hat{\sigma} = \hat{\rho}^{-1} \qquad (2a)$$

$$\hat{\varepsilon}'' = \hat{\pi}''/T = \hat{\varepsilon}'\hat{\gamma}^{-1}\hat{\rho}^{-1} \qquad (2b)$$

$$\hat{\lambda}'' = \hat{\gamma}^{-1}(1 + \hat{\varepsilon}'\hat{\varepsilon}''T) \approx \hat{\gamma}^{-1} = \hat{\lambda} \qquad (2c)$$

In equation (2), the kinetic tensors $\hat{\sigma}$ and $\hat{\lambda}''$ of electrical and thermal conductivity, and $\hat{\varepsilon}''$, the thermoelectric tensor, are expressed as functions of the experimental tensors $\hat{\rho}$ and $\hat{\gamma}$ of electrical and thermal resistivity and $\hat{\varepsilon}'$, an adiabatic thermoelectric tensor, respectively.

---
* This work was supported by the U.S. Atomic Energy Commission and the U.S. Army Research Office, Durham.
† More details can be obtained, for example, in C. G. Grenier.[1]

Geometry and symmetry of the sample result in the possibility of its basal plane being treated as a complex plane with the vector fields and fluxes becoming complex numbers, $\mathbf{E} = E_1 - iE_2$, etc. Similarly, the 2 × 2 tensors can be written as complex numbers $\hat{\rho} = \rho_{11} + i\rho_{12}$, etc. The complex representation gives an easy way to compute and represent the transport effects.

The Experimental Coefficients of the Transport Effects

A complete set of measurements was performed at each of the temperatures 4°, 3°, 2.1°, and 1.6°K in a field ranging from 0.1 to 17.8 kG. Using standard experimental procedures, we measured the following effects: The magnetoresistivity ρ_{11} and the Hall resistivity $\rho_{21}(= -\rho_{12})$; the thermomagnetic resistivity γ_{11} and the Righi–Leduc resistivity $\gamma_{21}(= -\gamma_{12})$; the thermoelectric coefficient ε'_{11}; and finally, the Nernst–Ettinghausen (N–E) coefficient $\varepsilon'_{21}(= -\varepsilon'_{12})$.

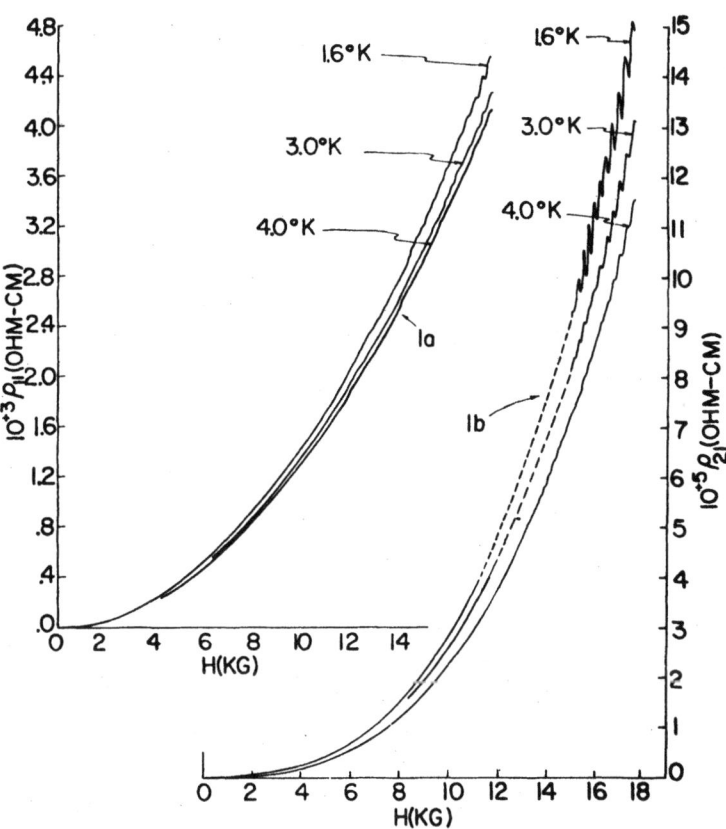

Fig. 1. The magnetoresistivity (a) and Hall resistivity (b) are plotted against applied field. Measurements were made over the range 133 G < H < 17.8 kG for each of four temperatures, but data for 2.1°K are omitted for reasons of their proximity to the 1.6°K curve. A dashed line indicates the presence of Shubnikov–de Haas oscillations of period $10.88 \times 10^{-7} \text{G}^{-1}$, too short to be shown on this reduced scale. Segments near the origin are also omitted in order to prevent crowding.

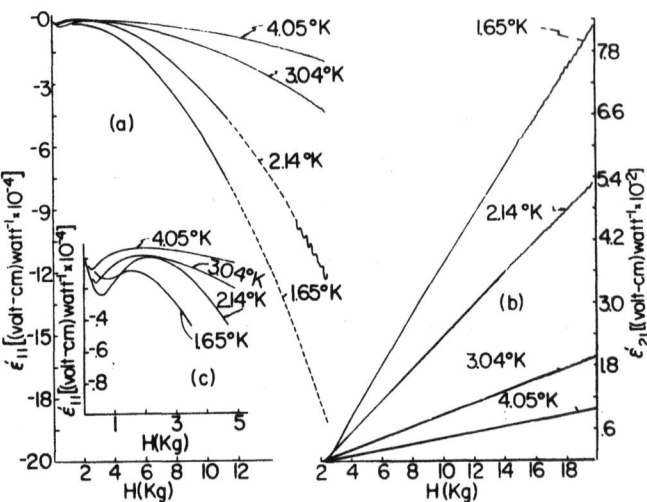

Fig. 2. The adiabatic [equation (1c)] thermoelectric effect (a) and Nernst–Ettinghausen effect (b) plotted vs. magnetic field, and the low-field detail of the thermoelectric effect (c) are shown at each of four temperatures for the range 133 G < H < 17.8 kG. Omitted or dashed segments have the same meaning as given in the caption to Fig. 1.

Results are shown in Figs. 1–3. The magnetoresistance, (Fig. 1a), shows a nearly quadratic field increase and small temperature dependence. The Hall resistivity (Fig. 1b) is positive, contrary to an earlier publication,[2] thus implying a majority of holes and holes which are more mobile than the electrons. The Hall resistivity is very small and exhibits nearly cubic field dependence rather than the linear dependence expected from the assumption $n_h = n_e$. Marked Shubnikov–de Haas oscillations appear on both effects.

The large positive N–E effect (Fig. 2b) is linear in field and strongly temperature dependent. The thermoelectric effect (Fig. 2a) is two orders of magnitude smaller and of opposite sign than the N–E effect. It exhibits strong temperature dependence and complicated field behavior (Fig. 2c). Electrons are the majority carrier in this effect, with a higher saturation field more than compensating the slightly larger density of state of the holes. Large oscillations appear on each effect.

The thermal resistivity, saturating at relatively low fields, as seen in Fig. 3a, indicates a nearly complete transport of heat by the lattice when the electronic component of heat transport is quenched by the magnetic field. The saturation value $\gamma_{11} \to \gamma_g$, interpreted as lattice thermal resistivity, is strongly temperature dependent ($\gamma_{11} \sim T^{-4.6}$), and at the lowest temperatures shows oscillations, Fig. 3c, of the Shubnikov–de Haas type. The Righi–Leduc thermal resistivity, positive and small like the Hall resistivity, passes through a low field maximum before a rapid decrease to zero at high field, Fig. 3b.

The Kinetic Coefficients of the Transport Effects

In the calculation of the kinetic coefficients by equation (2), corrections for the thermoelectric power of the constantan leads were neglected as was the difference between $\hat{\lambda}''$ and $\hat{\lambda}$. Much of the gross behavior of these coefficients can be interpreted in terms of a two-band model with holes more mobile than electrons. A Sondheimer–Wilson[3] type of theory is used, neglecting any size effect correction. With a rough

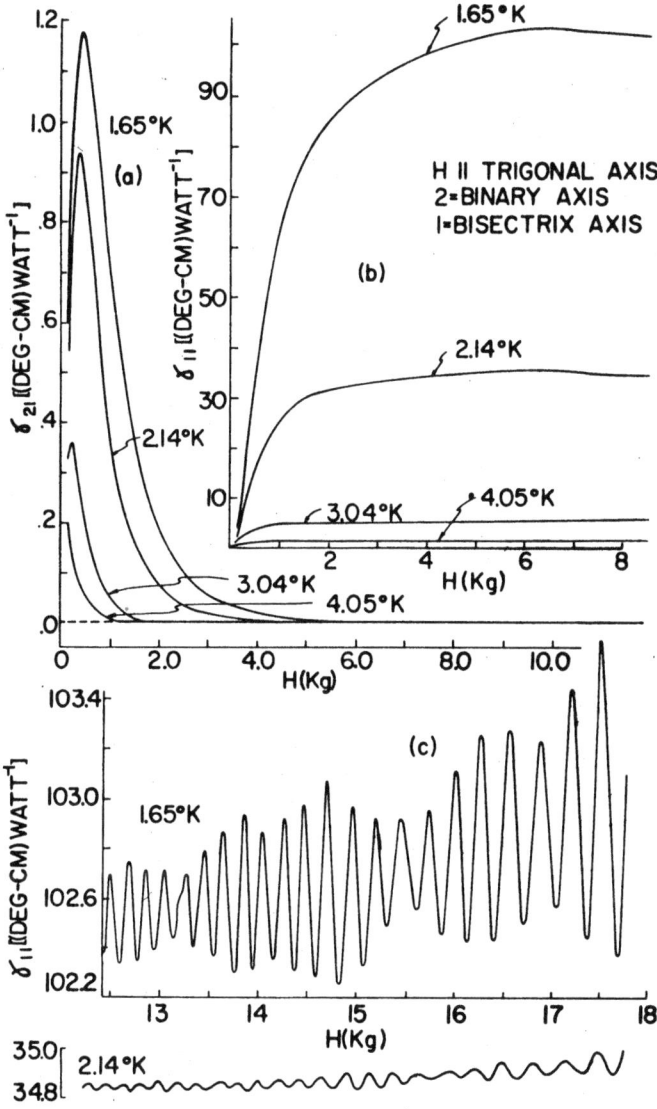

Fig. 3. The thermal magnetoresistivity is plotted against field (a) and is observed to saturate to its lattice value at high fields. The Righi-Leduc resistivity (b) is observed to decrease to zero after passing through a maximum. Shubnikov–de Haas type oscillations which are highly temperature dependent are shown (c) in the thermal resistivity at high field. These are explained as oscillations in a lattice conductivity limited by the scattering of phonons by electrons.

analysis of the high-field asymptotic behavior only, some of the transport effect parameters have been obtained and are shown in Table I.

Besides an expected behavior of the mobilities (as measured by α, the ratio of theoretical and experimental Lorentz numbers and the saturation fields $H_{e\sigma}$, $H_{h\sigma}$, $H_{e\lambda}$, and $H_{h\lambda}$), some specific points of interest can be stressed as follows:

Table 1 †

	1.6°K	2.1°K	3.0°K	4.0°K
(a) Experimental values				
$n_h - n_e$	8.1×10^{15}	7.9×10^{15}	8.0×10^{15}	7.6×10^{15}
$n(a_h H_{h\sigma} + a_e H_{e\sigma})$	4.33×10^{21}	4.38×10^{21}	4.61×10^{21}	4.76×10^{21}
$Z_h + Z_e$	3.05×10^{33}	4.45×10^{33}	7.48×10^{33}	10.9×10^{33}
$f(\alpha)(Z_e a_e H_{e\sigma} - Z_h a_h H_{h\sigma})$	4.84×10^{35}	6.72×10^{35}	10.5×10^{35}	13×10^{35}
$\alpha = \tau_\lambda/\tau_\sigma = L_n/L_1$	0.384	0.322	0.143	0.115
(b) Calculated values				
$H_{e\sigma} = c/\mu_{e\sigma}$	29.4	29.8	31.4	32.4
$H_{h\sigma} = c/\mu_{h\sigma}$	5.7	5.8	6.1	6.3
$H_{e\lambda} = (1/\alpha)H_{e\sigma}$	76	92	220	282
$H_{h\lambda} = (1/\alpha)H_{h\sigma}$	15	18	43	55
$\alpha(Z_h + Z_e)$	1.2×10^{33}	1.4×10^{33}	1.1×10^{33}	1.2×10^{33}

† Part (a) consists of experimental quantities in cgs Gaussian units. Indices σ and λ represent electrical and thermal processes, respectively, while e and h represent electrons and holes, respectively. τ_σ is the relaxation time for an electrical process and $H_{e\sigma} = m_e^* c/e\tau_\sigma$ is the saturation field of electrons in an electrical process with the associated mobility $\mu_{e\sigma}$. The coefficient $\alpha = H_{e\sigma}/H_{e\lambda} = \tau_\lambda/\tau_\sigma$ is a measure of the efficiency of scattering between electrical and thermal processes and is supposed to be the same for electrons and holes. It is the ratio of theoretical and measured asymptotic Lorentz numbers experimentally. Z_e is the density of electron states. Part (b) presents quantities calculated from part (a) with the aid of Rao's² values $n \approx n_e \approx n_h = 5.05 \times 10^{19}$ holes/cm³, and $H_{h\sigma}/H_{e\sigma} = 0.194$; chemical potentials $\zeta_e = 20 \times 10^{-14}$ ergs and $\zeta_h = 13 \times 10^{-14}$ ergs; $a_e = 2.6$, $a_h = 1.6$, where a_e is the ratio between linear and mean-square averages of electron mobilities in the basal plane.

(a) a 0.2% excess of holes over electrons; (b) a strong temperature variation of the apparent density of states $Z_e + Z_h$ with values larger than

$$3/2(n_e/\zeta_e + n_h/\zeta_h) = 0.96 \times 10^{33} \text{ erg}^{-1}/\text{cm}^3$$

obtained from Rao's data,[2] but of the same order as obtained from specific heat measurements[4] (this behavior is strongly reminiscent of results in the study of bismuth[5]); and (c) an apparent proportionality between $(Z_e + Z_h)$ and α^{-1}.

Analysis of the oscillatory components of the transport effects is in progress and will be published at a later date.

The Lattice Conductivity

One of the most striking results to come from the transport effects in antimony pertains to the asymptotic behavior of the thermal conductivity. As shown in Figs. 3a and 3b, $\gamma_{11} \to \gamma_g$ and $\gamma_{21} \to 0$. This implies that $\lambda_{11} = (\hat{\gamma}^{-1})_{11} \approx \gamma_{11}^{-1}$. Because $\hat{\varepsilon}'\hat{\varepsilon}''T \ll 1$, it is concluded that $\lambda''_{11} \approx \lambda_{11} \approx \gamma_{11}^{-1} \to \gamma_g^{-1}$. It should be mentioned however, that a slight increase of λ_g (decrease of γ_g) above 7 kG at the lower temperatures was found to be due to the term $\hat{\varepsilon}'\hat{\varepsilon}''T$ in equation (2c). The large oscillations appearing at high field in γ_{11} at 1.65° and 2.14°K (Fig. 3c) seem definitely associated with an oscillation of the lattice conductivity. Experimental evidence and theoretical considerations rule out other possible origins of this effect such as: (a) magnetocaloric effect in the sweeping fields; (b) effect of the Wiedemann–Franz law term, $\alpha^{-1} L_n T \sigma_{11}$ (L_n is the Lorentz number); (c) effect of the $\hat{\varepsilon}'\hat{\varepsilon}''T$ term; and (d) phonon drag effect.

Steele and Babiskin[6] observed oscillations of this nature in bismuth. They noted that the lattice resistivity is expected to contain a term proportional to the square of a number of carriers which is field dependent, and it was suggested, without analysis, that the oscillations were due to the scattering of lattice waves by electrons. It is shown below that these suggestions are probably correct for bismuth, and their correctness is quantitatively established for our data on antimony. In justification of the above conclusion, it is pertinent to note that the scattering of phonons must be preponderantly due to electrons. Such is the case, as the phonon mean free path is still order of magnitude less than the sample size or the expected mean free path for U-processes at the temperatures used. The number of charged carriers is expected to dominate the scattering with a detailed knowledge of the electron and phonon states expected to be necessary for a precise calculation. If the second alternative is neglected, an expression of the form $|\tilde{\gamma}|/\gamma = \beta|\tilde{n}|/n$ relates the relative amplitudes of the thermal resistivity oscillations to the relative amplitudes of the number of carrier oscillations as defined by the expression

$$\tilde{n} = -\partial\tilde{\Omega}/\partial\zeta$$

where $\tilde{\Omega}$ is the oscillatory part of the grand canonical potential. The value of β resulting from the relation may depend on many factors: (a) $\tilde{n}$ given by the grand canonical potential does not represent the effective variation Δn in the number of carriers. For example, in a two-band model one has

$$\Delta n_e = \Delta n_h = [Z_e/(Z_e + Z_h)]\tilde{n}_h + [Z_h/(Z_e + Z_h)]\tilde{n}_e$$

(b) Relative scattering efficiencies of electrons and holes on transverse and longitudinal phonons might be expected to introduce complications. If the simplest assumptions are made, namely that any phonon is scattered with equal efficiency by any hole or electron with a scattering proportional to the square of the number of each type of carrier,* the value $\beta = 1.20$ is obtained using Rao's[2] chemical potentials. Since the oscillations are due to the Shoenberg set of ellipsoids, $\tilde{n}/n$ can be computed[2] with the results $1.17 < \beta < 1.75$ at $1.65°K$, and $0.41 < \beta < 0.66$ at $2.14°K$, for a $12\,kG < H < 18\,kG$ field range. One is led to conclude that oscillations in the lattice conductivity are due to scattering of the lattice wave by Landau quantized electrons with the generally accepted n^2 proportionality.

Acknowledgment

The authors thank Dr. G. N. Rao for his assistance in preparing the experiment and taking the data.

References

1. C. G. Grenier, J. M. Reynolds, and N. H. Zebouni, *Phys. Rev.* **129**, 1088, 1963.
2. G. N. Rao, N. H. Zebouni, C. G. Grenier, and J. M. Reynolds, *Phys. Rev.* **133A**, 141, 1964.
3. A. H. Wilson, *Theory of Metals*, third edition, Cambridge University Press, New York (1953).
4. C. Nanney, *Phys. Rev.* **129**, 109, 1963.
5. C. G. Grenier, J. M. Reynolds, and J. R. Sybert, *Phys. Rev.* **132**, 58, 1963.
6. M. C. Steele and J. Babiskin, *Phys. Rev.* **98**, 359, 1955.
7. I. I. Hanna and E. H. Sondheimer, *Proc. Roy. Soc. (London) Ser. A* **239**, 247, 1957.
8. P. G. Klemens, *Solid State Physics*, Vol. 7 Academic Press, Inc., New York (1958), p. 1.

---

* General references are given in the articles by I. I. Hanna and E. H. Sondheimer[7] or P. G. Klemens.[8]

OSCILLATORY MAGNETOMORPHIC BEHAVIOR IN THE ELECTRON TRANSPORT EFFECTS OF A CADMIUM SINGLE CRYSTAL AT LOW TEMPERATURES*

J. M. Reynolds, K. R. Efferson, C. G. Grenier, and N. H. Zebouni

Louisiana State University
Baton Rouge, Louisiana

Introduction

The Hall effect and magnetoresistance in a cadmium crystal exhibit well-defined magnetomorphic oscillations periodic in the field over a wide field range.[1] Additional measurements of the Hall effect and magnetoresistance in a cadmium crystal as well as a study of the oscillatory behavior of the thermomagnetic effects are presented here. Oscillations periodic in H were found in all the measured effects. The results of an analysis of the amplitudes and periods of these oscillations are presented below.

Crystal and Experimental Procedure

The sample was a single crystal of zone refined Cd cut from a bar which contained three large crystals.† It was cut and planed with a spark cutter so that its dimensions were $1.8 \times 0.558 \times 0.109$ cm, and so that the hexagonal axis was perpendicular to the face of the slab. The resistance ratio of the sample was $\rho_{300°}/\rho_{4.2°} = 35,000$.

In the measurements of the transport effects, the crystal was oriented so that the x-direction was along its length, the y-direction was across its face, and the z-direction was perpendicular to its face and parallel to the hexagonal axis. The primary fluxes, i.e., the current density J and the heat current density W, flowed along the x-direction and the affinities, i.e., the electrical field E and gradient of temperature $\nabla T = -G$, were measured in the x- and y-directions, while the field H was in the z-direction.

Experimental Results

The experimental transport coefficients‡ were determined from the measurements described above. The magnetoresistivity coefficient ρ_{11} and the Hall resistivity coefficient ρ_{21} are shown in Figs. 1b and 1c at the temperatures 1.6°, 2.0°, 2.8°, and 4.0°K. They exhibit essentially the same features already observed by Zebouni et al.[1]

---

* This work was supported by the U.S. Atomic Energy Commission and the U.S. Army Research Office (Durham).

† The bar of cadmium was obtained from Cominco Products, Inc. It was grade 69 with a purity of 99.9999%.

‡ For definition of effects see C. G. Grenier et al.[2]

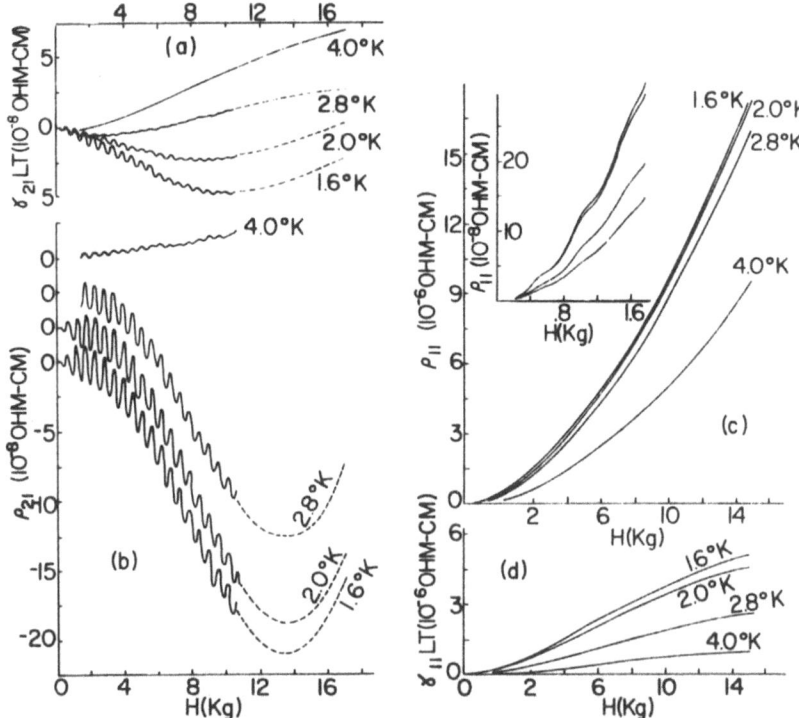

Fig. 1. Quantities plotted against the magnetic field H are (a) Righi–Leduc resistivity multiplied by T and L, the Lorentz number, (b) the Hall resistivity, (c) transverse magnetoresistivity, and (d) transverse thermal magnetoresistivity. Magnetomorphic oscillations are superposed on all these effects; however, in the ρ_{11} curves and the $\gamma_{11}LT$ curves they are small compared to the gross effect and must be viewed on a scale such as the one shown in the inset in (c). The two upper curves at 1.6° and 2.0°K are ρ_{11} curves and the two lower curves are $\gamma_{11}LT$ curves for 1.6° and 2.0°K.

The Hall effect is comparable to the effect in bulk specimens at 4.0°K since it is essentially positive. Below this temperature it shows a strong dip into negative values. The oscillations in the Hall effect are periodic in H with a period of about 570 G and were observed over the entire field range of 18 kG. The dotted lines in all cases represent oscillations which were observed but not analyzed. The transverse magnetoresistance ρ_{11} appears to behave very classically, however, as can be seen from the inset in Fig. 1c, oscillations are superposed on the gross effect, and are sizable at low field and low temperatures only.

The transverse thermal magnetoresistivity γ_{11} and the Righi–Leduc resistivity γ_{21} are both multiplied by the free-electron Wiedemann–Franz ratio and plotted in Figs. 1d and 1a, respectively, for easy comparison with the corresponding coefficients ρ_{11} and ρ_{21}. The appearance of a saturation effect in γ_{11} is due to the thermal conductivity of the lattice, and the generally smaller value for the γ's than the ρ's is in qualitative agreement with the more efficient scattering for the thermal effect. This implies $\alpha = \tau_\lambda/\tau_\sigma < 1$, where τ_λ and τ_σ are relaxation times for thermal and electrical processes, respectively. Morphic oscillations are very pronounced in the Righi–Leduc effect at the lowest temperatures, but no oscillations could be detected at 4.0°K. Oscillations in γ_{11} appear at low field and low temperature as shown by the inset in Fig. 1c.

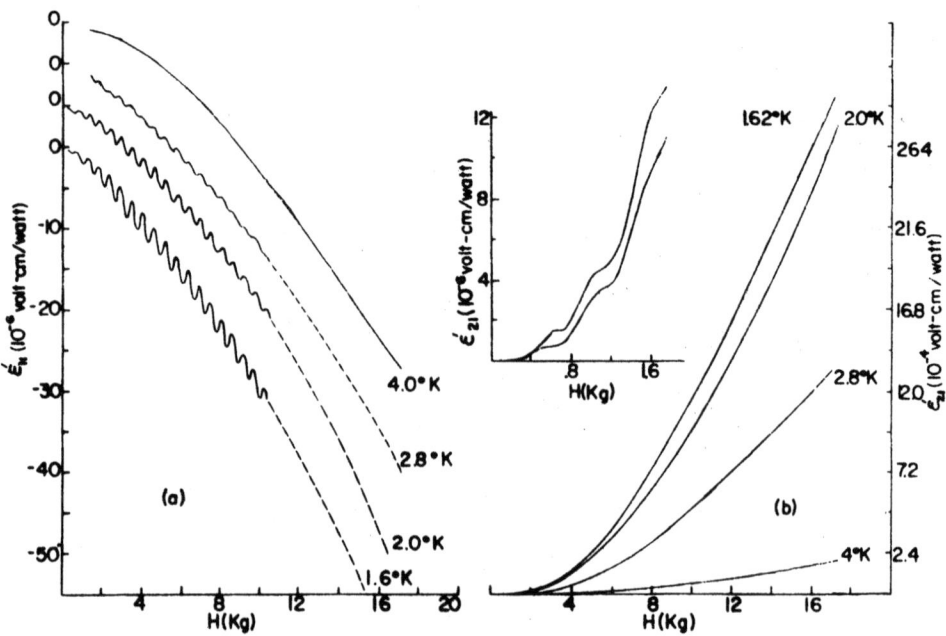

Fig. 2. (a) The adiabatic thermoelectric coefficient ε'_{11} and (b) the Nernst–Ettinghausen coefficient ε'_{21}, showing the magnetomorphic oscillations superposed on the gross effect. The inset in (b) shows low-field oscillations in ε'_{21} at 1.6° and 2.0°K.

The adiabatic thermoelectric coefficient (Th–E) ε'_{11} and the adiabatic Nernst–Ettinghausen (N–E) coefficient ε'_{21} are shown in Figs. 2a and 2b. The Th–E coefficient is practically independent of temperature as opposed to the N–E coefficient which is strongly temperature dependent. The oscillations are strongly marked at the lowest temperatures in the Th–E coefficient whereas none could be detected at 4.0°K. As shown by the insert in Fig. 2b, the oscillations are observable in the N–E coefficient at low field and low temperatures.

The Kinetic Coefficients

The measured coefficients are not usually compared directly to the theory, thus it is necessary to compute the kinetic coefficients as pointed out in the previous paper.* The magnetoconductivity and thermomagnetic conductivity can be obtained by simply inverting the corresponding resistivity tensors, or by taking the inverse of the complex numbers which are holomorphic with them, i.e., $(\sigma_{11} + i\sigma_{12}) = (\rho_{11} + i\rho_{12})^{-1}$ and $(\lambda_{11} + i\lambda_{12}) = (\gamma_{11} + i\gamma_{12})^{-1}$. If the thermocouple effect of the leads is neglected, the coefficient $\varepsilon'' = (\varepsilon''_{11} + i\varepsilon''_{12})$ is related to the adiabatic and isothermal coefficients through the equation

$$\varepsilon'' = (\varepsilon'_{11} + i\varepsilon'_{12})(\sigma_{11} + i\sigma_{12})(\lambda_{11} + i\lambda_{12}) \qquad (1)$$

The tensors $\hat{\sigma}$, $\hat{\lambda}$, and $\hat{\varepsilon}''$ so determined can be readily compared to Blatt's[3] theory, which is an extension of E. H. Sondheimer's theory[4-7] to thermal processes,

---
*For definition of effects see C. G. Grenier et al.[2]

for the case of free electrons. Blatt's theory can be used to express the tensor elements as follows:*

$$\sigma_0 = \sigma_{11} + i\sigma_{12} = \sigma_{b_0} \int_1^\infty (\tfrac{3}{2})(t^{-2} - t^{-4})\{1 + [st]^{-1}[\exp(-st) - 1]\} \, dt$$

$$\lambda_0 = \lambda_{11} + i\lambda_{12} \cong \lambda'' = \lambda''_{b_0} \int_1^\infty (\tfrac{3}{2})(t^{-2} - t^{-4})\{1 + [st]^{-1}[\exp(-st) - 1]\} \, dt \qquad (2)$$

$$\varepsilon''_0 = \varepsilon''_{11} + i\varepsilon''_{12} = \varepsilon''_{b_0} \int_1^\infty t^{-2}\{1 + [st]^{-1}[\exp(-st) - 1]\} \, dt$$

where the index b refers to the bulk effect, the index 0 refers to the case of free electrons, and

$$\sigma_{b_0} = (e^2\tau'/m_0)n_0$$
$$\lambda''_{b_0} = L_n T(e^2\tau'/m_0)n_0 \qquad (3)$$
$$\varepsilon''_{b_0} = \pi^2 k^2 T/3e(e^2\tau'/m_0)Z_0$$

where n is the number of electrons per cubic centimeter in the Fermi sphere, m_0 the free electron mass, Z_0 the density of states on the Fermi sphere, L_n the Lorentz number, $1/\tau' = 1/\tau_b + i\Omega_0$, τ_b the bulk time of relaxation, $e = |$electronic charge$|$, $s = a/v_f\tau' = \kappa + i\beta_0$, v_f the Fermi velocity, a the thickness of the sample, $\beta_0 = 2\pi H/P_0$, $P_0 = (2\pi c/ea)m_0 v_f$ is the period of the oscillation, $t = v_f/v_z$ is the parameter of integration, and v_z is the component of the electron velocity parallel to the field. In equation (2), only the case of diffuse scattering has been considered for electrons scattered at the surface of the sample.

The Asymptotic Behavior of the Oscillatory Effects

Limiting the actual study to the oscillatory phenomena and noting that the oscillations extend to a very wide field range, it was found that for high field behavior, i.e., $|s| \gg 1$ and $\beta_0 \gg \kappa$, a good approximation to the results could be obtained from the first term in the asymptotic expansion of equation (2) so that the oscillatory part is given by

$$\tilde{\sigma}_{11} + i\tilde{\sigma}_{12} = [i\sigma_{12_{b_0}}(i\beta_0)^{-3}e^{-\kappa}]e^{-i\beta_0} = (-3\sigma_{12_{b_0}}\beta_0^{-3}e^{-\kappa})\{\cos\beta_0 + i\cos[\beta_0 + (\pi/2)]\}$$

$$\tilde{\lambda}_{11} + i\tilde{\lambda}_{12} = L_n T e^{(\kappa - \kappa')}(\tilde{\sigma}_{11} + i\tilde{\sigma}_{12}) \qquad (4)$$

$$\tilde{\varepsilon}''_{11} + i\tilde{\varepsilon}''_{12} = [i\varepsilon''_{12_{b_0}}(i\beta_0)^{-2}e^{-\kappa}]e^{-i\beta_0} = (-\varepsilon''_{12_{b_0}}\beta_0^{-2}e^{-\kappa''})\{\cos[\beta_0 - (\pi/2)] + i\cos\beta_0\}$$

with

$$\sigma_{12_{b_0}} = -(n_0 ec)H^{-1}$$

and

$$\varepsilon''_{12_{b_0}} = -(\pi^2 k^2 TcZ_0/3)H^{-1}$$

The symbols κ' and κ'' are introduced to include the possibility of different relaxation times for the different processes, i.e., $\kappa/\kappa' = \alpha$, and $\kappa/\kappa'' = f(\alpha)$, where $f(\alpha)$ is an unknown function of α which most probably would be of the form $f(\alpha) = \alpha^\nu$, $0 < \nu < 1$.

---

* The authors are indebted to Dr. H. J. Mackey for helpful collaboration on this matter.

The following conclusions about phase and amplitude can be obtained from equation (4):

(a) The coefficients $\tilde{\sigma}_{11}$, $\tilde{\lambda}_{11}$, $\tilde{\varepsilon}''_{12}$ are in phase with maxima given directly by integral values of P_0, $H_{max} = nP_0$.

(b) The coefficients $\tilde{\lambda}_{12}$ and $\tilde{\sigma}_{12}$ are in phase, but in opposite phase with $\tilde{\varepsilon}''_{11}$. $\tilde{\sigma}_{12}$ leads $\tilde{\sigma}_{11}$ by $\pi/2$. Thus, $H_{max}(\tilde{\sigma}_{12}) = H_{min}(\tilde{\varepsilon}''_{11}) = (n - \tfrac{1}{4})P_0$.

(c) The amplitudes $|\tilde{\sigma}_{11}|$, $|\tilde{\sigma}_{12}|$, $|\tilde{\lambda}_{11}|$, and $|\tilde{\lambda}_{12}|$ should be proportional to H^{-4}, whereas the amplitudes $|\tilde{\varepsilon}''_{12}|$ and $|\tilde{\varepsilon}''_{11}|$ should be proportional to H^{-3}.

(d) Those effects for which the amplitudes should be equal are $|\tilde{\sigma}_{11}| = |\tilde{\sigma}_{12}|$, $|\tilde{\lambda}_{11}| = |\tilde{\lambda}_{12}|$, and $|\tilde{\varepsilon}''_{11}| = |\tilde{\varepsilon}''_{12}|$. Also, for the lowest temperatures with $e^{-\kappa} \approx e^{-\kappa'} \approx 1$, the Wiedemann–Franz law should be applicable, i.e., $|\tilde{\lambda}_{12}|(L_n T)^{-1} = |\tilde{\sigma}_{12}|$.

Comparison with Experiment

(1) The experimental asymptotic period of 570 G corresponds to $P_{exp} = 1.07 P_0$.†

(2) $\tilde{\sigma}_{11}$, $\tilde{\lambda}_{11}$, and $\tilde{\varepsilon}''_{12}$ are in phase and $H_{max} = nP$ is well fulfilled in the range of experimental accuracy, thus statement (a) holds very well.

(3) Statement (b) is fulfilled for $\tilde{\sigma}_{12}$ and $\tilde{\lambda}_{12}$, since they are in phase with the positions of maxima being given by $H_{max} = (n - \tfrac{1}{4})P$, but $\tilde{\varepsilon}''_{11}$ is also in phase with $\tilde{\sigma}_{12}$ instead of having the opposite phase.

(4) The field dependence of the amplitudes is given in Figs. 3a, 3b, and 3c. In the case of $\tilde{\sigma}_{12}$, $\tilde{\lambda}_{12}$ and $\tilde{\sigma}_{11}$, twice the amplitude multiplied by H^4 is shown, and in the case of $\tilde{\varepsilon}''_{11}$, twice the amplitude multiplied by H^3 is shown. It is readily seen that these quantities approach a constant value above about 3000 G. Thus the asymptotic conditions for the amplitudes are readily fulfilled for $|s| > \approx 30$.

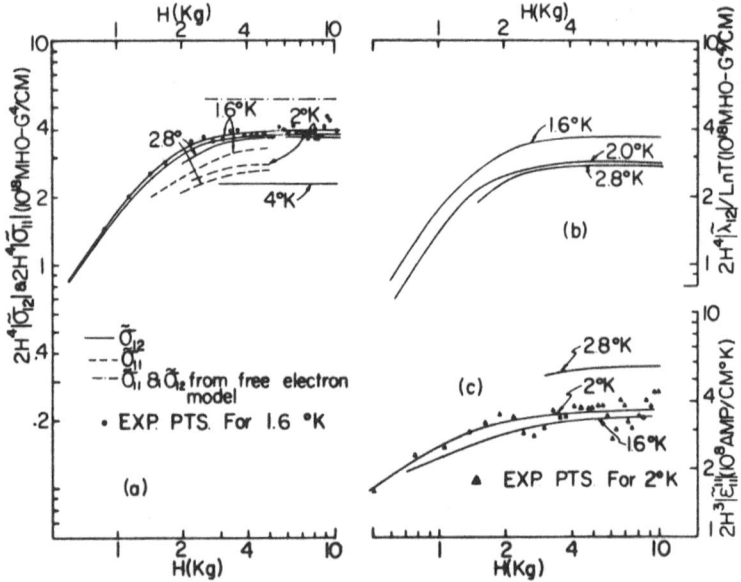

Fig. 3. The amplitudes of the magnetomorphic oscillations multiplied by appropriate powers of the field H are plotted against H to show field dependence. (a) $|\tilde{\sigma}_{12}|$ and $|\tilde{\sigma}_{11}|$ are shown to fall off as H^{-4} for high fields. (b) $|\tilde{\lambda}_{12}|$ falls off as H^{-4} for high fields. (c) $|\tilde{\varepsilon}''_{11}|$ falls off as H^{-3} for high fields.

† By use of a corrected value for the thickness, a value of $0.98 R_{f_0}$ was found.

(5) The precision of the $|\tilde{\sigma}_{11}|$ amplitudes is poor at large fields. Results shown in Fig. 4a indicate that the $|\tilde{\sigma}_{11}|$ amplitudes are slightly smaller than the $|\tilde{\sigma}_{12}|$ amplitudes by at most a factor of 25%. This is still in relatively good agreement with relation (d). The same discrepancy exists in $\tilde{\lambda}_{11}$ and $\tilde{\lambda}_{12}$. Oscillations in $|\tilde{\varepsilon}''_{12}|$ were very poor and no comparison could be made with $|\tilde{\varepsilon}''_{11}|$. Comparison of Figs. 3a and 3b shows excellent agreement with the Wiedemann–Franz law. The smaller values obtained for $\lambda(L_n T)$ are expected from the condition $\kappa' > \kappa$.

(6) The expected free-electron amplitude for $2H^4|\tilde{\sigma}_{12}|_0$ and similar quantities for σ_{11}, $\lambda_{11}(L_n T)^{-1}$, and $\lambda_{12}(L_n T)^{-1}$ would be $2H^4|\tilde{\sigma}_{12}|_0 = 5.5 \times 10^{18}$ G^4/Ω-cm for $e^{-\kappa} = 1$. This value is only slightly larger than the value obtained at 1.6°K.

(7) The only large discrepancy between these results and those expected for the free-electron case is in the amplitude of the $|\tilde{\varepsilon}''_{11}|$ oscillations. It is expected that $2H^3|\tilde{\varepsilon}''_{11}|_0 = 1.7 \times 10^8$ G^{3}A/cm-deg K at 1.6°K, while it is actually found that the amplitudes in $\tilde{\varepsilon}''_{11}$ are some 20 times larger than this. It is possible that multiplication of experimental errors, due to the tensor manipulation, is responsible for some or possibly all of the discrepancy. The matter is now being investigated.

The Lens-Shaped Fermi Surface in Cadmium

The relatively good fit of the experimental results of the free-electron theory indicates that the lens-shaped Fermi surface in the third Brillouin zone is responsible for the asymptotic effects. This can be concluded because the asymptotic oscillatory behavior is characteristic of apex properties of the Fermi surface in the hexagonal direction and the lens is known to nearly osculate the free-electron Fermi sphere in this direction.[8]

In the case of a lens which does not exhibit a quadratic energy distribution, the value $m^* v_z = (2\pi c/ea)^{-1} P$ would be characteristic of the cyclotron mass and velocity along the field for electrons at the apex of the lens. In this case, $m^* v_z$ can also be interpreted as $(m^* v_z)_{\text{apex}} = (1/2\pi)(\partial s/\partial p_z)_{\text{apex}}$, where s is the cross section of the lens cut by the plane $p_z = $ constant. If the lens is a surface of revolution, this quantity can be interpreted as the radius of curvature of the Fermi surface at the apex, i.e., $m^* v_z = R_f$. In the present case the results show that the radius of curvature $R_f = 1.07 R_{f_o}$† is slightly larger than the free-electron sphere. The lens model given by Ziman[9] and analyzed by Daniel and MacKinnon[8] would have $R_f < R_{f_o}$. The influence of the boundaries of the zones other than the (002) Brillouin plane would probably tend to modify Ziman's lens. The fact that the values obtained experimentally for the amplitude of $|\tilde{\sigma}_{12}|$ are only slightly smaller than the expected free-electron values tends to give qualitative agreement with the modification of the theory to fit the lens scheme. Calculations are under way to obtain this extension.

Acknowledgments

The authors wish to thank Dr. G. N. Rao and J. R. Long for assistance in making the measurements.

References

1. N. H. Zebouni, R. E. Hamburg, and H. J. Mackey, *Phys. Rev. Letters* **11**, 260, 1963.
2. C. G. Grenier, J. R. Long, J. M. Reynolds, and N. H. Zebouni, this volume, p. 802.
3. F. J. Blatt, *Phys. Rev.* **96**, 13, 1954.
4. E. H. Sondheimer, *Phys. Rev.* **80**, 401, 1952; *Advan. Phys.* **1**, 1, 1952.
5. G. Chambers, *Proc. Roy. Soc. (London) Ser. A* **202**, 378, 1950.
6. R. B. Dingle, *Proc. Roy. Soc. (London) Ser. A* **201**, 545, 1950.
7. D. K. C. MacDonald and K. Sarginson, *Proc. Roy. Soc. (London) Ser. A* **203**, 223, 1950.
8. M. R. Daniel and L. MacKinnon, *Phil. Mag.* **8**, 537, 1963.
9. J. M. Ziman, *Electrons and Phonons*, Oxford University Press, Fair Lawn, New Jersey (1960), p. 100.

† See footnote, preceding page.

THE THERMAL HALL EFFECT IN COPPER

S. G. Lipson

Magnetic Laboratory, Cavendish Laboratory
Cambridge, England

Introduction

In comparison to the interest given to electrical effects, very little interest has been shown in the thermal magnetoresistance and Righi–Leduc (thermal Hall) effect in metals. There is no *a priori* reason to suppose that the Wiedemann–Franz law relating the electronic contributions to the thermal and electrical conductivity tensors should not hold in a magnetic field, and the lattice contribution to the thermal conductivity should be negligible at 2°K; in fact, Azbel et al.[1] have stated explicitly that the Lorentz ratio for the transverse effects should be independent of Fermi surface and scattering mechanisms in the limit of high magnetic fields. There are a number of practical reasons, however, which weigh strongly in favor of the measurement of thermal effects. The first of these concerns noise levels; a typical sample carrying an electric current of a few amps will generate a transverse voltage of perhaps 10^{-7} V, which is not easy to measure accurately, especially when the magnetic field is not absolutely steady. A sample for thermal measurements can be considerably larger—about 5 mm square has been found satisfactory—and still generate temperature differences of 10^{-2} deg which can be measured reliably and accurately using carbon resistance thermometers. The effects of field fluctuations can be eliminated by using a thermal link between the sample and the thermometers, which can be placed outside the field. A second disadvantage of the electrical measurements is that the large electric current necessary may itself distort the applied field; the sample mentioned above could produce a field difference of about 100 G across it, and a corresponding alteration in direction of the total field of half a degree at 10 kG. This may be sufficient to smear out interesting details in the highly anisotropic magnetoresistance tensor of a single crystal.

It is for these reasons that we attempted to measure the features of the magneto-resistance and Hall effect in copper using the thermal technique, and we present here some of the results; they are not yet in any way conclusive.

Hall Coefficients in High Symmetry Directions

The values of the Hall coefficients in the three high symmetry directions (100), (110), and (111) are particularly interesting, as their electron-per-atom equivalents (Hall volumes) can be predicted accurately with very little knowledge of the Fermi surface dimensions. A geometrical calculation, involving subtraction of the numbers of electrons describing electron orbits (e.g., "belly" orbits in copper) from the number describing hole orbits (e.g., "dog's bone" orbits) turns out to involve only the following quantities:

1. The volume of the Brillouin zone (2 electrons per atom).
2. The volume of the Fermi surface (1 electron per atom).

3. The volume of the section of the zone which contains the hole orbits, itself involving the position and projected diameter of the Fermi surface necks only.

The neck diameter and shape have been carefully measured by Joseph and Thorsen,[2] and calculations based on their measurements lead to the following Hall volumes (all electron-like):

      (100)   0.54 electron per atom
      (110)   0.55 electron per atom
      (111)   0.08 electron per atom

The high-field limits of the Hall constants should be the reciprocals of these quantities.

The problems in measuring the constants in (100) and (110) are not great, and the accuracy depends mainly on the care taken in calibration and shielding from extraneous heat leaks to the thermometers. We have measured the constants and, using the free-electron Lorentz ratio, obtain the following values of the Hall volumes (electron-like):

      (100)   0.48 electron per atom
      (110)   0.40 electron per atom

with an estimated accuracy of $\pm 5\%$. The lattice contribution to the conductivity has been neglected, pending a reliable estimate of its magnitude.

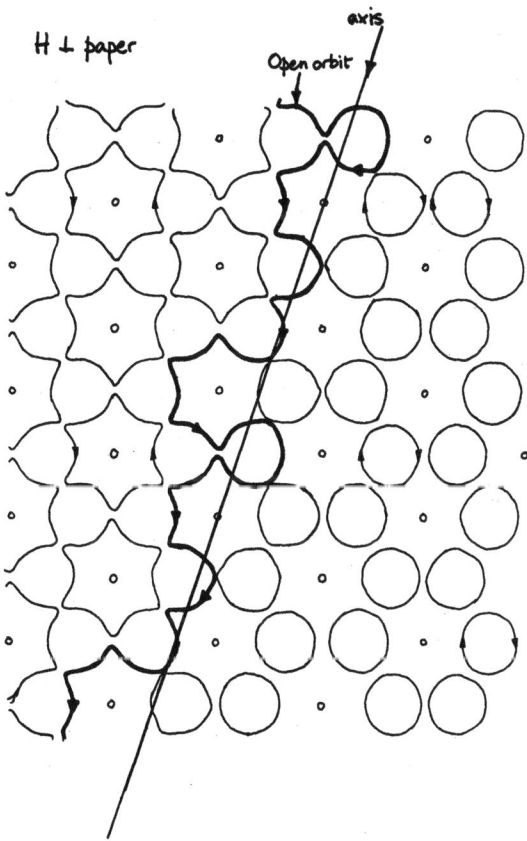

Fig. 1. Aperiodic open orbit near $\langle 111 \rangle$.

Aperiodic Open-Orbit Effects

The (111) experiment has produced some very interesting side effects which have effectively obscured the measurement originally intended. Near to the high-symmetry direction, aperiodic open orbits (Fig. 1) become possible, and the resultant conductivity has a peculiar angular dependence. In particular, the aperiodic open-orbit contribution to the transverse conductivity must vanish when the orbits conduct either exactly along or exactly across the sample. The direction of the open-orbit axis is a very sensitive function of the angular deviation from the high-symmetry direction, and we have studied the resultant resistivity, which has a quadratic field dependence, within a region extending about 1° in all directions from (111). Experimentally this has involved the addition of a tipping coil to the magnet, producing a field perpendicular to the main field and thus rotating it by angles up to 1°. Variation of the angle in the perpendicular direction is obtained by rotating the cryostat. The results, plotted out as a contour map, are shown in Fig. 2, and the extremely sensitive dependence on angle is clear. It is doubtful, in view of the effect of measuring current on the field direction, whether such variations would be so prominent in an electrical Hall effect experiment. Finally, having located the high-symmetry direction by plotting the contour map, we have measured the transverse temperature gradient there as a function of field. Although the small Hall volume

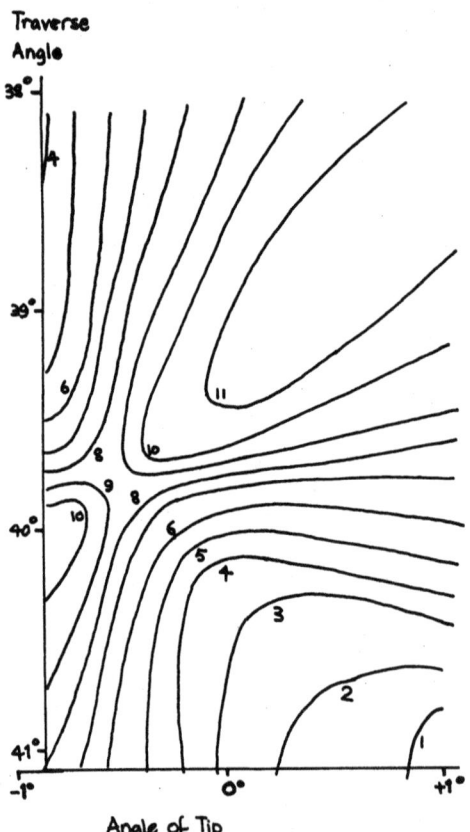

Fig. 2. Transverse temperature gradient in copper at 30 kG near ⟨111⟩.

calculated suggests a very large Hall coefficient, we do not think the coefficient observed is significantly different from zero. It should be noted that the magnetoresistance and Hall coefficient ought to saturate to their high-field values at about the same field, and in this direction the magnetoresistance is showing definite signs of saturation; it is not complete even though $\omega\tau$ is as large as 20 at the highest fields we have used.

Discussion

It is probably too early to discuss reasons for the disagreement, but there is a strong possibility that the Wiedemann–Franz law may not hold at very large fields, where $\hbar\omega$ is much greater than kT, in which case an electrical experiment would give different results. This possibility is shortly to be investigated.

References

1. M. Azbel, M. I. Kaganov, and I. M. Lifshitz, *Zh. Eksperim. i Teor. Fiz.* **32**, 1188, 1957. (*Soviet Phys. JETP (English Transl.)* **5**, 967, 1957.)
2. A. S. Joseph and A. C. Thorsen, *Phys. Rev.* **134**, A979, 1964.

THE SURFACE IMPEDANCE OF METALS IN A WEAK MAGNETIC FIELD*

J. F. Koch† and A. F. Kip
University of California
Berkeley, California

The surface impedance of a metal single crystal at low temperature and microwave frequency exhibits rapid variations with magnetic field. The resistance derivative maxima and minima that we observe (Fig. 1) occur in a range of magnetic fields below those necessary for the observation of cyclotron resonance. In this weak magnetic field regime the electron mean free path is insufficient for the completion of cyclotron orbits. The low field effect has been observed in tin,[1,2] aluminum, indium,[2] cadmium,[2] and tungsten,[3] but on the other hand, is notably absent in copper,[4] silver,[5] sodium,[6] and potassium.[6]

In a sense the experiment is a low field by-product of cyclotron resonance. The sample requirements and experimental arrangement are essentially those for the observation of cyclotron resonance. The sample forms the end wall of a cylindrical microwave cavity. In our apparatus the magnetic field can be rotated in the plane of the sample surface as well as in a plane perpendicular to the sample.[7] The RF currents are approximately linear and can be directed parallel or perpendicular to the magnetic field.

We have studied the absorption derivative signals extensively in single crystals of tin, indium, and aluminum. The crystals were cut with several different symmetry planes from single-crystal boules, and subsequently electropolished. The experiments were done at frequencies between 30 and 70 Gcps and at liquid-helium temperatures. We wish first to summarize our experimental results and observations.

1. As in Fig. 1, a typical signal consists of a number of distinct derivative maxima and minima at magnetic fields below 100 Oe. The most characteristic feature seems to be the pronounced derivative maxima, and for lack of a detailed theory we choose the position of these peaks as the relevant experimental parameter. The number of maxima observed, as well as their position, amplitude, and width varies with the orientation of H in the sample plane. The fact that the peak positions shift differently with orientation for the various peaks, as well as the fact that the range of angles over which they can be observed is different, lead us to believe that the signal is a superposition of individual, unrelated peaks. We do not observe the relationships between the fields of successive resistance maxima that Khaikin[2] has found. It is possible, though, that the observed peaks are fundamentals of series and that subsequent peaks at lower fields are not sufficiently resolved.

2. The effect shows a striking dependence on the sample surface in which it is observed. Figure 1 shows the effect with the magnetic field along the [001] axis

* Work supported by the Air Force Office of Scientific Research and the Advanced Research Projects Agency.
† Present address: University of Maryland, College Park, Maryland.

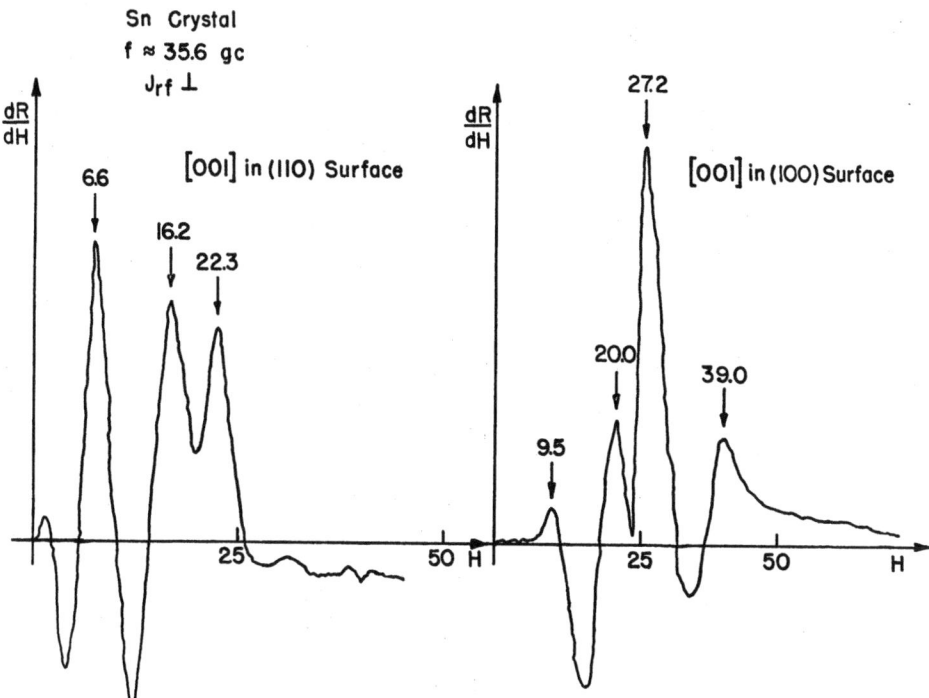

Fig. 1. Low-field effect observed with the magnetic field in the [001] direction in a tin crystal. The two traces show the signals observed in the (110) and (100) sample planes, respectively.

of tin, but observed in the (110) and (100) crystal planes. Relative amplitudes and peak positions are very different.

3. We find that the polarization of the RF current will affect the amplitude, but not the position of the derivative maxima. In fact, the dependence on the direction of the RF current relative to the magnetic field is the same as that observed for cyclotron resonance at higher fields. Along an axis of symmetry, where strong cyclotron resonance is observed only with the current perpendicular to the magnetic field, the low-field signals are observed with that mode only. For arbitrary directions of the magnetic field, signals can generally be seen with both current polarization modes, although amplitudes are different.

4. If the experimental frequency is increased, the peak positions shift to higher fields as $\omega_{RF}^{2/3}$. These measurements were carried out at several frequencies in the range of 30 to 70 Gcps. The measured slopes on a logarithmic plot of derivative maxima positions vs. frequency are 1.50 ± 0.05.

5. The low-field effect can readily be observed with the magnetic field tipped at arbitrarily large angles with respect to the sample surface. The signals observed are found to be almost identical with those for an equivalent direction in the same sample surface, except for a marked decrease in amplitude as the field is nearly perpendicular to the surface.

6. The preparation of a highly polished surface is essential to the observation of the effect. A tin crystal with an electropolished surface was exposed to dilute HCl vapor to lightly etch the surface. This crystal failed to show the low-field effect even though cyclotron resonance could still be seen.

7. The effect is independent of temperatures between 4.2°K and the superconducting transition temperature. Below this temperature and with the magnetic field less than the critical field, there is also observed structure in the microwave absorption. The dR/dH maxima observed in the superconducting state are different with respect to position in magnetic field, line shape, and number of peaks from those observed in the normal state. It is not certain if the superconducting signals are related to those in the normal state.

As a result of our observations, we are led to suggest that the observed signals are due to the interaction of conduction electrons, traversing the skin region on a segment of cyclotron orbit, with the RF electric field. While our simple model, and simple-minded calculation, can account for many of the features of our data, it is not without some serious shortcomings. Whereas we would quote our experimental results as plus or minus a few per cent, the probable error of the theoretical speculations is more or less 50%.

As in Fig. 2, we imagine an electron on a trajectory through the skin depth. It starts on this path at (Z_1, t_1) and is scattered at (Z_2, t_2), dissipating the energy that it has acquired in its transit of the skin region. Ignoring the complications of the anomalous skin effect (A.S.E.) to carry out our calculation, we write for the energy acquired by the electron[8]

$$\Delta E = \frac{e^2 E_0^2}{2m} \left[\int_{t_1}^{t_2} e^{-z(t)/\delta} \cos\left(\omega t - \frac{z(t)}{\delta} + \chi\right) dt \right]^2 \tag{1}$$

χ is a phase factor determining the phase of the electric field that the electron encounters at the top of its orbit. Since the major contribution to the integral arises from the section of the path immediately near the surface, we approximate the path

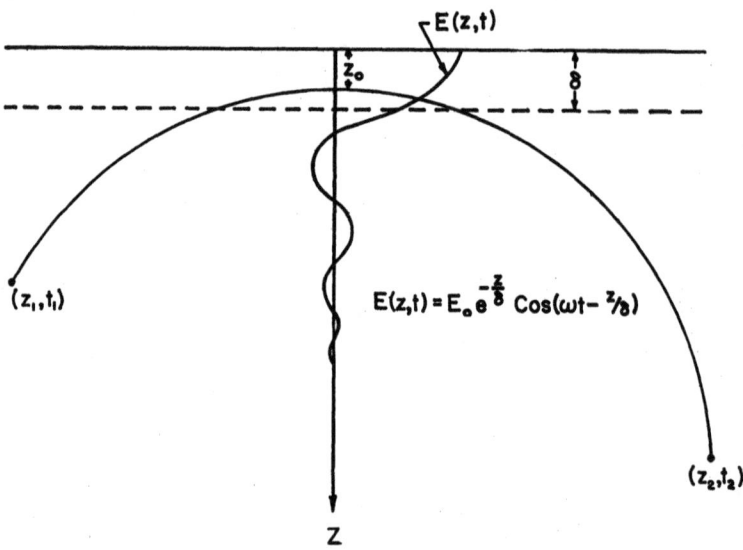

Fig. 2. Electron trajectory through the region of the skin depth. While it travels through the skin, the electron encounters an electric field $E(z, t)$ that changes both in time and space.

of the electron by a parabola

$$Z(t) = Z_0 + \tfrac{1}{2}\frac{v_F^2}{R}t^2 \qquad (2)$$

where R is the radius of curvature of the path and v_F the velocity of the electron for the segment of the path in the skin. For all but the very lowest fields ($H \ll 1$ G), the mean free path is very much longer than the section of the orbit that lies in the skin, so that each trajectory can be considered to start far away from the surface, and likewise to end far away. We extend the limits of integration to $\pm \infty$ and average over all values of χ from 0 to 2π to find the average energy absorbed per trajectory passing through Z_0

$$\overline{\Delta E} = \frac{\pi}{2\sqrt{2}}\frac{eE_0^2}{m}\frac{\hbar c \rho \delta}{v_F^2 H} e^{-(2Z_0/\delta)} e^{-(\hbar c \rho \delta \omega^2 / 2 e v_F^2 H)} \qquad (3)$$

ρ is the radius of curvature of the Fermi surface at the point where the electron is going through the skin. The total power absorbed from the microwave field will be the product of $\overline{\Delta E}$ with the number of electrons traversing an orbit through Z_0 per unit time, summed over all values of Z_0. Since this number of electrons is directly proportional to H, we must have for the power absorbed as a function of the magnetic field

$$P(H) = \text{constant} \cdot e^{-(\hbar c/2e)(\omega^2 \delta \rho / H v_F^2)} \qquad (4)$$

The absorption derivative has a maximum when

$$H = \frac{\hbar c}{4e}\omega^2 \delta \frac{\rho}{v_F^2} \qquad (5)$$

This expression does give the correct magnitude for H. With $\omega = 10^{11}$ sec^{-1}, $\delta = 10^{-5}$ cm, $\rho = 10^8$ cm^{-1}, and $v_F = 10^8$ cm/sec, we would estimate the required magnetic field as about 20 Oe. This value is in excellent agreement with our observations.

It is also possible to arrive at our criterion for the occurrence of a derivative peak by the following simple argument: The time that the electron spends in the surface is limited by the curvature of its trajectory in the magnetic field, rather than the mean free path. The variation in absorption is then associated with a variation of the time which the electron spends in the surface. To have maximum effect on the surface impedance the electron should remain in the surface $\tfrac{1}{2}$ RF cycle. For a longer time the electron would experience a reversal of the field and energy acquired would decrease; for shorter times it would not be exposed to the field sufficiently long. Except for an insignificant difference in the numerical factor, this argument leads to equation (5).

The value of ρ/v_F^2 will vary widely over the Fermi surface. We expect a signal for all those points on the Fermi surface for which this parameter has a stationary value. For the many sheets of the Fermi surface of the multivalent metals there should be several such points, to account for the many derivative maxima that are observed.

The explanations for the anisotropy of the signals and the polarization dependence are immediately obvious in terms of our model. Since we are observing an effect due to a point on the Fermi surface, the dependence of the signals on the crystal surface is expected. The frequency scaling leaves some doubt. With $H \propto \omega^2 \delta$, and $\delta \propto \omega^{-\frac{1}{2}}$ as in the normal skin effect, the $\frac{3}{2}$ scaling that is observed would be immediately obvious, but, in fact, such metal samples in which the effect is observed are strictly in the A.S.E. regime, and accordingly, at least in the absence of a magnetic field, $\delta \propto \omega^{-\frac{1}{3}}$. This would require a $\frac{5}{3}$ exponent for the scaling results.

Another, possibly more serious, shortcoming of our model is its failure to adequately explain the tipping result. One would expect that when the field is tipped to an angle θ, that the effective skin depth becomes $\delta/\cos\theta$ (at least for those electrons moving in a plane perpendicular to H) and that the peaks should shift correspondingly. We observe no significant shift.

The questions raised by the failure of the model to account for the tipping result and the frequency scaling point out the need for a more detailed calculation. In particular, the calculation should be done for the A.S.E. and should also include the effects of the magnetic field on the skin depth. In conclusion, we would like to note that our calculation is similar to that for the "RF kick" that the electron gets in its traversal of the skin in cyclotron resonance.[8,9] However, at low magnetic fields it is not possible to neglect the time variation of the electric field along the trajectory through the skin, as is done for the cyclotron resonance calculation. For long mean free path or high experimental frequency, it should be possible to observe cyclotron resonance at such low fields so that it coincides with the low-field effect. Under those circumstances there should be a variation in the amplitude of the resonance, such that the envelope of the resonance is the low-field effect signal.

References

1. A. F. Kip, et al., *Phys. Rev.* **108**, 494, 1957.
2. M. S. Khaikin, *Proceedings of the Eighth International Conference on Low Temperature Physics*, University of Toronto Press, Toronto, Canada (1961); *Zh. Eksperim. i Teor. Fiz.* **39**, 152, 1961.
3. E. Fawcett and W. M. Walsh, Jr., *Phys. Rev. Letters* **8**, 476, 1962.
4. A. F. Kip, D. N. Langenberg, and T. W. Moore, *Phys. Rev.* **124**, 359, 1961.
5. D. Howard, (to be published in *Phys. Rev.*).
6. C. C. Grimes and A. F. Kip, *Phys. Rev.* **132**, 1991, 1963.
7. J. F. Koch and A. F. Kip, *Phys. Rev. Letters*, **8**, 473, 1962.
8. M. Azbel and E. A. Kaner, *Soviet Phys. JETP (English Transl.)* **5**, 730, 1957.
9. A. B. Pippard, *Rept. Progr. Phys.* **23**, 176, 1960.

A RADIO FREQUENCY SIZE EFFECT IN GALLIUM*

D. M. Sparlin† and D. S. Schreiber

Northwestern University
Evanston, Illinois

The effects observed when the surface impedance of a metal in a magnetic field is measured at relatively low radio frequencies have been shown to be effective tools for the study of the Fermi surface of a number of metals.[1-3] Perhaps the most direct results of the various experiments were obtained by Gantmakher in measurements of size-dependent surface impedance anomalies in thin samples.[2]

When the mean free path is sufficiently long, closed electron orbits should give rise to a singularity in the surface impedance at the field value, H_i, for which they just fit within the boundaries of a thin parallel plate sample (H lying in the plane of the sample). The equation of motion

$$\dot{\mathbf{p}} = -(e/c)\mathbf{v} \times \mathbf{H}$$

shows that the orbits in momentum space and in real space are similar and are rotated relative to each other by 90°. The width of the electron orbit in momentum space, $2p_i$, can thus be related to the field value H_i and the sample thickness d by the equation

$$2p_i = (e/c)\, dH_i$$

Experimental Details

Three single-crystal plates, 0.165 mm × 8 mm × 8 mm in dimension, were grown from 99.999% gallium with the minor dimension in the **a**, **b**, and **c** direction, respectively. The crystals were grown between two Pyrex plates separated by a single thickness of Scotch No. 33 electrical tape. A $\frac{1}{16}$-in.-square cross section crystal grown simultaneously with each thin sample was used to determine a residual resistance ratio (38,000 for all samples).

The surface impedance of a sample was monitored through use of a standard NMR apparatus using a Pound–Knight marginal oscillator[4] as the detecting element. This oscillator was used with a radio frequency ranging between 10 and 15 Mcps and a field modulation frequency of 100 cps. The 100-cps modulation amplitude never exceeded 5 G.

The samples were sandwiched between two 8 mm × 8 mm pieces of glass microscope cover slide and inserted into the coil of the oscillator. The polarization of the RF field was initially set parallel to one of the two crystallographic directions in the plane of the sample and was varied by removal of the sample, rotation by

---

* Supported by ARPA and the National Science Foundation.
† Present address: Western Reserve University, Cleveland, Ohio.

90°, and reinsertion into the coil. The sample was held firmly in place at low temperatures with silicone stopcock grease.

The magnetic field calibration for this experiment relied on the accuracy of the Varian field dial system; the field dial was checked at three points for linearity and absolute value using a Rawson field probe. Fields lower than 100 G were inaccessible in this experiment.

Results

A plot of dR/dH vs. H was obtained at two degree intervals in the ac, ab, and bc planes for fields between 100 and 1000 G. The RF polarization was directed along each of the principal axes in the plane in turn, making a total of six separate sample orientations investigated in this work. Each of the curves was inspected for singularities of width $\Delta H/H \approx$ (skin depth/sample thickness).

A "typical" data curve is shown as Fig. 1 in which size effect signals below about 1 kG, quantum oscillations* in the conductivity of the Shubnikov–de Haas variety between 2 and 4 kG, and a NMR signal from the protons in the coil insulation and holder at 3 kG may be seen. While this variety of signals is of interest, only the size effect signals below 1000 G have been analyzed in detail.

Figure 2 presents the values of H_i at the singularities in dR/dH plotted in k-space ($\mathbf{k} = \mathbf{p}/\hbar = (ed/c\hbar)H_i$). The results are at the same time both promising and disappointing. Many line segments are formed by the data points, but only in a few cases is it possible to follow a signal over the entire plane.

The elliptical cross section mapped out in the bc plane is the only section which could be observed continuously with either **b** or **c** polarization of the RF field.

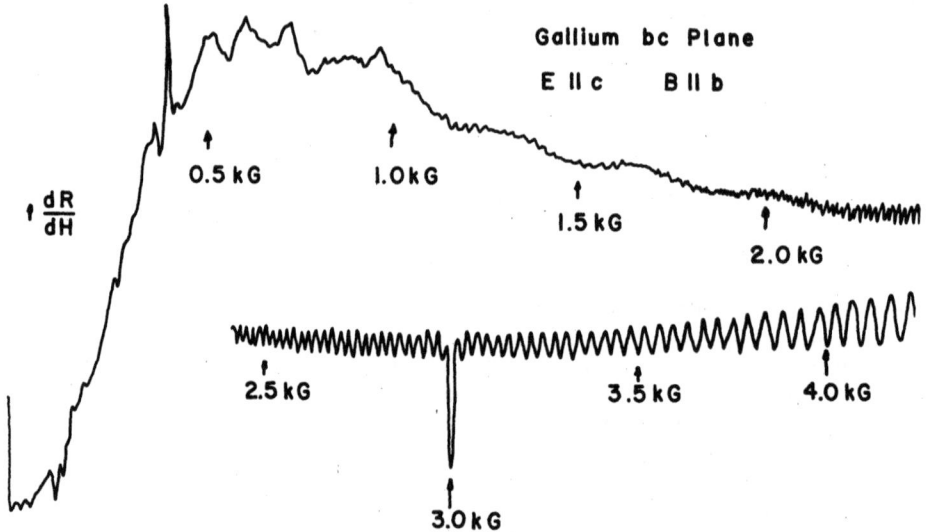

Fig. 1. "Typical" recording of dR/dH vs. H for gallium bc plane showing size effect signals, quantum oscillations, and proton NMR signal.

---

* Periods observed with $B \| b$ were $29.3 \times 10^{-7}\,G^{-1}$, $20.7 \times 10^{-7}\,G^{-1}$, $\sim 1 \times 10^{-7}\,G^{-1}$, and $\sim 0.5 \times 10^{-7}\,G^{-1}$.

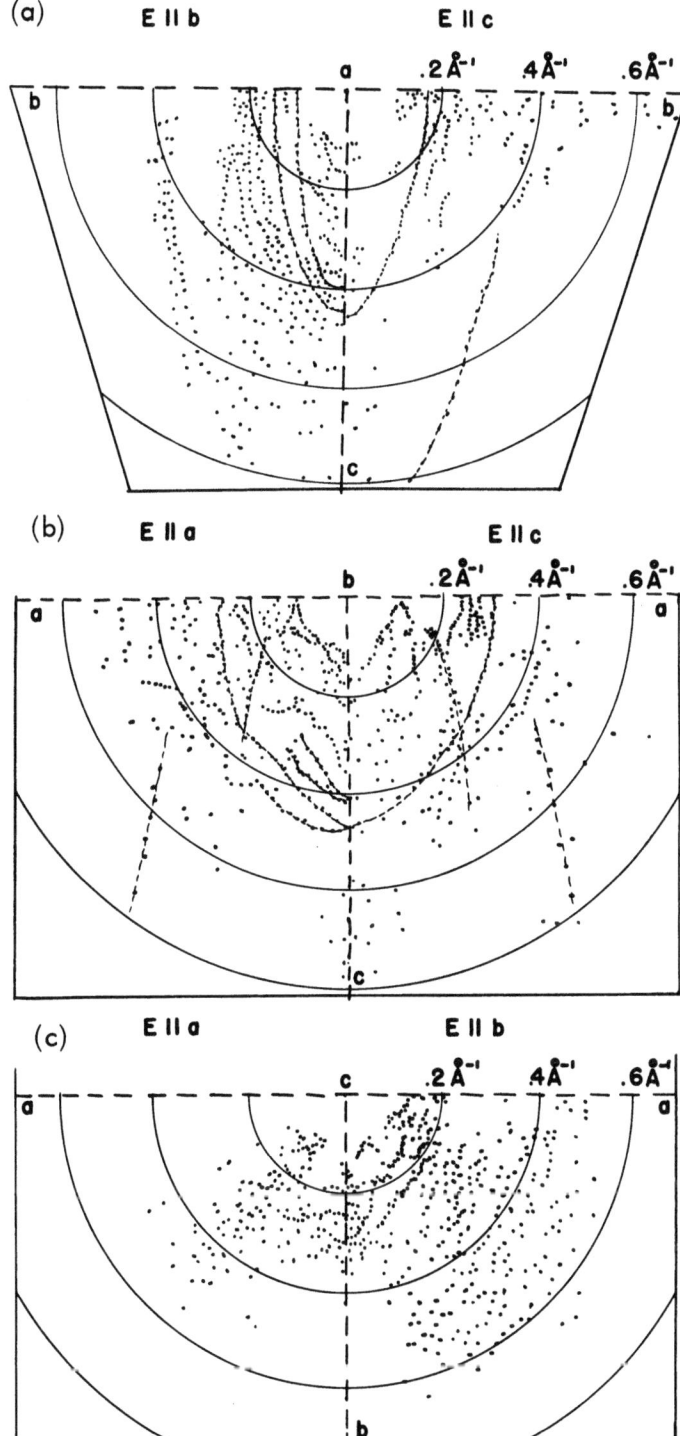

Fig. 2. Plot of the observed singularities in dR/dH in k space for $d = 0.165$ mm. (a) bc plane; (b) ac plane; (c) ab plane.

The area was determined with a planimeter and was found to be 0.207 Å^{-2} for the $\mathbf{E} \parallel \mathbf{b}$ polarization and 0.216 Å^{-2} for the $\mathbf{E} \parallel \mathbf{c}$ case. A second ellipse whose area was 0.134 Å^{-2} was mapped out for $\mathbf{E} \parallel \mathbf{b}$. These two elliptical cross sections have corresponding de Haas–van Alphen periods of 0.453×10^{-7} G^{-1} (average) for the larger and 0.716×10^{-7} G^{-1} for the smaller. These two periods have been observed for $\mathbf{H}$ along the $\mathbf{a}$ axis in a previous de Haas–van Alphen study of gallium.[5]

The only complete cross section mapped out in the ac plane was observed with $\mathbf{E} \parallel \mathbf{c}$. The area of this section is 0.358 Å^{-2}, corresponding to a de Haas–van Alphen period of 0.267×10^{-7} G^{-1}. This cross section was also observed in the de Haas–van Alphen study.[5] It is to be noted that the dimension of this section along the $\mathbf{c}$ direction in the ac plane agrees with that observed in the bc plane.

It is clear from the experience gained by the authors in this initial work that future field sweeps must be taken at one degree intervals in order to sort out the loci of points more completely.

References

1. M. S. Khaikin, *Soviet Phys. JETP (English Transl.)* **16**, 42, 1963.
2. V. F. Gantmakher, *Soviet Phys. JETP (English Transl.)* **17**, 549, 1963.
3. E. P. Vol'skii, *Soviet Phys. JETP (English Transl.)* **19**, 89, 1964.
4. R. V. Pound and W. D. Knight, *Rev. Sci. Instr.* **21**, 219, 1950.
5. J. H. Condon, *Bull. Am. Phys. Soc.* **9**, 239, 1964.

THE ELECTRONIC MEAN FREE PATH AND THE AREA OF THE FERMI SURFACE IN ALUMINUM AND INDIUM

G. Brändli, P. Cotti, E. M. Fryer,* and J. L. Olsen

Institut für kalorische Apparate und Kältetechnik, Swiss Federal Institute of Technology Zürich, Switzerland

There is a simple relation connecting the electronic mean free path in a metal and the area of the Fermi surface. This is found by considering the normal expression for the conductivity σ of a metal. This may be transformed into

$$\frac{\sigma}{\bar{l}} = \frac{e^2}{12\pi^3 \hbar} S \qquad (1)$$

where S is the total area of the Fermi surface, and $\bar{l}$ is a mean free path average defined by

$$\bar{l} = \frac{1}{S} \iint dS \tau(\mathbf{k}) \cdot |v(\mathbf{k})| \qquad (2)$$

The integral is taken over the free parts of the Fermi surface. $\tau(k)$ and $v(k)$ are the relaxation time and velocity of an electron of wave vector $\mathbf{k}$. Equation (1) holds for both single crystals and polycrystals of materials with cubic symmetry irrespective of the details of the anisotropy of $\tau(k)$ and $v(k)$ (Cotti[1]).

Clearly then the ratio of $\sigma/\bar{l}$ is a property of the metal investigated, and a knowledge of $\sigma/\bar{l}$ should permit us to estimate the area of the Fermi surface.

It is well known[2] that size effects in the thermal and electrical conductivity of metals may be used to determine the electronic mean free path. Until recently, such determinations of the free path have involved measurement of the resistivity of a range of samples of different thickness but with identical bulk free paths. This method has been used by various authors for a range of metals. Unfortunately, the requirement of identical bulk free paths for all specimens is difficult to fulfill and the results of such determination show very poor consistency. Sufficient information is, however, available to show that there is serious discrepancy between these measurements and the results expected from a free electron model or obtained from anomalous skin effect investigations. For this reason we have remeasured σ/l in indium and aluminum by an eddy current method suggested by Cotti.[1,3] In this method σ and l are found by comparing the DC resistivity ρ_F of a thin plate with a resistivity ρ_e defined by the decay time of eddy currents. This eliminates the uncertainties connected with the need for measurements on different samples.

However, the problem of estimating the fraction p of specular reflection at specimen boundaries remains. To investigate the effect of nonzero values of p,

---
* On leave from Pomona College, Claremont, California.

Cotti's calculation has been extended to $p \neq 0$. The results are plotted in Fig. 1, where ρ_F/ρ_0 and ρ_τ/ρ_0 are plotted as a function of l/d for various values of p. (ρ_0 is the bulk resistivity.) A detailed description of this calculation with numerical data will be published elsewhere. It can be seen from Fig. 1 that the eddy current size effect is much less dependent on p than the DC size effect. The value of σ/l derived by Cotti's method will therefore depend on the choice of p.

Values of σ/l obtained by this method, using four different values for the parameter p, are tabulated in Table I for 5 indium specimens and 2 aluminum specimens. It will be seen that the values of σ/l deduced from the data of Table I are only 50% of those expected theoretically, and that this discrepancy becomes larger as p departs from zero. Thus the largest Fermi surface area derivable from the present measurements corresponds to $p = 0$ and is 43 and 59% of the free electron Fermi surface for indium and aluminum, respectively.

In indium, σ/l, computed with $p = 0.5$ and 1, is seen to vary with specimen thickness. Since σ/l should be a constant, this may be taken as an indication that p must be less than 0.5 in this metal. The fact that the disagreement between size effect and other estimates of σ/l is smallest for $p = 0$ would suggest that this is the most probable value. This is, of course, in agreement with the generally accepted view that $p = 0$ (Sondheimer,[2] Chambers,[4] and Gaide and Wyder[5]).

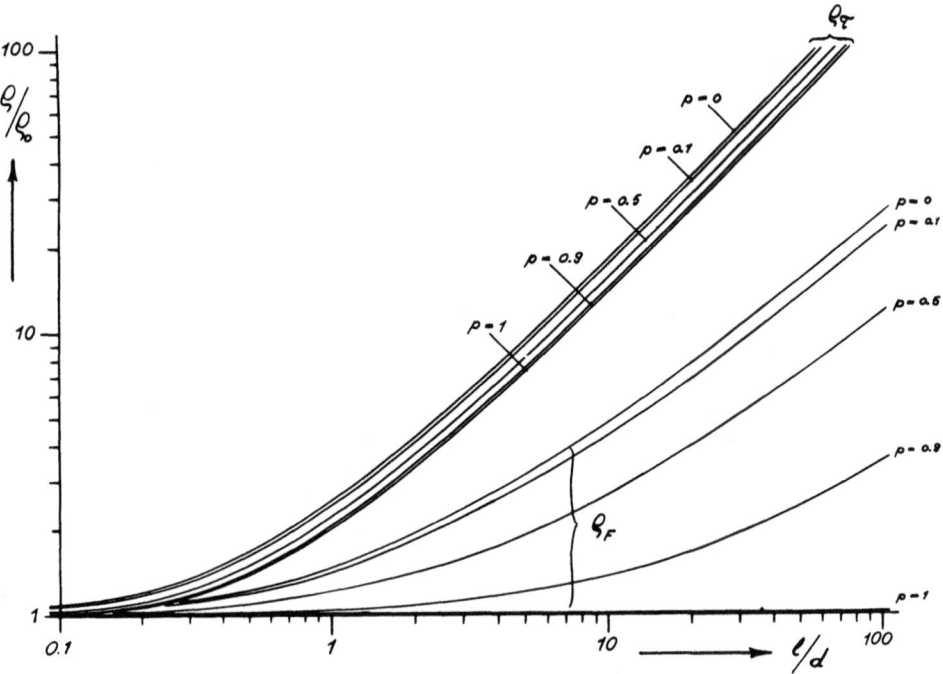

Fig. 1

Table I. σ/l (Indium at 3.4°K and Aluminum at 4.2°K)*

Probe	d, mm	$\sigma/l \times 10^{-10} \Omega^{-1}/cm^2$			
		$p = 0$	$p = 0.1$	$p = 0.5$	$p = 1$
Indium 1	0.47	7.52	7.24	6.21	5.23
Indium 2	0.35	7.57	7.30	6.54	6.05
Indium 3	0.17	7.30	7.09	6.66	6.21
Indium 4	0.11	7.57	7.40	7.00	6.45
Indium 5	0.077	7.40	7.35	6.85	6.45
Aluminum 1	0.193	14.9	14.5	12.5	11.1
Aluminum 2	0.106	15.2	14.6	12.7	11.2

*For free electrons, $\sigma/l = 18.5 \times 10^{10} \Omega^{-1}/cm^2$ in indium; $\sigma/l = 25.1 \times 10^{10} \Omega^{-1}/cm^2$ in aluminum.

There are very strong reasons for supposing that the area of the Fermi surface in both aluminum and indium is actually very close to that for a free electron model, and it is of interest to investigate the causes of the observed discrepancy. The most plausible explanation is found if it is assumed that the free path varies strongly over the Fermi surface. In this case the mean free path evaluated by comparing size effect investigations with the behavior calculated for a free electron model can no longer be regarded as the mean value defined by (2). This can be seen from the calculations of Sondheimer,[6] who assumed a Fermi surface consisting of two different uniaxial ellipsoids with different free paths. Sondheimer showed that for such a model the observed size effect in thin films may depend strongly on the orientation of the ellipsoids.

The simple case of a spherical two-band model has been discussed by one of us (Olsen[7]) for the case of thin wires. We consider that a fraction α_1 of the electrons has a free path l_1, and the remaining fraction α_2 has the free path l_2, where $l_1 > l_2$. The limiting form of Nordheim's relation $\rho = \rho_0(1 + l/d)$ now becomes

$$\rho = \rho_0[1 + (l_1/d)] \qquad (d \gg l_1) \qquad (3)$$

$$\rho = \rho_0[1 + (\alpha_1 l_1 + \alpha_2 l_2)/d] \qquad (l_2 \gg d) \qquad (4)$$

Clearly, work in the region of large d gives values for the free path very different from its average value $l = l_1\alpha_1 + l_2\alpha_2$ and tends to give a value of l close to the maximum value l_1. The resulting apparent Fermi surface area is the part of the total area where this value of l applies. An analogous effect may be expected in plates.

Wyder's[8] measurements on thin wires show no apparent change in l as a function of d up to $l/d \sim 10$. In terms of a two-band model, this means that while d is of the order of the observed l_1, it must still be much greater than l_2. We conclude then that the free paths for the unobserved parts of the Fermi surface must be several times shorter than those we observe.

The existence of such a spread of mean free paths at very low temperatures, where the scattering of electrons is due to impurities and lattice defects, may at first sight seem surprising, since it would appear that large-angle scattering should be the dominant mechanism for restoring the electron distribution to equilibrium. Small-angle scattering is by no means absent, however, and Klemens[9] has pointed out that such small-angle scattering is particularly effective in reducing the relaxation time for the strongly distorted parts of the Fermi surface in higher zones.

Taylor's work on copper has shown that this is so for the phonon case, where there are remarkable anisotropies in τ at low temperatures.

In our present two cases we find an apparent Fermi area somewhat smaller than the second zone area. We believe therefore that the free paths on the higher zone monsters and in regions close to the sharp edges of the second zone body must be very short. These areas therefore contribute negligibly to the electron transport phenomena.

The reason that anomalous skin effect measurements lead to values of the Fermi surface area close to the accepted one is that work so far has been carried out at frequencies which make the penetration depth shorter than the shortest free path on the Fermi surface. Investigations of the frequency dependence of the anomalous skin effect should lead to a spectrum of free paths.

References

1. P. Cotti, *Phys. Kondens. Materie* **3**, 40, 1964.
2. E. H. Sondheimer, *Phil. Mag. Suppl.* **1**, 1, 1952.
3. P. Cotti, *Phys. Letters* **4**, 114, 1963.
4. R. G. Chambers, *Proc. Roy. Soc. (London) Ser.* **215**, 418, 1952.
5. A. Gaide and P. Wyder, *Koninkl. Vlaam. Akad. Wetenschap.* **20**, 411, 1963.
6. E. H. Sondheimer, *J. Phys. Radium* **17**, 201, 1956.
7. J. L. Olsen, *Koninkl. Vlaam. Akad. Wetenschap.* **20**, 338, 1963.
8. P. Wyder, *Phys. Kondens. Materie* **3**, 263, 1965.
9. P. G. Klemens, private communication.

THE EFFECT OF TENSION ON THE FERMI SURFACES OF THE NOBLE METALS

D. Shoenberg and B. R. Watts

Royal Society Mond Laboratory
Cambridge, England

Introduction

An apparatus has been constructed to measure the effect of applying different uniaxial tensions to single metal crystals by observing the change in frequency of the de Haas–van Alphen oscillations and hence the change in area of the extreme cross sections of the Fermi surface. The magnetic field **H** used for measuring the de Haas–van Alphen effect is parallel to the direction of the tension, which in turn is along the axis of the cylindrical specimens. Experiments have been carried out so far on $\langle 111 \rangle$ specimens of the noble metals, since these exhibit two well-established frequencies which arise from the "belly" (a very high frequency) and the "neck" (a rather low frequency) of the Fermi surface. We will see that somewhat different techniques have to be used with high and low frequencies. However, before describing the details of measurements, the sizes of the effects to be expected will be discussed, because they bear directly upon the design of the apparatus. The amount of strain which may be produced in the specimens is limited to between 10^{-4} and 10^{-3} by the fact that only elastic effects are of primary interest in regard to the change in shape of the Fermi surface. If the lower limit of 10^{-4} is assumed to be the largest strain permissible, the corresponding changes in the Fermi surface would probably be of the same order (10^{-4}) for the large parts and usually something appreciably greater than this for the smaller parts. If $H = 50\,kG$, the phase of the belly oscillations is about 10^4 and that of the neck oscillations is about 5×10^2. Consequently a phase change of about 2π for the belly oscillations would be expected for maximum strain. It turns out in practice that owing to a much more rapid fractional change in area, the neck as well as the belly oscillations change phase by about 2π when the maximum tension is applied. Therefore the apparatus has been designed to measure phase changes in fixed **H** to an accuracy of better than 0.2π. When the specimen is being put under tension, three principal mechanisms which could produce spurious phase changes must be considered: (1) Variations in **H**; (2) longitudinal variations of specimen position in the solenoid, which has a nonuniform field profile; (3) bending of the specimen, and therefore changing its angle with respect to the magnetic field. It is straightforward to calculate the necessary stability in the three cases to prevent more than 0.2π phase change occurring.

1. If $H = 50\,kG$, the necessary field stability is $0.5\,G$ for the belly and $10\,G$ for the neck.
2. The specimen can be set to within 2 mm of the center of the solenoid, and the field profile is such that in these circumstances the maximum allowed longitudinal shifts of the specimen when applying the tension are 0.02 and 0.2 mm for the belly and neck, respectively.

3. The angular variation of frequency of the belly and neck allow a maximum angular shift of 0.03° and 0.3°, respectively, if the specimen axis is within 2° of the symmetry direction (as is most likely).

Experimental Technique

With such small effects to be expected, the way to measure the change in the de Haas–van Alphen frequency is to use a "steady field" method for detection, and measure the change in phase of the oscillations at constant field. The RF method of Shoenberg and Stiles[1] is used to pick up the oscillations, together with a 50 kG superconducting solenoid which can be used in the persistent mode when necessary to give very high field stability. Figure 1 shows the method of applying the tension to the specimens (of the order $1\frac{1}{2}$ mm diameter by 10 mm long), which are soldered into position. The overall contraction of the suspension when the tension is applied only shifts the specimen by 0.005 mm along the axis of the solenoid, which does not introduce any significant error for the belly or neck. The angular change of the specimen can be estimated or measured, as will be seen in the description of measuring the belly phase change, and it is unimportant for the neck, but it could be a serious source of error for the belly.

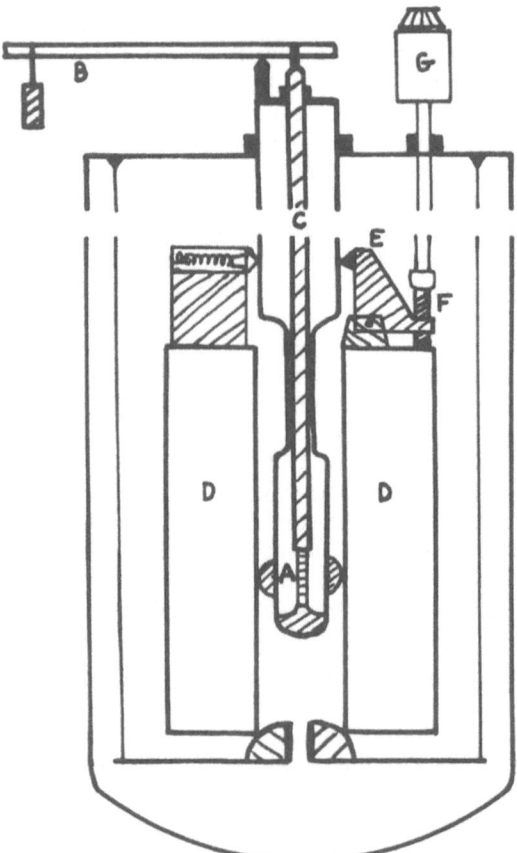

Fig. 1. Diagram of the principal mechanical parts of the apparatus. The tension is applied to the specimen A by means of the lever B and the rod C. The magnet D is rotated about an axis perpendicular to the plane of the diagram by the lever E (acting against a spring) operated by a screw F. This is controlled outside the dewar by the helipot G.

Owing to the relative insensitivity of the phase of the neck oscillations to these sources of error, providing the specimen is aligned nearly enough along $\langle 111 \rangle$, a simple method of determining the phase change is used. This consists of recording the de Haas–van Alphen oscillations on an X–Y recorder as shown in Fig. 2a, and running the field up through several oscillations for each successive value of the tension. Comparison of the phases for each tension can then be made at fixed values of x; a fixed value of x should correspond to a fixed value of $\mathbf{H}$, except when DC drifts in the recorder and other electrical circuits occur. These effects have been checked and found to be insignificant for the neck but not for the belly. The problem of field stability for the belly is most simply solved by using the solenoid in the persistent mode, and using a pair of auxiliary coils to slowly modulate the field over several periods of the belly oscillations. However, the results obtained were somewhat irreproducible and this was attributed to bending of the specimen.* In order to overcome the problem of bending, the magnet was designed so that it could rotate through $\pm 2°$ (to an accuracy of better than $0.04°$) about either of two perpendicular axes which are perpendicular to the direction of tension. This is done by

---

* This method is unsuitable for the neck because it is not possible to modulate the field sufficiently to cover several oscillations owing to the comparatively large periodicity in $\mathbf{H}$.

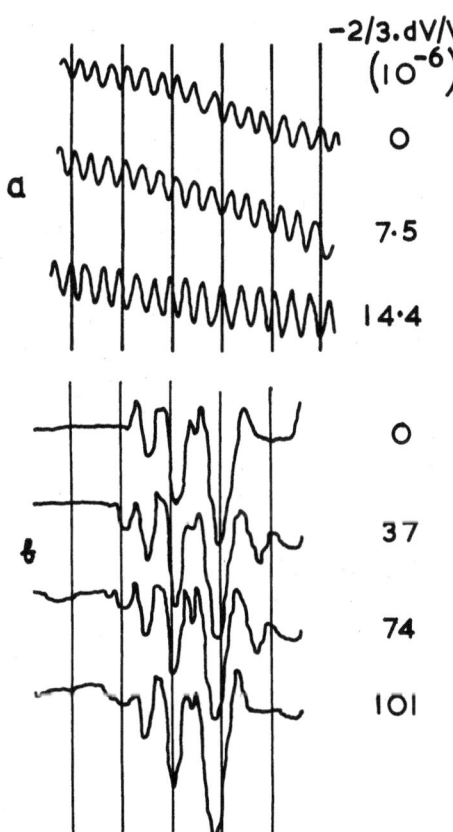

Fig. 2. (a) The neck oscillations in gold for three different strains. The lines of constant H help to show up the phase differences. (b) Rotation diagrams of the belly oscillations in silver. The phase change at the turning point is clearly very small, though there is a slight displacement of the curves corresponding to a bending of about 0.1°.

Table I. Ratios of (dA/A) **to** $(-2dV/3V)$

Element	Neck	Belly
Copper	-40 ± 10	—
Silver	—	0.0 ± 0.1
Gold	-60 ± 10	0.6 ± 0.1

a system of levers driven by two 10-turn helical potentiometers (see Fig. 1) so that the angle can be recorded as the x-coordinate on the X–Y recorder. The y-coordinate is the de Haas–van Alphen signal, and if the magnet is kept in the persistent mode, a rotation diagram of the kind shown in Fig. 2b is obtained. To a first approximation, directions of equal phase lie on circular cones round $\langle 111 \rangle$, and it is therefore possible to estimate the change in phase of the turning point in the rotation diagram and its change of position (if any) due to the bending of the specimen. If this is done for two rotations at right angles, it is possible to measure and allow for the effect of bending on the extreme phase.* It should be noted that this method may only be used for specimens oriented along a symmetry direction where the angular variation of frequency has a turning point.

Results

The results which have so far been obtained are given in Table I, where the fractional change of cross-sectional area of the Fermi surface (dA/A) with tension is expressed as a ratio to the fractional change in area $(-2dV/3V)$ of the free electron sphere due to the volume change dV of the specimen with tension. The effect of the tension is (1) to decrease the volume of the Brillouin zone, and (2) to make the Brillouin zone faces approach each other along the line of tension and recede from each other normal to the line of tension. This second effect, which distorts the Brillouin zone, should increase the distortion of the Fermi surface from sphericity near the "end" faces of the zone which approach the Fermi surface and reduce the distortion of the Fermi surface in regions which are nearer the "lateral" zone faces which recede from the Fermi surface. Consequently, the large negative values given for the necks in Table I, corresponding to a rapid increase in area, are attributed to the effect of the approaching zone faces. The values of less than unity for the bellies can be explained in the following way: Since the $\langle 111 \rangle$ extremum of the belly is a minimum with respect to $\mathbf{k_H}$ (the direction of $\mathbf{H}$ in k-space), the effect of reducing the distortion round this orbit (because it is nearer the lateral zone faces) would be to increase its area and so tend to compensate the overall volume reduction. This compensation appears to be approximately complete in silver.

Note Added in Proof

The entry for silver in Table I has since proved to be erroneous and should read 0.9 ± 0.2.

Reference

1. D. Shoenberg and P. J. Stiles, *Phys. Letters* **4**, 274, 1963.

---

* This method is also unsuitable for the neck because the angular separation of the oscillations is too large.

FERMI SURFACE OF SODIUM AND LITHIUM BY POSITRON ANNIHILATION*

J. J. Donaghy, A. T. Stewart, D. M. Rockmore, and J. H. Kusmiss

University of North Carolina
Chapel Hill, North Carolina

In recent years, the angular correlation of photons from the annihilation of positrons in metal single crystals has been used to obtain information about the shape of the Fermi surface. This paper presents an account of the results of such measurements on single crystals of sodium and lithium. The measurements yield information closely related to the areas of slices through the Fermi surface perpendicular to the (100), (110), and (111) directions.

Upon entering a metal, a positron is slowed by collisions with the lattice and electrons until it reaches thermal velocities. The thermalization time (10^{-12} sec)[1] is short compared to the lifetime (10^{-10} sec) of the positron, so that it can be assumed to be at rest when the annihilation takes place. Most of the annihilations result in the creation of two photons which are emitted at 180° to each other in the center-of-mass system of the two particles. The observed angle between the photons differs very slightly from 180° due to the velocity of the center-of-mass. For the velocities of interest, the small deviation of the photon directions from 180° is directly proportional to that component of momentum of the center-of-mass of the annihilating pair which is transverse to the photon direction; i.e., $\theta = p_z/mc$.

The apparatus used was similar to that described previously, but with improved angular resolution. The detector slits subtended an angle at the sample of 0.0002 rad. The coincidence counting rate for photon pairs is proportional to the number of such pairs for which the z-component of total momentum lies between p_z and $p_z + dp_z$, that is, the momentum distribution, $N(p_z)$, of annihilating electron-positron pairs is obtained directly as the angular correlation of annihilation photons.

The angular correlation data for a series of experiments on sodium are shown in Fig. 1. The center of the data has been taken as the 180° direction of the photons, and the data are symmetric about this direction. The broad portion of the curves (shown by the dashed line) arises mainly from annihilation with the core electrons of sodium. There is also a small contribution to this portion of the curve from annihilations in the thin oxide film on the surface of the sodium, as well as a slight background due to "chance" coincidences. The upper portion of each curve arises solely from annihilations with conduction electrons, and its shape for the different directions depends upon the shape of the Fermi surface.

The sodium data have been analyzed by subtracting the "background" of core annihilations and superimposing the data for the three directions. When this is done, it is found that the results for the three directions are the same within the experimental error of 1.5%

---

* Work supported by the National Science Foundation and the Advanced Research Projects Agency.

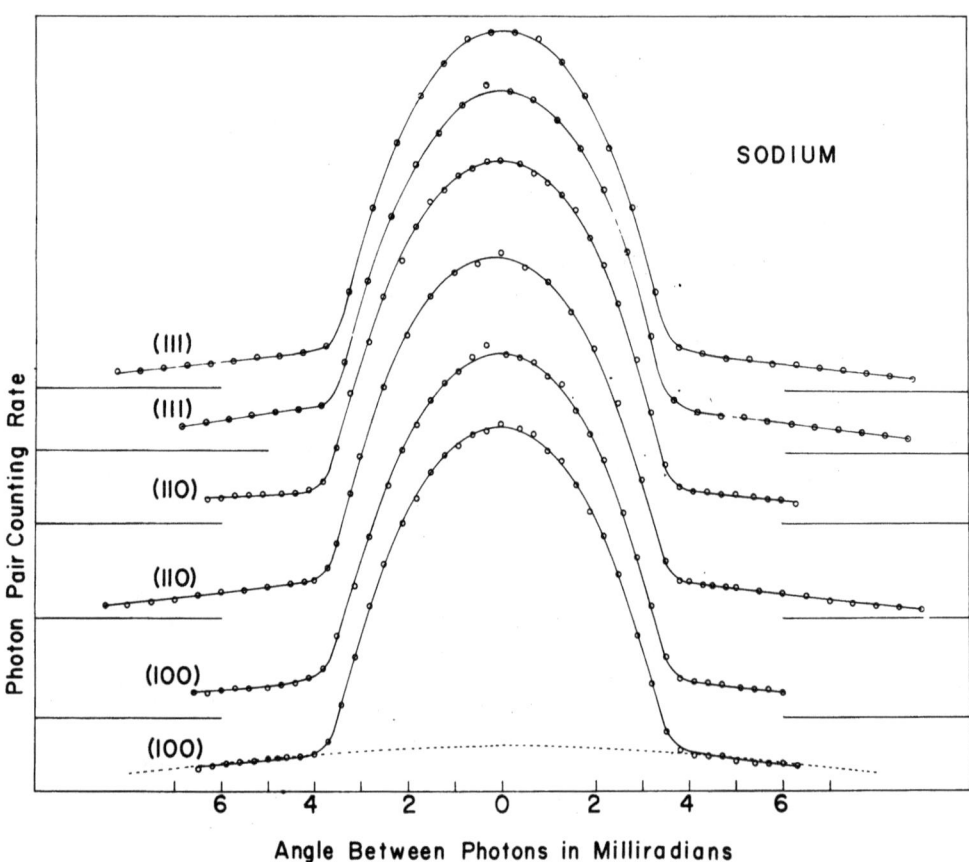

Fig. 1. The angular correlation of photons from positrons annihilating in oriented sodium crystals. The dashed line represents an approximation to background and core annihilation effects.

In addition to determining the shape of the Fermi surface in sodium, these data can yield some information concerning the velocity dependence of the annihilation cross section. If the cross section were independent of velocity, the sodium data, after subtraction of the background of core annihilations, would be parabolic. A calculation by Kahana[2] has shown that the annihilation probability increases with increasing electron velocity as

$$\varepsilon(\gamma) = a + b\gamma^2 + c\gamma^4 \qquad \gamma = k/k_f$$

where k_f is the free electron Fermi radius. The values of b/a and c/a which give a reasonable fit to the sodium data are 0.30 and 0.27, respectively. These values are slightly larger than the ones obtained from the calculation.

The angular correlation data for lithium, after subtraction of the background of core annihilations, are shown in Fig. 2. In this case there is a clear difference between the three directions which is well outside the experimental error of 1.5%. At $\theta = 0$, the difference between the (111) and (110) directions is about 4.5%; the difference between the (100) and (110) directions is about 2%. The zone boundary in the (110) directions is at 4.92 mrad. Thus, it can be seen directly from the

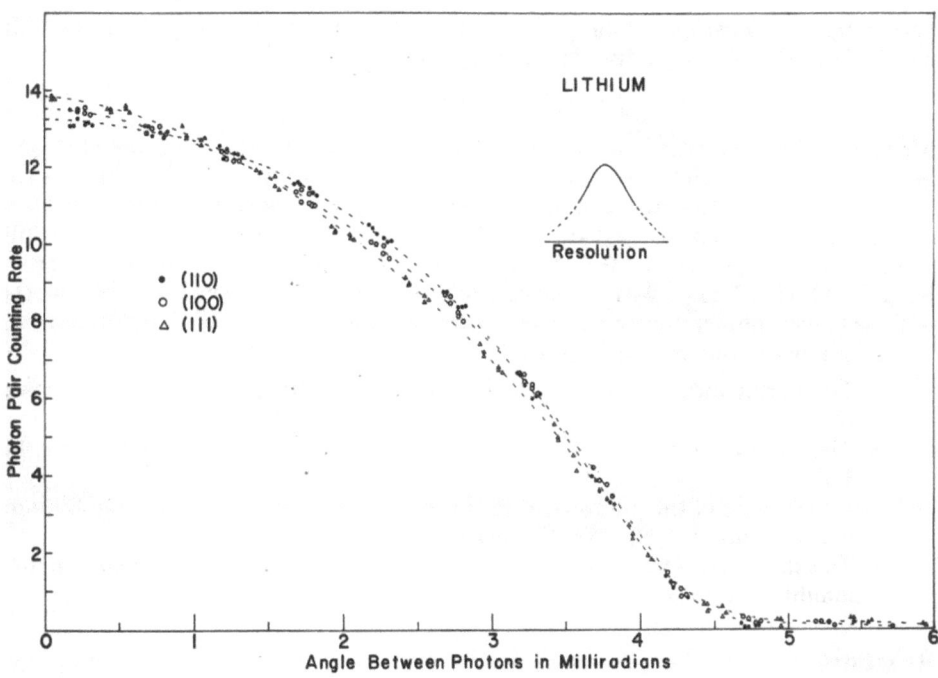

Fig. 2. The angular correlation data for lithium after folding about the center and subtracting a smoothed approximation to background and core effects. The dashed lines are visual fits to the data for the three directions.

(110) angular correlation data that the Fermi surface does not touch this zone boundary unless by an unrealistically narrow neck.

We have obtained estimates of the radii of the Fermi surface in the three principal directions by comparing the data with areas calculated from a phenomenological model of the Fermi surface. The model chosen consists of a sphere of radius r with bulges superimposed toward the nearest zone faces. The radius of the Fermi surface in the direction (β, ϕ) is given by this model as

$$k_F(\beta, \phi) = r + \Delta r = r + A \exp[-\alpha d(\beta, \phi)]$$

where β and ϕ are the polar and azimuthal angles, respectively, in spherical coordinates, and $d(\beta, \phi)$ is the distance to the nearest zone boundary from the point (β, ϕ) on the underlying sphere. The parameters r, A, and α are fixed by choosing values of the three principal radii to the Fermi surface—k_{100}, k_{110}, and k_{111}. Having determined these parameters, we calculated the areas of slices through the Fermi surface perpendicular to the three principal directions. These calculated areas have the same general appearance as the data in the interval from $\theta = 0$ to 2.5 mrad. For values of $\theta > 2.5$, the calculated curves differ appreciably from the experimental results. This discrepancy is probably due to the presence of higher momentum components in the wave functions of the conduction electrons.

Most of the annihilations take place in the interstitial region of the lattice. In this region, the electron wave function can be represented as a plane wave for most values of **k** inside the Fermi surface, so that most of the annihilations result in photon pairs with momentum $\mathbf{p} = \hbar\mathbf{k}$. At points near a zone boundary, the "smooth part" of the electron wave function becomes

$$\chi_\mathbf{k} = a_0(\mathbf{k})e^{i\mathbf{k}\cdot\mathbf{r}} + a_G(\mathbf{k})e^{i(\mathbf{k}-\mathbf{G})\cdot\mathbf{r}}$$

where $\mathbf{G}$ is the reciprocal lattice vector perpendicular to the zone boundary in question. This state gives rise to photon pairs of momentum $\hbar(\mathbf{k} - \mathbf{G})$ with relative probability $|a_G|^2$. The amplitude a_G has been estimated from the almost-free electron model for an energy gap of 3 eV. This model gives $|a_G|^2 = 0.16\,|a_0|^2$ for all values of $\mathbf{k}$ such that $(\mathbf{k}\cdot\mathbf{G})/G = k_f$; and $|a_G|^2 = 0.03\,|a_0|^2$ for all values of $\mathbf{k}$ satisfying $(\mathbf{k}\cdot\mathbf{G})/G = 0.8k_f$. When the calculated areas are corrected for the effects of these higher momentum components, a much improved fit to the data is obtained.

In summary, our conclusions are:

1. The Fermi surface of lithium is distorted, k_{110} being about 5 or 6% greater than k_{100}.
2. The Fermi surface of sodium is spherical within the experimental error of 1.5%.
3. An estimate of the intensity of the higher momentum components in lithium is consistent with the experimental data.
4. The data also yield some information on the velocity dependence of the annihilation cross section.

References

1. G. E. Lee-Whiting, *Phys. Rev.* **97**, 1557, 1955.
2. S. Kahana, *Phys. Rev.* **129**, 1622, 1963.

THE LORENZ NUMBER OF SINGLE CRYSTALS OF LEAD AND INDIUM IN TRANSVERSE MAGNETIC FIELDS

L. J. Challis, J. D. N. Cheeke, and P. Wyder*

Department of Physics, University of Nottingham
Nottingham, England

Since the theoretical work of Lifshitz[1] and his school, and the experimental measurements of Alekseevskii and Gaidukov,[1] the problem of the electrical magnetoresistance is solved in principle and there remains only the task of explaining the experimental results in terms of the topological structure of the Fermi surface. On the other hand, only a few measurements exist on the thermal magnetoresistance[2] of metals and, apart from the investigations of Grüneisen and co-workers[3] and de Nobel[4] in the liquid-hydrogen range and of Alers[5] and Wyder[6] in the liquid-helium range, no simultaneous measurements exist of resistivity and thermal conductivity at high fields in very pure metals where the heat is carried almost entirely by the electrons. Furthermore, most of the work that has been done was carried out on polycrystalline material. There is, however, a certain interest in investigations of this kind.[7] In this paper, we present preliminary results on the electrical and thermal magnetoresistance of a lead and an indium single crystal.

The crystals were supplied by Metals Research Ltd., Cambridge. They were grown from 99.9999% purity bulk material and were cut in the form of square rods about 40 mm long and 2 mm across. The specimen axis of the lead crystal was along the [111] direction and the indium crystal was randomly orientated, the specimen axis being about 19° off the [111] direction. The resistivity ratio of the specimens between room temperature and 4.2°K was for lead $\rho_{4.2}/\rho_{290} \sim 2.2 \times 10^{-4}$ and for indium $\rho_{4.2}/\rho_{290} \sim 0.47 \times 10^{-4}$. The zero-field values in the superconducting region were carefully extrapolated from the low-field measurements.

The measurements of the thermal conductivity were made using the conventional constant heat flow technique. The temperature differences were measured with carbon resistors ($\frac{1}{4}$W, 33 Ω L.A.B.), using a modified Wheatstone bridge arrangement with a phase sensitive detector operated at 1 Kcps. The temperature differences were calculated using a digital computer[8] after corrections had been made for the very small magnetoresistance of the thermometers. The vacuum can surrounding the specimen for the thermal measurements was removed for the electrical measurements so that the specimen was fully immersed in liquid helium. This reduced any heating of the specimen by the considerable currents used (up to 5 A). The voltage measurements were made using a galvanometer amplifier of the MacDonald type having a sensitivity of 10^{-9} V and the usual current reversing procedure was carried out to eliminate possible effects of Hall voltages. The field could be rotated in a horizontal plane and the specimen was mounted vertically;

---

* Present address: Swiss Federal Institute of Technology, Zurich, Switzerland.

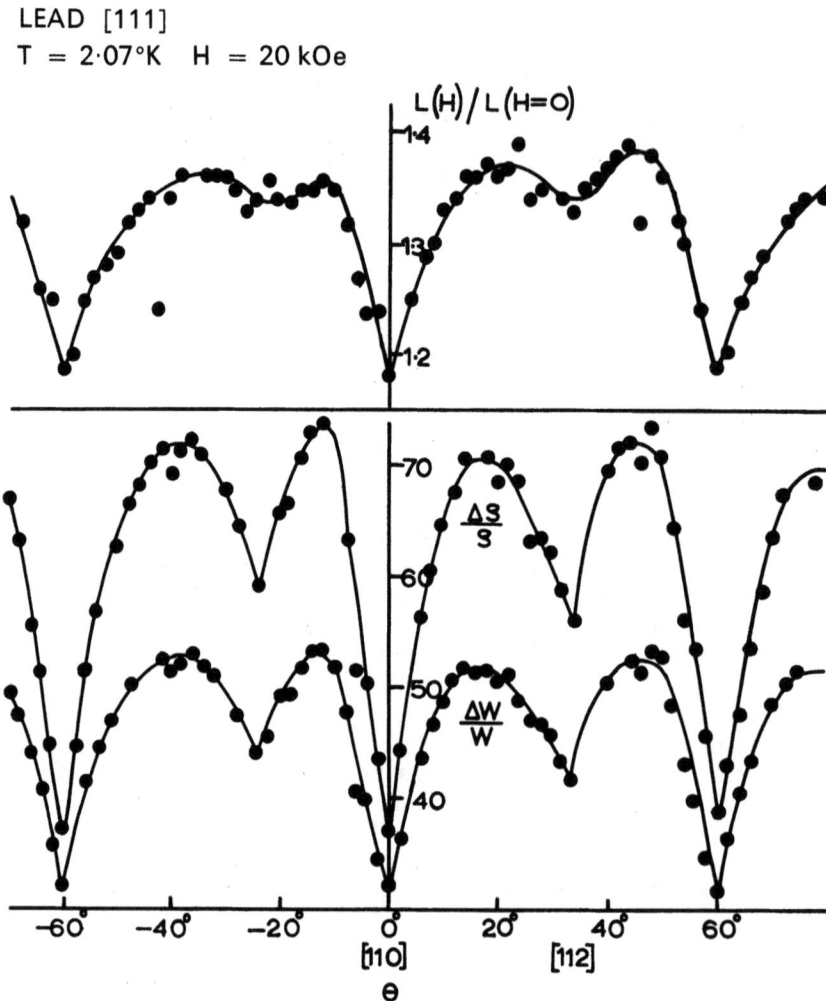

Fig. 1. The angular dependence of the Lorenz number, the electrical and the thermal resistance of a lead single crystal in a magnetic field.

the error in alignment could have been up to 5°. We note that since the thickness of the specimen was not vanishingly small compared to its length, the results in high fields could be in error due to the Hall effect,[9] in the electrical case and the Righi–Leduc effect[10] in the thermal case. It has been explained that this error was eliminated in the electrical case by reversing the current, but this was not possible in the thermal case. However, this error is thought to be small.

Metals can be divided into three classes by the behavior of their electrical resistivity in high magnetic fields, as indicated by Alekseevskii et al.[11]

1. If, for some direction of the magnetic field, H, there are no open sections of the Fermi surface but there are equal numbers of electrons and holes, $n_e = n_h$, then the resistance ρ increases quadratically with field $\rho \propto H^2$.

2. If for some direction of H there are no open sections of the Fermi surface but $n_e \neq n_h$, then the resistance reaches saturation in high magnetic fields.
3. If for some direction of H there is a layer of open sections of the Fermi surface then $\rho \propto H^2$. According to the careful investigation on the Fermi surface of lead made by Alekseevskii and Gaidukov[12] for the field directions used in our measurements on lead, we are observing the behavior of class 1. Borovik and Volotskaya[13] have shown that indium belongs to class 2.

The behavior of the thermal conductivity in a magnetic field can be described, to a very rough approximation, by making use of the Wiedemann–Franz law.[14] The components of the thermal conductivity can be transformed into the corresponding components of the electrical conductivity by dividing by $L_0 T$, where L_0 is the Lorenz number at 0°K defined by

$$L_0 = \frac{\rho_0}{(WT)_0} = \frac{\pi^2}{3}\left(\frac{k}{e}\right)^2.$$

(ρ is the electrical resistivity, W the thermal resistivity, and k, e, and T have their usual meanings.) This implies that the thermal conductivity should have the same behavior as the electrical conductivity. It can be seen in Fig. 1 that this is roughly true in the case of the angular dependence in 20 kOe. This approximate agreement can only be expected if the scattering of the electrons is elastic: the argument breaks down if the electrons are scattered inelastically by phonons since in this case the Wiedemann–Franz law is no longer valid.

The behavior of the thermal conductivity in a magnetic field has also been discussed in terms of a two-band model by Sondheimer and Wilson.[15] According to this model, at temperatures where the Wiedemann–Franz law is not obeyed, the Lorenz number should no longer be constant but should be field dependent. In general, $L(H, T)$ should be increased by a magnetic field. For measurements such that the behavior of the metal is that of class 1, as in the present measurements on lead, $L(H, T)$ should increase steadily from $L(0, T)$ to $L_0^2/L(0, T)$. For measurements corresponding to class 2, as in the measurements on indium, $L(H, T)$ should pass through a maximum as H is increased and then tend to $L(0, T)$ again. Azbel, Kaganov, and Lifshitz[16] have also examined the behavior of the Lorenz number in the high-field limit on the basis of the Lifshitz theory of galvanomagnetic effects in metals. They conclude that for the measurements in class 1 and class 2, the Lorenz number should reach a value independent of the field as in the two-band model although this value has not been determined.

Figure 1 shows the angular dependence of the electrical resistivity, the thermal resistivity and the Lorenz number of the lead specimen. An angular dependence of L is to be expected since L should be measured as a function of $\omega_c \tau$ (where ω_c is the cyclotron frequency and τ the relaxation time) which, while proportional to H, is known to vary over the Fermi surface. In contrast with this behavior, only a small change in the thermal resistance, $\sim 5\%$ of the total, was observed in a 180° rotation of the field around the indium specimen.

Figure 2 shows the Lorenz number of lead and indium as a function of magnetic field. At 4.1°K, where about 50% of the thermal resistivity of the lead specimen is due to inelastic electron–phonon scattering, the Lorenz number increases with field with no sign of saturation. At 1.9°K, where nearly all the phonons have been frozen out, and the Wiedemann–Franz law is therefore valid, the Lorenz number stays essentially constant in agreement with the theories. On the other hand,

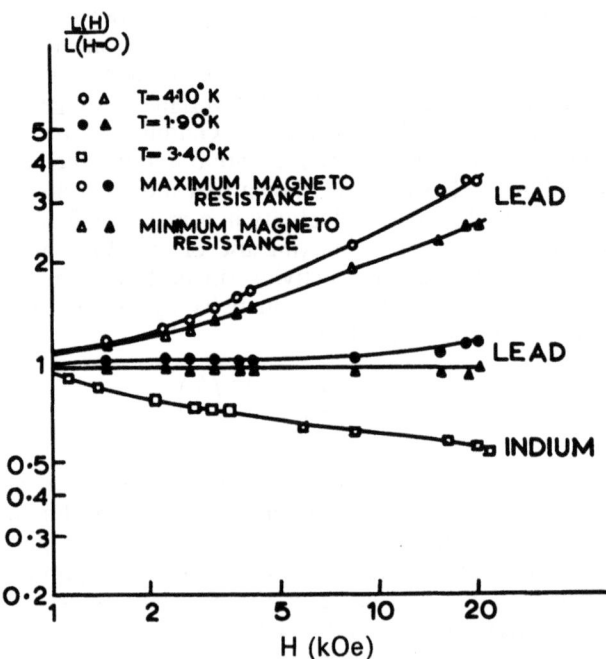

Fig. 2. The Lorenz number of lead and indium as a function of the magnetic field.

in indium at a temperature of 3.4°K, where about 80% of the thermal resistance is due to electron–phonon scattering, the Lorenz number decreases with increasing field. This was also observed in polycrystalline indium[6] and cannot be explained by either of the theories mentioned above.

Acknowledgments

We should like to thank Professor J. L. Olsen for drawing our attention to these problems. One of us (P.W.) is most grateful to O.E.C.D. for a Fellowship which made possible his stay in Nottingham.

References

1. For detailed references see, for example, W. A. Harrison and M. B. Webb (eds.). *The Fermi Surface*, John Wiley & Sons, Inc., New York (1960), p. 100.
2. K. Mendelssohn and H. M. Rosenberg, *Proc. Roy. Soc. (London) Ser. A* **218**, 190, 1953.
3. E. Grüneisen and H. Adenstedt, *Ann. Physik* **31**, (5), 714, 1938.
4. J. de Nobel, *Physica* **15**, 532, 1949.
5. P. B. Alers, *Phys. Rev.* **101**, 41, 1956.
6. P. Wyder, *Phys. Kondens. Materie*, **3**, 263, 1965.
7. J. L. Olsen, *Electron Transport in Metals*, Interscience Publishers, Inc., New York (1962).
8. L. J. Challis, *Cryogenics* **2**, 23, 1961.
9. P. Cotti, *Helv. Phys. Acta* **34**, 8, 1961.
10. J. Thorn and P. Wyder (to be published).

11. N. E. Alekseevskii, Yu, P. Gaidukov, I. M. Lifshitz, and V. G. Peschanskii, *Soviet Phys. JETP* (*English Transl.*) **12**, 837, 1961.
12. N. E. Alekseevskii and Yu. P. Gaidukov, *Soviet Phys. JETP* (*English Transl.*) **14**, 256, 1962.
13. E. S. Borovik and V. G. Volotskaya, *Soviet Phys. JETP* (*English Transl.*) **11**, 189, 1960.
14. A. B. Pippard, *Low-Temperature Physics*, Gordon & Breach Science Publishers, Inc., New York (1962).
15. A. H. Wilson, *Theory of Metals*, second edition, Cambridge University Press, Cambridge (1953).
16. M. Azbel', M. I. Kaganov, and I. M. Lifshitz, *Soviet Phys. JETP* (*English Transl.*) **5**, 967, 1957.

EFFECT OF COLLISIONS ON THE CONDUCTIVITY TENSOR OF A QUANTUM PLASMA IN A UNIFORM MAGNETIC FIELD

S. Tosima,* J. J. Quinn, and M. A. Lampert

RCA Laboratories
Princeton, New Jersey

The conductivity tensor $\sigma(\mathbf{q}, \omega)$ of a quantum plasma in the presence of a uniform magnetic field has been investigated recently by a number of authors.[1-3] In these papers the semiclassical theory of Cohen et al.[4] has been extended, in rather natural fashion, into the quantum realm. One difficulty which arises in making this extension is the treatment of the scattering of electrons by thermal phonons, point defects, and other lattice imperfections. In this paper we take collision effects into account in the simplest possible manner, namely, following the time-honored procedure of introducing a single phenomenological relaxation time τ.

In the usual semiclassical treatment of conductivity, collisions are taken into account by introduction of a term $[f(\mathbf{v}) - f_0(\mathbf{v})]/\tau$ into the Boltzmann equation. Here $\mathbf{v}$ is the electron velocity, $f(\mathbf{v})$ the Boltzmann distribution function under the influence of the imposed fields, $f_0(\mathbf{v})$ the thermal-equilibrium distribution function, and τ the phenomenological relaxation time. The natural generalization of this procedure in the density-matrix treatment of the quantum magnetoconductivity problem is the introduction of a term of the form $(\rho - \rho_0)/\tau$ into the equation of motion of the single-particle density matrix ρ.† If one naively takes the ρ_0 in this term, $\rho_0 = \rho_0(H_0)$, where H_0 is the Hamiltonian in the absence of the perturbing self-consistent field, then one obtains a physically nonsensical result for the DC conductivity tensor. In this paper we establish the correct form for the relaxation term, and present the results obtained by applying it in a calculation of the quantum conductivity tensor.

The equation of motion of the single-particle density matrix ρ, including the phenomenological damping term, is

$$\frac{\partial \rho}{\partial t} + \frac{\rho - \rho_0}{\tau} = -\frac{i}{\hbar}[H, \rho] \tag{1}$$

where $[A, B]$ denotes $AB - BA$, and H is the single-electron Hamiltonian in the presence of the electromagnetic field

$$H = H_K + H_P \tag{2}$$

---

* Present address: Laboratories RCA, Inc., CPO Box 219, Tokyo, Japan.
† The paper by Quinn and Rodriguez[3] studies ultrasonic attenuation for which problem there is an additional damping term not considered in the present paper, namely $(\tilde{\rho}_0 - \rho_0)/\tau$, where $\tilde{\rho}_0$ is the local value of the thermal-equilibrium density matrix referred to a system of coordinates which is moving with the local ionic velocity. The effects arising from this term have been studied in detail by earlier authors.

subscripts K and P denoting the kinetic and potential contributions, respectively.

$$H_K = \frac{1}{2m}\left[\mathbf{p} - \frac{e}{c}\mathbf{A}(\mathbf{r},t)\right]^2 \tag{3}$$

and

$$H_P = e\,\phi(\mathbf{r},t) \tag{4}$$

$\mathbf{A}(\mathbf{r},t)$ and $\phi(\mathbf{r},t)$ are the vector and scalar potentials, respectively, of the electromagnetic fields; the electric and magnetic fields, $\mathbf{E}$ and $\mathbf{B}$, respectively, are given by

$$\mathbf{E} = -\frac{1}{c}\frac{\partial \mathbf{A}}{\partial t} - \nabla\phi \tag{5}$$

$$\mathbf{B} = \nabla \times \mathbf{A} \tag{6}$$

The fundamental assumption of this paper is that the equilibrium distribution function ρ_0 appearing in (1) is $\rho_0[H_K(t)]$. We have written the kinetic Hamiltonian as $H_K(t)$ in order to emphasize the fact that it is explicitly time dependent. By $\rho_0[H_K(t)]$ is meant the equilibrium value of the single-particle density matrix appropriate to a system in which the Hamiltonian is held fixed for all time at the value which $H_K(t)$ has at time t.

The basic reason for this choice of ρ_0 stems from a consideration of gauge transformations. In a Hamiltonian description of dynamics, whether classical or quantum mechanical, the electromagnetic potentials rather than the fields appear in the formalism, but the potentials are not unique. If $\mathbf{A}, \phi$ yield the fields $\mathbf{E}, \mathbf{B}$, then it follows from (5) and (6) that so likewise do the potentials

$$\mathbf{A}' = \mathbf{A} + \nabla\chi \tag{7}$$

$$\phi' = \phi - \frac{1}{c}\frac{\partial\chi}{\partial t} \tag{8}$$

where χ is an arbitrary continuous function of $\mathbf{r}$ and t. Now ρ is not an invariant under the gauge transformation defined by (7) and (8), but transforms according to the equation

$$\rho' = \exp\left(\frac{ie}{\hbar c}\chi\right)\rho\exp\left(-\frac{ie}{\hbar c}\chi\right) \tag{9}$$

It is obvious from (1) that the choice of ρ_0 must be such that ρ_0 transforms precisely like ρ. The choice $\rho_0 = \rho_0(H_K)$ always satisfies this requirement because H_K itself transforms according to equation (9).

There is another cogent reason why the choice $\rho_0(H_K)$ and not $\rho_0(H_0)$ must be made in (1). Because of the coalescence of electronic energy levels into Landau levels, the quantum thermal-equilibrium properties of a plasma in a magnetic field are sensitive to the presence of the field, notable examples being the Landau diamagnetism and the de Haas–van Alphen effect. (Because the thermal-equilibrium Boltzmann distribution function is insensitive to the presence of static magnetic fields, this problem does not occur in the classical theory of conductivity.) However, the system cannot distinguish between unperturbed and perturbing magnetic forces, and hence both must be included in the thermal-equilibrium system toward which the system is trying to relax. This is obviously achieved by the choice $\rho_0(H_K)$.

We now consider a degenerate electron gas consisting of N-electrons together with a fixed, uniform background of positive charge providing electrical neutrality, contained in a cubic box of volume $\Omega = L^3$. A DC magnetic field of induction $\mathbf{B}_0$ and an electromagnetic disturbance that varies as $\exp(i\omega t - i\mathbf{q}\cdot\mathbf{r})$ act upon the system. The vector potential $\mathbf{A}(\mathbf{r}, t)$ is the sum of a time-independent part $\mathbf{A}_0 = (0, B_0 x, 0)$, which is responsible for the DC magnetic field, and a time-dependent part $\mathbf{A}_1(\mathbf{r}, t)$. The single-particle Hamiltonian is given by equation (2). The kinetic part of the Hamiltonian, which is given by equation (3), can be divided into a time-independent part

$$H_0 = \frac{1}{2m}\left(\mathbf{p} - \frac{e}{c}\mathbf{A}_0\right)^2 \tag{10}$$

and a time-dependent part

$$H_A = -\frac{e}{2c}(\mathbf{v}_0 \cdot \mathbf{A}_1(\mathbf{r}, t) + \mathbf{A}_1(\mathbf{r}, t) \cdot \mathbf{v}_0) + \frac{e^2}{2mc^2}A_1^2(\mathbf{r}, t) \tag{11}$$

In these equations, $\mathbf{v}_0 = m^{-1}(\mathbf{p} - (e/c)\mathbf{A}_0)$.

The equation of motion of the single-particle density matrix is equation (1), where of course $\rho_0 = \rho_0(H_K)$. To solve this equation, we let $\rho = \rho_0(H_K) + \rho_1$; substitution into equation (1) gives

$$\frac{\partial}{\partial t}\rho_0(H_K) + \frac{\partial}{\partial t}\rho_1 + \frac{\rho_1}{\tau} = -\frac{i}{\hbar}[H, \rho_1] - \frac{i}{\hbar}[H_P, \rho_0(H_K)] \tag{12}$$

One can see by inspection that the solution of equation (12) is

$$\rho_1(t) = -\int_{-\infty}^{t} dt'\, e^{-(t-t')/\tau} \left\{T\exp\left[-\frac{i}{\hbar}\int_{t'}^{t} H(s)\,ds\right]\right\} \left\{\frac{\partial}{\partial t'}\rho_0(H_K(t'))\right.$$

$$\left.+ \frac{i}{\hbar}[H_P(t'), \rho_0(H_K(t'))]\right\} \left\{T^{-1}\exp\left[\frac{i}{\hbar}\int_{t'}^{t} H(s)\,ds\right]\right\} \tag{13}$$

Here $H(t)$ is the full time-dependent Hamiltonian. The symbol T stands for the chronological operator, which orders all products of time-dependent operators such that later times stand to the left. By T operating on the exponential we mean that T operates on each term in the power series expansion of the exponential. The operator T^{-1} is the inverse of T, that is, it orders products of operators such that later times stand to the right.

The equilibrium density matrix $\rho_0(H_K)$ can formally be expanded in the form

$$\rho_0(H_K(t)) = \rho_0(H_0) + \sum_n Q_n(t)\,\phi_n(H_0) \tag{14}$$

where

$$Q_n(t) = \left\{1 - \int_0^{n\beta} d\lambda\, e^{-\lambda H_0} H_A(t) e^{\lambda H_0} + \cdots \right\} e^{-n\beta(\mu_0 - \mu)} - 1 \tag{15}$$

and

$$\phi_n(H_0) = (-1)^{n+1} \exp[-n\beta(H_0 - \mu_0)] \tag{16}$$

The unperturbed chemical potential μ_0 is determined by the condition Trace $\rho_0(H_0) = N$, while the instantaneous chemical potential μ is determined by the condition Trace $\rho_0(H_K) = N$. The current density at a point $\mathbf{r}_0$ at time t is given by

$$\mathbf{j}(\mathbf{r}_0, t) = \frac{e}{2} Tr\{[\mathbf{v}\delta(\mathbf{r} - \mathbf{r}_0) + \delta(\mathbf{r} - \mathbf{r}_0)\mathbf{v}][\rho_0(H_K(t)) + \rho_1(t)]\} \tag{17}$$

The velocity operator $\mathbf{v}$ is equal to $m^{-1}[\mathbf{p} - (e/c)\mathbf{A}_0 - (e/c)\mathbf{A}_1]$. By using (13)–(16) in (17), and linearizing one can obtain the following result for $\sigma(\mathbf{q}, \omega)$

$$\sigma(\mathbf{q}, \omega) = \frac{\omega_p^2}{4\pi i \omega}[\mathbf{1} + \mathbf{I}(\mathbf{q}, \omega)] \tag{18}$$

The symbol **1** stands for the unit tensor, and **I** is defined by

$$\mathbf{I}(\mathbf{q}, \omega) = \frac{m}{N} \sum_{v'v} \left[1 + \frac{\hbar\omega}{E_{v'} - E_v - \hbar(\omega - i/\tau)}\right]$$

$$\times \frac{\rho_0(E_{v'}) - \rho_0(E_v)}{E_{v'} - E_v} \langle v'|\mathbf{V}(\mathbf{q})|v\rangle \langle v'|\mathbf{V}(\mathbf{q})|v\rangle^* \tag{19}$$

The operator $\mathbf{V}(\mathbf{q})$ is defined by

$$\mathbf{V}(\mathbf{q}) = \tfrac{1}{2}(e^{i\mathbf{q}\cdot\mathbf{r}}\mathbf{v}_0 + \mathbf{v}_0 e^{i\mathbf{q}\cdot\mathbf{r}}) \tag{20}$$

where $\mathbf{v}_0 = m^{-1}[\mathbf{p} - (e/c)\mathbf{A}_0]$

A more detailed analysis of the effect of collisions on the conductivity tensor will be presented elsewhere.

References

1. P. S. Zyryanov and V. P. Kalashnikov, *Soviet Phys. JETP (English Transl.)* **14**, 799, 1962.
2. H. Stolz, *Phys. Stat. Sol.* **2**, 1029, 1962.
3. J. J. Quinn and S. Rodriguez, *Phys. Rev.* **128**, 2487, 1962.
4. M. H. Cohen, M. J. Harrison, and W. A. Harrison, *Phys. Rev.* **117**, 937, 1960.
5. J. J. Quinn and S. Rodriguez, *Phys. Rev.* **128**, 2494, 1962.

ELECTROMAGNETIC EXCITATION MODES IN PURE METALS IN HIGH MAGNETIC FIELDS, INCLUDING THEIR CONTRIBUTION TO THE SPECIFIC HEAT

Murray A. Lampert, John J. Quinn, and Soitiro Tosima*
RCA Laboratories
Princeton, New Jersey

The studies herein reported were undertaken in order to ascertain theoretically whether the electromagnetic modes of excitation (the so-called helicon modes) of a pure metal in a high magnetic field might make a measurable contribution to any low temperature properties of the metal. As an indicative example, we consider in particular their contribution to the specific heat. The theoretical analysis is confined to the free-electron gas which has a spherical Fermi surface, although the lines of generalization to real Fermi surfaces are indicated.

Theoretical studies of helicon propagation in metals are customarily based on the local theory of conductivity, which consists of replacing the full q-dependent conductivity tensor $\sigma(\mathbf{q}, \omega)$ corresponding to a disturbance of the form $\exp[i(\omega t - \mathbf{q} \cdot \mathbf{r})]$, by its q-independent limit, $\sigma(0, \omega)$. This procedure is justified under the conditions $q_y R < 1$ and $q_z l < 1$, where q_z, q_y are the components of $\mathbf{q}$ parallel and perpendicular, respectively, to the magnetic field, R is the cyclotron radius at the Fermi energy, $R = v_F/\omega_c$, with v_F the Fermi velocity and $\omega_c/2\pi$ the electron cyclotron frequency, and l is the electron mean free path at the Fermi surface, $l = v_F \tau$, with τ the electron collision time. With this approximation the dispersion equation relating $\mathbf{q}$ and ω for the helicon mode is[1]

$$\omega = \frac{c^2 \omega_c q_z q}{\omega_p^2 + c^2 q^2}\left(1 + i\frac{q}{\omega_c \tau q_z}\right) \qquad q = \sqrt{q_y^2 + q_z^2} \qquad (1)$$

where c is the velocity of light in vacuum and $\omega_p/2\pi$ is the electron plasma frequency.†

It is important to note that a damped mode qualifies as an independent, elementary excitation only if it is not strongly damped within one modal period; thus, only those helicon modes described by (1) for which $q \lesssim \omega_c \tau q_z$ are admissible for the specific heat calculation. In this sense, for a given q_z, the highest admissible q_y is $\omega_c \tau q_z$, approximately. The limit on q_z is set by the requirement that the propagation frequency lie below the threshold for Doppler-shifted cyclotron resonance, since heavy damping once again nullifies the concept of an independent excitation within the cyclotron resonance. (The situation concerning propagation "windows" between cyclotron resonances is touched upon briefly below.) This requirement is approximately: $q_z < q_{z,\,\text{max}} \simeq \omega_c/v_F$; correspondingly, $q_{y,\,\text{max}} \simeq \omega_c \tau q_{z,\,\text{max}}$.

---
* Present address: RCA Laboratories, Tokyo, Japan.
† The result quoted in Kaner and Skobov,[2] equation (7) of the translation, is in error in that the factor q/q_z is missing from the collision–damping term in parentheses in (1). The Landau damping in the same equation (2.7) is correct.

The electromagnetic specific heat must compete with the electronic specific heat and the phonon specific heat. For order-of-magnitude estimates it is sufficient to consider the respective heat contents per unit volume, that for electrons is $U_e \simeq kTN_e(T/T_F)$; at low temperatures that for phonons is $U_p \simeq 10^2 kTN_2(T/\theta_D)^3$, where N_e and N_2 are the number of electrons and atoms, respectively, per unit volume and T_F and θ_D are the Fermi and Debye temperatures, respectively. The crossover temperature T_x is given by $T_x/\theta_D \approx (\theta_D/T_F)^{\frac{1}{2}}/10 \approx 10^{-2}$ for $\theta_D \simeq 300°K$ and $kT_F \simeq 2.5\,eV$. Above T_x, U_p dominates U_e, and below T_x, conversely. The optimum temperature region over which to search for the electromagnetic contribution to the specific heat is the region around T_x, since the electromagnetic specific heat is not likely to vary faster with temperature than the phonon specific heat or slower with temperature than the electronic specific heat. A reasonable choice of ω_c is $kT_x/\hbar \approx 5 \times 10^{11}$ rad/sec for $T_x \simeq 3°K$. The corresponding magnetic field, for the free electron mass, is $B_0 \simeq 25{,}000$ Oe. With this choice the Dulong–Petit limit is reached at about T_x and the electromagnetic heat content is

$$T \approx T_x: \qquad U_{em} \approx \frac{kT}{4\pi^2} q_{y,\max}^2 q_{z,\max} \tag{2}$$

With the above choice of ω_c, $q_{z,\max} \simeq 5 \times 10^3$ cm^{-1} for $v_F \simeq 10^8$ cm/sec. If we take $\tau \simeq 4 \times 10^{-8}$ sec, as might be achieved with one or two metals (magnesium, gallium) under ultrapurification, $q_{y,\max} \simeq 10^8$ cm^{-1}, an upper limit under *any* conditions. Finally, from (2) it follows that $U_{em}/kT_x \approx 10^{18} \approx U_e/kT_x$. Unfortunately, this optimistic result can be given no weight because $q_{z,\max} l \approx 2 \times 10^4 \approx q_{y,\max} R$, and so the use of the q-independent conductivity $\sigma(0, \omega)$ on which the argument rests is grossly unjustified.

The starting point for a self-consistent calculation of the helicon dispersion relation at large q_y, q_z is the full q-dependent, semiclassical conductivity tensor $\sigma(\mathbf{q}, \omega)^2$

$$\sigma(\mathbf{q}, \omega) = \frac{3}{2}\sigma_0 \sum_{n=-\infty}^{\infty} \int_0^{\pi} d\theta \begin{pmatrix} i(\partial/\partial Y) \\ -n/Y \\ \cos\theta \end{pmatrix} J_n(Y\sin\theta) \begin{pmatrix} -i(\partial/\partial Y) \\ -n/Y \\ \cos\theta \end{pmatrix} J_n(Y\sin\theta)$$

$$\times \frac{\sin\theta}{1 + i(n\omega_c + \omega - q_z v_F \cos\theta)\tau} \tag{3}$$

where σ_0 is the DC conductivity, $\sigma_0 = \omega_p^2 \tau/4\pi$ and $Y = q_y v_F/\omega_c$. Since our attention is confined to the domain $\omega < \omega_c$, we have analyzed the $n = 0, +1$, and -1 terms in great detail and have summed the remaining infinity of terms via expansion of the denominator in the integrands in powers of $(n\omega_c)^{-1}$. In accordance with our interest in large Y, we have replaced the Bessel functions by their asymptotic expressions, e.g., $J_{0,1}^2(Y\sin\theta) \sim (\pi Y \sin\theta)^{-1}[1 \pm \sin(2Y\sin\theta)]$, and so on. The greatest analytical complications arise in the surviving $n = 0$ terms in (3), (namely σ_{xx}, σ_{zz}, and σ_{xz}) where integrals of the form

$$\int_0^{\pi} \sin(2Y\sin\theta)\sin^m\theta \cos^p\theta [1 + (\omega - q_z v_F \cos\theta)^2 \tau^2]^{-1} d\theta$$

with m and p nonnegative integers, appear. Significant contributions to this integral come from two narrow domains[3] in θ: (1) a region of stationary phase in the rapidly oscillating numerator, denoted by SP, of width $2Y^{-\frac{1}{2}}$ centered around $\theta = \pi/2$, and (2) a region of resonance in the denominator, denoted by R, of width $2(|q_z|l)^{-1}$ centered around $\theta = \arccos \omega(q_z v_F)^{-1}$. For $\omega(|q_z|v_F)^{-1} > 1$, the R region does not, of course, exist. A case of complete overlap of the SP and R regions, with the SP region much broader, namely corresponding to $\omega\tau < 1$ and $1 \ll Y < (q_z l)^2$, has been treated by Kaner and Skobov[2] and leads to propagation "spikes" at nearly perpendicular propagation ($q_y \gg q_z$). More germane to the specific heat problem are two cases: (1) with

$$\omega\tau < |q_z|l \quad \omega\tau < \omega_c\tau - |q_z|l$$

and nonoverlap of the SP and R domains assured by

$$\omega\tau > |q_z|lY^{-\frac{1}{2}} + 1$$

and (2) with

$$1 < |q_z|l < \omega\tau < \omega_c\tau - |q_z|l$$

so that there is no R domain.[4] The conductivity tensor has been evaluated in both cases and the helicon dispersion relation obtained numerically by solving the secular equation, $\det(\mathbf{\Gamma} - \mathbf{\sigma}) = 0$, on a digital computer. Here, $\mathbf{\Gamma}$ is the propagation tensor,[5] $ic^2q^2(4\pi\omega)^{-1}\mathbf{1} - ic^2(4\pi\omega)^{-1}\mathbf{\Lambda}$, where $\mathbf{1}$ is the unit diagonal tensor, $\Lambda_{\mu\nu} = q_\mu q_\nu$, and displacement currents have been neglected. For case (1) we find that there are no roots at large Y ($Y > 30$ for the cases studied) and for case (2) there is always a root at large Y (verified analytically), but it is infinitesimally displaced from the Doppler-shifted, $n = -1$ resonance and so will be heavily damped. In effect, even for $\omega_c\tau$ as large as 10^4, the largest admissible $q_{y,\max}$ in (2) is less than 10^6. We conclude that the electromagnetic contribution to the specific heat is far too small to be measurable even under optimum conditions. At the very least, two orders of magnitude improvement in purification, yielding $\tau \gtrsim 10^{-5}$ sec, is required and even then boundary scattering of electrons will presumably nullify such an achievement.

References

1. P. Cotti, P. Wyder, and A. Quattropani, *Phys. Letters* **1**, 50, 1962.
2. E. A. Kaner and V. G. Skobov, *Zh. Eksperim i Teor. Fiz.* **45**, 610, 1963 (*Soviet Phys. JETP* (English Transl.) **18**, 419, 1964).
3. M. H. Cohen, M. J. Harrison, and W. A. Harrison, *Phys. Rev.* **117**, 937, 1960.
4. A. B. Pippard, *Rept. Progr. Phys.* **23**, 176, 1960.
5. J. J. Quinn and S. Rodriguez, *Phys. Rev.* **128**, 2487, 1962.

COLLISION BROADENING OF GIANT QUANTUM OSCILLATIONS

John J. Quinn
RCA Laboratories
Princeton, New Jersey

The study of giant quantum oscillations in the attenuation of acoustic waves promises to become the most important tool yet developed for the study of the electronic properties of metals. In principle, the study of these giant quantum oscillations can yield the cross-sectional area of the Fermi surface, and the cyclotron effective mass and g-factor for electrons not only at the extremal cross sections, but for any cross-sectional area of the Fermi surface normal to the direction of the DC magnetic field. Previous treatments[1–5] of giant quantum oscillations have either neglected the effect of collisions completely or treated the effect incorrectly. The object of the present paper is to study the effect of collisions on the giant quantum oscillations. A detailed calculation has been performed by using the collision dependent quantum magnetoconductivity tensor of Tosima, Quinn, and Lampert[6] in the study of the attenuation of sound waves in a longitudinal magnetic field. The important results of the detailed calculation can be qualitatively understood in a very simple way. In this summary only the nonmathematical qualitative aspects of the problem will be presented.

In the presence of a uniform DC magnetic field of induction $\mathbf{B}_0$, the energy of a single electron is given by

$$E_n(k_z) = \hbar\omega_c(n + \tfrac{1}{2}) + \hbar^2 k_z^2/2m \tag{1}$$

In this equation, n is the Landau level quantum number, k_z the component of the wave vector in the direction of $\mathbf{B}_0$, and

$$\omega_c = \frac{|e|B_0}{mc}$$

is the cyclotron frequency. When a sound wave of wave vector $\mathbf{q}$ is absorbed by the electron gas, both energy and the component of wave vector parallel to the direction of the DC magnetic field must be conserved, that is, if an electron makes a transition from the initial state $|n, k_z\rangle$ to the final state $|n + \alpha, k_z + q_z\rangle$ in the process of absorbing the phonon, then the following equation must be satisfied:

$$E_{n+\alpha}(k_z + q_z) - E_n(k_z) = \hbar\omega_q \tag{2}$$

where ω_q is the frequency of the phonon. One can solve this equation for k_z and find that only electrons with certain specific values of k_z, namely, $k_z = K_\alpha$, can participate in the absorption process:

$$K_\alpha = \frac{m}{\hbar q_z}(\omega_q - \alpha\omega_c) - \tfrac{1}{2}q_z \tag{3}$$

The absorption processes in which the electrons do not change their Landau level quantum number were first considered by Gurevich, Skobov, and Firsov.[1] For these processes

$$K_0 = \frac{msq}{\hbar q_z} - \tfrac{1}{2}q_z \tag{4}$$

where s is the velocity of sound. Unless $\mathbf{q}$ is almost normal to the direction of the DC magnetic field, K_0 is very small compared to the Fermi wave vector k_F. For $\alpha \neq 0$, K_α can be of the same order of magnitude as k_F, even when $\mathbf{q}$ is parallel to $\mathbf{B}_0$.

The giant quantum oscillations result simply from the requirement that the initial state be at least partially occupied and the final state at least partly empty. Thus the absorption can occur only when either the initial or final states lie within kT of the Fermi level. (It is of course necessary that $\hbar\omega_c \gg kT$ in order that temperature broadening be small enough to make the quantum oscillations observable.) As one varies the magnetic field strength this situation occurs periodically, every time a Landau level passes through the Fermi surface at the plane $k_z = K_\alpha$. This result is quite reminiscent of the de Haas–van Alphen effect; however, here the period of the oscillations is a measure of the cross-sectional area of the Fermi surface at the plane $k_z = K_\alpha$. For $\alpha = 0$ this plane is close to the extremal cross-section, because $K_0/k_F \ll 1$ (unless of course q/q_z is of the order of v_F/s). For $\alpha \neq 0$ the position of the plane $k_z = K_\alpha$ is itself a function of B_0. In the metals, however, the position of the plane changes by an exceedingly small amount over one period of oscillation. Thus, for $\alpha \neq 0$ one can obtain the cross-sectional area of any plane normal to $\mathbf{B}_0$ from the period of the quantum oscillations. The cyclotron effective mass and the g-factor of the electrons at the plane $k_z = K_\alpha$ can be obtained from the width and spin splitting of the absorption lines.[7,8]

The effect of collisions is most simply understood by using the uncertainty principle. Due to collisions, the electrons have an energy uncertainty $\hbar/\tau$, where τ is the collision time. If one introduces this energy uncertainty into equation (2), the resulting solution $k_z = K_\alpha$ becomes uncertain by the amount $\Delta K_\alpha = m/\hbar q\tau$. This uncertainty in the initial wave vector results in an uncertainty in the energy of the initial state which is equal to $(\hbar^2/m)\Delta K_\alpha[K_\alpha + \tfrac{1}{2}\Delta K_\alpha]$. Unless this uncertainty in the initial energy is smaller than $\hbar\omega_c$, simultaneous absorption by more than a single Landau level is possible. This would destroy the quantum oscillations. Thus, we arrive at the condition

$$\frac{m}{q^2\tau^2}(1 + |\omega\tau - \alpha\omega_c\tau|) < \hbar\omega_c \tag{5}$$

in order to see quantum oscillations due to electronic transition in which Δn, the change in the Landau level quantum number, is equal to α. For $\Delta n = 0$ transitions, the frequency ω of the sound wave must satisfy the relation

$$\omega\omega_c[1 + (2\omega\tau)^{-1}]^{-1} > \frac{ms^2}{\hbar\tau} \tag{6}$$

while for $\Delta n \neq 0$ transitions, ω must satisfy the inequality

$$\omega^2 > \frac{ms^2}{\hbar\tau} \tag{7}$$

In writing down (7) we have assumed $\omega_c \gg \omega$. Obviously, the inequality (7) requires higher ultrasonic frequencies or purer materials than inequality (6).

The giant quantum oscillations due to $\Delta n = 0$ transitions have already been observed in a number of materials. Thus far, the much more interesting giant quantum oscillations due to $\Delta n \neq 0$ transitions have not been observed. These oscillations should be observable in a very pure metal ($\tau \approx 10^{-8}$ sec) at ultrasonic frequencies of the order of a few kilomegacycles.

References

1. V. L. Gurevich, V. G. Skobov, and Yu. A. Firsov, *Soviet Phys. JETP (English Transl.)* **13**, 552, 1961.
2. J. J. Quinn and S. Rodriguez, *Phys. Rev.* **128**, 2487, 1962.
3. J. J. Quinn, *Phys. Letters* **7**, 235, 1963.
4. D. N. Langenberg, J. J. Quinn, and S. Rodriguez, *Phys. Rev. Letters* **12**, 519, 1964.
5. S. V. Gantsevich and V. L. Gurevich, *Soviet Phys. JETP (English Transl.)* **18**, 403, 1964.
6. S. Tosima, J. J. Quinn, and M. A. Lampert, *Phys. Rev.* **137**, A883, 1965.
7. Y. Shapira and B. Lax, *Phys. Rev. Letters* **12**, 166, 1964.
8. Y. Shapira, *Phys. Rev. Letters* **13**, 162, 1964.

8.4. Magnetic Breakdown

MAGNETIC BREAKDOWN IN THE LONG DHVA PERIODS IN ZINC[†]

J. R. Lawson and W. L. Gordon
Case Institute of Technology
Cleveland, Ohio

In early studies of the de Haas–van Alphen effect in zinc,[1] it was found that the amplitude of the long period oscillations (which are now attributed to the needle-shaped electron portions in the third band) fell off unexpectedly with increasing magnetic fields. The solution of this puzzle was indicated by Stark's[2] identification of the occurrence of magnetic breakdown in magnetoresistance effects in zinc. The present work was initiated to further explore the influence of breakdown on these DHVA oscillations.

As discussed by Stark earlier in this conference, magnetic breakdown occurs between the second band hole surface, or monster, and the needle-shaped electron surface in the third band shown in Fig. 1a. Following Pippard's notation,[3,4] the probability of an electron tunneling from a needle orbit onto the monster can be written as $p^2 = e^{-H_0/H}$. Here a characteristic field is defined as $H_0 = \pi\varepsilon^2/(4\hbar e v_x v_y)$, with v_x and v_y the normal and tangential components, respectively, of the free-electron velocity at the Brillouin zone boundary, and ε the energy gap between bands.

Pippard,[4] in considering the case of the coupling of orbits on neighboring bands, has shown that under breakdown conditions Landau level broadening takes place as p^2 becomes important. In addition, he has treated the case of interest in this study of zinc where for magnetic field directions $(\theta) > 5°$ from [0001], orbits link three arms of the monster with the needles. For such orbits he found that the amplitude of the fundamental of the DHVA oscillations due to the needle should be reduced by the factor $(1 - e^{-H_0/H})^{\frac{3}{2}}$ and obtained a reasonable fit to the data of Dhillon and Shoenberg[1] with this expression.

In the absence of magnetic breakdown, the oscillatory torque is given by the usual DHVA expression:[5]

$$C = ATH^{\frac{1}{2}}f(\theta) \sum_l \frac{\cos(l\pi g m^*/2m) \exp(-2\pi^2 lkx/\beta H) \sin\left(\frac{lcS_m}{e\hbar H} - 2\pi l\gamma - \frac{\pi}{4}\right)}{\sqrt{l} \sinh[(2\pi^2 kTl)/\beta H]} \quad (1)$$

where S_m is the extremal cross section of the Fermi surface in momentum space cut by planes perpendicular to H, θ is the angular distance between the field direction and [0001], $\beta = e\hbar/m^*c$, with m^* the cyclotron mass, and x is the usual Dingle temperature. The $\cos(l\pi g m^*/2m)$ describes the effects of spin splitting of the Landau levels on the lth harmonic. For the usual g value of 2, this term has a value close to unity for the first few harmonics. However, for these needle orbits in zinc, Stark[6]

[†]Supported by the Air Force Office of Scientific Research.

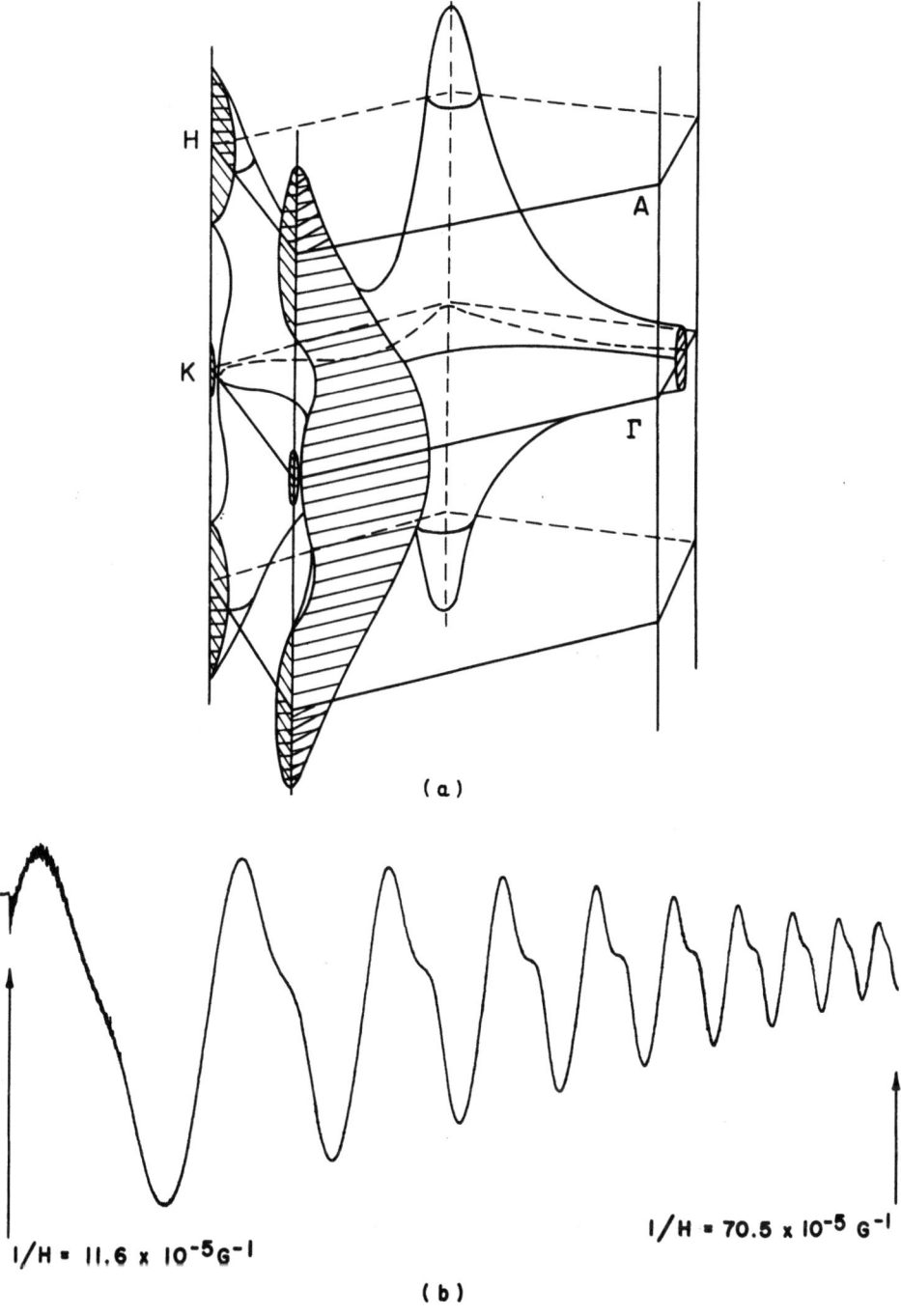

Fig. 1. (a) First and second band hole surface, or monster, in zinc together with the needle-shaped electron surfaces from the third band. (b) Oscillations in the torque on a zinc crystal at 1.95°K. The magnetic field lies 20° from [0001] in the (11$\bar{2}$0) plane.

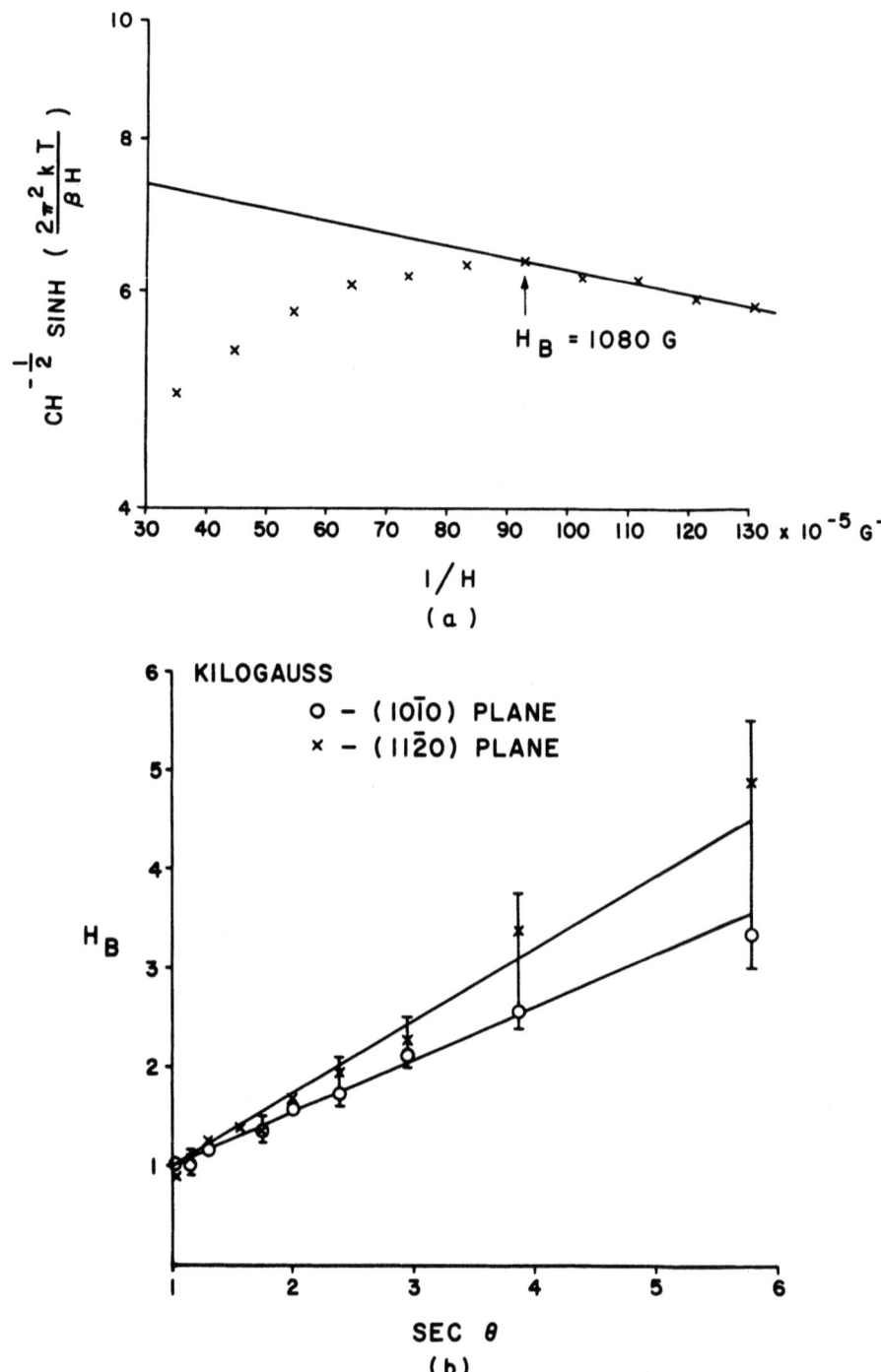

Fig. 2. (a) Typical amplitude plot obtained for $\theta = 5°$ in a $(11\bar{2}0)$ plane. (b) Angular dependence of the field at which breakdown becomes observable ($H_B \approx H_0/4$).

found the g- factor to be considerably larger, so that the term $\cos(l\pi gm^*/2m)$ becomes important.

Magnetic breakdown can be observed most readily on an amplitude plot as the deviation from a straight line as illustrated in Fig. 2a. Here only the amplitude of the fundamental is considered. Theory predicts a straight line, as observed at low fields, with a slope proportional to the Dingle temperature. In the low-field linear region, the value of 0.1°K which we calculate is consistent with the relatively large harmonic content shown in Fig. 1(b). The m^* values used in the calculations have been obtained from a temperature-dependent study of the oscillation amplitudes at sufficiently low fields to avoid breakdown effects.

H_0 has been calculated, using the Pippard factor to compute the reduction of the amplitude, ΔC, of the fundamental at $H/H_0 < 1$. We have limited ourselves to this region to avoid the large uncertainties involved in correcting for harmonic content at higher fields. In this limit the factor becomes $(1 - \frac{3}{2}e^{-H_0/H})$. Thus H_0 is obtained from the slope of a logarithmic plot of $\Delta C/C$ against $1/H$, independent of the fraction $\frac{3}{2}$, and has values increasing from 4100 g at $\theta = 5°$ to 14,500 g at $\theta = 70°$. The values of H_0 are uncertain to about $\pm 15\%$. The corresponding values of the fraction are less certain but lie near $\frac{3}{2}$ except in a few cases. These uncertainties are due to the number of corrections involved in obtaining the amplitude plots as well as noise in the low-field data.

From the definition of H_0, its angular dependence is determined by the velocity components v_x and v_y on the assumption of a constant energy gap between bands. However, for simplicity in representing the angular dependence we have chosen to plot H_B as in Fig. 2b against $\sec \theta$, since this represents the dominant term in the velocity variation and H_B can be obtained simply from the amplitude plots.

A clear reduction of harmonic content appeared at fields above about $H_0/3$ as illustrated in Fig. 1b. This we have attributed to a rapid increase in Landau level broadening caused by magnetic breakdown. It is described in terms of an additional field-dependent Dingle temperature, $y(H)$, obtained for a series of field values from the observed ratio of second harmonic to first. For several angular orientations, $y(H) \propto e^{-5H_0/2H}$.

The effect of spin splitting of the Landau levels as mentioned in connection with equation (1) is observed in the unexpectedly large amount of second harmonic present in these oscillations. A value for $gm^*/2m$ was found to be 0.37 from the ratio of second harmonic to fundamental at fields below H_B, in good agreement with Stark's[6] value of 0.34. No angular variation of this $gm^*/2m$ product was observed in the angular range of 5° to 70°. This is consistent with the calculations of Bennett and Falicov.[7]

References

1. J. S. Dhillon and D. Shoenberg, *Phil. Trans. Roy. Soc. London*, Ser. A **248**, 1, 1955.
2. R. W. Stark, *Phys. Rev. Letters* **9**, 482, 1962.
3. A. B. Pippard, *Proc. Roy. Soc. (London)* Ser. A **270**, 1, 1962.
4. A. B. Pippard, *Phil. Trans. Roy. Soc. London*, Ser. A **256**, 317, 1964.
5. I. M. Lifshitz and A. M. Kosevitch, *Zh. Eksperim. i Teor. Fiz.* **29**, 730, 1955. (*Soviet Phys. JETP* (*English Transl.*) **2**, 635, 1956.)

6. R. W. Stark, *Phys. Rev.* **135**, A1698, 1964.
7. A. J. Bennett and L. M. Falicov, *Phys. Rev.* **136**, A998, 1964.

EFFECT OF ALLOYING ON MAGNETIC BREAKDOWN IN ZINC[†]

R. J. Higgins,[‡] J. A. Marcus, and D. H. Whitmore

Northwestern University
Evanston, Illinois

In the course of torsion measurements[1] of the de Haas–van Alphen periods in zinc–copper alloys, it became apparent that the high-field amplitude of the "needle" period showed a nearly normal field dependence, in contrast to the anomalously low amplitude in pure zinc.[2] Computations have been made to interpret the changing field dependence in terms of magnetic breakdown[3] effects. Our results in pure zinc are in qualitative agreement with those of Lawson and Gordon,[2] with certain quantitative differences which are presumably due to differences in approach. Since breakdown effects become apparent at about 3 kG, our magnetic field range of 1 to 30 kG results in a wide range of breakdown probability, in an attempt to provide a stringent test of any theoretical interpretation. In the high-field region the hyperbolic sine term in the Lifshitz formula[4] may not be replaced by the exponential approximation. The computations are therefore unwieldy, and were carried out on the IBM 709 computer. The amplitude of the DHVA oscillations was measured at three temperatures and at about ten angles in the $(11\bar{2}0)$ and $(10\bar{1}0)$ planes in pure zinc and in the $(10\bar{1}0)$ plane in Zn–0.14 at.% Cu, and Zn–0.21 at.% Cu alloys. The computer first uses the temperature dependence of the amplitudes to calculate the effective mass, varying the mass iteratively[5] until equation (1) is satisfied.

$$\frac{Y[(1/H, T_1)]}{Y[(1/H, T_2)]} = \frac{T_1 \sinh[(\lambda(m^*/m_0)(T_2/H)]}{T_2 \sinh[(\lambda(m^*/m_0)(T_1/H)]} \tag{1}$$

where $\lambda = \pi^2 k/\mu_B$. After the effective mass calculated at a number of values of $1/H$ is suitably averaged, a least-squares fit is performed on the field dependence of the amplitudes. We write the Lifshitz expression[4] for the envelope of the fundamental torque oscillation $Y(1/H)$ as

$$Y\left(\frac{1}{H}\right) = \frac{T\sqrt{H}e^C}{\sinh[\lambda(m^*/m_0)(T/H)]} \exp\left[-\lambda \frac{m^*}{m_0} \frac{T_D}{H}\right] F\left(\frac{H_0}{H}\right) \tag{2}$$

where e^C is an unknown constant involving the shape of the needle, T_D is the unknown scattering temperature[6] and $F(H_0/H)$ is a function of the unknown breakdown field H_0. Following a suggestion of Pippard,[7] we have used a breakdown function

$$F\left(\frac{H_0}{H}\right) = (1 - P)^{\frac{3}{2}} \tag{3}$$

[†] Supported by the Advanced Research Projects Agency, through the Northwestern University Materials Research Center, and by the National Science Foundation.
[‡] National Science Foundation Predoctoral Fellow. Present address: University of Oregon, Eugene, Oregon.

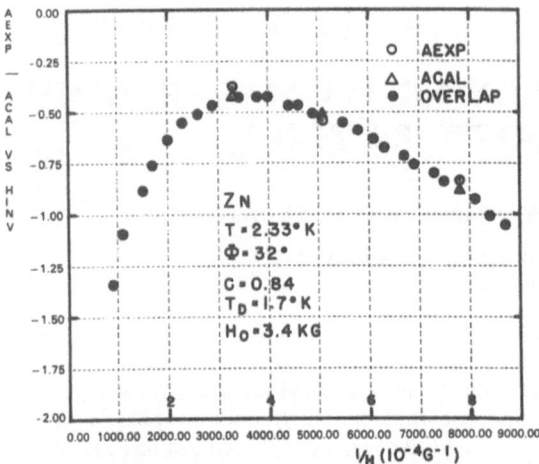
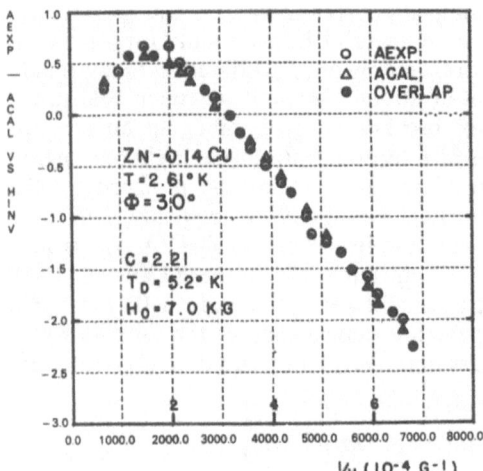
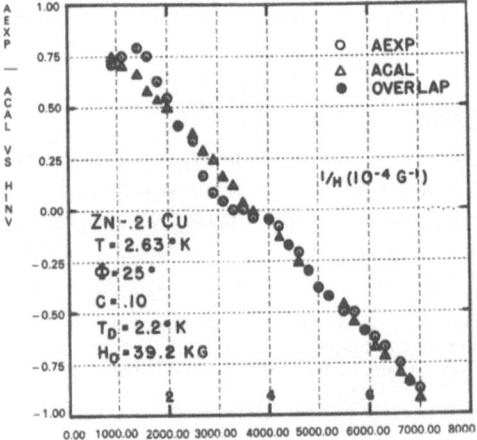

Fig. 1. Computer plotted amplitudes (arbitrary units), calculated from equations (4) and (5) as a function of inverse magnetic field $1/H$. Open circles are A_{EXP}, open triangles are A_{CAL}, and solid circles indicate points where the least-squares fit A_{CAL} overlaps the experimental amplitude A_{EXP} within the accuracy of the graph. Values of H_0 are: 3.4 kG (pure zinc); 7.0 kG (0.14% Cu); 39.2 kG (0.21% Cu). One unit on the x-axis corresponds to $1 \times 10^{-4} \, \text{G}^{-1}$ or $1000 \times 10^{-7} \, \text{G}^{-1}$.

where $P = \exp(-H_0/H)$. The computer program then varies the parameters C, T_D, and H_0 until the quantity

$$\sum_i \left[A_{CAL}\left(\frac{1}{H_i}\right) - A_{EXP}\left(\frac{1}{H_i}\right) \right]^2$$

is a minimum,[8] where

$$A_{EXP}\left(\frac{1}{H_i}\right) = \ln\left\{ \frac{Y(1/H_i)\sinh[\lambda(m^*/m_0)(T/H)]}{TH^{\frac{1}{2}}} \right\} \tag{4}$$

and

$$A_{CAL}\left(\frac{1}{H_i}\right) = C - \lambda\frac{m^*}{m_0}\frac{T_D}{H_i} + \tfrac{3}{2}\ln[1 - \exp(-H_0/H)] \tag{5}$$

Preliminary results of these calculations indicate that in pure zinc $H_0 \simeq 2\,\text{kG}$ for H nearly parallel to [0001]. H_0 rises slowly away from [0001], reaching about 5 kG 50° from [0001]. H_0 then rises rapidly, reaching about 18 kG 82° from [0001]. These numbers are somewhat sensitive to the value of the effective mass and to scatter in the experimental data, and further refinements are in progress.

Since other forms for the breakdown function have been suggested,† work is underway to determine from a purely empirical point of view whether the data can be fitted by $(1 - p)^n$, for various values of n. Such testing of functional dependence is quite tedious by hand but becomes quite simple in the computer calculation. It appears that to determine a unique fit requires amplitude data both below and above the breakdown field. Otherwise very different values of H_0 occur. For example, at H 12° from [0001], changing n from $\tfrac{3}{2}$ to 3 changes the value of the apparent breakdown field from 2 to 8 kG.

In spite of present uncertainties in the form of the breakdown function, it is quite clear that alloying with copper increases the breakdown field markedly. Figure 1 shows how the field dependence of amplitudes for H about 30° from [0001] changes upon addition of 0.14% Cu and 0.21% Cu. The linear portion of equation (5) extends to lower values of $1/H$ in the 0.14% Cu alloy (middle curve) than in pure zinc (upper curve). Within the scatter of preliminary data, the amplitude in the 0.21% Cu alloy appears linear over the whole region of $1/H$. Fitting these amplitudes to equation (5) results in the following values of H_0: 3.4 kG (pure zinc); 7.0 kG (Zn–0.14% Cu); 39.2 kG (Zn–0.21% Cu).

The rapid rise in H_0 with alloying suggests a rise in the relevant band gap V_g, since $H_0 \propto (V_g^2)$[11,12]. However, the interpretation of these results is complex. Further work is in progress both to test the functional dependence of breakdown and to calculate reliable values of breakdown field.

Acknowledgments

One of the authors (R. J. H.) is grateful to A. B. Pippard for suggesting the form of the breakdown expression [equation (3)], to L. M. Falicov and R. W. Stark for helpful discussions on the same subject, and to W. L. Gordon for discussing his experimental results prior to publication (preceding paper). The decision to undertake these experiments was in part influenced by a discussion with J. H. Condon, who demonstrated that the anomalous field dependence of amplitudes reported by Shoenberg[13] could be interpreted by a simple magnetic breakdown expression (unpublished).

† One equation has been suggested by L. M. Falicov;[9] another is in W. A. Harrison.[10]

References

1. J. H. Condon, *Phys. Rev.* **134**, A446, 1964.
2. J. R. Lawson and W. L. Gordon, this volume, p. 854.
3. Morrell H. Cohen and L. M. Falicov, *Phys. Rev. Letters* **7**, 231, 1961.
4. I. M. Lifshitz and A. M. Kosevich, *Zh. Eksperim. i. Teor. Fiz.* **29**, 730, 1955. (*Soviet Phys. JETP (English Transl.)* **2**, 635, 1956.
5. J. B. Scarborough, *Numerical Mathematical Analysis,* fifth edition, The Johns Hopkins Press, Baltimore (1962), p. 199.
6. R. B. Dingle, *Proc. Roy. Soc. (London) Ser. A* **211**, 517, 1952.
7. A. B. Pippard, private communication, later published in *Phil. Trans. Roy. Soc. London, Ser. A* **256**, 317, 1964.
8. J. B. Scarborough, *Numerical Mathematical Analysis,* fifth edition, The Johns Hopkins Press, Baltimore (1962), p. 539.
9. L. M. Falicov, private communication.
10. W. A. Harrison, *Phys. Rev.* **126**, 497, 1962.
11. E. I. Blount, *Phys. Rev.* **126**, 1636, 1962.
12. A. B. Pippard, *Proc. Roy. Soc. (London) Ser. A* **270**, 1, 1962.
13. J. S. Dhillon and D. Shoenberg, *Phil. Trans. Roy. Soc. London, Ser. A* **248**, 1, 1955.

POSSIBLE NEW OPEN ORBITS IN THE FERMI SURFACE OF ZINC*

J. E. Schirber

Sandia Laboratory
Albuquerque, New Mexico

A great deal of attention has been given recently to the galvanomagnetic properties of the hexagonal metals magnesium[1] and zinc,[2-4] particularly concerning the phenomena associated with magnetic breakdown. In line with a continuing investigation of the effect of pressure on Fermi surface topologies,[5,6] we have collected very careful magnetoresistance data on zinc at zero pressure.

Using a tipping technique described in detail elsewhere,[7] we have observed an additional set of regions of applied magnetic field which appear to support open orbits in the Fermi surface of zinc. These regions are found at angles (as measured from the [0001] direction) greater than the extremities of the short whiskers in the (11$\bar{2}$0) planes. (see Stark[2] for a magnetoresistance stereogram for zinc). The regions are marked by the appearance and subsequent disappearance of cusplike troughs in the "transverse" magnetoresistance *vs.* angle plots as the sample is tipped relative to the plane of the applied magnetic field. A series of sweeps of the "transverse" magnetoresistance as the crystal (current parallel to [10$\bar{1}$0]) is tipped in a (11$\bar{2}$0) plane is shown in Fig. 1a. The sharp cusplike trough disappears between the curves in Fig. 1a labeled 4.39 and 5.76° and is replaced by a distinct peak. A new trough has appeared by the 7.29° sweep which disappears before 15.99°. Figures 1b and 1c show more detailed sweeps near the ends of the short whiskers and ends of the new regions, respectively.

The samples ($\frac{1}{16} \times \frac{1}{16} \times \frac{5}{8}$ in.) were cut by acid erosion from a zone refined (five pass) ingot grown from 99.999 + % ASARCO zinc. The ratio of the resistance in zero field at 300°K to that at 4°K was between 10^4 and 2×10^4. The data described here were all taken in fields of 8 to 10 kOe.

We can relate these features in the magnetoresistance to dimensions of the Fermi surface in the following way: If we denote the length of the short whiskers by δ_1, the beginning and end of the new regions by δ_0 and δ_2 respectively, and the length of the long whiskers in the (10$\bar{1}$0) plane by δ_3, we can write the following relations for these angles: (Our measured values in degrees for these angles appear in parentheses.)

$$\delta_1(5.63 \pm 0.05) = \tan^{-1}[(h_N + d)/2r]$$

$$\delta_0(6.4 \pm 0.2) = \tan^{-1}[(H_N - d)/(2a - 2r)]$$

$$\delta_2(15.2 \pm 0.1) = \tan^{-1}[d/(2r - a)]$$

---

*This work was supported by the United States Atomic Energy Commission.

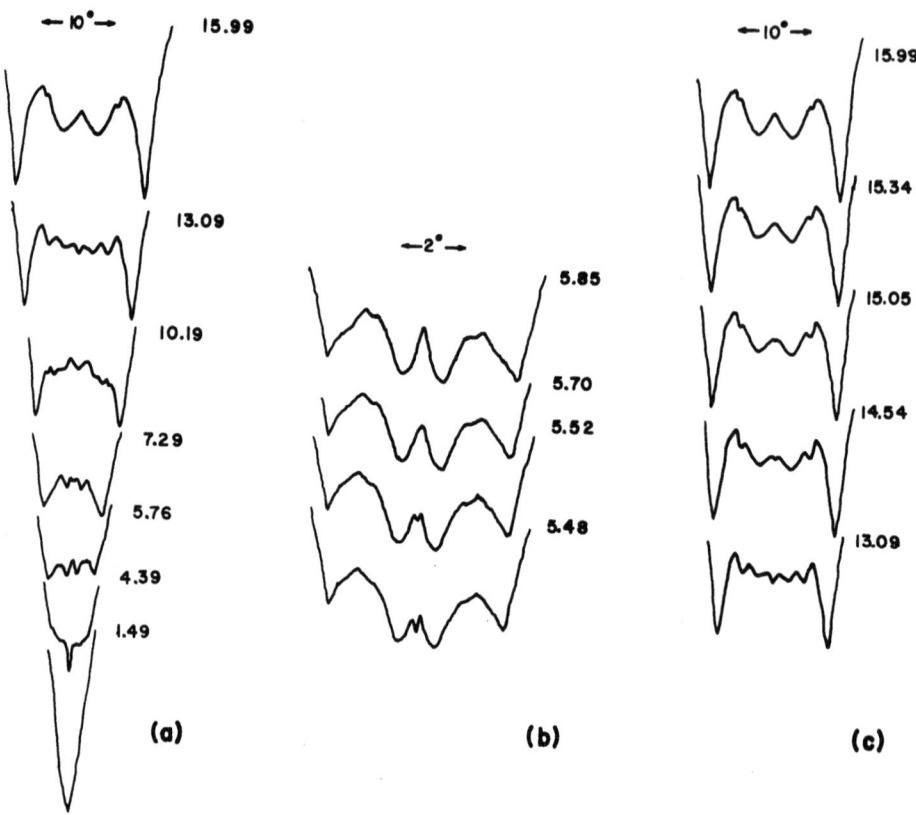

Fig. 1. (a) A reproduction of a recorder tracing of the magnetoresistance of zinc as the sample is tipped in the (11$\bar{2}$0) plane. Tip angle zero is defined when the field is along the c-axis and all traces are labeled in degrees of tip from the axis. (b) A detailed set of sweeps determining δ_1, the end of the short whisker in the (11$\bar{2}$0) plane. (c) Same as (b), determining δ_2, the end of the new region.

$$\delta_3(42.5 \pm 0.3) = \tan^{-1}[(\sqrt{3}h_N + \sqrt{3}\,d)/(3r - 2a + \sqrt{3}r_N)]$$

$$\approx \tan^{-1}[(\sqrt{3}h_N + \sqrt{3}\,d)/(3r - 2a)]$$

We have defined h_N as the height parallel to the c-axis of magnetic breakdown on the needles, d as the thickness in the c-direction of the waists of the combined first and second zone hole surface (crown or monster), r as the radius of the crown, $2a$ as the reciprocal lattice vector in the basal plane, r_N as the minor dimension of the needles, and H_N as the height of the notches in the crown into which the needles fit.

The dimensions and the angles defined are shown schematically in Fig. 2. The tip angle could be determined in two independent ways: primarily by the counter reading of the goniometer arc position,[7] and secondly, by the angular separation of troughs due to the long whiskers in the (10$\bar{1}$0) planes. These troughs are the outermost cusps in the tracings in Fig. 1. All features were measured for

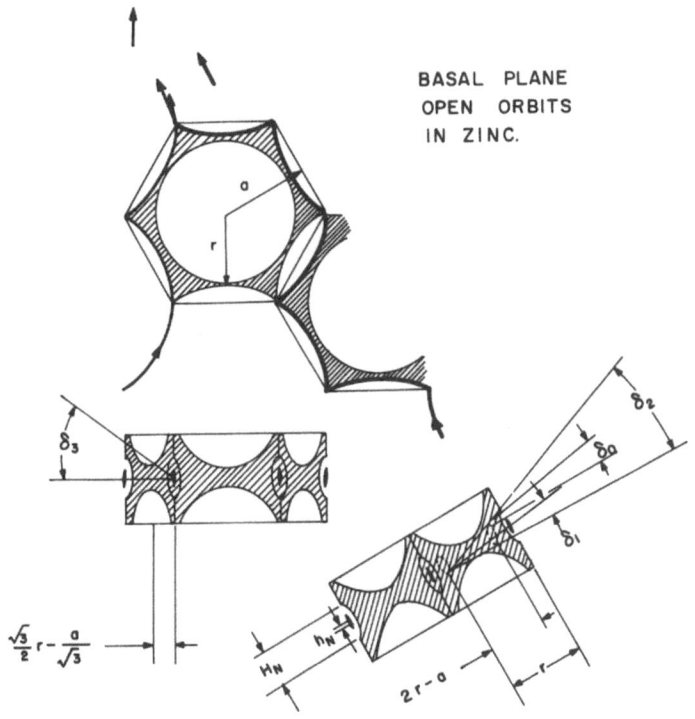

Fig. 2. Projections of the crown and needle portions of the Fermi surface of zinc showing the dimensions and angles discussed in text. The angles are exaggerated in the interest of clarity.

both directions of tip from the c-axis, and the angular determinations by the two methods agreed to within approximately 0.1°.

It is known from the DHVA measurements[8] that r_N is about 1% of r, so that the second formula for δ_3 is an excellent approximation. With these four equations, we can determine the values of d, h_N, H_N, and r without recourse to other experiments or to dimensions from the free-electron model. This is a practical necessity for the pressure studies for which the present work is a prerequisite.

The value for r of 1.04 Å^{-1} is consistent with the value from Harrison's band calculation[9] and with the magnetoacoustic result.[10] Our assignment requires that $h_N < d$ and that H_N be greater than the height of the needle (0.22 Å^{-1} from Joseph and Gordon[8]). We calculate $d = 0.19$ Å^{-1}, $h_N = 0.01 \pm 0.01$ Å^{-1}, and $H_N = 0.27$ Å^{-1}. Our value for d agrees quite well with the value of 0.17 Å^{-1} obtained by Stark.[11] Our value of h_N is about a factor of seven lower than Reed and Brennert's[4] value and a factor of 70 lower than that of Alekseevskii and Gaidukov,[3] both as determined from the Hall coefficient, and leads to a prediction that the giant orbit[1] should be seen for an angular deviation of only about $\frac{1}{2}°$ from the c-axis. Thorsen et al.[12,13] observe this orbit for somewhat larger angles than this. We cannot explain this discrepancy at the present time.

Acknowledgments

The author is indebted to D. D. Sand for his help throughout the investigation and to Dr. R. W. Stark and Dr. A. C. Thorsen for discussions of their work prior to publication.

References

1. M. G. Priestley, L. M. Falicov, and Gideon Weisz, *Phys. Rev.* **131**, 617, 1963.
2. R. W. Stark, *Phys. Rev. Letters* **9**, 482, 1962.
3. N. E. Alekseevskii and Yu. P. Gaidukov, *Soviet Phys. JETP (English Transl.)* **16**, 1481, 1963.
4. W. A. Reed and G. F. Brennert, *Phys. Rev.* **130**, 565, 1963.
5. D. Caroline and J. E. Schirber, *Phil. Mag.* **8**, 72, 1963.
6. J. E. Schirber, *Bull. Am. Phys. Soc.* **9**, 98, 1964.
7. J. E. Schirber, *Phys. Rev.* **131**, 2459, 1963.
8. A. S. Joseph and W. L. Gordon, *Phys. Rev.* **126**, 489, 1962.
9. W. A. Harrison, *Phys. Rev.* **126**, 497, 1962.
10. D. F. Gibbons and L. M. Falicov, *Phil. Mag.* **8**, 177, 1963.
11. R. W. Stark, *Phys. Rev.* **135,** A1698, 1964.
12. A. C. Thorsen, A. S. Joseph, and L. E. Valby, this volume, p. 867.
13. A. C. Thorsen, private communication.

GIANT ORBITS IN ZINC*

A. C. Thorsen† and L. E. Valby†

*Atomics International Division of North American Aviation, Inc.
Canoga Park, California*

and

A. S. Joseph

*North American Aviation Science Center
Thousand Oaks, California*

Recent measurements of the de Haas–van Alphen effect in magnesium[1,2] and zinc[3] have shown the Fermi surfaces of these metals to be similar and consistent with those predicted by the nearly free-electron model. Magnetic breakdown phenomena in these metals have been found in a number of crystallographic directions by both de Haas–van Alphen effect[4] and magnetoresistance[5,6] experiments. In particular, with the magnetic field oriented along the hexagonal axis of magnesium, magnetic breakdown results in a giant orbit having an area larger than the cross-sectional area of the hexagonal face of the first Brillouin zone. In this paper, we report on the existence of a high-frequency oscillation in zinc which can be associated with a similar orbit. In addition, beating effects in these oscillations are possibly indicative of other orbits of comparable cross section. A preliminary survey of longer periods observed with the magnetic field near the [0001] axis shows a number of oscillations heretofore unreported.

The data were taken with a conventional pulsed field apparatus described elsewhere.[7] Two single-crystal zinc samples were spark-cut in the form of 0.036 in.-diameter cylinders from a large crystal having a resistivity ratio of 24,000. The samples were then etched down to approximately 0.020 in. diameter with dilute HNO_3. Great care was taken in handling and orienting the samples to minimize strains. The accuracy in orienting the [0001] axis with respect to the magnetic field is estimated to be $\pm 0.5°$.

Figure 1 shows an oscilloscope trace of the oscillations which we associate with the giant orbit in zinc. The horizontal traces are calibration lines and the diagonal trace is proportional to the magnetic field. The field changes in value from 174.6 to 176.0 kG in a time of approximately 0.5 msec. The dominant oscillations have a period of $1.175 \times 10^{-9} \, G^{-1}$ ($\pm 2\%$) corresponding to an extremal cross-sectional area in k-space of $8.12 \times 10^{16} \, cm^{-2}$. This is to be compared with the hexagonal cross-sectional area of the Brillouin zone of $6.44 \times 10^{16} \, cm^{-2}$ and an extremal cross section of the free-electron sphere of $7.88 \times 10^{16} \, cm^{-2}$. These oscillations were observed over a total angular range of $2.4° \pm 0.2°$. This angular range is in excellent agreement with that found from magnetoresistance studies.[6]

---

* Supported in part by the U.S. Atomic Energy Commission.
† Presently at NAA Science Center, Thousand Oaks, California.

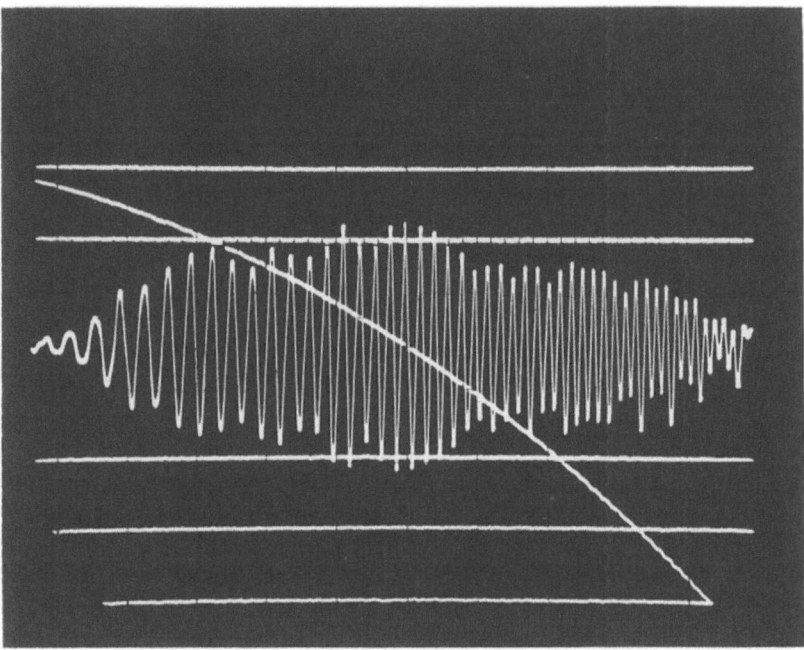

Fig. 1. Oscilloscope trace of giant orbit oscillations in zinc showing beats due to the presence of other oscillatory terms. The magnetic field is approximately 175 kG and the temperature of the sample is 1.1°K.

The beating effects in the dominant oscillations are quite pronounced and definitely indicate at least one set of oscillations having a period smaller than that associated with the giant orbit. At least four oscillations appear to be present in most of the data; their values are summarized in Table I. When there is some question as to whether the periods of the subsidiary oscillations are above or below the period of the dominant oscillations, both values are given in parentheses. All of these periods may be associated with orbits on the hexagonal face of the Brillouin zone (shown schematically in Fig. 2) if one invokes the possibility of multiple magnetic breakdown.[8,9] The dominant oscillation corresponds to complete breakdown to free-electron behavior and results in an orbit shown in Fig. 2a. The oscillations having the next strongest amplitude appear to have a period of $0.91 \times 10^{-9}\,\text{G}^{-1}$ which would result from an orbit even larger than the free-electron sphere. A possible explanation for these oscillations is shown in Fig. 2b. This orbit traces out six small loops in addition to the orbit around the free-electron sphere. The total area found from the nearly-free-electron model is $11.25 \times 10^{16}\,\text{cm}^{-2}$, compared to the experimental value of $11.0 \times 10^{16}\,\text{cm}^{-2}$. The long beats seen in Fig. 1 are evidence of the period 1.12 (or 1.25) $\times 10^{-9}\,\text{G}^{-1}$ given in Table I. The smaller period could possibly arise from an orbit shown in Fig. 2c. Here, only one loop is completed in addition to the orbit around the free-electron sphere yielding an area of $8.5 \times 10^{16}\,\text{cm}^{-2}$. The nearly-free-electron model predicts a corresponding

Table I. Experimentally Determined Period Values Compared with the Theoretical Values Found from the Nearly Free-Electron Approximation.

Experiment period, $\times 10^9$ G	Theoretical period ($\times 10^9$ G) (nearly-free-electron approximation)	Orbit
$1.175 \pm 2\%$	1.21	"Giant" orbit
$0.91 \pm 3\%$*	0.888	Multiple breakdown (nonlinearity)
$(1.12 \pm 5\%)(1.25 \pm 5\%)$*	1.14	Multiple breakdown
$(0.83 \pm 5\%)(2.02 \pm 7\%)$*	1.89	Outside monster (nonlinearity)
$3.99 \pm 2\%$	3.77	Lens
$3.06 \pm 2\%$*		Inside monster

*Derived from beat pattern.

area of 8.78×10^{16} cm^{-2}. The last, and weakest, oscillation that is detected by virtue of its beat pattern has a period of either 0.83 or 2.0×10^{-9} G^{-1}. We tentatively associate the latter value with a hole orbit around the outside of the "monster" shown in Fig. 2d. The theoretical value of period for this orbit is 1.89×10^{-9} G^{-1}.

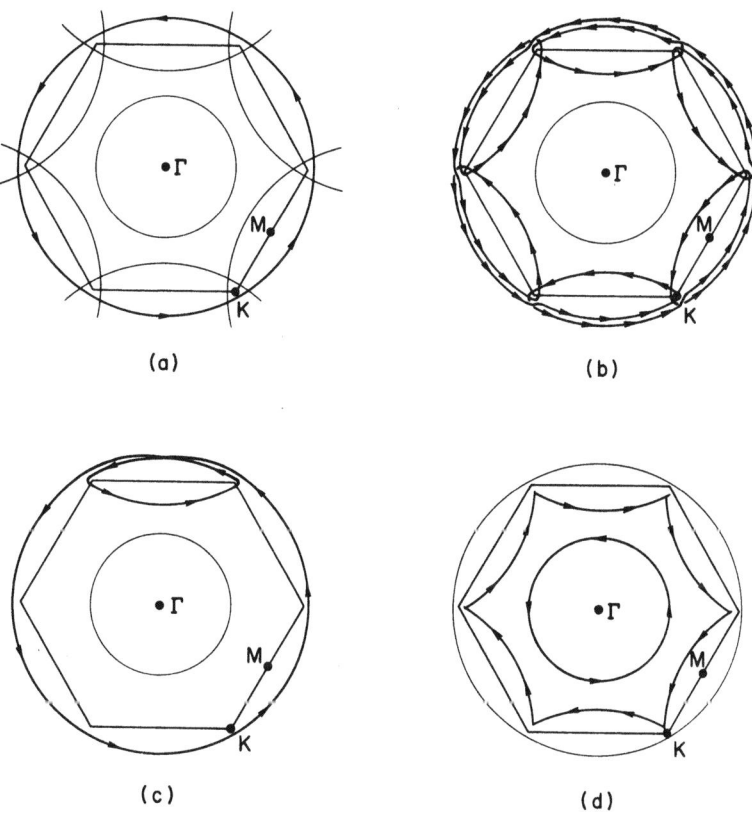

Fig. 2. Some possible electron and hole orbits in zinc in the ΓKM plane.

Oscillations with somewhat larger periods have also been investigated near the [0001] axis. The dominant one of these has a period of $3.99 \times 10^{-9} \, G^{-1}$ and is a consequence of an orbit around the "lens" (the inner circle in Fig. 2d). There are also strong beats in the lens oscillations from which we deduce another oscillatory term having a period of $3.06 \times 10^{-9} \, G^{-1}$. This term, resulting from an orbit slightly larger than the lens, can be associated with the inside of the monster. The area of this orbit in the nearly-free-electron approximation has a common value with that of the lens. When the energy gap due to the lattice potential is taken into account, the area of the lens decreases while the area of the inner portion of the monster increases. The net separation of these two bands is a measure of the energy gap which may be estimated from the measured areas and the corresponding effective masses according to[10]

$$Eg = \tfrac{1}{2}\left(\frac{\hbar^2}{2\pi}\right)\left(\frac{A_M}{m_M^*} - \frac{A_L}{m_L^*}\right) \approx \frac{\hbar^2}{4\pi m_L^*}(A_M - A_L)$$

where A_M is the area of the inner portion of the monster, A_L the area of the lens, and m_M^* and m_L^* are the associated effective masses. Taking $m_L^* = 1.2 \, m_0$ (from cyclotron resonance measurements) and assuming $m_L^* \approx m_M^*$, we have $Eg = 0.37$ eV.

Although the nearly-free-electron model provides an explanation for the above results, we must not rule out the possibility that the beats in the giant orbit oscillations are due to nonlinear effects of the sort envisaged by Shoenberg[11] and Pippard[12]. If the amplitudes of two oscillatory terms are sufficiently large, it is possible to obtain oscillations which have a frequency equal to the sum of the two separate frequencies. Supposing that such a situation occurs in zinc, then a frequency could be present having the sum of the lens and giant orbit frequencies. The corresponding period is $0.906 \times 10^{-9} \, G^{-1}$, in good agreement with the value in Table I. The orbit around the inside of the monster could similarly produce a sum frequency with the giant orbit resulting in a period of $0.848 \times 10^{-9} \, G^{-1}$. Nonlinear effects could thus account for two of the three subsidiary oscillations which have been observed.

Acknowledgments

The authors wish to thank R. K. Willardson and W. P. Allred of Bell and Howell Research Center for supplying the high-purity zinc crystal, T. G. Eck for communicating the results of his cyclotron resonance measurements previous to publication, T. G. Berlincourt and W. M. Lomer for many helpful discussions and suggestions, and D. G. Swarthout and P. C. Romo for orienting the samples.

References

1. W. L. Gordon, A. S. Joseph, and T. G. Eck, *The Fermi Surface*, W. A. Harrison and M. B. Webb (eds.), John Wiley & Sons, Inc., New York (1960), p. 84.
2. M. G. Priestley, *Proc. Roy. Soc. (London) Ser. A* **276**, 258, 1963.
3. A. S. Joseph and W. L. Gordon, *Phys. Rev.* **126**, 489, 1962.
4. M. G. Priestley, L. M. Falicov, and G. Weisz, *Phys. Rev.* **131**, 617, 1963.
5. R. W. Stark, T. G. Eck, W. L. Gordon, and F. Moazed, *Phys. Rev. Letters* **8**, 360, 1962.
6. Magnetoresistance measurements on Zn indicating magnetic breakdown effects have been carried out by R. W. Stark (private communication).
7. A. C. Thorsen and A. S. Joseph, *Phys. Rev.* **131**, 2078, 1963.
8. A. B. Pippard, *Proc. Roy. Soc. (London) Ser. A* **270**, 1, 1962.
9. A. S. Joseph and A. C. Thorsen, *Phys. Rev.* **133**, A1546, 1964.
10. A. H. Wilson, *The Theory of Metals*, Cambridge University Press, London (1958), p. 36.
11. D. Shoenberg, *Phil. Trans. Roy. Soc. London, Ser. A* **255**, 85, 1962.
12. A. B. Pippard, *Proc. Roy. Soc. (London) Ser. A* **272**, 192, 1963.

9
MAGNETISM

COMPLEX ANTIFERROMAGNETISM

J. van den Handel and H. C. Meijer

Kamerlingh Onnes Laboratorium, University of Leiden
Leiden, The Netherlands

In the classical examples of ferromagnetism and antiferromagnetism, such as iron, MnO_2, $CuCl_2 \cdot 2H_2O$, all dipoles together undergo a transition into a long range order phase when the temperature passes the Curie or Néel point. In the course of time, several salts have been found in which the transition from a state of weak short range interaction into a state of complete long-range order is more complicated and is realized in two or more steps. In the preceding low-temperature conference, some examples have already been given.[1,2] The principal results were:

$CuSO_4 \cdot 5H_2O$ and $CuSeO_4 \cdot 5H_2O$ (triclinic, single crystals). The suggestion of Geballe and Giauque that the magnetic dipoles could be divided into two groups was confirmed and considered in more detail in the Kamerlingh Onnes Laboratory.[1] Each group contains 50% of the dipoles. The first group shows a short-range order at about 1°K, the second becomes antiferromagnetic at $T = 0.046°K$ in the case of $CuSeO_4 \cdot 5H_2O$ and at 0.029°K for $CuSO_4 \cdot 5H_2O$.

Azurite [$Cu_3(CO_3)_2(OH)_2$; monoclinic, single crystal].[2] The results can be understood when it is assumed that two systems of dipoles exist, containing two-thirds and one-third of the total number of magnetic ions. The first group is short-range ordered at about 40°K, the second remains paramagnetic till 1.85°K, where it becomes antiferromagnetic.

Similar results were now obtained for nickeliodate ($Ni(IO_3)_2 2H_2O$; powder). RF measurements of the susceptibility by Burgiel, Jaccarino, and Schawlow[3] showed that nickeliodate has a transition at about 3.1°K. A sharp maximum in χ was interpreted as an indication of the onset of ferromagnetism. Our susceptibility measurements[4] in a static field also demonstrated the transition, and, though a maximum in χ was found, this has not the special character of that, found by Burgiel et al. Our results are in agreement with measurements of Williams and Sherwood.[3]

With an AC field (ν between 5 and 500 Hz), however, Miss Blom[5] could reproduce this special maximum which in a simultaneous weak static field gradually shrunk; at about 80 Oe it had completely disappeared.

Below 3°K the character of the σ–H curves, measured in a DC field changed, again indicating the ferromagnetic behavior, mentioned above. After having split σ into a ferromagnetic and a paramagnetic part and extrapolating to $T = 0$, the ferromagnetic part proved to be only 4% of the saturation magnetization of all dipoles in the salt. A change in the slope of the magnetization curve (for weak fields at about 2.5°K) indicates another transition.

Specific heat measurements were performed by Du Chatenier, Boerstoel, and De Nobel.[6] The c/T vs. T curves for several values of H (Fig. 1) indicated two transitions, at about 3.0° and 2.4°K. A closer examination of the curves leads to a combination of two λ curves, each corresponding roughly to a change in entropy

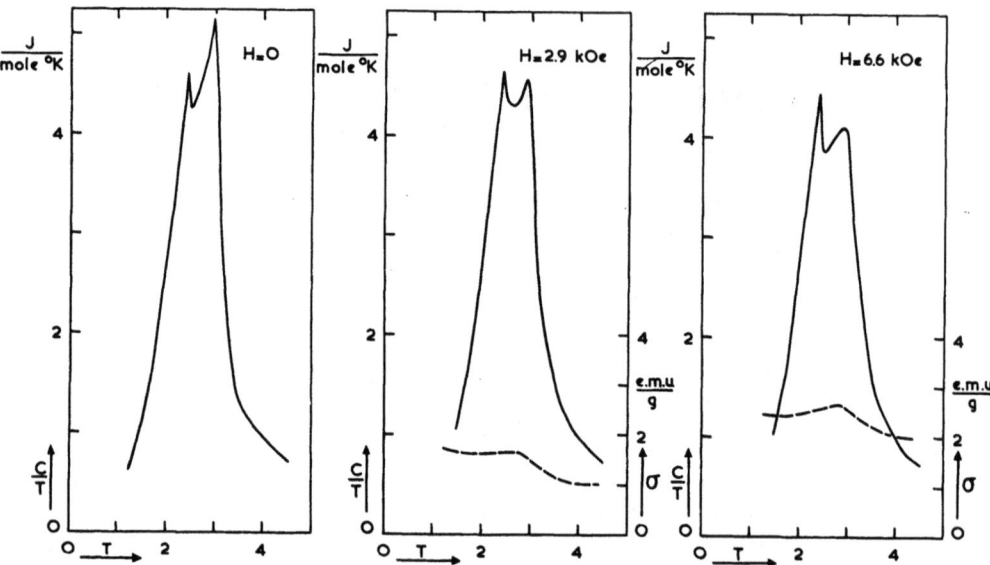

Fig. 1. Specific heat over temperature (———) and magnetization (- - -) of nickeliodate.

of $\frac{1}{2}R\ln 3$ per mole. Because of these results, a picture was accepted according to which there are two groups of nickeldipoles which order separately. The group with a long-range order transition at 3.1°K shows, moreover, the above-mentioned weak ferromagnetism, probably caused by the small rotation of the spontaneous magnetizations of the two sublattices with respect to each other, as predicted by Dzialoshinskii[7] and by Moriya.[8] This results in a residual magnetic moment.

Vivianite [$Fe_3(PO_4)_2 \cdot 8H_2O$; monoclinic, single crystal]. Specific heat,[9] magnetization,[10] and nuclear resonance measurements[11] were carried out. The values of c and σ as functions of T are given in Fig. 2. The c–T curve can be decomposed into three curves with maxima at about 11°, 8.8°, and 4°K. The total entropy under the curve c/T vs. T is 3 R ln 5 per mole in agreement with an ordering of the Fe^{++} ions.

An assumption about the shape of the curve belonging to the lowest maximum in the c/T vs. T diagram, which seems to us to be reasonable, gives an entropy connected with this lowest transition of about one-third of the total ordering entropy. This indicates that only one-third of the copper dipoles is ordered here. It is not sure which type of ordering takes place. The curve has the shape of a Schottky curve, but at low temperatures only antiferromagnetic nuclear resonance lines are found so that the possibility exists that the detailed shape of the subcurve is hidden by the main curve. The supposition of the splitting of the group of the dipoles into two groups containing one and two-thirds of the total number is supported by the crystal structure, where a system of single ions and a system of pairs of ions can be seen.[11] The system of single ions will thus remain "free" when the first system is ordered. This ordering takes place in two steps. At about 11°K the pairs show their preference for the parallel state,[11] thus forming gradually a system of dipoles with $S = 4$. This system undergoes a long-range order at 8.8°K.

The NMR and the susceptibility measurements support this view. The latter ones show that the preferred directions for both systems are in a plane normal to the

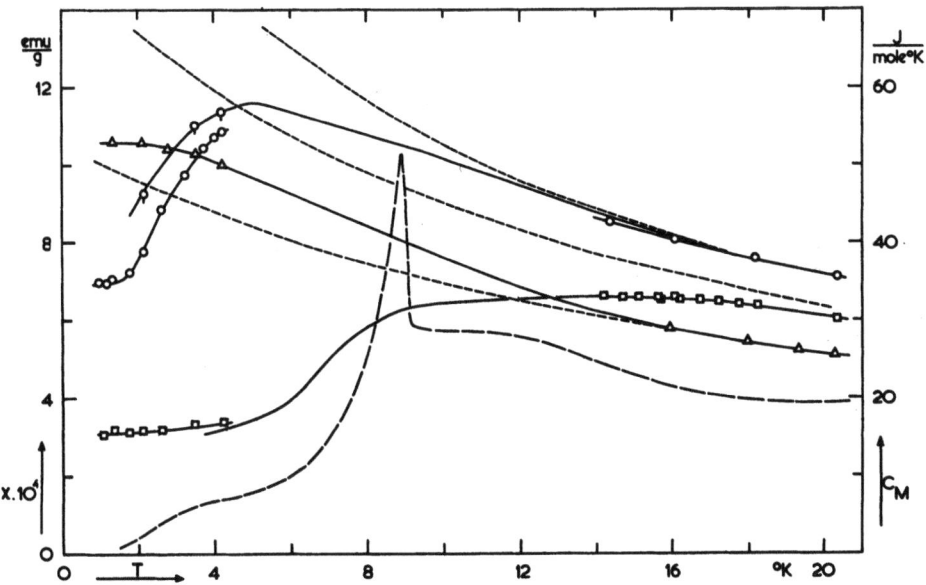

Fig. 2. Specific heat C_M and magnetization σ of vivianite. ———, experimental σ values; ---- extrapolated Curie–Weiss law; ———, C_M.

b axis, parallel and perpendicular to the a axis. For independent antiferromagnetic systems, one has $\chi_\perp = [Ng^2\mu_B^2 S(S+1)]/6kT_N = A[S(S+1)]/T_N$.[12] For two systems with equal N, with $S = 2$ and $S = 4$, and with $T_N = 4°K$ and $T_N = 8.8°K$, the values of $\chi_\perp$ are 1.5 A and 2.3 A. With the preferred directions as indicated, one would expect to find for χ the values 1.5, 2.3, and 3.8 A in the three perpendicular directions. The values 3.1, 7.0, and 10.6 × 10^{-4} emu/g which were really found do not seem to be in contradiction with the general picture, taking into consideration that neither an influence of the crystalline field nor that of a mutual magnetic interaction of the two systems were taken into account and that the formula is based on isotropic exchange interaction.

A remark, also applicable to other salts, must be made about the position of the magnetic axes. One of them was in general fixed (along the b axis), but in several cases the magnetic axes, in a plane normal to the b axis, are rotating, sometimes over a small angle (about 5°) such as in azurite, sometimes over a large angle (about 38°), as in vivianite.

Note Added in Proof

The supposition about the ordering of the dipoles in vivianite is based on Van der Lugt's remark that at about 11°K pairs are formed with $S = 4$. There has, however, arisen some doubt about the conclusion.

References

1. W. T. Duffy, Jr., J. Lubbers, H. van Kempen, T. Haseda, and A. R. Miedema, *Proceedings of the Eighth International Conference on Low Temperature Physics*, Butterworths, London, 1963, p. 245; A. R. Miedema, H. van Kempen, T. Haseda, and W. J. Huiskamp, *Physica* **28**, 119, 1962, and *Commun. Kamerlingh Onnes Lab.* No. 331a (1962).

2. E. Frikkee, W. van der Lugt, and J. van den Handel, *Proceedings of the Eighth International Conference on Low Temperature Physics*, Butterworths, London, 1963, p. 226; W. van der Lugt and N. J. Poulis, *Physica* **25**, 1313, 1959, and *Commun. Kamerlingh Onnes Lab*. No. 318c (1959); W. van der Lugt, thesis University of Leiden (1961); E. Frikkee and J. van den Handel, *Physica* **28**, 269, 1962; and *Commun. Kamerlingh Onnes Lab*. No. 332b.
3. J. C. Burgiel, V. Jaccarino, and A. L. Schawlow, *Phys. Rev.* **122**, 429, 1961.
4. H. C. Meijer and J. Van den Handel, *Physica* **30**, 1633, 1964; *Commun. Kamerlingh Onnes Lab*. No. 340b.
5. J. C. Blom (unpublished results).
6. F. J. Du Chatenier, B. M. Boerstoel, and J. De Nobel, *Physica* **30**, 1625, 1964; *Commun. Kamerlingh Onnes Lab*. No. 340a.
7. J. E. Dzialoshinskii, *Soviet Phys. JETP* (*English Transl.*) **5**, 1259, 1957.
8. T. Moriya, *Phys. Rev.* **117**, 635, 1960; *Phys. Rev. Letters* **4**, 228, 1960.
9. D. Onderdelinden, J. J. Kimmel, and Z. Dokoupil (unpublished results).
10. H. C. Meijer and J. Van den Handel (unpublished results).
11. W. Van der Lugt and N. J. Poulis, *Physica* **27**, 733, 1961, and *Commun. Kamerlingh Onnes Lab*. No. 327c; W. Van der Lugt, thesis University of Leiden (1961).
12. J. H. Van Vleck, *J. Chem. Phys.* **9**, 85, 1941.

ANTIFERROMAGNETIC TRANSITIONS IN SOME THIOUREA COORDINATED COMPOUNDS*

R. Au, J. A. Cowen, R. D. Spence, and H. Van Till†

*Department of Physics and Astronomy, Michigan State University
East Lansing, Michigan*

Introduction

The low-temperature study of the transition metal chlorides and bromides coordinated with thiourea was undertaken in order to determine the nature of any observable magnetic ordering transition. Each of the three salts listed here exhibits an antiferromagnetic transition in the behavior of both the proton magnetic resonance and the magnetic susceptibility in the temperature range 0.4°K to 4.2°K. This preliminary report will summarize the data taken up to the present time and give some indications of the nature of the ordered spin states of these materials.

Experimental Methods

Temperatures from 1.1° to 4.2°K were obtained by a standard ^{4}He cryostat. For lower temperatures a glass ^{3}He cryostat was used.[1] Proton magnetic resonance data were obtained with a marginal oscillator and phase-sensitive detection system, the oscillator coil being wrapped around the outside of the ^{3}He dewar. Magnetic susceptibility was measured with a set of mutual inductance coils and a commercial 17-cps mutual inductance bridge.

All crystals were grown from an aqueous solution of thiourea and the transition metal chloride or bromide.[2] A preliminary X-ray analysis has been done to determine the space group, unit cell dimensions, and number of chemical formula units per unit cell. A more complete analysis will be reported later.

Co[(NH$_2$)$_2$CS]$_4$Cl$_2$

Co(tu)$_4$Cl$_2$ forms dark blue bipyramidal crystals with tetragonal point symmetry $4/m$. From the X-ray data it is found that the space group is P4$_2$/n, the unit cell dimensions are $a = 13.52$ Å and $c = 9.10$ Å, and there are four formula units per unit cell.

The proton magnetic resonance diagram in the (110) plane is shown in Fig. 1 with a representative number of lines traced out. At this temperature (0.47°K) the resonance pattern is antiferromagnetic as shown by the 360° period and pairing of high and low field lines. As the temperature is raised the lines collapse toward the free proton line position, widen, and disappear with the appearance of the paramagnetic lines of period 180° and much smaller splitting. Measurements in zero applied field below 0.93°K show the presence of a number of true internal magnetic fields.

---

* This work was supported by the Air Force Office of Special Research and the Army Research Office (Durham).
† National Science Foundation Predoctoral Fellow.

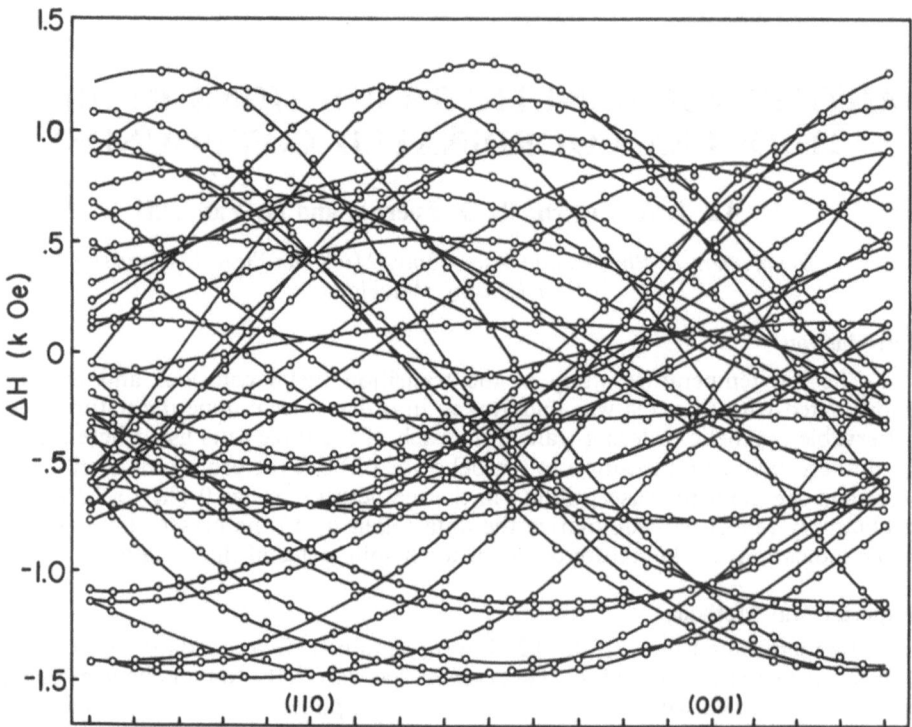

Fig. 1. Proton magnetic resonance diagram of Co(tu)$_4$Cl$_2$ in the (110) plane at $T = 0.47°$K. ΔH is the difference between the actual resonance field and the resonance field for the free proton at 9.71 Mcps.

Magnetic susceptibility measurements have been made on both single-crystal and powder samples. Typical data are shown in Fig. 2. In the limited temperature range between 2° and 4°K, the susceptibility obeys a Curie–Weiss law with $\theta \approx 3.7°$K, the Curie constant for the powder being $C \approx 2.5$/mole. The maximum in $d\chi/dT$ occurs at $T \approx 0.93°$K. The data shown in Fig. 2 indicate that the sublattice magnetization is along the c-axis.

Mn[(NH$_2$)$_2$CS]$_4$Cl$_2$

Mn(tu)$_4$Cl$_2$ forms very pale green bipyramidal crystals, similar in habit to Co(tu)$_4$Cl$_2$, with tetragonal point symmetry $4/m$. X-ray data indicate the space group $P4_2/n$ and four formula units per unit cell of dimensions $a = 13.76$ Å and $c = 9.07$ Å.

Proton magnetic resonance has been observed as a function of temperature. Above 0.56°K the resonance is typical of the paramagnetic state, while below this temperature the observed resonance lines indicate an antiferromagnetic state. A full investigation of the angular variations has not yet been completed.

The magnetic susceptibility follows a Curie–Weiss law in the paramagnetic region with $\chi_{001} < \chi_{110}$. Below 0.56°K, χ_{110} decreases very rapidly while χ_{001} falls off more slowly. Thus, it appears that the sublattice magnetization is not along the c-axis as in Co(tu)$_4$Cl$_2$.

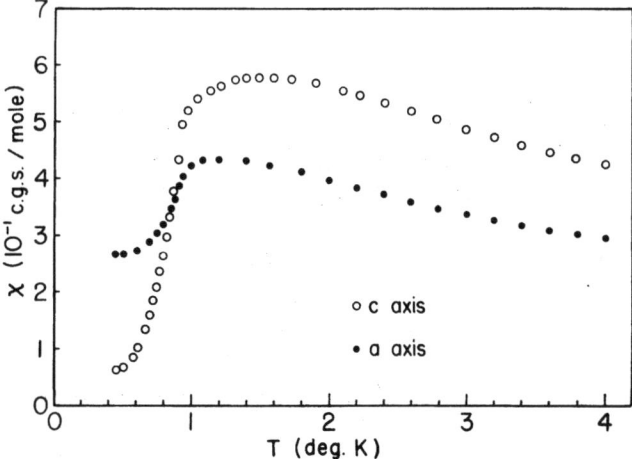

Fig. 2. Magnetic susceptibility data on single-crystal samples of Co(tu)$_4$Cl$_2$.

Ni[(NH$_2$)$_2$CS]$_6$Br$_2$

Ni(tu)$_6$Br$_2$ forms dark yellow–green prismatic crystals with monoclinic point symmetry $2/m$. X-ray data indicate unit cell dimensions $a = 24.30$ Å, $b = 8.91$ Å, $c = 16.79$ Å, and $\beta = 137.02°$, with space group $C2/c$ and four formula units per unit cell.

From proton resonance data a transition temperature of 2.28°K is found. Below this temperature the resonance pattern is similar to that shown in Fig. 1. Resonance experiments in zero applied field indicate at least fourteen local fields between 200 and 700 G. At present, the data are being analyzed for the purpose of identifying the magnetic symmetry group.

References

1. J. A. Cowen, R. D. Spence, H. Van Till, and H. Weinstock, *Rev. Sci. Instr.* **35**, 914, 1964.
2. M. Nardelli, L. Cavalca, and A. Braibanti, *Gazz. Chim. Ital.* **86**, 867, 1956.

THERMAL CONDUCTIVITY OF CRYSTALS WITH MAGNETIC LINEAR CHAINS AT VERY LOW TEMPERATURES

A. R. Miedema, J. N. Haasbroek, and F. W. Gorter

Kamerlingh Onnes Laboratory
Leiden, The Netherlands

The thermal conductivity of dielectric hydrated crystals has been studied using the steady-state method. At temperatures above 1.5°K, the liquid-helium bath serves as a cooling reservoir and the temperature gradient along the specimen is measured by means of two carbon resistors. Two heaters are used: One of them produces the temperature gradient while the other heater raises the temperature of the specimen as a whole above the temperature of the surrounding helium bath.

Figure 1 describes the apparatus used for measurements below 1°K. Principally the temperature gradient is determined by one thermometer (A) only, which measures two temperatures in turn by means of two thermal switches (lead wires in a coil of niobium wire, D_I and D_{II}). The cooling reservoir (F) is made of slabs of chrome–alum single crystal, glued with grease between brass plates. The plates are screwed to a 3-mm copper rod to which the lower end of the specimen (C) is connected.

The coils of niobium wire, which produce the 800 Oe field needed for the thermal switches, are mounted on top of the cooling salt. For the heater (B), we used a brass strip with a resistance of $1 \times 10^{-2}\,\Omega$. The heater serves both for creating a temperature gradient and raising the temperature of the specimen as a whole. The steady-state condition was realized to a good approximation, as long as the specimen temperature was considerably higher than that of the cooling salt, while the latter must be in the region of large heat capacity ($T < 0.25°K$). In some experiments an extra thermal resistance was mounted between the low-temperature end of the specimen and the cooling salt, so that it was possible to have a steady-state condition even at relatively high temperatures ($T \approx 1°K$).

The temperature is derived from the susceptibility of a cerium magnesium nitrate single crystal. Because of uncertainties in the Curie constant calibration in the liquid–helium temperature range and of small parasitic susceptibilities (glass thermal shield), the experimental accuracy may be about 2% for the temperature itself, but since we use only one thermometer, the temperature gradient is measured with practically the same accuracy.

Thermal contact between the specimen and cooling salt, thermal switches, and heater had been obtained by soldering them to copper foil, glued to the specimen with Apiezon grease. The specimens were cut from large single crystals and ground to dimensions of about 3×3 mm. Their length was about 20 mm and the distance between the thermometer positions was about 10 mm.

Generally, one expects a T^3 dependence for the thermal conductivity of dielectric crystals at very low temperatures, according to the formula:

$$\lambda = \tfrac{1}{3} C_v v_0 l \tag{1}$$

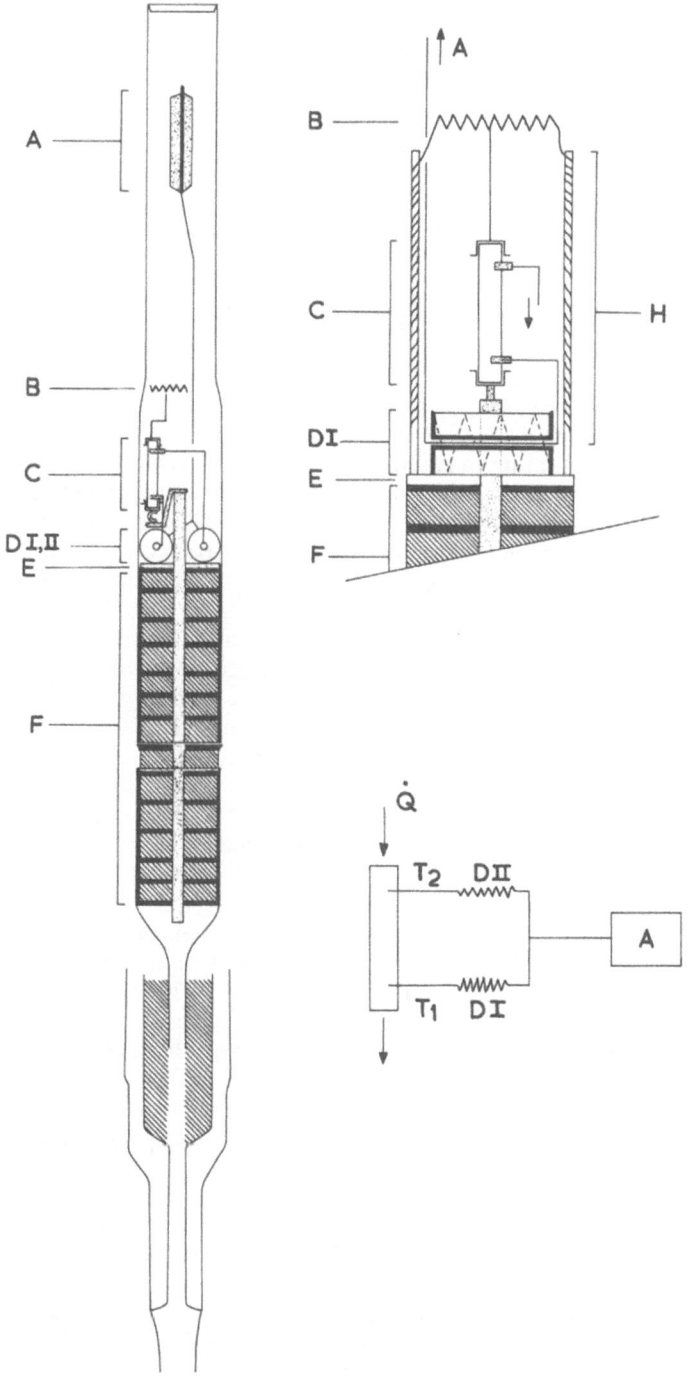

Fig. 1. Apparatus for measuring thermal conductivities of dielectric crystals below 1°K. A: magnetic thermometer; B: heater; C: specimen; D_I, D_{II}: coils and thermal switches; E: perspex isolating plate; F: cooling salt; H: connection to heater (lead wires on nylon).

where C_v is the heat capacity of the lattice heat waves. v_0 is the sound velocity, and l the phonon mean free path. At very low temperatures, l will be independent of temperature, so that $\lambda \sim c_v \sim T^3$. The T^3 law has been found experimentally in many cases; from the proportionality constant a value for l can be estimated. The values for l may be different for different specimens of the same salt (see Berman et al.[1]), but we have found that in optically perfect single crystals of

$$Co(NH_4)_2(SO_4)_2 \cdot 6H_2O$$

and

$$Ce_2Mg_3(NO_3)_{12} \cdot 24H_2O$$

the phonon mean free path equals the smallest dimension of the specimen. For $Co(NH_4)_2(SO_4)_2 \cdot 6H_2O$, the T^3 dependence is reached below 2°K and λ shows its maximum (3 W cm/deg K) at $T = 10°K$.

For crystals containing magnetic atoms, some influence on λ may be expected at temperatures where T is of the order of magnitude of the magnetic energy splittings, provided the spin lattice relaxation times are sufficiently short. On one hand, the thermal conductivity may be reduced, as l may be reduced, while on the other hand, λ may be enhanced if energy is transferred by magnetic effects (spin diffusion, spin waves).

The first effect has been clearly demonstrated in experiments of Donaldson and Edmonds[2] on $CuCl_2 \cdot 2H_2O$, while both effects seem to be important for the thermal conductivity of the ferromagnetic crystal $Cu(NH_4)_2Cl_4 \cdot 2H_2O$.[3]

From our experiments on salts, in which the magnetic atoms behave as nearly isolated linear chains, we hope to be able to distinguish between the two effects mentioned.

Figure 2 shows the thermal conductivity of $CuSO_4 \cdot 5H_2O$. Three samples have been investigated, all being cut from the same single crystal. The long axes of the specimen makes an angle of 0°, 45°, and 90°, respectively, with the crystallographic a-axis. The figure shows that especially near 1°K the thermal conductivity is very anisotropic, the conductivity for the 0° specimen being larger by more than a factor 10. One may notice that λ for the 45° sample is not in between that for the other two directions for temperatures above 0.5°K. The existence of a very large anisotropy in the thermal conductivity of $CuSO_4 \cdot 5H_2O$ has been verified by measuring the temperature gradient perpendicular to the direction of the heat flow for the 45° sample. In case of extreme anisotropy [$\lambda(0°) \gg \lambda(90°)$], one expects the temperature gradients existing parallel and perpendicular to the direction of the heat flow to be nearly equal. This has been found to be the case near 1°K.

Similar effects have been observed for $Cu(NH_3)_4SO_4 \cdot H_2O$, for which we were unable to study different specimens cut from the same single crystal. Near 1°K, the conductivity parallel to the c-axis, the direction of the chain, is higher than that perpendicular to the chains. At temperatures below 0.4°K, where the interactions between the chains establish long-range order, λ is not different for the 0° and 90° directions and attains a T^3 dependence below 0.25°K. This indicates that the spin lattice relaxation times become long below T_N, so that at $T = 0.25°K$ the spins no longer influence the thermal conductivity.

In the temperature region just above T_N (0.4° < T < 0.8°K) we found that $\lambda(0°)$ is definitely smaller than $\lambda(90°)$, while the average is about 3 times smaller than the value obtained by extrapolation of the T^3 term.

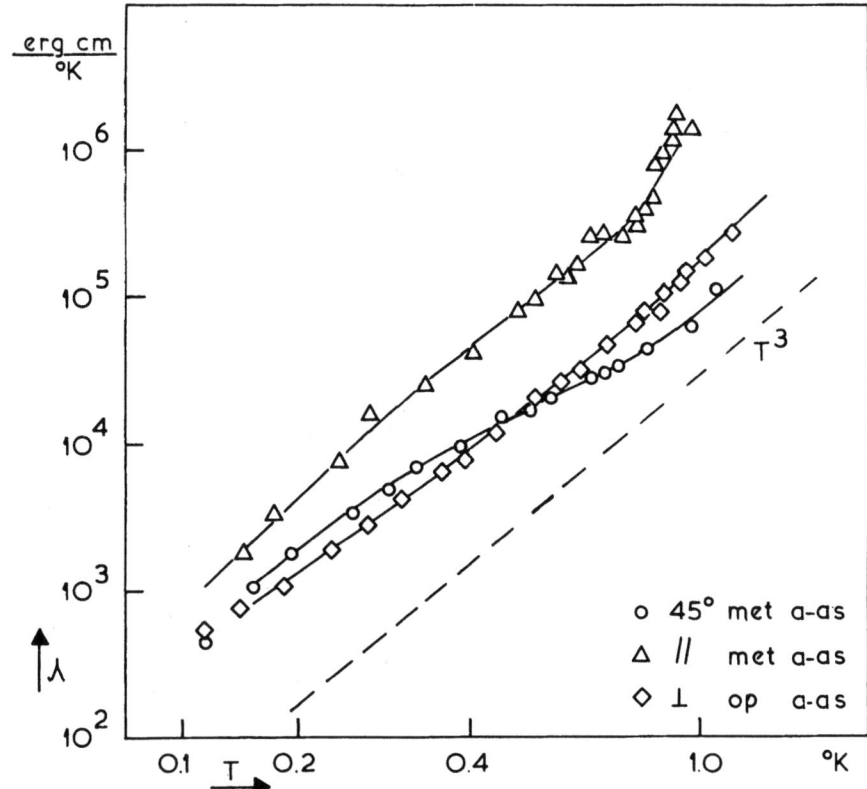

Fig. 2. Thermal conductivity of $CuSO_4 \cdot 5H_2O$ below 1°K for three orientations with respect to the crystallographic a-axis.

From the experimental data which at this moment are still incomplete, some important conclusions may be drawn:

1. In copper salts (isotopic exchange), magnetic effects in the thermal conductivity may occur in a wide temperature range.
2. Energy transfer by magnetic interactions may be effective even if only short-range magnetic ordering exists. Magnetic conductivities of at least 10^{-1} W cm/deg K are found.
3. In $CuSO_4 \cdot 5H_2O$, the chains, in which half the number of copper ions are arranged magnetically, run parallel to the crystallographic a-axis.
4. For a magnetic linear chain, the spin-phonon interaction is angular dependent. For phonons perpendicular to the chain direction, the interaction with the magnetic spins may be a minimum. Furthermore, this interaction seems to be no longer angular dependent, if the chains attain long-range order.

References

1. R. Berman, J. C. F. Brock, and D. J. Huntley, *Phys. Letters* **3**, 310, 1963.
2. R. H. Donaldson and D. T. Edmonds, *Phys. Letters* **2**, 130, 1962.
3. E. A. Van Kempen, thesis, Leiden (1965).

THERMAL BEHAVIOR OF THE ANTIFERROMAGNET CoCl₂ · 6H₂O AT ITS NÉEL POINT*

J. Skalyo, A. F. Cohen, and S. A. Friedberg

Carnegie Institute of Technology
Pittsburgh, Pennsylvania

Several years ago,[1] the heat capacities of the antiferromagnetic salts $CoCl_2 \cdot 6H_2O$ and $NiCl_2 \cdot 6H_2O$ were found to vary with $|T - T_N|$ in essentially a logarithmic way as one approached within 0.07°K of their Néel points from either side. This paper reports briefly new thermal and magnetic measurements performed with improved temperature resolution on one of these salts, $CoCl_2 \cdot 6H_2O$, near its Néel point, $T_N = 2.3°K$. The cobaltous salt was chosen for this work primarily for its convenient T_N. It also has other properties recommending it for such study. The effective spin of the Co^{++} ion at low temperatures is $S = \frac{1}{2}$ and the moment is quite anisotropic so that the Ising approximation may be applicable in treating its behavior. In addition, an analysis[1] of earlier c_p data reveals that the spin entropy gained above T_N, 0.36R, is more than half that associated with the complete disordering of the spin system, namely, $R \ln(2S + 1) = 0.69R$. This fact and the structure of the material are consistent with the idea that $CoCl_2 \cdot 6H_2O$ may resemble a two-dimensional antiferromagnet whose microscopic treatment is tractible in the Ising approximation.

The heat capacities of two different specimens of $CoCl_2 \cdot 6H_2O$ have been measured in a vacuum calorimeter by a discontinuous heating technique. Temperature intervals $\leqslant 10^{-4}$ °K were covered in heating periods near T_N. This temperature resolution was made possible by the use of carbon resistance thermometers in an audio frequency bridge circuit (8 cps) employing synchronous detection. Temperature measurements had an absolute accuracy of ± 0.002°K and a relative accuracy of $\pm 1.0 \times 10^{-6}$ °K. Measured c_p values were found to rise rapidly as T_N was approached but not without limit. The λ-like anomaly is rounded off over an interval of $\sim 10^{-2}$°K, which includes T_N. It is in this interval that Sawatzky and Bloom,[2] using proton resonance techniques, found evidence of the coexistence of para and antiferromagnetic phases. It thus appears possible that the observed truncation of the λ-anomaly is associated with inhomogeneity of the specimen crystals through chemical impurity, physical imperfection, or strain. These would act to reduce the effective specimen size,[3] limiting the range over which long-range order might propagate and with it the magnitude of c_p at T_N. Imperfections could also cause local variations in the strengths of the interactions among spins and thus broaden the λ-anomaly.

Rounding-off of the anomalous peak in c_p makes the identification of T_N not obvious. We find, however, that a particular choice of T_N, namely, 2.2890°K,

---
* Work supported by the National Science Foundation and the Office of Naval Research. Some of these results are described in *Phys. Rev. Letters* **13**, 133, 1964.

causes a plot (Fig. 1) of c_p vs. $\ln|T - T_N|$ outside the rounded region to exhibit the kind of behavior seen to much smaller values of $|T - T_N|$ for ^{4}He at its λ-point.[4] Over significant intervals, the curves for $T > T_N$ and $T < T_N$ on this diagram are well fitted by parallel straight lines given by $(c_p/R) = -0.015 - 0.271 \ln|T - T_N| + \Delta$, where $\Delta = 0$ for $T > T_N$ and $\Delta = 0.574$ for $T < T_N$. Other choices of T_N tend to destroy this symmetry. Interestingly enough, such symmetrical logarithmic behavior is expected for two-dimensional Ising models[3,5] with which, as was mentioned above, $CoCl_2 \cdot 6H_2O$ may show some common features. Furthermore, the coefficient of the logarithmic term, $0.27R$, is comparable in magnitude with the values calculated for two-dimensional Ising models. It can actually be reproduced quantitatively with Onsager's results[3] for a square lattice with suitably chosen anisotropy of the interactions, although the actual lattice may have somewhat different symmetry.

We are inclined to believe that T_N as chosen above is rather close to the "actual" value, i.e., the value for which c_p measured for an ideal crystal of this material would assume its maximum value. At the same time it appears possible that c_p for such an ideal crystal might vary with temperature in the way indicated by extrapolation with the fitted formula given above to smaller values of $|T - T_N|$. In other words, outside the interval in which truncation of the heat capacity anomaly occurs, our specimens may approximate rather well the equilibrium behavior of an ideal crystal of this salt. Let us consider an observation consistent with this idea.

In the course of their resonance studies on $CoCl_2 \cdot 6H_2O$, Sawatzky and Bloom[2] remeasured its parallel susceptibility at a frequency of 21 Mcps at temperatures rather close to T_N. Since these data show features not seen in earlier measurements at low audio frequencies,[6] it was felt to be of interest to repeat the latter measurements with better temperature resolution. This has been done for several specimens at temperatures within $\approx 10^{-2}$ °K of T_N. Most of the data were taken at 275 cps, although checks at higher and lower frequencies yielded similar values. The results confirm the earlier low-frequency values, $\chi_\parallel$ varying with T as expected and

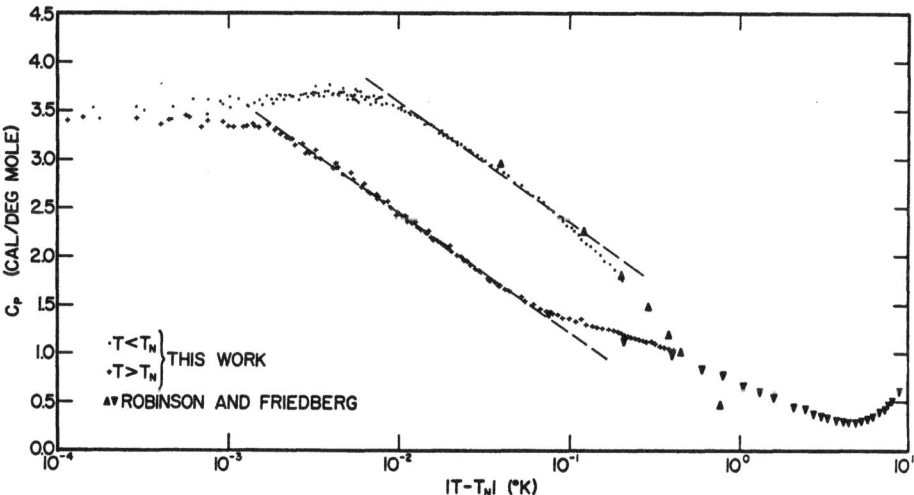

Fig. 1. The heat capacity of $c_p(H = 0)$ of $CoCl_2 \cdot 6H_2O$ vs. $\ln|T - T_N|$ for $T_N = 2.2890$°K.

showing its maximum temperature derivative quite close to T_N in accord with theory. Fisher[7] has shown for both Ising and Heisenberg models that $C \propto A[\partial(\chi T)/\partial T]$, where A varies slowly with T near T_N. A plot of $\int_0^T C\, dT$ vs. $\chi_\parallel T$ for $2.19° < T < 2.39°K$ yields a good straight line as this relation requires.

As did Sawatzky and Bloom, we may adapt a result from Buckingham and Fairbank's[8] thermodynamic treatment of λ points to a magnetic system and write $(C_H/T) = (dH/dT_N)^2 \chi(T) +$ terms varying slowly with T near T_N. If $\chi(T) = \chi_\parallel(T, H = 0)$, then dH/dT_N is the initial slope of the antiferro-paramagnetic phase boundary in the $H-T$ plane for H parallel to the preferred spin direction. A plot of $c_p(H = 0)/T$ vs. $\chi_\parallel$, using only data outside the region in which the λ-anomaly is truncated, yields curves approaching asymptotically, from above and below, the line $\chi = \chi(T_N)$. An ideal crystal might thus be expected to exhibit a rather sharp phase boundary with infinite slope at $H = 0$, as is reasonable. The state of the real crystal in the interval $\Delta T \sim 10^{-2}$ °K about T_N does not appear to have a simple description. Thus, even were values of $\chi_\parallel$ available in this interval, it is not clear if their inclusion in the above plot would be meaningful. Note that this analysis of the new data removes the paradox suggested by Sawatzky and Bloom for $T > T_N$.

References

1. W. K. Robinson and S. A. Friedberg, *Phys. Rev.* **117**, 402, 1960.
2. E. Sawatzky and M. Bloom, *Can. J. Phys.* **42**, 657, 1964.
3. L. Onsager, *Phys. Rev.* **65**, 117, 1944.
4. W. M. Fairbank, M. J. Buckingham, and C. F. Kellers, *Proceedings of the Fifth International Conference on Low Temperature Physics and Chemistry*, J. R. Dillinger (ed.), University of Wisconsin Press, Madison (1958), p. 50.
5. See, for example: M. E. Fisher, *J. Math. Phys.* **4**, 278, 1963.
6. R. B. Flippen and S. A. Friedberg, *J. Appl. Phys.* **31**, 338S, 1960; T. Haseda, *J. Phys. Soc. Japan* **15**, 483, 1960.
7. M. E. Fisher, *Phil. Mag.* **7**, 1731, 1962.
8. M. J. Buckingham and W. M. Fairbank, *Progress in Low Temperature Physics*, Vol. III, (1961), p. 189.

MAGNETIC SUSCEPTIBILITY OF SINGLE CRYSTAL FeCl$_2$ PARALLEL AND PERPENDICULAR TO THE TRIGONAL AXIS*

J. W. Stout, Carl L. Brandt, and Charles Trapp

Department of Chemistry and Institute for the Study of Metals
University of Chicago, Chicago, Illinois

The magnetic susceptibilities of single crystals of FeCl$_2$ have been measured parallel and perpendicular to the trigonal axis over the temperature range from 1.2° to 300°K. The method of measurement is a refinement of the null coil method described by Arrott and Goldman.[1] The single-crystal sample, in the form of a right circular cylinder encased in a polyethylene capsule, is placed in a uniform magnetic field of approximately 100 G. The magnetic field arising from the uniform magnetization of the sample is cancelled by means of a direct current through a null coil wound on the cylindrical surface of the sample. Demagnetization effects are accurately taken into account. The accuracy is a few tenths of one percent in the susceptibility.

Both the parallel and perpendicular susceptibilities show peak values at 24°K, in qualitative agreement with the results of earlier investigators, although there are large quantitative differences between the present results and earlier data. The peak value of the parallel susceptibility is 1.05 cm^3/mole and that of the perpendicular susceptibility is 0.2406 cm^3/mole. At 1.3°K, the parallel susceptibility is 0.0104 cm^3/mole and the perpendicular susceptibility is 0.178 cm^3/mole. The unusual drop in the perpendicular susceptibility below the Néel temperature, as well as the other features of the susceptibility curve, are quantitatively explained by a calculation in which the positions of the one-atom electronic levels are determined by the spin-orbit coupling constant and a trigonal field parameter which splits the t_{2g} orbital states. The ferromagnetic interactions between Fe^{++} ions in a layer and the antiferromagnetic interactions between ions in adjacent layers are approximated by molecular fields.

Reference

1. Arrott and Goldman, *Rev. Sci. Instr.* **28**, 99, 1957.

---

* Supported in part by the National Science Foundation and by the Office of Naval Research.

MAGNETIC AND CALORIC PROPERTIES OF K_3MoCl_6

P. A. van Dalen and H. M. Gijsman
Technological University
Eindhoven, The Netherlands

and

N. Love and H. Forstat*
Physics Department, Michigan State University
East Lansing, Michigan†

The susceptibility of powdered K_3MoCl_6 has been measured in the temperature range of 1°–300°K. The salt was commercially obtained and had a purity of about 98%. Calorimetric experiments were made on a sample of the same origin.

The results of the susceptibility experiments, performed with a magnetic balance, can be summarized as follows: Above 15°K, the susceptibility follows a Curie–Weiss law

$$\chi = \frac{Ng^2 S(S+1)\beta^2}{3k(T-\theta)}$$

with $S = \frac{3}{2}$, $g = 1.87$, and $\theta = -5°K$, the magnetic moment of the Mo^{3+} ion being 3.61 Bohr magnetons. The experiments were made at eight different temperatures between 15°K and room temperature. The g-value of the salt, obtained from resonance experiments at room temperature, was found to be 1.92. These measured values are in rather good agreement with those found by Epstein and Elliot,[1] who found a Curie law between 77° and 300°K with $g = 2$, and with resonance experiments on a dilute sample, quoted by Griffiths,[2] leading to a g-value at room temperature of 1.93–1.96.

The results in the temperature region of liquid helium are shown in Fig. 1. Figure 1a plots the magnetization vs. magnetic field strength. Below about 1000 Oe, the magnetization rises sharply with increasing field; in higher fields the magnetization increases linearly with the applied field. The slope of the line decreases with decreasing temperature.

If we suppose that the sharp rise in magnetization is due to a ferromagnetic contribution which is saturated at about 1000 Oe, and plot this saturation value as a function of temperature, a curve as shown in Fig. 1b is obtained.

If we write $\sigma(H, T) = \sigma_0(T) + \chi(T)H$ for fields higher than 1000 Oe, the susceptibility $\chi(T)$ proves to be of an antiferromagnetic character, as is shown by Fig. 1c.

A mutual inductance method (operating frequency 515 Hz) was applied in order to obtain susceptibility values above the normal boiling point of liquid

---

* Fulbright Fellow at Trinity College, Dublin, Ireland (1963–64).
† Part of this work was performed under a grant from the U.S. Air Force Office of Scientific Research.

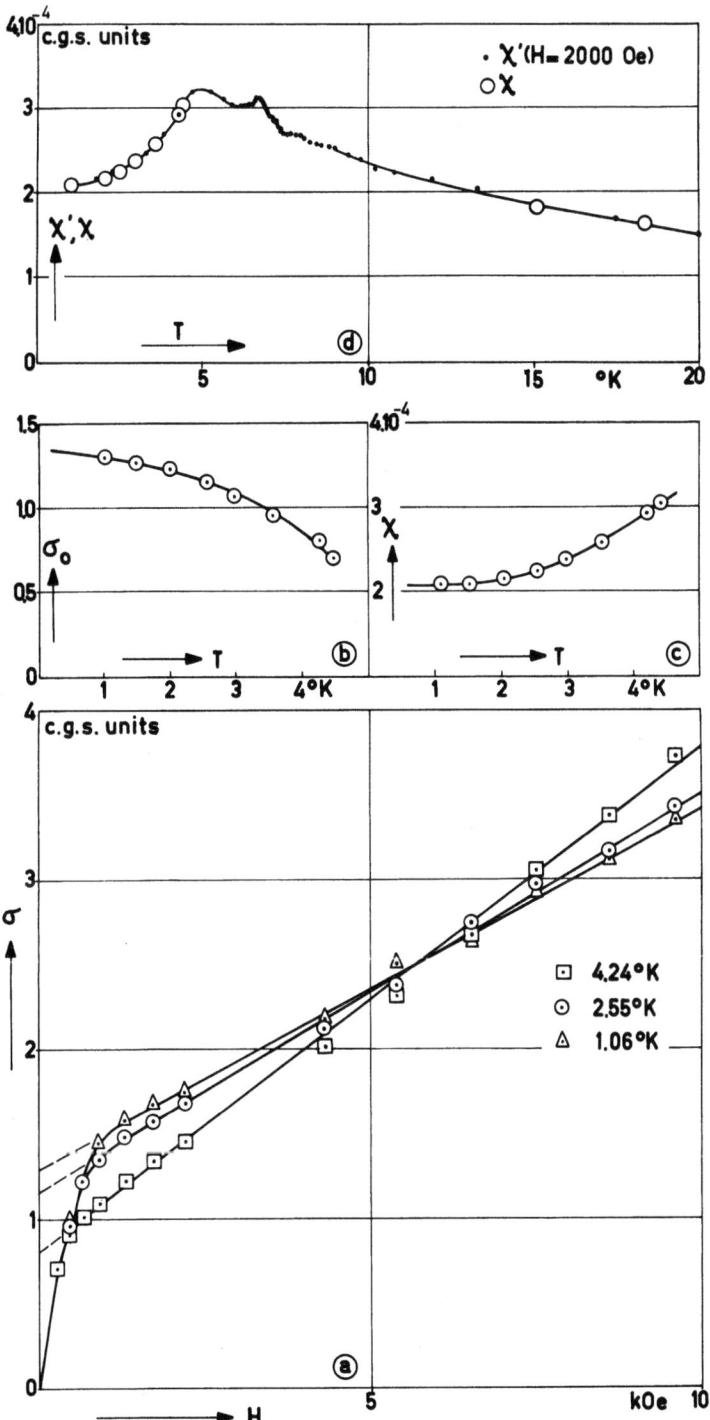

Fig. 1. Magnetization and susceptibilities per gram of K_3MoCl_6. (a) Magnetization as a function of the magnetic field in the helium region. (b) Temperature dependence of the ferromagnetic contribution to the magnetization. (c) The antiferromagnetic susceptibility below 5°K. (d) The AC susceptibility in an external field of 2000 Oe. The values of the static susceptibility are also given.

helium. The temperature measurements were performed with a carbon resistor, calibrated in the liquid-helium range. The heating occurred also with this resistor. Preliminary values of the susceptibility χ' in a transverse external field of about 2000 Oe are shown in Fig. 1d. The values of χ' were obtained by fitting the results of the AC measurements in the hydrogen temperature range to the values following from the Curie–Weiss law. In the helium region the values of χ' thus obtained and χ_{static} proved to be in agreement. It was found that maxima in the susceptibility occur at 5° and 6.7°K. Measurements in zero external field showed that the temperature at which the second maximum was situated is independent of the strength of the applied field; the maximum at 5°K seems to move to a lower temperature with decreasing fieldstrength.

The specific heat vs. temperature curve gives some more information about the complicated behavior of the material in the temperature region of 4°–10°K. A calorimetric method was used with a triple can calorimeter; heating occurred with a manganin wire heater; the temperatures were measured with a carbon resistor, calibrated against the vapor pressure of liquid helium. Four separate runs were made on the sample; three runs were made on the empty container for the corrections. The final results are shown in Fig. 2, where a λ-type transition which occurs at approximately 6.8°K, a temperature corresponding with one of the maxima in the susceptibility, can be seen. Assuming that the specific heat above 6.8°K could be written as $C_p = a/T^2 + bT^3$, we obtained the constants a and b by plotting $C_p T^2$

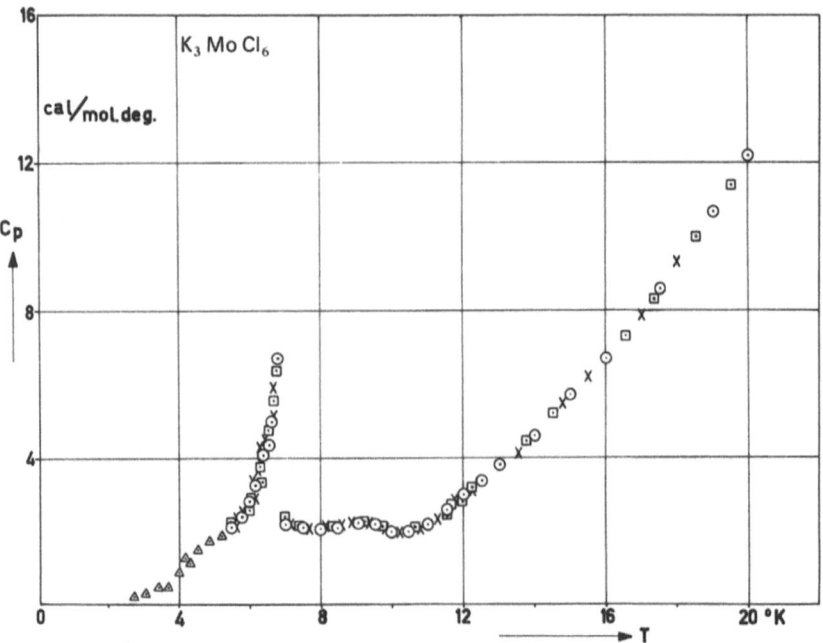

Fig. 2. The specific heat of K_3MoCl_6 between 1° and 20°K.

vs. T^5. The lattice term was found to be $1.5 \times 10^{-3} \ T^3$— corresponding with a Debye temperature of about 70°K—the paramagnetic term $80/T^2$.

The total entropy change associated with the transition was obtained by measuring the area under the C_p(magn)$/T$ vs. T curve, and amounted to approximately 2.20 cal/mole-deg. This value has to be compared with the theoretical value of $R \log(2S + 1) = 2.78$ cal/mole-deg for $S = \frac{3}{2}$. The entropy above the transition temperature amounts to about 40% of the total entropy change. It may be noted that a small rounded peak occurs at about 9°K. There is also evidence for some small irregularity in the susceptibility at this temperature (Fig. 1d). We might try to explain the previously mentioned results as a combined antiferromagnetic–ferromagnetic behavior, as was recently described by Moriya[3] and earlier by Dzyaloshinsky,[4] i.e., a weak ferromagnetism existing in a mainly antiferromagnetic material. This behavior has been found in several other salts.[5,6]

On the basis of a molecular field model, we can understand the ferromagnetic contribution, assuming that the magnetization of the sublattices are not exactly antiparallel but slightly canted. In fact, the ferromagnetic contribution in this salt is small; if we calculate the effective moment p associated with the ferromagnetic magnetization at zero temperature (Fig. 1b), we find a value of $p = 0.1\beta$. For further explanations and more detailed comparison with Moriya's theory, it should be necessary to perform experiments on single crystals. The growing of single crystals, however, seems to be rather impossible; crystallographic data are lacking, too. This applies also to two other investigated salts $K_2MoCl_5 \cdot H_2O$ and $K_3MoCl_6 \cdot 2H_2O$, which show similar magnetic properties.

With respect to the χ' vs. T curve (Fig. 1d) we could possibly explain the maximum in the susceptibility, occurring at 5°K, by considering how the magnetization varies with the applied field at different temperatures. Further investigations on the field dependence of the susceptibility are necessary in order to give a more complete description of the magnetic behavior. Also, measurements of the specific heat in magnetic fields may be worthwhile.

References

1. C. Epstein and N. Elliot, *J. Chem. Phys.* **22**, 634, 1954.
2. J. H. E. Griffiths, J. Owen, and I. M. Ward, *Proc. Roy. Soc. (London) Ser. A* **219**, 526, 1953.
3. T. Moriya, *Phys. Rev.* **120**, 91, 1960; in: *Magnetism, Vol. I*, G. T. Rado and H. Suhl (eds.), New York (1963), p. 85.
4. I. Dzyaloshinsky, *J. Phys. Chem. Solids* **4**, 241, 1958.
5. C. Domb and A. R. Miedema, in: *Progress in Low Temperature Physics, Vol. IV*, C. J. Gorter (ed.), North Holland Publishing Co., Amsterdam (1964), p. 296.
6. R. B. Flippen and S. A. Friedberg, *Phys. Rev.* **121**, 1591, 1961.

HYPERFINE STRUCTURE COUPLING IN FERRIC AMMONIUM SULFATE AS A FUNCTION OF MAGNETIC FIELD AND TEMPERATURE*

F. E. Obenshain, L. D. Roberts, C. F. Coleman,† and D. W. Forester‡

Oak Ridge National Laboratory
Oak Ridge, Tennessee

and

J. O. Thomson

University of Tennessee
Knoxville, Tennessee

In the study of the hyperfine structure coupling in ferromagnetic or antiferromagnetic materials by means of the Mössbauer effect the concept of an effective magnetic field H_{eff} has been widely used.[1] A precise definition of H_{eff} may be given in terms of the spin Hamiltonian $\mathcal{H}_i$ associated with the ith magnetic ion of spin S_i in the magnetic solid:

$$\mathcal{H}_i = A\mathbf{I}_i \cdot \mathbf{S}_i + g_n\beta_n\mathbf{I}_i \cdot \mathbf{H} + g_e\beta_e\mathbf{S}_i \cdot \mathbf{H}$$
$$+ \sum_{j \neq i} J_{ij}\mathbf{S}_i \cdot \mathbf{S}_j + g^2\beta^2 \sum_{j \neq i} \left[\frac{\mathbf{S}_i \cdot \mathbf{S}_j}{r_{ij}^3} - 3\frac{(\mathbf{S}_i \cdot \mathbf{r}_{ij})(\mathbf{S}_j \cdot \mathbf{r}_{ij})}{r_{ij}^5}\right] \quad (1)$$

Here, H is an externally applied magnetic field, $\mathbf{I}_i$ is the nuclear spin of the ith ion, the fourth term is the exchange interaction, the fifth term is the dipole–dipole interaction, and the remaining symbols have their usual meaning. In materials such as metallic iron and Fe_2O_3 the exchange interaction is much larger than the other terms in $\mathcal{H}_i$, including in particular the hyperfine structure coupling. Correspondingly, the spin–spin relaxation time τ_2 will be very short compared to the nuclear precession time τ_L. When the temperature T is below the Curie temperature T_c, magnetic ordering effects suppress the off-diagonal terms in $A\mathbf{I}_i \cdot \mathbf{S}_i$. If $\tau_L/\tau_2 \gg 1$, the hyperfine interaction may then be described approximately as

$$H_{\text{hfs}} = g_n\beta_n I_{iz} H_{\text{eff}} \quad (2)$$

where

$$H_{\text{eff}} = A\langle S_{iz}\rangle/g_n\beta_n \quad (3)$$

and

$$\langle S_{iz}\rangle = Tr S_{iz} e^{-\mathcal{H}_i/kT}/Tr e^{-\mathcal{H}_i/kT} \quad (4)$$

The above effective field model gives a good description of Mössbauer measurements of the hfs splitting occurring in metallic iron[2] and Fe_2O_3.[3] It would be of

---

* Research sponsored by the U.S. Atomic Energy Commission under contract with the Union Carbide Corporation.
† Visiting scientist from the Atomic Energy Research Establishment, Harwell, England.
‡ Oak Ridge Graduate Fellow from the University of Tennessee, Knoxville, Tennessee.

interest to explore magnetically dilute materials in which the relative magnitudes of the terms in $\mathcal{H}_i$ may differ markedly from those for Fe or Fe_2O_3.

One substance of interest is the dilute paramagnetic material $FeNH_4(SO_4)_2 \cdot 12H_2O$. Here the exchange interaction[4] is many orders of magnitude smaller than in metallic iron, and the dipole–dipole interaction is roughly one order of magnitude smaller. In this salt, the dipole–dipole interaction is thought to be much larger than the exchange term in $\mathcal{H}_i$, which now contains an additional term $D[S_z^2 - \tfrac{1}{3}S(S+1)]$. The coefficient D has not been measured, but may be of comparable magnitude[5] to the dipole–dipole term. Magnetic ordering in this material for $H = 0$ is presumably due to the D term and dipole–dipole interaction and has a Néel temperature T_N near $0.04°K$.[4] If we apply a magnetic field of about 10^4 Oe to the salt at temperatures in the liquid-helium range, a value for $\langle S_{iz} \rangle$ near magnetic saturation may be produced. In the measurements described below we have chosen to make $g_e\beta_e S_{iz}H_z$ about one order of magnitude larger than the dipole–dipole and D terms, which in turn are large compared to the other terms in $\mathcal{H}_i$. The magnetic ordering is now due primarily to the external field, in contrast to metallic iron, where the spin ordering is almost entirely due to the exchange interaction. We may then inquire how far the hfs observed in these measurements may be correlated with the predictions of the effective field model. An additional interest in this measurement lies in the fact that ferric alum has been used for many years in magnetic cooling studies.

A radioactive source was prepared by diffusing about 70 mCi of ^{57}Co into a 0.0003-in.-thick copper foil. The diameter of the active area was approximately 2.2 cm. The absorber was a polycrystalline sample of $FeNH_4(SO_4)_2 \cdot 12H_2O$ containing 20 mg/cm^2 of natural iron, pressed into the form of a disk 1 in. in diameter. Both source and absorber were placed in the same liquid-helium bath, but the magnetic field was applied only to the absorber. The direction of the magnetic field was normal to the plane of the absorber and parallel to the γ-ray propagation direction. Measurements were made at H/T values of 14.10, 10.65, 8.40, 6.50, 5.72, 4.30, 2.86, and 0 kOe/deg K. A selection from these results is shown in Fig. 1.

At the largest H/T value of 14.10 kOe/deg K (Fig. 1a), a hyperfine structure spectrum of four lines was observed in which the relative spacings of the lines corresponded to those observed in ferromagnetic iron, and all the lines showed widths ~ 0.85 mm/sec. The line width found for this source at $4.2°K$ using a $K_4Fe(CN)_6 \cdot 3H_2O$ absorber at $300°K$ was 0.47 mm/sec. The relatively small line broadening in this spectrum compared to that found for the smaller H/T values discussed below results from the fact that here $\langle S_{iz} \rangle$ is near saturation, i.e., the electron spins spend most of the time in the energetically most favored orientation $S_z = -\tfrac{5}{2}$. For a thin single crystal of the salt, we would expect the lines to approach their natural width as H/T approaches infinity. Assuming for the present that H_{eff} is opposite in sign to the applied field H, and using equations (2–4) to extrapolate to saturation magnetization, we see that the observed splitting corresponds to $H_{eff} = 591 \pm 20$ kOe at $\langle S_z \rangle /S = 1$. This may be compared to the value 552 ± 10 kOe measured at $4.2°K$ for antiferromagnetic Fe_2O_3, in which the iron atoms are also in the ferric state.

The other spectra, of which Fig. 1b is a typical example, show a very marked progressive asymmetric broadening and decrease of amplitude of the outer lines as H/T decreases. In sharp contrast with this, the amplitudes and widths of the inner lines are very nearly independent of H/T, within statistical error, over the range from 14.10 to 2.86 kOe/deg K. The arrows in Fig. 1 show where the peaks of the outer lines would appear if their positions followed equations (2–4).

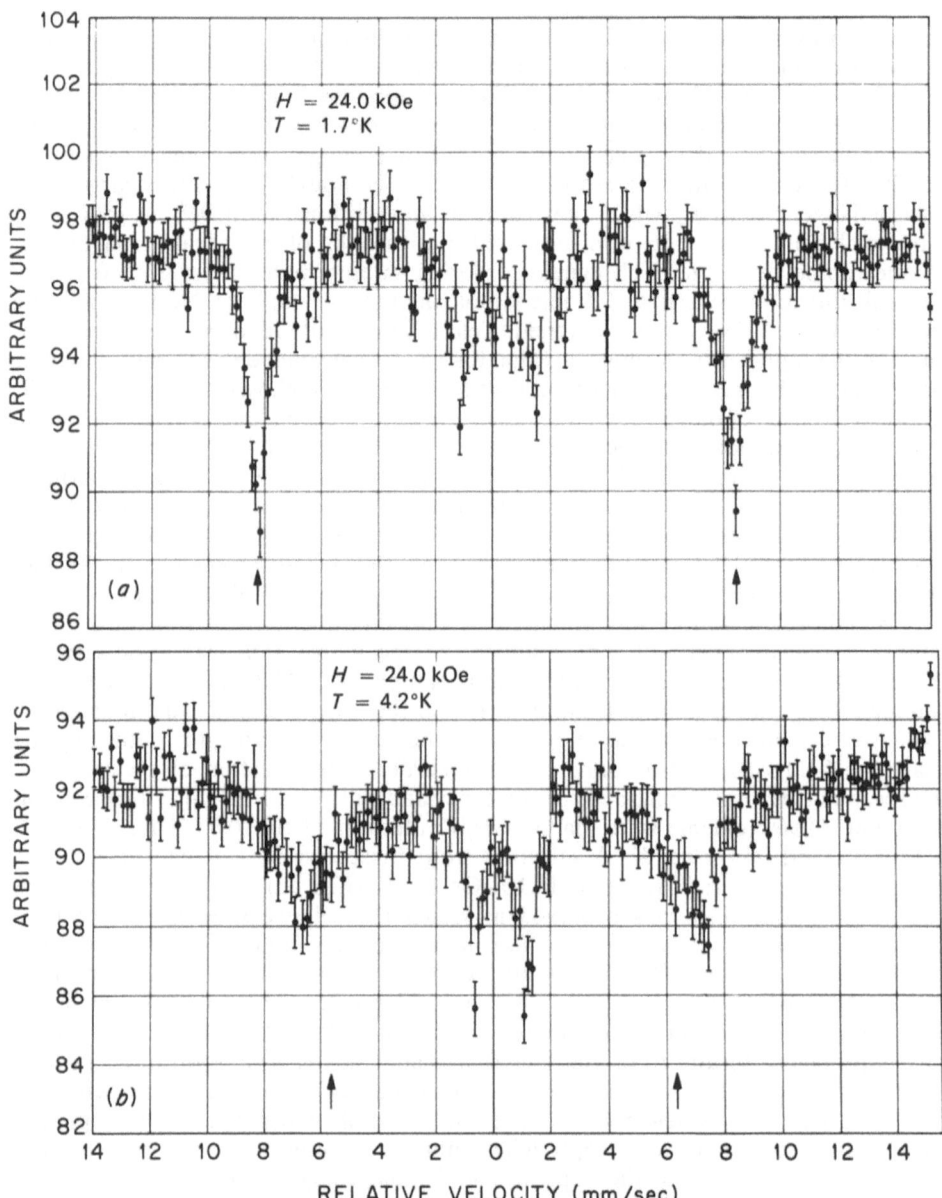

Fig. 1. Hyperfine structure spectra of ^{57}Fe in ferric ammonium alum. (a) $H = 24.0$ kOe, $T = 1.7°$K. (b) $H = 24.0$ kOe, $T = 4.2°$K.

The separations of the outer line pair and those of the inner pair for all of our data are shown as a function of $\beta H/kT$ in Fig. 2. The solid curves are a plot of $\langle S_z \rangle$ given by (4), normalized in each case to the splitting of the corresponding line pair at the largest value of H/T. The spacing of the inner pair appears to follow the Brillouin function (4) within the statistical uncertainties of the measurements, but

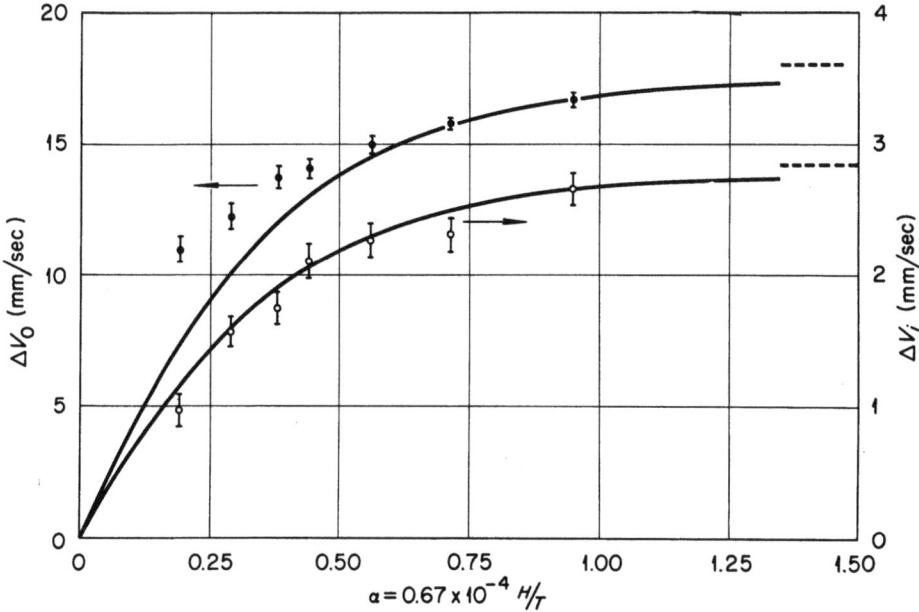

Fig. 2. Peak separations in hyperfine structure spectrum of ferric ammonium alum as a function of $\beta H/kT$. Solid circles: separation of outer peaks; open circles: separation of inner peaks. See text for explanation of curves.

the peak separation of the outer lines clearly falls above the calculated curve as H/T decreases.

In our discussion of these measurements, three characteristic times are of importance. These are the nuclear life time τ_N, the spin–spin relaxation time τ_2, and the nuclear precession time τ_L. Clearly, τ_L/τ_N must be small compared to unity in order to have a resolved spectrum, and τ_L/τ_2 must be large compared to unity so that fluctuations of the electron spin magnetization may have a negligible effect on the line widths. As mentioned above, these conditions are well satisfied in metallic iron. In contrast, τ_L/τ_2 in ferric ammonium alum is much smaller and possibly comparable with unity. Here fluctuation effects may be expected to have a strong influence on the observed Mössbauer spectrum. In this context we may note that the broadening effects are much more pronounced for the outer lines, for which $|(m_e\mu_e/I_e - m_g\mu_g/I_g)|$ is large, than for the inner lines, and that as the outer lines become broader they also become progressively more asymmetric. The origin of this asymmetry may arise from the detail electronic spin distribution, but there seems to be no theoretical treatment of this effect at the present. A possible contribution to the asymmetry may come from $\Delta m_I = 0$ transitions. Because of the presence of off-diagonal terms in the spin Hamiltonian, these transitions will not be completely forbidden.

The hfs spectrum of ferric ammonium alum was also observed under field-free conditions at temperatures of 1.7°, 2.25°, 4.2°, and 300°K. In each case the spectrum showed a single rather broad line, the width at 4.2°K being 1.8 mm/sec. This is to be contrasted with the resolved multiline spectra observed under field-free conditions in magnetically very much more dilute materials by Wertheim and Remeika.[6]

It is clear that under the conditions of this experiment the effective field model does not give an adequate description of the behavior of ferric ammonium alum. The discrepancies are quantitative, in their failure to account for the positions of the outer peaks, and qualitative, in that the model assumes that the lines will show only the natural width associated with the nuclear lifetime. Both types of discrepancy may well be removed by a theory which takes into account the fluctuations in the electron spin magnetization.

References

1. H. Frauenfelder, *The Mössbauer Effect*, W. A. Benjamin, Inc., New York (1962).
2. R. S. Preston, S. S. Hanna, and J. Heberle, *Phys. Rev.* **128**, 2207, 1962.
3. K. Ono, Y. Ishikawa, A. Ito, and E. Hirahara, *J. Phys. Soc. Japan, Suppl. B-I* **17**, 129, 1962.
4. A. H. Cooke, *Proc. Phys. Soc. (London) Ser. A* **62**, 269, 1949.
5. K. D. Bowers and J. Owen, *Rept. Progr. Phys.* **18**, 304, 1955.
6. G. K. Wertheim and J. P. Remeika *Bull. Am. Phys. Soc.* **9**, 464, 1964.

SPECIFIC HEAT OF FOUR RARE-EARTH COMPOUNDS BETWEEN 0.4° AND 5°K*

David G. Onn, Roland Gonano, and Horst Meyer

Department of Physics, Duke University
Durham, North Carolina

We have measured the specific heat of lanthanum and holmium ethylsulfate (LaES and HoES) and of lutetium and gadolinium iron garnet (LuIG and GdIG) by means of a cryostat using liquid ^{3}He or ^{4}He. The sample, in the form of single crystals or a sintered block, is attached to a support which is made up of a quartz frame and thin copper strips in order to have a low specific heat. This support can be brought in contact with the helium refrigerator by means of a heat switch. Two carbon resistors are used: a Speer 450-Ω resistor covers the region from 0.35° to 0.9°K, and a 40-Ω $\frac{1}{10}$ W Allen–Bradley is used at higher temperatures. The resistor operating below 0.9°K is calibrated by means of the vapor pressure of liquid ^{3}He and the susceptibility of three single crystals of cerium magnesium nitrate which obeys Curie's law in this temperature region. The samples are cooled from room temperature entirely by means of the heat switch and without any helium exchange gas. This is to prevent adsorption of the gas on the specimen, in particular on the sintered samples, which have been found to be porous.

Lanthanum Ethylsulfate

This compound has already been investigated before by Meyer and Smith[1] between 1.7° and 20°K. These authors found that below about 3°K the ratio C/T^3 increased with decreasing temperature. This behavior was interpreted by Orbach[2] as the effect of a low-lying lattice mode responsible for the anomaly in the heat conductivity of holmium ethylsulfate,[3] a salt isomorphous to LaES. The previous specific heat results were obtained with crystals sealed in a thin-walled copper calorimeter with a small amount of exchange gas. Since the specific heat of the diamagnetic salt is small, it could have been artificially increased by adsorbed gas, and a check was therefore important. In the present experiments three large crystals with a total weight of 18 g were used. Below 3°K the new results were found to disagree with the previous ones, and below 2°K the ratio C/T^3 tends to $(6.1 \pm 0.2) \times 10^{-3}$ J/mole-deg^4. The scatter becomes appreciable ($\pm 10\%$) below about 0.8°K, mainly because of the smallness of the specific heat in relation to that of the addenda. The results, however, show clearly that the lattice mode is not detected and that the former results must have been in error below 3°K, presumably because of exchange gas adsorption. The results above 3°K, however, seem correct because the effect of the gas becomes unimportant. Also, the previous results[1] on salts with a large Schottky anomaly were probably not affected by exchange gas.

---

* Work supported by a grant from the Army Research Office (Durham) and a contract from the Office of Naval Research.

Holmium Ethylsulfate

Paramagnetic resonance experiments[4] on the salt diluted with LaES have shown that the resonance spectrum could be interpreted by assuming a spin Hamiltonian which in zero field is

$$\mathcal{H} = D\left[S_z^2 - \frac{S(S+1)}{3}\right] + AS_zI_z + B(S_xI_x + S_yI_y) + P\left[I_z^2 - \frac{I(I+1)}{3}\right]$$

with $A = 0.168 \text{ cm}^{-1}$, $B = 0.1 \pm 0.02 \text{ cm}^{-1}$, and $P = 3 \times 10^{-4} \text{ cm}^{-1}$. The meaning of the symbols is explained by Baker and Bleaney.[4] The splitting D between the non-Kramers doublet and the singlet was found to be $D = -5.8 \text{ cm}^{-1}$. Data of optical absorption on the concentrated salt[5] gave $A = 0.235 \text{ cm}^{-1}$ and $D = -6.01 \text{ cm}^{-1}$. Our specific heat results are presented in Fig. 1. Below about 0.5°K the measurements of this large heat capacity were made difficult by a very low thermal diffusivity in the crystals and very long drift rates had to be taken, up to 30 min at 0.35°K. By solving the energy eigenvalues of the Hamiltonian given above and using the parameters given by Baker and Bleaney, we calculated the specific heat. This is shown in Fig. 1 (curve BB). Adding the calculated dipole–dipole contribution $CT^2/R = 0.0617$, one obtains good agreement with the experimental results.

Lutetium Iron Garnet

The specific heat of this garnet in the temperature range considered is theoretically expected to be given by

$$C = \alpha T^3 + \delta T^{3/2} + \varepsilon T^{-2}$$

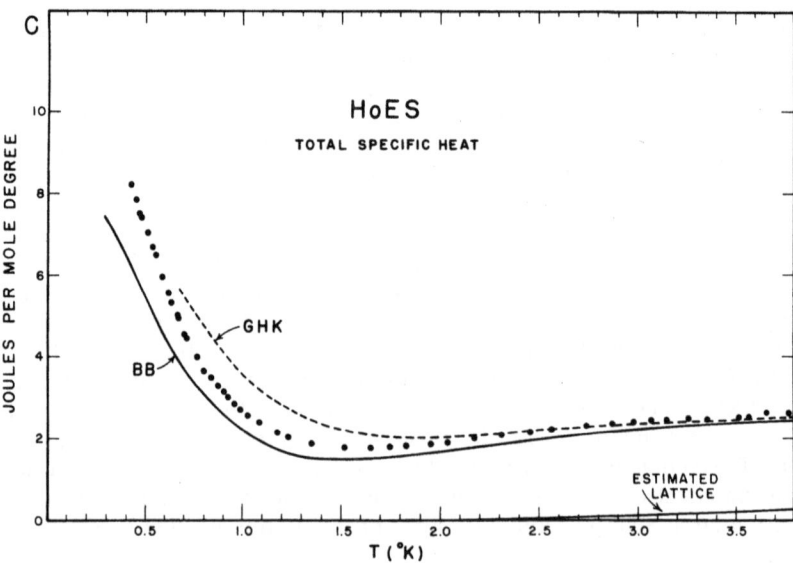

Fig. 1. The total specific heat of holmium ethylsulfate. The solid and dashed curves are those calculated from the data of Baker and Bleaney[4] and of Grohmann, Hellwege, and Kahle.[5]

where the terms are respectively those of the lattice, the spin wave contribution, and the nuclear contribution of the lutetium isotopes. This last term is expected to be caused mainly by the coupling of the magnetic field created by the a and the d Fe^{3+} sublattices to the nuclear spins of ^{175}Lu $(I = \frac{7}{2})$ and ^{176}Lu $(I = 7)$. Neglecting (provisionally) the effect of the nuclear quadrupole interaction, and assuming the local magnetic field to be the same at the nuclei of both isotopes, we can derive this field from the experimental value of the constant ε. However, the presence of other rare-earth impurities with a high nuclear specific heat, such as HoIG, limits the accuracy of the magnetic field determination in LuIG. By subtracting the maximum estimated specific heat due to these impurities, we have estimated the local field within about $\pm 12\%$.

The sample was a cylindrical bloc of sintered material. The lutetium was 99.9% pure, the chief impurity being ytterbium. The specific heat results shown in Fig. 2 are in agreement within the combined experimental error with previous data[6] extending down to 1.7°K. We found $\varepsilon = 2.5$ mJ-deg/mole garnet. The analysis of the results gives $H_{loc} = 130 \pm 15$ kG, and the splittings Δ between the consecutive nuclear levels in both isotopes were estimated to be $\Delta_{175} = 64 \pm 6$ Mcps and $\Delta_{176} = 46 \pm 4$ Mcps.

Experiments are in progress to measure these splittings directly using a frequency-modulated Robinson spectrometer. It may be difficult to observe the line because of the expected broadening due to the nuclear quadrupole term. This broadening is expected to be of the order of several megacycles. Mössbauer experiments, such as those carried out on other garnets[7] would be of interest in this compound.

Gadolinium Iron Garnet

This compound has a spin wave contribution due principally to the Gd^{3+} in the exchange field of the iron ions. Below about 2.5°K, this contribution is expected to be approximately[8]

$$\frac{C}{R} \approx \frac{0.113}{4} \left(\frac{kT}{D^*}\right)^{\frac{3}{2}} (1 + 0.11T) \exp(-g\beta H_{anis}/kT)$$

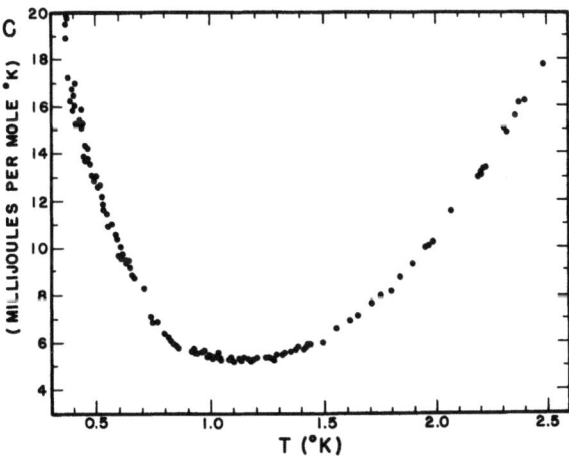

Fig. 2. Specific heat of lutetium iron garnet.

where D^* depends mainly on the exchange interactions between the a and d sublattices,[8] g is the Lande factor of Gd^{3+}, and β is the Bohr magneton. H_{anis} is the anisotropy field, approximately 500 G, as determined from microwave experiments.[9] At temperatures above 2.5°K, the magnetic specific heat is more complicated to express, and has been computed numerically by Harris.[8] Our results are found to be in good agreement with the calculations of Harris, provided that D^* is taken to be 7.25 cm^{-1}. This is somewhat smaller than the D^* of 9.4 cm^{-1} predicted for GdIG from experiments on YIG.[6] At temperatures below about 1.2°K, the experimental specific heat becomes systematically larger than the calculated one. This can be accounted for by the nuclear contributions of small traces of TbIG, SmIG, and HoIG, as well as by that of GdIG.[10]

We are pursuing a program of study on the nuclear parameters in several rare-earth iron garnets. We are also studying the magnetic exchange interactions in rare-earth gallium garnets between 0.4° and 4°K. From high-resolution specific heat measurements, we have found the transition temperatures of these garnets to be: SmGaG (two samples) 0.966°K and 0.922°K; ErGaG, 0.788°K; DyGaG, 0.36°K; NdGaG, 0.515°K; YbGaG, $T_N < 0.35°K$.

Acknowledgments

The authors are greatly indebted to Dr. A. B. Harris for his interest in these experiments. They are grateful to Dr. James Schooley of the N.B.S. in Washington, D.C. for the loan of the LaES and HoES crystals.

References

1. H. Meyer and P. L. Smith, *J. Phys. Chem. Solids* **9**, 296, 1959.
2. R. Orbach, *Phys. Rev. Letters* **8**, 394, 1962.
3. I. P. Morton and H. M. Rosenberg, *Phys. Rev. Letters* **8**, 200, 1962.
4. J. M. Baker and B. Bleaney, *Proc. Roy. Soc. (London) Ser. A* **245**, 156, 1958.
5. I. Grohmann, K. H. Hellwege, and H. G. Kahle, *Naturwissenschaften* **12**, 277, 1960.
6. A. B. Harris and H. Meyer, *Phys. Rev.* **127**, 101, 1962.
7. S. G. Cohen, I. Nowick, and S. Ofer, *Rev. Mod. Phys.* **36**, 378, 1964; P. Kienle, *Rev. Mod. Phys.* **36**, 372, 1964.
8. A. B. Harris, *Phys. Rev.* **132**, 2398, 1963.
9. G. P. Rodrigue, H. Meyer, and R. V. Jones, *J. Appl. Phys.* **31**, 3768, 1960.
10. J. I. Budnick, *Bull. Am. Phys. Soc.* **9**, 464, 1964.

THE SPECIFIC HEAT OF YTTERBIUM, TERBIUM, AND DYSPROSIUM METALS BETWEEN 3° AND 25°K*

O. V. Lounasmaa†

Argonne National Laboratory
Argonne, Illinois

Introduction

Most of the rare earths are ferro- or antiferromagnetic at low temperatures. Due to this, their heat capacity has usually four different components: the lattice specific heat C_L, the electronic specific heat C_E, the magnetic specific heat C_M, and the nuclear specific heat C_N. A fairly large amount of work has recently been done for establishing the heat capacity of lanthanides in the liquid-helium region and, in particular, below 1°K. As a result, the nuclear specific heat, which is large below 1°K, is well understood.

Much less is known about the magnetic specific heat. The reason is, at least partly, that C_M is usually the dominant contribution in the observed C_p, and thus most easily separated from the other components, between 3° and 25°K, a temperature range characterized by lack of experimental data.

The magnetic structures of rare-earth metals are very interesting and quite complicated. Neutron diffraction studies have revealed spiral spin arrangements in the higher lanthanides. In these metals, C_M is caused by exchange interaction between the $4f$-electronic spins. Several theories have been proposed recently for calculating the magnetic specific heat of rare-earth metals. For a ferromagnetic metal, the simple spin wave theory[1] predicts $C_M = CT^{\frac{3}{2}}$. For the more complicated spiral spin structures, $C_M \propto \exp(-E_g/kT)$,[2,3] where E_g is the energy gap at the bottom of the spin wave spectrum. The physical reason for E_g is that it always takes a finite energy to turn a spin against the anisotropy field. Energy gaps of the order of $E_g/k \sim 20°$–$30°$K have been predicted for terbium and dysprosium. By combining the exponential temperature dependence of C_M with the prediction of the simple spin wave theory, one obtains for a ferromagnetic metal $C_M = CT^{\frac{3}{2}} \exp(-E_g/kT)$. More complicated formulas have also been proposed.[2]

Impurities seem to have a rather large effect in the specific heat of several rare-earth metals; good examples are terbium and dysprosium.[4,5] At least two different mechanisms are possible. If there is a magnetic transformation in the impurity, the resulting heat capacity peak is appreciable, particularly at low temperatures, as compared to the heat capacity of the metal itself, even when the impurity content is quite small. The specific heat peak in terbium, observed by some investigators at 2.4°K, appears to be caused by an antiferromagnetic transition in Tb_2O_3.[6] Besides this direct effect, impurities may also alter E_g and thus cause a profound change in C_M. Small impurity contents might be of importance here.

---

* Based on work performed under the auspices of the U.S. Atomic Energy Commission.
† Present address: Institute of Technology, Otaniemi, Finland.

We have started a program of measuring the heat capacity of all rare-earth metals between 3° and 25°K. Our cryostat has been described earlier[5]; some modifications were made, the most important one being the use of a germanium resistance thermometer.* In this paper, we present our data on ytterbium, terbium, and dysprosium. Due to lack of space and of sufficient time for analysis, a more thorough discussion of the results must be postponed to a later publication.

Results

Ytterbium. Ytterbium has a full $4f$-shell and thus, contrary to most other lanthanides, this metal is nonmagnetic and its heat capacity should have only the usual lattice and electronic contributions. We found earlier,[7] from measurements between 0.4° and 4°K, that $C_E = 2.90T$ mJ/mole-deg K and $C_L = 1.180T^3$ mJ/mole-deg K, the lattice specific heat corresponding to a Debye temperature $\theta = 118.1$°K. Our new results are in good agreement with the earlier data and, by assuming the above value of C_E, we find the θ's given in Table I. The θ vs. T curve thus has a familiar appearance. Our specific heats are about 2–3% lower than those by Gerstein *et al.*[8] in the overlapping region from 15° to 25°K.

* We are indebted to D. W. Osborne and H. E. Flotow for supplying us with a calibration of this thermometer.

Table I. The Debye θ of Ytterbium

T, °K	θ, °K	T, °K	θ, °K	T, °K	θ, °K
4	117.6	10	109.9	16	109.2
6	114.7	12	109.1	20	109.8
8	111.5	14	108.9	24	109.8

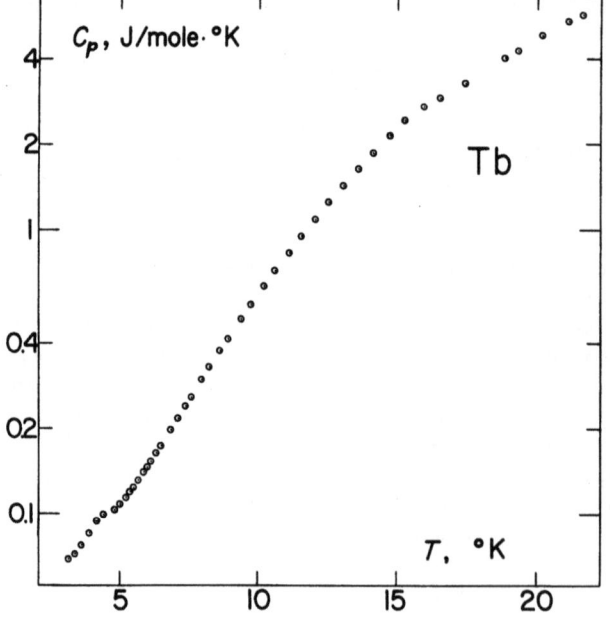

Fig. 1. The specific heat of terbium metal.

Terbium. Our results on terbium are shown in Fig. 1. A rather small anomaly was observed around 4.4°K and an irregularity found in the heat capacity curve at about 16°K. In view of the low-temperature anomaly, which was unsuspected on the basis of measurements below 4°K,[4] the Debye temperature calculated by Lounasmaa and Roach, $\theta = 150°K$, is obviously too low.* The value $\theta = 181°K$ given by Lounasmaa in a later paper[9] seems to be closer to the truth. The specific heat of terbium has been measured above 15°K by Jennings et al.[10] At 15°K, our results are about 15% higher than theirs, but the discrepancy, which is probably caused by the anomaly at 16°K, almost disappears by the time 25°K is reached.

An analysis of our results is incomplete at this time. However, values in Table II were calculated by assuming $C_E = 10.5T$ mJ/mole-deg K for terbium,[9] by subtracting this amount from the observed C_p, and by expressing the remaining specific heat with a Debye temperature. The θ's thus represent $C_L + C_M$.

Dysprosium. Our specific heat data on dysprosium are shown in Fig. 2. The anomaly around 5°K is probably caused by impurities and its existence is not surprising in view of our earlier results for dysprosium below 4°K.[5] Our new data are in excellent agreement with the measurements by Flotow and Osborne[11] (the

_____
* The ytterbium, terbium, and dysprosium samples used in these experiments were the same as those measured by us below 4°K. [4,5,7]

Table II. The Debye θ Corresponding to $C_L + C_M$ for Terbium

T, °K	θ, °K	T, °K	θ, °K	T, °K	θ, °K
4	140	10	159	16	143
6	170	12	153	20	143
8	166	14	147	24	140

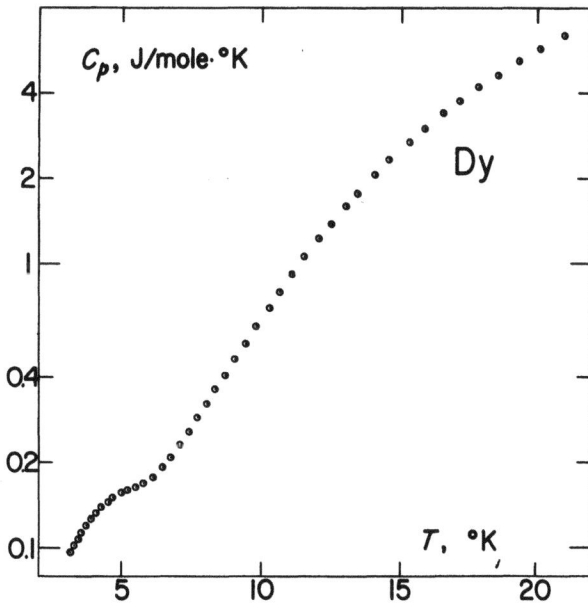

Fig. 2. The specific heat of dysprosium metal.

Table III. The Debye θ Corresponding to $C_L + C_M$ for Dysprosium

T, °K	θ, °K	T, °K	θ, °K	T, °K	θ, °K
4	112	10	154	16	137
6	155	12	146	20	132
8	162	14	141	24	128

same sample was used) above 8°K; below this temperature their platinum resistance thermometer was not sufficiently sensitive for observing the anomaly, since too large temperature increments had to be used.

The Debye temperature of dysprosium, again corresponding to $C_L + C_M$ and calculated in the same way as for terbium, is given in Table III. Since C_L should correspond to a θ not smaller than 180°K,[5,9] the magnetic specific heat of dysprosium is appreciable. A preliminary analysis gives support to an exponential temperature dependence of C_M.

References

1. J. van Kranendonk and J. H. van Vleck, *Rev. Mod. Phys.* **30**, 1, 1958.
2. B. R. Cooper, *Proc. Phys. Soc. (London) Ser. A* **80**, 1225, 1962.
3. A. R. Mackintosh, *Phys. Letters* **4**, 140, 1963.
4. O. V. Lounasmaa and P. R. Roach, *Phys. Rev.* **128**, 622, 1962.
5. O. V. Lounasmaa and R. A. Guenther, *Phys. Rev.* **126**, 1357, 1962.
6. B. C. Gerstein, F. J. Jelinek, and F. H. Spedding, *Phys. Rev. Letters* **8**, 425, 1962.
7. O. V. Lounasmaa, *Phys. Rev.* **129**, 2460, 1963.
8. B. C. Gerstein, J. Mullaly, E. Phillips, R. E. Miller, and F. H. Spedding, *J. Chem. Phys.* **41**, 883, 1964.
9. O. V. Lounasmaa, *Phys. Rev.* **133**, A219, 1964.
10. L. D. Jennings, R. M. Stanton, and F. H. Spedding, *J. Chem. Phys.* **27**, 909, 1957.
11. H. E. Flotow and D. W. Osborne, Proceedings of the Third Rare Earth Conference, Clearwater (1963).

SPECIFIC HEAT OF SCANDIUM BETWEEN 0.15° AND 3°K*

P. Lynam, R. G. Scurlock, and E. M. Wray

Department of Physics, University of Southampton
Southampton, Great Britain

The magnetic properties of scandium metal are not well understood. Chechernikov et al.[1] have observed a weak temperature-dependent paramagnetic susceptibility with a Curie–Weiss temperature of about 1000°K, suggesting the occurrence of antiferromagnetism. However, NMR measurements[2] have shown that no magnetic ordering occurs in scandium above 1.7°K.

We have measured the specific heat of scandium metal between 0.15° and 3°K. High-purity scandium is difficult to obtain, and measurements were made on two specimens of about 99.8% purity, having different impurity contents (Table I). Measurements on specimen I, weighing 17 g, were made in a ^{3}He cryostat between 0.5° and 3°K; specimen II, weighing 9 g, was measured in an adiabatic demagnetization cryostat between 0.15° and 1°K. The results, plotted as C/T vs. T^2 in Fig. 1, show a large electronic specific heat with an anomalous deviation from the straight line behavior of a normal metal below 1°K. Values for the electronic and lattice specific heat parameters for specimen I are $\gamma = 10.9 \pm 0.1$ mJ/deg^2-mole and $\Theta_D = 344 \pm 25$°K. These values are in reasonable agreement with previous measurements made on the same specimen between 1.7 and 4.2°K.[3]

The anomaly below 1°K is much larger for specimen II than for specimen I, and may be explained in various ways.

1. *Nuclear Schottky specific heat.* The anomalies may be described as the high-temperature tail of a Schottky specific heat, arising from a nuclear hyperfine interaction and varying as $C = BT^{-2}$. If the anomalous specific heats are fitted to

---
* This work was supported by research and maintenance grants by the Department of Scientific and Industrial Research.

Table I. Impurity Contents of Specimens from Mass Spectrometer Analysis

Impurity	Proportion, parts per 10^6	
	I	II
Tantalum	0.3	1000
Gadolinium	40	400
Dysprosium	40	8
Silver	600	0.8
Copper	1000	30
Iron	60	60
Aluminum	300	100
Fluorine	300	100

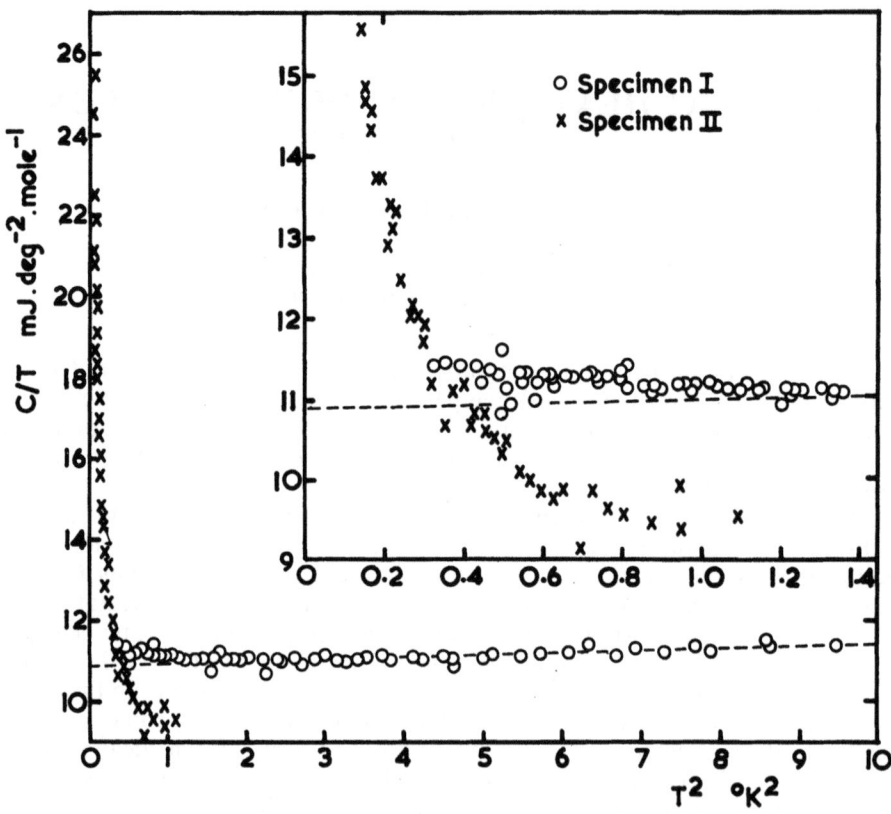

Fig. 1. Specific heat of scandium between 0.15° and 3°K, plotted as C/T vs. T^2. Inset: results in the region of 1°K in greater detail. The dotted line corresponds to $C = 10.9T + 0.048T^3$ mJ/deg-mole.

this term, values of B of 0.05 ± 0.03 and 0.23 ± 0.05 mJ deg/mole are obtained for specimens I and II, respectively. The reason for the disparity between these two values is not clear, but it is possible that any hyperfine interaction at the ^{45}Sc nucleus might depend on the impurities present in the specimens. There are far too few impurity atoms present for a hyperfine interaction at their nuclei to produce values of B such as are observed.

2. *Long-range magnetic ordering.* Both specimens contain the magnetic impurities gadolinium, dysprosium and iron. Specific heat anomalies, similar to those reported here, have been observed in several alloys containing magnetic impurities of the 3d-transition group,[4] and have been attributed to long-range magnetic ordering. The difference in the magnitude of the anomalies in the two specimens suggests that in this case the magnetic coupling may be between the gadolinium atoms, since specimen II contains ten times as much of this impurity as I.

3. *Superconducting transition.* The anomaly could arise from the onset of a superconducting transition. Montgomery[5] has found that the electrical resistivity of scandium at 0.3°K is 30% less than at 1°K. This decrease is removed on the application of a magnetic field. The possibility that high-purity scandium becomes superconducting cannot therefore be ruled out.

In view of the uncertainty in interpreting these results, it is clear that further measurements on scandium specimens of higher purity are needed before any definite conclusion can be reached.

Acknowledgment

We would like to thank Dr. H. Montgomery of A.E.R.E., Harwell, for communicating his results to us and for the loan of specimens.

References

1. V. I. Chechernikov, I. Pop, and O. V. Naumkin, *Soviet Phys. JETP* (*English Transl.*) **17**, 1228, 1963.
2. W. P. Blumberg, J. Eisinger, V. Jacarino, and B. T. Matthias, *Phys. Rev. Letters* **5**, 52, 1960.
3. H. Montgomery and G. P. Pells, *Proc. Phys. Soc.* **78**, 622, 1961.
4. See, for example, F. J. du Chatenier and J. De Nobel, *Physica* **28**, 181, 1962.
5. H. Montgomery, private communication (1964).

MAGNETO-OPTICAL BEHAVIOR OF FERROMAGNETIC METALS

S. Doniach and D. H. Martin

Queen Mary College
London, England

Magneto–optical properties can be used to investigate the band structures and Fermi surfaces of ferromagnetic metals. This will be illustrated by the results of theoretical and experimental studies of nickel and iron.

MAGNETIC SUSCEPTIBILITY OF Mn^{++} IN CdS AND EFFECTS OF ANTIFERROMAGNETIC EXCHANGE

Marshall M. Kreitman*

Aerospace Research Laboratories
Wright-Patterson AFB, Ohio

Frederick J. Milford

Battelle Memorial Institute
Columbus, Ohio

and

J. G. Daunt†

Department of Physics, Ohio State University
Columbus, Ohio

This paper reports low-temperature measurements of the magnetic susceptibility of single crystals of CdS (in the hexagonal modification) with manganese impurity added. The data provide clear evidence for isotropic antiferromagnetic exchange interactions between next-nearest neighbor and in fact even more distant pairs of Mn^{++} ions. Antiferromagnetic effects in other dilute paramagnetic insulating crystals have been observed previously, as, for example, in susceptibility measurements on Cr^{+++} in Al_2O_3 by Daunt et al.,[1] on Mn^{++} in ZnS by Brumage et al.,[2] in resonance measurements on platinum-diluted ammonium chloriridate by Griffiths and co-workers,[3–5] on Mn^{++} in MgO by Coles, Orton and Owen,[6] on Cr^{+++} in Al_2O_3 by Rimai et al.,[7] and by Gill,[8] and in optical observations on Cr^{+++} in Al_2O_3 by Schawlow, Wood, and Clogston,[9]‡ and on Mn^{++} in ZnS by McClure.[11]

Measurements of the magnetic susceptibility of ½-in.-diameter meltgrown§ spherical single crystals of CdS, doped with manganese, were made in the temperature range 1° to 4.2°K, using a mutual inductance technique.[12] Measurements were made on nine different samples of mole fractions, x, of Mn^{++} ions ranging from 0.026 to 5.0% and with the specimens orientated with their c-axis at various angles to the magnetic field. It was found that the susceptibilities were isotropic within experimental errors, which ranged from ±3% at the lowest manganese concentrations to ±1% at the highest concentration. The mole fraction of manganese was determined by destructive wet chemical tests which yielded concentrations quite close to the nominal concentrations estimated by the premelt ratios of the constituents.

Plots of our observed susceptibilities, χ_{obs}, vs. $1/T$ showed the curves to be approximately linear at the lowest temperatures (approximately $1° < T < 2°K$)

---
* Cryotronics Field Engineering Division at Wright-Patterson Air Force Base, Ohio.
† Work supported in part by the National Science Foundation.
‡ See P. Kisliuk and W. F. Krupe[10] for further details and subsequent work.
§ Supplied by the Eagle-Picher Co., Miami, Oklahoma.

and here we evaluated a constant C^*_{obs} given by $C^*_{obs} = \partial \chi_{obs}/\partial(1/T)$ for all our specimens. Figure 1 gives C^*_{obs} as a function of x. Also shown in Fig. 1 are various calculated Curie constants, C_1, C_2, and C_3. C_1 is the independent ion Curie constant, $C_1 = xNg^2\beta^2 S(S+1)/3k$ for $S = 5/2$. C_2 is calculated assuming that clusters of two or more nearest neighbors contribute nothing to the susceptibility and consequently $C_2 = C_1(1-x)^{12}$. C_3 is calculated similarly to C_2, but assuming that only ions without either nearest or next-nearest neighbors contribute and thus $C_3 = C_1(1-x)^{18}$. C_2 and C_3 then are lower bounds on the effective Curie constant if only nearest neighbor interactions are important or if nearest and next-nearest neighbor interactions are important, respectively. It is clear that at low concentrations ($x \gtrsim 0.25\%$) the free ion theory is applicable, but at the highest concentrations not only nearest and next-nearest neighbor interactions but also some more distant pairs are important.

If clusters of three or fewer ions interacting in pairs through one or the other of two distinct exchange interactions are considered, then one must recognize and treat separately isolated ions, two kinds of pairs, and seven kinds of triples. The susceptibility of any cluster, with the exception of one kind of triple, can be readily computed using the vector model for the composition of two or more angular momenta.[13,14] The remaining triple can be treated to an adequate approximation by a thermodynamic perturbation theory. Four of the resulting susceptibilities involve only one or the other of the exchange interactions, while the other three involve both. On the basis of previous work, it is expected that the nearest

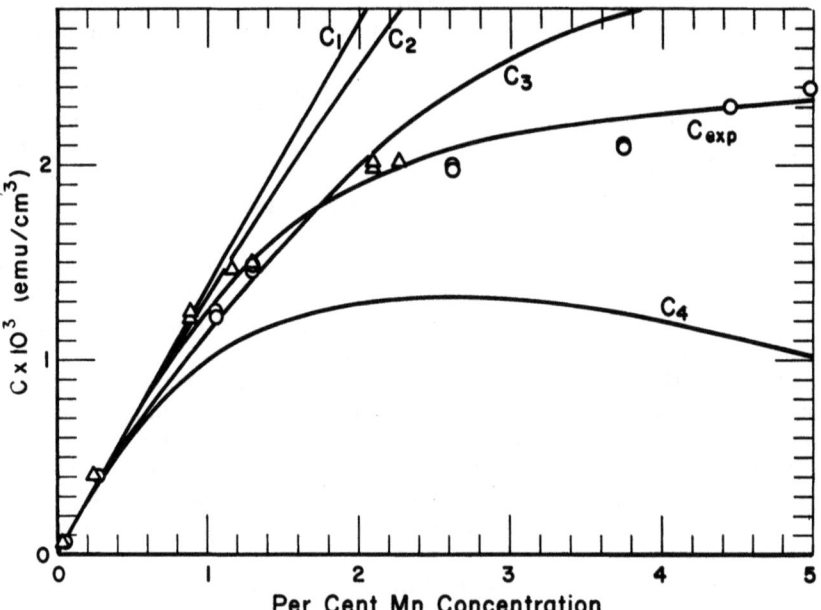

Fig. 1. Effective Curie constants vs. concentration. Free ions, C_1; ions without nearest neighbors, C_2; ions without nearest or next-nearest neighbors, C_3. Low-temperature experimental values are indicated and labeled C_{exp}.

neighbor exchange interaction will be of the order of 10°K and consequently that for the range of temperatures of interest here, $T/j_{nn} \gtrsim 0.4$. For such temperatures, the susceptibility due to a cluster involving only nearest neighbor interactions has its (constant) low-temperature limiting value to a very good approximation. The dependence on j_{nn} may also be removed in the mixed cases by noting that T/j_{nn} is extremely small. These susceptibilities have been evaluated in detail and provide a set of susceptibilities which depend only on T/j_{nnn}, which can be used for comparison with experimental.

Using the results of these detailed considerations, the total susceptibilities χ_{nn} and χ_{nnn}, which are the sums of four and ten cluster terms, respectively, have been evaluated for a concentration of 5%. In the case of χ_{nnn}, the low-temperature limit with respect to T/j_{nn} has been used. In Fig. 2, χ_{nn}, χ_{nnn}, $C_{nn} = \partial\chi_{nn}/\partial(1/T)$, $C_{nnn} = \partial\chi_{nnn}/\partial(1/T)$ and the experimental results C^*_{obs} and χ_{obs} are shown in the form of ratios to C_1 or χ_1. For the experimental data a true temperature scale has been used, while for the calculated quantities a reduced temperature T/j_{nn} or T/j_{nnn} is more convenient. With respect to the theoretical curves, it is striking that there is so little variation of either C/C_1 or χ/χ_1 over a range of τ from 0 to 4 or 5. In fact, an examination of extended calculations shows that the significant changes in these quantities occur over a range of τ from 0 to 40 or more.

Taking these comments into consideration, we can readily understand the results presented in Fig. 2. At temperatures between about 2° and 4°K, the nearest neighbor exchange interactions completely dominate thermal effects, consequently, ions with nearest neighbors contribute essentially nothing to the susceptibility.

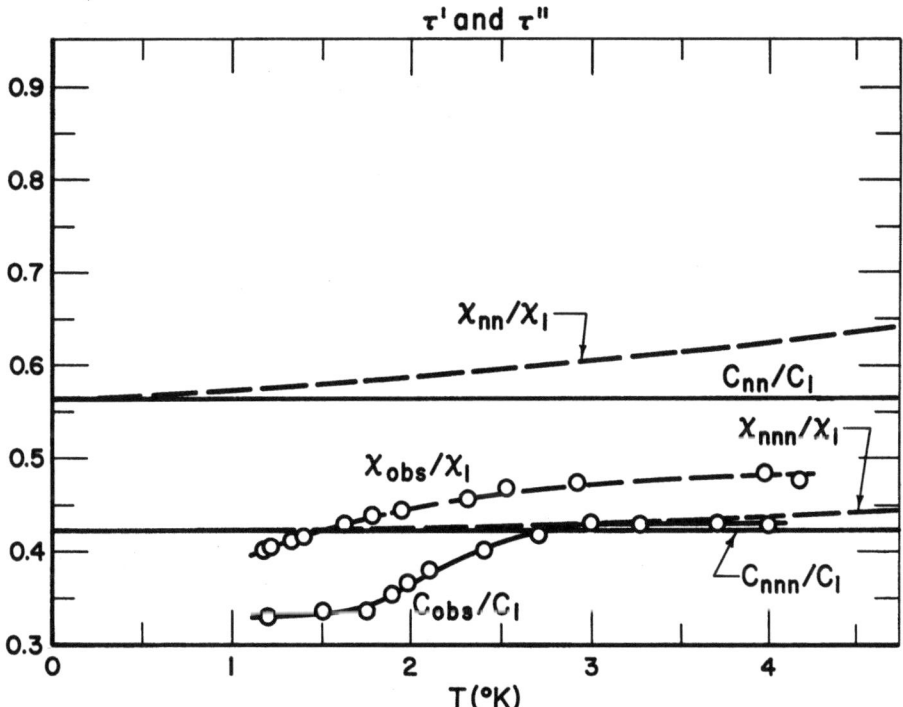

Fig. 2. Susceptibility and effective Curie constant vs. temperature. Experimental points and theoretical curves for nearest neighbor and nearest and next-to-nearest neighbor theories.

For $x = 5\%$, this reduces the susceptibility to $\sim 0.57\chi_1$. In addition, there is a next-nearest neighbor exchange of the order of 1°K which causes small changes in χ and C^* over the temperature range considered. Finally, there is an interaction between more distant neighbors of the order of a few hundredths of a degree Kelvin, which reduces the susceptibility still further in the temperature range below 2°K.

The essential points of the argument are:

1. C^*_{obs}/C_1, starting from a value of about 0.32 at 1.2°K, rises monotonically with increasing temperature, and in the range 3° to 4°K approaches very closely to the theoretical value for C^*_{nnn}/C_1, which over a very broad range of τ'' remains at 0.423. This conclusion is considered to be of significance, since the very slow variation of C^*_{nnn}/C_1 with τ'' permits a meaningful comparison to be made between theory and experiment despite our lack of knowledge of j_{nnn}. It is concluded therefore that our experimental observations indicate the importance of nnn interactions in the description of the susceptibility of Mn^{++} in CdS in the helium temperature range.

2. χ_{obs}/χ_1, starting from a value of about 0.40 at 1.2°K, rises monotonically with increasing temperature and reaches a value of about 0.475 at 4.2°K; χ_{nnn}/χ_1, starting from a value of 0.421 at $\tau'' = 0$, also rises monotonically with increasing τ, reaching a value of about 0.44 at $\tau'' \approx 3.5$. Unfortunately, it is not possible to use these data to determine j_{nnn} with any precision. Within the range of uncertainty, however, in our knowledge of the precise values of χ_{obs}/χ_1 and χ_{nnn}/χ, we may deduce that $j_{nnn} \approx 1°K$. The data at other values of x are also in agreement with this evaluation.

3. The observed facts that between 1° and 4°K the values of χ_{obs}/χ_1 and C^*_{obs}/C_1 lie so far below those of χ_{nn}/χ_1 and C^*_{nn}/C_1, respectively, for all values of τ' indicate that the nearest neighbor exchange interactions must have "locked-in" at temperatures well above 4°K and that therefore $j_{nn} > 4°K$. Unfortunately, the measurements of χ do not extend to sufficiently high temperatures to permit an evaluation of j_{nn}.

4. The results evident in Fig. 2—that at the lowest temperatures of measurement both χ_{obs}/χ_1 and C^*_{obs}/C_1 fall well below the lowest values that χ_{nnn}/χ_1 and C^*_{nnn}/C_1, respectively, can have for any value of τ''—lead one to conclude that at these temperatures further antiferromagnetic ordering, beyond that imposed even by next-nearest neighbor exchange interaction, is becoming important. This, as has been indicated above, is to be expected theoretically.

Acknowledgments

We wish to thank Mr. Dan Cooper and Miss Jean Clement for their help in the measurements, and Dr. Richard P. Kenan for assistance with the calculations.

References

1. Daunt, Edwards, Kreitman, Pandorf, and Snider, *Proceedings of the Eighth International Conference on Low Temperature Physics*, University of Toronto Press, Toronto, Canada (1960), p. 360.
2. W. H. Brumage, C. R. Yarger, and C. C. Lin, *Phys. Rev.* **133**, A765, 1964.
3. J. H. E. Griffiths, J. Owen, J. G. Park, and M. F. Partridge, *Phys. Rev.* **108**, 1345, 1957 and *Proc. Roy. Soc. (London) Ser.* A **250**, 84, 1959.
4. Bleaney and Bowers, *Proc. Roy. Soc. (London) Ser.* A **214**, 451, 1952.
5. J. Owen, *J. Appl. Phys.* **33**, 355, 1962.
6. B. A. Coles, J. W. Orton, and J. Owen, *Phys. Rev. Letters* **4**, 116, 1960, E. A. Harris and J. Owen, *Phys. Rev. Letters* **11**, 9, 1963, J. Owen, *J. Appl. Phys.* **32**, 355, 1962.
7. Rimai, Statz, Weber, de Mars, and Koster, *Phys. Rev. Letters* **4**, 125, 1960; *J. Appl. Phys.* **32**, 218, 1961.

8. J. C. Gill, *Proc. Phys. Soc. (London) Ser. A* **79**, 58, 1962.
9. A. L. Schawlow, D. L. Wood, and A. M. Clogston, *Phys. Rev. Letters* **3**, 271, 1959; A. L. Schawlow, *J. Appl. Phys.* **33**, 395, 1960.
10. P. Kisliuk and W. F. Krupe, *Appl. Phys. Letters* **3**, 215, 1963.
11. D. S. McClure, *J. Chem. Phys.* **39**, 2850, 1963.
12. J. G. Daunt and D. O. Edwards, *Temperature, Its Measurement and Control in Science and Industry*, Vol. 3, Part I, p. 133, 1962.
13. K. Kambe, *J. Phys. Soc. Japan* **5**, 48, 1950.
14. J. S. Smart, *Magnetism, Vol. III*, G. T. Rado and H. Suhl (eds.), Academic Press, Inc., New York (1963).

MAGNETORESISTANCE MEASUREMENTS ON CdS WITH MANGANESE IMPURITY

Marshall M. Kreitman*

Aerospace Research Laboratories (ARX)
Wright-Patterson AFB, Ohio

Measurements have been carried out on the magnetoresistance of a number of melt-grown CdS single crystals containing manganese impurity in amounts from 0.01 to 1.9 wt. %. The crystals investigated were cleaned in acetone, and copper leads tinned with indium were attached with the aid of an ultrasonic soldering iron. A large electromagnet provided the static magnetic fields which were used in the measurements. Samples were placed in a bath of liquid helium and temperatures were obtained from the readings of oil and mercury vapor pressure manometers. A simple manostat[1] was used to keep the temperature constant while the resistivity was measured in different magnetic fields. In all the measurements the current was along the [0001] direction which was perpendicular to the magnetic field. A study of the current–voltage characteristics of the crystals was utilized to ensure ohmic conditions during the measurements.

Zero-field dark resistivity values for the samples at nitrogen temperature ranged from several ohm centimeters to a thousand. As the temperature was reduced from 4.2° to 1.1°K, the resistivities were observed to increase from 10^5 to 10^7 Ω-cm.

Some typical results are shown in Fig. 1, where the ratio of the change in resistivity to the resistivity in zero field is plotted as a function of magnetic field for several representative temperatures. These data were taken for a crystal having 0.4 wt. % Mn, and an $A/1$ ratio of 0.071 cm. The widths of maxima observed in the magnetoresistance ratio in Fig. 1 were found to increase with an increase in temperature. It was also observed that the magnetic fields of the maxima were, within the limits of experimental error, directly proportional to the absolute temperature.

The results have indicated that the magnetic field dependence of the maxima could be influenced by other effects in addition to the temperature. For example, a small increase in manganese concentration caused the maxima to occur at a lower magnetic field. A sixfold increase in electric field also caused a small shift of the maximum to a lower magnetic field. Furthermore, the crystals were found to be photoconductive and a shift of the maximum to a higher magnetic field could be effected by illuminating the crystals with tungsten light during the measurements.

The Hall coefficient, R, the magnetoresistive ratio, $\Delta\rho/\rho_0$, and the resistivity, ρ, are shown as a function of magnetic field in Fig. 2 for a temperature of 2.983°K. These measurements were taken with the crystal in darkness and with an applied

---
* Cryotronics Field Engineering Division, at Aerospace Research Laboratories, Wright-Patterson Air Force Base, Ohio.

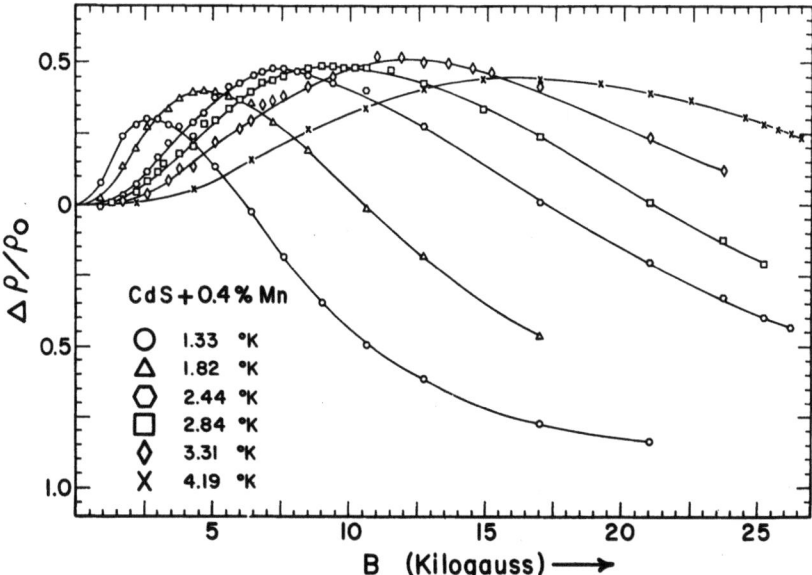

Fig. 1. Magnetoresistive ratio vs. magnetic field.

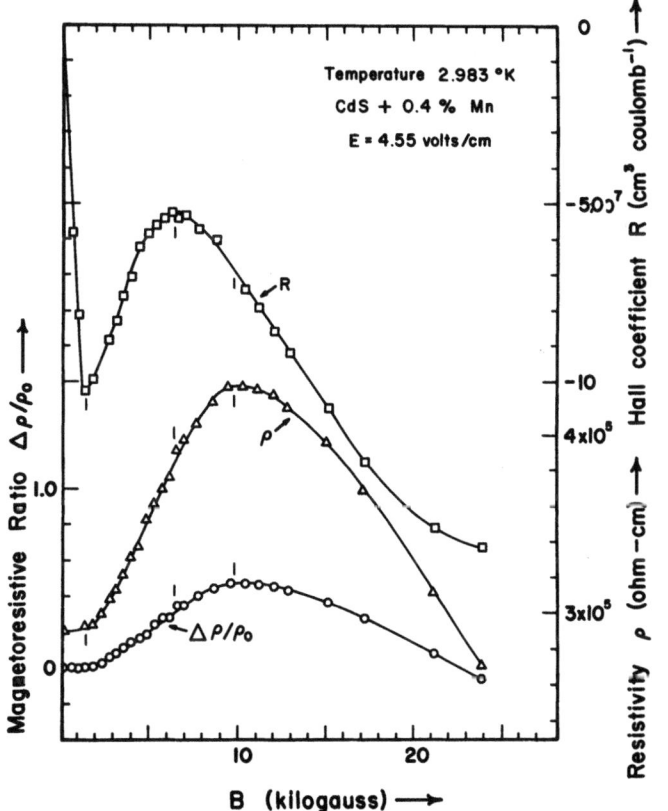

Fig. 2. Magnetoresistive ratio, Hall coefficient, and resistivity vs. magnetic field.

electric field of 4.55 V/cm. Figure 2 illustrates, for a representative temperature, that a maximum observed in the Hall coefficient occurs at a lower field than the maximum in the resistivity. The Hall coefficient maxima were also observed to vary directly as the absolute temperature, yielding a B/T ratio of 2.6 kG/deg K.

The Mn^{++} ion enters the CdS lattice substitutionally and has, in the cadmium site, a $3d^5$-configuration with a ^6S-ground state. Neglecting the hyperfine interaction[2] of the d-electrons with the nuclear moment of ^{55}Mn, and the small splittings[3] caused by the crystalline field, we may write for the Zeeman separation of the energy levels in a magnetic field, B,

$$\Delta E = g\beta\sqrt{s(s+1)}\,B$$

where $\beta = -0.927 \times 10^{-20}$ ergs/Oe, $S = \frac{5}{2}$, and[3] $g \approx 2.0$. Setting $\Delta E = kT$ and solving for B/T yields 2.51 kG/deg K, in fair agreement with the value obtained from the temperature dependence of the magnetic field corresponding to the Hall coefficient maximum.

Since the Hall coefficient varies inversely as the number of carriers, these results imply that the number of carriers in the conduction band is a minimum when the level splitting is equal to kT.

References

1. M. M. Kreitman, *Rev. Sci. Instr.* **35**, 749, 1964.
2. G. A. Slack and S. Galginaitis, *Phys. Rev.* **133**, A253, 1964.
3. P. B. Dorain, *Phys. Rev.* **112**, 1058, 1958.

ANTIFERROMAGNETIC STRUCTURE OF COBALT FLUOSILICATE BELOW 1°K

Akio Ohtsubo and Eizo Kanda

The Research Institute for Iron, Steel, and Other Metals, Tohoku University Sendai, Japan

Electron spin resonance research on cobalt fluosilicate has been done by Bleaney et al.[1] with zinc fluosilicate, which keeps rhombohedral structure down to helium temperatures, as host. The ground state of Co^{2+} ion at helium temperatures can be described by an effective spin $S' = \frac{1}{2}$ and its g-values are highly anisotropic between directions parallel and perpendicular to the trigonal field axis ($g_\parallel = 5.80$, $g_\perp = 3.44$).

Cobalt fluosilicate has also been thought to have a simple crystal structure like that of zinc fluosilicate. Tsujikawa and Couture[2] found by optical observation a low-temperature modification in cobalt and manganese fluosilicates, single crystals of which shattered below about $-50°C$ into hybrid column crystals consisting of three parts with lower symmetry, each having extinction angle differing by 120°. X-ray analysis on the low-temperature modification of cobalt salt was done by Prof. Watanabe and his group at Osaka University. According to their private communication, it has monoclinic or triclinic symmetry and in the unit cell there are two kinds of $Co(H_2O)_6^{2+}$ ions (A and B), each of which forms sheet alternately along the hexagonal axis of the room-temperature form.

A weak ferromagnetism will occur generally, if antiferromagnetic interaction exists between these dissimilar ions with different crystalline field axis. Weak ferromagnetism of manganese and cobalt Tutton salts[3] are examples of such a case. Uryu[4] derived quite generally the canted spin structures using Hamiltonians including isotropic exchange interactions, dipole interaction, and DS_z^2 term for the manganese salt and dipole and anisotropic exchange interactions for cobalt salt. The fluosilicate of low-temperature modification might be a similar case.

The theoretical magnetization curve of an antiferromagnet has been derived but the real magnetization curve measurement has seldom covered the whole field region. This is due to the fact that even a molecular field of an antiferromagnet with T_n at helium temperatures exceeds the conventional magnetic field at, say 20 kG. A measurement of the whole magnetization curve on an antiferromagnet with T_n below 1°K, by a recording fluxmeter as stated in a previous paper,[5] would give a clear example of the molecular field theory and also give an understanding of magnetic structure of the individual substance. Adiabatic magnetization curve measurements were undertaken in order to obtain knowledge of the spin structure. Discussion will be given (1) on the antiferromagnetism and (2) on the weak ferromagnetism separately, as the weak ferromagnetism is thought to be small due to the small deviation of the crystalline field axis from the c-axis. Existence of the weak ferromagnetism, although small, will play a role to eliminate all the possible antiferromagnetic arrangements but the real one.

Magnetic measurements were made parallel and perpendicular to the column direction of the hybrid crystal which coincided with the hexagonal axis of the room-temperature form. Each will be described by suffix ∥ and ⊥. From the susceptibility measurements between 4.2° and 1.3°K, $g_{\|c} = 6.8 \pm 0.05$, $g_{\perp c} = 2.5_5 \pm 0.05$ (mean values); $\theta_\| = -0.19° \pm 0.03°K$, $\theta_\perp = -0.15° \pm 0.04°K$; temperature-independent terms were very small. Adiabatic magnetization curves ($M_s \sim H$) were measured for several entropy values. With decreasing field from an initial field of isothermal magnetization, M_s remains constant in the paramagnetic region and decreases as transforming into antiferromagnetic state. In parallel direction, M_s decreased in S-shape for entropies below a value S_n (= 0.44R). The field at the steepest gradient became the higher for the lower entropy, for example, about 1000 G for $S = 0.01R$. In perpendicular direction, M_s decreased monotonically, for example, below about 6000 G for $S = 0.17R$. For entropies below S_n, weak ferromagnetism, showing abrupt change of magnetization near zero field, appeared. The weak ferromagnetic moment was 3.5% of the saturation moment in the perpendicular direction at the lowest entropy. The corresponding temperature T_n to S_n was determined to be 0.20°K. Specific heat measurement was made using a CMN thermometer and eddy current heating in metal plates for conduction. A specific heat maximum was observed at 0.19°K. (Fig. 1). Its tail can be expressed as $C_m T^2/R = 1.5_2 \times 10^{-2}$ deg². Benzie, Cooke, and Whitley[6] obtained 1.85×10^{-2} deg² by the dispersion method. In Fig. 1 susceptibilities which derived from the initial gradient of magnetization curves are also shown against temperature. The $\chi_\perp$ value after subtraction of the weak ferromagnetic part is lower than $\chi_\|$ even below T_n.

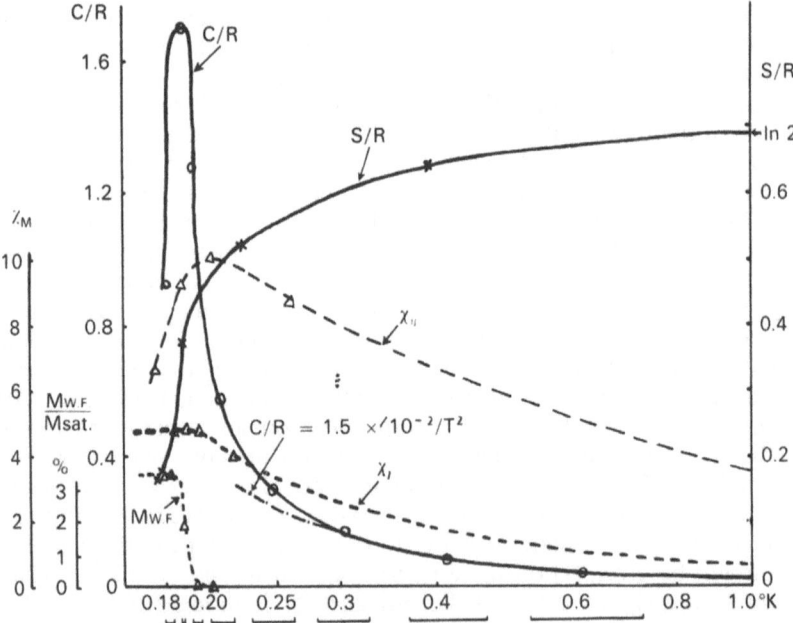

Fig. 1. Entropy, specific heat, susceptibilities, and weak ferromagnetic moment against temperature. The weak ferromagnetic moment is shown as the ratio to the saturation moment in the perpendicular direction. Length of lines below temperature axis indicates a heating interval for each specific heat measurement.

Magnetization curves show antiferromagnetic character with the preferred axis nearly along the hexagonal axis. From the present isentropics, the $M_{s=0} \sim H$ curve for the sphere was drawn by rough extrapolation. By a demagnetizing field correction $[H_d = (4\pi/3) \times (M/V)$, where $M_\parallel = 18{,}900$ G-cm^3/mole, $M_\perp = 7090$ G-cm^3/mole, and $V = 148$ cm^3/mole], the $M_{s=0} \sim H$ curve for an infinitely long specimen was obtained, the spin ordering of which will be discussed. According to the theory, when an anisotropic field is much larger than a molecular field, the magnetization in a direction along the preferred axis ($M_\parallel$) remains zero until H_{ext} reaches an antiferromagnetic molecular field H_e, where $M_\parallel$ steeply increases to the saturation value. In this process both parallel and antiparallel spins are in line with H_e and no flopping takes place. The present $M_\parallel \sim H$ curve is just this case and $H_e = 700$ G. This is nearly the same as the value from the relation $kT_n = \mu H_e$ ($H_e = 880$ G, $\mu = \frac{1}{2} \times 6.8 \times \mu_B$, $T_n = 0.20°$K). "No flopping" can also be explained by the fact that $\chi_\parallel > \chi_\perp$ even below T_n and this anisotropy is due to the anisotropic exchange interaction which also appears in $M_\perp \sim H$ curve as a large saturation field (about 6000 G). The exchange contribution in the total specific heat ($C_m T^2/R = 1.5 \times 10^{-2}$ deg^2) is determined to be 1.3×10^{-2} deg^2 by subtracting the dipole part (1×10^{-3}) and the hfs part (1×10^{-3}). The exchange interaction is dominant and the magnetic ordering due to this exchange shows a specific heat maximum at $S = 0.38R$ remaining 45% of total entropy due to short-range order.

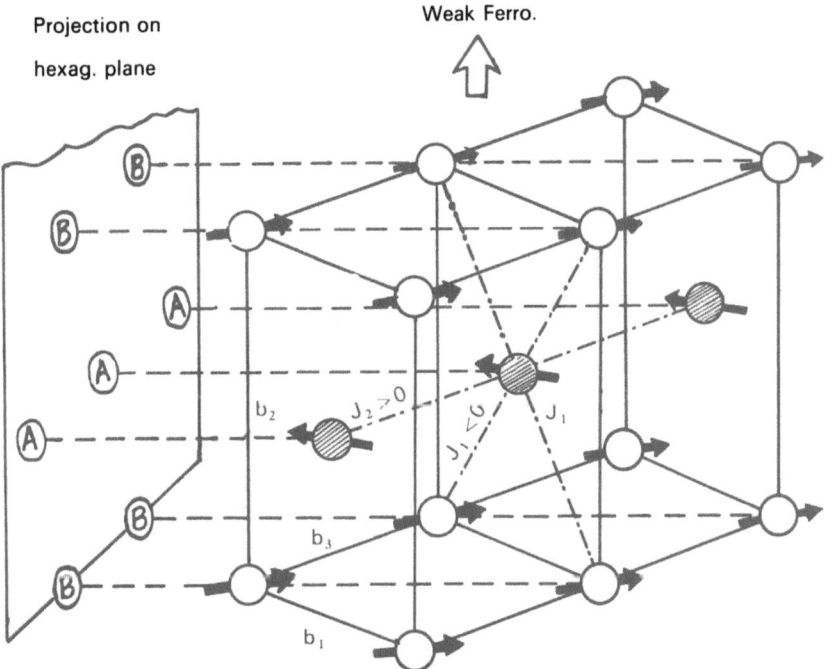

Fig. 2. Proposed spin structure for cobalt fluosilicate.

As for the weak ferromagnetism, quantitative discussion is impossible because neither perfect X-ray analysis such as tilt angle of crystalline field axis nor ESR data such as g-values referred to the crystalline field axis have been obtained. However, the crystal structure known so far and the appearance of the weak ferromagnetism lead us to the following spin structure. A and B ions which have different directions in the crystalline field of the water octahedron form ferromagnetic sheets along the hexagonal axis of the room-temperature form, and spins in the A-sheets are antiparallel to those in the B-sheets, accordingly each spin has four antiparallel and two parallel neighbors (Fig. 2). The tilt angle of spins is determined to be about $0°\ 50'$, considering the hybrid crystal consisting of nearly the same amount of three equivalent single crystals with extinction angles different by $120°$. The crystalline field axis must cant a little more than this value. The crystalline field is thought to be trigonal because of small angle deviation of the axis from the hexagonal axis.

Acknowledgments

The authors would like to thank Prof. T. Watanabe and Drs. Ohsaki and E. Kishida at Osaka University for their kindness in undertaking the research on the crystal structure and the private communication on the result.

References

1. B. Bleaney and D. J. E. Ingram, *Proc. Roy. Soc. (London) Ser. A* **208**, 143, 1951.
2. I. Tsujikawa and L. Couture, *J. Phys. Radium* **16**, 430, 1955.
3. A. R. Miedema, J. van den Broek, H. Postma, and W. J. Huisskamp, *Physica* **25**, 1177, 1959.
4. N. Uryu, *J. Phys. Soc. Japan* **16**, 2139, 1961.
5. A. Ohtsubo and E. Kanda, *Cryogenics* **2**, 339, 1962; *J. Phys. Soc. Japan, Suppl. B-1* **17**, 497, 1962.
6. R. J. Benzie, A. H. Cooke, and S. W. Whitley, *Proc. Roy. Soc. (London) Ser. A* **232**, 277, 1955.

SUPPRESSION OF NUCLEAR DYNAMIC POLARIZATION BY RF RADIATION

G. E. G. Hardeman and G. Gerritsen

Philips Research Laboratories, N.V. Philips' Gloeilampenfabrieken
Eindhoven, The Netherlands

The nuclear spins surrounding the centers in a solid containing paramagnetic centers are subjected to shifts of their Zeeman frequencies. In such a system the nuclear spin diffusion is the essential mechanism for the distribution of any excess nuclear polarization over the lattice. However, the inhomogeneous frequency shifts result in diffusion barriers around the centers.

In the case of nuclear dynamic polarization, the nuclear spin temperature around the centers is lowered by microwave radiation H_e. Strong gradients of spin temperature can then be expected due to the diffusion barriers. This proves that the presence of a perturbing RF field $H^* = H_1^* \cos 2\pi v^* t$ together with H_e hampers the enhancement of the nuclear polarization of the majority of nuclear spins outside the barriers.

The effect can be understood as the formation by H^* of "walls" of saturated nuclear spins resonating on frequency v^* within the barriers.

The phenomenon was observed on phosphorus-doped silicon at $T = 4.2°K$. The attenuation was only observed for v^* within 100 kcps separated from the pure ^{29}Si nuclear resonance frequency. This fact indicates that the ^{29}Si nuclear dynamic polarization is only generated on the peripheral parts of the donor electron density distribution around the phosphorus. Consequently one can expect the nuclear spin-lattice relaxation of the majority spins to take place on the periphery too.

The nuclear dynamic polarization in a solid containing paramagnetic impurity centers is usually described as follows: Due to the magnetic dipolar coupling between the electron spins and the surrounding nuclear spins in the lattice, one can induce combined electron-nuclear spin transitions with an alternating magnetic field H_e. The frequency v_e, which has to obey the relation $v_e = v_s \pm v_I$ where v_s and v_I are the electron and nuclear resonance frequency, respectively. With sufficient intensity of H_e, the double transitions can be induced at such a rate that the polarization of the interacting nuclei is enhanced.

For a very small concentration of paramagnetic centers, the excess nuclear spin polarization is transported and distributed over the total nuclear spin system by means of nuclear spin diffusion.

In the case of a relatively long electron spin-lattice relaxation time, the Zeeman frequencies v_l of the nuclei on the lattice sites l near to the paramagnetic center differ appreciably from each other due to inhomogeneous frequency shifts Δv_l relative to the value v_0 of the majority nuclei I_∞ far from the centers. With a constant $v_I = v_0$, the nuclear spin diffusion can easily take place by simultaneous nuclear spin flips with conservation of energy. If, however, the frequency difference due to Δv_l between two neighboring nuclei exceeds their intrinsic resonance line width due to their mutual dipole–dipole interaction, these spin flips become difficult. Thus a diffusion barrier can be expected around the paramagnetic centers within which the nuclei are more or less unable to carry off any deviation of polarization to the majority nuclei with $v_l = v_0$. The influence of the diffusion barrier on the nuclear spin-lattice relaxation has been investigated by Blumberg[1] and Khutsishvili.[2]

It is the purpose of this paper to show that the diffusion barrier makes it possible to inhibit the polarization process by the application of a perturbing radiation $H^* = H_1^* \cos 2\pi v^* t$, together with H_e and the external field $\bar{H}_0$.

Phosphorus-doped silicon was chosen for the experiments. The various frequency shifts of the ^{29}Si lattice nuclei at the positions $\bar{r}_l$ relative to the donor impurity are given by[3]

$$\Delta v_l = \pm [\tfrac{1}{2}a_l - \tfrac{1}{2}b_l(\bar{r}_l, \theta)]$$

$$a_l = \tfrac{8}{3}\gamma_I \mu_B |\psi(\bar{r}_l)|^2$$

(1)

The term a_l gives the scalar contribution due to a finite electron spin density $|\psi(\bar{r}_l)|^2$, while b_l represents the dipolar contribution, which is a function of $\bar{r}_l$ and the orientation θ of $\bar{H}_0$. For $T = 4.2°K$, a donor concentration of $7 \times 10^{16}/\text{cm}^3$ and a microwave frequency $v_e \approx 10^{10}$ cps, the following experiments were carried out: (1a) With a given H_0 and v_e, the ^{29}Si nuclear system was polarized during a time t_p (about 20 min) starting from the condition of zero polarization. The NMR signal was then recorded. This signal is proportional to the polarization P_0 obtained for the majority spins (I_∞). (1b) The same experiment was carried out, but with the RF field H^* also present, together with H_e. Then the polarization P_1 was determined. (2a) As in case 1a, the polarization P_0 was built up, but the NMR signal was recorded after a time t_r taken from the moment that the microwave field H_e was switched off. The polarization P_0' was now slightly smaller than P_0, owing to the nuclear spin-lattice relaxation time, which time is 4 hr for the studied sample. (2b) The experiment 2a was repeated but with H^* present during the relaxation phase t_r. The polarization P_1' was then registered. (3) The polarization P_0 was built up during the time t_p. Then the radiation H^* was switched on while the microwave radiation was continued for another time t_p. Afterwards the polarization P_1'' was recorded.

Figure 1 shows the attenuation factors P_1/P_0 and P_1'/P_0' plotted as a function of v^*, together with the part of the Endor[3] spectrum in the same frequency range. No resolved Endor structure could be found in the attenuation curves. The vanishing of P_1 and P_1' for $v^* \approx v_0$ is trivial because of direct saturation of the majority spin system. However, the width of the dip in the P'/P_0' curve is considerably larger than the observed NMR linewidth of about 300 cps. The NMR line probably has broad wings below noise level due to the shifts (1) corresponding to large r_l. From a comparison of the two attenuation curves we conclude that a frequency range exists where H^* has negligible influence on an existing polarization but strongly hampers the enhancement of polarization when H_e is also present.

Experiment 3 was carried out for $|v^* - v_0| = 15$ kcps, where $P_1'/P_0' \approx 1$ but $P_1/P_0 \ll 1$. It appeared that $P_1'' \approx P_0$. Thus with H_e present the radiation H^* does not destroy a finite polarization of the I_∞ spins by a competing saturation process, but only blocks further enhancement.

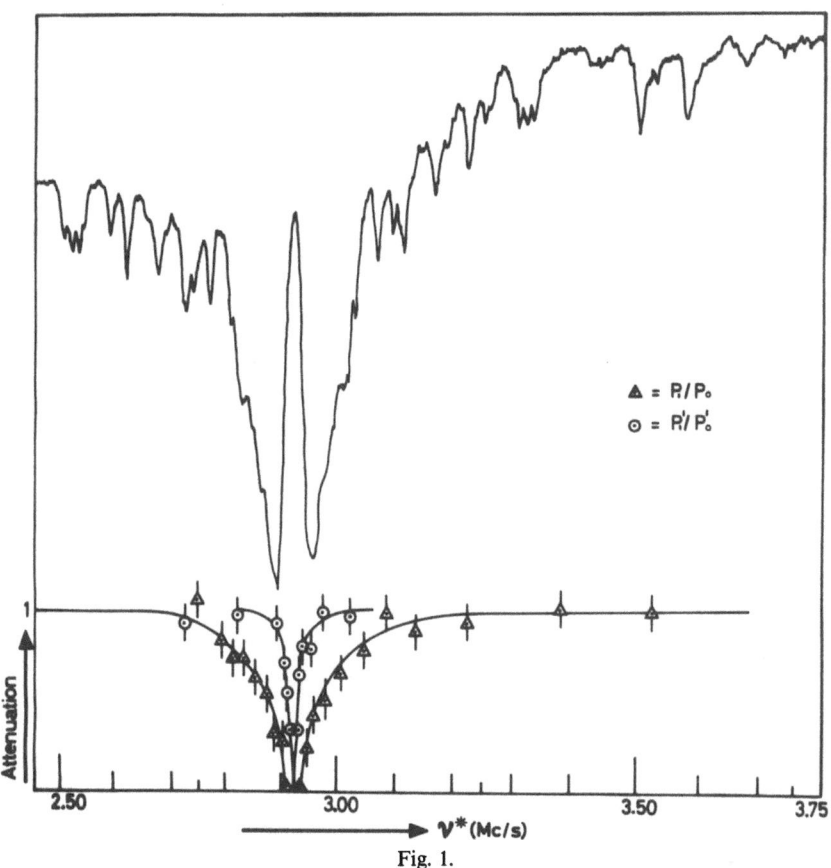

Fig. 1.

The absence of any resolved Endor structure in the attenuation curves indicates that the I_l spins close to the paramagnetic centers have practically no spin diffusion coupling with the I_∞ spin system and thus do not contribute to the polarization.

The observed effects can be ascribed to the saturation of nuclear spins in the outer part of the diffusion barrier, where a limited spin diffusion can take place due to low Δv_l values ($\Delta v_l \ll v_0$). The excess nuclear polarization is now carried off along a steep spin temperature gradient to the saturated spins instead of the I_∞ spins.

We propose the following mechanism for the hampered spin–spin interaction between two neighboring nuclei I_l and I_{l+1}, whose frequency differs by the value $\delta v_l = \Delta v_l - \Delta v_{l+1}$.

As a condition for a hampered but not impossible exchange of spin orientation, we have

$$v_I \gg \delta v_l > T_2^{-1}$$

where T_2 is the transversal relaxation time of the nuclear spin system in the absence of free-electron spins. Assuming $I = \frac{1}{2}$, we denote the nuclear spin states by $+$ and $-$. For the two-spin system we have to consider the probability of the transition $+ - \leftrightarrows - +$. The nuclear dipole interaction admixes the adjacent state

$+-$ to the state $-+$ with a coefficient $\alpha_{\text{II}} \approx 1/\delta v_l T_2$. Due to the admixture, the time dependent part of the scalar interaction of the nuclei with the electron

$$H_{\text{SII}} = A_l \bar{I}_l \cdot \bar{S} + A_{l+1} \bar{I}_{l+1} \cdot \bar{S}$$

can now result in the necessary transition $+- \leftrightarrows -+$ with the low energy $h\delta v_l$. Experiment 3 shows that if there is any moderate polarization present in the I_x spin system, the diffusion barrier protects the I_∞ spins against a speedy loss of the polarization to the saturated spins in the barrier.

References

1. W. E. Blumberg, *Phys. Rev.* **119**, 79, 1960.
2. G. R. Khutsishvili, *Soviet Phys. JETP* (*English Transl.*) **15**, 909, 1962.
3. G. Feher, *Phys. Rev.* **114**, 1219, 1959.

INTERNAL LOCAL FIELDS ON ANTIMONY AND YTTRIUM NUCLEI IN IRON

B. N. Samoilov, V. N. Agureev, V. D. Gorobchenko, V. V. Sklyarevskii, and O. A. Chilashvili

Institute for Atomic Energy
Moscow, USSR

Our earlier preliminary experiments[1] indicated that the effective magnetic field acting on antimony nuclei in a dilute alloy of antimony in iron was positive, although the results obtained so far were not amenable to an exact analysis in view of an insufficient energy resolution of the β-electrons counting channel.

The present paper summarizes the final results of our measurements of the local magnetic field sign on antimony nuclei and those of the measurements of the sign and magnitude of the field on yttrium nuclei in dilute solutions of these elements in iron. The isotope ^{122}Sb decays by β-emission with a half-life of 2.7 days into ^{122}Te mainly through two β branchings $2^-(\beta)2^+(\gamma)0^+$ and $2^-(\beta)0^+$. The first transition, with an intensity 69% and an end-point energy $W_0 = 1.4$ MeV, is a first-forbidden transition in which case the angular distribution of β-electrons, emitted by oriented nuclei, depends on six nuclear matrix elements, the exact values of which are unknown. The second transition, with an intensity 27% and an end-point energy $W_0 = 2.0$ MeV, is a unique first-forbidden β-transition and depends only on a single nuclear matrix element which enters the angular distribution as an unimportant constant factor. The isotope of yttrium, ^{90}Y, decays with an intensity 100% into the ground state of ^{90}Zr with a half-life of 64 hr. In the latter case, the transition is also a unique first-forbidden $2^-(\beta)0^+$ with an endpoint energy $W_0 = 2.3$ MeV. In view of the simple dependence of the β-electrons angular distribution on the nuclear matrix element, we have made our measurements of the asymmetry of the β-radiation from ^{122}Sb and ^{90}Y nuclei on their unique transitions, separating the latter in the case of antimony against the background of a more intensive $2^-(\beta)2^+$ transition.

A silicon surface-barrier semiconductor counter of n–p type, mounted ~ 2 cm above the radioactive sample and cooled to the temperature of liquid nitrogen, was used to detect β-electrons. The temperature of the sample was determined from measurements of the γ-radiation anisotropy of ^{60}Co nuclei polarized in 1% alloy of cobalt in iron. Two samples, one made of such alloy and one to be investigated, were soldered side by side on the end of a copper heat link, being thus in equal temperature conditions. The value of 289.7 kOe was accepted for the magnitude of the local field on cobalt nuclei.[2] The magnetization of the sample was provided with coils, wound from a superconducting wire (a niobium–zirconium alloy). The remaining experimental devices, as well as the experimental procedure, were essentially the same as in our previous communications.[3]

Making use of the angular distribution function of β-electrons from oriented nuclei, found by M. Morita and R. S. Morita,[4] we can readily show that for the unique $2^-(\beta)0^+$ transition of antimony and yttrium the angular distribution has the form

$$N(\theta) = |M(iB_{ij})|^2(p^2 + q^2)\left[1 - \frac{q^2 + \frac{3}{5}p^2}{p^2 + q^2}\frac{P}{W}f_1\mathscr{P}_1(\cos\theta) - \frac{2p^2}{p^2 + q^2}f_2\mathscr{P}_2(\cos\theta)\right] \quad (1)$$

Here, $M(iB_{ij})$ is the nuclear matrix element which determines the transition, W is the energy of the electron and $p = \sqrt{W^2 - 1}$ its momentum, $q = W_0 - W$, the f_K are parameters describing the degree of orientation of the nuclei (we have neglected the parameter of orientation f_3, in view of its smallness in equation (1)); the $P_K(\cos\theta)$ are Legendre polynomials, and θ is an angle between the axis of nuclear polarization and the direction of the β-electron emission.

In experiment, the quantities

$$\varepsilon_\beta^\uparrow = \frac{N(30°) - N_0}{N_0} \quad \text{and} \quad \varepsilon_\beta^\downarrow = \frac{N(150°) - N_0}{N_0} \quad (2)$$

were directly measured. Here, $N(30°)$ and $N(150°)$ are the β-counting rates from oriented nuclei at the angles correspondingly 30° and 150° relative to the axis of orientation, and N_0 is the isotropic counting rate in the absence of orientation. It is easy to show that in the case of the unique transition and for not very low temperatures, the dependence of $\varepsilon_\beta^\uparrow$ and $\varepsilon_\beta^\downarrow$ upon magnetic field at the nucleus can be written in the form

$$\varepsilon_\beta^\uparrow = -a\left(\frac{\mu H}{\kappa T}\right) - b\left(\frac{\mu H}{\kappa T}\right)^2$$

$$\varepsilon_\beta^\downarrow = +a\left(\frac{\mu H}{\kappa T}\right) - b\left(\frac{\mu H}{\kappa T}\right)^2 \quad (3)$$

where a and b are some positive constants, depending on the end-point energy of the β-spectrum, as well as on the setting of the analyzer thresholds selecting the registered portion of the spectrum, μ is the magnetic moment of the nucleus, H is the effective magnetic field at the nucleus; κ is the Boltzmann constant, and T is the temperature of the sample.

The experimentally measured dependences of the quantities $\varepsilon_\beta^\uparrow$ and $\varepsilon_\beta^\downarrow$ on the temperature for the cases of ^{122}Sb and ^{90}Y are plotted correspondingly in Figs. 1 and 2. In so far as the magnetic moments of the ^{122}Sb and ^{90}Y nuclei are negative, a simple joint analysis of equation (3) and the curves in Figs. 1 and 2 gives evidence that the effective magnetic fields on antimony and yttrium nuclei in iron are positive. A value ~ 205 kOe has been obtained in both cases for the magnitudes of the fields. The solid curves in Figs. 1 and 2 represent theoretical dependences $\varepsilon_\beta(1/T)$ for this value of the fields. A statistical error in the value of the fields indicated is of the order of 5%, but real error may be somewhat greater due to the fact that no corrections were made for a contribution of scattered electrons to the β-counting rate, for inaccurate separation of $2^-(\beta)2^+$ electrons in the case of ^{122}Sb and for some uncertainty in solid angle, in which electrons were detected.

The positive sign of the internal local field at antimony nuclei in iron is in good agreement with the theory developed by Daniel and Friedel,[5] who suggested the conduction electrons, polarized by s–d exchange coupling, to contribute the

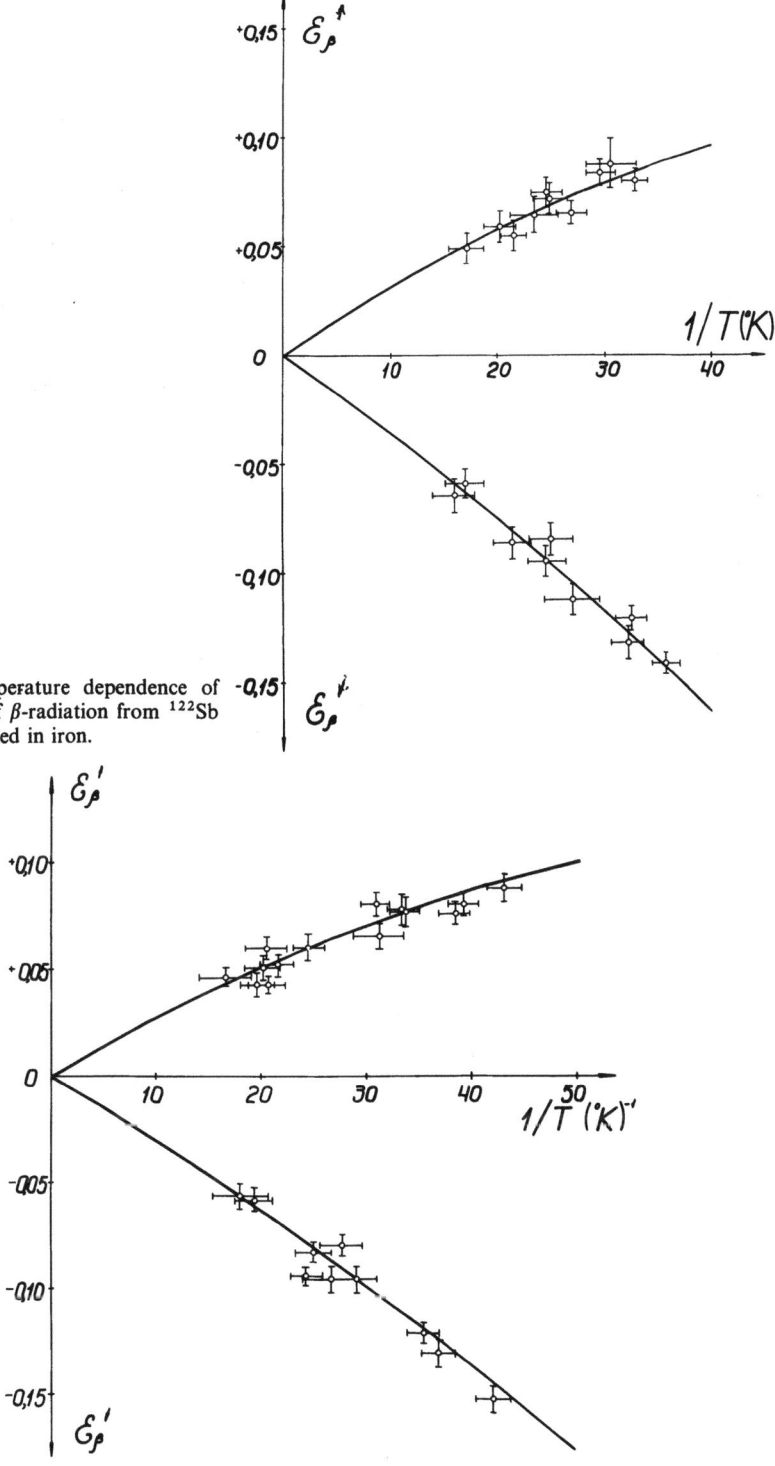

Fig. 1. Temperature dependence of asymmetry of β-radiation from ^{122}Sb nuclei dissolved in iron.

Fig. 2. Temperature dependence of asymmetry of β-radiation from ^{90}Y nuclei dissolved in iron.

bulk of the hyperfine field on nuclei of impurity nonmagnetic elements in ferromagnets. Due to the action of a perturbing potential introduced by the impurity, the conduction electrons' polarization around the impurity nucleus is locally altered and may be either positive or negative depending on the valence difference between the impurity atom and the atom of the host metal. According to the theory of Daniel and Friedel, the internal local field on yttrium nuclei in iron should be negative, while our experiments give quite opposite results.

We have also made an attempt to determine the sign of the effective magnetic field at copper nuclei in a dilute alloy of copper in iron. The existence of the field and its magnitude had been established by the method of nuclear magnetic resonance in a work by Japanese physicists.[2] Preliminary results of the investigation of the asymmetry of the β^+ and β^- radiations from ^{64}Cu nuclei, polarized in iron, give indication that the field on copper nuclei is negative, if we assume (in accord with the nuclear shell model) a negative value for the nuclear magnetic moment of ^{64}Cu.

References

1. B. N. Samoilov, V. V. Sklyarevskii, and V. D. Gorobchenko, *Proceedings of the Eighth International Conference on Low Temperature Physics,* Butterworths, London (1963), p. 265.
2. T. Kushida, A. H. Silver, Y. Koy, and A. Tsujimura, *J. Appl. Phys.* **33**, 1079, 1962.
3. B. N. Samoilov, V. V. Sklyarevskii, V. D. Gorobchenko, and E. P. Stepanov, *Zh. Eksperim. i Teor. Fiz.* **40**, 1871, 1961. (*Soviet Phys. JETP (English Transl.*) **14**, 1267, 1962.) B. N. Samoilov, V. V. Sklyarevskii, and V. D. Gorobchenko, *Zh. Eksperim. i Teor. Fiz.* **41**, 1783, 1961. (*Soviet Phys. JETP (English Transl.*) **14**, 1314, 1961.)
4. M. Morita and R. S. Morita, *Phys. Rev.* **109**, 2048, 1958.
5. E. Daniel and J. Friedel, *J. Phys. Chem. Solids* **24**, 1601, 1963.

PARAMAGNETISM OF HEMOGLOBIN AND MYOGLOBIN DERIVATIVES AT LOW TEMPERATURE

M. Kotani
University of Tokyo
Tokyo, Japan

Some of the hemoglobin and myoglobin derivatives have high spin $S = \frac{5}{2}$, and some others low spin $S = \frac{1}{2}$, in their ground states. According to electron paramagnetic resonance experiments, the high-spin sextet seems to be split into three Kramers doublets with anomalously large intervals ($\sim 10 \text{ cm}^{-1}$). Furthermore, some derivatives are believed to be thermal mixtures of high-spin and low-spin molecules. From information provided by electron paramagnetic resonance and other experiments, the temperature dependence of effective Bohr magneton number at low temperature is calculated and the possibility of determining values of parameters from low-temperature susceptibility measurements is discussed. A preliminary experiment on paramagnetic anisotropy at low temperatures of myoglobin single crystals in the high-spin state gives $D = 12 \text{ cm}^{-1}$, where D is the coefficient in the spin Hamiltonian DS_z^2.

10
DILUTE ALLOYS

10.1. General

LOCALIZED STATES IN DILUTE ALLOYS

E. Daniel
Institut de Physique
Strasbourg, France

and

J. Friedel
Physique des Solides-Faculté des Sciences
Orsay, France

Introduction

Most of the characteristic properties of pure metals are rather satisfactorily described by treating their conduction electrons as almost free electrons, the wave functions of which extend throughout the crystal. If a dilute alloy is made by dissolving an atom X in a metallic matrix M, some localization of electronic charge must occur around X, because of the necessity for electrons to screen out the extra ionic charge of X over M, in order to satisfy the classical law of electrical neutrality of conductors in equilibrium. It turns out that, even in a free electron gas, the screening radius of a point charge is of the order of atomic dimensions for an electronic density corresponding to the number of conduction electrons in an ordinary metal.

Now, in a real solid solution, the wave function of a conduction electron must retain something of its atomic behavior around each nucleus. When the atomic valence wave functions of both X and M are of the same symmetry type and quite overlapping, as is the case for instance with the s states in monovalent or divalent metals, one expects the screening charge to be built up continuously from each state of the conduction band until it reaches the Fermi level, like around a point charge in a free electron gas. The situation is quite different if X is a transition element and M a noble metal. This kind has an essentially s-type conduction band, while X has both s and d valence electrons with nearly the same energy. The d radial wave function does not extend very much out of the X atomic cell, especially near the end of the transition series, so it must keep an important part of its atomic character in the alloy. Nevertheless, since its energy is higher than the bottom of the conduction band, it gets broadened by mixing with the nearly free electrons having approximately the same energy in the conduction band. If this broadening is not too great, the electronic charge of X is not spread out in the conduction band but remains strongly localized in the X cell, in a "virtual d-bound state" which looks very much like an atomic d-state. Another equivalent way of describing the same phenomenon is as follows: One first thinks of X as giving up all its d and s valence electrons to the conduction band of the alloy; then, the extra ionic charge of X produces a perturbing potential which, because of the atomic properties of X, must be almost strong enough to have d-bound states. As a result, the d components of the nearly free conduction electron wave functions are strongly attracted in the atomic cell of X when their energy is close to an average value somewhat higher than the bottom of the conduction band. One still gets a "virtual d-bound" state, localized on X and appearing here as a resonance for conduction electrons in the potential of the atomic X cell.

In the following discussion, we shall first demonstrate in detail the equivalence of these two descriptions of virtual bound states. Then, we shall give an account of the effects of these states on the physical properties of dilute alloys: electronic specific heat, residual resistivity, size effect, and the diffusion coefficient of solute atoms. Special attention will be devoted to the case where the virtual bound states become nondegenerate with respect to spin, giving rise to localized magnetic moments in the alloys.

Screening of a Point Impurity and Phase Shifts[1-4]

Let us first recall some important ramifications of the theory of alloys. As is well known, the extra ionic charge Z associated with a foreign atom X in a metal M is screened out over a distance of the order of an atomic radius, apart from some long-range oscillations of small amplitude in the electronic density. This means that the perturbing potential of the impurity is small to the point of vanishing out of a sphere of atomic dimensions. In a first approximation, one can assume that this potential is spherically symmetric and so describe the conduction electrons by plane waves. (This means replacing the Bloch conduction electron wave functions by their plane wave part and taking an adequate pseudopotential with spherical symmetry instead of the real potential of the impurity.)

With these simplifications, it is possible to make a partial wave analysis of the scattering of the conduction electrons by the impurity. At large distances from the impurity, its effect is to produce a phase shift η_l in each spherical wave of orbital quantum number l. The screening condition gives Friedel's sum-rule:

$$Z = \frac{2}{\pi} \sum_l (2l+1)\eta_l \tag{1}$$

According to elementary scattering theory, the differential cross-section for an electron with wave vector k is:

$$\frac{d\sigma}{d\Omega} = \frac{1}{k^2} \left| \sum_l (2l+1) e^{i\eta_l} \sin \eta_l P_l(\cos\theta) \right|^2 \tag{2}$$

where θ is the scattering angle and P_l the Legendre polynomial of order l. The total cross-section is:

$$\sigma = \frac{4\pi}{k^2} \sum_l (2l+1) \sin^2 \eta_l \tag{3}$$

but the effective cross section for electrical resistivity:

$$2\pi \int_0^\pi \frac{d\sigma}{d\Omega}(1-\cos\theta)\sin\theta\, d\theta$$

makes the residual resistivity due to an atomic concentration c of impurities in a metal with A conduction electrons per atom equal to (in Hartree atomic units, $e = m = \hbar = 1$):

$$\rho = \frac{4\pi c}{A k_F} \sum_l l \sin^2(\eta_l - \eta_{l-1}) \tag{4}$$

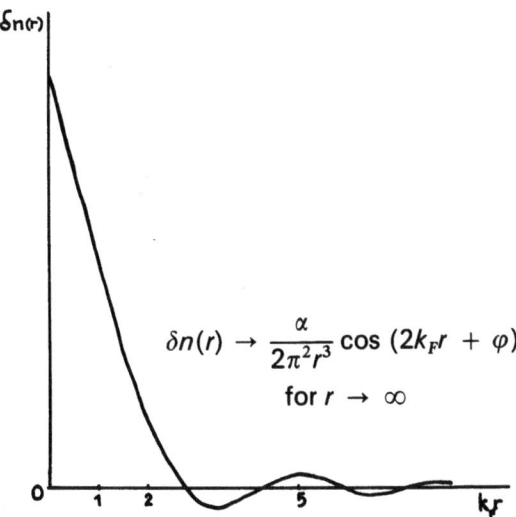

Fig. 1. Screening of a point charge and long-range oscillations in electronic density.

with the phase shifts computed at the Fermi level

$$\mathscr{E}_F = \frac{\hbar^2 k_F^2}{2m}$$

An important consequence of the phase shifts is to produce some oscillations of the electronic density in the matrix, which extend far from the impurity, with, asymptotically, half the wavelength of electrons at the Fermi level and an amplitude decreasing like r^{-3} at large distance r from the impurity (Fig. 1):

$$\delta n(r) \simeq -\frac{\alpha}{2\pi^2 r^3} \cos(2k_F r + \varphi) \tag{5}$$

with
$$\begin{cases} \alpha \sin \varphi = \sum_l (-)^l (2l+1) \sin^2 \eta_l(k_F) \\ \alpha \cos \varphi = \sum_l (-)^l (2l+1) \sin \eta_l(k_F) \cos \eta_l(k_F) \end{cases}$$

In fact, equation (5) is valid in the free electron approximation. The true change in local density could be roughly deduced from it by multiplication by the square of the periodic part of the wave function if this one is real.

Virtual Bound State

Let X be a transition element dissolved in a monovalent normal matrix M, for instance a noble metal. Let us assume that X has an atomic d level, the energy of which lies within the bandwidth of the s conduction electrons (Fig. 2). For $k \neq 0$, the conduction electron wave functions are not purely s, even in a cubic crystal, and their d part gets mixed with the atomic d level of X with which they are degenerate.

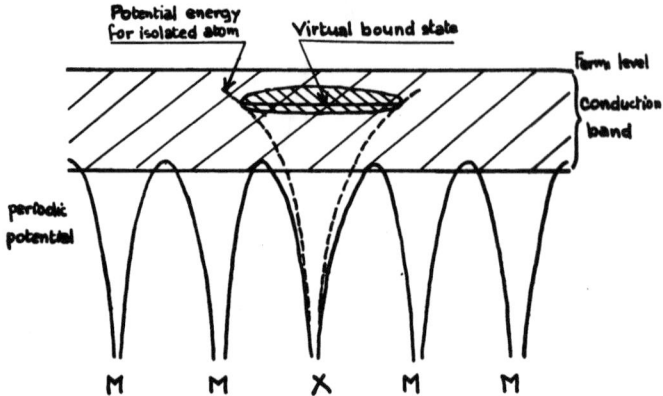

Fig. 2. Virtual d level on X in MX.

In order to give a simple account of this phenomenon, let us first assume that X has but a single level $|0\rangle$ with energy $\mathscr{E}_0$ which may give rise to a virtual bound state; let k be an unperturbed Bloch state for an electron with energy k in the conduction band. If we neglect the s–d interaction V between the conduction electrons and the atomic state $|0\rangle$, both kind of states can be considered as eigenstates of an unperturbed Hamiltonian H_0 such that:

$$H_0|0\rangle = \mathscr{E}_0|0\rangle \quad \text{and} \quad H_0|\mathbf{k}\rangle = \mathscr{E}_\mathbf{k}|\mathbf{k}\rangle \tag{6}$$

We assume that the $|\mathbf{k}\rangle$ states are normalized in such a way that:

$$\langle \mathbf{k}'|\mathbf{k}\rangle = \delta(\mathbf{k}' - \mathbf{k})$$

By taking into account the s–d interaction, we obtain the true Hamiltonian $H = H_0 + V$. Because of the degeneracy of $|0\rangle$ with states in the conduction band, we look for eigenstates $|\psi\rangle$ of H as wave packets built from $|0\rangle$ and $|\mathbf{k}\rangle$ states close to it in energy:

$$|\psi\rangle = c|0\rangle + \int a(\mathbf{k})|\mathbf{k}\rangle\, d^3\mathbf{k} \tag{7}$$

such that:

$$H|\psi\rangle = (H_0 + V)|\psi\rangle = \mathscr{E}|\psi\rangle \tag{8}$$

By taking the scalar products of $H|\psi\rangle$ with $|0\rangle$ and $|\mathbf{k}\rangle$, we get the set of two coupled equations for c and $a(\mathbf{k})$:

$$\begin{cases} c(\mathscr{E} - \mathscr{E}_0) = \int a(\mathbf{k})\langle 0|V|\mathbf{k}\rangle\, d^3\mathbf{k} \\ a(\mathbf{k})(\mathscr{E} - \mathscr{E}_\mathbf{k}) = c\langle \mathbf{k}|V|0\rangle \end{cases} \tag{9}$$

if we assume that a proper choice of H_0 makes $\langle 0|V|0\rangle = 0$ and that the matrix elements $\langle \mathbf{k}'|V|\mathbf{k}\rangle$ are negligible.

The solution for a state which tends adiabatically to the incoming Bloch wave k_0 when the interaction vanishes is given by:

$$a(\mathbf{k}) = \delta(\mathbf{k} - \mathbf{k}_0) + c\frac{\langle \mathbf{k}|V|0\rangle}{\mathscr{E} - \mathscr{E}_\mathbf{k} + i\lambda} \quad (10)$$

where $\mathscr{E} = \mathscr{E}_{\mathbf{k}_0}$, and $\lambda \to +0$.

The corresponding solution for c is, according to the first equation (9),

$$c = \frac{\langle 0|V|\mathbf{k}_0\rangle}{\mathscr{E} - \mathscr{E}_0 - \Delta + i(\Gamma/2)} \quad (11)$$

where

and

$$\begin{aligned}\Delta &= \text{p.p.} \int \frac{|\langle 0|V|\mathbf{k}\rangle|^2}{\mathscr{E} - \mathscr{E}_\mathbf{k}} d^3k \\ \Gamma &= 2\pi \int |\langle 0|V|\mathbf{k}\rangle|^2 \delta(\mathscr{E}_\mathbf{k} - \mathscr{E}) d^3k \\ &= 2\pi \overline{|\langle 0|V|\mathbf{k}\rangle|^2} g(\mathscr{E}) \end{aligned} \quad (12)$$

with $g(\mathscr{E})$ being the density of states with energy $\mathscr{E}$ in the conduction band and $\overline{|\langle 0|V|\mathbf{k}\rangle|^2}$ an average value of the s–d interaction between $|0\rangle$ and the electron states on the $\mathscr{E}$ energy shell in $\mathbf{k}$ space.

Going back to equation (10) with this result, we finally get for $a(\mathbf{k})$:

$$a(\mathbf{k}) = \delta(\mathbf{k} - \mathbf{k}_0) + \frac{\langle 0|V|\mathbf{k}_0\rangle\langle \mathbf{k}|V|0\rangle}{\mathscr{E} - \mathscr{E}_0 - \Delta + i(\Gamma/2)} \frac{1}{\mathscr{E} - \mathscr{E}_\mathbf{k} + i\lambda} \quad (13)$$

For a narrow level, Δ must not depend very strongly upon the energy, so that we shall take it as a constant in equations (11) and (13). The amount of d state inside the state $|\psi\rangle$ with initial wave vector $\mathbf{k}_0$ is then given by:

$$|c|^2 = \frac{|\langle 0|V|\mathbf{k}_0\rangle|^2}{(\mathscr{E} - \mathscr{E}_0 - \Delta)^2 + (\Gamma^2/4)} \quad (14)$$

This shows that the initial atomic state $|0\rangle$ is broadened in energy into a Lorentz-shaped line shifted by Δ from $\mathscr{E}_0$ and having Γ for line width. This is the virtual bound state, broadened by resonance with the continuum of extended electronic states of the conduction band (Fig. 3). It is clear from equations (12) and (14) that:

$$\int_{-\infty}^{\infty} |c|^2 g(\mathscr{E}) \, d\mathscr{E} = 1$$

so that this virtual bound state can accommodate the total amount of charge initially included in the atomic level $|0\rangle$; of course, this charge is strongly located in the atomic X cell.

It must be noticed that, according to (7), the initial d state $|0\rangle$ is now incorporated, by mixing with the Bloch states $|\mathbf{k}\rangle$, into many extended states of the conduction band of the alloy, the energies of which are close to $\mathscr{E}_0$, apart from the shift Δ. It is convenient to label each $|\psi\rangle$ state with the initial wave vector $\mathbf{k}_0$. As the amount of charge in the conduction states does not change, one can take into account

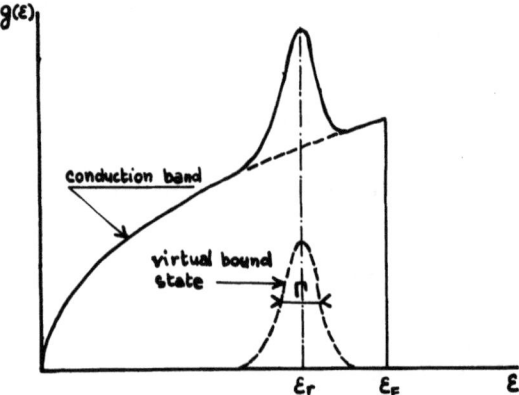

Fig. 3. Effect of a virtual bound state on the density of states versus energy.

the extra electrons accommodated in the d part of $|\psi\rangle$ by adding to the initial density of states in $\mathbf{k}$ space a term proportional to $|c|^2$, as given by equation (14) times the number of X atoms per unit volume in the alloy.

It is clear that the analysis just given here of a virtual bound state in an alloy is nothing but an adaption to solid state physics of the well-known theory of resonances and metastable states in atomic and nuclear physics (see any modern book on quantum mechanics or quantum theory of wave scattering). Obviously, this remains the case for the partial wave analysis that we shall now make.

Partial Wave Analysis

In order to demonstrate in a simple way the equivalence of the previous approach to virtual bound levels with the description by scattering of the conduction electrons in the potential of the substituted X atom, we now make some additional assumptions needed for mathematical simplicity, which are not likely to modify anything fundamental in the physical results. We reduce the Bloch wave functions for conduction electrons to their plane-wave part and we assume spherical symmetry for the interaction V and the surfaces of constant energy in $\mathbf{k}$ space. These hypotheses allow us to perform a partial wave analysis of the wave functions in the alloy.

We now have:

$$|\mathbf{k}\rangle = \frac{1}{(2\pi)^{3/2}} e^{i\mathbf{k}\mathbf{r}}$$

$$= \frac{4\pi}{(2\pi)^{3/2}} \sum_l i^l j_l(kr) \sum_m Y_{l,m}^*(\Omega_\mathbf{k}) Y_{l,m}(\Omega_\mathbf{r})$$

where j_l is the spherical Bessel function of order l and $Y_{l,m}$ the corresponding spherical harmonics, while $\Omega = (\theta, \varphi)$ indicates the angular coordinates of the vectors.

Instead of limiting ourselves to a single state $|0\rangle$, we can take into account the usual orbital degeneracy by considering the $2l + 1$ atomic states $|m\rangle = Y_{l,m}(\Omega_\mathbf{r}) f(r)$ with energy $\mathscr{E}_0$ if we neglect any crystalline field splitting. We have, of course, $l = 2$ for a d level. We now write for the wave function instead of equation (7)

$$|\psi\rangle = \sum_m C_m |m\rangle + \frac{4\pi}{(2\pi)^{\frac{3}{2}}} \sum_{l,m} i^l \int j_l(kr) b_{l,m}(k) Y_{l,m}(\Omega_r) k^2 \, dk \qquad (15)$$

Equations (9) are now replaced by the following set:

$$\begin{cases} (\mathscr{E} - \mathscr{E}_0)C_m = -\int b_{2,m}(k)\langle d|V|k\rangle k^2 \, dk \\ (\mathscr{E} - \mathscr{E}_k)b_{l,m}(k) = + C_m \delta_{l,2}(-i)^l \langle k|V|d\rangle \end{cases} \qquad (16)$$

with

$$\langle k|V|d\rangle = \frac{4\pi}{(2\pi)^{\frac{3}{2}}} \int j_2(kr) V(r) f(r) r^2 \, dr$$

From equations (16), it is seen that the atomic d state of X mixes only with the d part of the conduction wave functions. (This is nothing but a consequence of our assumption of spherical symmetry.)

The solution of the set of equations (16) gives

$$C_m = \frac{-\langle d|V|k_0\rangle Y^*_{2,m}(\Omega_{k_0})}{\mathscr{E} - \mathscr{E}_0 - \Delta + i(\Gamma/2)} \qquad (17)$$

where

$$\left.\begin{array}{l} \Delta = \text{p.p.} \int \frac{|\langle d|V|k\rangle|^2}{\mathscr{E} - \mathscr{E}_k} k^2 \, dk \\[2mm] \Gamma = 2\pi |\langle d|V|k_0\rangle|^2 k_0^2 \dfrac{\partial k}{\partial \mathscr{E}} \\[2mm] = \tfrac{1}{2}|\langle d|V|k_0\rangle|^2 g(\mathscr{E}) \end{array}\right\} \qquad (18)$$

and

Obviously, we get here the same Lorentzian broadening of the virtual d level as in the previous description.

According to equation (15), outside of X, as far as the $l = 2$ component is concerned, the wave function is just the d part of the incoming plane wave with wave vector k_0 plus a scattered wave which becomes, at large distances from X:

$$-\frac{4\pi}{(2\pi)^{\frac{3}{2}}} \sum_m Y^*_{2,m}(\Omega_{k_0}) Y_{2,m}(\Omega_r) \frac{\pi|\langle d|V|k_0\rangle|^2 m \hbar^{-2}}{\mathscr{E} - \mathscr{E}_0 - \Delta + i(\Gamma/2)} \frac{e^{ik_0 r}}{r}$$

$$= -\frac{4\pi}{(2\pi)^{\frac{3}{2}}} \sum_m Y^*_{2,m}(\Omega_{k_0}) Y_{2,m}(\Omega_r) \frac{\Gamma/2}{\mathscr{E} - \mathscr{E}_0 - \Delta + i(\Gamma/2)} \frac{e^{ik_0 r}}{k_0 r}$$

$$= +\frac{4\pi}{(2\pi)^{\frac{3}{2}}} \sum_m Y^*_{2,m}(\Omega_{k_0}) Y_{2,m}(\Omega_r) e^{i\eta_2} \sin \eta_2 \frac{e^{ik_0 r}}{k_0 r}$$

where

$$\tan \eta_2 = \frac{\Gamma}{2(\mathscr{E}_0 + \Delta - \mathscr{E})} \qquad (19)$$

Let us call θ the angle between the initial wave vector k_0 and the distant point given by **r**. The scattered wave at this point can be written:

$$\frac{1}{(2\pi)^{\frac{3}{2}}} \frac{5}{k_0} P_2(\cos\theta) e^{i\eta_2} \sin\eta_2 \frac{e^{ik_0 r}}{r} \quad (20)$$

In formula (20), $P_2(\cos\theta)$ is the Legendre polynomial of order 2; the quantity $(5/k_0)P_2(\cos\theta)e^{i\eta_2}\sin\eta_2$ is exactly the contribution from the $l = 2$ spherical wave to the scattering amplitude:

$$f(k_0, \theta) = \sum_l \frac{(2l+1)}{k_0} e^{i\eta_l} \sin\eta_l P_l(\cos\theta)$$

when a plane wave of vector k_0 is scattered by a spherically symmetrical potential of finite range which produces the phase shift η_l in the lth spherical component. From this result it is clear that the two approaches, that is to say, s–d mixing and scattering by a potential, must lead to the same conclusions concerning any physical properties of a dilute alloy. We shall now show this in detail.

Instead of describing the impurity atom X in the alloy by a quasi-bound state d which gets mixed with the states of the conduction band, we could have used a second approach in which we represent the impurity by a localized potential acting on the conduction electrons. Apart from having a short range of the order of the atomic volume, this potential must be almost strong enough to bind a state with d-like symmetry, taken out of the conduction band. That is to say, there must be a resonance in this potential for some mean energy $\mathscr{E}_r$ in the conduction band. As is well known from scattering theory, near such a resonance the η_2 phase shift increases rapidly from a value close to zero (if we assume no bound state) to almost π in a narrow range of energy, according to the law (Fig. 4):

$$\eta_2(\mathscr{E}) = \arctan\frac{\Gamma}{2(\mathscr{E}_r - \mathscr{E})} \quad (21)$$

Γ being the width of the resonance. This is exactly the same law as that given by formula (19). If Γ is small, that is to say, if the resonance is sharp, the total charge

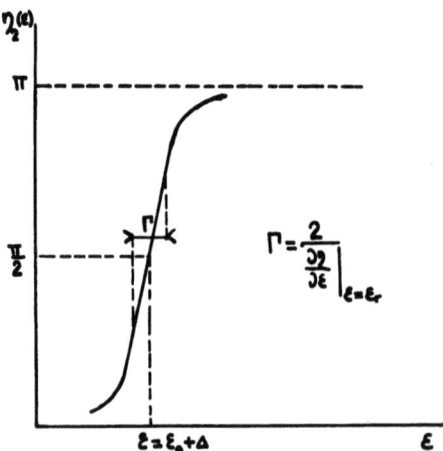

Fig. 4. Phase shift for a virtual d-bound state of width Γ.

attracted by the potential is strongly concentrated in the volume of the impurity so that one gets a virtual bound state in this volume, with a narrow spread in energy around $\mathscr{E}_r$. The electronic structure of the alloy is then seen to be fully equivalent to that which we obtained previously by mixing a localized d state with the conduction band. We note that:

$$\Gamma = \frac{2}{\left.\frac{\partial \eta}{\partial \mathscr{E}}\right|_{\mathscr{E}=\mathscr{E}_r}}$$

In the s–d mixing approach, we get for the total screening charge up to energy $\mathscr{E}$ [according to equations (14), (17), and (18)] when we include spin degeneracy:

$$\begin{aligned}
Z(\mathscr{E}) &= 2 \sum_{m=-2}^{2} \int^{\mathscr{E}} |C_m(k_0)|^2 \, d^3k_0 \\
&= 10 \int^{\mathscr{E}} \frac{|\langle d|V|k_0\rangle|^2}{(\mathscr{E} - \mathscr{E}_0 - \Delta)^2 + (\Gamma^2/4)} k_0^2 \frac{\partial k_0}{\partial \mathscr{E}} \, d\mathscr{E} \\
&= \frac{10}{\pi} \int^{\mathscr{E}} \frac{\Gamma/2}{(\mathscr{E} - \mathscr{E}_0 - \Delta)^2 + (\Gamma^2/4)} \, d\mathscr{E} \\
&= \frac{10}{\pi} \eta_2(\mathscr{E})
\end{aligned} \qquad (22)$$

as $\eta_2(0)$ can be made equal to zero, because Γ is small and there is no potential to subtract a bound state from the conduction band. This result is identical to that given by Friedel's sum-rule,[1] by neglecting phase shifts other than $l = 2$ but assuming spin degeneracy in the scattering approach to virtual bound states.

Finally, in the scattering formalism, we have a long-range oscillating variation of the electronic density due to the interference between the incident and scattered waves. According to formula (5), in which we keep only the $l = 2$ component in the case of a virtual bound state, this variation in density is given asymptotically by:

$$\delta n(\mathbf{r}) \simeq -\frac{(2l+1)}{2\pi^2 r^3} \sin \eta_2(k) \cos[2kr + \eta_2(k)] \qquad (23)$$

where $\eta_2(k)$ is written for $\eta_2(\mathscr{E})$ with $\mathscr{E} = \hbar^2 k^2/2m$.

One gets exactly the same result from the s–d mixing. Starting from the expression (15) for the mixed wave function corresponding to an incident plane wave with wave vector $\mathbf{k}$, one obtains for the difference in electronic densities in the region of space in which the d wave function is vanishingly small:

$$\frac{2}{\pi} \sum_{m,m'} \sum Y^*_{2,m}(\Omega_\mathbf{k}) Y_{2,m'}(\Omega_\mathbf{k}) Y^*_{2,m'}(\Omega_\mathbf{r}) Y_{2,m}(\Omega_\mathbf{r}) \sin \eta_2(k)$$

$$\times \{[n_2^2(kr) - j_2^2(kr)] \sin \eta_2(k) - 2j_2(kr) n_2(kr) \cos \eta_2(k)\}$$

because, as a consequence of equation (19):

$$\frac{\Gamma^2/4}{(\mathscr{E}_0 + \Delta - \mathscr{E}) + \Gamma^2/4} = \sin^2 \eta_2 \quad \text{and} \quad \frac{(\Gamma/2)(\mathscr{E}_0 + \Delta - \mathscr{E})}{(\mathscr{E}_0 + \Delta - \mathscr{E})^2 + \Gamma^2/4} = \sin \eta_2 \cos \eta_2$$

By integrating over the angles for **k**, at given energy, this gives:

$$\frac{\partial}{\partial \mathscr{E}}[\delta n(\mathbf{r})] = \frac{(2l+1)}{2\pi^2} k^2 \frac{\partial k}{\partial \mathscr{E}} \sin \eta_2(\mathscr{E}) \times [(n_2^2 - j_2^2) \sin \eta_2 - 2 j_2 n_2 \cos \eta_2]$$

The asymptotic values of the spherical Bessel and Neumann function j_2 and n_2 for large values of their argument give, if we take into account spin degeneracy:

$$\frac{\partial}{\partial \mathscr{E}}[\delta n(\mathbf{r})] \xrightarrow[r \to \infty]{} \frac{5}{\pi^2 r^2} \frac{\partial k}{\partial \mathscr{E}} \sin \eta_2(\mathscr{E}) \sin[2kr + \eta_2(\mathscr{E})]$$

This result is what has already been obtained from scattering theory, so that one gets expression (23) again by integrating on the energy. So the two descriptions of a virtual bound state are equivalent.

The previous analysis is, of course, an oversimplification of the real state of affairs. By neglecting the matrix elements $\langle \mathbf{k}'|V|\mathbf{k}\rangle$, only the resonance scattering for the $l = 2$ partial wave of an electron has been taken into account, the smooth change of phase shifts due to potential scattering being neglected. This approximation is a good one for a very narrow virtual bound level, as may happen when l is large. For an s state, that is to say $l = 0$, the width would be so large that the concept of virtual bound level would lose its meaning. The situation improves with increasing values of l and to speak of virtual d-bound states is certainly meaningful. Nevertheless, estimates by Blandin and Friedel[5] of the width of such a level by numerical computation of the η_2 phase shift in a square-well potential give Γ of the order of $\mathscr{E}_r/3$, that is to say, several electron volts nearer the Fermi level of an ordinary metal, the energy $\mathscr{E}_r$ of the resonance being counted from the bottom of the conduction band. The conclusion is that the simple Lorentz shape of the level must not be taken too literally in the most interesting case of a resonance close to the Fermi level, because its width reaches several electron volts and potential scattering may not be negligible on such a range.

Physical Effects of Virtual Bound States

When a virtual bound state occurs in an alloy, the number of electrons per unit of volume which can be accommodated from the bottom of the conduction band up to the Fermi level increases. In fact, one gets an extra density of states per energy unit given by:

$$|C(\mathscr{E})|^2 g(\mathscr{E}) = \frac{1}{\pi} \frac{\Gamma/2}{(\mathscr{E} - \mathscr{E}_0 - \Delta)^2 + \Gamma^2/4}$$

or, in case of spherical symmetry with orbital and spin degeneracy:

$$\sum_m 2|C_m(\mathscr{E})|^2 g(\mathscr{E}) = 2 \frac{(2l+1)}{\pi} \frac{\Gamma/2}{(\mathscr{E} - \mathscr{E}_0 - \Delta)^2 + \Gamma^2/4}$$

$$= \frac{10}{\pi} \frac{\partial \eta_2}{\partial \mathscr{E}}$$

for a d state. This gives rise to an extra electronic specific heat $\gamma' T$, where:

$$\gamma' = c \frac{10\pi}{3} k_B^2 \left. \frac{\partial \eta_2}{\partial \mathscr{E}} \right|_{\mathscr{E} = \mathscr{E}_F}$$

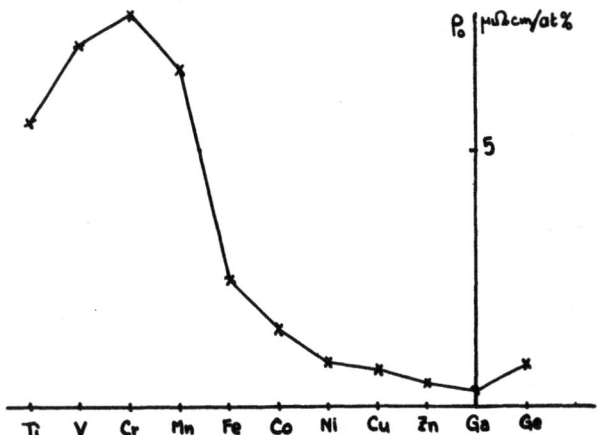

Fig. 5. Residual resistivities of dilute alloys of aluminum.[7]

if there are c impurities per unit of volume and if the energy $\mathscr{E}_0 + \Delta$ of the virtual level is close to the Fermi level $\mathscr{E}_F$.

The validity of a direct check of this prediction is somewhat uncertain for the time being, due to a lack of experimental data on aluminum base alloys, for instance. Measurements have mainly been made on alloys of transition elements in noble metal matrix,[6] but in this case, as will be seen later, the main contribution to low temperature specific heat has a magnetic origin with a similar temperature dependence, that is to say, proportionality to T, so that one has first to remove the magnetic anomaly. This anomaly is not present in copper–nickel, for instance, and the present theory accounts satisfactorily for the low temperature specific heat of these alloys.

Along with excess electronic specific heat, a similar increase of the Pauli spin susceptibility is expected when a virtual d-bound state crosses the Fermi level of the alloy, but the diamagnetism of the d shell must overcome it when the virtual d-bound state sinks into the Fermi sea.

Simultaneously with these effects, a virtual level gives rise to a large residual resistivity. According to equations (4) and (19) this is equal to:

$$\rho_0 \simeq \frac{20\pi c}{Ak_F} \sin^2 \eta_2(\mathscr{E}_F)$$

$$= \frac{20\pi c}{Ak_F} \frac{(\Gamma/2)^2}{(\mathscr{E}_F - \mathscr{E}_0 - \Delta)^2 + (\Gamma/2)^2} \tag{24}$$

if the effect of the other phase shifts is negligible. ρ_0 reaches a maximum value $\rho_{max} \simeq 20\pi c/Ak_F$ independent of Γ when the virtual bound state crosses the Fermi level. ρ_{max} may be of the order of $10\mu\Omega$ cm %, that is to say, ten times larger than the residual resistivity due to an ordinary impurity atom.

The relation between the electrical resistivity and the thermoelectric power:

$$Q = -LeT\frac{d\log\rho}{d\mathscr{E}_F}$$

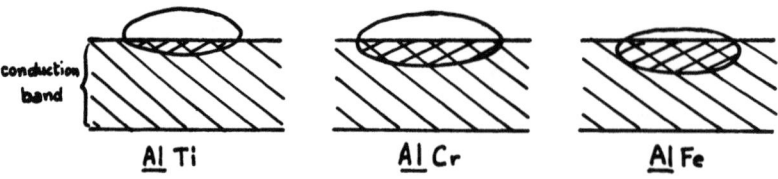

Fig. 6. Virtual d levels of transition elements dissolved in aluminum.

where $L = \pi^3 k_B^2/3e^2$ is the Lorentz number (k_B being the Boltzmann constant) shows that a virtual d-bound state crossing the Fermi level may give rise to a large thermoelectric power with positive or negative sign, depending on the virtual level being more or less than half filled.

The model of virtual bound state that we just described accounts satisfactorily for the electrical properties of dilute alloys of aluminum with transition elements of the iron group, for instance, as can be seen from Fig. 5. When the atomic number of the solute increases, from titanium to copper the corresponding virtual bound state sinks into the Fermi sea of the aluminum conduction band (Fig. 6), so that the atomic residual resistivity increases up to chromium, where the resonance energy coincides with the Fermi level; then it decreases as expected from formula (24). Simultaneously, the predicted anomaly in thermoelectric power is observed.

As we shall see later, the situation is different when the matrix is a noble metal, with only one valence electron instead of the three for aluminum. There, the exchange energy becomes sufficient for lifting the spin degeneracy of the virtual d-bound states, which then splits into two localized states with a net magnetic moment giving rise to Curie–Weiss type magnetic susceptibility and to two successive maxima for the residual electrical resistivity versus the atomic number of the solute atom.

Magnetic Localized States

Up to now, we assumed that a virtual bound state occurred at the same energy for both spin directions. In fact, this is a situation which seems to be realized in aluminum, for instance, as we have just seen. However, as the electrons in such a state are almost as much localized as in an atom, we expect exchange and correlation forces to be very effective, especially for a narrow level, with a tendency to favoring spin alignment, according to Hund's rule. When these forces prevail, the virtual bound states have different energies for opposite spin directions so that they accommodate unequal numbers of electrons up to the Fermi level and a localized magnetic moment results (Fig. 7).

A simple variational argument may be used, which gives a prescription for the occurrence of such a lifting of spin degeneracy.[5] Let us imagine that, starting with a spin-degenerate virtual bound state, we shift a small number δn electrons from the half level with, say, spin-up into the half level with spin-down. The energy of the electrons moving in the average Hartree potential is then increased by

$$\delta E_H = \delta n \left(\frac{\partial \mathscr{E}}{\partial n} \delta n \right) \bigg|_{\mathscr{E} = \mathscr{E}_F}$$

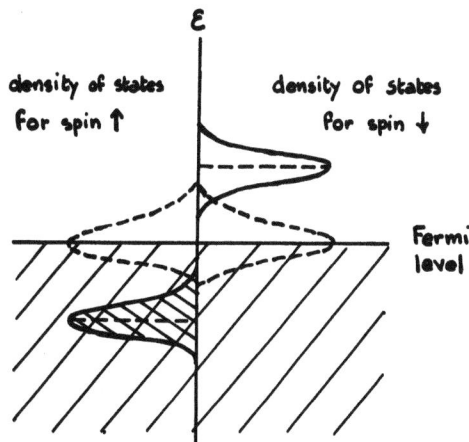

Fig. 7. Localized magnetic moments arising from a virtual bound state.

if n is the number of electrons up to energy $\mathscr{E}$ in a half level with fixed spin direction. On the other hand, the exchange energy is lowered by

$$\delta E_{ex} = -U(\delta n)^2$$

if U is the effective exchange energy per electron pair. The spin-degenerate state is unstable, that is to say, the spin degeneracy must be lifted, when the total change in energy is negative:

$$\delta E_H + \delta E_{ex} = \left[\left(\frac{\partial \mathscr{E}}{\partial n}\right)_{\mathscr{E}_F} - U\right](\delta n)^2 < 0$$

From

$$n(\mathscr{E}) = \frac{2l+1}{\pi}\eta_l(\mathscr{E}) = \frac{2l+1}{\pi}\arctan\frac{\Gamma}{2(\mathscr{E}_r - \mathscr{E})}$$

the condition for occurrence of a localized magnetic moment can be written, in the case of a narrow Lorentzian virtual d-bound state

$$\frac{2l+1}{\pi} U \frac{\Gamma/2}{(\mathscr{E}_r - \mathscr{E}_F)^2 + (\Gamma^2/4)} > 1 \qquad (25)$$

For a given value of U, this condition is most easily satisfied the closer $\mathscr{E}_r$ is to the Fermi level $\mathscr{E}_F$. When $\mathscr{E}_r = \mathscr{E}_F$, the virtual bound state is just half filled, so that it accommodates the right screening charge for elements in the middle of the transition series. One then expects localized magnetic moments to occur more easily for elements of the middle of the series, such as chromium and manganese, than for titanium or nickel, which lie at the beginning or at the end of the period. The general tendency does in fact agree with experimental data on the magnetic susceptibility of dilute alloys of a noble metal with iron-group transition elements. For $\mathscr{E}_r = \mathscr{E}_F$, the condition (25) takes the simple form

$$2U\frac{2l+1}{\pi} > \Gamma \qquad (26)$$

As the virtual bound state broadens when the resonance energy increases, the width may become too large for the splitting to occur if the Fermi energy of the matrix is too high. This seems to be the case for aluminum, which has three conduction electrons per atom, giving $\mathscr{E}_F \simeq 12$ eV instead of 7 eV for copper and 5.5 eV for silver and gold. According to the condition (26) and assuming Γ to be as large as 4 eV in aluminum, it would be sufficient for U to be a little larger than 1.2 eV to produce spin splitting of a dissolved chromium atom virtual d level. The fact that this splitting does not occur indicates that the effective exchange energy U is not as great as the 10 eV of the Coulomb self-energy used by Anderson[8] in his model with no orbital degeneracy. As a matter of fact, when orbital degeneracy is taken into account for a d level, the mean interaction between an electron in a given d orbital which reverses its spin and the other electrons in the shell corresponds to an average between the self-energy for the same orbital and ordinary exchange energy for the electron in other orbitals. A calculation made by Kanamori[9] for pure transition metals suggests that the magnitude of U is strongly reduced by electron correlations. Finally, a value $U \simeq 1$ eV, of the order of magnitude of an ordinary atomic exchange energy, seems reasonable for elements in the first transition series. Tentative predictions of spin splitting of virtual bound states have been made on such a basis by Friedel and Blandin.[5,7] In Fig. 8, one can see the general agreement of predicted splitting with experimental evidence for localized magnetic moments in different alloys.

When such localized magnetic moments are present in the alloy, they give rise to a Curie–Weiss type of paramagnetism at temperatures higher than some critical

	Pd H	Au	Ag	Cu	Mg	Zn	Al
Sc							
Ti		−					−
V		+					−
Cr	+	+	+	+		+	−
Mn	+	+	+	+	+	+	−
Fe	+	+		+	+	−	−
Co	+	+		?			−
Ni	−	−		−		−	

Fermi energy increases →

Fig. 8. Occurrence of localized magnetic moments ($\mathscr{E}_F$ increases palladium (H) to aluminum).

temperature T_c, approximately proportional to the concentration of the alloy, below which some spin ordering is established, usually antiferromagnetic, because of the interaction between localized states. A typical case is that of dilute copper–manganese alloys. When a virtual d-bound state crosses the Fermi level, it generally accommodates a noninteger number of electrons, so that there is some theoretical difficulty in determining the exact value of the magnetic moment in order to ascribe it to a Curie–Weiss law.

Apart from these magnetic properties, the main effect of spin splitting of a virtual d-bound state is that one now has two such states with different energies, which are expected to cross the Fermi level for different values of the atomic number of the solute when the elements of a transition series are dissolved in a noble matrix. One especially expects two peaks of residual resistivity instead of one. This is indeed observed experimentally, as shown in Fig. 9, where the measured residual resistivities of copper alloys are compared to those that we computed, assuming virtual d levels and deducing their occupation from the screening condition and experimental data on magnetic susceptibilities.

As for any localization of electronic charge in an alloy, the occurrence of magnetic localized states gives rise to long-range oscillations of charge and, in this

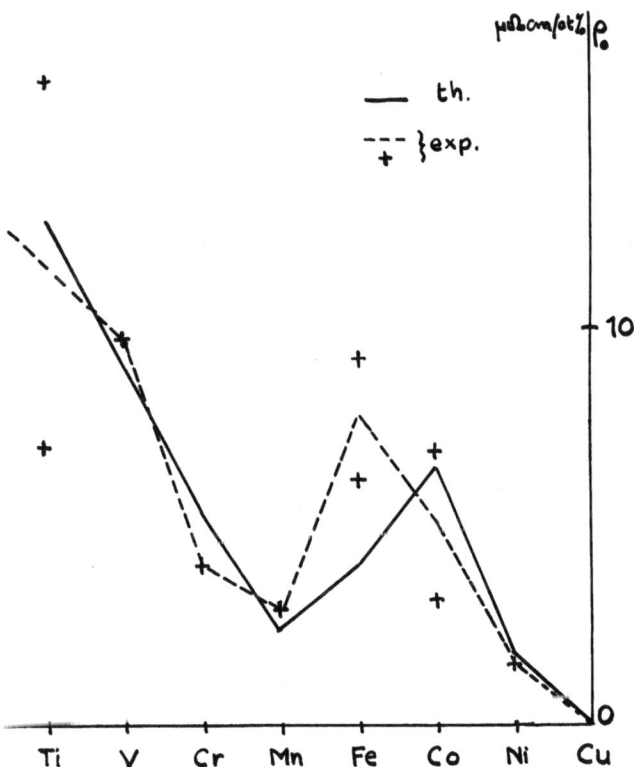

Fig. 9. Residual resistivity of copper-containing transition elements as impurities.[10]

case, of spin density around such a state. One expects these perturbations to induce a coupling of the magnetic moments at low temperatures, with antiferromagnetic spin order for a disordered solid solution, as first shown by Blandin and Friedel.[5] We shall now look at the way in which localized states interact with each other.

Interactions in Dilute Alloys

From the long-range oscillations in electronic density arising around a virtual bound state according to formula (23), the corresponding impurity is expected to interact with other atoms in the matrix or other impurities even in a very dilute alloy. In trying to give an analysis of the resulting effects, it must be kept in mind that until now an accurate description of the individual noninteracting virtual bound states has been given for only very simple models, such as a single nondegenerate state, which is rather unphysical for a d level in a cubic crystal, or a free electron approximation for the conduction band of the matrix, which allows the simple phase-shifts analysis. Even with these simplifying assumptions, the problem of the coupling between virtual bound states remains so difficult that only qualitative or semiquantitative results have been reached. A simple description of the mechanism of interaction can be given by looking at the perturbation produced by the solute element at the site occupied by any atom in the matrix or the alloy. We shall proceed mostly with this method in describing chemical interactions and coupling between localized magnetic states.

Chemical Interactions. According to formula (5) for ordinary screening or formula (23) for a virtual bound state, the localization of screening charge on an impurity X produces an oscillating change $\delta n(\mathbf{r})$ of electronic density in the matrix, the amplitude of which decreases as r^{-3} at large distances from X:

$$\delta n(\mathbf{r}) \simeq \frac{-\alpha}{2\pi^2 r^3} \cos(2k_F r + \varphi)$$

From the Poisson equation, this corresponds to an electrostatic potential

$$V(r) \simeq \frac{\alpha}{2\pi k_F^2} \frac{\cos(2k_F r + \varphi)}{r^3}$$

i.e., to a radial electric field of magnitude

$$E(r) \simeq \frac{\alpha}{\pi k_F} \frac{\sin(2k_F r + \varphi)}{r^3}$$

This electric field which arises from X acts on the atoms of the matrix. Blandin and Déplanté have used this approach to compute the relative change in lattice parameter with concentration $1/a \, da/dc$ for dilute alloys with a noble metal matrix. Figure 10 shows a comparison of experimental data for this size effect and the theoretical predictions of Déplanté, according to whom:

$$\frac{1}{a}\frac{da}{dc} \simeq \frac{\chi p A_e}{9\Omega} R_1 E(R_1)$$

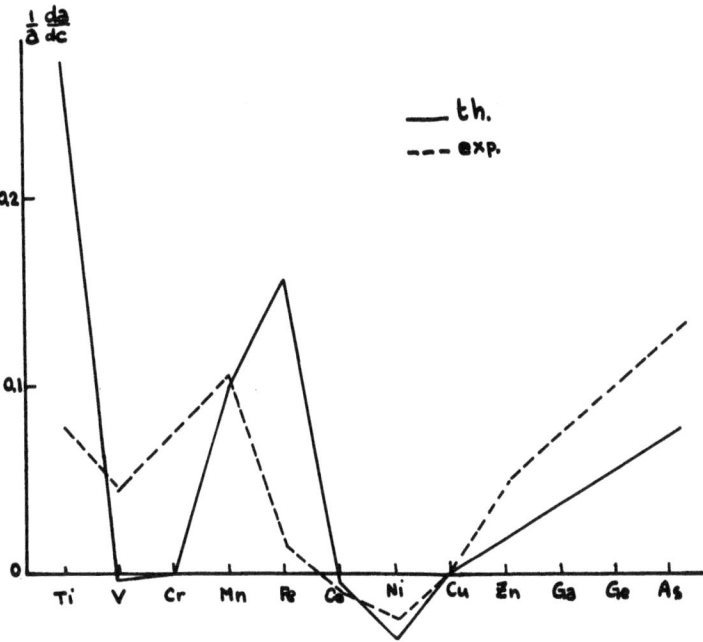

Fig. 10. Size effect of impurities in noble metals.

if χ is the compressibility of the matrix, Ω its atomic volume, A its valency, p the number of nearest neighbors of a given atom, $E(R_1)$ the electric field at the distance R_1 of the nearest neighbors from an impurity. The experimental values for silver and gold have been shifted in such a way as to give zero in copper in order to eliminate the period effect present besides the pure valency effect.

The energy of interaction W of an impurity with a vacancy can be computed along the same lines, taking $-A$ as the effective charge of the vacancy. In Fig. 11, the theoretical results of Déplanté are compared to the difference ΔQ of activation energy for impurity diffusion and self-diffusion in copper. As for the change in lattice parameter, the general behavior of dilute alloys is remarkably well reproduced, even if the agreement between theory and experiment is far from quantitative in each specific case.

Magnetic Coupling. When the spin degeneracy is lifted in a virtual bound state, the phase shifts $\eta_\uparrow$ for spin-up and $\eta_\downarrow$ for spin-down electrons are different, so that the oscillating charge density in the matrix is then associated with an oscillating spin density of the same wavelength (Fig. 12). For a virtual d level, assuming that only the partial wave with $l = 2$ is perturbed, asymptotically:

$$\delta n_\uparrow(r) - \delta n_\downarrow(r) \simeq \frac{-5}{2\pi^2 r^3} \{\sin \eta_\uparrow(k_F) \cos [2k_F r + \eta_\uparrow(k_F)] - \sin \eta_\downarrow(k_F) \cos [2k_F r + \eta_\downarrow(k_F)]\}$$

(27)

where the subscript $l = 2$ for the phase shifts has been dropped and only the spin

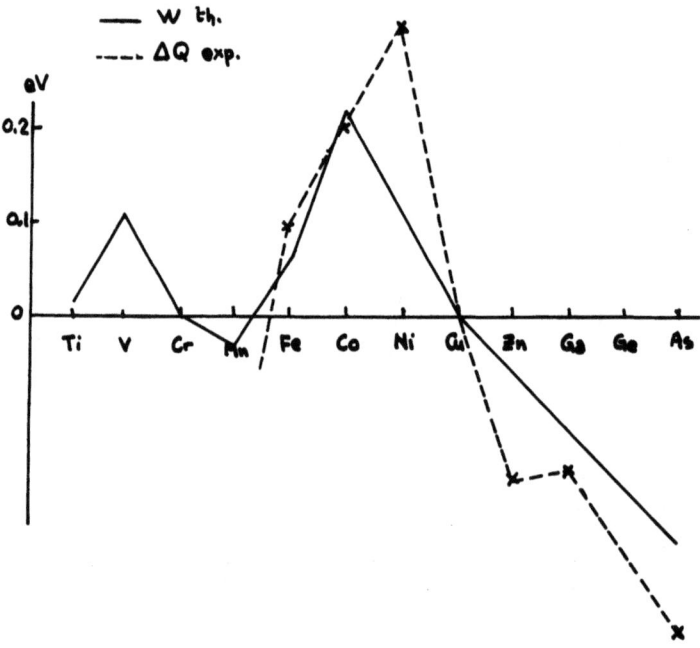

Fig. 11. Vacancy–impurity interaction energy in copper.

index is kept. Through the s–d exchange interaction, such an excess or defect of spin density at the site of a second localized magnetic state is equivalent to a positive or negative local molecular field, so that the magnetic moment of the second atom tends to line up parallel or antiparallel to the first one. In a perfectly random dilute solution, one then expects on average an antiferromagnetic behavior of the solute atoms at low temperature. This model was first devised by Blandin and Friedel[5] to account for the magnetic properties of copper–manganese alloys. The extension of their theory [12-14] explains most of the properties of this kind of dilute alloy; a Néel temperature proportional to the concentration c, an excess specific heat and spin disorder resistivity increasing linearly from very low temperature toward the

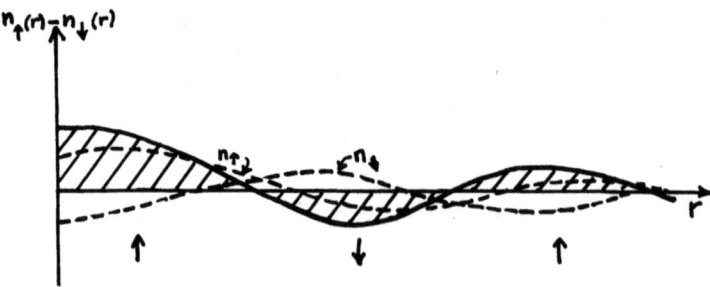

Fig. 12. Oscillations of spin density around a virtual bound state.

Néel point (with a slope independent of c), and a remanent magnetization when cooled in a magnetic field. This behavior of very dilute alloys is a consequence of the fact that for such a solid solution, the distribution function $p(H)$ for the effective local magnetic field on an atom at 0°K is peaked like c^{-1} at $H = 0$, with a width proportional to c (Fig. 13), so that the total number of atoms which experience a field lower than a small given value is approximately independent of c. When the temperature increases from very low values, the number of disordered spins increases as T, as do the spin disorder resistivity and the specific heat. In fact, the shape of the distribution function $p(H)$ changes with increasing temperatures, so that the simple linear temperature dependence of resistivity and specific heat may not be valid up to the Néel point. By detailed calculation of $p(H)$ at several temperatures and concentrations, Brout and Klein[14] have been able to account satisfactorily of the thermal behavior of copper–manganese, for instance. Finally, the distribution function $p(H)$ becomes unsymmetrical with respect to positive and negative values of the local effective field if an external magnetic field is applied, so that a remanent magnetization arises when the alloy is cooled in an external field.[15,16]

It must be stressed that the type of coupling between magnetic states assumed by Blandin and Friedel is somewhat hybrid, being of resonance type on one atom and s–d exchange on the second. The resonance coupling is larger than the exchange coupling by the ratio $\mathscr{E}_F/J$ of the Fermi energy so the s–d exchange integral, that is to say, something of the order of 10, so that this hybrid coupling is one order of magnitude stronger than the double s–d exchange mechanism of Yoshida.[17] However, it may well be that the true mechanism of interaction is through resonance on both localized states. At first sight, it seems that this double resonance mechanism introducing another factor $\mathscr{E}_F/J$ would be an order of magnitude too strong, and give a too high Néel temperature for copper–manganese alloys, for instance. This is not necessarily true, because the effective interaction depends on the exact values of the phase shifts, and it is doubtful that an accurate estimate of the strength of the coupling has ever been made. Recent work by Caroli[18] seems to indicate

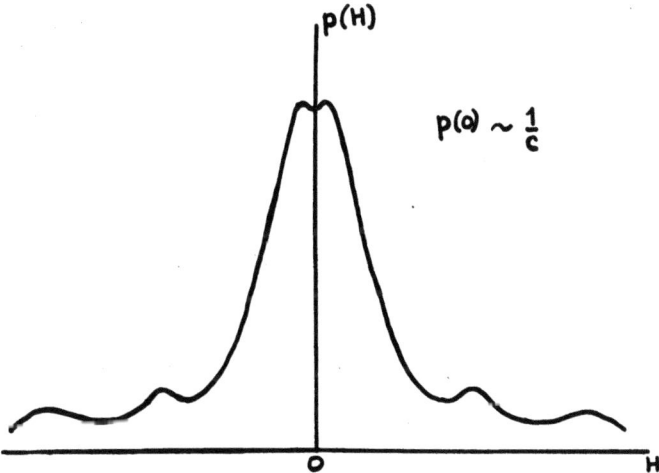

Fig. 13. Distribution of local effective fields in a dilute alloy with localized magnetic moments.

that the double resonance mechanism must be retained, especially to account for the strong coupling of pairs of transition atoms close to each other in the alloy and the possibility for such a pair to become magnetic even when the spin degeneracy would not be lifted if the two atoms were far away from each other, as may well be the case for copper–cobalt alloys. According to Caroli, the interaction energy between two identical nondegenerate virtual bound states at a distance R from each other can be written

$$E = \frac{\mathscr{E}_F}{\pi} \sin^2 \eta \frac{\cos(2k_F R + 2\eta)}{(k_F R)^3}$$

$\mathscr{E}_F$ being the Fermi energy of the metal and η the phase shift produced by each impurity. It must be noted that the sum of the two phase shifts is involved in the argument of the cosines, instead of one phase shift only in the resonance exchange coupling of Blandin and Friedel.

Finally, the occurrence of localized magnetic states seems always to be related to anomalous low temperature resistivities of dilute alloys of transition or rare earth metal in ordinary metals.[19] Usually, a minimum in electrical resistivity is observed at low temperature and, very often, a maximum at still lower temperature (Fig. 14). It is likely that, when the maximum has not been observed experimentally, it is because measurement should have been performed at still lower temperatures in order to reach it.

An explanation of these anomalies could be as follows: With increasing temperature from absolute zero, the electrical resistivity first increases from its true residual value to a maximum value at about the Néel point of the alloys by progressive disappearance of long-range spin order, as described above. The subsequent decrease might be linked to short-range order involving pairs of magnetic localized states above the Néel temperature. By making explicit use of the Rudermann–Kittel type of coupling between localized magnetic moments in the formal theory of Brailsford and Overhauser[20] Béal[21] was able to show that the short-range spin

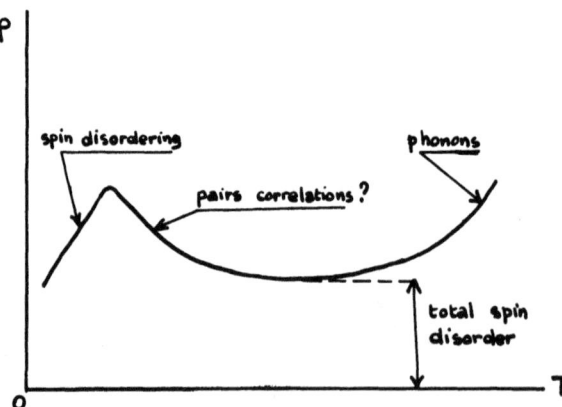

Fig. 14. Anomalous low temperature resistivity in a dilute alloy with localized magnetic moments.

order resistivity should decrease with increasing temperature because of the same spatial dependence in $r^{-3}\cos 2k_F r$ of both the interaction and the interference function. This resistivity decreases like c^2/T down to the perfect spin disorder value, but the increasing contribution from phonon scattering leads to the appearance of a minimum on the experimental curves. Quantitative agreement with experimental data is not very good, which may be partly due to the fact that the calculations were made using Born approximations.

More recently, Kondo[22] has found, by calculating the scattering of conduction electrons by individual localized magnetic states in the second Born approximation, that the s–d interaction could lead to a term varying like $c \log T$ in the low-temperature resistivity, down to some cut-off value T_0 corresponding to the Zeeman splitting of levels of magnetic moments. When added to the phonon contribution, this would give a minimum in resistivity at a temperature proportional to $c^{\frac{1}{5}}$ with a depth proportional to c, if the s–d interaction is negative. Although a rather good fit was found for gold-iron data, the question of what mechanism leads to the resistivity minimum still remains, but the experimental results presented at this conference by M. P. Sarachik[25] indicate that pair correlation effects are not preponderant.

Conclusion

There is no doubt that the concept of virtual bound state is physically meaningful in the theory of electronic structure of dilute alloys involving transition elements. The two intuitive pictures that one may give of such a state, either a d-localized state broadened by mixing with conduction electron wave function of the same energy or a state almost bound in a strongly attractive local potential, turn out to be equivalent. The second point of view may be more convenient when considering alloys of transition elements with themselves.[1,23,24] In this case, a virtual d-bound state may be thought of as extracted from the d band itself. On the other hand, an element to the left of the matrix in the periodic table may give rise to a virtual d-bound hole taken out of the d band. Several properties of alloys of transition metals might be explained in this way.

References

1. J. Friedel, *Phil. Mag.* **43**, 153, 1952; *Advan. Phys.* **3**, 446, 1954; *Nuovo Cimento (Suppl. 2)* **7**, Ser. 10, 287, 1958.
2. A. Blandin and E. Daniel, *J. Phys. Chem. Solids* **10**, 126, 1959.
3. J. S. Langer and S. H. Vosko, *J. Phys. Chem. Solids* **12**, 196, 1959.
4. A. Blandin, *J. Phys. Radium* **22**, 507, 1961.
5. A. Blandin and J. Friedel, *J. Phys. Radium* **20**, 160, 1959.
6. J. E. Zimmerman and F. E. Hoare, *J. Phys. Chem. Solids* **17**, 52, 1960; L. T. Crane and J. E. Zimmerman, *Phys. Rev.* **123**, 113, 1961.
7. J. Friedel, *Can. J. Phys.* **34**, 1190, 1956; *J. Phys. Radium* **23**, 692, 1962; *Metallic Solid Solutions*, W. A. Benjamin, Inc., New York (1963).
8. P. W. Anderson, *Phys. Rev.* **124**, 41, 1961.
9. J. Kanamori, *Progr. Theoret. Phys. (Kyoto)* **30**, 275, 1963.
10. E. Daniel, *J. Phys. Chem. Solids* **23**, 975, 1962.
11. A. Blandin and J. L. Déplanté, *J. Phys. Radium* **23**, 609, 1962; in: *Metallic Solid Solutions*, W. A. Benjamin, Inc., New York (1963); and article to be published.
12. A. Blandin, Thesis, Paris (1961).
13. W. Marshall, *Phys. Rev.* **118**, 1519, 1960.

14. M. W. Klein, *Phys. Rev. Letters* **11**, 408, 1963; and article to be published.
15. R. W. Schmitt and I. S. Jacobs, *J. Phys. Chem. Solids* **3**, 324, 1957.
16. J. S. Kouvel, *J. Phys. Chem. Solids* **21**, 57, 1961.
17. K. Yoshida, *Phys. Rev.* **106**, 893, 1957.
18. B. Caroli (to be published).
19. G. L. Van den Berg and J. de Nobel, *J. Phys. Radium* **23**, 665, 1962; in: *Metallic Solid Solutions*, W. A. Benjamin, Inc., New York (1963). T. Sugawara, R. Soga, and I. Yamase, *J. Phys. Soc. Japan* **19**, 780, 1964.
20. A. D. Brailsford and A. W. Overhauser, *J. Phys. Chem. Solids* **15**, 140, 1960; **21**, 127, 1961.
21. M. T. Béal, Thesis, Paris (1963); M. T. Béal and J. Friedel, *Phys. Rev.* **135**, A446, 1964.
22. J. Kondo (to be published).
23. P. A. Wolff, *Phys. Rev.* **124**, 1030, 1961.
24. A. M. Clogston, *Phys. Rev.* **125**, 439, 1962; A. M. Clogston, B. T. Mathias, M. Peter, H. J. Williams, E. Corenzwit, and R. O. Sherwood, *Phys. Rev.* **125**, 541, 1962.
25. M. P. Sarachik, this volume, p. 1044.

ANOMALIES IN DILUTE METALLIC SOLUTIONS OF TRANSITION METALS

G. J. van den Berg

*Kamerlingh Onnes Laboratorium
Leiden, The Netherlands*

The purpose of this paper is to present a survey of the experiments on dilute alloys of transition elements both in noble or other normal metals, and in transition metals. Not all aspects of this problem can be treated here. The following properties will be presented: the electrical and thermal resistivity as a function of temperature and of a magnetic field, the Hall effect, the thermoelectric power, the magnetization and susceptibility, and the specific heat.*

The Electrical Resistivity

Apart from the experiments on this property of "pure" metals, which show, in some cases,[2] a minimum in the graph representing the electrical resistivity as a function of temperature at low temperatures, we must mention that the detailed investigations were initiated by Gerritsen and Linde[3] in 1949. In Fig. 1, the results of their measurements of the electrical resistivity ratio of dilute alloys of manganese in silver are plotted as a function of temperature. For about 0.1 at.% Mn, the curve shows a minimum followed by a maximum at lower temperatures. Gerritsen and Linde,[3] from the point of view of an $s-d$ scattering, expected and also observed in fields up to 20 kOe an anomaly in the magnetoresistance of silver–manganese alloys, which was negative for the lowest temperatures and for not too small concentrations. The value amounted in some cases to -30%. In certain magnetic fields, this value depends on the temperature and the concentration of Mn[4] (Fig. 2). Similar results had already been found by Nakhimovich,[5] in 1941, in the case of gold–iron alloys.

Gerritsen and, later on, other investigators extended the measurements to the transition metals chromium, manganese, iron, cobalt and nickel in the noble metals gold, silver, and copper as far as the solubility permitted. Anomalies were found in the $R-T$ curves. Manganese,[6-8] chromium,[6,7] and iron[7-11] as an impurity in gold gave a maximum and a minimum in the curve (Fig. 3) for a concentration of about 0.1 at.% alloy with a decrease of the resistance in a magnetic field[5,9,11-14] (negative magnetoresistance). Manganese in silver[4] and copper[13,15,16] gave similar results. Alloys of iron and chromium in copper[6,8,17,18] show no maximum in the $R-T$ curve (Fig. 4).

The negative magnetoresistivity seemed to be coupled to the maximum in the $R-T$ curve, but recently Mutô, Noto, and Hedgcock[19] measured the magnetoresistivity of some dilute copper–iron alloys, a copper–manganese and copper–zinc alloys in fields up to 100 kOe. The authors did not find a negative value for

---

* A more extensive review can be found in *Low Temperature Physics*, Vol. IV.[1]

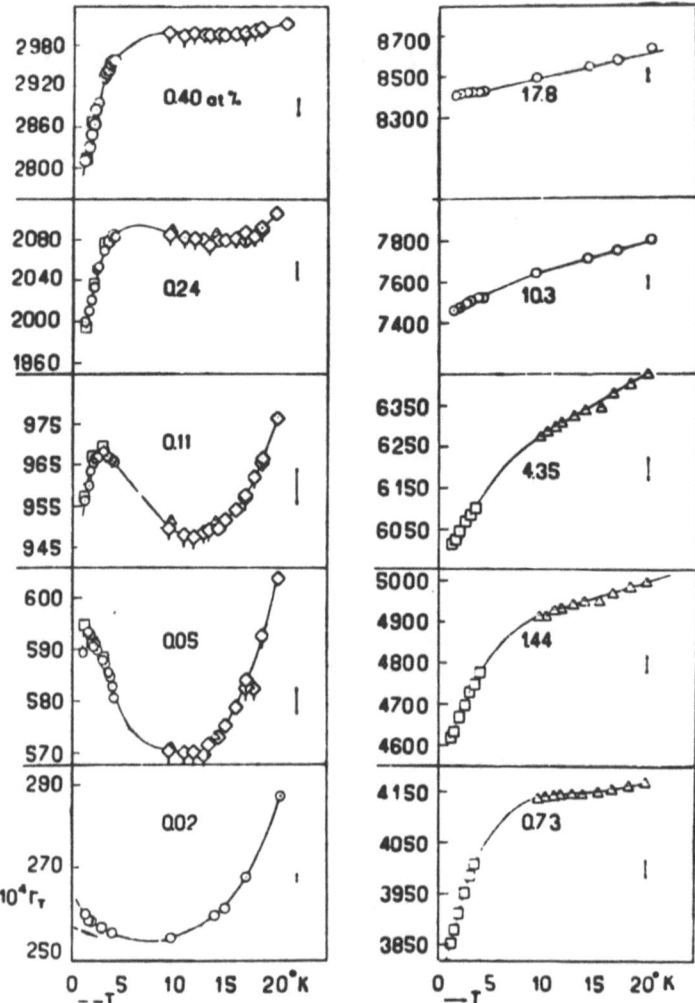

Fig. 1. The resistance ratio, $r_T = R_T/R_{273}$, as a function of T for silver–manganese alloys with atomic concentrations as indicated in the graph (Gerritsen and Linde[3]).

copper–zinc, as was expected for alloys of nontransition metals. Copper–iron alloys containing more than 0.04 at.% Fe and a copper–manganese alloy with 0.007 at.% Mn exhibited a negative magnetoresistivity at 4.2°K. Mutô et al. proposed to represent the magnetoresistivity by a sum of a positive (obeying Kohler's rule) and a negative (magnetic spin) component. In the case of the 0.007 at.% Cu–Mn, they succeeded in establishing this separation (Fig. 5) totally.

Nickel and cobalt[7,20,21] as impurities do not cause anomalies in the R–T curve. The accidental introduction of iron as an impurity during the treatment, e.g., during annealing in a reductive atmosphere[22] or by reducing by means of the added metal,[23] may sometimes simulate another result, but the random behavior of the R–T curve with, e.g., an increasing cobalt content[24] demonstrates this simulation.

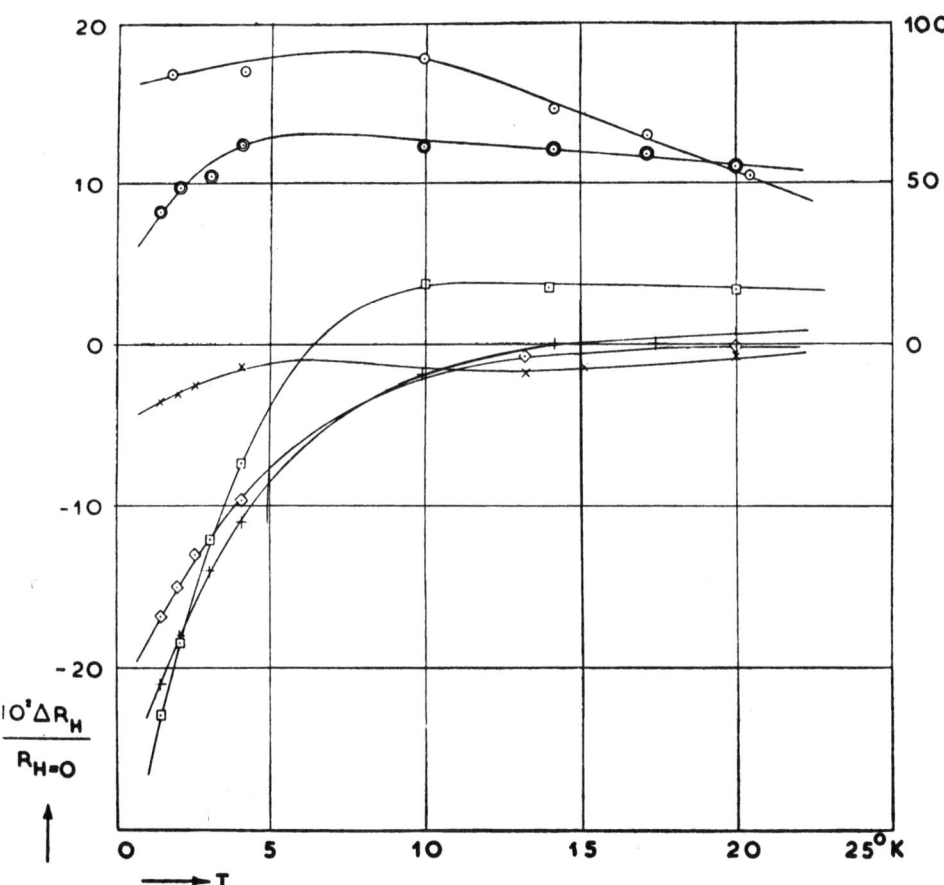

Fig. 2. The magnetoresistance of silver–manganese alloys in a transverse magnetic field of 20 kOe. ⊙: $c = 0.00$; ◎: $c = 0.02$; □: $c = 0.05$; +: $c = 0.15^5$; ◇: $c = 0.24$; ×: $c = 0.61$. The ordinate scale at the right is only for $c = 0.00$ (Gerritsen and Linde[4]).

A normal metal as an impurity in a pure noble metal does not cause a minimum in the R–T curve. Knook et al.,[17]* showed this directly by measuring the R–T curves of the dilute alloys: copper–tin, copper–iron, and copper–tin–iron. The base metal copper has to contain less than 0.8 ppm of iron. Earlier, Schmitt[25] had demonstrated that zinc did not cause a minimum in the R–T curve of copper.

The anomalous effect due to small quantities of iron as an impurity in copper and gold can be eliminated by annealing in an atmosphere with a small amount of oxygen, as was demonstrated by MacDonald and Pearson,[26] Coltman et al.,[27] Dominicali and Christenson,[28] and Anderson and Nielsen.[29] The residual resistance of the dilute alloy treated in this way has a much lower value than that of such an alloy annealed in high vacuum. By reducing with hydrogen,[22] graphite,[26] or carbon monoxide,[27] the anomalous behavior of the electrical resistance as well as the higher value of the residual resistance returns.†

* See also *Progress in Low Temperature Physics*, Vol. IV, p. 202.[1]
† Attention can be drawn to the gas permeation through fused silica capsules during high temperature heat treatment in vacuum.[141]

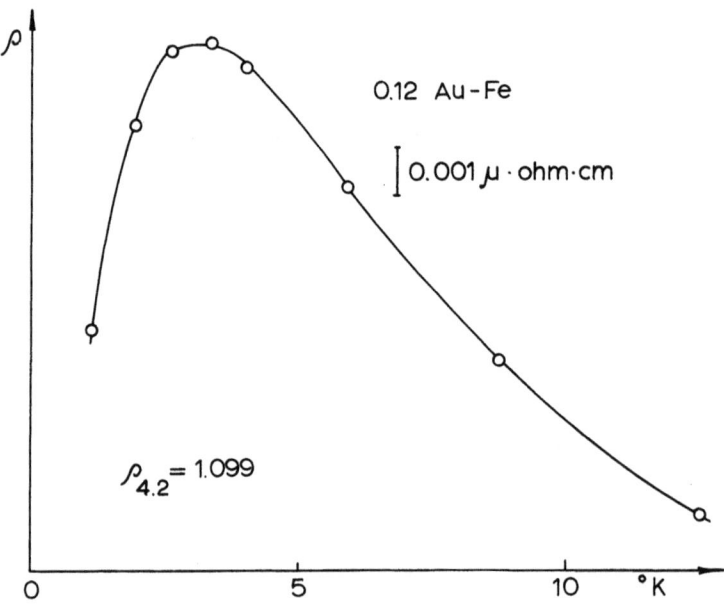

Fig. 3. The electrical resistance of 0.12 at. % Au–Fe as a function of temperature (Linde[10]).

After mentioning the alloys of the first long period with the noble metals as base metals, we must focus our attention now on the experimental results on alloys of manganese and chromium in zinc,[16,30] manganese and iron in aluminum and magnesium,[31] and manganese in cadmium.[32,33] Hedgcock et al.[31] investigated the alloys magnesium–manganese, magnesium–iron, aluminum–iron, and aluminum–manganese and succeeded in demonstrating maxima and minima in the $R-T$ curve of mangesium–manganese, minima in that of magnesium–iron and a normal $R-T$ curve for the aluminum alloys. This last mentioned result is in agreement with experiments by Boorse and Niewodniczanski[34] and by Thomas and Mendoza.[35] As shown in Fig. 6, the results on the magnesium–manganese alloys show, e.g., the behavior of the $R-T$ curve some degrees below the maximum at 6°K in this curve for 0.6 at. % Mg–Mn. For a 0.1 at. % Cd–Mn[33] nominal a maximum as well as a minimum could be detected. Deaton[36] estimated a minimum around 3°K in a cadmium sample by means of magnetic acoustic waves.

Concerning the alloys with zinc as the base metal, Mutô et al.[16,30] investigated the alloys zinc–manganese and zinc–chromium and compared the results with the normal one for a 0.05 at. % Zn–Sn alloy and with the abnormal one for a 5.4 at. % Cu–Mn alloy, respectively. For the 0.10 at. % Zn–Cr, a negative temperature coefficient was found, while the magnetoresistance was positive. In the $R-T$ curve, for 0.12 at. % Zn–Mn there appeared a minimum and the decrease of the negative slope below the temperature of the minimum suggested the approach of a maximum at or below 1°K.[3,6] Other experiments did not give any maximum above 1.8°K for the whole range of solubility (to 0.42 at. %).[37] The presence of a negative magnetoresistance for a 0.12 at. % Zn–Mn alloy in fields between 0 and 60 kOe (above 60 kOe the value is positive), though previously mostly found in the case of the

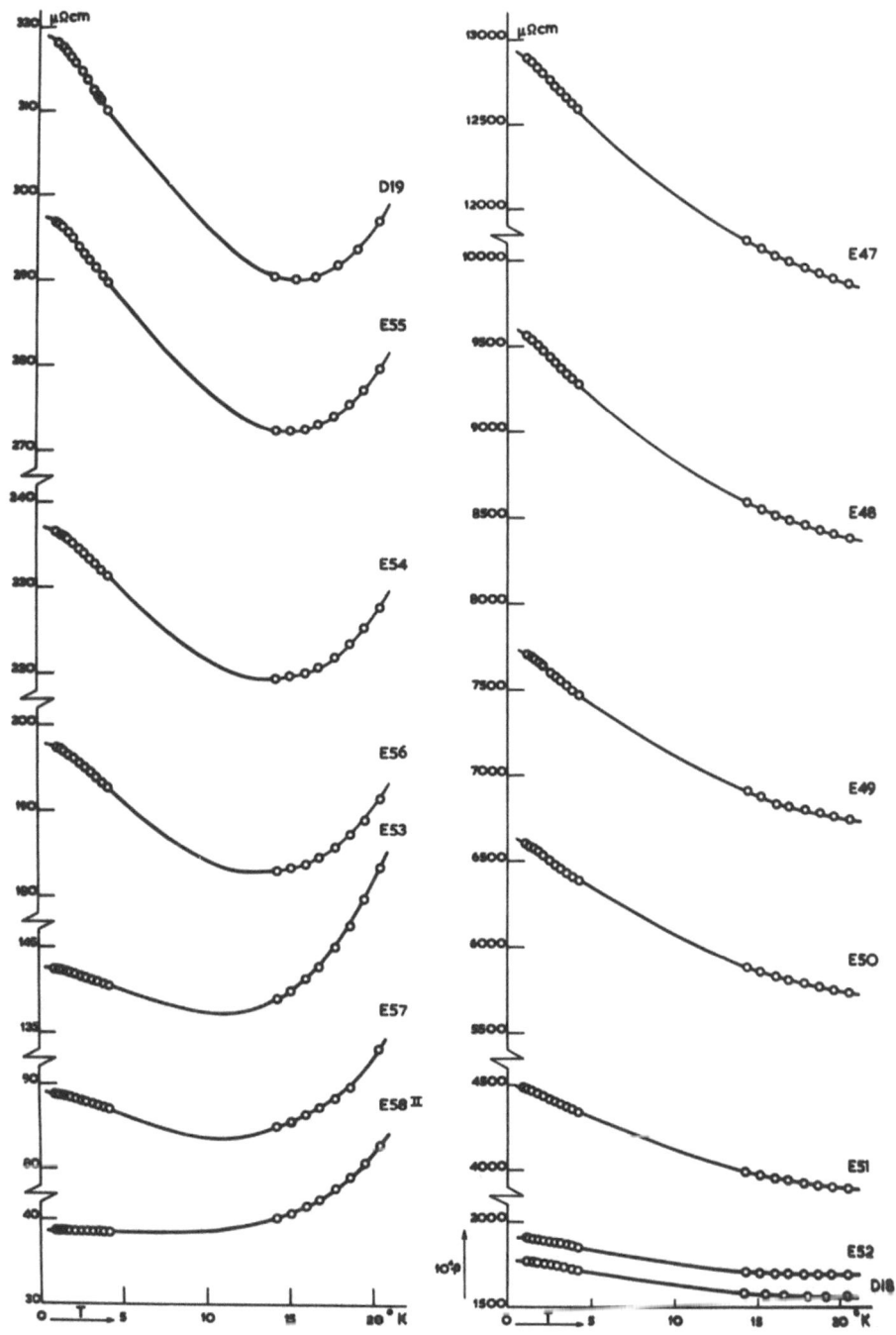

Fig. 4. ρ–T curves for a series of copper–iron wires. The concentration increases from 0.0005 to 0.123 at. %. Note that the scales of the two ordinates are different (Knook et al.[17]).

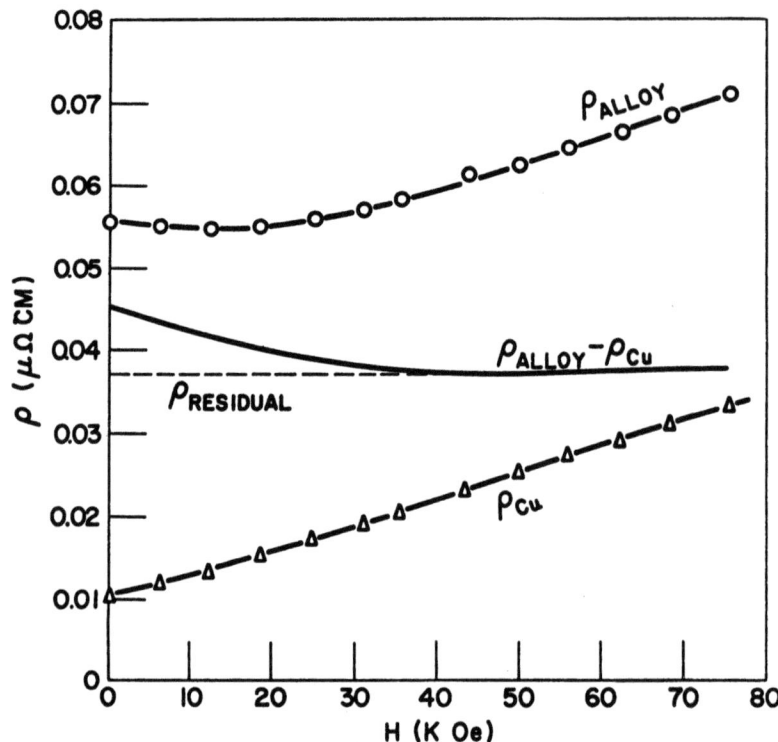

Fig. 5. Graphical analysis of the transverse magnetoresistivity of a 0.007 at.% Cu–Mn alloy at 1.3°K (Mutô, Noto, and Hedgcock[19]).

existence of a maximum, is no proof for the occurrence of such a maximum in the R–T curve[19] (Fig. 7); this has been previously mentioned.

Collings, Hedgcock, and Mutô[38] calculated the value of the exchange integral J for spin $\frac{3}{2}$ to be about 1.5×10^{-12} ergs for zinc–manganese alloys. The authors based their calculations on theoretical equations which assume a scalar s–d interaction between paramagnetic ions and conduction electrons and used values of the Néel temperature, the magnetoresistivity as a function of the field, and the suppression of the superconducting critical temperature, respectively. They concluded from the fact that the value of J derived in three different ways is approximately the same, that the nature of the magnetic interaction leading to these phenomena may be of the same physical origin. The value of J for zinc–manganese is of the same order of magnitude as that for copper–manganese, 2.1×10^{-12} ergs, calculated for spin $\frac{5}{2}$ by Yosida,[39] but much larger than that calculated for aluminum–manganese from the suppression of the transition temperature of superconductivity.[40] This value is 0.23×10^{-12} ergs for an estimated spin value of $\frac{1}{2}$ for manganese,[41] which possesses a very small moment, if any, when dissolved in aluminum.

The alloys 0.045 at.% Al–Mn and 0.02 at.% Al–Fe[42] behave normally with respect to the influence of the magnetic field on the electrical resistance, as also do magnesium–cadmium and magnesium–aluminum alloys.[42]

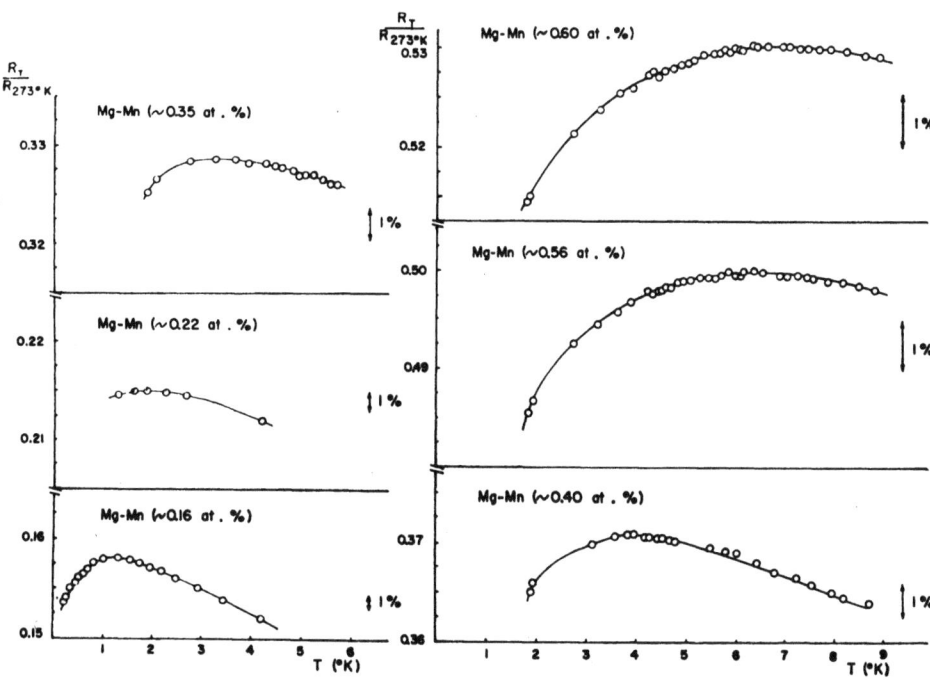

Fig. 6. The electrical resistance ratio $r_T = R_T/R_{273}$, as a function of temperature T for magnesium–manganese alloys with concentrations between 0.16 and 0.60 at.% (Gaudet et al.[31]).

As mentioned above[19,38] the Philadelphia group proposed to consider the magnetoresistivity of a dilute alloy containing paramagnetic impurities as the sum of a normal term independent of temperature in the impurity scattering range ($\Delta\rho_n$) and an anomalous term which is negative and decreases with decreasing temperature ($\Delta\rho_s$). Using Kasuya's[43] expressions for the magnetoresistivity under various field and temperature conditions, Hedgcock and Mutô[42] calculated for magnesium–manganese alloys an average value of 0.48×10^{-12} ergs for J and 2.78×10^{-12} ergs for the Coulomb interaction constant A, which are independent of concentration. By using A and J for the calculation of $\Delta\rho_s$, we could determine $\Delta\rho_n$ from the experiments at 4.2° and 1.5°K. The value for $\Delta\rho_n$ at 4.2°K obeys the Kohler rule, but at 1.5°K does not. Only if the ratio A/J was altered, could agreement with experiment for a particular magnetic field be found.

Summarizing the extensive series of experiments mentioned above concerning the alloys of the normal metals with small amounts of transition metals of the first long period, one can state that the transition metals with the largest numbers of unpaired spins, like chromium, manganese, and iron, preferably cause anomalies in the magnetoresistivity of both the noble metals, and magnesium, zinc and cadmium, details depending on the host metals. According to the concept of the virtual bound state by Friedel and co-workers,[44,45]* there will be a localized magnetic moment present,[45-47] when in first approximation the product of the number of electrons or holes in the d-shell, p, and the average difference in energy, ΔE, between two d-electrons of parallel and antiparallel spins in an atom is larger

* See E. Daniel.[142]

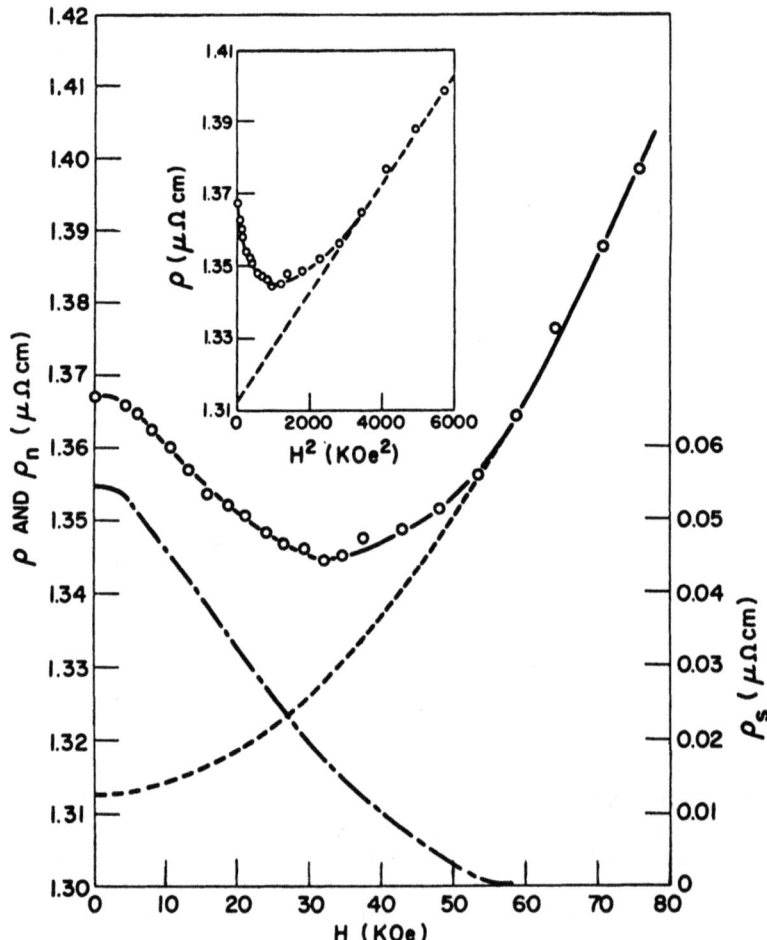

Fig. 7. Resistivity as a function of magnetic field strength for a 0.11 at.% Zn–Mn alloy. The inset is the resistivity as a function of H^2, showing the normal magnetoresistivity component, ρ_n, as a dashed line. This normal component is also shown as a dashed line in the resistivity vs. H plot. The dotted line is the difference $(\rho - \rho_n)$ and hence represents ρ_S (Collings et al.[38]).

than the width of the states, w, which is approximately one-third of the Fermi energy of the host metal ($\Delta E \approx 0.8$ eV). In accordance with this rule of thumb, alloys with aluminum ($E_F = 13$ eV) as a host metal must not show anomalies, as confirmed by the experimental results. Nickel and cobalt are not expected to cause anomalies, even in gold ($E_F = 5.5$ eV). Linde[48] found a minimum in the R–T curve of a gold–vanadium alloy at room temperature and no maximum at lower temperatures. A further investigation is in progress.

The presence of a localized moment on iron which is coupled to the host lattice copper was an explanation for the temperature dependence of the negative internal field found by Taylor et al.[49] by means of the Mössbauer hyperfine spectra of ^{57}Fe in applied fields up to 61 kOe and at temperatures between 1.1° and 310°K.

Experiments are in progress on the influence of an addition of a normal metal on the magnetic moment of the transition metal ion in a dilute alloy. Knook et al.[17]

reported an increasing influence of an iron impurity in copper by adding tin. Collings et al.[50] announced the disappearance of the minimum in the R–T curve of 0.025 at.% Mg–Mn by adding 0.93, 2.59, and 4.99 at.% Al, respectively, in agreement with susceptibility data. The results of experiments by van Rongen et al.[20] on gold–iron–tin show, as described in the article on p. 1041 of this volume, that tin does not have an influence similar to that in both cases mentioned above. More experiments are needed for a reliable statement concerning this effect.

With elements of the second long period, the following series of dilute alloys has been investigated: gold–palladium, silver–palladium, copper–palladium, gold–rhodium, copper–rhodium, copper–ruthenium, gold–technetium,[20] and gold–molybdenum.[1,17,51] For these investigations, it is very important that the base metal be free from impurities of the first period, especially iron, manganese, and chromium. The resistance and magnetoresistance were anomalous only in the case of gold–molybdenum[1,51] (Fig. 8). At 1.3°K and in a magnetic field of 21 kOe, the magnetoresistance amounted to -12%. The anomalies for gold–technetium are not well established at present.

The following elements of the third transition series have been studied:[17] gold–tantalum, gold–iridium, gold–osmium, gold–rhenium, silver–rhenium, and copper–rhenium. A minimum in the R–T curve was observed in gold–osmium, gold–rhenium, silver–rhenium, and copper–rhenium. Also from these groups of transition elements, those with the largest number of impaired spins appear to cause anomalies.

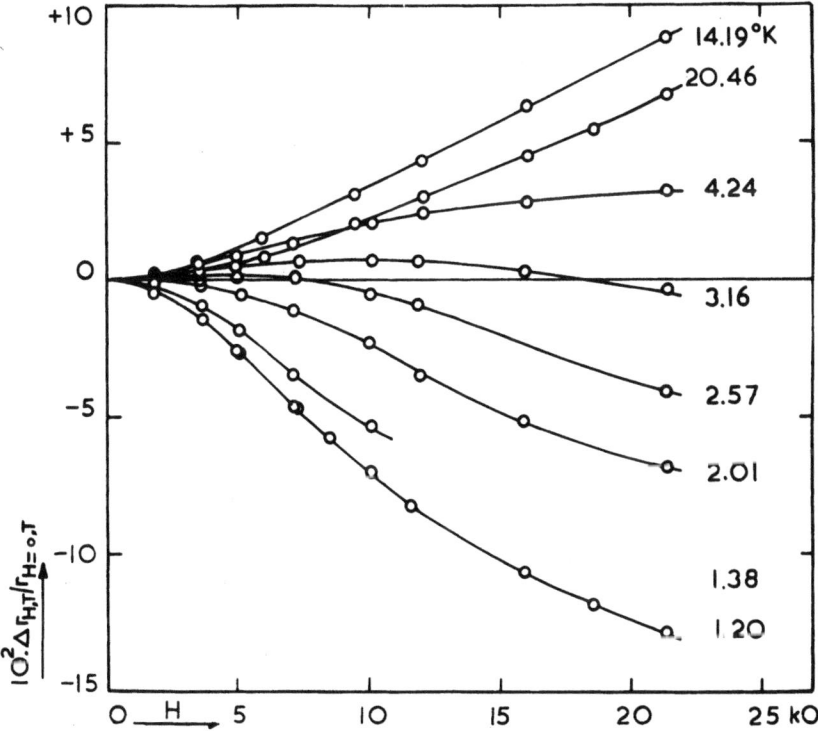

Fig. 8. The magnetoresistance of gold–molybdenum as a function of the transverse magnetic field for temperatures below 21°K (Knook[51]).

Thomas and Mendoza[35] measured the electrical resistance of molybdenum, cobalt, and tungsten, metals which were not very pure. Iron, nickel, and chromium were present in the molybdenum wires, which had a minimum in the R–T curves, while one of these showed an anomalous magnetoresistive behavior at 1.2°K (in high fields the magnetoresistance became negative). An extensive investigation was carried out recently on alloys, with titanium as solvent and manganese, chromium, iron, cobalt, nickel, niobium, aluminum, and zirconium as solutes, by Berlincourt et al.[52] The 99.92 wt.% Ti contained inter alia 0.03 wt.% Zr, 0.003 wt.% Mn, 0.002 wt.% Fe and Cr. The R–T curve showed a minimum at 14.1°K with a difference of 0.84% with the value of the resistance at 4.2°K. The titanium–manganese alloys had a deeper minimum and a negative magnetoresistance of the order of -30% at 1.2°K in a field of 120 kOe for the 1 at.% alloy and -5% for the 0.1 at.%. The saturation for the last mentioned alloy appeared to be complete below 100 kOe (Fig. 9). The R–T curves for the alloy 1.15 at.% Ti–Cr and 0.96 at.% Ti–Fe showed a minimum of 0.84 and 1.66%, respectively. The magnetoresistivity for these alloys, as well as for those with cobalt, nickel, niobium, zirconium, and aluminum as solutes, is small and positive. Of all the transition elements (chromium, manganese, iron, cobalt, and nickel), only manganese displays strong localized-moment behavior in dilute solution in hcp titanium, again demonstrating

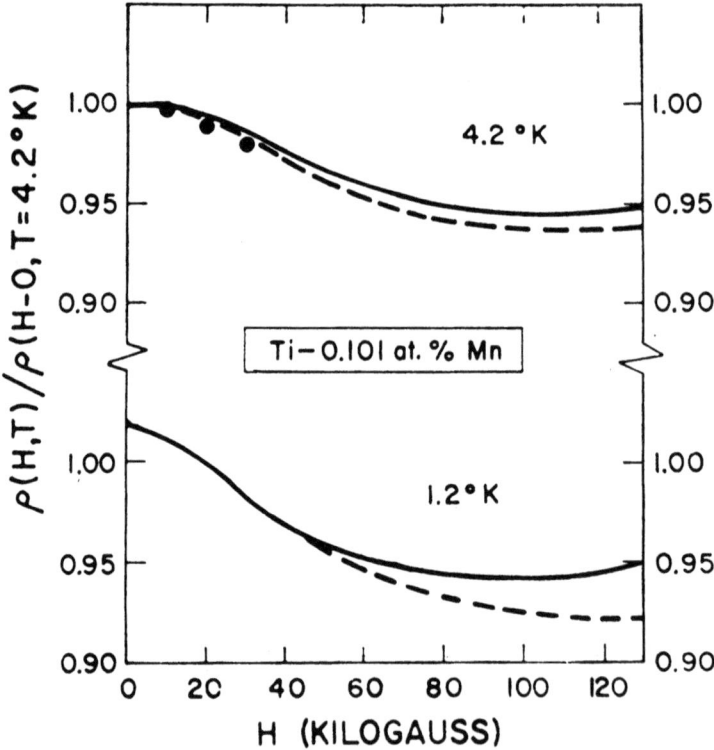

Fig. 9. The electrical resistance ratio $\rho(T)/\rho(4.2°K)$ as a function of magnetic field for 0.101 at.% Ti–Mn at 4.2° and 1.2°K. The solid and dashed curves correspond, respectively, to transverse and longitudinal fields. The black dots represent transverse steady-field data (Berlincourt et al.[52]).

the crucial importance of the electronic character of the solute in determining localized-impurity-state characteristics.

Addition of about 1 at. % Mn in zirconium[53] produces a negative magnetoresistance at 4.2°K, but that of chromium and iron does not.[54] The same result was obtained in hafnium for "dilute" addition of both manganese and chromium, but not of iron. At 1.2°K, however, negative magnetoresistance was also observed[54] in dilute hcp zirconium–chromium and hafnium–iron.

Coles[55] investigated dilute alloys with iron as a solute and molybdenum, niobium, and palladium as solvents. The iron atoms appeared to carry localized moments in molybdenum, as can be concluded from a minimum and a maximum in the R–T curve, from a negative magnetoresistance, and from susceptibility measurements, but in niobium these iron atoms did not (Fig. 10). The 0.10 at. % Pd–Fe alloy showed a decrease in resistivity below about 2°K, but no sign of a minimum above it, contrary to gold–iron and molybdenum–iron alloys of similar concentrations, but comparable with silver–manganese of higher concentration (Fig. 1). Magnetic data[56,57] indicate a strong interaction between the iron and palladium atoms, which may remove the special significance of the nearest neighbor iron pairs needed for the appearance of a minimum in the R–T curve.[58] Experiments with iron in rhodium produced new results.[59] The resistance decreases rather steeply at decreasing temperatures. The resistivity increment vs. temperature curve

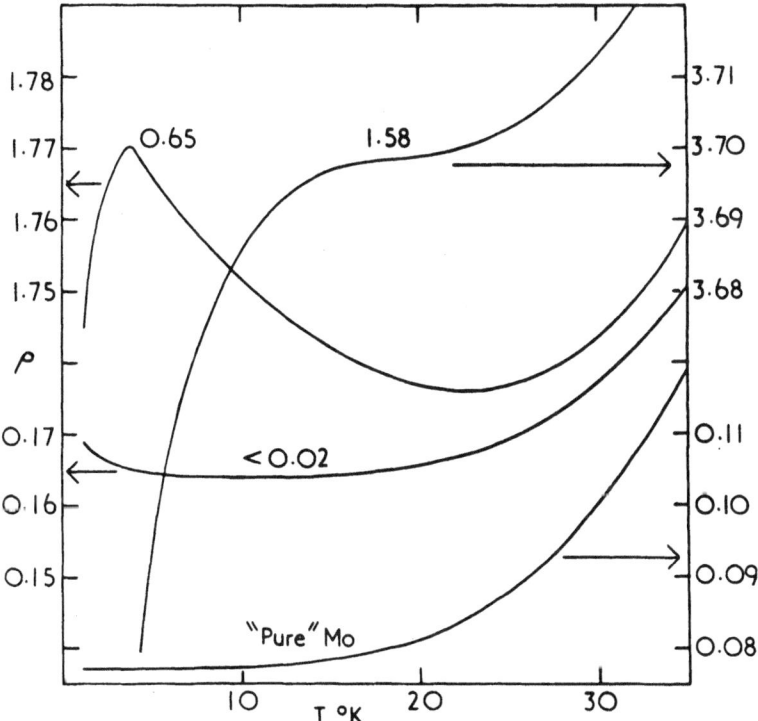

Fig. 10. The specific resistivity of dilute molybdenum–iron alloys as a function of temperature. The iron contents in atomic percentages are given in the plot. "Pure" molybdenum contains less than 10 ppm iron or manganese (Coles[55]).

is steep and approximately composition independent. No abnormal magnetoresistance was observed. Sarachik et al.[57] showed the strong correlation between a localized moment of 1 at. % Fe in the niobium–molybdenum–rhenium system and the anomalies in the electrical resistance and magnetoresistance at low temperatures.

Finally, attention will be paid to dilute alloys with rare-earth elements both as solutes and as solvents. Magnesium with less than 0.25 at. % Nd and Gd, respectively, did not show any resistance anomaly[60] at low temperatures. However, gadolinium introduced in silver[61] and palladium[62] with concentrations of the order of 0.5 at. % resulted in R–T curves with a sharp maximum at about 4°K. At higher concentrations (> 1 at. %), the R–T curve of silver–gadolinium shows a sharp drop. A concentration of 1 at. % Ce in silver did not result in any anomaly in the resistivity.[62] The R–T curves of yttrium–cerium alloys with concentrations between 0.25 and 2 at. % show a minimum, which temperature is proportional to $C^{0.29}$. This means a stronger relation between T_{min} and the concentration than, e.g., for copper–iron.[63,17] One at. % Ce in lanthanum does not cause an anomaly above 4.5°K, the transition temperature to superconductivity; for 2 at. % a minimum exists above 2.9°K. The magnetoresistance of yttrium–gadolinium has a behavior similar to that of copper–manganese, that of yttrium–terbium is large and positive, while the resistance in zero field drops like that of yttrium–gadolinium in two stages, of which the lower one is due to the antiferromagnetic ordering. Further investigations are in progress.[62]

The Thermal Resistivity

This property, as far as it is due to the electrons, shows a similar anomaly as the electrical resistivity.[64] At liquid-helium temperatures, the electronic thermal

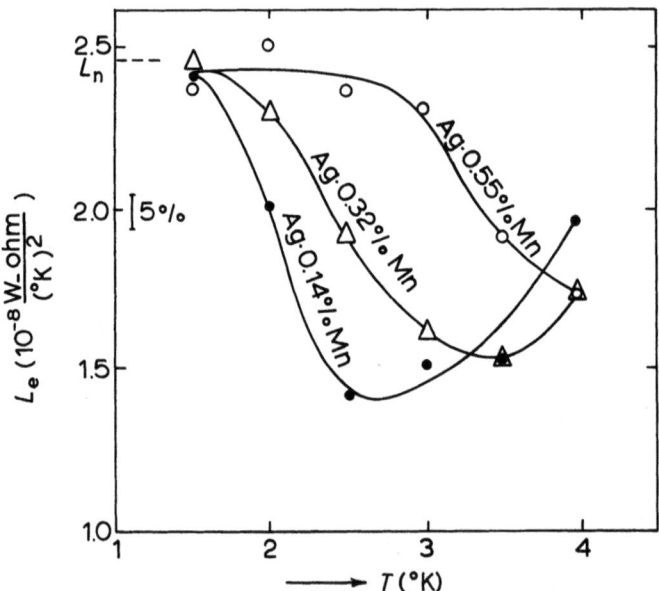

Fig. 11. The electronic Lorenz parameter L_e as a function of temperature for three silver–manganese alloys (Chari[66]).

resistivity may decrease instead of increasing when a magnetic field is applied. Recently, Berman et al.[11] found such a result for 0.03 at. % Au–Fe. Previously, De Nobel and Chari[65] performed measurements on some silver–manganese alloys with concentrations 0.55, 0.32, and 0.14 at. % Mn, the last alloy having the largest decrease (28%) in resistance at 1.5°K in a magnetic field of 25 kOe. The results were recently reanalyzed by Chari,[66] who paid a great deal of attention to the Wiedemann–Franz–Lorenz parameter in the temperature range of the anomaly. Chari found an extra dip in the curve of this parameter as a function of temperature, which shifted to higher temperature for increasing manganese concentration (Fig. 11). Anderson and Nielsen[29] found very recently, in agreement with Fenton et al.,[67] a parameter value of $(2.46 \pm 0.02) \times 10^{-8}$ WΩ/degK^2 for oxidized gold strips (iron ineffective), while for a vacuum-annealed sample this value is 20–30% higher.

The Hall Effect

For 0.03 and 0.05 at. % Au–Cr, Teutsch and Love[68] found a monotonic increase of the coefficient at very low temperatures and they did not find anomaly corresponding to the resistance minimum in gold and the mentioned alloys. This is in contradiction to the results of Fukuroi and Ikeda[69], who have made measurements on an unannealed gold wire. An explanation for this contradiction has been given by Gaidukov,[70] from the results for a gold wire with a normal R–T curve and a field-independent Hall coefficient, to the results for a gold wire with an anomalous resistance curve, whose Hall coefficient remained field independent below 8 kOe. Teutsch et al. measured in a field strength smaller than 8 kOe, Fukuroi et al. in a 20 kOe magnetic field.

Franken et al.[71] investigated silver–manganese and copper–manganese alloys, the pure metals, and normal alloys for comparison. The results were complicated. Those for 1.0 and 4.2 at. % Ag–Mn were interesting. The Hall coefficient could be represented by a sum of a normal term and an anomalous one. The first term increased from room temperature to helium temperature by 10 and 0%, respectively. The second one appeared to be proportional to the magnetization. This could be derived from measurements of the magnetoresistance, which is proportional to the square of the magnetization[72,15] (Fig. 12).

Alekseevski et al.[73] found a field dependence for dilute gold–iron alloys for magnetic fields higher than 8500 Oe, the value at which also the magnetization changed and the value of the field needed to eliminate the minimum in the R–T curve (Fig. 13).

Blue[74] carried out measurements on the Hall effect of not very dilute alloys (3 and 5 at. %) of manganese, iron, cobalt and nickel in gold between 4° and 300°K. He found a very large temperature influence. This could be attributed to the "magnetic" term of the above-mentioned sum for the Hall coefficient.

The results of the measurements on 1.0 and 2.0 at. % Ti–Mn could also be represented by a compound coefficient. The first part, not depending on the magnetization, had double the value of that for the base metal titanium at temperatures below 4.2°K.[52]

The Thermoelectric Power

The thermoelectric power (T.P.) is a property which is very sensitive to impurities in metals and to the inhomogeneity of the specimen. Direct absolute

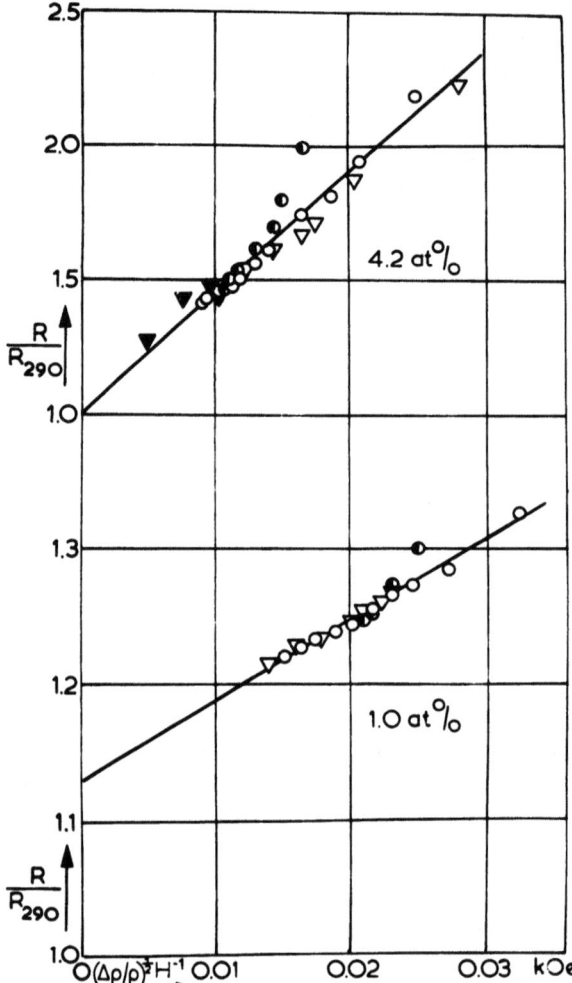

Fig. 12. The relative Hall coefficient, R/R_{290}, as a function of $|\Delta\rho/\rho|^{\frac{1}{2}}H^{-1}$ for the alloys 1.0 and 4.2 at.% Ag–Mn (Franken et al.[71]). $\triangledown$: 4.2°K, decreasing field; $\bigcirc$: 1.3°K, decreasing field; $\blacktriangledown$: 4.2°K, increasing field; ◐: 1.3°K, increasing field.

measurements against a superconductor are preferable to indirect ones. The early measurements by Borelius et al.[75] and by Keesom and Matthijs[76] on dilute alloys of gold and copper with some of the transition metals were in general performed against "silver normal." These investigators found very large values, of the order of -10μV/degK (at present called "giant" T.P. after Bailyn[77]) in comparison with the "normal" value of the order of some hundredth of a μV/degK.

MacDonald and Pearson[26] found a linear relation between the T.P. at 15°K and the "depth" of the minimum, which means an approximately linear relation between the T.P. and the concentration.[17] This is in disagreement with Sommerfeld's[78] theory of electrons. According to this theory, the T.P. should be proportional to the temperature and independent of the concentration of the impurity. As described by de Vroomen,[79] these "transitional" impurity atoms like iron cannot be considered "static" in their exchange with the electrons because they possess an internal degree of freedom, as shown by the peculiar effect in the specific heat.

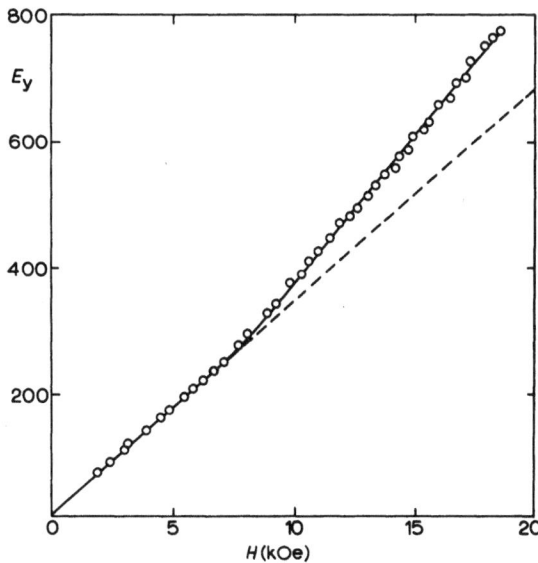

Fig. 13. The Hall field, E_y, as a function of magnetic field at $T = 1.45°K$ (Alekseevskii et al.[73]).

Application of the conventional theory is risky or even unpermitted. Experiments on the T.P. of alloys 0.01–0.49 at.% Ag–Mn demonstrated a large negative value for T.P. at a certain temperature. The curve of T.P. vs. concentration had a peak for a certain concentration.[80] The value of this concentration depended on the thermal treatment. Above a concentration of 0.3 at.%, the T.P. at 3°K was positive. Japanese research[81] on copper–manganese gave a positive value below 20°K for atomic concentrations of 0.03 and 5.4%.

Gold et al.[23] studied the influence of different impurities on the T.P. of copper. The negative value of T.P. increased with iron concentration. Using a modified Nordheim–Gorter[82] relation, Gold et al.[23] could calculate the T.P. of several introduced metals. The value for iron appeared to be the largest negative value of $-16.2\ \mu V/degK$ (Fig. 14). The influence of oxygen on copper could be determined (Fig. 15).

Recently, Anderson and Nielsen[29] investigated the influence of an air (oxygen) annealing at 950°C on some transport properties of gold. The T.P. appeared to be positive above the lowest temperature of 2.2°K and was of the order of 0.01 $\mu V/degK$ at 4°K, contrary to the value of about $-14\ \mu V/degK$ for the T.P. of a vacuum-annealed gold strip of 40 μ. The Ottawa N.R.C. group studied the T.P. and the electrical resistivity of the same specimen. The material consisted of alloys of chromium, manganese, iron, and cobalt in gold[7] and copper[21] and manganese and iron in platinum and palladium. The values for the T.P. of the alloys were "giant," except for the alloys iron and manganese in palladium, though in the case of cobalt we have to be cautious. The T.P. of manganese in magnesium, zinc and aluminum were studied by Hedgcock and Muir.[83] Surveying the results, we conclude:

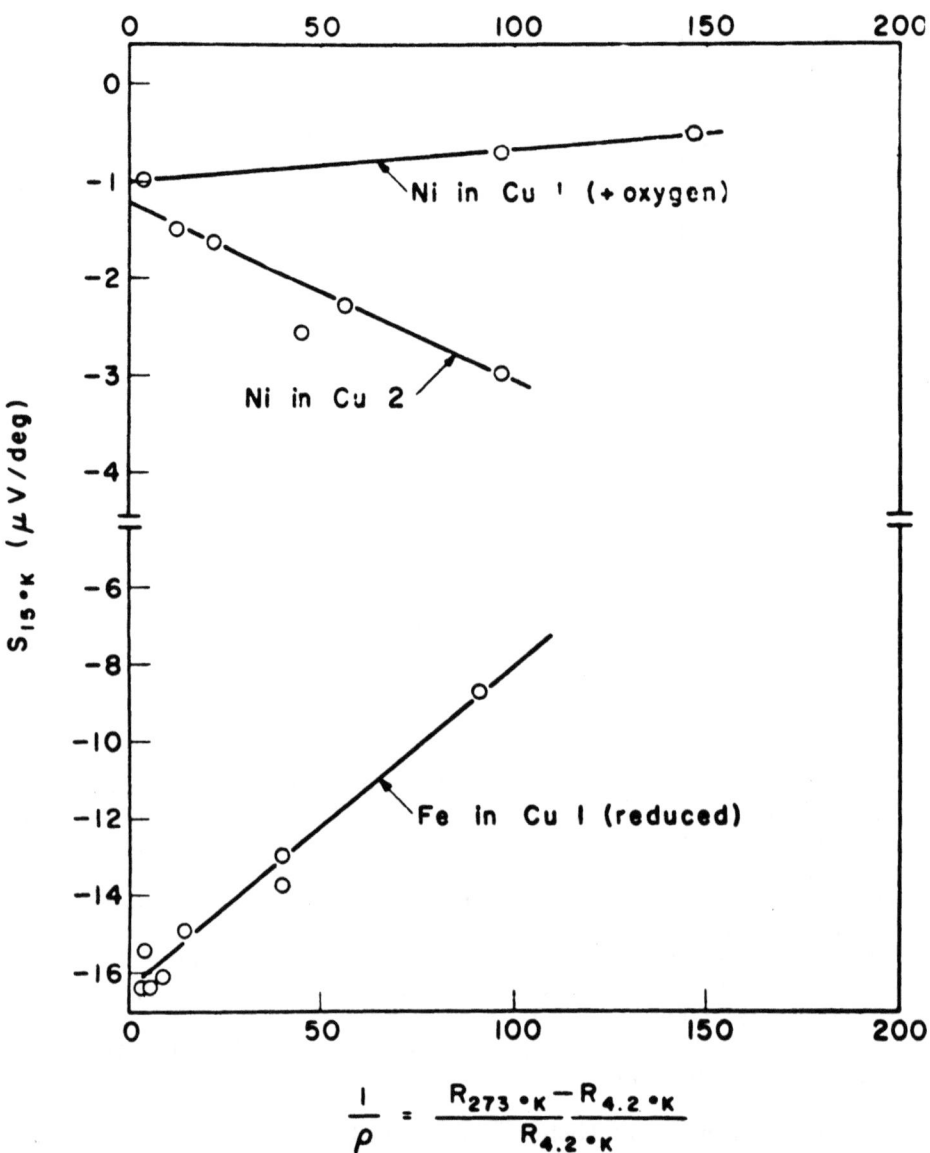

Fig. 14. The thermoelectric power, S_{15}, at 15°K, as a function of the inverse residual resistance ratio for dilute copper–nickel and copper–iron alloys (Gold et al.[23]).

1. In many investigations the starting material was not pure enough nor did it remain so during annealing and after the treatment necessary for the mounting. This has even greater importance for the T.P. than for the electrical resistivity.
2. The temperature dependence is still questionable.
3. For the alloys with a localized magnetic moment, the T.P. is giant. This is never the case in aluminum.

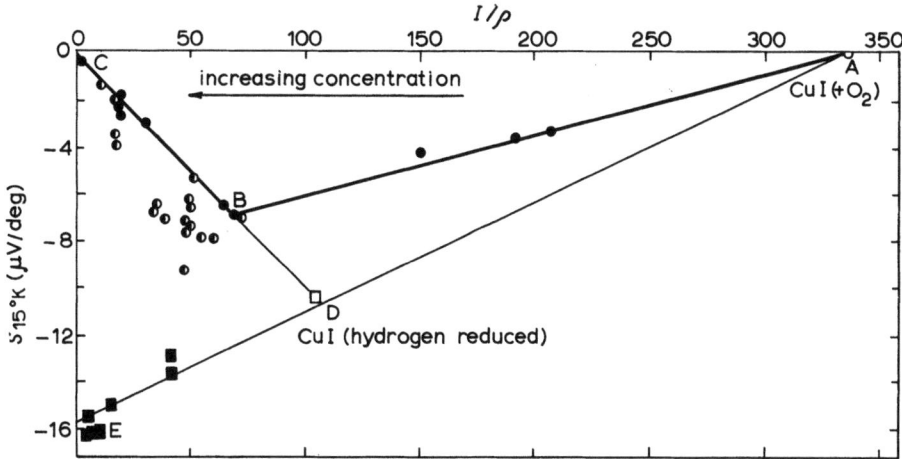

Fig. 15. The thermoelectric power, at 15°K, S_{15}, as a function of the inverse residual resistance for dilute alloys of iron and tin in J.T.H. (oxygen containing) copper (Gold et al.[23]). ⊙: copper; ●: copper–tin; □: copper (reduced); ■: copper–iron (reduced); ◐: copper–tin (reduced).

Berman, Brock, and Huntley[11] studied gold–iron alloys with respect to their use as thermoelements. The best result was obtained for a 0.03 at.% Au–Fe alloy. The thermopower appeared to saturate (20% changed) at 1.3°K in a magnetic field of 20 kOe, while the electrical and thermal resistivity showed no sign of doing so.

The Magnetic Susceptibility and Remanence

Measurements by Bowers[84] on pure copper indicate that the susceptibility, χ, of copper is substantially independent of temperature. The contribution to χ in the form of a $1/T$ term was partly due to the nuclear moment and partly to 3 parts in 10^7 of paramagnetic ions (Fe^{++}); this level of impurity was plausible for the copper used. As shown by Sonder and Sekula,[85] the annealing atmosphere (for pure copper, 99.999%) is of great influence. An anneal of 2 hr under 15 μ of CO at 900°C created a resistance minimum. The susceptibility was field independent but below 100°K an increase in paramagnetism occurred which could be accounted for by assuming approximately 5×10^{17} magnetic centers per cm³, each having 2 Bohr magnetons. Such an anneal of 4 hr under 20 μ of air removed the resistance minimum, while the susceptibility data suggested a precipitation of 2×10^{17} iron atoms (2 ppm). The same experiments carried out on a 0.1 at.% Cu–Fe showed that almost all the iron was in a paramagnetic state. After an anneal in air, the sample was strongly ferromagnetic.

Nontransition elements, as a solute, will not cause a temperature-dependent susceptibility. When such a susceptibility is found, it must be ascribed to a reduction of, e.g., iron oxide present in the base metal.[86] Several investigators studied the susceptibility of manganese in copper,[15,87–89] gold[90] (Fig. 16), silver,[88,89] magnesium,[88,41] aluminum,[41] and zinc.[38] The $1/\Delta\chi$ vs. T curves (Fig. 17) show a minimum in the cases of the higher concentrations ($\approx$ 1 at.%). We assume that the place of this minimum (maximum in $\Delta\chi$ value) mainly determines the Néel temperature, though criticism in some cases is possible.[91] The spin susceptibility

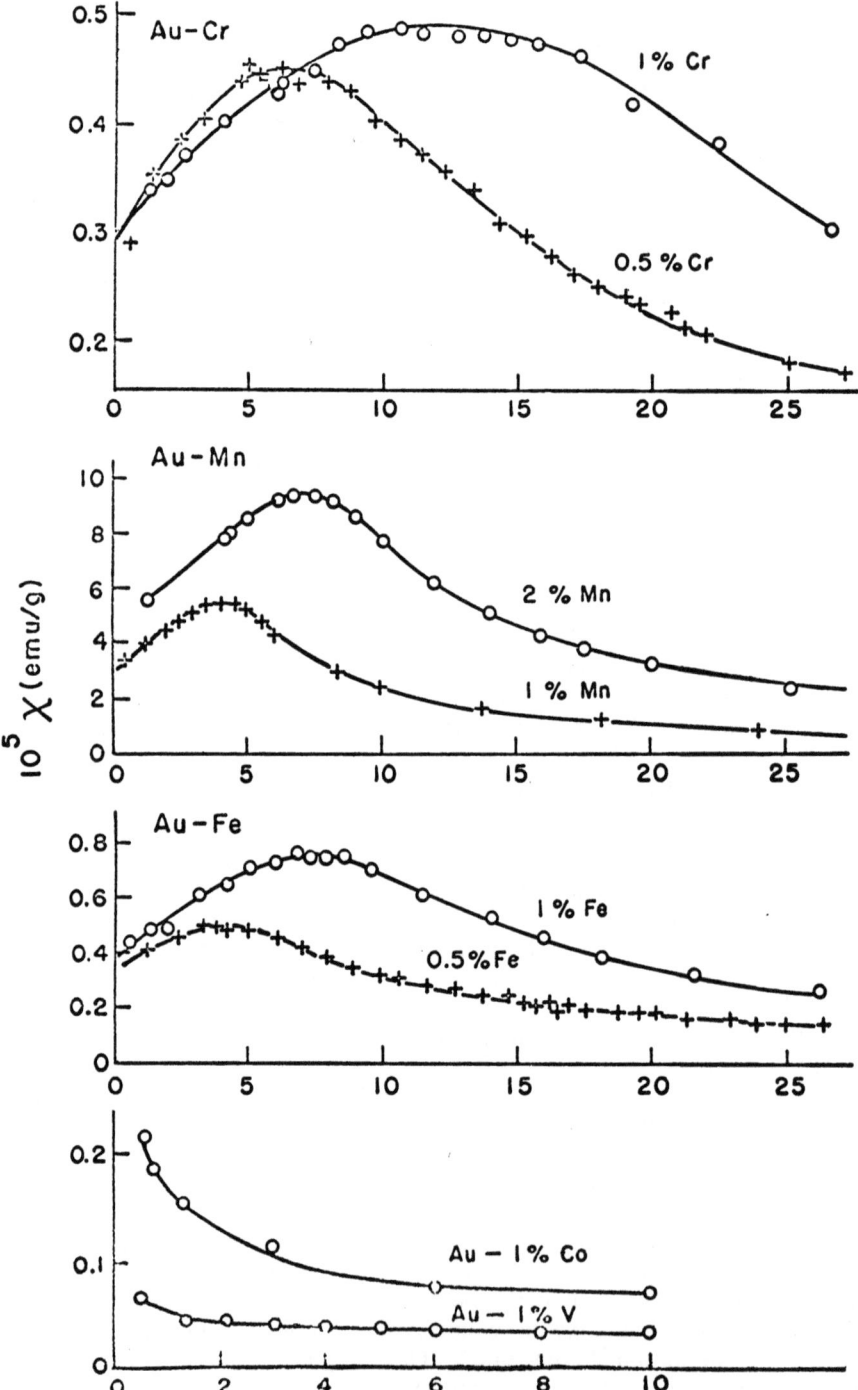

Fig. 16. Magnetic susceptibility vs. temperature. χ is the susceptibility per gram of alloy (Lutes and Schmit[90]).

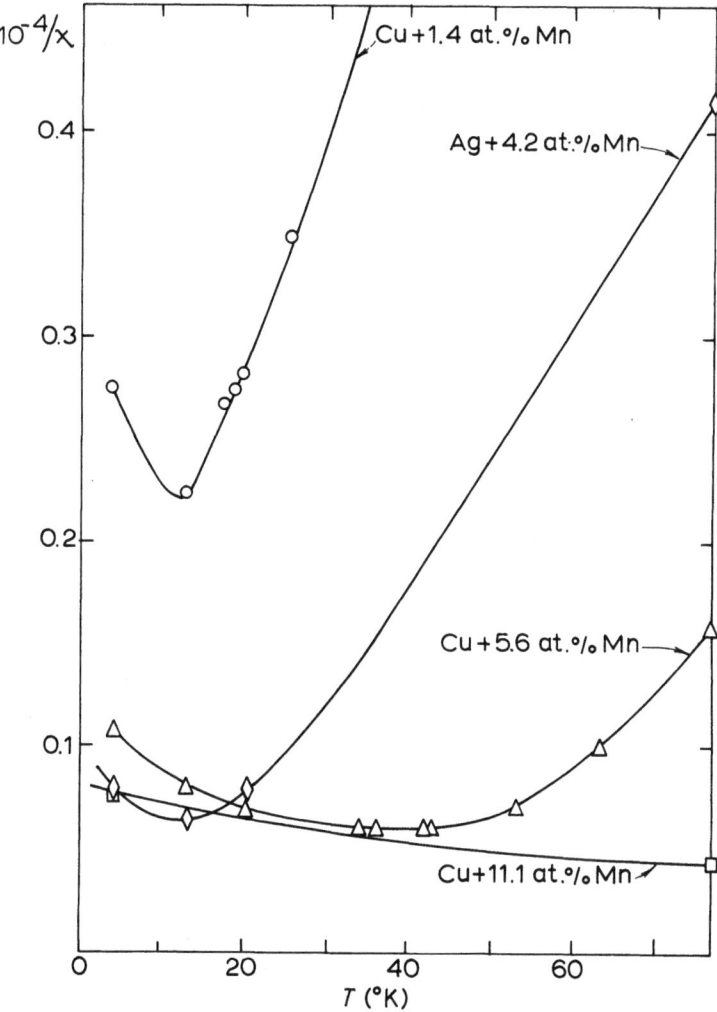

Fig. 17. Inverse of volume susceptibility, χ, as a function of temperature (Owen et al.[88]).

of the conduction electrons of the alloy is not allowed to be greater than that of the matrix. The effective magnetic moment of manganese in copper[92] is of the order of 4.9 (assuming $g = 2$, the spin value is 2), in gold about 5.9 ($s = \frac{5}{2}$), in silver[92] approximately 5.9 ($s = \frac{5}{2}$), in magnesium 4.9–3.9 ($s = 2$ to $\frac{3}{2}$), depending on the temperature range of the data. In aluminum, the spin value, if any, for manganese will be of the order of $\frac{1}{2}$ and in zinc this value is constant, $\frac{3}{2}$, between 5° and 273°K,[38] again assuming no spin-orbit coupling. Collings et al.[50] added manganese to magnesium–aluminum alloys in order to study the influence of aluminum (higher electron to atom ratio than magnesium) on the localized magnetic moment of 0.025 at. % Mn in magnesium ($s = 2$). The determination of susceptibility will always be a product of the concentration of paramagnetic ion and the square of the effective magneton number (nP_{eff}^2). The susceptibility data showed a disappearance of the localized magnetic moment of manganese at 5 at. % Al, while at 0.93 at. % Al this was present, causing a pronounced resistance minimum. However, the results of magnetoresistivity measurements, together with calculations based on Kasuya's theoretical expression[43] for the negative magnetoresistance component, indicate that the concentration of manganese in the magnesium–aluminum alloys is dependent on the aluminum concentration.

Not many experiments have been conducted on dilute alloys with the normal metals as host metals and with titanium, vanadium, chromium, iron, cobalt, and nickel as solutes at low temperatures. Very recently 0.5 and 1 at. % Au–Cr were studied and a susceptibility maximum at temperatures approximately proportional to the concentration was found. The effective moment indicated a spin value of $\frac{3}{2}$, assuming the gyromagnetic factor to be 2.[90] Hedgcock[86] found an increase of the paramagnetism below the temperature of the resistance minimum for copper–iron alloys. At temperatures higher than 90°K, Scheil et al.[93] calculated an average effective magnetic moment of 4.9 for 0.57–2.16 at. % Fe in copper. For gold–iron alloys,[94] the same sort of experiments also procure a value of 4.9 for the smallest concentrations. For the temperature range 0.5°–35°K, Lutes and Schmit[90] found an effective moment for 0.5 and 1 at. % Fe in gold of 3.6 and 3.3, respectively, comparable with the values 3.0 and 3.4 calculated from susceptibility data for 0.63 at. % Fe in the range 14°–77°K and 77°–300°K by Kaufmann et al.[95] Vogt and Gerstenberg[96] measured the susceptibility of gold–titanium and gold–vanadium alloys, with 1 at. % as the smallest concentration. For the gold–titanium alloy, a temperature-independent paramagnetism was found above 90°K. However, for gold–vanadium alloys, the susceptibility increased with decreasing temperature and a part of it obeyed a Curie–Weiss law with a negative Curie temperature. The effective number of magnetons increased so steeply with decreasing concentration of vanadium that Vogt and Gerstenberg[96] expected the more dilute alloys ($\ll 1$ at. %) to have $3d$-electrons responsible for the magnetic moment. At higher concentrations of vanadium, the behavior may be compared with that of titanium in gold. For 1 at. % V, Lutes and Schmit[90] did not find a Curie–Weiss law nor a magnetic transition at low temperatures (0.5–30°K). The same behavior that was found for gold–titanium can be expected for gold–nickel, the former having a small number of unpaired spins and the latter a small number of open places in the $3d$-band. Experiments by Pugh et al.[97] on copper–nickel and also on silver–palladium showed that one has to be careful with conclusions drawn from the representation of the susceptibility by $aT + b + (c/T)$. An impurity like manganese or iron may play a confusing role.

Another group of experiments was carried out on alloys with transition metals as host metals. Solutions of chromium,[98] manganese,[98] iron,[98,99] and cobalt[98,99] in titanium were studied (0.1–4.0 at. %). Only hcp titanium–manganese alloys exhibited a strong temperature dependence at low temperatures. Taking the Landé factor to be 2, Cape[98] derived, using the nominal concentration of manganese atoms, a spin value of approximately 1.5 (effective moment $\approx 3.5\, \mu_B$).

In vanadium none of the transition elements titanium, manganese, iron, cobalt, and nickel appeared to have a localized moment[100] for the concentrations smaller than 5.8 at.%, just as for bcc 14 at. % Ti–Mn.

In hcp zirconium, only 1 at. % Mn[101] and not iron[50] produced a Curie–Weiss type temperature dependence of the 4°–300°K magnetic susceptibility. Susceptibility measurements on alloys with hcp hafnium as host metal resulted in a magnetic moment of 1.7, 3.8, and 0.7 for chromium, manganese, and iron, respectively. Gadolinium exhibited an effective moment of 7 in zirconium and 6.2 in hafnium.[101]

One atomic percent iron exhibited a localized moment in molybdenum, rhodium, and palladium[56] as a host metal. In binary alloys of adjacent elements, the gradual onset of such a moment could be found, e.g., between 50 and 80 at. % Mo in niobium, from 70 at. % Rh in ruthenium.

In a 50–50 alloy of titanium and zirconium, 0.2 at. % Mn caused a Curie–Weiss type behavior of the susceptibility[102] with an effective moment of 1.1 Bohr magneton per manganese ion.

The magnetization experiments by Schmitt and Jacobs[15] on copper–manganese alloys with concentrations of about 0.05, 0.2, 0.4, 1.0, and 1.8 at. % showed clearly that those samples with a concentration of 0.4 at. % and more, exhibited hysteresis in the liquid-helium range. The susceptibility of those samples goes through a maximum at a transition temperature,[88] below which the hysteresis first appears. The existence of the maximum was interpreted as a gradual transition to antiferromagnetism (Berkeley group), but the evidence linking the maximum in the susceptibility to the presence of hysteresis and remanence can also be interpreted

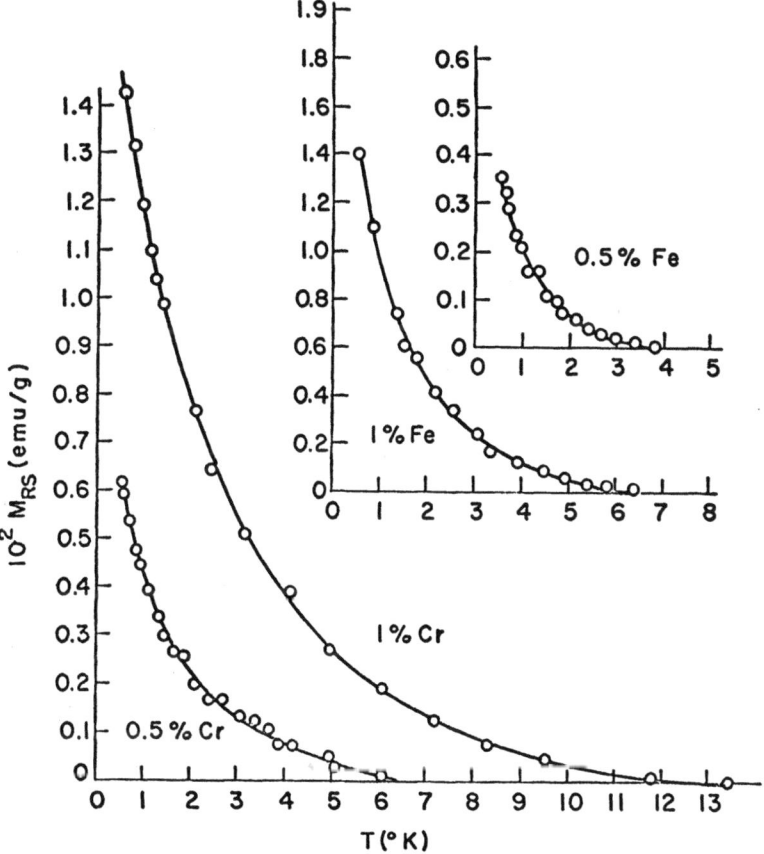

Fig. 18. Temperature dependence of saturated remanence for gold–chromium and gold–iron alloys (Lutes and Schmit[90]).

by the short-range order picture of small "ferromagnetic" domains coupled "antiferromagnetically."[103]

Lutes and Schmit[90] studied gold-based alloys with about 1 at.% V, Cr, Mn, Fe, and Co between 0.5° and 35°K. They found maxima in the susceptibility vs. temperature (χ–T) curve for gold–chromium, gold–manganese, and gold–iron (Fig. 16). Above the maxima, these alloys exhibit temperature-dependent paramagnetism. Below the maxima, remanent magnetization was observed, in agreement with earlier experiments on copper–manganese.[104] Estimates of the saturated magnetic remanence at 0°K were obtained by fitting the low temperature remanence data to paramagnetic expressions and resulted in a few percent of the available total moment of the solute atoms (Fig. 18). The linear dimension of a ferromagnetic domain for copper–manganese was calculated to be of the order of 10^{-6} cm for 1 at.% Mn.[104] There is no direct correlation with the particular form of magnetic transition exhibited at much higher concentrations.[95,105,106] For the gold–chromium, gold–manganese, and gold–iron alloys, the transition temperatures are roughly 12°, 4°, and 8°K per at.%, respectively. The magnetic transitions did not occur in cobalt and vanadium alloys of 1 at.% concentration.

More extensive information will be found in a report by Arrott.[91]

The Specific Heat

Measurements of this property of the dilute alloys mentioned before have been started for two reasons: (1) the determination of the "magnetic" contribution to the specific heat; (2) the measurement of the change in the electronic specific heat term as an indicator of the change in the density of states. After the first measurements carried out on a 0.09 at.% Ag–Mn specimen by De Nobel and du Chatenier[107] (Fig. 19), the following long series of alloys was studied:

vanadium,[109] chromium,[108] manganese,[108] iron,[109] and cobalt[110] in gold
chromium[108] and manganese[107,108] in silver
chromium,[108] manganese,[107,108,111] iron,[108,112,113] cobalt,[109,114] and nickel[108,115] in copper
chromium,[108] manganese,[108] and iron[108] in zinc
chromium,[109] manganese,[116,117] and iron[116] in magnesium
manganese[33] in cadmium
chromium,[108] manganese,[119] and iron[108] in aluminum
manganese,[102] iron,[120] and cobalt[102] in titanium
manganese[102] in 50 at.% Ti–Zr
cobalt[118] and iron[118a] in palladium

(Liu[121] pointed out that lanthanum–gadolinium[122] behaved like copper–manganese.)

It was impossible to represent the specific heat of the alloy silver–manganese by the same formula, $C = \alpha(T/\theta)^3 + \gamma T$, that was used for a pure metal. The first term represents the lattice contribution and the second the electronic contribution to the specific heat. The hump in the C/T vs. T curve could be described roughly as of a Schottky type, assuming the ^6S level for manganese to be split up into equidistant levels. An internal field of the order of about 25 kOe must be assumed for an alloy with 0.28 at.%. The area between the curve for the 0.09 at.% alloy and that for the host metal was approximately $0.09R \ln 6$ or $cR \ln (2s + 1)$, indicating a spin value of $\frac{5}{2}$ for the manganese ion. Experiments at temperatures below 1°K are needed for a reliable extrapolation of the curves. With increasing magnetic field the hump in the C/T vs. T curve became lower, while the abovementioned area remained

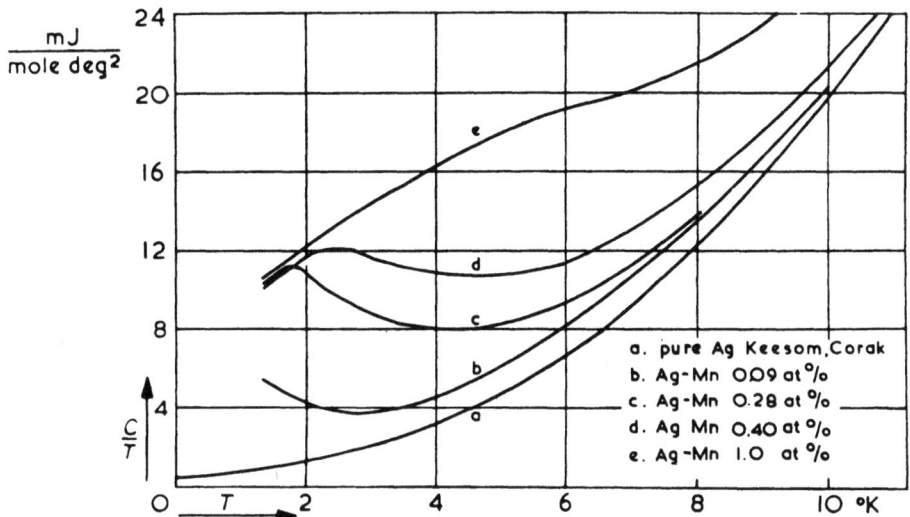

Fig. 19. The ratio of specific heat to temperature, C/T, as a function of T for silver–manganese alloys (De Nobel et al.[107]).

approximately constant. In this way manganese showed a localized magnetic moment in silver.

Considering the list of alloys, we can remark that vanadium and cobalt did not show an anomaly in the specific heat *vs.* temperature curve. The coefficient γ of the electronic part of the specific heat for gold–vanadium alloys increased as a function of concentration from 0.80 to 1.28 mJ/mole-deg K^2 for $c = 0.1$ to 1.4 at.%, while γ had a value of 0.74 for pure gold. The anomalies found[110] for higher concentrations of cobalt may be due to impurities like iron or, in comparison with experiments by Friedberg et al.[115] on copper–nickel alloys and by Weil et al.[123] on copper–cobalt and gold–cobalt alloys, due to fluctuations[124] in the concentration.* Overhauser[125] gave an argument on the nonlocalization of cobalt impurities in copper (see also Klein and Brout[126]).

The alloys, copper–nickel, copper–cobalt, zinc–iron, aluminum–iron, and aluminum–chromium, because of their normal behavior, offered the possibility of comparing the values of γ and θ_D with those of the pure metal. The value of γ for a 0.6 at.% Cu–Ni alloy is about 1% higher than that for pure copper, while this difference for a 0.09 at.% Cu–Co amounts to 4%. The values of γ and θ_D for 0.1 at.% Zn–Fe, 0.1 at.% Al–Fe, and 0.1 at.% Al–Cr are equal to those of the host metals, while the value for the 0.045 at.% Al–Mn alloy does not differ from the aluminum value. For a small concentration (0.013%) of iron in magnesium,[116] no anomaly was found above 3°K. Neither was it found for 0.043% Mn in magnesium,[116] but Martin[117] showed the existence between 0.4° and 1.5°K. The specimen 0.1 at.% Mg–Cr appeared to have a concentration of 0.02% at the top (spectrographic analysis) and of 3.3% at the bottom, though microhardness experiments suggested a rather good homogeneity. The γ and θ_D values were those of pure magnesium. However, we cannot come to any conclusions about anomalies in magnesium–chromium.

---

* For details, see *Low Temperature Physics, Vol. IV*,[1] p. 243.

Experiments on 1 at. % Fe^{120} and Co^{102} in titanium did not cause an anomalous behavior in the specific heat at low temperatures.

All the other investigated dilute alloys had an anomalous C/T vs. T curve (Fig. 20), the place where the maximum occurred probably being dependent on the probability distribution of the effective field.[109] Calculating the magnetic entropy from this graph, we can deduce a spin value for the solute. Manganese as a solute in gold, silver, and copper has roughly a spin value of $\frac{5}{2}$. In zinc, however, this value is $\frac{3}{2}$, in magnesium it is unknown, while in titanium a spin of $\frac{3}{2}$ is plausible. This value decreases when zirconium is added to the titanium. For chromium in gold the spin value points to $\frac{3}{2}$. In silver the value should be 1.4 and 0.7 for the concentrations 0.099 and 0.28, respectively. For increasing concentration (0.07–0.56), the spin value in copper seems to decrease from 0.8 to 0.2, the first value being the more reliable one. In zinc the value for chromium is 0.3, the accuracy being rather small due to the extrapolation of the $\Delta C/T$ vs. T curve to 0°K. The spin value derived from the data of a 0.092 at. % Au–Fe alloy is 0.65 and from those of copper–iron alloys is $\frac{1}{2}$. For cobalt and iron in palladium, the values 1.5 and 1.1 ± 0.3, respectively, were found, in disagreement with magnetic data.[56]

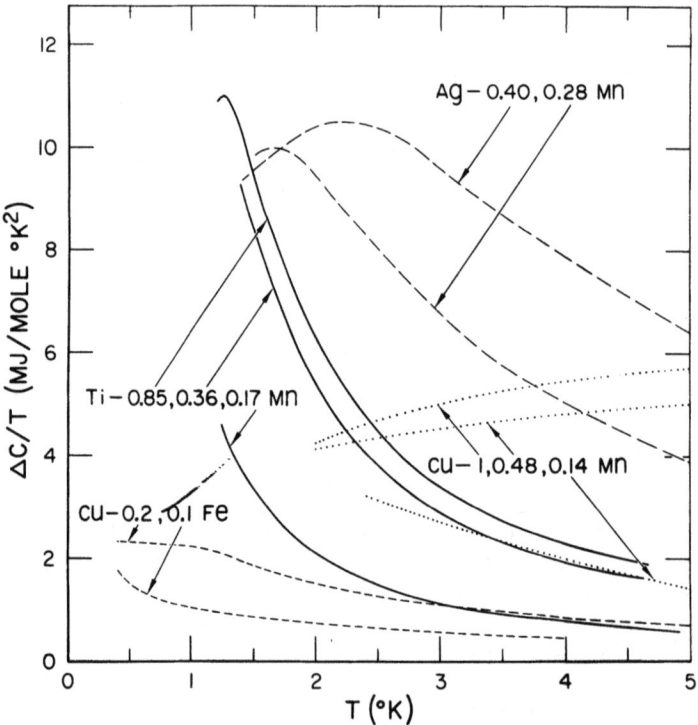

Fig. 20. $\Delta C/T$ vs. T for dilute hcp titanium–manganese alloys, silver–manganese,[107] copper–manganese,[111] and copper–iron[113] alloys. ΔC is the difference between the specific heat of the alloy and that of the solvent, approximately the "anomalous" specific heat contribution (Hake and Cape[102]).

Zimmerman and Hoare[111] measured the specific heat of copper–manganese alloys with rather high concentrations (0.5 to 10 at. % Mn) and found no systematical influence of the concentration. They concluded that the extra specific heat at low temperatures is nearly linear in T and independent of manganese concentration (Fig. 21). Overhauser[125] gave an explanation based on his concept of spin density waves. Marshall,[127] who had objections to this concept of antiferromagnetism, used Yosida's[128] interaction (Ruderman and Kittel[129]) for another explanation, and Blandin and Friedel[45] started from the concept of the virtual bound state. Franck et al.[113] found indications of a magnetic term proportional to temperature for copper–iron alloys between 0.1° and 1.5°K. Du Chatenier and Miedema[108,130] carried out experiments on gold–manganese, gold–iron, silver–manganese, copper–chromium, and copper–manganese, below 1°K. The contribution of the hyperfine structure (interaction between electronic and nuclear moments of the manganese ions) hindered the exact determination of the linear term. Between 0.06° and 0.4°K, the specific heat for gold–manganese could be described by $C = 6.6 \times 10^{-3}/T^2 + 7.8T + C_{Au}$ mJ/mole-deg K for 0.083 at. % Mn and by $C = 10.6 \times 10^{-3}/T^2 + 7.8T + C_{Au}$ for 0.16 at. % (Fig. 22). For copper–manganese alloys such a relation is valid between 0.06° and 0.3°K. A silver–manganese alloy was measured down to 0.03°K and $\gamma_m T$ was 4.6T mJ/mole-deg K. Better confirmation of the existence of the linear term was obtained by measuring 0.1 and 0.6 at. % Cu–Cr and 0.09 at. % Au–Fe. Only 9% of the natural chromium possesses a nuclear spin (^{53}Cr) and only 2.2% of the iron (^{57}Fe). These alloys do not demonstrate a hyperfine-structure effect in the specific heat. Between 0.04° and 1.0°K, both chromium alloys have a magnetic term with $\gamma_m = 1.8$ mJ/mole-deg K^2, while between 0.1° and 1.0°K, the gold–iron alloy has $\gamma_m = 2.2^6$ mJ/mole-deg K^2.

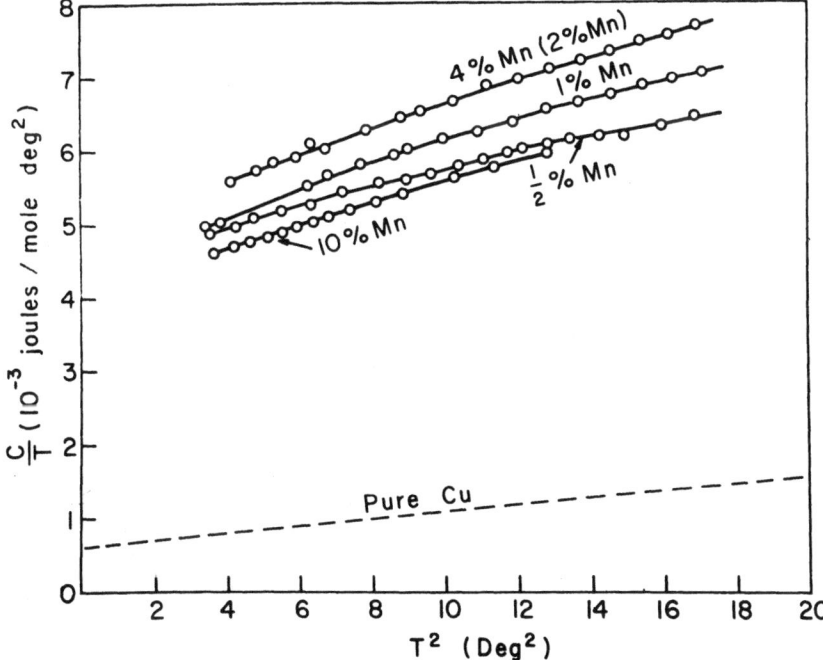

Fig. 21. The ratio, C/T, as a function of T^2 for copper–manganese alloys (Zimmerman et al.[111]).

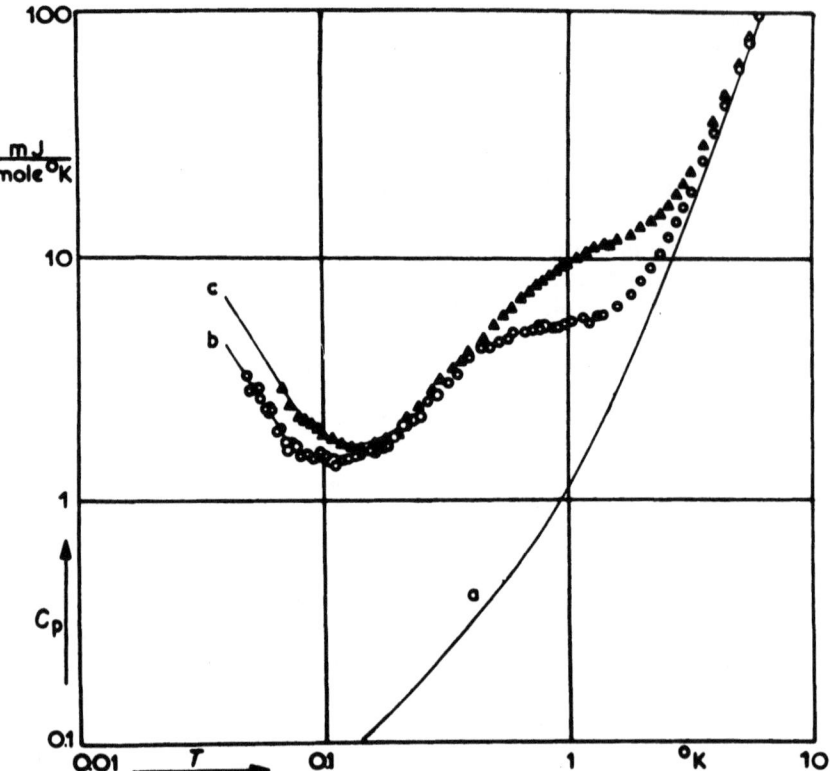

Fig. 22. The specific heat of gold–manganese as a function of temperature (du Chatenier et al.[108]). a: pure gold; ○ b: 0.083 at.%; △ c: 0.16 at.%

To conclude, we mention that the extensive investigations on the specific heat have shown that Friedel's scheme approaches reality rather well. The possibility of the presence of a localized moment is largest for the transition metal with the largest number of unpaired spins. The extra magnetic contribution at very low temperatures is independent of the concentration of the transition metal and proportional to the temperature.

Theoretical Considerations

Since the suggestion of a resonance hypothesis by Korringa and Gerritsen in 1953,[72] this hypothesis was reconsidered by Domenicali,[131] the energy dependent collision time τ being zero in a small interval and being τ_0 at other energies. The suggestions by other authors may be divided into two groups: those starting from a molecular field concept (Schmitt,[132] Schmitt and Jacobs,[15] Owen, Browne, Knight, and Kittel,[87] Yosida,[39,128] Marshall,[127] de Vroomen and Potters,[133] Kasuya,[43] Bailyn,[77] Guénault and MacDonald[134]), and those starting from the interaction between pairs of ions (Dekker[135] and Brailsford and Overhauser[136]).

The molecular field model may explain the anomaly of the thermopower in a reasonable way, while the anomaly of the R–T curve is better understood with the other model. However, Friedel and his co-workers[44,45] discussed the concept of the virtual bound states for the ions introduced in the base metal. The localized magnetic moment was also studied by Anderson[46] and Wolff.[47] More recent publications pertaining to electrical resistivity, by Mrs. van Peski–Tinbergen and Dekker,[137] Liu,[138] Miss Béal,[139] and Kondo,[140] may be mentioned. Professor Daniel's paper[142] may be referred to for a better representation of the several models and calculations.

References

1. *Progress in Low Temperature Physics, Vol. IV*, C. J. Gorter (ed.), North Holland Publishing Co., Amsterdam (1964), p. 194.
2. W. Meissner and G. Voigt, *Ann. Phys.* **7**, 761, 892, 1930; W. J. de Haas, J. de Boer, and G. J. van den Berg, *Physica* **1**, 1115, 1933/34; *Commun. Kamerlingh Onnes Lab.*, Leiden, No. 233b; G. J. van den Berg, thesis, Leiden University (1938); *Proceedings of the Seventh International Conference on Low Temperature Physics*, University of Toronto Press, Toronto, Canada (1961), p. 246 (review).
3. A. N. Gerritsen and J. O. Linde, *Physica* **17**, 573, 1951; *Commun. Kamerlingh Onnes Lab.*, Leiden, No. 285c.
4. A. N. Gerritsen and J. O. Linde, *Physica* **17**, 584, 1951; *Commun. Kamerlingh Onnes Lab.*, Leiden, No. 285d.
5. N. M. Nakhimovich, *J. Phys. (USSR)* **5**, 141, 1941.
6. A. N. Gerritsen and J. O. Linde, *Physica* **18**, 877, 1952; *Commun. Kamerlingh Onnes Lab.*, Leiden, No. 290d.
7. D. K. C. MacDonald, W. B. Pearson, and I. M. Templeton, *Proc. Roy. Soc. (London), Ser. A* **266**, 161, 1962.
8. C. A. Domenicali and E. L. Christenson, *J. Appl. Phys.* **32**, 2450, 1961.
9. A. N. Gerritsen, *Physica* **23**, 1087, 1957; *Commun. Kamerlingh Onnes Lab.*, Leiden, No. 309d.
10. J. O. Linde, *Proceedings of the Fifth International Conference on Low Temperature Physics*, University of Wisconsin Press, Madison (1958), p. 402.
11. R. Berman, J. F. Brock, and D. J. Huntley, *Clarendon Lab. Report* No. 140/64 (1964).
12. E. Mendoza and J. G. Thomas, *Phil. Mag.* **42**, 291, 1951.
13. A. N. Gerritsen, *Physica* **19**, 61, 1953; *Commun. Kamerlingh Onnes Lab.*, Leiden, No. 291c.
14. W. Meissner and H. Scheffers, *Physik. Z.* **30**, 827, 1929; W. F. Giauque, J. W. Stout, and C. W. Clark, *Phys. Rev.* **51**, 1108, 1937; W. F. Giauque and J. W. Stout, *J. Am. Chem. Soc.* **60**, 388, 1938; J. W. Stout and R. E. Barieau, *J. Am. Chem. Soc.* **61**, 238, 1939.
15. R. W. Schmitt and I. S. Jacobs, *J. Phys. Chem. Solids* **3**, 324, 1957.
16. Y. Mutô, *Sci. Rep. Res. Inst. Tohoku Univ. Ser. A* **13** (1), 1961, *J. Phys. Soc. Japan* **15**, 2119, 1960.
17. B. Knook, thesis, Leiden University (1962); B. Knook and G. J. van den Berg, *Proceedings of the Eighth International Conference on Low Temperature Physics*, Butterworths, London (1963), p. 239; B. Knook, W. M. Star, H. J. M. van Rongen, and G. J. van den Berg, *Physics* **30**, 1124, 1964; *Commun. Kamerlingh Onnes Lab.*, Leiden, No. 339a.
18. W. B. Pearson, *Phil. Mag.* **46** (7), 911, 1955; G. K. White, *Can. J. Phys.* **33**, 119, 1955; J. S. Dugdale and D. K. C. MacDonald, *Can. J. Phys.* **35**, 271, 1957.
19. Y. Mutô, K. Noto, and F. T. Hedgcock, *Can. J. Phys.* **42**, 15, 1964.
20. H. J. M. van Rongen, B. Knook, and G. J. van den Berg (to be published in *Physica*).
21. A. Kjekshus and W. B. Pearson, *Can. J. Phys.* **40**, 98, 1962.
22. W. B. Pearson, *Phil. Mag.* **45** (7), 1087, 1954.
23. A. V. Gold, D. K. C. MacDonald, W. B. Pearson, and I. M. Templeton, *Phil. Mag.* **5**, 765, 1960.
24. A. N. Gerritsen, *Physica* **25**, 489, 1959; *Commun. Kamerlingh Onnes Lab.*, Leiden, No. 315c.
25. R. W. Schmitt, *Phys. Rev.* **96**, 1446, 1954.
26. D. K. C. MacDonald and W. B. Pearson, *Acta Met.* **3**, 392, 403, 1955.
27. R. R. Coltman, T. H. Blewitt, and S. T. Sekula, *Proc. colloq. int. métaux de très haute pureté*, C.N.R.S., Paris (1960), pp. 73, 77; S. T. Sekula, *Phys. Rev. Letters* **3**, 416, 1959.
28. C. A. Domenicali and E. L. Christenson, *J. Appl. Phys.* **31**, 1730, 1960.
29. H. H. Anderson and M. Nielsen, *Riso Report* 77, 1964; *Phys. Letters* **6**, 17, 1963.
30. Y. Mutô, Y. Tawara, Y. Shibuya, and T. Fukuroi, *J. Phys. Soc. Japan* **14**, 380, 1959.
31. F. T. Hedgcock, W. B. Muir, and E. Wallingford, *Can. J. Phys.* **38**, 376, 1960; G. Gaudet, F. T. Hedgcock, G. Lamarche, and E. Wallingford, *Can. J. Phys.* **38**, 1134, 1960.
32. W. B. Muir, *J. Phys. Soc. Japan* **16**, 2598, 1961.
33. Metalgroup, Leiden (to be published).

34. H. A. Boorse and H. Niewodniczanski, *Proc. Roy. Soc. (London), Ser. A* **153**, 463, 1936.
35. J. G. Thomas and E. Mendoza, *Phil. Mag.* **43**, 900, 1952.
36. B. C. Deaton, *Phys. Letters* **7**, 7, 1963.
37. F. T. Hedgcock and W. B. Muir, *Phys. Rev.* **129**, 2045, 1963.
38. E. W. Collings, F. T. Hedgcock, and Y. Mutô, *Phys. Rev.* **134**, A1521, 1964.
39. K. Yosida, *Phys. Rev.* **107**, 396, 1957.
40. G. Boato, G. Gallinare, and I. Rizzuto, *Phys. Letters* **5**, 20, 1963; Proc. Colgate Conf., *Rev. Mod. Phys.* **36**, 162, 1964.
41. E. W. Collings and F. T. Hedgcock, *Phys. Rev.* **126**, 1654, 1962.
42. F. T. Hedgcock and Y. Mutô, *Phys. Rev.* **134**, A1593, 1964.
43. T. Kasuya, *Progr. Theoret. Phys. (Kyoto)* **22**, 227, 1959.
44. J. Friedel, *Nuovo Cimento, Suppl.* **7**, 287, 1958 (Varenna Conf. 1957); *J. Phys. Radium* **19**, 573, 1958; *J. Phys. Radium* **23**, 692, 1962.
45. A. Blandin and J. Friedel, *J. Phys. Radium* **20**, 160, 1959.
46. P. W. Anderson, *Phys. Rev.* **124**, 41, 1961.
47. P. A. Wolff, *Phys. Rev.* **124**, 1030, 1961.
48. J. O. Linde, *Physica* **24**, S109, 1958.
49. R. D. Taylor, T. A. Kitchens, D. E. Nagle, W. A. Steyert, and W. E. Millet (to be published in *Solid State Commun.*).
50. E. W. Collings, F. T. Hedgcock, W. B. Muir, and Y. Mutô, *Phil. Mag.* **10**, 159, 1964.
51. B. Knook, *Proc. Tenth Int. Congress Refrigeration*, Copenhagen 1, 69, 1959; B. Knook and G. J. van den Berg, *Proceedings of the Seventh International Congress on Low Temperature Physics*, University of Toronto Press, Toronto, Canada (1961), p. 257.
52. T. G. Berlincourt, *Phys. Rev.* **114**, 969, 1959; R. R. Hake, D. H. Leslie, and T. G. Berlincourt, *Phys. Rev.* **127**, 170, 1962; T. G. Berlincourt, R. R. Hake, and A. C. Thorsen, *Phys. Rev.* **127**, 710, 1962.
53. R. R. Hake and D. H. Leslie (unpublished).
54. R. R. Hake, J. A. Cape, and D. H. Leslie (unpublished).
55. B. R. Coles, *Phil. Mag.* **8**, 335, 1963.
56. A. M. Clogston, B. T. Matthias, M. Peter, H. J. Williams, E. Corenzwit, and R. Sherwood, *Phys. Rev.* **125**, 541, 1962; J. Crangle, *Phil. Mag.* **5**, 335, 1960.
57. M. P. Sarachik, E. Corenzwit, and L. D. Longinotti, *Phys. Rev.* **135**, A1041, 1964; M. P. Sarachik, this volume, p. 1044.
58. A. D. Brailsford and A. Overhauser, *J. Phys. Chem. Solids* **15**, 140, 1960.
59. B. R. Coles, *Phys. Letters* **8**, 243, 1963.
60. J. Bijvoet, B. de Hon, J. A. Dekker, and G. W. Rathenau, *Solid State Commun.* **1**, 237, 1963.
61. T. Sugawara, R. Soga, and I. Yamase, *J. Phys. Soc. Japan* **19**, 780, 1964.
62. T. Sugawara, private communication.
63. W. B. Pearson, *Phil. Mag.* (7) **46**, 920, 1955.
64. W. R. G. Kemp, A. K. Sreedhar, and G. K. White, *Proc. Phys. Soc. (London) Ser. A* **66**, 1077, 1953; D. A. Spohr and R. T. Webber, *Phys. Rev.* **105**, 1427, 1957.
65. J. De Nobel, *Suppl. Bull. Inst. Intern. Froid, Annexe* **2**, 97, 1956; M. S. R. Chari and J. De Nobel, *Physica* **25**, 60, 84, 1959; *Commun. Kamerlingh Onnes Lab.*, Leiden, No. 313b and Suppl. No. 114b.
66. M. S. R. Chari, *Proc. Phys. Soc. (London) Ser. A* **78**, 1361, 1961.
67. E. W. Fenton, J. S. Rogers, and S. B. Woods, *Can. J. Phys.* **41**, 2026, 1963.
68. W. B. Teutsch and W. F. Love, *Phys. Rev.* **105**, 487, 1957.
69. T. Fukuroi and T. Ikeda, *Sci. Rept. Res. Inst. Tohoku Univ. Ser. A* **8** (3), 1956.
70. Yu. P. Gaidukov, *Zh. Eksperim. i Teor. Fiz.* **34**, 836, 1958. (*Soviet Phys. JETP (English Transl.)* **7**, 577, 1958.)
71. B. Franken and G. J. van den Berg, *Physica* **26**, 1030, 1960; *Commun. Kamerlingh Onnes Lab.*, Leiden, No. 324a.
72. J. Korringa and A. N. Gerritsen, *Physica* **19**, 357, 1953; *Commun. Kamerlingh Onnes Lab.*, Leiden, Suppl. No. 106.
73. N. E. Alekseevskii and Yu. P. Gaidukov, *Zh. Eksperim. i Teor. Fiz.* **32**, 1589, 1957 (*Soviet Phys. JETP (English Transl.)* **5**, 1301, 1957.)
74. M. D. Blue, *J. Appl. Phys.* **33**, 3060, 1962.
75. G. Borelius, W. H. Keesom, C. H. Johansson, and J. O. Linde, *Proc. Koninkl. Akad., Amsterdam* **33**, 32, 1930; **35**, 15, 25, 1932. *Commun. Kamerlingh Onnes Lab.*, Leiden, No. 206b, 217d, and 217e.
76. W. H. Keesom and C. J. Matthijs, *Physica* **2**, 623, 1935; *Commun. Kamerlingh Onnes Lab.*, Leiden, No. 238b; C. J. Matthijs, thesis, University of Leiden (1939); see also G. Borelius, *Physica* **19**, 333, 1953.

77. M. Bailyn, *Westinghouse Res. Rept. 029-B000-P1* (1961).
78. A. Sommerfeld, *Z. Physik* **47**, 1, 43, 1928; A. Sommerfeld and H. Bethe, *Handbuch der Physik* 24/2, 1933; N. F. Mott and H. Jones, *Theory of the Properties of Metals and Alloys*, Clarendon Press, Oxford, England (1936), Dover Publications, Inc., New York (1958); A. H. Wilson, *Theory of Metals*, Cambridge University Press, New York (1954).
79. A. R. de Vroomen, C. van Baarle, and A. J. Cuelenaere, *Physica* **26**, 19, 1960: *Commun. Kamerlingh Onnes Lab.*, Leiden, No. 319d; A. R. de Vroomen, thesis, University of Leiden (1959).
80. G. J. van den Berg, *Proceedings of the Fifth International Conference on Low Temperature Physics and Chemistry*, University of Wisconsin Press, Madison (1958), p. 489.
81. S. Tanuma, *J. Phys. Soc. Japan* **14**, 541, 1959.
82. L. Nordheim and C. J. Gorter, *Physica* **2**, 383, 1935.
83. F. T. Hedgcock and W. B. Muir, *J. Phys. Soc. Japan* **16**, 2599, 1961.
84. R. Bowers, *Westinghouse Res. Rept. 60-94469-2-P10* (1955); *Phys. Rev.* **102**, 1486, 1956.
85. E. Sonder and S. T. Sekula, *J. Phys. Chem. Solids* **21**, 315, 1961.
86. F. T. Hedgcock, *Phys. Rev.* **104**, 1564, 1956.
87. J. Owen, M. E. Browne, W. D. Knight, and C. Kittel, *Phys. Rev.* **102**, 1501, 1956.
88. J. Owen, M. E. Browne, V. Arp, and A. F. Kip, *J. Phys. Chem. Solids* **2**, 85, 1957.
89. A. van Itterbeek, R. Polluntier, and W. Peelaers, *Appl. Sci. Res., Sect. B* **7**, 329, 1958; A. van Itterbeek, W. Peelaers, and F. Steffens, *Appl. Sci. Res., Sect. B* **8**, 337, 1960.
90. O. S. Lutes and J. L. Schmit, *Phys. Rev.* **134**, A676, 1964.
91. A. Arrott, *Report Sci. Lab.*, Ford Motor Company, Dearborn, Michigan, p. 122.
92. See *Low Temperature Physics Vol. IV*, C. J. Gorter (ed.), North Holland Publishing Co., Amsterdam (1964) p. 226, Figs. 26 and 27.
93. E. Scheil, E. Wachtel, and A. Kalkuhl, *Ann. Phys.* **4** (7), 58, 1959.
94. E. Scheil, H. Specht, and E. Wachtel, *Z. Metallk.* **49**, 590, 1958.
95. A. R. Kaufman, S. T. Pan, and J. R. Clark, *Rev. Mod. Phys.* **17**, 87, 1945.
96. E. Vogt and D. Gerstenberg, *Ann. Phys.* **4** (7), 145, 1959.
97. E. W. Pugh, B. R. Coles, A. Arrott, and J. E. Goldman, *Phys. Rev.* **105**, 814, 1957; E. W. Pugh and F. M. Ryan, *Phys. Rev.* **111**, 1038, 1958.
98. J. A. Cape, *Phys. Rev.* **132**, 1486, 1963.
99. B. T. Matthias, *I.B.M. J. Res. Develop.* **6**, 250, 1962.
100. B. G. Childs, W. E. Gardner, and J. Penfold, *Phil. Mag.* **8**, 419, 1963.
101. J. A. Cope, this volume, p. 353.
102. R. R. Hake and J. A. Cape, *Phys. Rev.* **135**, A1151, 1964.
103. J. C. Fisher, reported in ref. 88 and in: R. W. Schmitt and I. S. Jacobs, *Can. J. Phys.* **34**, 1285, 1956; see also C. J. Gorter, G. J. van den Berg, and J. De Nobel, *Can. J. Phys.* **34**, 1281, 1956.
104. O. S. Lutes and J. L. Schmit, *Phys. Rev.* **125**, 433, 1962.
105. E. Wachtel and V. Vetter, *Z. Metallk.* **52**, 525, 1961.
106. J. Cohen, G. Quezel, and G. Rimet, *Berichte der Arbeitsgemeinschaft Ferromagnetismus*, Verlag Stahleisen G.M.B.H., Dusseldorf, 1960, p. 74.
107. J. De Nobel and F. J. du Chatenier, *Physica* **25**, 969, 1959; *Commun. Kamerlingh Onnes Lab.*, Leiden, No. 317c.
108. F. J. du Chatenier and J. De Nobel, *Physica* **28**, 181, 1962; G. J. van den Berg and J. De Nobel, *J. Phys. Radium* **23**, 665, 1962; J. De Nobel and F. J. du Chatenier, *Proceedings of the Eighth International Congress on Low Temperature Physics*, Butterworths, London (1963), p. 241.
109. F. J. du Chatenier, thesis, University of Leiden (1964); and articles to be published in *Physica*.
110. L. T. Crane, *Phys. Rev.* **125**, 1902, 1962.
111. J. E. Zimmerman and F. E. Hoare, *J. Phys. Chem. Solids* **17**, 52, 1960.
112. G. J. van den Berg, *Proceedings of the Seventh International Conference on Low Temperature Physics*, University of Toronto Press, Toronto, Canada (1961), p. 246.
113. J. P. Franck, F. D. Manchester, and D. L. Martin, *Proc. Roy. Soc. (London) Ser. A* **263**, 494, 1961.
114. L. T. Crane and J. E. Zimmerman, *Phys. Rev.* **123**, 113, 1961.
115. G. L. Guthrie, S. A. Friedberg, and J. E. Goldman, *Phys. Rev.* **111**, 45, 1959.
116. J. K. Logan, J. R. Clement, and H. R. Jeffers, *Phys. Rev.* **105**, 1435, 1957.
117. D. L. Martin, *Can. J. Phys.* **39**, 1385, 1961.
118. B. M. Boerstoel, F. J. du Chatenier, and G. J. van den Berg, this volume, p. 1071.
118a. B. W. Veal and J. A. Rayne, *Phys. Rev.* **115**, A442, 1964.
119. D. L. Martin, *Proc. Phys. Soc. (London) Ser. A* **78**, 1489, 1961.
120. F. J. Morin and J. P. Maita (unpublished).
121. S. H. Liu, *J. Phys. Chem. Solids* **24**, 475, 1963.
122. N. E. Phillips and B. T. Matthias, *Phys. Rev.* **121**, 105, 1961.

123. R. Tournier, J. J. Veyssié, and L. Weil, *J. Phys. Radium* **23**, 672, 1962; R. Tournier and L. Weil, *J. Phys. Radium* **23**, 522, 1962; J. le Guillerm, R. Tournier, and L. Weil, *Proceedings of the Eighth International Conference on Low Temperature Physics*, Butterworths, London (1963), p. 236.
124. R. Smoluchowski, *Phys. Rev.* **84**, 511, 1951; other references in H. C. van Elst, B. Lubach, and G. J. van den Berg, *Physica* **28**, 1297, 1962; *Commun. Kamerlingh Onnes Lab.*, Leiden, No. 334b.
125. A. W. Overhauser, *J. Phys. Chem. Solids* **13**, 71, 1960.
126. M. W. Klein and R. Brout, *Phys. Rev.* **132**, 2412, 1963.
127. W. Marshall, *Phys. Rev.* **118**, 1519, 1960.
128. K. Yosida, *Phys. Rev.* **106**, 893, 1957.
129. M. A. Ruderman and C. Kittel, *Phys. Rev.* **96**, 99, 1954.
130. F. J. du Chatenier and A. R. Miedema, this volume, p. 1029.
131. C. A. Domenicali, *Phys. Rev.* **117**, 984, 1960.
132. R. W. Schmitt, *Phys. Rev.* **103**, 83, 1956.
133. A. R. de Vroomen and M. L. Potters, *Physica* **27**, 1028, 1961.
134. A. M. Guénault and D. K. C. MacDonald, *Phil. Mag.* **6**, 1201, 1961; D. K. C. MacDonald, *Thermoelectricity and Introduction to Principles*, John Wiley & Sons, Inc., New York (1962).
135. A. J. Dekker, *Physica* **24**, 697, 1958; **25**, 1244, 1959. *J. Phys. Radium* **23**, 702, 1962.
136. A. D. Brailsford and A. W. Overhauser, *J. Phys. Chem. Solids* **15**, 140, 1960.
137. T. van Peski-Tinbergen and A. J. Dekker, *Physica* **29**, 917, 1963.
138. S. H. Liu, *Phys. Rev.* **132**, 589, 1963.
139. M. T. Béal, *J. Phys. Chem. Solids* **25**, 543, 1964.
140. J. Kondo, *Progr. Theoret. Phys.* (*Kyoto*) **32**, 37, 1964.
141. T. J. Norton and A. V. Seybolt, *Trans. Met. Soc. AIME* **230**, 595, 1964.
142. E. Daniel, this volume, p. 933.

A CORRELATION OF THE MÖSSBAUER ISOMER SHIFT AND THE RESIDUAL ELECTRICAL RESISTIVITY FOR ^{197}Au ALLOYS*

Louis D. Roberts, Richard L. Becker, F. E. Obenshain

Oak Ridge National Laboratory
Oak Ridge, Tennessee

and

J. O. Thomson

University of Tennessee
Knoxville, Tennessee

Introduction

In the theoretical discussion of metals, it is often convenient to describe the metallic sample by using a single potential well for the conduction electrons. The Bloch wave functions which extend throughout the entire sample give an approximate solution to the corresponding quantum mechanical problem. When an impurity is dissolved in the metal, the electron wave functions again will, in general, extend throughout the sample. We may then expect a correlation between different physical phenomena associated with the impurity which may each depend predominantly on the different regions of the wave function. For a suitable host and impurity, we should observe a correlation between the charge density at the impurity nucleus, $\rho(0)$, and the transport cross section per impurity atom, σ_{tr}, which the impurity atoms present to the host conduction electrons. For impurity atoms whose nuclei have resonant γ-ray transitions suitable for Mössbauer effect studies,[1] a measurement of the isomer shift will give information[2] about $\rho(0)$. For suitable dilute alloys, the measurement of the residual electrical resistivity per at. %, $\Delta R/c$, will give σ_{tr}. Thus, we may expect a correlation between the isomer shift, v_I, of the resonance γ-ray energy, E_γ, of the nucleus of an impurity in a dilute alloy with $\Delta R/c$ due to that impurity in the alloy.

The isotope ^{197}Au has a suitable γ-ray for recoilless radiation (Mössbauer) investigations[3] and, since gold is a noble metal, it is particularly suitable for a study of the correlation of v_I with $\Delta R/c$. Thus, if we assume that only the 6s-shell of gold is appreciably modified when gold is dissolved in a host metal, measurements of v_I and of $\Delta R/c$ may be used in conjunction to give information about the s-band wave function in the gold alloy system. In the following we present measurements of v_I and of $\Delta R/c$ for gold alloyed in a variety of host metals and show a correlation between these measured quantities, using a simple theoretical model which is based on the use of an impurity potential.[4] Information is obtained about the impurity potential and about the s-band filling.

* Research sponsored by the U.S. Atomic Energy Commission under contract with the Union Carbide Corporation.

The isomer shift v_I arises as a result of the Coulomb (electric monopole) interaction between the nucleus and the atomic s-electrons which penetrate it. In first-order perturbation theory, the Coulombic energy shift of a nuclear state x produced by a single Dirac s-electron is $D|\psi_s(0)|^2 \langle x|r^{2\sigma}|x \rangle$. Here, D is a calculable constant, $|\psi_s(0)|^2$ is the nonrelativistic electron density at the origin, and $\sigma = (1 - \alpha^2 z^2)^{\frac{1}{2}}$, where z is the nuclear charge and α is the fine structure constant. If the matrix elements $\langle x|r^{2\sigma}|x \rangle$ differ for the nuclear ground and excited states, E_γ will be modified by $D\rho_s(0)[\langle r^{2\sigma} \rangle_e - \langle r^{2\sigma} \rangle_g]$, where e and g refer to the excited and ground nuclear states and where $\rho_s(0) = \Sigma_i |\psi_{si}(0)|^2$ is the total nonrelativisitic charge density at the origin due to all of the s-electrons. Under the circumstances that $\rho_s(0)$ is different for the γ-ray source and for the absorber in the Mössbauer measurement, there will be an energy ΔE_{abs} which must be added to the source γ-ray energy to bring about resonance with the nuclei in the absorber, namely

$$\Delta E_{abs} = D[\rho_{s,abs}(0) - \rho_{s,source}(0)][\langle r^{2\sigma} \rangle_e - \langle r^{2\sigma} \rangle_g] \tag{1}$$

This energy increment may be supplied to E_γ as a Doppler energy. The required Doppler velocity will be $v_{abs} = c\Delta E_{abs}/E_\gamma$, where c is the velocity of light. It is convenient to refer these shifts and the equivalent Doppler velocities to pure gold as the absorber. The isomer shift for a gold alloy as absorber referred to pure metallic gold as the absorber and expressed as a Doppler velocity v_I is then

$$v_I = (v_{alloy} - v_{Au}) \tag{2}$$

$$= g[\rho'_{Au}(0) - \rho_{Au}(0)] \tag{3}$$

where $\rho_{Au}(0)$ is the charge density at a gold nucleus in metallic gold, $\rho'_{Au}(0)$ is that charge density at a gold nucleus for a gold atom in an alloy, and g is a constant. In general, primes will be used to designate quantities which describe alloy properties.

Experimental Results

Experimental values for v_I in mm/sec have been determined at 4.2°K as a function of gold concentration in the range from near zero to 100 at.% Au in binary alloys of gold with copper, silver, palladium, platinum, or nickel. Table I gives values for these alloys of $v_I(0)$, the isomer shift extrapolated to zero gold concentration. The experimental techniques used for these isomer shift measurements have been previously described.[3]

Experimental values for $\Delta R/c$ in microhm centimeters per atomic percent were measured at 4.2°, 77°, and 300°K for gold concentrations of 0.50, 1.00, and 2.00 at.% Au in each of the above metals as host. The $\Delta R/c$ values were found to be independent of concentration over the range investigated and of temperature for gold in copper, silver, palladium, and platinum. For nickel, $\Delta R/c$ was also independent of concentration for the three concentrations measured, but was found to rise nearly linearly and quite strongly with temperature from 0.38 $\mu\Omega$-cm/at.% at 4.2°K to 0.7 $\mu\Omega$-cm/at.% at 300°K. This temperature dependence of $\Delta R/c$ must be related to the ferromagnetism of nickel and gold–nickel.[5] Table I gives the values of $\Delta R/c$ measured at 4.2°K for the above alloy systems. Because of the ferromagnetism of nickel and gold–nickel, it may not be possible to interpret $\Delta R/c$ for the nickel–gold system in the same context with the alloys of gold with copper, silver, palladium, and platinum.

Table I. Parameters Used in the Correlation of the Isomer Shift with the Residual Resistance for Gold. The Phase Shifts δ_0 and δ_1 Are Given for the Case of a Potential Attractive for s-Waves and for the Valence η_A, also Given Below, which Results in the Best Correlation (Fig. 1). The Quantity $k_0^2/2$ Is the Corresponding Well Depth in Electron Volts.

Host metal	Experimental residual resistivity $\Delta\rho/c$, $\mu\Omega$-cm/at.%	Experimental isomer shift v_I, mm/sec	$\dfrac{\delta a}{a}$	Poisson's ratio σ	$1 + \dfrac{3}{\gamma}\dfrac{\delta a^h}{a}$	Host valence used η_A	Phase shifts δ_0	δ_1	Well depth $k_0^2/2$, eV
Copper	0.52 ± 0.02	4.4 ± 0.2	0.157^a	0.364^f	1.337	1.0	0.0562	-0.1951	0.305
Silver	0.36 ± 0.02	2.1 ± 0.2	-0.00735^b	0.37^f	0.984	1.0	0.2155	-0.06343	1.07
Nickel	0.38 ± 0.02	5.4 ± 0.2	0.209^c	0.31^f	1.397	0.6	0.2358	$+0.00598$	1.15
Palladium	0.70 ± 0.02	2.4 ± 0.2	$0.0528^{b,d}$	0.39^g	1.120	0.67	0.337	$+0.016$	1.61
Platinum	1.55 ± 0.02	1.4 ± 0.2	0.040^e	0.39^f	1.091	0.50	0.461	$+0.084$	2.07

[a] L. Vegard and A. Kloster, Z. Krist. **89**, 560, 1934; C. S. Smith, Min. and Met. **9**, 458, 1928.
[b] G. Sachs and J Weerts, Z. Phys. **50**, 481, 1930.
[c] E. C. Ellwood and K. Q. Bagley, J. Inst. Metals **80**, 617, 1951/52.
[d] V. G. Kuznecov, Izv. Sektora Platiny Akad. Nauk. SSSR **20**, 5, 1946; see Structure Reports, **10**, 54, 1945/46.
[e] A. S. Darling, R. A. Mintern, and J. C. Chaston, J. Inst. Metals **81**, 125, 1952/53.
[f] From the American Institute of Physics Handbook, McGraw-Hill Book Co., Inc, New York (1957).
[g] Assumed the value for platinum.
[h] $\gamma = 3(1 - \sigma)/(1 + \sigma)$.

Theoretical Discussion of $v_1(0)$ and of $\Delta R/c$

Basically, the γ-ray isomer shift of an impurity in an alloy may be viewed as an aspect of the screening of the impurity. In the case where the gold impurity is dissolved in another noble metal, the screening will presumably be predominantly by the conduction s-band of the host. In the case where the host is a transition metal, however, the situation may be more complex. We shall take the usual view that the eigenstates of the outer electrons in a transition metal host may be described by two bands, one of dominantly s and the other of dominantly d-character. For dilute gold alloys with the transition metals, we further assume (1) that the states of gold other than the $6s$-state, in particular the $5d$-states, are not sufficiently modified, relative to pure gold, to significantly alter their screening of the gold s-shells, (2) that only the s-band contributes to the isomer shift, e.g., we neglect any hybridization of gold s and $p_{\frac{1}{2}}$ states into the d-band, (3) that in a gold cell, the hybridization of gold p and d functions into the s-band is nearly the same as in pure gold, and (4) that the residual resistivity is attributable entirely to the s-band, because of the high effective mass of the d-band holes.

The interaction of the impurity atom with the electrons of the s-band of the host metal will be described by associating a perturbing pseudopotential with the impurity atom. The pseudopotential will give rise to a transport cross section σ_{tr} for the conduction electrons and will thus produce a residual electrical resistivity per atomic percent, $\Delta R/c$. From a measurement of $\Delta R/c$, σ_{tr} for electrons at the Fermi level may be obtained. Then, from σ_{tr} and the requirement that the impurity atom must be screened in an electrically conducting alloy as described through the Friedel sum rule[6] [see equation (11)], the effective potential for the $l = 0$ component of Bloch waves in the s-band may be estimated. Using this well, we may obtain a calculated value of $\Sigma_i |\psi_{si,\text{abs}}(0)|^2$, and thus of the isomer shift to within a constant which depends on the nuclear size change.

We consider the substitution of an impurity atom B for an atom A of the host metal at the origin of the coordinate system. The wave function Φ_k for the s-partial wave of an s-band electron in the alloy with energy E_k will be written as

$$\Phi_k(r) = \phi_k(r) U(r) \tag{4a}$$

where

$$U(r) = \begin{cases} U_B(r) & r \leq R \\ U_A(r) & r > R \end{cases} \tag{4b}$$

is assumed to be independent of k. Here, $U_A(r)$ or $U_B(r)$ is taken to be the wave function at the bottom of the s-band and within the Wigner–Seitz cell of the pure host metal A or of the pure solute metal B. The radius R will be taken as the radius, r_B, of the Wigner–Seitz cell of the pure metal B. It will be assumed that U_A and U_B have the same amplitude at the surface of their respective Wigner–Seitz spheres, i.e., $U_A(R) = U_B(R)$. The charge density $\rho'_{Bk}(0)$ at the origin, i.e., at the impurity nucleus B in the alloy, arising from the s-partial wave of a given $\mathbf{k}$ state is then

$$\rho'_{Bk}(0) = |U_B(0)|^2 \cdot |\phi_k(0)|^2 \tag{5}$$

where $\mathbf{k}$ is the electron crystal momentum vector. For pure gold,

$$\rho_{Bk}(0) = |U_B(0)|^2 \tag{6}$$

and thus the isomer shift arising from a given **k** state $v_{I\mathbf{k}}$ may be written

$$v_{I\mathbf{k}} = g|U_B(0)|^2(|\phi_k|^2 - 1) \tag{7}$$

To obtain the isomer shift, $v_{I\mathbf{k}}$ must be multiplied by the density of states $n(\mathbf{k})$ and integrated over filled **k**-states. Assuming $n(\mathbf{k})$ to be proportional to k^2, and where k_F is the s-band electron momentum at the Fermi surface of the host metal, we have

$$v_I = g_1|U_B(0)|^2 \int_0^{k_F} (|\phi_k(0)|^2 - 1)k^2 \, dk \tag{8a}$$

$$= g_2|U_B(0)|^2(\bar{P} - 1) \tag{8b}$$

where g_1 and g_2 are constants.

It is now necessary to obtain an estimate of $|\phi_k(0)|^2$. In the asymptotic region

$$\phi_k(r) = \sin[kr + \delta_0(k)]/kr \tag{9}$$

Here $\delta_0(k)$ is the phase shift of the s-partial wave arising from the effect of the impurity atom on a conduction s-band electron state of energy E_k. It will be assumed that at the Fermi surface only $\delta_0(k_F)$ and $\delta_1(k_F)$ (where δ_1 is the phase shift of the p-partial wave) are appreciably different from zero. Then an estimate of $\delta_0(k_F)$ and $\delta_1(k_F)$ may be obtained using the measured $\Delta R/c$ and the Friedel sum rule.

For infinitely dilute alloys, Huang[7] has given the result

$$\Delta R/c = (r_A/0.703\eta_A^{1/3}) \sum_l (l+1) \sin^2[\delta_l(k_F) - \delta_{l+1}(k_F)] \tag{10}$$

which describes $\Delta R/c$ in terms of the phase shifts $\delta_l(k_F)$ of the lth partial wave produced at the Fermi surface by the impurity. Here, r_A is the Wigner–Seitz cell radius in atomic units and η_A is the number of s-band electrons per atom for pure metal A. The Friedel sum rule

$$Z = \frac{2}{\pi} \sum_l (2l+1)\delta_l(k_F) \tag{11}$$

where Z is the difference of charge between impurity and host atoms, gives a second condition on the phase shifts. If it is assumed that only δ_0 and δ_1 are appreciable, and given the value for Z, the simultaneous solution of equations (10) and (11) will give an estimate of $\delta_0(k_F)$.

The difference of ionic charge Z between impurity and host, equation (11), is assumed to have the form

$$Z = (\eta_B - \eta_A) - \left(\frac{1+\sigma}{1-\sigma}\right)\left(\frac{\delta a}{a}\right)\eta_A \tag{12}$$

Here, η_B, like η_A which was defined earlier, is the number of s-band electrons per atom in the pure metal B, and $(\eta_B - \eta_A)$ gives the ionic charge on the impurity to be screened by the s-band of the host metal A. The second term on the right in (12) gives the contribution to Z which arises from the dilation of the host lattice A by the impurity B and is of the form used by Blatt.[8] Here, σ is Poisson's ratio and $\delta a/a$ is the change of lattice parameter in percent of metal A per atomic percent of metal B dissolved (Table I).

Then, to obtain an estimate of $|\phi_k(0)|^2$ we assume that $\delta_0(k_F)$ is produced by a square well of radius R and of a suitably adjusted depth $k_0^2/2$. Using this well, we may then obtain the result

$$|\phi_k(0)|^2 = \left[1 - \frac{k_0^2 R^2}{(k_0^2 + k^2)R^2} \sin^2(\sqrt{k_0^2 + k^2}R)\right]^{-1} \quad (13)$$

a form given previously by Daniel[9] in obtaining an estimate of the Knight shift in dilute alloys. The use here of (13) amounts to assuming only that, in its effect on s-waves over a limited range of energy, the pseudopotential may be replaced by a roughly equivalent square well. By substituting (13) into (8a) and performing the indicated integration, we may obtain an estimate of v_I from $\Delta R/c$.

There are in general two sets of solutions in obtaining $\delta_0(k_F)$ and $\delta_1(k_F)$ from $\Delta R/c$ and the sum rule, because equation (10) involves the squares of the sine functions. We present the results of the calculations for both sets of solutions. For the cases we have treated, both positive and negative values of $\delta_0(k_F)$ may occur. If the potential is weak, a positive $\delta_0(k_F)$ corresponds to an attractive potential for s-waves, and a negative $\delta_0(k_F)$ arises from a repulsive potential. Consequently, the two types of solution give strikingly different results for the isomer shift.

Comparison of Theory with Experiment

As shown in equation (8b), the isomer shift v_I is proportional to $(\bar{P} - 1)$. This $(\bar{P} - 1)$ is estimated from $\Delta R/c$ by the calculation outlined in the preceding section, where, in the calculation, there are two free parameters, the s-band filling in the host metal η_A, and the choice between the possible solutions of (10), (11), and (12), from which the impurity potential is obtained.

Because the experimental isomer shifts are all of the same sign, it may be seen from equation (3) that for all of the alloys studied $\rho'_{Au}(0) > \rho_{Au}(0)$ or for all of the alloys $\rho'_{Au}(0) < \rho_{Au}(0)$. In Fig. 1 the calculated $(\bar{P} - 1)$ is presented for gold alloyed in copper, silver, palladium, and platinum as a function of s-band filling for the case where the impurity potential is attractive (i.e., $\rho' > \rho$) and in Fig. 2 for the case where $\rho' < \rho$. The proportionality constant $G = g_2|U_B(0)|^2$ of (8b), which relates v_I and $(\bar{P} - 1)$, is obtained by normalizing the theoretical value of $\bar{P} - 1$ at $\eta_A = 1$ for gold in copper to the experimental value $v_I = 4.4 \pm 0.2$ mm/sec. This must be done separately for the attractive well solutions, $\rho' > \rho$, yielding a constant $G_A = 8.0$ mm/sec, and for the solutions with $\rho' < \rho$, giving a constant $G_R = -31$ mm/sec. The quantities $G_A(\bar{P} - 1)$ are also given as functions of η_A in Fig. 1, and the quantities $G_R(\bar{P} - 1)$ are given similarly in Fig. 2. For copper and silver, η_A is taken to be unity so that the experimental isomer shifts are plotted as points at $\eta_A = 1$ with the experimental errors indicated. For the three transition metal alloys, the value of η_A is less certain, so that the experimental isomer shifts are plotted as cross-hatched bands with heights equal to the experimental errors.

An examination of Fig. 1 shows a good measure of agreement between theory and experiment for the attractive potential case. In detail, experiment and theory agree for gold in palladium with a palladium s-band filling in the range 0.65 to 0.68. This is in quite reasonable agreement with the usual interpretation of the results of magnetic measurements on palladium. For the case of platinum as host, the s-band filling at which experiment and calculation agree (Fig. 1) is in the range 0.48 to 0.51. There have been several previous papers on the subject of the s-band filling in platinum. Some years ago, Wohlfarth[10] gave an interpretation of measurements

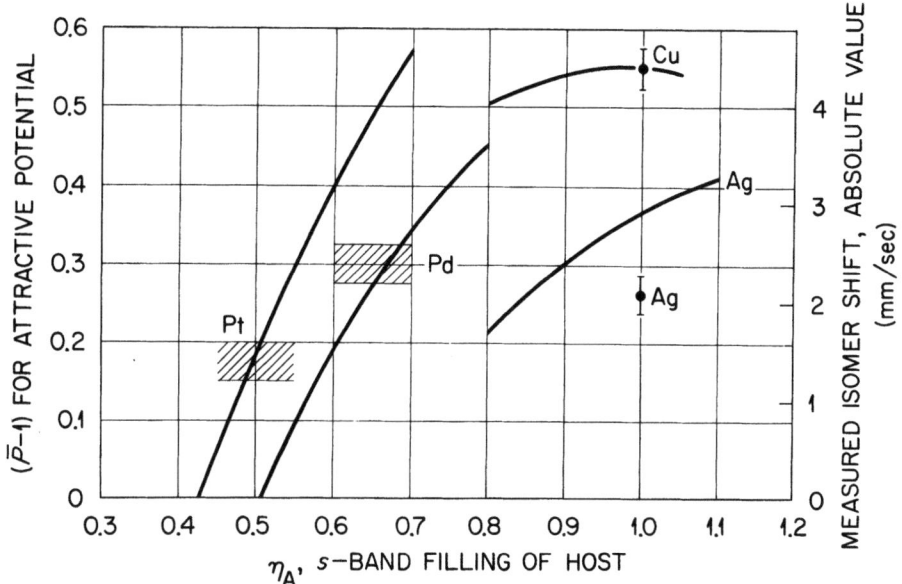

Fig. 1. Comparison between the calculated isomer shift $v_I = g_A(\bar{P} - 1)$ using an attractive impurity potential and the experimental isomer shift, Table I. The experimental values for v_I are represented by points or crosshatched areas. The calculated v_I are given by the solid curves where $g_A = 8.0$ mm/sec has been chosen to give agreement between theory and experiment for copper as host. Closer overall agreement is found here than for a repulsive potential (see Fig. 2).

of the magnetic susceptibility and of the specific heat of pure platinum in which it was suggested that the number of holes in the d-band was in the vicinity of 0.2 to 0.3. The number of electrons in the s-band would then be expected to lie in a similar range. A recent treatment of the Knight shift in platinum by Clogston et al.[11] also indicates an s-band filling in this range. A much higher value of 0.58 for this s-band filling has been suggested by Budworth, Hoare, and Preston[12] from a rigid band model interpretation of their specific heat measurements on platinum–gold alloys.

For the case of gold in silver (Fig. 1), our experimental isomer shift lies about 30% below the theoretical curve. The agreement between theory and experiment for the case of gold and silver could be made to appear better by changing the normalization constant to a somewhat lower value near $G_A = 7.0$ mm/sec. The experimental results for both gold in silver and gold in copper would then lie almost within their experimental errors of the theoretical curves, and the results for the s-band filling in palladium and platinum would be very little changed.

For the case of gold in nickel, at $\eta_A = 0.6$ the experimental result lies about 60% higher than the theoretical curve. In our calculation of the isomer shift from the electrical resistance, ferromagnetism of the host was not taken into account in any way. We find a reasonable agreement between theory and experiment for the paramagnetic or nonmagnetic cases but not where ferromagnetism occurs. As was observed earlier, $\Delta R/c$ for gold in nickel is strongly temperature dependent.[5] While this temperature dependence must be due mainly to magnetic scattering, it is not absolutely clear that $\Delta R/c$ at $T = 0°K$ gives a useful measurement of the transport cross section associated with the gold impurity in the context of our model.

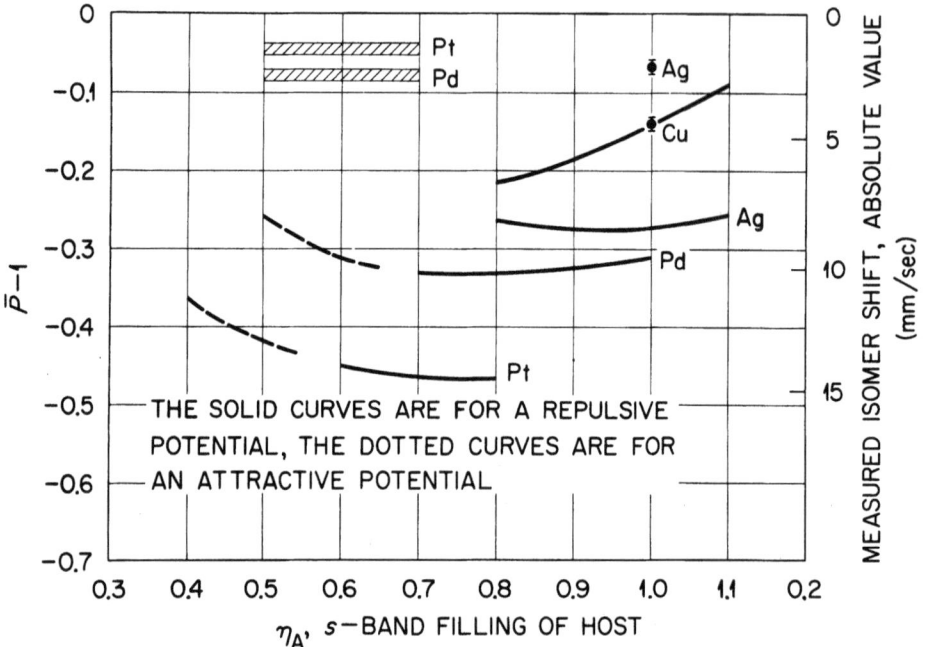

Fig. 2. Comparison between isomer shifts $v_I = -g_R(\bar{P} - 1)$ calculated using a repulsive (or, in the cases of palladium and platinum for small values of η_A, a slightly attractive) s-wave impurity potential, and the experimental isomer shifts of Table I. The experimental values for v_I are represented by points or crosshatched areas. The calculated v_I are given by the solid curves where $g_R = -31.0$ mm/sec has been chosen to give agreement between theory and experiment for copper as host. Little agreement is found, whereas in the case of an attractive potential a close overall agreement is found (see Fig. 1).

Referring to Fig. 2, we may now investigate the comparison between experiment and the theoretical model for the case where $\rho' < \rho$. With theory and experiment again normalized for gold in copper we find $G_R = -31$ mm/sec. Here, however, for gold in silver, palladium, and platinum, the theoretical values for $|v_I|$ are all greater than for gold in copper while the experimental values are all less. There is no agreement for these cases between theory and experiment for any value of the s-band filling η_A. Considering all of these cases, we realize that the use of a repulsive well in the above calculation is not consistent with experimental results.

The results presented in Fig. 1 thus indicate that in gold alloys or compounds where v_I is greater than zero, the electron charge density near the gold nucleus is greater than the density in pure gold, $[\rho'_{Au}(0) - \rho_{Au}(0)] > 0$. This result is in agreement with the conclusion of Mott[13] that gold when alloyed in silver should present an attractive potential to valence band electrons. From a study of the correlation between the isomer shift of gold in a variety of alloys with the electronegativity difference between gold and the host metal, Shirley et al.[14] have also concluded that $v_I > 0$ corresponds to $[\rho'_{Au}(0) - \rho_{Au}(0)] > 0$. We note from equation (1), in agreement with Shirley et al., that with $[\rho'_{Au}(0) - \rho_{Au}(0)] > 0$ we have

$$(\langle r^{2\sigma}\rangle_e - \langle r^{2\sigma}\rangle_g) > 0$$

This result is in agreement as to sign with the description of ^{197}Au given by Zeldes,[15] using the shell model.

In the use of the theoretical model given above to describe the isomer shift, the range of the pseudopotential was taken to be the radius of a gold atom, r_B. Strictly, this assumption is not required in the above theoretical model. We have thus repeated the above calculation for an R of $0.25r_B$, $0.50r_B$, $0.75r_B$, and $1.25r_B$. The best correlation between isomer shift and residual resistance was found for the case presented in detail, namely, $R = r_B$.

References

1. R. L. Mössbauer, *Z. Physik* **151**, 124, 1958.
2. O. C. Kistner and A. W. Sunyar, *Phys. Rev. Letters* **4**, 412, 1960.
3. L. D. Roberts and J. O. Thomson, *Phys. Rev.* **129**, 664, 1963. L. D. Roberts, H. Pomerance, J. O. Thomson, and C. F. Dam, *Bull. Am. Phys. Soc.* **7**, 565, 1962.
4. L. D. Roberts, R. L. Becker, and J. O. Thomson, *Bull. Am. Phys. Soc.* **8**, 42, 1963; R. L. Becker, L. D. Roberts, and J. O. Thomson, *Bull. Am. Phys. Soc.* **8**, 558, 1963; L. D. Roberts, R. L. Becker, F. E. Obenshain, and J. O. Thomson (to be published).
5. L. D. Roberts, F. E. Obehshain, R. L. Becker, and J. O. Thomson, *Bull. Am. Phys. Soc.* **9**, 398, 1964.
6. J. Friedel, *Phil. Mag.* **43**, 153, 1952; *Advan. Phys.* **3**, 446, 1954. P. deFaget de Casteljau and J. Friedel, *J. Phys. Radium* **17**, 27, 1956. J. Friedel, *Nuovo Cimento* **7**, Ser. X, 287, 1958.
7. K. Huang, *Proc. Phys. Soc. (London) Ser. A* **60**, 161, 1948.
8. F. J. Blatt, *Phys. Rev.* **108**, 285, 1957; **108**, 1204, 1957.
9. E. Daniel, *J. Phys. Chem. Solids* **10**, 174, 1959.
10. E. P. Wohlfarth, *Proceedings of the Leeds Philosophical and Literary Society, Sci. Sect.* (1948), pp. 89–101.
11. A. M. Clogston, V. Jaccarino, and Y. Yafet, *Phys. Rev.* **134A**, 650, 1964.
12. D. W. Budworth, F. E. Hoare, and J. Preston, *Proc. Roy. Soc. (London) Ser. A* **257**, 250, 1960.
13. N. F. Mott, *Proc. Camb. Phil. Soc.* **32**, 281, 1936.
14. D. A. Shirley, *Rev. Mod. Phys.* **36**, 339, 1964; P. H. Barrett, R. W. Grant, M. Kaplan, and D. A. Shirley, *J. Chem. Phys.*, **39**, 1035, 1963.
15. N. Zeldes, *Nucl. Phys.* **2**, 1, 1956/57.

10.2. Magnetic Scattering, Localized Moments, Hyperfine Splitting, Other Magnetic Phenomena

ON THE QUESTION OF MAGNETIC ORDERING IN DILUTE ALLOYS

A. M. de Graaf and R. Luzzi

Department of Physics, University of São Paulo
São Paulo, Brazil

Dilute copper–manganese alloys show an apparent antiferromagnetic behavior at low temperatures. Marshall tried to explain this phenomenon on the basis of the familiar Ruderman Kittel–Yosida indirect exchange interaction between the magnetic ions. Overhauser,[1] on the other hand, suggested that in these alloys, the conduction electrons system is not in the paramagnetic plane wave Hartree–Fock state, but instead develops a static spin density wave (SDW) of the form

$$\bar{P}_0 \cos(\bar{q}_m \cdot \bar{r})$$

The exchange interaction between the magnetic impurities and the SDW should overcome the SDW excitation energy

$$\mu^2 \frac{n^2 p_0^2}{4\chi(q_m)}$$

where $\chi(q)$ is the spin susceptibility of the electron gas in momentum space and q_m is the wave number for which $\chi(q)$ has a maximum or a singularity. In this paper we calculate $\chi(q)$ of an interacting electron gas, using the generalized random phase approximation. The susceptibility is determined by the solutions of an integral equation, which describes the scattering of quasi-particles. We show, using a variational expression for the susceptibility, that, for electron densities typical of normal metals, $\chi(q)$ has a maximum if not a singularity for a finite value of q. If $\chi(q)$ has a singularity, it probably occurs for values of q at least two times the Fermi momentum. We thus show that a SDW may indeed represent a stable state.

Reference

1. A. W. Overhauser, *Phys. Rev.* **128**, 1437, 1962.

MAGNETIC SCATTERING IN DILUTE METALS*

F. T. Hedgcock and D. Mathur

Franklin Institute Laboratories
Philadelphia, Pennsylvania

Recently, Toyozawa[1] has presented a theory for the experimentally observed negative magnetoresistance in heavily doped semiconductors exhibiting metallic impurity conduction.[2,3] This theory requires a localized spin in the impurity band which interacts with the free charge carriers to give rise to the same kind of magnetic interaction which results in resistive and magnetoresistive anomalies observed in dilute paramagnetic metal alloys. The statistical nature of the donor or acceptor impurity distribution in the semiconductor results in a spin-dependent potential which under conditions of metallic conductivity may give rise to an electron state of sufficient lifetime to have a spatially localized spin.

Thermoelectric measurements in the liquid-helium range have been made on n-type samples of germanium doped with antimony in the concentration limits of 5×10^{16} and 5×10^{18} charge carriers. Thermoelectric anomalies have been observed in these samples and the magnitude of the anomaly is two to three times the magnitude of the normal electron diffusion term

$$\pi^2 k^2 T/|e|\zeta \sim -(3 \to 4)\mu V/\text{degK at } 4.2°\text{K}$$

If the magnetoresistance data on the same samples are used, it is possible to estimate a value of the ratio of the Coulomb scattering integral to the s–d exchange integral ($A/J \approx 36$). With this information, a theoretical estimate of the thermoelectric power based on the theory of Kasuya[4] can be made, and the theoretical curve normalized to unit concentration of localized spin carriers is found to agree qualitatively with the experimentally determined temperature dependence of the thermoelectric power. The number and moment of the localized spin systems necessary to produce the magnitude of the experimental thermoelectric anomaly agree with the theoretical estimates of Toyozawa.

References

1. Toyozawa, *J. Phys. Soc. Japan* **17**, 986, 1962.
2. W. Sasaki, C. Yamamouchi, and G. M. Hatoyama, International Conference on Semiconductors, 1960, Prague.
3. W. Sasaki and R. de Bruyn Ouboter, *Physica* **27**, 877, 1961.
4. T. Kasuya, *J. Phys. Soc. Japan* **17**, 630, 1962.

---

* This research was supported by the Aeronautical Systems Division under contract AF33(657)–8744.

INELASTIC IMPURITY SCATTERING OF ELECTRONS IN GOLD ALLOYS AT LOW TEMPERATURES*

D. H. Damon and P. G. Klemens

Westinghouse Research Laboratories
Pittsburgh, Pennsylvania

The electrical resistivity of some binary gold alloys with copper and platinum was measured from 1.5° to 40°K, in order to study deviations from Matthiessen's rule and to investigate a possible contribution of phonon-assisted impurity scattering to the resistance. This process was originally suggested by Koshino,[1] who ascribed it to the perturbing effect of the displacement of the impurity potential by a lattice wave. This was criticized by Taylor,[2,3] who showed that the displacement perturbation could be transformed away, leaving only a small residual effect on the resistivity. However, there is a further effect arising from the strain on the impurity site due to a lattice wave, which should also lead to phonon-assisted scattering, and to an additional resistance of the form[4]

$$\Delta\rho = A\rho_0 \langle \varepsilon^2 \rangle = \beta(T)\rho_0 \tag{1}$$

where ρ_0 is the residual resistivity, A a numerical constant, and $\langle \varepsilon^2 \rangle$ the mean-square thermal strain, so that $\beta(T) \propto T^4$ at low temperatures.

Annealed wires of 0.7 to 2.5×10^{-2}-cm diameter and about 20-cm length were loosely wound on a lava sample holder. The mounted samples were sealed in a copper chamber filled with helium gas at a pressure of a few centimeters of mercury, and their resistance measured in terms of the voltage developed by a known current between two potential probes spot-welded to the wire. The absolute resistivity (accurate to $\pm 0.3\%$) was determined by measuring the voltage drop per unit length, and weighing measured lengths for each sample in a separate experiment at room temperature. Between 1.5° and 4.2°K, and between 14° and 20.4°K, measurements were made with the sample chamber immersed in liquid helium or hydrogen. In order to reach other temperatures above 4.2°K, the sample chamber, fitted with a cold finger, was raised above a liquid-helium bath. The temperature was controlled by balancing the heat produced by a heater wound on the sample chamber against the heat conducted to the bath. The chamber had a styrofoam jacket to reduce internal thermal gradients, and temperatures were then measured by means of gold–cobalt thermocouples, with their reference junction at 4.2°K. Measurements were also made this way from 14° to 20.4°K and agreed quite well with measurements in liquid hydrogen. In each run an alloy specimen was measured together with a pure gold wire ($\rho_0 < 10^{-8}$ Ω-cm).

---

* Sponsored by the Air Force Office of Scientific Research of the Office of Aerospace Research under Contract No. AF 49(638)–1165.

Table I. Solute Content and Residual Resistivity of the Gold Specimens and Alloys

Sample	Solute	$\rho_0 \times 10^6$ Ω-cm
1	—	0.00779
2	—	0.00910
3	Platinum	0.2230
4	Copper	0.2715
5	Platinum	0.4943
6	Platinum	0.9239
7	Copper	1.136
8	Platinum	1.576

Table I describes the samples used. Figure 1 shows the measured values of $\rho - \rho_0$ as a function of T. For the pure gold specimens these values are in good agreement with previous work. The total resistivity of the alloys behaved regularly except for sample No. 3, which showed a slight minimum, so that its ρ_0 is somewhat uncertain. The measured values of $\rho(T) - \rho_0$ for gold may be identified with the ideal resistivity $\rho_i(T)$. It is clear that $\rho - \rho_0$ of the alloys is substantially larger than ρ_i, and increases with ρ_0.

However, we cannot attribute the entire change to an additional scattering mechanism, since positive deviations from Matthiessen's rule must be expected whenever two or more groups of electrons are involved in conduction and are affected differently by the scattering mechanisms. This is exemplified by a two-band model;[5] in the case of gold, the role of the two bands is played by different parts of the Fermi surface. The resulting deviations can be expressed as

$$\rho - \rho_0 - \rho_i(T) = \alpha(\rho_0, T)\rho_i(T) \tag{2}$$

and it can be shown readily that α increases rapidly with ρ_0, reaching a saturation value of α^0 in the limit $\rho_i \ll \rho_0$ which is independent of temperature if the ideal resistivity of both bands has the same temperature dependence, but which in general should show some temperature dependence. Together with the resistance of the phonon-assisted impurity scattering, which should affect the two bands in the same ratio as the elastic impurity scattering, we would expect deviations from Matthiessen's rule of the form

$$\rho - \rho_0 - \rho_i(T) = \alpha(\rho_0, T)\rho_i(T) + \beta(T)\rho_0 \tag{3}$$

where α attains a value $\alpha^0(T)$ in the limit $\rho_0 \gg \rho_i$. Thus $(\rho - \rho_0)$ should tend toward a linear dependence on ρ_0 for large values of ρ_0, though this limit is rapidly pushed toward high values of ρ_0 as T is increased.

Figure 2 shows two examples of such plots for $T = 14°K$ and $T = 17°K$. Between 10° and 15°K with ρ_0 between 0.2×10^{-6} and 1.6×10^{-6} Ω-cm, the linear relation is indeed found; from the slope and intercept we can deduce $\beta(T)$ and $\alpha^0(T)$. At higher temperatures the analysis of the deviations into the two components becomes increasingly uncertain. It is remarkable that both copper and platinum solutes yield the same deviations as function of ρ_0 at the lower temperatures, though divergences are noted at higher temperatures. It is found that α^0

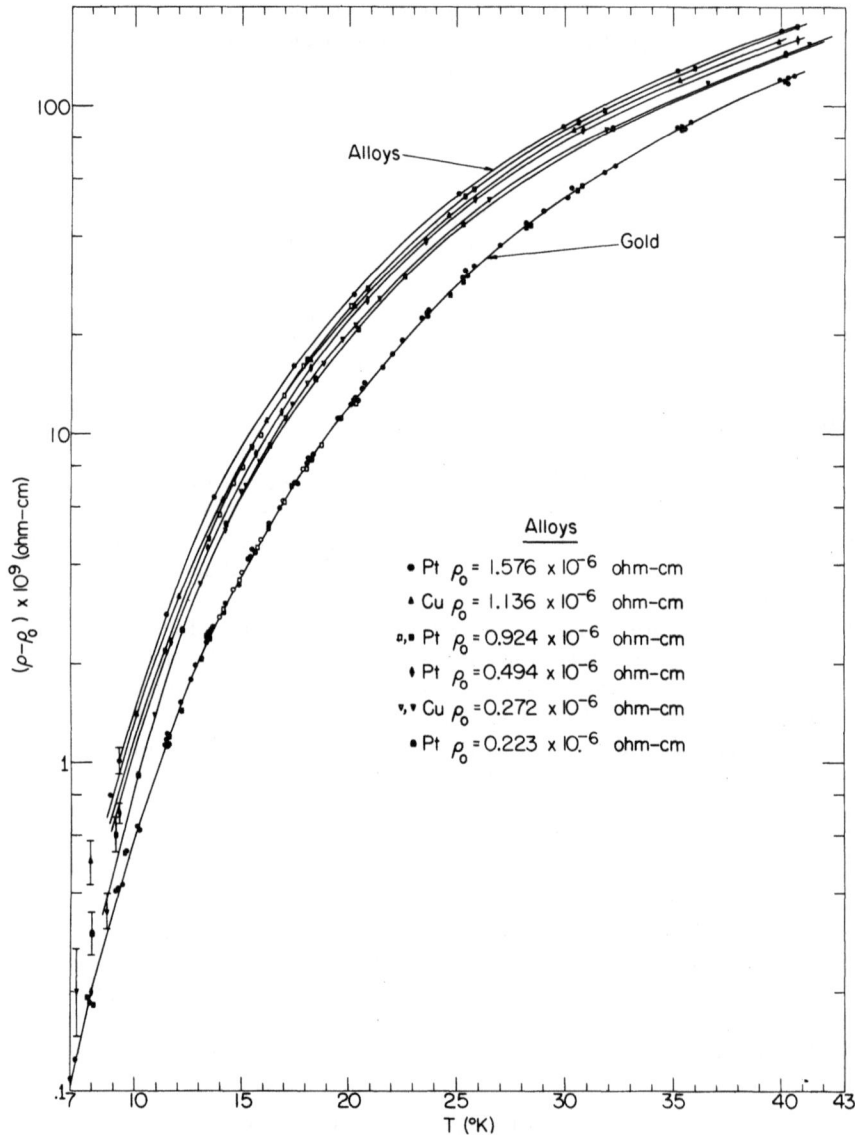

Fig. 1. Values of $\rho - \rho_0$ vs. T. A log scale is used for $\rho - \rho_0$ and a linear scale for T only for convenience in displaying the results. The lower curve shows the behavior of the pure gold specimens (samples 1 and 2). The upper curves show the results for the alloys and are distinguished in the legend. Open symbols are points determined by hydrogen vapor pressure, full symbols by thermocouples.

increases with temperature from 0.4 at 10°K to 0.6 at 16°K. This is not unexpected, although a detailed theory has not yet been worked out. It must also be recognized that a two-band model cannot fully describe the real situation, and that its deficiencies should become more apparent, the larger ρ_i, i.e., the higher T.

The values of $\beta(T)$ thus found are shown in Fig. 2 as functions of T. Below about 15°K, $\beta(T)$ varies at T^4, as expected theoretically.[4] Equating β with $A\langle\varepsilon^2\rangle$ at these

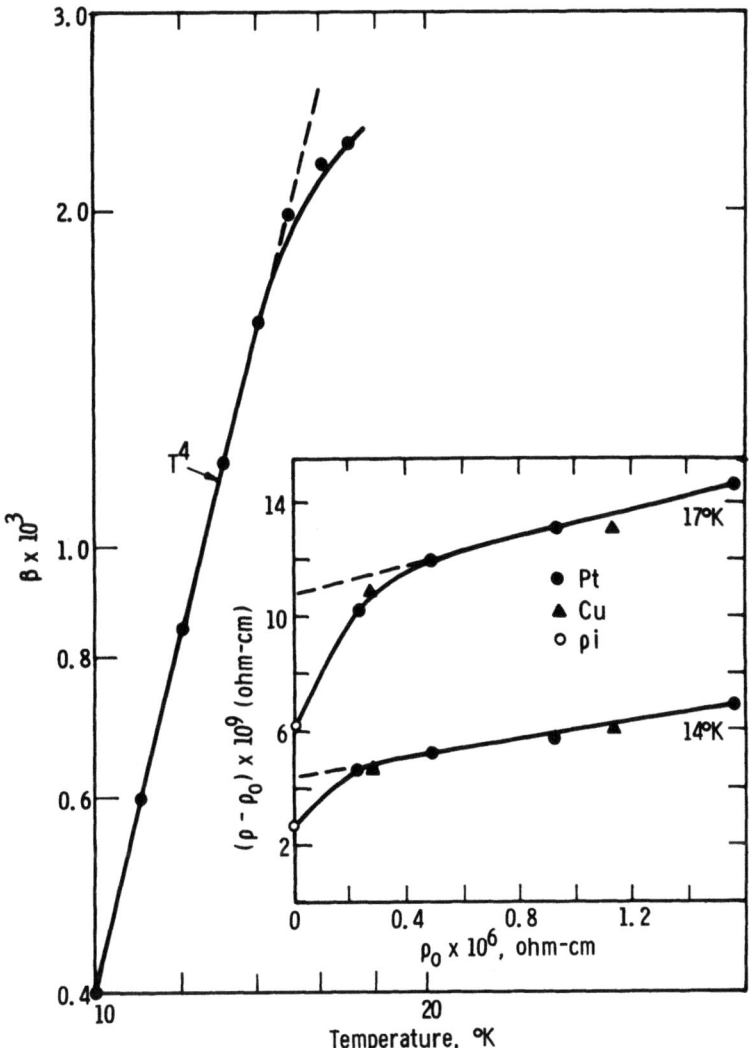

Fig. 2. Values of β vs. T. α and β are defined by $\rho - \rho_0 - \rho_i = \alpha\rho_i + \beta\rho_0$. The inset shows values of $\rho - \rho_0$ plotted against ρ_0 at 14° and 17°K. The slope (of each plot) of $\rho - \rho_0$ vs. ρ_0 at large values of ρ_0 is β, and α is obtained from the extrapolated intercept at $\rho_0 = 0$.

temperatures, we find $A \simeq 100$, and is thus somewhat larger than the theoretical estimate of 20. This implies that the impurity potential is very sensitive to the lattice strain. At higher temperatures, β seems to increase more slowly than $\langle\varepsilon^2\rangle$. Again, this may be a fault of the present method of separating the deviation from Matthiessen's rule into two components, but may be, at least in part, the result of a decrease of A with phonon frequency.

References

1. S. Koshino, *Progr. Theoret. Phys. (Kyoto)* **24**, 484, 1049, 1960.
2. P. L. Taylor, *Proc. Phys. Soc. (London) Ser. A* **80**, 755, 1962.
3. P. L. Taylor, *Phys. Rev.* (in press).
4. P. G. Klemens, *J. Phys. Soc. Japan* **18**, suppl. II, 77, 1963.
5. E. H. Sondheimer and A. H. Wilson, *Proc. Roy. Soc. (London) Ser. A* **190**, 435, 1947.

ANOMALOUS ELECTRON SCATTERING IN DILUTE MAGNETIC ALLOYS

S. H. Liu*

IBM Watson Research Center
Yorktown Heights, New York

Recently Kondo explained the resistance minimum in dilute alloys by the s–d scattering model. He showed that the second-order Born approximation to the scattering cross section contains a log T term which gives a good fit to the observed resistance at low temperatures. This term arises as a consequence of the dynamical properties of the spin operators and the sharp variation in electron population at the Fermi level. In this paper the Kondo theory is extended to show that the resistance anomaly is independent of other scatterings in the system and that the magnetoresistance contains a term like log H in strong enough fields. The effect of higher-order scattering on superconductivity is also studied. It is found that the inclusion of the Kondo term does not modify the qualitative conclusions of the existing theory.

We study the properties of a dilute alloy described by the following simplified model Hamiltonian:

$$H = H_0 + H' + H'' \qquad (1)$$

where H_0 is the free-electron term, H' is the s–d exchange term, and H'' represents phonon and nonmagnetic scattering interactions. In units with $\hbar = 1$, we may write

$$H_0 = \sum_\kappa \varepsilon_k c_\kappa^* c_\kappa \qquad (2)$$

$$H' = \sum_{\kappa\kappa'} H'_{\kappa'\kappa} c_{\kappa'}^* c_\kappa \qquad (3)$$

where $\kappa = \mathbf{k}, s$ is an index for both the momentum and the spin state of an electron, $\varepsilon_k = k^2/2m$, and $H'_{\kappa'\kappa}$ contains the spin operator of the impurity ions, e.g.,

$$H'_{\mathbf{k}'\uparrow\mathbf{k}\uparrow} = -N^{-1}J \sum_j S_j^z e^{i(\mathbf{k}-\mathbf{k}')\cdot \mathbf{R}_j} \qquad (4)$$

The quantity $2J$ is the s–d exchange constant, N the total number of ions in the sample, $\mathbf{R}_j, \mathbf{S}_j$ are the position and spin of the jth impurity, respectively, and the sum is over all the impurities. We define a Green function in the familiar manner

$$G_\kappa(\tau) = \langle T c_\kappa(\tau) c_\kappa^*(0) \rangle \qquad (5)$$

where

$$c_\kappa(\tau) = e^{-\tau(\mu\eta - H)} c_\kappa e^{\tau(\mu\eta - H)} \qquad (6a)$$

$$\eta = \sum_\kappa c_\kappa^* c_\kappa \qquad (6b)$$

* Present address: Department of Physics, Iowa State University, Ames, Iowa.

μ is the Fermi energy, and T is the ordering operator for τ. The bracket denotes thermal average. This Green function can be Fourier-analyzed

$$G_\kappa(\tau) = \frac{1}{\beta} \sum_\kappa G_\kappa(\omega_n) e^{-i\omega_n \tau} \tag{7}$$

where $\omega_n = (2n + 1)\pi/\beta$ and n is an integer. In general, $G_\kappa(\omega_n)$ has the form

$$G_\kappa(\omega_n) = [\varepsilon_\kappa - \mu - \Sigma(\omega_n) - i\omega_n]^{-1} \tag{8}$$

where $\Sigma(\omega_n)$ is the self-energy. In the present problem we may write

$$\Sigma(\omega_n) = \Sigma'(\omega_n) + \Sigma''(\omega_n) \tag{9}$$

where Σ' comes from H' interaction and Σ'' from H''. The diagrams for the first three orders of Σ' are shown in Fig. 1. The lines denote electron Green functions and the crosses represent the impurities. The cross terms between different impurities are ignored. The diagrams may be evaluated to give, after averaging over the random orientations of the impurity spins,

$$\Sigma'(\tau) = N^{-1} J^2 cS(S+1) \sum_{\mathbf{k}'} G_{\mathbf{k}'}(\tau) - N^{-2} J^3 cS(S+1) \sum_{\mathbf{k}'\mathbf{k}''} \int_0^\tau G_{\mathbf{k}''}(\tau - \tau') G_{\mathbf{k}'}(\tau') \, d\tau'$$

$$+ N^{-2} J^3 cS(S+1) \sum_{\mathbf{k}'\mathbf{k}''} \int_\tau^\beta G_{\mathbf{k}''}(\tau - \tau') G_{\mathbf{k}'}(\tau') \, d\tau' \tag{10}$$

where c is the impurity concentration, and the spin index on the Green functions has been suppressed because of complete degeneracy. The self-energy due to H'' is supposed to be known. Equations (10) and (8) may be solved simultaneously for G_κ and Σ. If we take the discontinuity of the self-energy across the imaginary frequency axis, we find the total level width

$$\Gamma(\omega) = \Gamma'(\omega) + \Gamma''(\omega) \tag{11}$$

Again, $\Gamma'(\omega)$ comes from H', and $\Gamma''(\omega)$ from $H''(\omega)$. The detail of this calculation will be published elsewhere.

The conductivity is related to the level width by

$$\sigma = -\frac{ne^2}{m} \int \frac{1}{\Gamma(\omega)} \frac{\partial f(\omega)}{\partial \omega} \, d\omega \tag{12}$$

where n is the electron density, and $f(\omega)$ is the Fermi distribution function. The result of this calculation is that the resistivity

$$\rho = \rho_0' \left(1 + \frac{3zJ}{\mu} \log \frac{kT}{\mu}\right) + \rho'' \tag{13}$$

where ρ_0' is the resistivity due to s–d interaction in the first Born approximation, z is the valence of the solvent metal, and ρ'' is the resistivity due to H''. This result

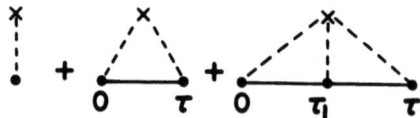

Fig. 1. The diagrams for the magnetic scattering contribution to the self-energy of conduction electrons.

is in complete agreement with Kondo.[1] However, this calculation shows that the resistance anomaly is independent of other scattering mechanisms that may be present in the alloy. In fact, the anomaly depends not on the sharpness of the Fermi surface as Kondo predicts, but rather on the abrupt change in electron population at the Fermi level.

The same procedure may be carried out to discuss the magnetoresistance of the material. We need only to add a Zeeman term for the impurity spins

$$-\omega_0 \sum_j S_j^z \tag{14}$$

in the Hamiltonian. The quantity ω_0 is the Zeeman splitting. The result may be written in the form

$$\rho(H) = \rho_0' \left(1 + \frac{3zJ}{\mu} \log \frac{kT}{\mu}\right)[1 - (S + 1)^{-1} B_s(\beta\omega_0) \tanh \tfrac{1}{2}\beta\omega_0]$$

$$- \rho_0'[S(S + 1)]^{-1} \frac{3zJ}{\mu} \int_{-\infty}^{\infty} G(u, \beta\omega_0) \, du + \rho''(H) \tag{15}$$

where $B_s(\beta\omega_0)$ is the Brillouin function and $G(u, \beta\omega_0)$ is a complicated function.

For $\omega_0 \gg kT$, the second term approaches

$$-\rho_0' \frac{6zJ}{\mu} \frac{\langle (S^z)^2 \rangle}{S(S + 1)} \log \beta\omega_0 \tag{16}$$

This term depends on the field like $\log H$ and is roughly 5% of ρ_0' for a copper–manganese alloy at 1°K and 20kG of field. Therefore, its effect should be experimentally observable.

We may also discuss the effect of anomalous scattering on superconductivity by using the matrix Green function for the electrons. The result shows that all the results of the Abrikosov and Gor'kov[2] paper are valid except that we should replace the spin flip scattering time τ_s by τ_s' where

$$\frac{1}{\tau_s'} = \frac{1}{\tau_s}\left(1 + \frac{3zJ}{\mu}\log\frac{kT}{\mu}\right) \tag{17}$$

Since the dependence of τ_s' on the temperature is a weak one, the inclusion of anomalous scattering does not produce any new effect.

In rare-earth ions there is a strong spin-orbit coupling in the 4f-shell. As a result the effective s–f coupling constant is $(g - 1)J$, where g is the gyromagnetic ratio. Since the resistance anomaly depends on the sign of the effective coupling constant and the quantity $(g - 1)$ has different sign for light and heavy rare earths, a systematic study of the resistivity of dilute rare earth alloys should give a critical test of the anomalous scattering theory. As a by-product, it also gives a direct measurement of the sign of the s–f coupling constant.

Acknowledgment

The author wishes to thank Dr. Jun Kondo for communicating to him his theory of anomalous scattering prior to its publication.

References

1. J. Kondo (to be published in *Progr. Theoret. Phys. (Japan)*).
2. A. A. Abrikosov and L. P. Gor'kov, *Zh. Eksperim. i Teor. Fiz.* **39**, 1781, 1960. (*Soviet Phys. JETP* (English Transl.) **12**, 1243, 1961.)

ANOMALOUS PROPERTIES OF DILUTE MAGNETIC ALLOYS

Jun Kondo

Electrotechnical Laboratory
Tanashi, Tokyo, Japan

Dilute alloys of transition-metal elements in noble metals show a number of anamolous behaviors at low temperatures, of which the most striking may be the resistance minimum and the giant thermoelectric power. These are striking in that they occur at sufficiently high temperatures compared with the exchange interaction between spins of transition-metal atoms. Thus, theories[1-3] which assumed an important part played by correlation or ordering of localized spins in accounting for these anomalies are incorrect. We have demonstrated[4] that the cross section of scattering due to impurity atoms, when calculated to the second Born approximation, involves a term which depends on the energy of the conduction electron in a singular way. The singularity of the dependence comes from the sharpness of the Fermi surface and the dynamical character of spin operators, but is not a consequence of correlation of spins. This energy-dependent cross section could account for the resistance minimum when combined with that due to lattice scattering.

Thermoelectric power is also a consequence of energy dependence of the cross section. The dependency required to account for it, however, is quite different from that required for the resistance minimum. Let us take zero-energy of the conduction electron at the Fermi surface. Then the cross section may be split into the even part and the odd part. The even part, unless it is a constant, gives rise to a temperature dependence of resistivity, while the odd part to thermoelectric power. We shall show that an odd function is obtained in the second Born approximation when the impurity potential contains an imaginary part. This function, which will turn out to be the Fermi distribution function $f(\varepsilon)$, has a large derivative at the Fermi surface and accounts for many features of the giant thermoelectric power. We shall shortly see that this term is a consequence of the Fermi statistics for the conduction electrons and the imaginary part of the impurity potential, but the exchange scattering is not essential, as it was to the resistance minimum. Thus the giant thermoelectric power is also expected in nonmagnetic alloys, which is actually the case. The characteristic temperature at which the thermoelectric power ceases to be of such a large value is determined in our case by the condition that the lattice scattering becomes appreciable, compared with the impurity scattering. In the old theories,[2,3] this temperature had to be the Néel or Curie temperature, which is actually much lower.

We shall consider the matrix element of the transition from the state k to the state k'. It is given in the first Born approximation by

$$H_{k'k} = \int \Psi_{k'}^*(r) \sum_n V(r - R_n) \Psi_k(r) \, dv$$
$$= N^{-1} \sum_n e^{i(k-k')R_n} V \qquad (1)$$

In the second Born approximation we have[4]

$$U_{k'k} = H_{k'k} + \sum_q \frac{H_{k'q}H_{qk}}{\varepsilon_k - \varepsilon_q + is}(1 - f_q) - \sum_q \frac{H_{qk}H_{k'q}}{\varepsilon_q - \varepsilon_{k'} + is} f_q$$

$$= H_{k'k} + \sum_q \left[P\left(\frac{1}{\varepsilon_k - \varepsilon_q}\right) + i\pi\delta(\varepsilon_k - \varepsilon_q)(2f_q - 1)\right] H_{k'q}H_{qk} \quad (2)$$

The principal part, when integrated over q, has a very little dependence on $\varepsilon_k (d/d\varepsilon_k$ gives $\sim 1/\varepsilon_F$). Thus we shall neglect it. We shall assume that the impurity atoms are distributed at random so that the coherence of waves scattered from different impurities may be neglected. The summation over q may be replaced by an integral $\int \ldots N(\varepsilon_q) d\varepsilon_q$. Then, schematically we may write

$$U_{k'k} = N^{-1} \sum_n e^{i(k-k')R_n} [V + i\pi\rho(\varepsilon_k)V^2(2f_k - 1)] \quad (3)$$

where $\rho(\varepsilon_k) = N(\varepsilon_k)/N$.

As stated above, this expression involves the Fermi distribution function. This does not immediately mean, however, that we shall have a large thermoelectric power. The transition probability is given by the absolute square of (3). The imaginary part, when squared separately, gives rise to an even term. Here we shall assume, and later actually show, that the impurity potential V contains an imaginary part. We shall replace V by $V + iK$. Then the absolute square of (3) becomes

$$|U_{k'k}|^2 = N^{-2} \sum_n |V + iK + i\pi\rho(\varepsilon_k)V^2(2f_k - 1)|^2$$

$$= (c/N)[V^2 + K^2 + 2\pi K\rho(\varepsilon_k)V^2(2f_k - 1) + \pi^2\rho(\varepsilon_k)^2 V^4(2f_k - 1)^2] \quad (4)$$

where c is the concentration of impurity.

The relaxation time τ_k is given by

$$\tau_k^{-1} = (2\pi/\hbar)|U_{k'k}|^2 N(\varepsilon_k) + W_l$$

$$= W_0 + W_1(2f_k - 1)^2 + W_2(2f_k - 1) + W_l \quad (5)$$

where W_l is the transition probability due to lattice scattering, which we assume is independent of ε_k but depends on T in order to reproduce the temperature dependence of the lattice resistivity. Following a standard theory[5] of transport phenomena in metals, we have for the resistivity R and thermoelectric power S

$$R = (e^2 K_1)^{-1} \quad (6)$$

$$S = -(K_2 - \varepsilon_F K_1)/eK_1 T \quad (7)$$

where

$$K_n = -\tfrac{2}{3} \int_0^\infty \tau(\varepsilon) v^2 \varepsilon^{n-1}(df/d\varepsilon) N(\varepsilon) d\varepsilon$$

Substituting from (5), assuming W_1 and W_2 smaller than $W_0 + W_l$, we have

$$R = \tfrac{3}{2} e^2 v_F^2 N(\varepsilon_F)[W_0 + W_l + (W_1/3)] \quad (8)$$

$$K_2 - \varepsilon_F K_1 = [2v_F^2 N(\varepsilon_F) W_2/3(W_0 + W_l)^2] \int (2f - 1)(\varepsilon - \varepsilon_F)(df/d\varepsilon) d\varepsilon$$

$$= [2v_F^2 N(\varepsilon_F) W_2/3(W_0 + W_l)^2] kT \quad (9)$$

Then, we have

$$S \cong -(k/e)[W_2/(W_0 + W_l)]$$
$$= -(k/e)(W_2/W_0)(R_{\text{residual}}/R)$$
$$\cong -(k/e)2\pi K\rho(\varepsilon_F)(R_{\text{residual}}/R) \tag{10}$$

From (10) we see that the thermoelectric power is independent of c and constant at low temperatures and tends to zero when the lattice resistance dominates over the residual resistance. These seem to agree with the observations made on copper–manganese[6] and gold–iron[7] alloys.

We can easily see why (10) is independent of T at low temperatures. The derivative of $f(\varepsilon)$ at the Fermi surface is of the order of $1/kT$, whereas in the normal metals the relevant derivative is of the order of $1/\varepsilon_F$, so that in our case we may expect $S \cong S_0(\varepsilon_F/kT)$. Since $S_0 \cong (k/e)(kT/\varepsilon_F)$, we have $S \cong k/e$, which is 86 $\mu V/\text{deg}$, quite large. For copper–manganese alloys we see $S = -5\,\mu V/\text{deg}$ at low temperatures,[6] so that $K \sim 0.09$ eV. This is much smaller than V, which must be larger than 1 eV in order to account for the residual resistivity of the alloy.

The most feasible way to obtain the imaginary part K is to use Anderson's model[8] of the s–d covalent mixing. The Hamiltonian is the sum of H_0 for the free electron and H', the s–d mixing effect. Nonperturbative scattering matrix is given by

$$H' \frac{1}{\varepsilon + is - H_0 - H'} H'$$

of which the k'–k element is

$$\frac{V_{sd}}{N^{\frac{1}{2}}} \left\langle d \left| \frac{1}{\varepsilon + is - H_0 - H'} \right| d \right\rangle \frac{V_{sd}}{N^{\frac{1}{2}}}$$
$$= \left(\frac{V_{sd}^2}{N}\right) \frac{1}{\varepsilon - \varepsilon_d + i\Delta}$$

where Δ is the width of the d-level. A more detailed account of the theory for copper–manganese alloys will be published elsewhere. Our derivation of the giant thermoelectric power, however, does not depend on a special model, provided the impurity potential has an imaginary part.

References
1. A. D. Brailsford and A. W. Overhauser, *J. Phys. Chem. Solids* **15**, 140, 1960.
2. T. Kasuya, *Progr. Theoret. Phys. (Kyoto)* **22**, 227, 1957.
3. A. R. de Vroomen and M. L. Potters, *Physica* **27**, 1083, 1961.
4. J. Kondo, *Progr. Theoret. Phys. (Kyoto)* **32**, 37, 1964.
5. A. H. Wilson, *The Theory of Metals*, Cambridge University Press, New York (1953).
6. A. Kjekshus and W. B. Pearson, *Can. J. Phys.* **40**, 98, 1962.
7. D. K. C. MacDonald, W. B. Pearson, and I. M. Templeton, *Proc. Roy. Soc. (London) Ser. A* **266**, 161, 1962.
8. P. W. Anderson, *Phys. Rev.* **124**, 41, 1961.

FERROMAGNETISM OF DILUTE SOLUTIONS OF COBALT IN PALLADIUM*

R. D. Dunlap, J. G. Dash, P. M. Higgs, D. G. Howard, and J. D. Siegwarth

University of Washington
Seattle, Washington

Previously reported measurements of ferromagnetic transitions in solid solutions of the cobalt–palladium system cover the concentration range from 100 to 0.1% Co.[1] The most dilute sample was reported to have a Curie temperature of 7°K and the extrapolation of the T_c vs. concentration curve seemed to indicate that a transition would occur for arbitrarily small concentrations of cobalt. The present work is an extension of these results in an attempt to determine whether there is a critical concentration below which ferromagnetism does not exist at $T = 0$.

Previous studies[2] have demonstrated that the Mössbauer spectra of ^{57}Fe impurities are indicators of the intensity of magnetization of ferromagnetic materials. In order to improve the sensitivity of detection of the ferromagnetic transition temperatures, we developed modifications of the conventional Mössbauer-effect techniques. The principal element in the new method is the use of a resonant absorber of greater energy width than the natural width of the zero-phonon radiation, which has been described elsewhere.[3] This absorber is interposed between the sample containing ^{57}Co and the detector. When the source is paramagnetic, the Mössbauer spectrum consists of a single line having an energy within the energy limits of the wide absorber, and, accordingly, most of the sharp radiation is absorbed. Below the Curie temperature the spectrum becomes split into six sharp lines having a maximum overall separation much greater than the width of the absorber, which therefore transmits a much higher fraction of sharp radiation.[4]† Thus the total transmitted intensity is strongly dependent on the saturation magnetization of the sample. Determination of the Curie temperature in this manner has several inherent advantages: The transition is observed directly rather than depending on an extrapolation procedure; the necessity of applying an external magnetic field, which can produce large changes in the magnetization near the Curie temperature, is avoided; the method is a local field measurement and so is independent of domain effects. This technique has particular utility for the study of dilute cobalt–palladium solutions, since the ^{57}Fe is produced by the electron capture decay of ^{57}Co, and the required activity implies very low concentrations of ^{57}Co. If absorbers of different energy widths are used, it is possible to emphasize different portions of the magnetization vs. temperature curve, and in particular it is possible to emphasize the region in the immediate neighborhood of the Curie temperature so that it may be located with great precision. The technique is suitable for quantitative determinations of the magnetization by analysis of the transmitted intensity in terms of the detailed

---
* Work supported by the Air Force Office of Scientific Research, Grant Nos. AF-AFOSR-62-298 and AF-AFOSR-594-64.

† This method has also been used for thermocouple calibration, using the Curie temperature as a fixed point.

shape of the absorber spectrum, and the intensity ratios and relative positions of the hyperfine spectral components.

Samples having concentrations varying from 4.5 to 0.07% were prepared by melting 99.999% Co with 99.999% Pd in an evacuated quartz capsule. The pellets obtained were rolled into sheets and portions of them were sputtered to a thickness of 0.2 mil in an attempt to ensure the homogeneity of the material. The Curie temperatures of the melted and rolled samples were observed to be the same as that of the sputtered samples; however, the transition for the rolled samples is much broader, indicating that the sputtering process does increase the homogeneity. Approximately 0.1 mC of ^{57}Co was diffused into the samples for 1 hr at 1000°C. This amount of activity is estimated to give an impurity concentration of from 0.005 to 0.01% and should be the largest impurity in the samples.

We have calculated an intensity curve which would arise from a sample whose magnetization follows a Brillouin function and which is filtered by the LiFeF$_4$ absorber used in all these experiments. The result of this calculation is shown by the dashed line in Fig. 1. The solid points give actual data taken for a sample containing 0.19% Co and show the result, typical for all our samples, that the transition region is much broader than the calculated curve. Such behavior might be expected for a spatially disordered alloy where some regions of concentration higher than the average will begin to magnetize at a relatively high temperature, and other regions of concentration lower than the average will begin to magnetize only after a lower temperature has been reached.

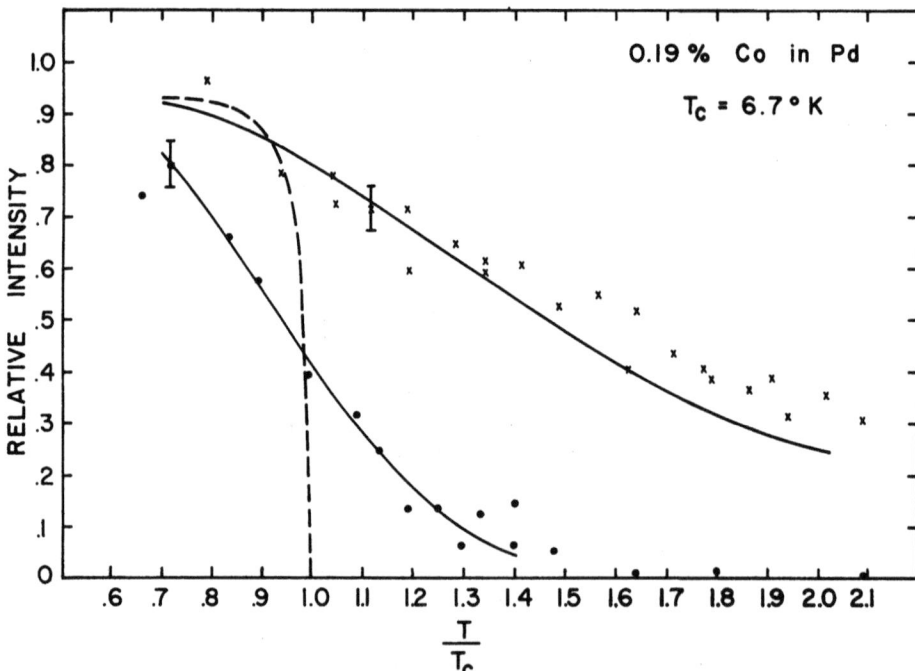

Fig. 1. Mössbauer intensity vs. temperature. The solid points give data taken in the absence of an external field and the crosses give data taken in the presence of 1.6 kG. The dashed line is the intensity anticipated from a magnetization that grows according to a $J = \tfrac{13}{2}$ Brillouin function. The solid lines are theoretical curves obtained as described in the text.

To see if this could be the case, calculations have been made on a highly simplified model, but one which should be at its best for the very dilute materials. We consider the sample to be divided into cells whose individual magnetic properties are determined only by their own concentrations and are not affected by interaction with the other regions. If the sample is statistically disordered, then the probability of finding a particular concentration in some cell is given by a Gaussian centered on the average value. If the T_c vs. concentration curve does not have an excessively high curvature and if the width of the Gaussian is not too large, we can approximate a linear relation between concentration and Curie temperature. In such a case the probability of finding a particular transition temperature in some cell will also be given by a Gaussian centered on the Curie temperature of the average concentration. If the magnetization of each cell varies according to a Brillouin function, then intensity curves similar to the dashed line of Fig. 1 will be obtained from each cell. The total intensity at any temperature is obtained by adding up the appropriate intensity contributions for that temperature from each cell, or what is equivalent, by integrating over all intensities appropriately weighted by the Gaussian. The dependence of the hyperfine field on temperature and on applied field has been measured for iron impurities in palladium,[5] and these results indicate that the magnetic moment associated with the iron atoms is localized and has a spin of about $\frac{13}{2}$. We used these results for our case of iron impurities in dilute cobalt–palladium, which means taking $J = \frac{13}{2}$ in each cell, and performed the numerical integration and fitted the data by adjusting the width of the Gaussian and the mean Curie temperature. The results obtained for the 0.19% sample are shown by the lower solid line in Fig. 1. The statistical widths obtained for all samples vary as (cobalt concentration)$^{-\frac{1}{2}}$, which is the result that should be obtained from samples which are only statistically disordered.

The mean Curie temperatures are given in Table I, as well as in Fig. 2. The most dilute sample thus far investigated has a cobalt concentration of 0.07% and shows no transition above 1.5°K, as has been indicated in the figure. The critical concentration, below which ferromagnetism does not exist, given by the extrapolated line is 0.05%, which corresponds to an average cobalt spacing of ~ 13 lattice sites.

Data similar to the above have been taken in the presence of permanent magnetic fields of a few hundred Gauss. In this case one expects the magnetization for a homogeneous ferromagnet to vary according to

$$\frac{M}{M_0} = B_J \left[\frac{3J}{(J+1)} \frac{M}{M_0} \frac{T_c}{T} + \frac{M_0 H}{NkT} \right] \quad (1)$$

where H is the applied field, N is the number of magnetic atoms present, and $B_J(x)$ is the appropriate Brillouin function. In the case of strong ferromagnets the extra

Table I. Mean Curie Temperatures of the Alloys

Cobalt concentration, at.%	Mean Curie temperature, °K
4.5	183 ± 1
1.91	87 ± 1
0.49	18.7 ± 3
0.19	6.7 ± 0.5
0.07	<1.5

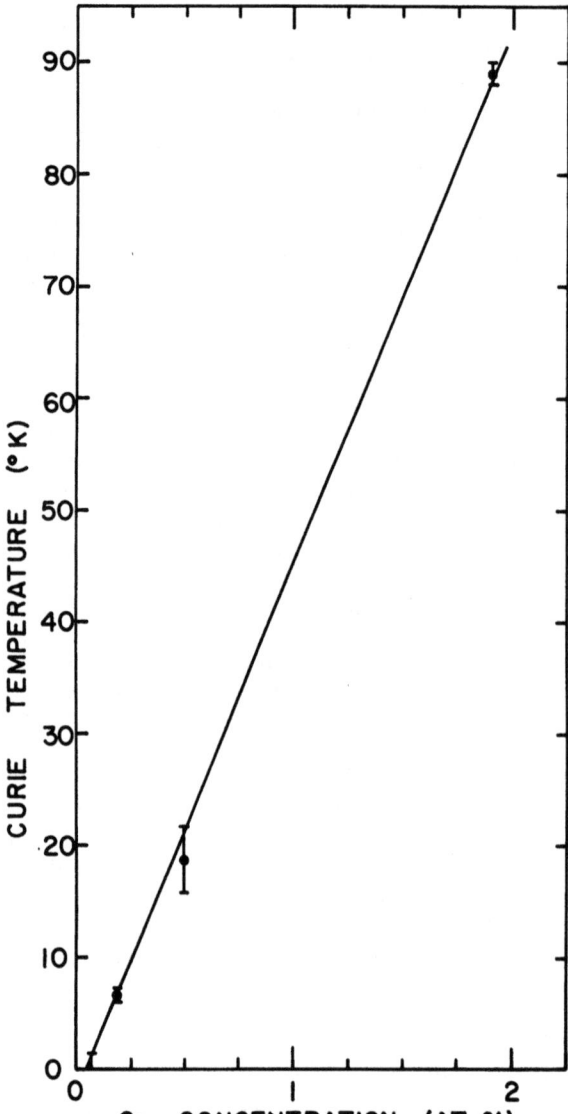

Fig. 2. Mean Curie temperature for alloys.

term will be observable only near the Curie temperature where the magnetization is small, or when H is very large. However, in the present case of dilute ferromagnets, M is always small enough so that a rather small field can produce a substantial change. According to equation (1), the applied field acts such as to widen the transition region into a long tail showing no sharp increase in magnetization. Data taken for the 0.19% sample in 1.6 kG are given by the crosses in Fig. 1. A calculation like the one previously described has been done using equation (1) for the temperature dependence of M with $J = \frac{13}{2}$, and using the mean Curie temperature and Gaussian width obtained for the case of zero applied field. In this calculation, M_0/N is a parameter to be adjusted, and the curve obtained is shown in Fig. 1 by the upper solid line. There has been no normalization adjustment of the intensity in the plot. The value of the moment obtained is 13 ± 2 magnetons, which agrees well with the value of 12.6 ± 0.4 obtained in a pure palladium matrix.[5]

If the same experiment is performed in two different fields, it is possible to estimate the Curie temperature of a sample whose transition has not been directly observed. This has been done for the 0.07% sample and we anticipate that $T_c \simeq 1.2°$K, which agrees with the extrapolation in Fig. 2. An empirical check of this value is now in preparation.

For all samples below ~ 2%, the Curie temperatures which we obtain are substantially lower than those obtained by Bozorth et al.[1] For example, the previous temperature for 0.19% was $\sim 12°$, while we obtain $6.7°$. Using reasonable values of the parameters in a calculation such as the one described above, we have attempted to reproduce the magnetization curves given by Bozorth et al.,[1] and find that the published magnetization curves are considerably broader than those we calculate. For this reason, we believe the discrepancies in the Curie temperatures to be due to a lack of sufficient homogeneity in those samples.

References

1. R. M. Bozorth, P. A. Wolff, D. D. Davis, V. B. Compton, and J. H. Wernick, *Phys. Rev.* **122**, 1157, 1961.
2. D. E. Nagle, H. Frauenfelder, R. D. Taylor, D. R. F. Cochran, and B. T. Matthias, *Phys. Rev. Letters* **5**, 364, 1960.
3. R. M. Housley, N. E. Erickson, and J. G. Dash, *Nucl. Instr. Methods* **27**, 29, 1964.
4. R. S. Preston, S. S. Hanna, and J. Heberle, *Phys. Rev.* **128**, 2207, 1962.
5. P. P. Craig, D. E. Nagle, W. A. Steyert, and R. D. Taylor, *Phys. Rev. Letters* **9**, 12, 1962.

LOCALIZED MOMENTS OF VERY DILUTE IRON IMPURITIES IN METALS FROM MÖSSBAUER EFFECT STUDIES OVER THE RANGE 0.5° TO 300°K*

R. D. Taylor, T. A. Kitchens, and W. A. Steyert

*Los Alamos Scientific Laboratory, University of California
Los Alamos, New Mexico*

Susceptibility measurements of various alloys of the second row transition metals containing 1% iron have demonstrated the presence of a localized moment on the iron atom having a magnitude of up to $12.7\mu_B$ (Bohr magnetons) in the neighborhood of palladium.[1] Clogston et al.[1] further showed a good correlation between the average number of electrons n outside of the last closed shell of the host alloy and the observed localized moment; a smaller localized moment occurs in the neighborhood of molybdenum ($n = 6$) and is zero in regions $4 < n < 5.5$ and $7 < n < 8.2$. We have utilized the Mössbauer effect to determine the localized moment on very dilute ^{57}Fe (generally less than 0.1%) in palladium,[2] copper,[3,†] and several other elements reported here. Our method is to study the hyperfine spectra at the iron impurity nucleus as a function of applied magnetic field H and temperature T. The magnitude of the splitting determines an effective field H_{eff} at the nucleus, a field composed of H plus an internal field H_i, which is proportional to the time average of the atomic spin vector $\mathbf{J}$ on the iron atom.[2]

From a complete dependence of H_i on H and T we can determine H_{sat}, the value of H_i for a fully aligned moment; μ, the localized magnetic moment on the iron; J, the spin characteristic of iron in its environment; and s, a parameter which represents a coupling between J and an assumed electronic field moving randomly in the host material.[3,6]

Sources were usually prepared by depositing high-purity ^{57}CoCl$_2$ in HCl (^{57}Co is the parent of the Mössbauer ^{57}Fe) on the host material to be studied. The deposit was then dried, reduced in hydrogen–argon at 300°–600°C, and finally diffusion-annealed into the host in a hydrogen or a hydrogen-inert gas atmosphere (usually 3 hr at a temperature of $\sim\frac{2}{3}$ of the melting temperature of the host) followed by a high-vacuum treatment at some reduced temperature. The source was washed to remove any activity on the surface and the average depth diffusion took place was determined by comparing the intensity ratio of 14.4 keV γ rays to the 124 keV γ rays (resulting from the ^{57}Co decay) before and after diffusion took place. All sources prepared as outlined gave narrow unsplit lines in zero external field down to 0.5°K;

---

* Work performed under the auspices of the U.S. Atomic Energy Commission.
† Brief preliminary results for iron in copper, titanium, niobium, gold, and platinum have been presented elsewhere.[4] N. Blum et al.[5] have made similar measurements generally using alloys containing 1% ^{57}Fe as absorbers.

thus no spontaneous magnetic ordering occurs for the impurity concentrations involved.

The sources were mounted on a heater–thermometer assembly located in a superconducting solenoid operating in liquid helium and capable of producing 62 kOe. By proper control of exchange gas and heater current, the sample could be maintained at any temperature between 1.1° and 300°K. A different holder allowed the use of liquid ^{3}He to achieve a temperature as low as 0.42°K.

The 14.4 keV γ-radiation was detected by a xenon-filled proportional counter and was analyzed by a ^{57}Fe-enriched $K_4Fe(CN)_6 \cdot 3H_2O$ (unsplit) absorber Doppler-shifted by a feedback loudspeaker system used in conjunction with a 400-channel analyzer to give directly an undistorted Mössbauer counts vs. velocity spectrum.[2]

The data for the magnetic behavior of iron in various hosts fall into several broad categories: (1) no localized moment on the iron; (2) pure paramagnetism associated with a large localized moment on the iron; and (3) a localized moment which is coupled to some randomly moving electronic field of the host metal.

Case (1) is realized for iron in vanadium, niobium, and tantalum, all of which are superconductors in the fifth column ($n = 5$). We find $H_{eff} \cong H$ at all temperatures studied implying that $\mu = 0$ for these materials. The concentration of magnetic impurities in one vanadium source is estimated to be as high as 0.4%, in the other, 0.05%. Susceptibility measurements gave $\mu = 0$ for 1% iron in niobium.[1]

Case (2) has been reported in detail for iron in palladium.[2] It is found that H_i is a function only of H/T and can be well represented by a Brillouin function B_J characteristic of a system of isolated spins. The uppermost curve of Fig. 1 is such a theoretical B_J curve for $J = 2$. H_{sat} is determined from the limiting behavior at

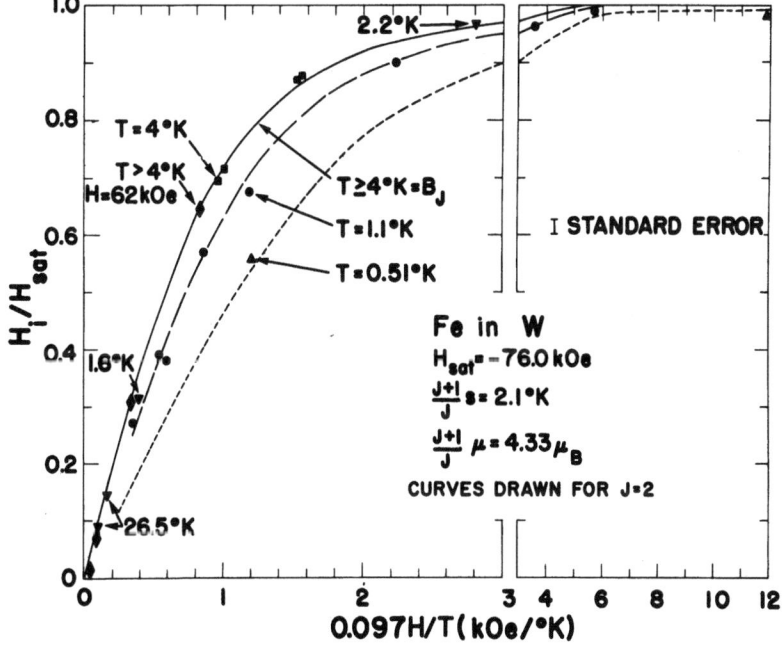

Fig. 1.

high H/T; the initial slope at low H/T determines the quantity $\mu(J+1)/J$, and J is determined by finding the B_J curve best fitting the data at all H/T. From the best fit of the palladium data, Craig et al.[2] found $\mu = 12.6 \pm 0.4\mu_B$, a local moment in good agreement with the susceptibility results. Measurements for iron in platinum (also $n = 10$) show a similar behavior giving a large localized moment, $\mu = 5.9\mu_B$ if $J = 2$. The magnitude of this moment is in agreement with the idea that the electron concentration is the important parameter.

In the other cases studied, some modification of the pure paramagnetism (B_J) model is necessary. The supposition is that in addition to the iron spin interacting with H, the spin is coupled to an electronic field of the host, a field which is assumed to be varying slowly compared to the thermal relaxation times, but rapidly compared to the lifetime of the ^{57}Fe excited state.[3,6] It is found that the experimental results are adequately described by taking the strength of coupling s of the iron spin-host field to be independent of T and dependent principally upon the host material. The calculated curves of Fig. 1 show how H_i is depressed relative to the B_J behavior at temperatures low compared to the coupling s. We note that the determination of $\mu(J+1)/J$ usually depends but little on a knowledge of the precise way that the iron spin is coupled to the host field. In fact, if sufficient magnetic fields were available to align the spins at temperatures high compared to s, at $T \gtrsim s$ only the B_J behavior would be evident.

The derived parameters for the best fit of the data shown for iron in tungsten are also shown in Fig. 1. Although the curves are drawn for the $J = 2$ case, the quantity $\mu(J+1)/J$ and the quality of the fit are not very sensitive to this choice of J; hence, the actual value of μ is defined only to the extent that we know J. There is some evidence that the choice[1] of $\mu/J \equiv g = 2$ may not be valid.[3] The magnitude of μ for iron in tungsten ($n = 6$, $\mu = 2.6\mu_B$ if $J = \frac{3}{2}$) is very similar to that found for iron in molybdenum ($n = 6$, $\mu = 1.9\mu_B$ if $J = 1$). Susceptibility results[1] for 1% Fe in molybdenum give $\mu = 2.1\mu_B$ if $g = 2$. Two molybdenum sources containing magnetic impurities of about 0.15% and about 0.005% gave consistent results. The only case for which we found a dependence on impurity concentration is that for iron in gold, where for an impurity concentration as high as 0.4% we obtained results[3] somewhat different from those given below for a very dilute source.

Our preliminary results are tabulated in Table I.

Table I

Source	n	$\mu(J+1)/J$ (Bohr magnetons)*	$s(J+1)/J$, °K	H_{sat}, kOe*	Probable spin
Vanadium, niobium, tantalum	5	0	—	—	
Molybdenum	6	3.71	2.6	-115	$\frac{1}{2} \leq J \leq 1$
Tungsten	6	4.33	2.1	-76	$\frac{1}{2} \leq J \leq 3$
Rhodium	9	5.30	20	-103	?
Palladium[2]	10	14.5	0	-295	$5 \leq J \leq 8$
Platinum	10	8.9	1 ± 1	-315	$1 \leq J \leq 3$
Copper	11	$\begin{cases} 8.8 \\ 2.2 \end{cases}$	33	$\begin{cases} -40 \\ -160 \end{cases}$	?
Silver	11	4.0	8.8	-40	$\frac{1}{2} \leq J \leq 3$
Gold	11	3.93	3.6	-195	$\frac{1}{2} \leq J \leq \frac{3}{2}$

*Typical error $\pm 2\%$ except for silver and rhodium, where the error may be as great as $\pm 50\%$; however, for silver, rhodium, and copper the quantity $H_{sat}\mu(J+1)/J$ is defined to better than $\pm 10\%$.

We have shown the presence of a localized magnetic state about dilute iron impurities in a number of host elements. The results are in good agreement with those from susceptibility measurements, where they exist, and are in agreement with the electron concentration correlation.[1] Comparison of these results with present theories should lead to a better understanding of these localized states.

References

1. A. M. Clogston, B. T. Matthias, M. Peter, H. J. Williams, E. Corenzwit, and R. C. Sherwood, *Phys. Rev.* **125**, 541, 1962; includes references to prior work.
2. P. P. Craig, D. E. Nagle, W. A. Steyert, and R. D. Taylor, *Phys. Rev. Letters* **9**, 12, 1962.
3. R. D. Taylor, T. A. Kitchens, D. E. Nagle, W. A. Steyert, and W. E. Millett, *Solid State Commun.* **2**, 209, 1964.
4. R. D. Taylor, W. A. Steyert, and D. E. Nagle, *Rev. Mod. Phys.* **36**, 406, 1964.
5. N. Blum, A. J. Freeman, and L. Grodzins, *ibid.*, p. 406.
6. R. M. Housley and J. G. Dash, *Rev. Mod. Phys.* **36**, 409, 1964; *Phys. Letters* **10**, 270, 1964.

POLARIZATION OF SILVER NUCLEI IN FERROMAGNETIC METALS*

G. A. Westenbarger† and D. A. Shirley

*Department of Chemistry and Lawrence Radiation Laboratory
University of California, Berkeley, California*

The effective magnetic field at the nucleus of a diamagnetic atom dissolved in a ferromagnetic transition-metal lattice has been measured experimentally for several impurity–host combinations. These hyperfine fields may exceed 10^6 Oe, as is the case for gold in iron. The fields are, in general, found to be directed oppositely to the bulk magnetization of the ferromagnetic host. While the data are still quite meager, certain systematic trends may be noted which make at least empirical estimation of the internal field possible in certain cases. A full quantitative theoretical prediction is not yet feasible.

There is strong indication that a major contribution to the field is produced by polarization of the outer electrons of the impurity atom by the conduction electrons of the host metal. If these electrons have appreciable s character, they may then polarize the nucleus by means of the Fermi contact interaction. While this conduction-electron-polarization (CEP) mechanism is not unique, other effects such as polarization of the atomic core are believed to be small when there are unpaired polarizable outer electrons. This is probably most true for the alkali metals and for the Ib group elements copper, silver, and gold dissolved in iron.

It is not possible to assess the relative importance of the contributions to the field nor to measure them separately in cases for which the field is of complex origin. For the Ib group elements, where there is nominally one outer unpaired s-type atomic electron, assumption of a simple model and a CEP mechanism leads to an estimation of both the sign and magnitude of the field in the case of silver in iron.[1] This field has been measured by us[2] and is given below.

The internal field at gold in iron, cobalt, and nickel has been determined by the Mössbauer effect[3] and at copper in these hosts by NMR.[4,5] Neither of these experimental techniques is amenable to measurement of the field at silver in transition metals; other methods are limited by a variety of experimental difficulties. The sign and magnitude of the field for silver in iron and nickel has been determined using an improved modification of the nuclear polarization technique originated by Samoilov.[6]

The 1-hr half-life isotope ^{104}Ag was dissolved in iron and the sample cooled to 0.01°K by contact with a magnetically cooled paramagnetic salt. After the iron was magnetized to saturation, both positron and γ-radiation were detected. The positron counting rate along the magnetizing field was smaller than the isotropic value. Nuclear theory allows one to predict that this will be true only if the magnetic

---

* This work done under the auspices of the U.S. Atomic Energy Commission.
† Present address: Ohio University, Athens, Ohio.

field acting to polarize the nucleus is directed oppositely to the external magnetizing field. Reversal of the external field increased the positron counting rate to a value greater than the isotropic value. Thus, the internal field at the silver nucleus is negative. The magnitude of the field was determined by measurement of the degree of anisotropy of γ-radiation emitted by the polarized nuclei. The 920-keV γ-ray from ^{104}Ag was used for this. The field was found to be -350 ± 100 kOe in iron.

More accurate data were obtained by polarizing the long-lived isotope ^{110m}Ag in iron. In this case, however, the nuclear magnetic moment is not known, so that only the product of the moment and the field could be determined. This isotope was polarized in both iron and nickel and the ratio of these products yielded the field at silver in nickel. The best value was -110 ± 80 kOe in nickel. The ^{110m}Ag data will give considerably more accurate values for these fields, but this must wait until the magnetic moment of this nucleus is accurately measured.

Figure 1 shows the s-electron contribution to the magnetic field at the nucleus in the free atom as a function of the atomic number. The open squares are for the monovalent cations. The smooth variation of the fields is striking. In Fig. 2, several hyperfine fields measured in iron are plotted *vs.* atomic number. The curves shown are those of Fig. 1 but reduced to 7% of the free-atom values. The correlation should be especially significant for those cases in which unpaired s-type electrons are present, as in the Ib group elements.

The major support for the CEP mechanism is that the fields observed for the Ib elements in iron are in fact roughly proportional to the fields attributed to the outer s-electrons in the free atoms. In particular, the Ib elements in iron have hyperfine fields which are about 7% of the s-electron contribution to the free-atom field. For these elements there is on the average one outer s-electron per impurity atom and hence about 7% polarization of the conduction electrons is expected.

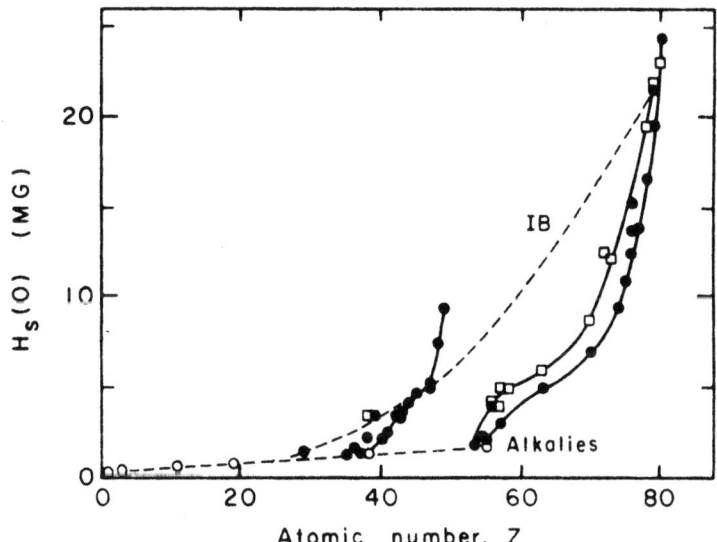

Fig. 1. Plot of s-electron contribution to hyperfine field as function of atomic number.

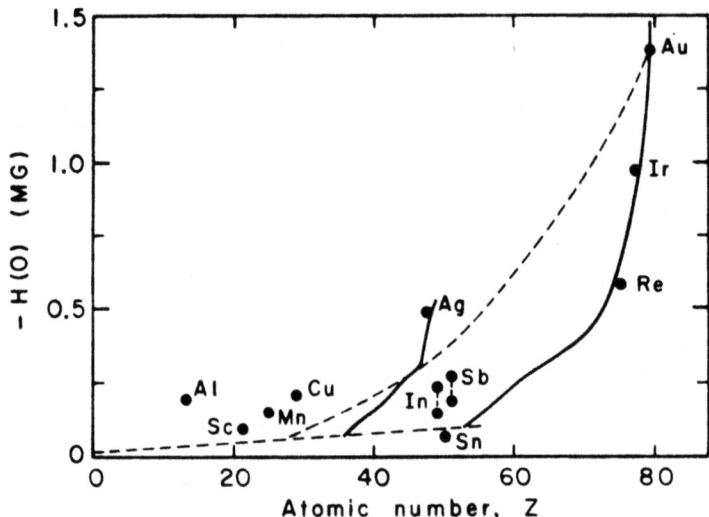

Fig. 2. Experimental hyperfine fields in iron plotted against atomic number. Curves are from Fig. 1, fitted to the experimental value of gold in iron.

This trend may be extended to the alkali metals as impurities in iron, but difficulties with both magnitude and sign are to be expected when the band structure of the alloy becomes highly altered by the impurities. If the value of 7% polarization is retained, we predict small negative hyperfine fields for these elements in iron.

A recent theoretical calculation[7] indicated that an iron atom has a spin density in the outer region that is opposite to the spin of the 3d-electrons. The magnitude of this polarization was only slightly greater than 7%.

References
1. D. A. Shirley and G. A. Westenbarger, *Phys. Rev.* **138**, A170, 1965.
2. G. A. Westenbarger and D. A. Shirley, *Phys. Rev.* **138**, A161, 1965.
3. R. W. Grant, *Study of the Nuclear Zeeman Effect in Au^{197}*, Lawrence Radiation Laboratory Report UCRL-10649, January 1963.
4. T. Kushida, A. H. Silver, Y. Koi, and A. Tsujimura, *J. Appl. Phys. Suppl.* **33**, 1079, 1962.
5. K. Asayama, S. Kobayashi, and J. Itoh, *J. Phys. Soc. Japan* **18**, 458, 1963.
6. B. N. Samoilov, V. V. Sklyarevskii, and E. P. Stepanov, *Soviet Phys. JETP (English Transl.)* **9**, 1383, 1959.
7. R. E. Watson and A. J. Freeman, *Phys. Rev.* **123**, 2027, 1961.

MÖSSBAUER STUDIES OF A DILUTE IRON IN PALLADIUM ALLOY IN EXTERNAL MAGNETIC FIELDS*

Romeo Segnan,† Paul P. Craig, and Robert C. Perisho‡

Brookhaven National Laboratory
Upton, New York

Iron and cobalt, when dissolved in palladium, form a series of random substitutional ferromagnetic alloys with Curie temperatures ranging over several orders of magnitude. Susceptibility and magnetization measurements[1–4] on these alloys indicate that the magnetic moment associated with each iron atom increases from $2.2\mu_B$ in pure iron to about $10\mu_B$ at zero iron concentration. Mössbauer measurements[5] have demonstrated the paramagnetism of exceedingly dilute iron in palladium, and have yielded a moment of $12.6 \pm 0.4\mu_B$ and a g-factor of 2.0 ± 0.4 corresponding to a spin $J = 6.5 \pm 1.5$ associated with the iron impurities. The large moments presumably arise from polarization of surrounding palladium atoms.

The Mössbauer technique permits study of the hyperfine magnetic field at the iron nucleus. In pure iron, the hyperfine field H_i is known to vary with temperature in the same way as the bulk moment M.[6] Thus $H_i/H_{sat} = M/M_{sat}$, where the subscript "sat" denotes the saturation value at $T = 0°K$. This result is thought to have general validity. However, the value of H_{sat} rarely bears any simple relation to that of M_{sat}, as sample composition is varied.[7,8]

We have undertaken a study of the hyperfine fields at ^{57}Fe nuclei in $Fe_{2.65}Pd_{97.35}$ both above and below the Curie temperature θ and in externally applied magnetic fields. The saturation magnetization in this system is about $7.6\mu_B$ per iron impurity atom.[2] Our measurements indicate that the relation $H_i/H_{sat} = M/M_{sat}$ is not obeyed in this system. Our zero external field results may be described in terms of a Brillouin function model with $\theta = 90°K$ and a spin $J = 1$. This spin is substantially lower than the value $J = \frac{13}{2}$ obtained from dilute alloy, bulk magnetization[2] and paramagnetic state Mössbauer results.[5] In the presence of external magnetic fields, we find complex behavior which approaches that predicted on a molecular field model with $J \sim 1$ at low temperature and fields, and approaches high spin ($J \sim \frac{13}{2}$) behavior in the high-temperature limit ($T \gg \theta$).

A resonant absorber ($Fe_{2.65}Pd_{97.35}$) was prepared by repeated arc melting of 76% ^{57}Fe-enriched metallic iron and five-ninths purity palladium, followed by rolling to a thickness of about 0.001 in. and annealing for 24 hr at 850°C. This absorber, together with a ^{57}Co in a copper unsplit source, was mounted in a 30 kOe superconducting solenoid. The source temperature was about 4.2°K and the variable absorber temperature could be maintained constant to about 0.010°K for many hours. The 14 keV γ-radiation emerged from the cryostat through beryllium

---
* This work was performed under the auspices of the United States Atomic Energy Commission.
† Now at the University of Pennsylvania, Philadelphia, Pennsylvania.
‡ Summer Research Assistant from Haverford College, Haverford, Pennsylvania.

windows and was detected by an argon–methane-filled proportional counter which was insensitive to stray magnetic fields from the superconducting solenoid.

The measurements were performed using parabolic source motion, linear velocity to pulse height conversion, and a multichannel analyzer. In order to correct for possible distortions in the velocity drive, the resonant velocity spectra were divided by nonresonant spectra recorded simultaneously in a separate subgroup of the analyzer memory. Hyperfine magnetic fields at the absorber nuclei result in six line unpolarized hyperfine spectra in the absence of an external field and in four line polarized spectra when the axial field is applied.

Figure 1 displays in normalized form the hyperfine fields deduced from the Mössbauer spectra vs. temperature for zero external field (experimental points). (Near $T = \theta$, the Curie temperature, the spectra collapsed into single broadened peaks from the widths of which the hyperfine fields were deduced.) The (negative) saturation field of -314 kOe at $T = 0°$K is in agreement with the results of Segnan et al.[9] The Curie temperature of $90 \pm 0.5°$K is somewhat below the value of $106°$K interpolated from Crangles' measurements.[2] Crangles' magnetization measurements on $Fe_{15.8}Pd_{84.2}$ ($\theta = 377°$K) and on $Fe_{3.2}Pd_{96.8}$ ($\theta = 122°$K) are also shown. Since magnetization results[2,4] on alloys more dilute than $Fe_{2.65}Pd_{97.35}$ show even flatter curves than those illustrated, the curve for our composition should lie below that for $Fe_{3.2}Pd_{96.8}$. (Magnetization measurements on our sample are in progress.) Hence we conclude that the temperature dependence of the magnetization for our sample is in striking disagreement with that of the hyperfine field. Zero-field molecular field calculations spins $J = 1$ and $J = \frac{13}{2}$ are also shown in Fig. 1. The calculation for $J = 1$ is seen to yield excellent agreement with experiment, while

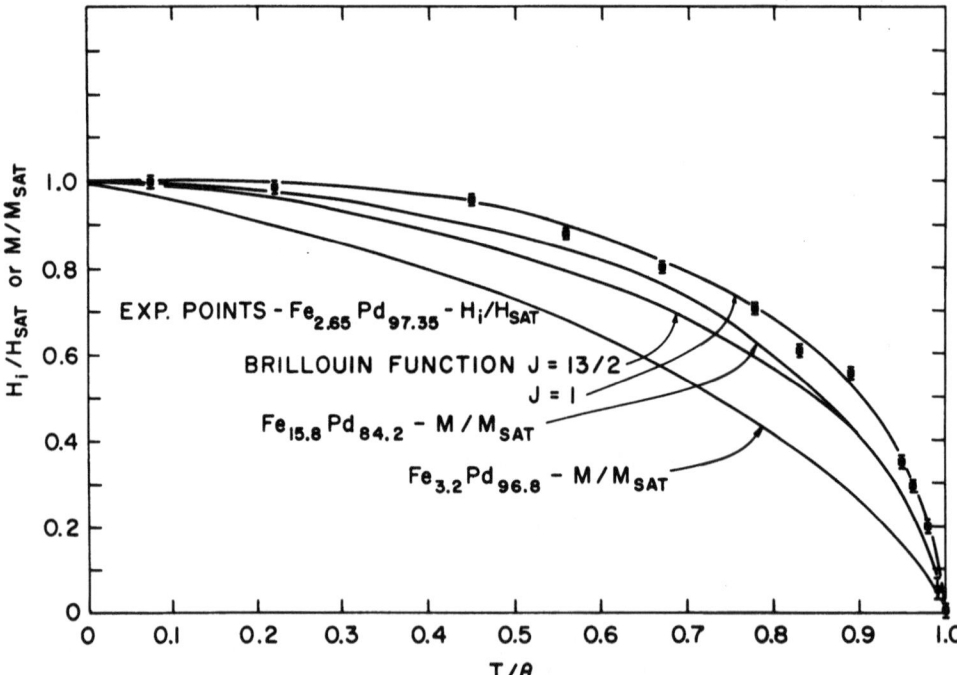

Fig. 1. The magnetic hyperfine field at iron sites vs. temperature for $Fe_{2.65}Pd_{97.35}$ in zero external field. The measurements are consistent with a spin $J = 1$ and are in contrast to the bulk magnetization temperature dependence.

for $J = \frac{13}{2}$, the spin expected on the basis of dilute alloy paramagnetic Mössbauer measurements,[5] the agreement is poor. (The spin expected for our composition is about four,[2] but the improvement in the fit with this spin is negligible.) This result suggests that the large moment per iron atom appearing in bulk magnetization measurements is only partially localized upon the iron atoms, and that spin localized upon the iron is not rigidly coupled magnetically to the palladium system. Rather, the iron impurities are a magnetically independent subsystem. This subsystem, with the same Curie temperature as the total system, is characterized by a moment of two to three Bohr magnetons, values which are comparable to the moment of $2.86\mu_B$ known from neutron diffraction measurements[10] on the ordered alloy Pd_3Fe to be localized spatially on the iron sites.

Figure 2 presents hyperfine fields obtained in various external fields above and below the Curie temperature θ. At fixed temperature, the external field produces additional polarization of the sample and hence modifies the hyperfine fields. Because the effective field at the iron nucleus is the sum of the positive external and negative polarization terms, the measurements may lie either above or below the zero field results.

While the molecular field calculations for nonzero external fields do not fit the experimental results over the entire range of H and T, we do find that in the low-temperature limit a spin $J = 1$ provides a reasonable fit. Furthermore, at the highest temperatures ($T > 130°$) the data approach the molecular field calculations for $J \sim \frac{13}{2}$. Thus in this high-temperature (Curie–Weiss) limit, our results are consistent with bulk susceptibility and paramagnetic Mössbauer conclusions. The details of the crossover phenomena will be discussed elsewhere. The measurements may be understood qualitatively by assuming that an external field produces a hyperfine field behavior characteristic of a moment of about $12\mu_B$, while spontaneous magnetization produces a hyperfine field characterized by a moment of about $2\mu_B$. No explanation for the nonequivalence of exchange polarization and externally produced polarization is known.

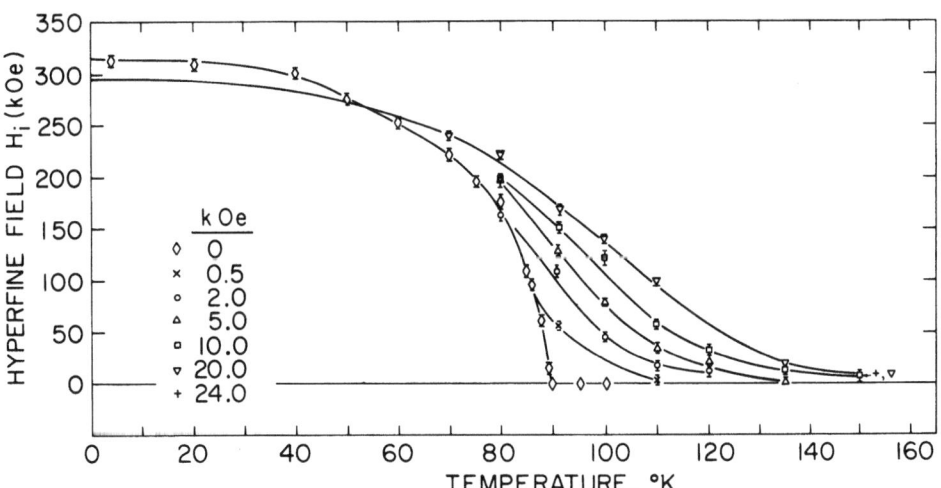

Fig. 2. Hyperfine magnetic fields at iron sites in $Fe_{2.65}Pd_{97.35}$ in the presence of external magnetic fields. At high temperatures, $T \gg \theta$, the measurements agree with bulk magnetization results, while at low temperatures complex behavior is found.

A problem of special interest because of recent theoretical predictions on susceptibility behavior is the region near $T = \theta$.[11-15] In order to test these predictions, we are currently extending our measurements close to the Curie temperature.

Note Added in Proof

Subsequent work has shown that the discrepancy between the temperature variation of the magnetization and the hyperfine field is an artifact and results from the extrapolation method[2] used to determine the Curie temperature from the bulk magnetization. A detailed discussion of the computation of magnetization and hyperfine field is given in ref. 16.

References

1. A. M. Clogston, B. T. Matthias, M. Peter, H. J. Williams, E. Corenzwit, and R. C. Sherwood, *Phys. Rev.* **125**, 541, 1962.
2. J. Crangle, *Phil. Mag.* **5**, 335, 1960.
3. J. Crangle and D. Parsons, *Proc. Roy. Soc. (London) Ser. A.* **255**, 509, 1960.
4. R. M. Bozorth, P. A. Wolff, D. D. Davis, V. B. Compton, and J. H. Wernick, *Phys. Rev.* **122**, 1157, 1961.
5. P. P. Craig, D. E. Nagle, W. A. Steyert, and R. D. Taylor, *Phys. Rev. Letters* **9**, 12, 1162.
6. D. E. Nagle, H. Frauenfelder, R. D. Taylor, D. R. F. Cochran, and B. T. Matthias, *Phys. Rev. Letters* **5**, 364, 1960.
7. W. Marshall and C. E. Johnson, *J. Phys. Radium* **23**, 733, 1962.
8. D. E. Nagle, P. P. Craig, P. Barrett, D. R. F. Cochran, C. E. Olsen, and R. D. Taylor, *Phys. Rev.* **125**, 490, 1962.
9. R. Segnan, B. Mozer, and P. P. Craig (submitted to *Phys. Rev.*).
10. G. Shirane, R. Nathans, S. J. Pickart, and H. A. Alperin, International Conference on Magnetism, Nottingham, England, September 7-11, 1964.
11. W. Marshall, *Proceedings of the Eighth International Conference on Low Temperature Physics*, R. O. Davies (ed.), Butterworths, London (1963), p. 215.
12. G. A. Baker, *Phys. Rev.* **124**, 768, 1961.
13. J. W. Essam and M. E. Fisher, *J. Chem. Phys.* **38**, 802, 1963.
14. C. Domb and M. F. Sykes, *J. Math. Phys.* **2**, 63, 1961; *Proc. Roy. Soc. (London) Ser. A* **240**, 214, 1957.
15. R. Brout, this volume, p. 623; C. Domb, this volume, p. 637.
16. P. P. Craig, R. C. Perisho, R. Segnan, and W. A. Steyart, *Phys. Rev.* **138A**, 1460, 1965.

CONTRIBUTION OF MAGNETIZATION (DOWN TO 0.05°K) AND SPECIFIC HEAT MEASUREMENTS TO THE STUDY OF SEGREGATION IN A GOLD–IRON ALLOY

O. Béthoux, Y. Ishikawa,* J. Souletie, R. Tournier, and L. Weil

Centre de Recherches sur les Très Basses Températures
Laboratoire d'Electrostatique et de Physique du Métal
Grenoble, France

Several studies of gold–iron alloys have recently been published. Some disagreement appears[1-3] in the interpretation of the magnetic properties because of the simultaneous ferromagnetic and antiferromagnetic aspects of these alloys. Ferromagnetism is characterized by a positive paramagnetic Curie temperature and a strong remanence. However, this remanence has precisely the same behavior as in diluted antiferromagnets: High fields are necessary to reduce it to zero as well

* The Institute for Solid State Physics, The University of Tokyo, Azabu Minato-Ku, Tokyo, Japan.

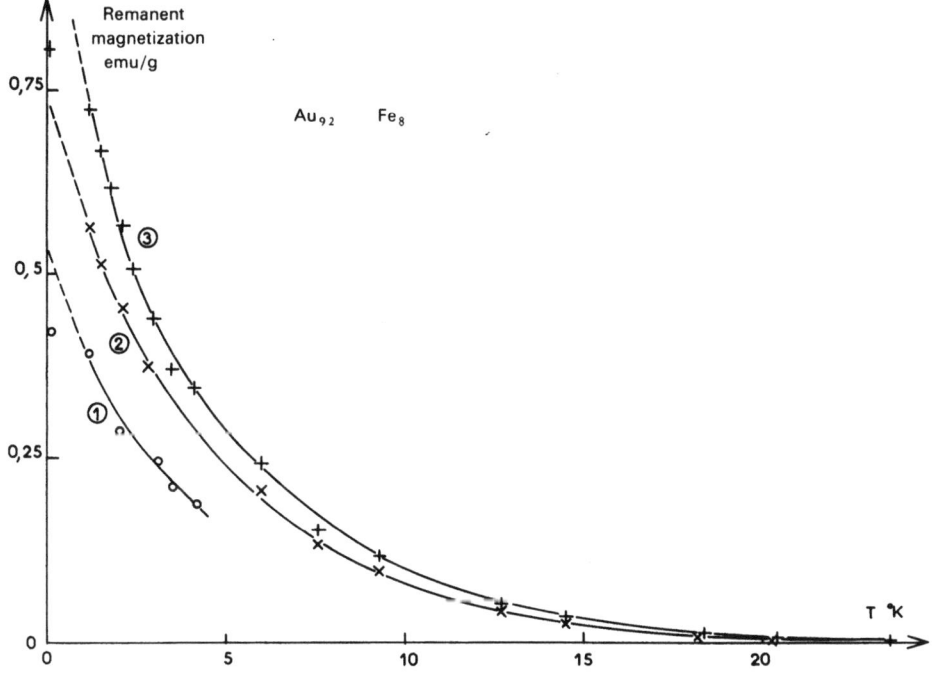

Fig. 1. Remanence σ_R as a function of temperature.

as to saturate it, and moreover it appears at temperatures proportional to concentration, at least at small values of the latter.[4,5]

In a recent paper,[6] it has been shown that it is possible to explain the high-field and rather high-temperature properties of gold–iron by attributing ferromagnetism to segregated areas with a predominant short-range order. Above 25°K, these regions are superparamagnetic. Below about 25°K, their "giant" magnetic moments are blocked through long-range interaction with a now antiferromagnetically ordered matrix; the size of these regions has been estimated at 40 iron atoms. The suggested antiferromagnetic matrix explains why, even in a field of 95 kOe at 1.5°K[2] or 20 kOe at 0.05°K, saturation is far from being reached.

This segregation should of course vary with the heat treatment. We will discuss observations made in this laboratory by magnetic and specific heat measurements, and elsewhere[1] by. Mössbauer effect measurements.

Magnetization vs. field of the same sample (8 at.% Fe) has been measured in three different conditions: (1) quenched from 900°C to room temperature; (2) aged a few days at room temperature and cooled several times to helium temperatures; (3) additionally annealed at 100°C for 14 hr. In the last condition, magnetization reaches 2.5 emu/g in 25 kOe, which should be compared to an estimated saturation of 4.8 emu/g on the basis of $2\mu_B$ per iron atom. Between (1) and (3), the magnetization in any field as well as at remanence has strongly increased. Aging and annealing have emphasized the ferromagnetic aspects, in good agreement with the hypothesis of an increase in segregation.

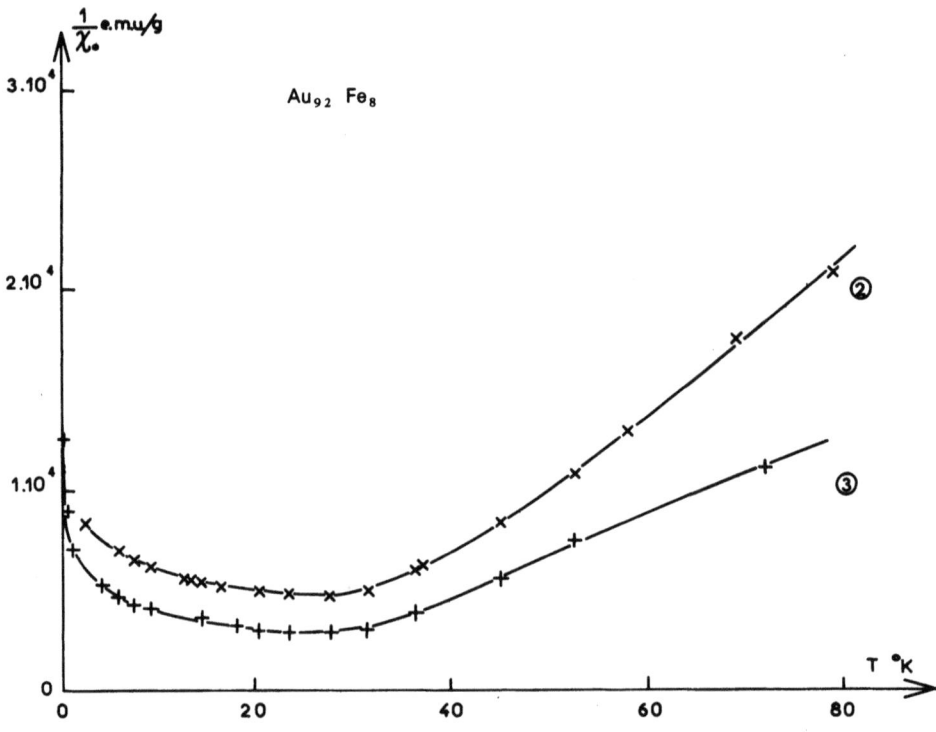

Fig. 2. Inverse initial susceptibility vs. temperature.

Figure 1 gives a detailed account of the remanence σ_R as a function of temperature; the dispersion of σ_R is attributable to aftereffects above 1°K and other phenomena below 1°K.[7] Figure 2 shows the inverse initial susceptibility; between (2) and (3) in the superparamagnetic region, the apparent Curie constant has been increased, probably because of the increase of the giant moments.

The specific heat of the same sample is not changed in the range 1.3° to 4.2°K, between conditions (2) and (3). This could be explained if the specific heat increase over that of gold, $\Delta c = aT$, where $a = 7 \pm 0.5$ mJ/mole of alloy, is mainly attributable to the matrix and if the specific heat of this matrix is, as found in copper–manganese,[8,9] concentration independent over a large range.

The Mössbauer effect (BBV)[1] gives apparently contradictory results, suggesting no segregation for an alloy of 7.38 at.% Fe. As has already been suggested,[10] at room and liquid-nitrogen temperatures, H_n is averaged out because of the high-rotation frequency of the giant moments: As long as the segregations are small enough (concentrations low enough, for instance) to be superparamagnetic at nitrogen or room temperatures, they remain invisible with Mössbauer effect.

References

1. R. J. Borg, R. Booth, and C. E. Violet, *Phys. Rev. Letters* **11**, 464, 1963.
2. W. E. Henry, *Phys. Rev. Letters* **11**, 468, 1963.
3. J. Crangle and W. R. Scott, *Phys. Rev. Letters* **12**, 126, 1963.
4. O. S. Lutes and J. L. Schmit, *Phys. Rev.* **134**, 3A, A676–A683, 1964.
5. O. S. Lutes and J. L. Schmit, *Phys. Rev.* **125**, 433–439, 1962.
6. R. Tournier and Y. Ishikawa, *Phys. Letters* **11**, 280, 1964.
7. B. Dreyfus, Y. Ishikawa, R. Tournier, and L. Weil, this volume, p. 1026.
8. J. E. Zimmerman and F. E. Hoare, *J. Phys. Chem. Solids* **17**, 52–56, 1960.
9. W. Marshall, *Phys. Rev.* **124**, 1030, 1961.
10. Y. Ishikawa, *J. Phys. Soc. Japan* **17**, 1835, 1962.

MAGNETIZATION DISCONTINUITIES BELOW 1°K IN ALLOYS

B. Dreyfus, Y. Ishikawa,* R. Tournier, and L. Weil

*Centre de Recherches sur les Très Basses Températures
Laboratoire d'Electrostatique et de Physique du Métal
Grenoble, France*

The magnetic properties of alloys below 1°K have been measured with a previously described apparatus.[1-3] The classical extraction method is used: The sample is cooled by means of a paramagnetic salt at a distance of about 40 cm. During the displacement, the sample remains in a nearly constant field. By means of a special device, the movement of the sample is very smooth, and it is possible to ascertain that the extraction is made without shock; it was verified that in this device the sample is not heated up during extraction. This has been done by a series of measurements of remanence in the case of alloys where σ_R varies rapidly

* The Institute for Solid State Physics, The University of Tokyo, Azabu Minato-Ku, Tokyo, Japan.

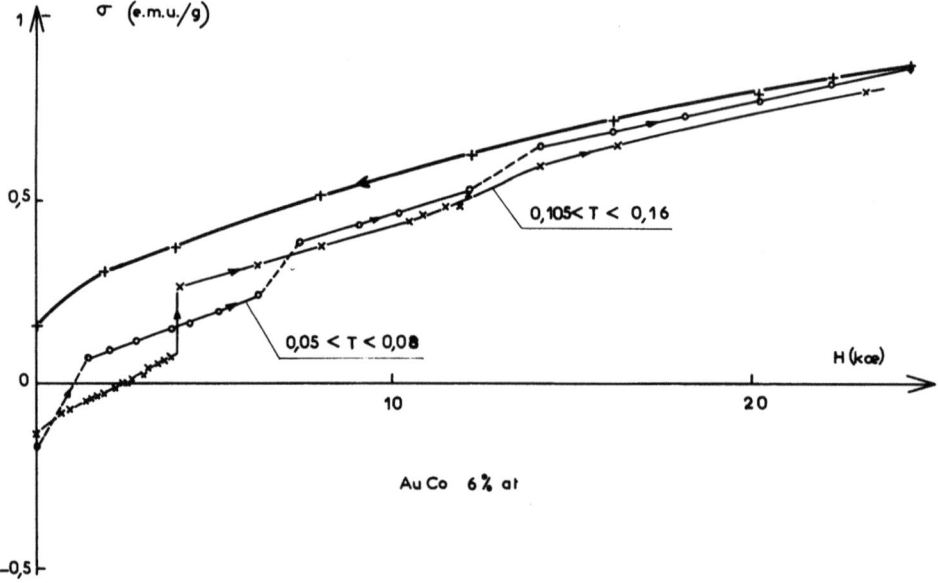

Fig. 1.

with temperature (copper–cobalt, for instance). The same method proves that thermal equilibrium is reached at least down to 0.1°K within a minute or so.

Discontinuities have been observed in the case of several alloys when their magnetization curves were plotted. Figure 1 gives results for increasing field in 6 at. % Au–Co. Similar behavior has been noticed as well in increasing as decreasing field on 3% Cu–Co,[4] 15% Cr–Fe,[3] 8% Au–Fe, and 2% Cu–Mn.[5] These discontinuities were found neither in 40 and 44 at.% Cu–Ni[5] nor in 1 and 2% Cu–Co and 2, 3, and 4% Au–Co.

To provide a better understanding for these effects, we decided to separate the reversible magnetization from the irreversible part. This can be done by measuring the hysteresis of the remanence,[6] i.e., the remanence σ_R as a function of the negative field previously applied. Figure 2 shows the jumps in 8% Au–Fe. The starting field of the jump depends critically upon the initial value of the remanence. The values of the initial remanence are determined by previous jumps, either at positive or negative fields. Figure 2 shows that the starting points are well aligned on a curve, in spite of the fact that the temperature range is rather extended (0.06° to 0.40°K). A smooth curve may be drawn through the "landing" points. For comparison the remanence hysteresis curve at 4.2°K, a temperature at which no discontinuities exist, has been plotted on Fig. 2.

A common feature of the alloys displaying jumps is a large hysteresis loop. A field H_c of some thousands of oersteds is necessary to reduce the remanence to

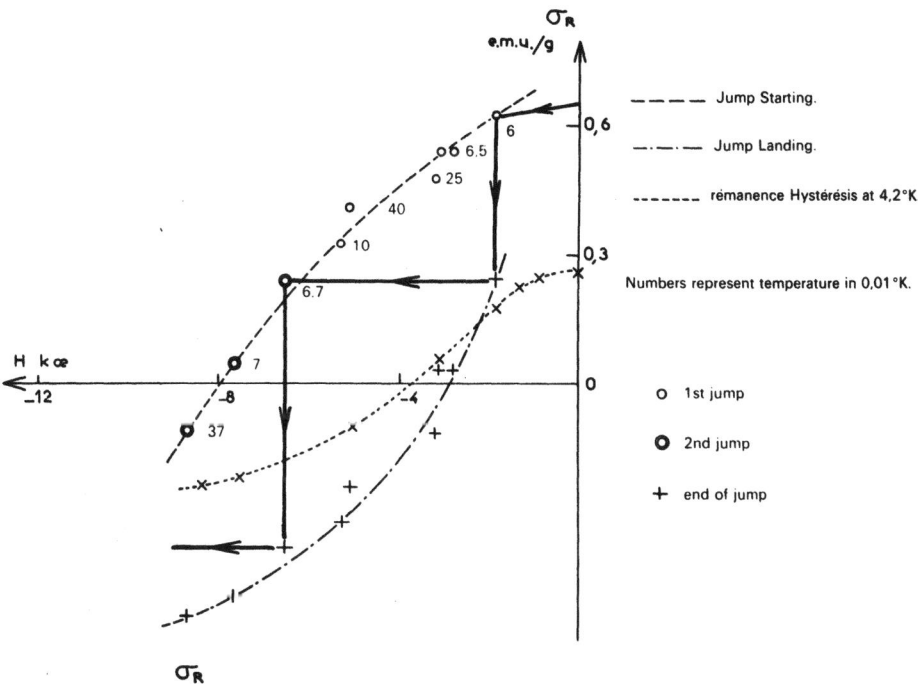

Fig. 2. σ_R as a function of the negative field previously applied.

zero. The heat dissipated by the reversal of magnetization, even when this occurs in a small region only, may be large. The local increase in temperature, assuming adiabatic reversal, may reach a few degrees due to the extremely low specific heat. As the heat diffuses, the temperature in the surrounding region also rises, leading to a magnetization reversal (H_c decreases with increasing T) and consequently an additional dissipation of heat. A real avalanche occurs (Fig. 2), capable of reversing a large part of the remanence.

The arresting of this avalanche effect becomes easier the lower the value of the thermal diffusivity and the higher the critical field H_c. If the whole sample had the same critical field, the jump starting curve would be a vertical straight line. The observed curve, however, is a dispersion curve of the critical field as a function of volume. Only a small fraction of the sample volume reaches the value of 8 kOe.

References

1. R. Tournier and L. Weil, *J. Phys. Soc. Japan* **17**, Suppl. BI, 118, 1962.
2. R. Tournier and Y. Ishikawa, *Phys. Letters* **11**, 280, 1964.
3. R. Tournier and Y. Ishikawa (to be published in *Compt. Rend.*).
4. R. Tournier and L. Weil (to be published in *J. Phys. Radium*).
5. J. A. Careaga and M. Bourrières, private communication.
6. L. Néel, *Ann. Geophys.* **5**, 99, 1949.

HEAT CAPACITY BELOW 1°K: OBSERVATION OF THE LINEAR TERM AND THE HFS CONTRIBUTION IN SOME DILUTE ALLOYS

F. J. du Chatenier and A. R. Miedema

Kamerlingh Onnes Laboratory
Leiden, The Netherlands

In the present paper, we will report on experiments on the heat capacities of dilute binary alloys of chromium, iron and manganese in copper, silver, and gold in the temperature region between 0.03° and 1°K.

The apparatus used for determining heat capacities below 1°K is shown in Fig. 1. The calorimeter, consisting of the sample (E), the thermometer (F), and the heater (D), is brought to the required low temperature by thermal conduction through a thermal switch (B) to a cooling salt (A).

The cooling salt is made of chrome–alum single crystals. A tin or lead wire is used for the thermal switch, the lead wire being used for experiments at temperatures above 0.25°K. The field of 800 Oe, required to bring the lead wire in the normal state, is supplied by a small iron magnet. For the heater, two copper rings are used in which heat can be generated by a 225–Hz alternating field. The temperature is derived from the susceptibility of a cerium magnesium nitrate single crystal, which follows Curie's law quite accurately.

At temperatures near 0.05°K, the equilibrium time between thermometer and metal sample has been found to be determined by spin lattice relaxation effects in the cerium magnesium nitrate thermometer. For a pure cerium magnesium nitrate crystal, the equilibrium time is about 5 min; if 1% copper ions are added to the cerium salt, the time is reduced to 1 min at 0.05°K. The heat leak to the calorimeter (thermometer–sample–heater) is of the order of 1 erg/min, the residual heat being due mainly to vibrations. The small leak is attained, using a glass thermal shield which by thermal conduction to the lower part of the cooling salt attains a temperature of about 0.15°K. The experimental accuracy depends mainly on the calibration of the heater and the Curie constant of the thermometer. It is estimated to be 2%.

Earlier results on the heat capacity of silver–manganese and copper–manganese alloys have led to the conclusion that the magnetic moments of the manganese ions are coupled by means of the conduction electrons. From the shape of the heat capacity vs. temperature curve, it could be concluded that the coupling (or internal field) is not the same for all magnetic ions, but may vary considerably.

From heat capacity data on copper–manganese alloys obtained between 1.5° and 40°K, Zimmerman[1] suggested the presence of a low-temperature contribution to the heat capacity, linearly dependent on temperature and independent of concentration. A theoretical background of Zimmerman's suggestion was given by Overhauser,[2] Marshall,[3] and Blandin.[4] The main feature of their theories is the existence of a continuous distribution of the internal fields in the alloy, giving a nonzero probability of ions with small energy splittings.

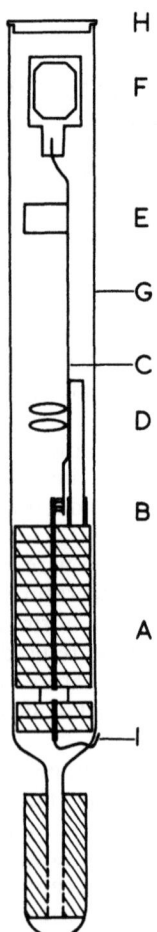

Fig. 1. Apparatus for determining heat capacities at temperatures below 1°K. A: chromium–alum cooling salt; B: thermal switch; C: copper wire; D: heater; E: sample; F: Curie's law thermometer; G: glass thermal shield; H: perspex top; I: guard salt.

The conclusion of Zimmerman was based on experiments on rather concentrated copper–manganese alloys (1–10%). However, data on more dilute alloys, taken by du Chatenier and de Nobel,[5] show deviations from Zimmerman's value ($C/T = 4.4$ mJ/mole-degK) for the linear term; thus experiments at temperatures below 1°K become quite interesting.

In the experiments on copper–manganese, silver–manganese, and gold–manganese alloys, the presence of the linear term could not be shown so well because of the large nuclear contribution (hfs) to the specific heat. The hfs specific heat causes a minimum in the heat capacity near 0.3°K. For that reason, a gold–iron and two copper–chromium alloys (0.1 and 0.6%) have also been investigated, the influence of hfs terms being negligible in these cases. The results on the latter alloys definitely reveal the existence of the linear term. The results are collected in Table I. It may be seen that the linear term (independent of concentration) is rather different for the three manganese alloys studied and is largest for gold–manganese.

Table I. The Magnetic and Hyperfine Structure Contributions to the Heat Capacity of Some Dilute Alloys

Alloy	Linear term C/T, mJ/mole-degK	hfs parameter AS/k, °K	Concentration, at.%
Gold–manganese	7.8	0.0173	0.083, 0.16
Silver–manganese	4.6	0.0161	0.078, 1.01
Copper–manganese	2.3	0.0222	0.15, 1.15
Copper–chromium	1.8		0.073, 0.56
Gold–iron	2.3		0.092

From deviations of the linear temperature dependence at higher temperatures, some conclusions may be drawn for the shape of the distribution function of the internal fields. It is suggested that at zero field this function has a minimum for the manganese alloys and a maximum in the case of copper–chromium 0.1% and Au–Fe 0.1%.

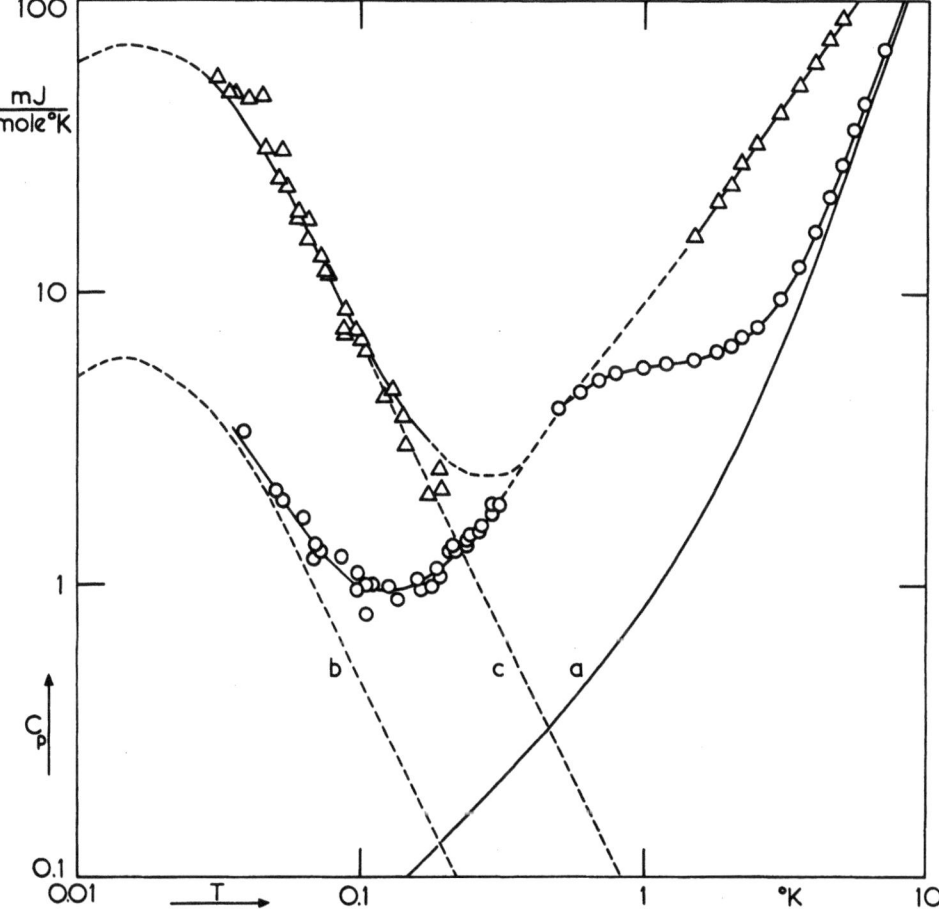

Fig. 2. Specific heat vs. temperature for two alloys of manganese in silver. △: 1.01% Ag–Mn; ○: 0.078% Ag–Mn; a: pure silver; b and c: Schottky curves, calculated using $AS/k = 0.0161°K$.

Values for the hfs parameter for manganese in copper, silver, and gold have been obtained from the heat capacity data obtained below 0.3°K. The values for AS/k, as derived from the T^{-2} term, are given in Table I. It may be seen that the nuclear splitting is largest for manganese in copper and smallest for manganese in silver.

It may be seen from Fig. 2 that the hfs parameters are not very different for different manganese ions in the same solvent metal. The data for each concentration can be described by a Schottky curve, assuming one single parameter for all manganese ions, while at 0.03°K the heat capacity is only slightly smaller than the theoretical maximum.

The values obtained for AS/k are considerably smaller than those for Mn^{++} in many ionic salts ($AS/k \approx 0.033°K$). According to entropy evaluations, this cannot be attributed to a value for S lower than $\frac{5}{2}$. Furthermore, nuclear resonance data by Van der Lugt[6] preclude the existence of nonlocalized d-electrons.

Another possibility might be the degree of freedom for the electron moments, which according to the linear term in the heat capacity, will exist even near 0.1°K. However, this would cause a reduction of the hfs splittings, which is different for different manganese ions. This disagrees with the experimental results.

Rather small hfs parameters may be found, if the conduction electrons contribute to the hfs interaction with a sign opposite to that of the inner s-electrons. The conduction electrons are polarized because of the Yosida interaction, in which case the degree of polarization will be inversely proportional to the Fermi energy. An attractive consequence of this suggestion is that it explains the differences between the hfs splittings for manganese in copper, gold, and silver, since they can be correlated to the differences in the Fermi energy.

References

1. J. E. Zimmerman and F. E. Hoare, *J. Phys. Chem. Solids* **17**, 52, 1960.
2. A. W. Overhauser, *J. Phys. Chem. Solids* **13**, 71, 1960.
3. W. Marshall, *Phys. Rev.* **118**, 1519, 1960.
4. A. Blandin, thesis, Paris (1961).
5. J. de Nobel and F. J. du Chatenier, *Physica* **25**, 969, 1959; *Commun. Kamerlingh Onnes Lab.*, Leiden, No. 317c.
6. W. Van der Lugt, N. J. Poulis, and W. P. A. Hass, *Physica* **25**, 97, 1959; *Commun. Kamerlingh Onnes Lab.*, Leiden, No. 314a.

HYPERFINE FIELDS IN DILUTE TRANSITION METAL ALLOYS*

J. A. Cameron, I. A. Campbell, J. P. Compton, M. F. Grant,
R. W. Hill, and R. A. G. Lines

The Clarendon Laboratory
Oxford, England

Ferromagnetic Alloys

An assembly of nuclei in a magnetic field H and at temperature T achieves at equilibrium a polarization with respect to the field axis which is determined by the parameter $\mu H/IkT$, μ being the magnetic moment and I the spin of a nucleus. All nuclei within one domain of a ferromagnetic material are polarized in this way, H being then the hyperfine field. If the ferromagnet is magnetized to saturation, all the domain magnetizations are brought parallel to the applied field, which then becomes the axis of alignment of all the nuclei in all the domains. The degree of alignment achieved, and hence the hyperfine field, can be determined by measuring the angular distribution of γ-rays emitted by suitable radioactive nuclei which form one constituent of the alloy. Thus in the present experiments the fields on manganese and vanadium, dissolved at very low concentration in iron, nickel, and cobalt, were measured using the radiation from ^{54}Mn and ^{48}V.

The experiments were carried out in an adiabatic demagnetization cryostat, the working substance being chromium potassium alum mixed with glycerol. A number of copper foils, to which the specimens were soldered, were embedded in this. The specimens cooled to temperatures close to that of the alum, yet were far enough away from it that polarizing fields of up to 15 kOe could be applied to the specimen without appreciably heating the salt. Thermometry was by means of a simultaneous nuclear alignment measurement; an iron plate containing ^{60}Co activity was also soldered to the foils. All the relevant parameters, including the nuclear moment and the hyperfine field, are well known for this case; thus the anisotropy of the ^{60}Co radiation provides a measure of the temperature.

The specimens were prepared by plating from carrier-free solution onto the host metal, followed by prolonged furnacing to allow the activity to diffuse into the host. Some trouble was experienced in preparing satisfactory specimens, but was always overcome by increasing either the diffusion time or the temperature. No result was considered reliable unless it could be reproduced in different specimens having had somewhat different heat treatments. Failure to take this precaution had previously led to the publication of an incorrect result for ^{54}Mn in iron.[1]

Magnetic saturation of the specimen was achieved by means of a superconducting magnet of niobium–zirconium wire. A three-element winding, with one coil carrying a reversed current, was used; this arrangement permitted the cancellation of the field in the region of the salt pill, while giving fields up to 15 kOe at the

---

* Supported by a research grant from the United Kingdom Atomic Energy Authority for which grateful acknowledgment is made.

specimen. The high current densities which are possible in superconducting windings make them especially suitable for the production of sharply profiled magnetic fields of this sort.

These fields are large enough to virtually ensure complete saturation in iron and nickel, but not in cobalt, because cobalt specimens invariably include regions of hexagonal structure which is magnetically much harder than material of cubic structure. Since these fields are not always negligible compared to the hyperfine field to be measured, extrapolation of the result to zero applied field is sometimes necessary. The extrapolation is not entirely straightforward, because the sign of the hyperfine field is not known; the applied field may add to or subtract from it. Moreover, the available fields are not large enough to permit a simple determination of this sign by using a wide variation of applied field in a range where saturation of the magnetization is effectively complete. However, it is known[2] that the magnetization near saturation varies with field as

$$\frac{M}{M_0} = 1 - \frac{a}{H} - \frac{b}{H^2}$$

and the constants a and b are also known.[2,3] Further, a small departure from saturation can be related to the consequent small change in the anisotropy of the γ-rays and hence to the reduction in the deduced value of the hyperfine field, giving a relationship between this latter quantity and the applied field which still depends on the assumed sign. It can be seen that, in favorable cases, only one sign is compatible with the data; the extrapolation to zero applied field is straightforward thereafter.

Analysis of the results for ^{54}Mn in nickel by this method leads to the result that $H = -316$ kOe; an earlier result[4] of ± 295 kOe is incomplete and slightly in error because complete saturation at too small an applied field was incorrectly assumed. None of the ^{48}V experiments have been analyzed in this way since the measured anisotropies are small, and the deduced field values have relatively large uncertainties. A similar determination of the sign of the effective field then becomes impossible.

These experiments measure the product μH, so that before the hyperfine field can be found, the nuclear magnetic moment μ must be known. In the case of ^{54}Mn, the moment is known from the dynamic polarization experiments of Jeffries.[5] No such measurement exists for ^{48}V, but the field on stable vanadium in cobalt is known from the nuclear magnetic resonance experiments of La Force and Day.[6] Assuming this value to be applicable to the present experiment, we can deduce the nuclear magnetic moment of ^{48}V from the alignment measurements; its value is 0.64 ± 0.1 mm. This value can then be used for the calculation of the hyperfine fields in the other vanadium alloys. The results, in kilo-oersteds, are summarized below:

	Iron	Cobalt	Nickel
^{48}V*	110 ± 30	183	<60
^{54}Mn	270 ± 15	130 ± 15	-316 ± 10

It may be seen that there is no simple trend in these results according to the host material. This is in contrast with the behavior of dilute alloys both of whose components lie on the right-hand side of the Slater–Pauling curve, where, for any given nucleus, the hyperfine field changes linearly with the magnetic moment per atom of the host material. In the case of iron, which falls slightly to the left of the maximum of the Slater–Pauling curve, the trend remains but the slope is different

* See note added in proof.

from that of the other metals. Manganese and vanadium fall well to the left, and show no systematic behavior at all. This is in line with other magnetic properties which behave much more predictably for materials on the right-hand side of the curve.

Nonferromagnetic Alloys

This section treats preliminary work on ^{54}Mn in copper and in gold. Specific heat measurements on manganese in copper[7] show that there is a hyperfine term in zero applied field, indicating the presence of a hyperfine field at the manganese nuclei. However, the effect cannot be attributed purely to the manganese nuclei, as part could arise from copper nuclei. Copper ions in the vicinity of a manganese neighbor could acquire some magnetic moment[8] and it is reasonable to suppose that these ions would contribute to the hyperfine specific heat. A complete analysis and deduction of the hyperfine fields cannot be made, however, because it is not known what fraction of the hyperfine specific heat is due to manganese ions. At Professor Phillip's suggestion, a nuclear alignment experiment was undertaken, since this would allow a direct determination of the field on the manganese nuclei.

The experiment was performed in the same way as those previously described. Polarizing fields of a few kilooersteds were used to provide the alignment axis, but the apparent hyperfine field was found to be independent of the applied field in the range used (4 to 10 kOe). The value of the hyperfine field was 280 ± 10 kOe. In contrast to this, the experiments on manganese in gold showed that an external field of about 8 kOe is necessary to produce the saturation effect; the hyperfine field was found to be 350 ± 10 kOe.

After the conclusion of these experiments, the heat capacity data of du Chatenier and Miedema[9] became known to us. These data lead to a hyperfine field on manganese in gold which is in good agreement with the one reported here, while the value for manganese in copper is considerably greater. This is to be expected if all the heat capacity is attributed to the manganese when part is due to neighboring nuclei. The effect should be negligible in gold, where the nuclear moments are small, but in copper it may well be important.

A great deal is known about the magnetic and electrical behavior of manganese in copper for manganese concentrations of around 1%. In particular, it is known that these alloys become magnetically ordered with transition temperatures proportional to the manganese concentration and of the order of a few degrees absolute. Manganese–gold alloys have not been studied to the same extent, but their behavior does not seem to be fundamentally different. The alloys used in the present work have extremely small manganese concentration (around 1 in 10^7), though other transition impurities were present with concentration ~ 1 in 10^6. A point of immediate interest is whether these very dilute alloys form a magnetically ordered phase at the temperatures reached in this work (0.01°K). If they do not, the application of a field H such that $\mu H \sim kT$, μ being the moment of the magnetic unit, should suffice to produce the full nuclear alignment. This criterion leads to values of H an order of magnitude smaller than those observed to be necessary in the gold–manganese work. It is concluded that some degree of magnetic order must exist in this case; for the copper–manganese alloy, no conclusion is possible until experiments have been conducted with smaller external fields.

Although these experiments are only of a preliminary nature, they suggest that nuclear alignment experiments are of value in studying transition metal alloys of very low concentration. Both ^{54}Mn and ^{60}Co are ideal nuclei for this purpose, giving large anisotropies in reasonable fields and free from any difficulty of interpretation from purely nuclear causes. An active program of investigation is under way.

Note Added In Proof

The value of 183 kOe for the field on vanadium in cobalt[6] has been questioned and is almost certainly incorrect.[10] Also, continuing experiments on ^{48}V in this laboratory have revealed further errors arising from second-order effects which are important in this case because of the extremely small measured anisotropy. All values given here for ^{48}V should therefore be ignored; they will be replaced in a forthcoming publication.

References

1. J. A. Cameron, R. A. G. Lines, B. G. Turrell, and P. J. Wilson, *Phys. Letters* **4**, 323, 1963.
2. R. M. Bozorth, *Ferromagnetism*, D. Van Nostrand Co., Inc., Princeton, New Jersey (1951).
3. K. Hoselitz, *Ferromagnetic Properties of Metals*, Oxford University Press, London (1952).
4. J. A. Cameron, R. A. G. Lines, B. G. Turrell, and P. J. Wilson, *Phys. Letters* **6**, 167, 1963.
5. C. D. Jeffries, *Prog. Cryogenics* (*London*), **131**, (1961).
6. R. C. La Force and G. F. Day, *Bull. Am. Phys. Soc.* **9**, 212, 1964.
7. N. E. Phillips, private communication.
8. K. Yosida, *Phys. Rev.* **106**, 893, 1957.
9. F. J. du Chatenier and A. R. Miedema, this volume, p. 1029.
10. Itoh, Asayma, and Kobayashi, *Proc. Internat. Conf. Magnetism,* Nottingham, England, 1964.

SPECIFIC HEAT OF $V_{10}Fe_{90}$ ALLOY BETWEEN 0.4° AND 7°K

D. F. Brewer, D. R. Howe, and B. G. Turrell

University of Sussex
Sussex, England

We have measured the specific heat of an alloy of iron with 9.0 ± 0.2 wt.% vanadium in the temperature region 0.4° to 7°K in order to determine the hyperfine contribution to the specific heat. Measurements were made in a ^{3}He cryostat, whose details will be published elsewhere. The specimen, whose mass was 439.1 g, was taken from the casting, machined, and used in the experiment without homogenizing or annealing. The concentration quoted above was obtained by microanalysis, after machining, of specimens from four different regions of the alloy.

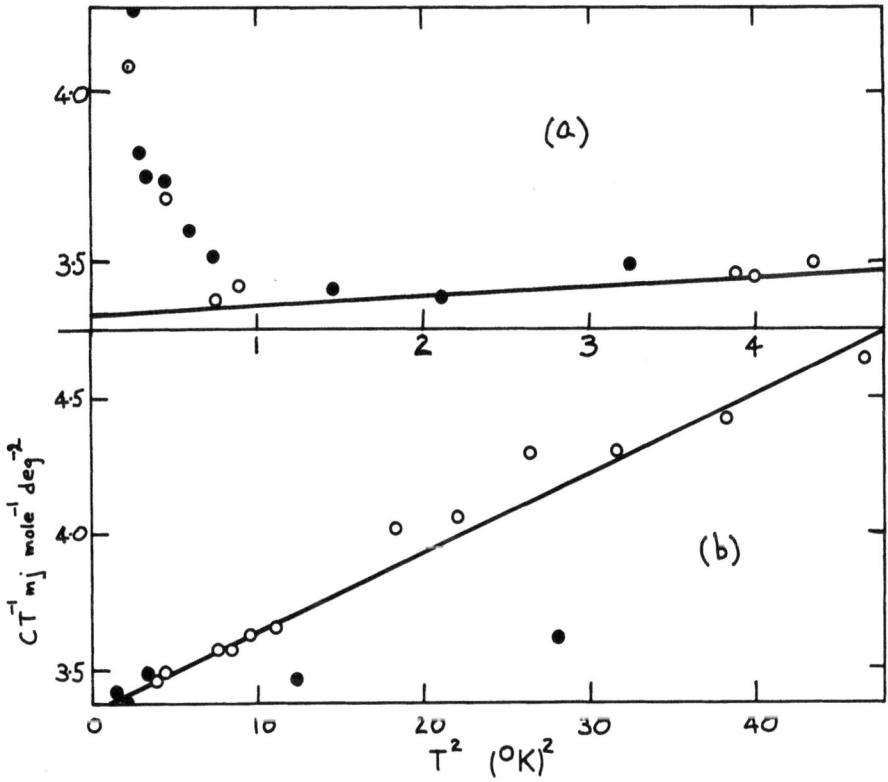

Fig. 1. C/T against T^2 for $V_{10}Fe_{90}$ alloy, in two temperature ranges. Different symbols refer to results obtained on different days.

Temperatures were measured with an "old Speer" 470-Ω carbon resistor, calibrated against ^{3}He and ^{4}He vapor pressures, with the use of a two-constant calibration equation with correction terms below 0.9°K.

The specific heat data are plotted in Fig. 1 as C/T against T^2. The straight line in the low temperature region of Fig. 1a is drawn with a slope and intercept taken from Fig. 1b at higher temperatures. This line gives the electronic and lattice contributions, $aT + bT^3$, with $a = 3.35$ and $b = 0.0291$, in units of millijoules/mole-degree, which fit in with previous measurements on iron alloys with higher vanadium content.[1,2] The remaining contribution in Fig. 1a, attributed to the vanadium hyperfine specific heat, fits a T^{-2} dependence from which we infer an effective field on the vanadium nucleus of 80 ± 15 kG. This value, and the error limits, are to be checked in repeat experiments.

From specific heat measurements on alloys with higher vanadium content, Cheng et al.[1] placed an upper limit of 61 kG on the effective field, rather lower than we find. More recently, nuclear resonance measurements have been made on vanadium–cobalt alloys,[3] and nuclear orientation measurements on vanadium–iron alloys.[4,5] The latter give a value of 90 kG, with fairly large error limits, in agreement with our specific heat measurements.

References

1. C. H. Cheng, C. T. Wei, and P. A. Beck, *Phys. Rev.* **120**, 426, 1960; *Phys. Rev.* **122**, 1129, 1961.
2. R. G. Scurlock and E. M. Wray, *Phys. Letters* **6**, 28, 1963.
3. G. F. Day and R. C. La Force, *Bull. Am. Phys. Soc.* **9**, 212, 1964.
4. J. A. Cameron, I. A. Campbell, J. P. Compton, M. F. Grant, R. W. Hill, and R. A. G. Lines, this volume, p. 1033.
5. R. W. Hill, private communication.

10.3. Electrical Resistivity, Thermoelectric Power, Electronic Specific Heat

MAGNETIC SCATTERING IN DILUTE METALS*

F. T. Hedgcock and D. P. Mathur

Eaton Laboratory, McGill University
Montreal, Canada

The observance of negative magnetoresistance at low temperatures in heavily doped n-type germanium, exhibiting metallic conductivity, was first reported by Sasaki et al.[1] at the 1960 Prague Semiconductor Conference. Since that time, numerous authors have reported similar behavior in other types of degenerate semiconductors. Toyozawa,[2] in 1962, suggested that there was a strong similarity between the behavior of the magnetoresistance of a heavily doped semiconductor at low temperatures and that of dilute paramagnetic metal alloys. It is experimentally known that in order to induce a negative magnetoresistance in these alloys, a localized magnetic moment must be present. Toyozawa suggested that the presence of localized moments in semiconductors could arise from the statistical nature of the distribution of impurity atoms, resulting in localized electron states in the continuum of conduction electron states in the impurity band, the lifetime of electrons in these quasi-localized states being sufficiently long to scatter magnetically the more mobile electrons associated with the impurity band.

At low magnetic fields, the magnetoresistivity in heavily doped n-type germanium at low temperatures is proportional to the square of the magnetic field; the magnetoresistivity then saturates and on further increase of field again varies with the square of the field strength. With an appropriate experimental analysis of the magnetoresistivity results and making use of the theoretical expressions of Yosida[3] and Kasuya,[4] we can calculate a value for the ratio of the s–d exchange interaction constant (J) to the Coulomb scattering constant (A) and solve for a value of the magnetic quantum number (j) of the magnetic scattering center. If these values are used, it is possible to calculate the spin-drag component of the thermoelectric power as a function of the reduced parameter H_0/kT, where H_0 is the pseudo-Wiess field. This curve is shown in Fig. 1, where the experimentally determined spin-drag component of the thermopower is plotted as a function of T^{-1}. (Corrections

---

* This research was supported by the Aeronautical Systems Division until May 31, 1964, under contract AF33(657)–8744 at the Franklin Institute Laboratories, Philadelphia, Pennsylvania.

Table I

Sample	Nf/cm^3 (charge carrier density)	N_i/cm^3	H_0, kOe	j	J/A
B	6.8×10^{17}	$2.0 \pm 0.2 \times 10^{17}$	18	11 ± 1	$3.9 \pm 0.4 \times 10^{-2}$
C	1.3×10^{18}	$3.8 \pm 0.4 \times 10^{17}$	19	13 ± 1	$4.3 \pm 0.4 \times 10^{-2}$
D	3.0×10^{18}	$7.3 \pm 0.7 \times 10^{17}$	27	14 ± 1	$3.5 \pm 0.4 \times 10^{-2}$
E	6.2×10^{18}	$4.5 \pm 0.5 \times 10^{18}$	29	17 ± 1	$2.2 \pm 0.3 \times 10^{-2}$

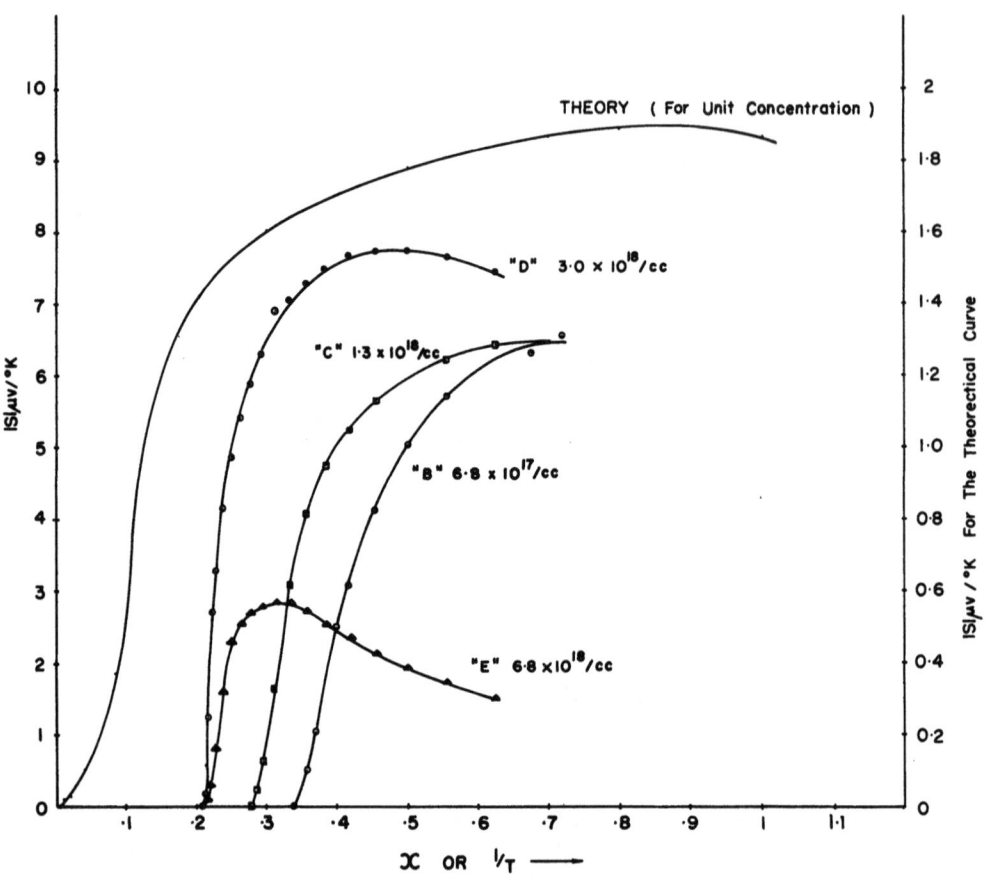

Fig. 1. Theoretical and experimental curves for the spin-drag component of the thermopower as a function of x or T^{-1}. The theoretical curve is calculated for unit concentration of localized spins and the experimental curves have been corrected for the normal electron diffusion term $(\pi^2 k^2 T/|e|\zeta_0)$. It has been assumed in calculating the theoretical curve that H_0 is finite at $T = 0$.

have been made for the normal electron diffusion thermopower.) By adjustment of coordinate scales it is possible, from the data in this figure, to solve for H_0 and the number of magnetic scattering centers, N_i, for each of the samples studied. Table I collects the pertinent parameters derived from the experimental results; it will be noted that J/A is concentration independent while N_i, j, and H_0 are concentration dependent. It would appear that the theory of Kasuya for spin-drag thermopower is adequate to explain the nature of the observed thermoelectric anomalies in heavily doped n-type germanium at low temperatures. This conclusion, if correct, lends further support to Toyozawa's speculation that spin scattering is taking place in the materials.

References
1. W. Sasaki, *et al.*, International Conference on Semiconductors, Prague, 1960.
2. Y. U. Toyozawa, *J. Phys. Soc. Japan* **17**, 986, 1962.
3. K. Yosida, *Phys. Rev.* **107**, 396, 1957.
4. T. Kasuya, *Progr. Theoret. Phys.* (*Kyoto*) **22**, 227, 1959.

THE ELECTRICAL RESISTANCE OF SOME DILUTE GOLD ALLOYS

H. J. M. van Rongen, B. Knook, and G. J. van den Berg

*Kamerlingh Onnes Laboratory**
Leiden, The Netherlands

Measurements of the resistivity of dilute alloys of noble metals with transition elements gave rise to the supposition that only transition elements with 5 or 6 electrons in the d-shell would cause anomalies in the electrical resistance at low temperatures. Therefore a number of gold–rhenium and gold–technetium alloys were investigated. Rhenium caused deep minima in the electrical resistivity, along with negative values for the magnetoresistance. From the results on the gold–technetium alloys it could be concluded that the solid solubility of technetium in gold at 1000°C is very small, and therefore no conclusions could be drawn about technetium as a possible cause of anomalies in the resistance.

Measurements on gold–cobalt alloys showed that cobalt ($3d^7 4s^2$) does not give anomalies below 0.1 at.-%.

Previous measurements on copper–tin–iron alloys showed a stimulating effect of tin on the anomalies due to the iron, but no such effect was found in gold–tin–iron alloys.

---

* Metal group F.O.M.–T.N.O.

THE RESISTANCE OF DILUTE ALLOYS OF MAGNESIUM AND GADOLINIUM OR NEODYMIUM*

A. N. Gerritsen and S. B. Das

Department of Physics, Purdue University
Lafayette, Indiana

In the search for anomalous resistance behavior in dilute paramagnetic alloys, attempts have been made to investigate dilute alloys of magnesium with gadolinium or neodymium as solutes.

For both rare-earth metals the solubility is supposedly less than 0.2 at.%. The values for the resistivity increase due to alloying are known within this solubility range; in this range no anomalous behavior has been observed[1] down to 1.2°K.

In the case of magnesium–gadolinium alloys, the authors find that with the proper melting technique, concentrations as high as 1.38 at.% solute exhibit solid solution properties.

For the atomic resistivity increase, the value 7.5 $\mu\Omega$-cm/at.% is within 10% of the value 8.2 $\mu\Omega$-cm/at.% as reported[1] for the increase in the solid solution range. Comparing this result with the strikingly different results for the magnesium–neodymium alloys, we may conclude that not much of the gadolinium is present as a precipitate. Changes in the slope of the low temperature resistance–temperature curves have been observed similar to those occurring at the onset of magnetic order. Antiferromagnetic transitions have been observed in yttrium–gadolinium alloys[2] and even though yttrium is a paramagnetic metal, the observation with the magnesium–gadolinium alloys indicates a behavior similar to that of the yttrium–gadolinium alloys. The ordering temperature varies proportionally with the gadolinium concentration for the two alloys for which the Néel temperature was observed (0.43 at.% Gd: $T_N \approx 4°K$; 1.38 at.% Gd: $T_N = 13.5°K$; rates 9° and 9.8°K/at.%, respectively).

The magnetoresistance for a 0.10 at.% Gd alloy (for which the Néel temperature would be below 1°K) was normal. For a 0.43 at.% Gd alloy, the magnetoresistance at 4°K was lower than that at 78°K; the value at 20°K was certainly not larger than that at 78°K. A clear-cut anomalous behavior was found for the 1.38 at.% Gd alloy, for which at 4°K the resistance in a magnetic field was lower than that in zero field by 0.01%. This small effect could easily be observed because of a relatively large sample resistance ($R_{4°} = 0.72 R_{273°}$).

When we compare the present results with the rather large negative magnetoresistance of magnesium–manganese alloys (for example $\Delta R_H = -0.05 R_0$ for a 0.7 at.% Mn alloy[3]), we conclude that the interaction between the strongly magnetic f-states and the conduction electrons is weak. Considering that the free moment of the gadolinium ion is approximately three times that of the free manganese ion, we

---

* This work is sponsored by a grant from the National Science Foundation.

wonder whether the magnetoresistance effect is the consequence only of the single 5d-electron at the gadolinium ion.

The results with the magnesium–neodymium alloys can be considered to support this tentative conclusion. Even though the presence of precipitates of the Mg_9Nd phase greatly confuses the results, no magnetoresistance anomalies do occur. We observe, however, for alloys rich enough in solute (0.5 to 1 at.% Nd) a transition temperature, independent of the concentration and hence presenting presumably the Mg_9Nd transition temperature.

In a discussion[4] on resistance anomalies based on a statistical evaluation of the number of electron spins seeing the molecular field created by the rare-earth localized states, it is found that the Néel temperature should vary linearly with the concentration of the rare-earth solute. The proportionality constant α can be expected to be of the order $\alpha = 3°K/at.\%$, though various values can be expected for different solvents. It is interesting to note that the ratio of the α's ($\alpha_{Mg}/\alpha_Y = 9.4/3.4 = 2.8$) for magnesium alloys and the yttrium alloys[2] is very close to the ratio between the coupling intensities λ that are the factors that can be expected to differ dominantly for different types of solvents.

Using the expression for λ from Sugarawa and Yamase,[2] we find

$$\lambda_{Mg}/\lambda_Y = (E_f)_{Mg}/(E_f)_Y \times (V_Y/V_{Mg})^2 \approx 1 \times (18.9/14.2)^2 = 1.8$$

In this crude comparison the Fermi energies E_f have been taken as the same for magnesium and yttrium; V is the atomic volume.

Using the same assumptions, we would arrive at the ratio $\lambda_{Ag}/\lambda_Y \approx 0.9$ (in this case the difference in valence has to be taken into account), or $\alpha \approx 3°K/at.\%$ for silver–gadolinium alloys. This does not agree with the experimental result: $\alpha \approx 7°K/at.\%$.[5] Considering the simplification made in the theoretical calculations, the uncertainty in the choice of the statistical distribution of aligned spins and the uncertainty in fundamental quantities, the exact agreement between the theory and the experimental results for the magnesium and yttrium alloys may be accidental. However, we can also find an estimate for the gadolinium spin interaction energy by using the Yoshida expression[6] for the total resistance drop between the Néel temperature and 0°K. This expression is less sensitive than that for λ for the differences in the solvent material, but it still is satisfying to see that the estimate for the exchange integral for the magnesium alloys turns out to be 0.11 eV,* which is the same value as was found for the magnesium–gadolinium alloys.

References

1. J. Bijvoet, B. de Hon, A. J. Dekker, and G. W. Rathenau, *Solid State Comm.* **1**, 237, 1963.
2. T. Sugawara and I. Yamase, *J. Phys. Soc. Japan* **18**, 1101, 1963.
3. F. T. Hedgcock and Y. Muto, *Phys. Rev.* **134** (A), 1593, 1964.
4. M. T. Béal, *J. Phys. Chem. Solids* **25**, 543, 1964.
5. T. Sugawara and R. Soga, *J. Phys. Soc. Japan* **19**, 780, 1964.
6. K. Yoshida, *Phys. Rev.* **107**, 396, 1957.

---

*From the graph in reference 5 for 0.5 at.% gadolinium in silver, the estimated exchange energy would be 0.09 eV.

RESISTIVITY AND MAGNETORESISTANCE OF DILUTE ALLOYS OF IRON IN $Mo_{0.8}Nb_{0.2}$

M. P. Sarachik

Bell Telephone Laboratories
Murray Hill, New Jersey

and

The City College
New York, New York

The resistivity *vs.* temperature and resistivity *vs.* magnetic field have been measured for nine dilute alloys of iron in $Mo_{0.8}Nb_{0.2}$. Iron concentrations ranged from 0.2 to 2.3%. Minima are observed in the resistivity *vs.* temperature curves in the range 28°K to 38°K, depending on iron concentration. It is found that both the temperature-independent impurity scattering and the spin scattering are linear functions of iron concentration for low concentrations, and the spin resistivity is a logarithmic function of temperature. The magnetoresistance is negative in all cases, and follows the approximate law $\Delta\rho = AH^x$, where the exponent x tends toward the value 2 for infinite dilution, and is smaller for finite values of the concentration. These results lend support to a recent theory by J. Kondo, which predicts both the linear dependence of the spin scattering on concentration, and its logarithmic dependence on temperature. If this theory is used, the s–d exchange integral is found to be negative and equal to -0.2 eV.

RESISTIVITY AND THERMOELECTRIC POWER OF SILVER–PALLADIUM ALLOYS

P. R. F. Simon

National Research Council
Ottawa, Canada

The resistivity and the thermoelectric power of the entire system of silver–palladium alloys have often been studied by different authors.[1] In fact, it is a system which, from the electronic point of view, is relatively simpler than the other alloys of noble metals with transition metals because it is free from ferromagnetism. The results of these studies are generally in good qualitative agreement with Mott's theoretical computations.[2] Nevertheless, the majority of the experimental results are given at relatively high temperatures, where influence of lattice vibrations is appreciable. We thought it might be interesting to make a study at very low temperature (between 0.1° and 4.2°K) of resistivity and thermoelectric power variations in these alloys as a function of concentration. These variations are caused by electron diffusion only. The Debye temperatures Θ_D for silver and palladium are relatively high (223° and 270°K). Consequently, it is unlikely that the electron–phonon interaction is important at low temperature. Experimental results have confirmed this hypothesis. The thermoelectric power in relation to temperature is linear up to approximately 2°K.

The samples, wire-shaped of 1 mm in diameter and 10 cm in length, were made of spectrographically pure metals and provided by Messrs. Johnson, Matthey and Co. Ltd. They were annealed under vacuum for 36 hr at 680°C before use. The resistivity was measured between 1.2° and 4.2°K and at 0°C. The thermoelectric power was measured between 0.1° and 3.2°K, making use of adiabatic demagnetization methods.[3] The specimen was mounted with one end in contact with a liquid-helium bath and the other end with a paramagnetic "pill." The potential difference E as a function of temperature T was measured during the warming up of the pill, with a superconducting reversing switch and a galvanometer amplifier.

Figures 1 and 2 represent the variations with alloy concentration of resistivity at 4.2°K and of thermoelectric power at 1°K, respectively.

Resistivity

It is known that the deviation of the resistivity from Nordheim's rule in alloys containing transition metals can be explained by the presence of an incomplete d-band in the transition metal (0.65 holes per atom for palladium). This permits s–d transitions which increase the resistivity. We have to consider two terms: (1) the residual resistivity ρ_i, and (2) the temperature-dependent term $\rho - \rho_i$. The basic features of these two terms with the concentration can be explained, with a very good quantitative agreement, by the rigid band theory.[4] Both the lattice scattering and the residual resistivity can be regarded as a sum of two terms, one from s–s scattering and one from s–d scattering when holes exist in the palladium d-band, the latter being proportional to the density of states in the d-band.

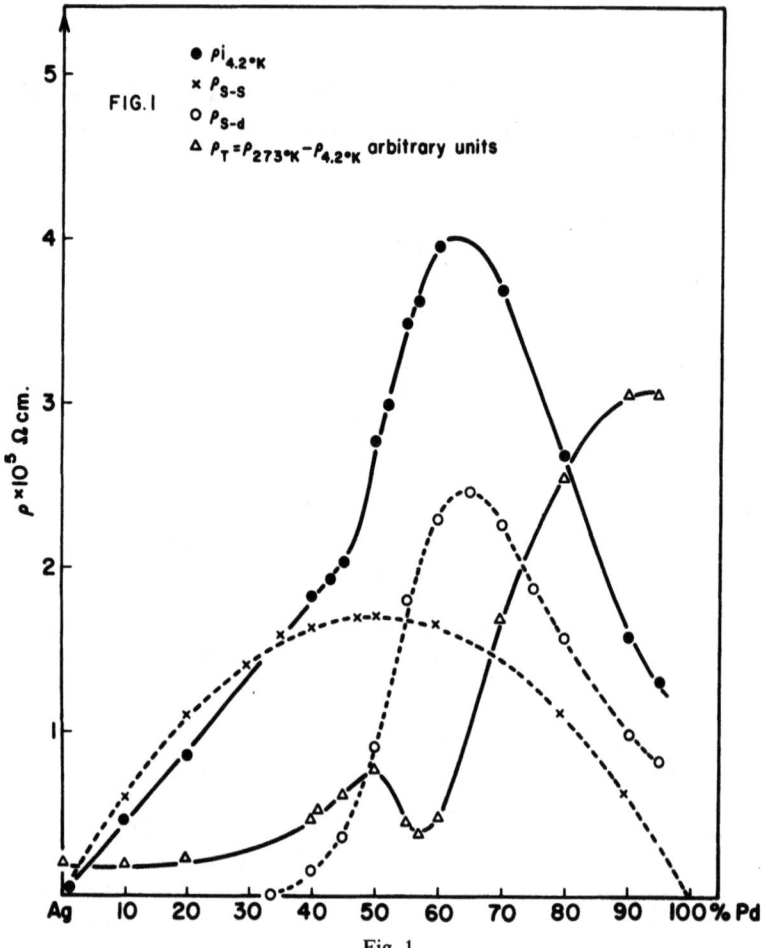

Fig. 1.

Residual Resistivity. In the silver-rich part, Nordheim's rule is obeyed, and the residual resistivity varies as $\rho_i = Kc(1-c)$, where c is the atomic fraction of the palladium component, but at 35% Pd, the s–d scattering begins to appear and for $c > 0.35$, ρ_i is the sum of two terms, ρ_{s-s} and ρ_{s-d} where $\rho_{s-s} = Kc(1-c)$ as before and $\rho_{s-d} = Bc(1-c)^2$, where B is directly proportional to the density of states in the d-band.[2–4]

We have plotted the different components ρ_{s-s} and ρ_{s-d} in Fig. 1.

Temperature-Dependent Resistivity $\rho_T = \rho - \rho_i$. The experimental points for 273°K are shown by triangles in the lower portion of Fig. 1. If the density of states in the d-band does not vary any more rapidly with the energy than that in the s-band, then

$$\rho_T \propto \gamma - \gamma_{0.35} + (\rho - \rho_i)_{0.35}$$

It is, however, well known that[2] the high temperature resistivity of pure palladium is modified by a factor $(1 - AT^2)$ due to the strong energy dependence of $N_d(E)$.

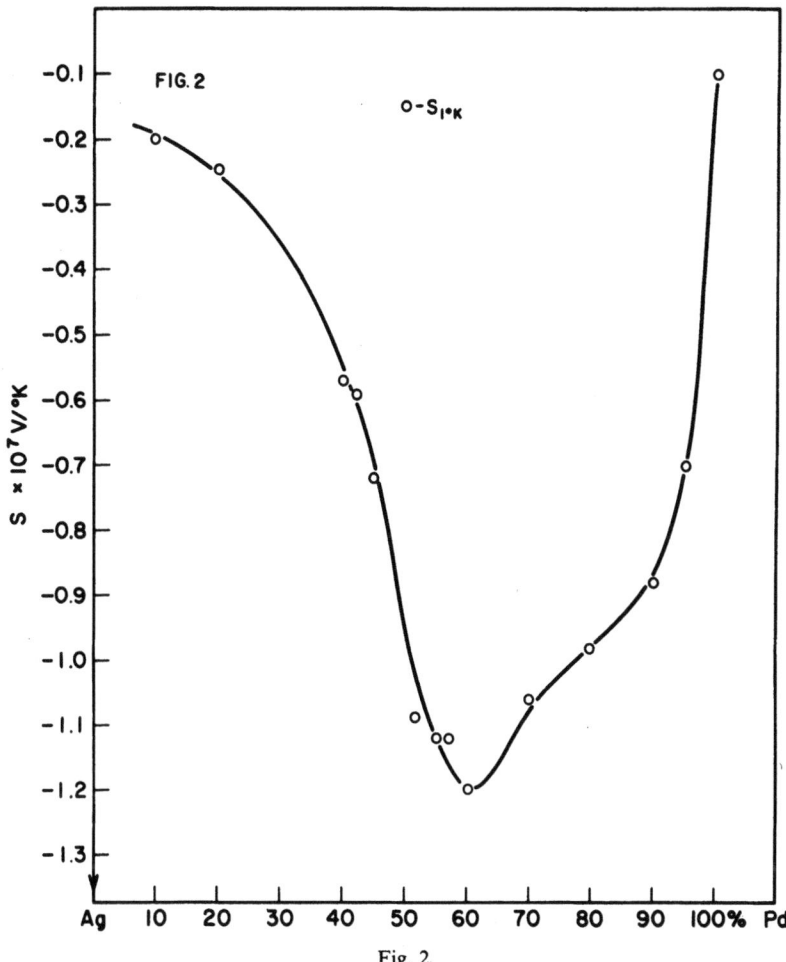

Fig. 2.

The additional effect in alloys is that the portion of ρ_i coming from s–d scattering is also modified by this $(1 - AT^2)$ factor as was pointed out by Coles and Taylor.[4] The large negative term thus introduced into the temperature-dependent part of the resistivity results in the apparent anomalies of Fig. 1 in the region of large residual resistivity.

Thermoelectric Power

It is well known that the thermoelectric power S of a metal due to electron diffusion can be written[5]

$$S = \frac{\pi^2 k^2 T}{3eE_F} \left[\frac{d \log N(E)}{d \log E} + \frac{d \log v(E)}{d \log E} + \frac{d \log l(E)}{d \log E} \right]_{E_F} \qquad (1)$$

where $N(E)$ is the density of states, $v(E)$ the electron velocity, and $l(E)$ the electron mean free path.

This formula is valid for pure metals at temperatures $T > \Theta_D$ and for alloys at low temperature provided that the term ρ_i is greater than ρ_T.

The silver-rich part, having a full d-band and 0.65 to 1.0 electrons per atom in the s-band, should be discussed in terms of "normal" scattering behavior by impurity centers. Presumably the introduction of palladium atoms reduces the number of electrons in the first Brillouin zone so that the Fermi surface tends toward a spherical shape, and we might hope that the assumption of "free electrons" ($n(E) \propto E^{\frac{1}{2}}, v \propto E^{\frac{1}{2}}$) is justifiable. We can then use equation (1) to deduce a rough value for $d \log l(E)/d \log E$. For 10 at.% Pd, $S = -2 \times 10^{-8}$ V/degK at 1°K and so we have, from equation (1), $(d \log l(E)/d \log E) \simeq 3.6$. This comparatively modest value indicates the absence of anomalous effects due to magnetic impurities. Above $c = 0.3$–0.4, the magnitude of the thermoelectric power is rapidly increased, presumably indicating that the s–d transitions begin to appear, reaches a maximum at $c = 0.6$ and then decreases again.

If we consider the contribution of the s–d transitions and if we neglect the s–s contribution, we get

$$S = \frac{\pi^2 k^2 T}{3e} \left[\frac{d \log N_d(E_F)}{dE} \right]_{E_F}$$

The results given by Montgomery[6] on the electronic specific heat are such that $(\gamma - \gamma_{0.40})/\gamma_{0.40}$ is linear in concentration between $c = 0.40$ and $c = 1$. Or, because

$$\frac{1}{N(E)} \cdot \frac{dN(E)}{dE} = 2 \frac{dN(E)}{dc} \quad \text{and} \quad \gamma = \tfrac{2}{3}\pi^2 k^2 N(E)$$

we obtain

$$S = +\frac{T}{e} \frac{d\gamma}{dc} \tag{2}$$

This should give rise to a constant (i.e., independent of concentration) and negative thermoelectric power in this concentration range, of the order of magnitude -1.5×10^{-7} V/degK at 1°K. This is far from being the case. Perhaps this is because the contribution of the d-band holes is not negligible. Even though conduction by d-band holes appears to give only a negligible contribution to the total conductivity, it is quite possible that because of the narrowness of the d-band, these can play a fairly important role in the thermoelectric power. If we take a value of the order of magnitude of 0.15 eV for the empty d-band width in pure palladium, we get approximately

$$S_d = +\frac{\pi^2 k^2 T}{3eE_F} \cdot \frac{d \log \sigma_d}{d \log E} \sim +1.5 \, 10^{-7} \text{ V/deg K at 1°K}$$

The Hall effect data[7] for a dilute alloy of silver in palladium can be interpreted as indicating a ratio of mobility of the d- to s-band of 0.33. Assuming $n_s = n_d$, this leads to a value of S for the dilute alloy given by: $S = S_s + (\sigma_d S_d/\sigma_s) = -0.7 \cdot 10^{-7}$ V/degK at 1°K in agreement with the experimental data. When silver is added, the d-band is progressively filled up and σ_d decreases toward zero. This gives rough agreement with the observed concentration dependence of the thermoelectric power.

References

1. J. M. Ziman, *Electrons and Phonons*, Oxford University Press, Fair Lawn, New Jersey (1960).
2. R. F. Mott, *Proc. Phys. Soc. (London)* **47**, 571, 1935; *Proc. Roy. Soc. (London) Ser. A* **156**, 368, 1936; *Phil. Mag.* **22**, 287, 1936.
3. D. K. C. MacDonald, W. B. Pearson, and I. M. Templeton, *Proc. Roy. Soc. (London) Ser. A* **256**, 334, 1960.
4. B. R. Coles and J. C. Taylor, *Proc. Roy. Soc. (London) Ser. A* **267**, 139, 1962.
5. See, for example, D. K. C. MacDonald, *Thermoelectricity*, John Wiley & Sons, Inc., New York (1962).
6. H. Montgomery, private communication.
7. A. I. Schindler, NRL Report 4788, 1956.

THE THERMOELECTRIC POWER OF NICKEL AND DILUTE NICKEL–COPPER ALLOYS

D. Greig and J. P. Harrison

Physics Department, University of Leeds
Leeds, England

Although many broad features of the variation of the thermoelectric power of metals with temperature are now understood, a number of anomalies remain. Included among these is the so-called "giant" thermoelectric power, which occurs at very low temperatures when small concentrations of impurity, and in particular ferromagnetic impurity, are added to a number of metals.[1] Several suggestions have been made regarding the exact role played by the solute magnetic ions in this process, but none of these is entirely satisfactory.[2] It therefore seems of interest to study the thermoelectric power of an actual ferromagnetic metal over a wide temperature range, and to investigate how this changed on alloying.

We have recently carried out a series of experiments designed to measure the electrical and thermal conductivity of nickel and dilute nickel–copper alloys in the temperature range 1° to 120°K. It was quite simple to extend these experiments to include the measurement of data on the absolute thermoelectric power, S. The results of measurements made on pure nickel and five alloys are shown in Fig. 1.

The pure nickel specimen was a 2-mm-diameter rod obtained from Messrs. Johnson, Matthey and Co. Ltd., while the alloys were prepared from J.M. 890 nickel and J.M. 30 copper from the same supplier. Each of the specimens, including the pure nickel, was annealed at 850°C before being mounted in the cryostat. The composition of alloys C and D was determined by X-ray microanalysis.* The concentration of copper in the other alloys was then deduced from these results and from measurements of the residual resistivity, as this is known to increase by 0.75 $\mu\Omega$-cm per added atomic percent copper.[3] The X-ray technique was also applied to specimen B, but failed to detect any copper, even though the limit of detectability was given as $\pm 0.05\%$ copper. Since one of the interesting features in the overall results is the pronounced maximum in alloy B, we checked that this was not due to some accidental impurity by making another specimen of similar nominal composition. This additional specimen was alloy F.

A recent theoretical model of nickel proposed by Ehrenreich, Philipp, and Olencha,[4] following magnetoresistance measurements by Fawcett and Reed,[5] suggests that the Fermi surface for the $\downarrow$ electrons is topologically the same as the Fermi surface of copper. They further suggest that it is those very copperlike s-electrons which dominate the electrical conductivity, and that their relaxation time "is quite similar to that for the corresponding electrons in copper." Since, however, the measured values of S in nickel are negative at all temperatures (Fig. 1), in contrast to the positive values found in the noble metals, it is clear that thermoelectric

---
* This analysis was performed by Metals Research Ltd., Cambridge, England.

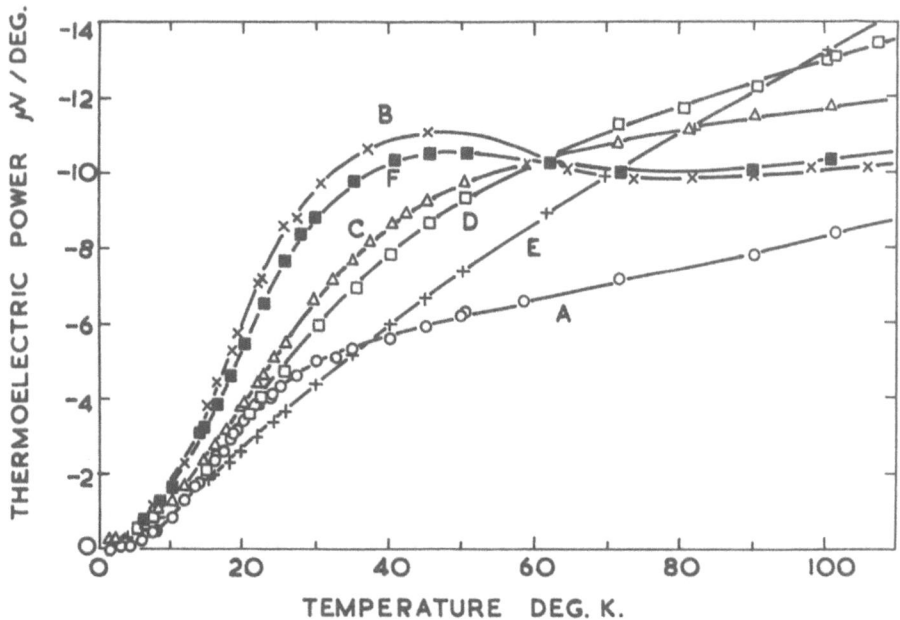

Fig. 1. Variation of absolute thermoelectric power with temperature for pure nickel (specimen A) and 5 nickel–copper alloys. Copper content of various specimens: B, 0.1 at.%; C, 0.6 at.%; D, 1.5 at.%; E, 4.9 at.%; F, 0.16 at.%.

phenomena cannot be discussed in terms of such a simple model, and electrons of both spin directions must be considered.

For two independent groups of charge carriers in a single conductor, the net diffusion thermopower is given by[6]

$$S_d = (\sigma_1 S_1 + \sigma_2 S_2)/(\sigma_1 + \sigma_2) \quad (1)$$

where σ_1 and σ_2 are the electrical conductivities of each of the groups, and S_1 and S_2 the thermopowers which would arise if each group were acting alone. Let us consider the ↓ electrons as group 1 and the ↑ electrons as group 2. On the proposed model $\sigma_\downarrow \gg \sigma_\uparrow$, so that the first term in (1) will be $\sim S_\downarrow$ and therefore probably similar in magnitude and sign to the thermopower of copper. The large negative S_d found in nickel must therefore originate in the second term, $\sigma_\uparrow S_\uparrow/\sigma_\downarrow$, even though the ratio $\sigma_\uparrow/\sigma_\downarrow$ is small.

At high temperatures, or at very low temperatures, the normal diffusion thermopower for negative charge carriers is given by[7]

$$S_d = \frac{-\pi^2 k^2 T}{3|e|} \left[\frac{\partial \ln \mathscr{A}}{\partial \varepsilon} + \frac{\partial \ln l}{\partial \varepsilon} \right]_{\varepsilon = \varepsilon_F} \quad (2)$$

where $\mathscr{A}$ is the effective area of the Fermi surface and l is the electronic mean free path. Although $\partial \ln \mathscr{A}/\partial \varepsilon$ is certainly positive for the ↑ electrons, there seems no reason why it should be unusually large. The thermopower of nickel is therefore probably determined by the second term in (2), the scattering term relating to the ↑ carriers.

At low temperatures the thermopower is the sum of S_d and a phonon drag term S_g, which has its maximum value in the temperature range 20° to 60°K. There is clear evidence of such an additional term in many of the curves shown in Fig. 1, but contrary to expectation, the maxima are more pronounced in the two most dilute alloys than in the pure metal. We have examined this anomaly by attempting to separate the S_d and S_g terms.

At the lowest temperatures, S_g has the same T^3 temperature dependence as the lattice specific heat. We have found that below 15°K a plot of S/T vs. T^2 gives a series of straight lines, so that at these temperatures the separation of S_d and S_g is easily effected.* At ~ 100°K, however, the temperature dependence of S_g is uncertain, and there is evidence to suggest that by room temperature this component of S is quite negligible.[9] We have therefore tried to estimate S_d over the whole temperature range by linking up earlier high temperature data† with our data from below 15°K, making use of the Kohler relationship,[11]

$$S_d = \sum_i W_i S_i / \sum_i W_i$$

(S_i and W_i represent the thermopower and electronic thermal resistivity, respectively, arising from the ith scattering process.) The values of S_g which are finally obtained when S_d is subtracted from the measured S are shown in Fig. 2.

* This technique was first used by A. R. de Vroomen et al.[8]
† For a summary of earlier work, see J. Nyström and Landolt-Börnstein.[10]

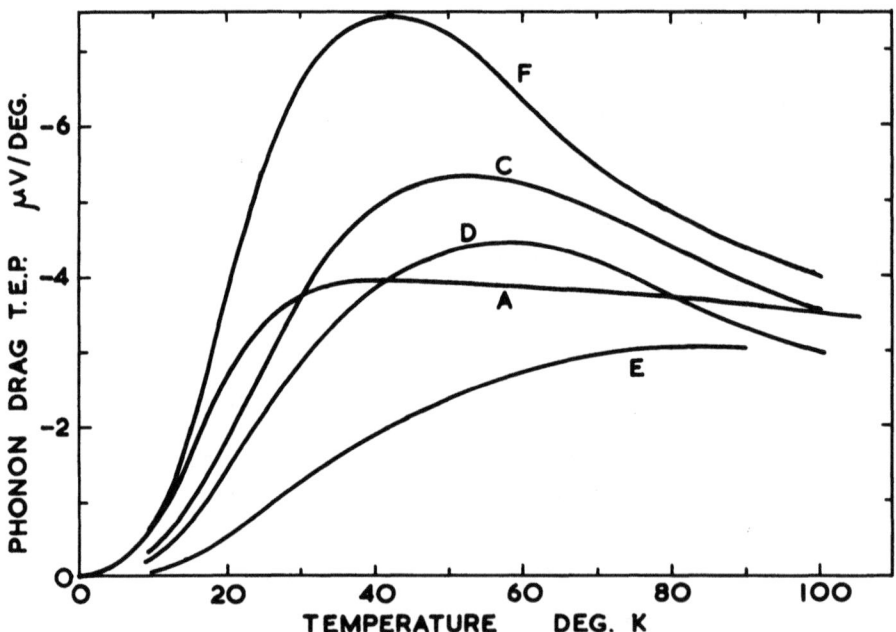

Fig. 2. Estimated variation of the phonon drag component of thermoelectric power with temperature. Lettering on curves refers to specimens in Fig. 1. The results from the later specimen F were chosen for detailed analysis in preference to those from B, but both sets of measurements would give almost identical curves on this diagram.

It can be seen that whereas in the alloys the value of S_g falls when the copper concentration is increased, the curve for the pure metal is out of order. Furthermore, although the decrease through the alloy series is quite expected owing to the increased scattering of phonons by defects, it is surprising that the decrease is so marked at temperatures below the maxima.

One explanation of the initial of S_g on alloying might be that whereas in the pure metal S_g is reduced by a phonon–magnon interaction, in the alloys this additional process disappears. Alternatively, the increase might be associated with the enhanced electron–phonon interaction arising from inelastic scattering at impurity atoms,[12,13] although Klemens[13] has predicted that this effect would actually reduce S_g.

Acknowledgments

This work was carried out while J.P.H. was in receipt of a D.S.I.R. Maintenance Grant, for which he acknowledges thanks. The cryostat was constructed by Mr. J. Palmer, who also kept us supplied with liquid helium.

References

1. G. J. van den Berg, *Progress in Low Temperature Physics*, Vol. *IV*, C. J. Gorter (ed.), North Holland Publishing Co., Amsterdam (1964), p. 215.
2. D. K. C. MacDonald, *Thermoelectricity*, John Wiley & Sons, Inc., New York (1962), p. 69.
3. E. I. Kondorskii, O. S. Galkina, and L. A. Chernikova, *Soviet Phys. JETP (English Transl.)* **7**, 741, 1958.
4. H. Ehrenreich, H. R. Philipp, and D. J. Olencha, *Phys. Rev.* **131**, 2469, 1963.
5. E. Fawcett and W. A. Reed, *Phys. Rev.* **131**, 2463, 1963.
6. D. K. C. MacDonald, *Thermoelectricity*, John Wiley & Sons, Inc., New York (1962), p. 115.
7. J. M. Ziman, *Electrons and Phonons*, Oxford University Press, Fair Lawn, New Jersey (1960), p. 397.
8. A. R. de Vroomen, C. van Baarle, and A. J. Cuelenaere, *Physica* **26**, 19, 1960.
9. F. J. Blatt, *Proc. Phys. Soc. (London) Ser. A* **83**, 1065, 1964.
10. J. Nyström and Landolt-Börnstein, *Zahlenwerte und Funktionen*, Springer-Verlag, Berlin (1959), Vol. 2(6), p. 929.
11. M. Kohler, *Z. Physik* **126**, 481, 1949.
12. S. Koshino, *Progr. Theoret. Phys. (Kyoto)* **24**, 484, 1960; **24**, 1049, 1960; **30**, 415, 1963.
13. P. G. Klemens, *Proceedings of International Conference on Crystal Lattice Defects*, Vol. *II* (1963), p. 77.

SPECIFIC HEAT MEASUREMENTS ON A SERIES OF GADOLINIUM–PRASEODYMIUM ALLOYS

B. Dreyfus, J. C. Michel, and A. de Combarieu

Centre de Recherches sur les Très Basses Températures
Grenoble, France

and

Centre d'Etudes Nucléaires de Grenoble
Grenoble, France

Introduction

In recent years, many specific-heat measurements have been performed on pure metallic rare earths, especially in the region below 4°K, aimed at determining the hyperfine coupling constant, through the identification of the tail in AT^{-2} of the Schottky anomaly which becomes important in this region.[1-10] The fact that rare-earth ions in metals seem to keep a definite $4f$ shell allows a comparison with the hyperfine coupling constants determined in salts, by the technique of electron paramagnetic resonance. This has been done extensively[5,11] and successfully on the whole series, including some information on quadrupoles coupling. In most cases in the metal, the exchange field is strong enough to draw the $4f$ shell in its maximum $J_z = J$ value. We then have a simple formula which connects the constants A_m and A_s for the metal and the dilute salt, respectively, the second being calculated from E.P.R. results:

$$\frac{A_m}{A_s} = \frac{4g^2 J^2}{g_\parallel^2 + 2g_\perp^2}$$

where (g, J) is related to the $4f^n$ shell and $g_\parallel g_\perp$ is related to E.P.R. results on salt of axial anisotropy.

The values of A obtained, using praseodymium, seem to be in strong disagreement with the others. The experimental A[5,8,9] lies between 20 to 35 mJ-°K/mole, much too low compared to the theoretical value ($\sim$1200 mJ-°K/mole). To face the situation, it has been advanced[5,12] that the exchange field was unable to overcome the crystalline field, so that instead of being in its maximum $J_z = J$ value, the praseodymium ion was in a nonmagnetic singlet ground state, with no hyperfine coupling. In order to check this idea, we have undertaken a series of measurements on gadolinium–praseodymium alloys, with the hope that the exchange field created by gadolinium (which is a good ferromagnetic when pure) would increase the magnetic moment of praseodymium to its maximum value. It must be recalled that gadolinium itself has no appreciable hyperfine coupling, due to its lack of orbital moment.

Recently, neutron diffraction experiments[13] brought more information. It has been shown that pure praseodymium was in fact in an ordered antiferromagnetic state, but with rather small magnetic moments per ion of the order of 0.7 to $1\mu_b$, much below the theoretical maximum value, which for $J = 4$, $g = \frac{4}{5}$ is $3.2\mu_b$, so that the purpose of our experiment is essentially unchanged.

Results

The experiments, in a preliminary step, which we present here, were made above 1.3°K, as the expected Schottky anomaly is rather large. The calorimeter is a classical one; it needs no further comment. The five samples were ingots, cylindrical in form ($\phi = 12$ mm), of weight about 30 g each. They were provided by Johnson, Matthey and Co.; their concentration is directly deduced from the weight of the components; no attempt has been made to check their homogeneity or actual concentration.

Figure 1 presents our results below 4°K; we have added the results on pure praseodymium obtained by Dreyfus et al.,*[5] which show no important deviation from Lounasmaa's.[8] The results were fitted to an equation of the form $AT^{-2} + \gamma T + \beta T^3$, where no attempt was made to take into account a more appropriate term for the magnetic contribution. This fact, added to the limitation of measurements down to 1.3°K, lowers the precision of A and makes it difficult to estimate. However for concentration smaller than 50%, we think it is not worse than 10%. Our results for A are reported in Table I. On the first line, they have been listed in terms of per mole of alloy. On the second line, the effect has been attributed to praseodymium atoms only, the same value has been listed as per mole of praseodymium.

Our results up to 80°K are reported in Fig. 2. They were obtained, for the whole series, by assuming a constant electronic contribution $\gamma = 10.5$ mJ/mole-degK^2 and a constant Debye temperature, which has been chosen to be 150°K.

* The value of A (35 ± 3) deduced from this curve[5] was in error. Our corrected value is 20 ± 1 closer to that of Lounasmaa (20.9) than that of Dempesy (37.5).

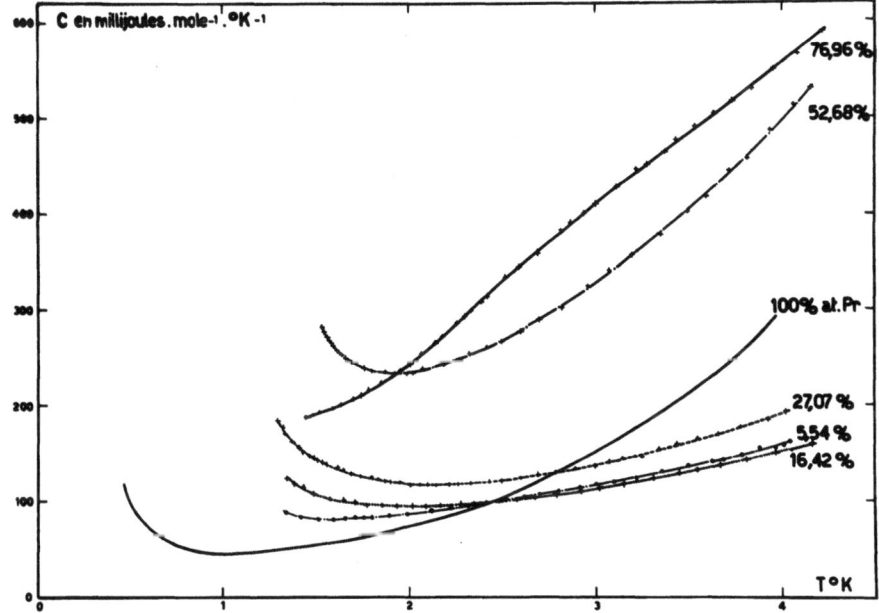

Fig. 1. The molar specific heat of praseodymium–gadolinium alloys.

Table I

C at.% Pr	5.54	16.42	27.07	52.68	76.96	100
A in mJ-degK/mole of alloy	62	154	228	437		20
A in mJ-degK/mole of praseodymium	1120	940	840	830	<100	20

Discussion

The most striking feature of our results is the increase by a factor of more than 50 of the constant A related to 1 mole of active atom. It is the first time, to our knowledge, that such a variation has been observed. The closeness of the

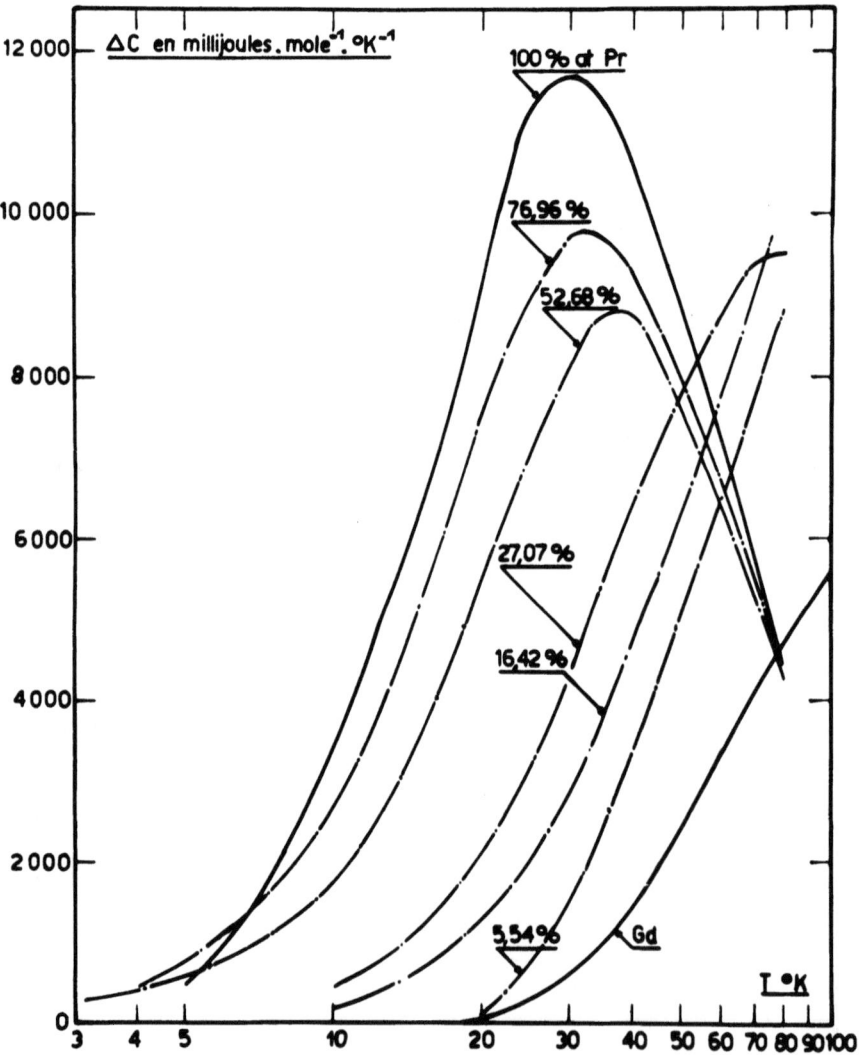

Fig. 2. Experimental molar specific heat of praseodymium–gadolinium alloys after subtraction of a constant electronic contribution $10.5T$ and a constant lattice contribution to θ_D (150°K).

experimental value for the most dilute alloy ($A = 1120$) to the predicted value in the presence of a strong exchange field ($A \sim 1200$) is a strong argument in favor of the possibility to use for praseodymium the general considerations which have been successful for the other rare-earth metals. It is clear that the low value observed for pure praseodymium is due to the relative weakness of the exchange interaction which is not able to develop a $J_z = 4$ fundamental level.

It would be interesting to find a model for the position of the $2J + 1 = 9$ levels of Pr^{+++} in the presence of both crystalline and exchange fields; however, this task is complicated by several facts:

1. In pure praseodymium, there exist two types of sites in equal number, half of them having hcp neighboring, the others fcc neighboring. Pure gadolinium has only sites of the hcp type, and when alloying occurs, we do not know exactly what happens. It is probable that the relative proportion of hcp sites is increasing with the content in gadolinium.
2. The form of the crystalline field is unknown. The most precise proposed until now (Bleaney[12]), based upon point charge model, seems somewhat inadequate since the more recent neutron diffraction experiment.
3. Presently the magnetic structure of praseodymium is still lacking in precision, and we must determine whether the magnetization is borne by both sites with a maximum value of $0.7\mu_b$ per atom, or only by the hcp sites, with a maximum value of $1\mu_b$.

Both possibilities are not in disagreement, as was pointed out by Cable et al.,[13] with the hyperfine coupling. Since A varies as $\langle J_z^2 \rangle$, if we admit that, for example, only the hcp sites have a magnetization of $1\mu_b$, after division by 2 due to sine modulation of the magnetic structure, we would find for $A \sim 30$ mJ-degK/mole not too far from the experimental value.

With our present results we can conclude that somewhere between 75 and 50% Pr, the lowest electronic level undergoes a drastic change of J_z from 1 to 4 under the action of an increasing exchange field. It seems to us that this change could affect only one type of site (probably hcp), the regular increase of A for concentration between 50 and 5% Pr being attributed to the relative increase of the proportion of hcp sites. Such an important change of the ground state is usually connected with a crossing, or a near crossing, of two lines of the scheme of levels plotted against the exchange field, the first excited level coming closer to the ground state. This is in accordance with the higher specific heat (Fig. 1) observed for the two samples in the critical region.

However, at the same time the other excited states do not appear to be strongly affected as can be seen in Fig. 2, where the ΔC curves vary smoothly for 100,[14] 75, and 50% Pr.

For a lower content of Pr (0,[15] 5, 15, and 25%), the ΔC curves of Fig. 2 have been plotted on a log scale against $1/T$, giving straight lines whose slopes are proportional to the exchange field (assuming for simplicity that all ions have the same set of equally spaced levels). This procedure gives a crude estimation of the mean exchange field, expressed relative to that of pure gadolinium in Table II.

Table II

C at.% Pr	(pure gadolinium)	5	15	25
H_{ex}	1	0.91	0.65	0.45

Note Added in Proof

In their letter dated April 5, 1965, Johnson Matthey and Co., who furnished the samples, informed us that as a result of an error in sample fabrication, the samples contained important percentages of yttrium (up to 30%), in addition to gadolinium and praseodymium. Since yttrium is magnetically inert and does not contribute to hyperfine coupling, we think that the main conclusions of this work remain valid, at least semiquantitatively.

References

1. N. Kurti and R. S. Safrata, *Phil. Mag.* **3**, 780, 1958.
2. R. M. Stanton, L. D. Jennings, and F. H. Spedding, *J. Chem. Phys.* **32**(2), 63, 1960.
3. J. G. Dash, R. D. Taylor, and R. P. Craig, *Proceedings of the Seventh International Conference on Low Temperature Physics*, University of Toronto Press, Toronto, Canada (1961), p. 705.
4. B. Dreyfus, B. B. Goodman, G. Trolliet, and L. Weil, *Compt. Rend. Acad. Sci.* **252**, 1743, 1961; **253**, 1085, 1961.
5. B. Dreyfus, B. B. Goodman, A. Lacaze, and G. Trolliet, *Compt. Rend. Acad. Sci.* **253**, 1764, 1961.
6. O. V. Lounasmaa and R. A. Guenther, *Phys. Rev.* **126**, 1357, 1962.
7. O. V. Lounasmaa and P. R. Roach, *Phys. Rev.* **128**, 622, 1962.
8. O. V. Lounasmaa, *Phys. Rev.* **126**, 1352, 1962; **128**, 1136, 1962; **129**, 2460, 1963; **113**, 211, 1964; **133**, 219, 1964; **133**, 502, 1964; **134**, 1620, 1964.
9. C. W. Dempesy, J. E. Gordon, and T. Soller, *Bull. Am. Phys. Soc.* **7**, 309, 1962.
10. H. Van Kempen, A. R. Miedema, and W. J. Huiskamp, *Physica* **30**, 229–236, 1964.
11. B. Bleaney, *J. Appl. Phys.* **34**(4), 1024, 1963.
12. B. Bleaney, *Proc. Roy. Soc. (London) Ser. A* **276**, 39, 1963.
13. J. W. Cable, R. M. Moon, W. C. Koehler, and E. O. Wollan, *Phys. Rev. Letters* **12**, 553, 1964.
14. D. H. Parkinson, F. E. Simon, and F. H. Spedding, *Proc. Roy. Soc. (London) Ser. A* **207**, 137, 1951.
15. M. Griffel, R. E. Skochdopole, and F. H. Spedding, *Phys. Rev.* **93**(4), 657, 1954.

LOW TEMPERATURE SPECIFIC HEAT OF SOLID SOLUTIONS OF THE THIRD TRANSITION SERIES*

E. Bucher,† F. Heiniger,† and J. Muller†
Swiss Federal Institute of Technology
Zurich, Switzerland

The purpose of this paper is to present new measurements of the electronic specific heat of alloys belonging to the 5d-transition series and to compare these with corresponding data in the case of body-centered cubic 3d-, 4d-, and 5d- alloys. The work on the tantalum–tungsten and tungsten–rhenium systems has been undertaken primarily with the aim of studying the occurrence of superconductivity. This aspect will be discussed in a separate paper.[1]

Details concerning the experimental methods have been given elsewhere.[2] The new calorimetric data on the 5d-series, obtained in the temperature range 1.2° to 20°K, are summarized in Table I. The plot of the electronic specific heat coefficients, γ, vs. electron concentration (Fig. 1), is completed by the reproduction of results of Beck and co-workers[3,4] on 3d alloys and by some of the γ-values found

Table I. Low Temperature Calorimetric Data of Alloys of the 5d-Elements

Alloy	Structure			γ mJ/degK^2-mole	θ_0 °K
		a, Å	c, Å		
$Hf_{0.30}Ta_{0.70}$	A2	3.373		8.30 ± 0.2	209 ± 10
$Ta_{0.84}W_{0.16}$	A2	3.280		4.36 ± 0.1	265 ± 10
$Ta_{0.60}W_{0.40}$	A2	3.243		3.08 ± 0.1	291 ± 10
$Ta_{0.40}W_{0.60}$	A2	3.212		1.63 ± 0.05	317 ± 10
$Ta_{0.20}W_{0.80}$	A2	3.188		0.88 ± 0.05	354 ± 10
$Ta_{0.10}W_{0.90}$	A2	3.176		0.92 ± 0.05	368 ± 10
W	A2	3.165		0.95 ± 0.05	396 ± 10
$W_{0.95}Re_{0.05}$	A2	3.160		1.14 ± 0.05	380 ± 10
$W_{0.925}Re_{0.075}$	A2	3.158		1.40 ± 0.05	378 ± 10
$W_{0.90}Re_{0.10}$	A2	3.157		1.63 ± 0.1	375 ± 10
$W_{0.85}Re_{0.15}$	A2	3.151		2.10 ± 0.05	365 ± 10
$W_{0.80}Re_{0.20}$	A2	3.148		2.20 ± 0.05	359 ± 10
$W_{0.75}Re_{0.25}$	A2	3.144		2.30 ± 0.05	351 ± 10
$W_{0.50}Re_{0.50}$	D8$_b$	9.63	5.01	2.69 ± 0.05	327 ± 10
$W_{0.12}Re_{0.88}$	A3	2.76	4.48	3.76 ± 0.1	332 ± 10
$Re_{0.70}Os_{0.30}$	A3	2.75	4.41	2.05 ± 0.1	351 ± 10
$Re_{0.30}Os_{0.70}$	A3	2.74	4.35	1.86 ± 0.1	382 ± 10

* This work has been supported by the Swiss National Science Foundation.
† Present address: Institut de Physique Expérimentale, Université de Genève, Geneva, Switzerland.

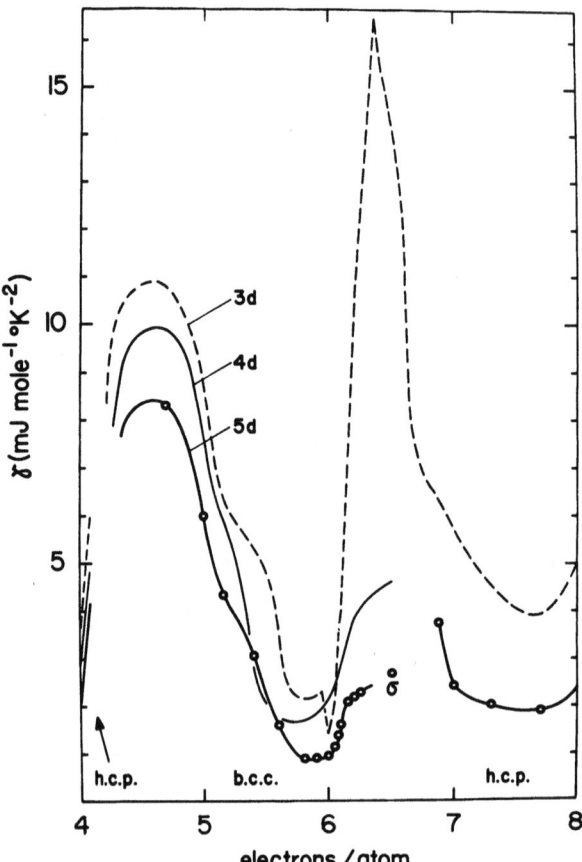

Fig. 1. Coefficient of electronic specific heat, γ, vs. electron concentration. Alloys of the 5d series (———O———) compared to 4d (—————) and 3d (- - - - - -) solid solutions. Alloy components are neighboring elements in the periodic table, except in the case of chromium–iron.[3]

by Morin and Maita[5] and Blaugher et al.[6] Shimizu et al. have made calculations involving the band structure derived from specific-heat measurements. However, their work on the 4d-series[7] is based on low temperature electronic specific heats of niobium and niobium–zirconium given by Morin and Maita,[5] which are considerably in error. The latter authors observed a discontinuous change of the linear specific-heat term of niobium and some of its alloys at 9.5°K, interpreting this behavior as a rapid change of the density of states with temperature. We have carefully remeasured the specific heat of very pure niobium (Fig. 2) and some alloys, and also compared the superconducting and normal entropies at the transition temperature T_c. The resulting value for the electronic specific-heat coefficient of pure niobium, $\gamma = 7.80 \times 10^{-3}$ J/degK^2-mole, is in agreement with very recent work by Van der Hoeven and Keesom[8] and Leupold and Boorse.[9] Moreover, magnetic susceptibility measurements of niobium in the normal state from 7° to 20°K showed no evidence of a change in the density of states at the Fermi level.*

One important feature of the comparison made in Fig. 1 is that, between 6 and 7 electrons per atom, electronic specific heats comparable in magnitude to those of chromium–iron (used in Fig. 1) and chromium–manganese[3] are found neither in the

---

* The susceptibility measurement, carried out by D. Bender, is gratefully acknowledged.

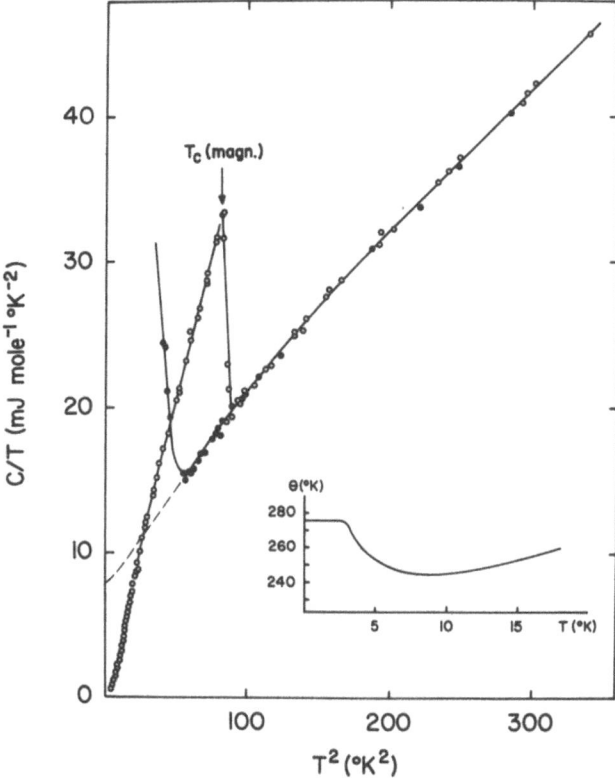

Fig. 2. Superconducting and normal specific heat of niobium. ○: magnetic field $H = 0$, ●: $H = 1000$ Oe. The measured superconducting entropy at $T_c = 9.32°K$ is consistent with $\gamma = 7.80 \times 10^{-3}$ J/mole-deg K^2 and the θ-variation reproduced in the inset (θ_0 is taken from Van der Hoeven and Keesom[8]).

4d- nor in the 5d-series. Although it is true that only in the first transition series, the body-centered cubic crystal structure persists up to more than 7 electrons per atom, it seems likely that the outstanding behavior of the 3d-alloys cannot be interpreted on the basis of a noninteracting electron gas.

In the body-centered cubic region of alloys with less than 6 electrons per atom, the electronic specific heat varies similarly in all three transition series. If the simple one-electron picture is adopted and a rigid band assumed, a conversion of the electron concentration to the energy scale leads to the corresponding density of states curves for one spin orientation reproduced in Fig. 3. These curves represent the total density of states, without distinction between d- and s-like electrons. It is obviously not permissible to subtract a constant γ_s from the measured electronic specific heat, since such a correction would result in too large energy separations between only slightly overlapping sub-bands of d-character. A comparison of the band shapes indicates a progressive broadening in the sequence of the 3d-, 4d- and 5d-periods. It must be pointed out, however, that such an analysis neglects the

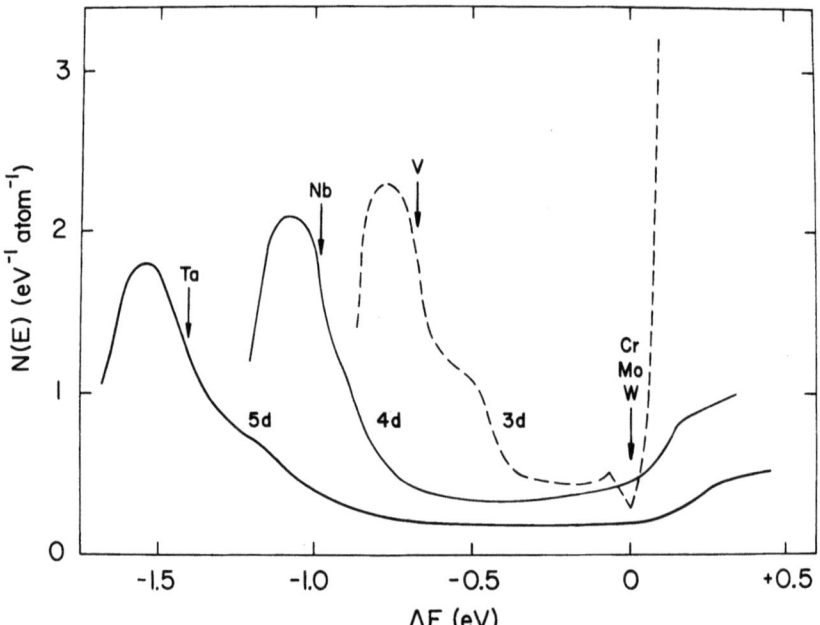

Fig. 3. One electron density of states (for one spin orientation) of bcc alloys of the transition metals. The arrows mark the position of the Fermi level of the corresponding elements (arbitrary zero of energy scale). The curves are calculated from the specific heat data, assuming $F_p + F_c = 0$ (see text).

influence of electron interactions on the electronic specific heat.[10] If such interactions were considered, the density of states at the Fermi level, as deduced from specific heat data, would become

$$N(E) = (\tfrac{2}{3}\pi^2 k^2)^{-1}(1 + F_p + F_c)\gamma$$

where F_p and F_c characterize the electron–phonon and the Coulomb interaction, respectively. The energy scale, in turn, as found by an integration over the electron concentration n, should be taken as

$$E - E_0 = \tfrac{1}{3}\pi^2 k^2 \int_{n_0}^{n} (1 + F_p + F_c)^{-1} \gamma^{-1} \, dn$$

As Clogston[11] has suggested for related substances with high densities of states, one might expect that the main correction arises from the electron–phonon interaction. In this case ($F_p < 0$), the maxima of $N(E)$ shown in Fig. 3 would be reduced and the corresponding width of the sub-band increased.

We have already mentioned the different behavior of first, second, and third row alloys at electron concentrations greater than 6. A study of the "diagonal" system vanadium–ruthenium (Fig. 4) indeed shows that the rise of the electronic specific heat in this region is particularly pronounced in the presence of a 3d-component, even in the case of nonmagnetic alloys. However, the energetic separation of the two sub-bands, calculated on the assumption $F_p + F_c = 0$, appears to be of the order of 0.6 eV larger in vanadium–ruthenium than in the isoelectronic system

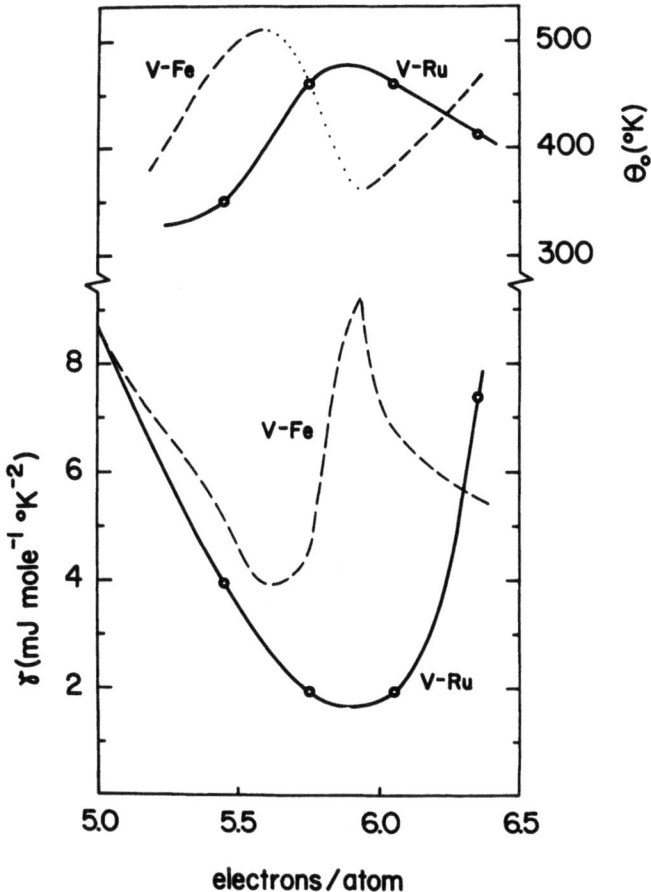

Fig. 4. Electronic specific heat, γ, and Debye temperature, θ_0, of the vanadium–ruthenium and vanadium–iron[3] systems.

vanadium–iron. The low temperature specific-heat data of vanadium–ruthenium (and vanadium–iron) further illustrate the fact, generally observed in transition metal alloys, that the variation of the Debye temperature, θ_0, is opposite to that of the electronic specific heat. θ_0 of vanadium–ruthenium exhibits a maximum at the composition of minimum density of states. There is no doubt that the effect is of electronic origin, although CsCl-type ordering sets in near 6 electrons per atom, since the influence of ordering would be expected to slightly increase θ_0. The relative variation of γ and θ is in qualitative agreement with the simple theoretical model,[12] according to which the electronic contribution to the bulk modulus should depend inversely on the density of states.

A comparison of the electronic specific heats of alloys of the sixth column elements (chromium, molybdenum, tungsten) with rhenium reveals an interesting anomaly exhibited by the chromium alloys. In Fig. 5, we consider the binary alloys of chromium, molybdenum, and tungsten with vanadium, niobium and tantalum,

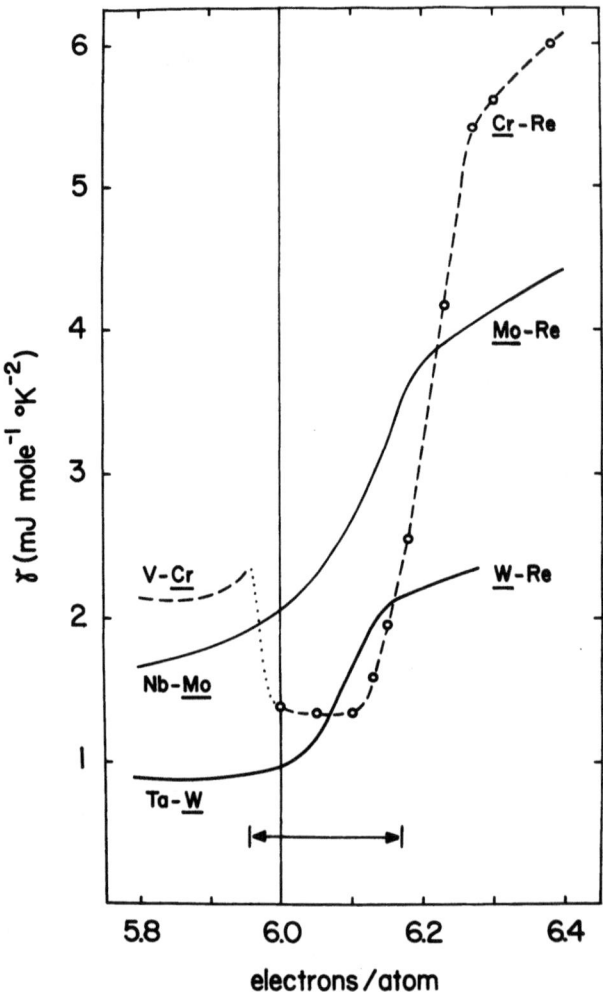

Fig. 5. Electronic specific heat, γ, of binary systems near 6 electrons per atom. The arrows at the bottom mark the limits of the antiferromagnetic region of the chromium alloys.

respectively, on one side, and with rhenium on the other. It clearly appears that a regular behavior of the density of states is shown in the series niobium–molybdenum–rhenium and tantalum–tungsten–rhenium, while the vanadium–chromium–rhenium systems are characterized by deep depression centered around a mean electron concentration of 6.07. This concentration just corresponds to the antiferromagnetic region with high Néel temperatures, the occurrence of which has been interpreted in terms of large amplitude spin density waves.[13] Outside the range of magnetic anomalies, the behavior of the alloys considered again supports the qualitative validity of the rigid band approximation in all transition metal series.

Acknowledgment

We would like to thank Prof. Martin Peter for valuable discussions.

References

1. E. Bucher, F. Heiniger, J. Muller, and J. L. Olsen, this volume, p. 616.
2. E. Bucher, F. Heiniger, and J. Muller, *Phys. Kondens. Materie* **2**, 210, 1964.
3. C. H. Cheng, C. T. Wei, and P. A. Beck, *Phys. Rev.* **120**, 426, 1960.
4. C. H. Cheng, K. P. Gupta, E. C. van Reuth, and P. A. Beck, *Phys. Rev.* **126**, 2030, 1962.
5. F. J. Morin and J. P. Maita, *Phys. Rev.* **129**, 1115, 1963.
6. R. D. Blaugher, J. K. Hulm, J. A. Rayne, B. W. Veal, and R. A. Hein, *Proceedings of the Eighth International Conference on Low Temperature Physics*, Butterworths, London (1963), p. 147.
7. D. O. Van Ostenburg, D. J. Lam, M. Shimizu, and A. Katsuki, *J. Phys. Soc. Japan* **18**, 1744, 1963.
8. B. J. C. Van der Hoeven and P. H. Keesom, *Phys. Rev.* **134**, A1320, 1964.
9. H. A. Leupold and H. A. Boorse, *Phys. Rev.* **134**, A.1322, 1964.
10. D. Simkin (to be published).
11. A. M. Clogston (to be published).
12. J. A. Rayne, *Phys. Rev.* **118**, 1545, 1960.
13. J. Muheim and J. Muller, *Phys. Kondens. Materie* **2**, 377, 1964.

PRECISION MEASUREMENT OF ELECTRONIC SPECIFIC HEAT IN DILUTE ALLOYS

S. Shinozaki and A. Arrott

Scientific Laboratory, Ford Motor Company
Dearborn, Michigan

In order to measure the changes of the electronic contribution to the specific heat of dilute alloys accurate enough for a small change in composition, we developed a technique of simultaneous calorimetry. The calorimetry uses three samples (one pure and two alloys) and a mechanical heat switch, which are arranged in a symmetric manner.

Simultaneous calibration of the carbon thermometers directly in a helium vapor bulb improves the precision of the thermometry. The agreement of the three thermometers among themselves is a factor of 10 better than the fit obtained with the usual analytical formula. The high precision obtained for the calibration using the standard formula and for the specific-heat results was achieved because the calibration of the thermometers and the specific-heat measurements are carried out at the same temperature for each sample. The accuracy of the relative change of the electronic specific heat thus deduced is at least an order of magnitude better than those obtained by individual measurements.

This technique was applied to measure the electronic contribution to specific heat, γT, of the iron alloys with $\frac{1}{2}-2$ at.% Ti, V, Cr, and Mn. The results show that the relative rate of change of γ with respect to solute concentration $(1/\gamma)(\Delta\gamma/\Delta c)$ is 2 for iron–manganese, while for iron–titanium, iron–vanadium, and iron–chromium, the rates of change are -1, -2, and -2, respectively. These results are to be compared with measurements of saturation magnetization, Curie temperature, Mössbauer effect, and small-angle neutron scattering in order to clarify the electronic structure of iron.

ELECTRONIC SPECIFIC HEAT OF DILUTE NOBLE METAL ALLOYS*

M. L. Glasser

Battelle Memorial Institute
Columbus, Ohio

The picture of a simple metal which has been emerging over the last few years is highly reminiscent of that current during the thirties. Most characteristic metallic properties can be explained in terms of weakly interacting particles, somewhat like electrons, but having, perhaps, a different mass, moving in a background of positive ions. The present view is that the interaction with the ions is also weak. It is reasonable to consider this electronic system as constituting a nearly free-electron gas; that is, the particles are looked on as a noninteracting Fermi gas in the presence of a weak crystalline potential which is to be treated by second-order perturbation theory.

It is interesting to investigate whether this model can be extended to simple alloy systems. Here, the rather puzzling behavior of the low-temperature electron specific heat of the α-brasses is considered from this point of view.

It has been thoroughly verified experimentally that the Fermi surface of pure copper is in substantial contact with the surface of the Brillouin zone. The Fermi level, therefore, lies beyond the corresponding maximum in the density of states. On the basis of the rigid band model, if the electron concentration is increased, as by alloying, the density of states at the Fermi level, and thus the electronic specific heat, should decrease. Rayne[1] found, however, that when zinc and germanium were added to copper the specific heat increased. Similar effects have now been reported in silver based alloys.[2,3] An explanation of this behavior has been recently advanced by Jones,[4] whose view is that, due to virtual impurity scattering, the Fermi surface is smeared out sufficiently for the peak in the density of states to be effective. This theory entails a relation between the specific heat increase and the residual resistivity which implies that the former should vary as the square of the valence difference (Linde's rule). This behavior is not observed for silver and is questionable for copper. The approach presented here is to try to account for the observed increase in specific heat purely in terms of band effects.

The partition function and free energy for a nearly free-electron gas under a variety of conditions has been derived elsewhere by the author.[5] From these, the following expression for the electronic specific heat coefficient of a pure metal can be obtained:

$$C_v = \gamma T$$
$$\gamma/\gamma_0 = 1 + \tfrac{1}{2} \sum_{k \neq 0} (V_K/2\zeta_0)^2 F[a_0(K)] \tag{1}$$

---

* This study was supported in part by Air Force Office of Scientific Research Grant No. AF-AFOSR-262-63.

where
$$F(x) = x/(1-x) - \sqrt{x}\tanh\sqrt{x}$$

$$a_0(K) = 4\zeta_0/\varepsilon(K) \qquad \varepsilon(K) = \hbar^2 K^2/2m \qquad \gamma_0 = k^2\pi^2 n/2\zeta_0$$

ζ_0 is the free-electron chemical potential corresponding to electron concentration n and $V_{\mathbf{K}}$ is the Kth Fourier coefficient of the lattice potential.

Let a concentration q of metal atoms (of atomic number Z_0) be replaced by point charges Ze and Z additional electrons. To make things as simple as possible, incoherent impurity effects are neglected by a random averaging, which effectively puts the same averaged potential at each lattice site. Then (1) is applicable to this system with $V_{\mathbf{K}}$ replaced by $qV_{\mathbf{K}}^i + (1-q)V_{\mathbf{K}}$, where $V_{\mathbf{K}}^i$ is the Fourier component of a neutralized lattice of point charges. After a straightforward calculation, the expression

$$M = \partial \gamma(q)/\partial q|_{q=0}$$
$$M = (\gamma_0/3)\{\delta + \sum_{k \neq 0}(V_{\mathbf{K}}/2\zeta_0)^2[3F(a_0)(\Delta V_{\mathbf{K}}/V_{\mathbf{K}} - \delta/2) + a_0\delta F'(a_0)]\} \qquad (2)$$

where
$$\delta = (Z - Z_0)/Z_0 \qquad \Delta V_{\mathbf{K}} = V_{\mathbf{K}}^i - V_{\mathbf{K}}$$

is obtained. (The small change in lattice spacing with alloying has been neglected.) Although the impurity average is taken at the wrong stage, equation (2) may not suffer from the neglect of impurity broadening since it refers to the limit of zero concentration where this broadening is small.

Since $V_{\mathbf{K}}$ is roughly proportional to ΔZ, M will behave linearly with change in valence of the solute, as Houghton[6] pointed out to be the case for a number of alloys. Similarly, the logarithmic derivative of γ with respect to the electron per atom ratio will be independent of ΔZ as observed by Green and Culbert. The expression for M is composed of three principal terms. The first and third correspond to a change in the density of states on the free-electron Fermi sphere due to changing electron concentration and the second due to changing band gaps. The physical interpretation of this expression is not clear, although it appears to bear on the Hume-Rothery interpretation of phase boundaries. The density of states which enters is that at the free-electron Fermi sphere, which does not contact the zone surface in the noble metals. We may therefore expect that a small increase in electron concentration might raise the specific heat.

A pseudo-potential is adopted, similar to that used by Ziman[7] in studying the transport properties of the noble metals, for the evaluation of this expression. The crystal potential is assumed to have only a (111) Fourier coefficient whose value for the solvent metal is chosen to give agreement with experiment. We use simply the linearized Thomas–Fermi potential for a point charge to treat the impurities. The neglect of the remaining Fourier coefficients of V^i is questionable, but the ratios $V_{\mathbf{K}}^i/\zeta_0$ will be small. The values for the Fourier coefficients are shown in Table I.

The results obtained are shown in Fig. 1. The experimental points are based on mean values and have generous error bounds due to the close cancellation of results good to about 1%. A careful effort to ensure the purity of the samples, and in particular to exclude paramagnetic impurities, dropped the experimental result for copper–zinc from 1.4 to 0.2 as shown by the point X.[8] A similar effort for copper–

Table I. Solute and Impurity Pseudo-Potential Fourier Coefficients. Values in Parentheses are Ziman's Estimates (see ref. 5)

	V_{111}*	V^i_{111} $z = 2$*	$z = 3$*	$z = 4$*
Cu	−2.83	−2.49	−3.74	−4.97
	(−3.5)			
Ag	−0.80	−2.14	−3.22	−4.29
	(−2.3)			
Au	−1.42	−2.14	−3.22	−4.29
	(−2.5)			

*Units are expressed in electron volts.

germanium would undoubtedly have the same effect and result in improved agreement with the theory.

A number of objections, apart from the limitations of the model, can be raised against the theory. In particular, (2) has been derived by truncating a series expansion for the free energy in powers of V_K/ζ_0, which converges only for $a_0(K)$ not too near unity. For the noble metals this value is 0.814, which may be dangerously

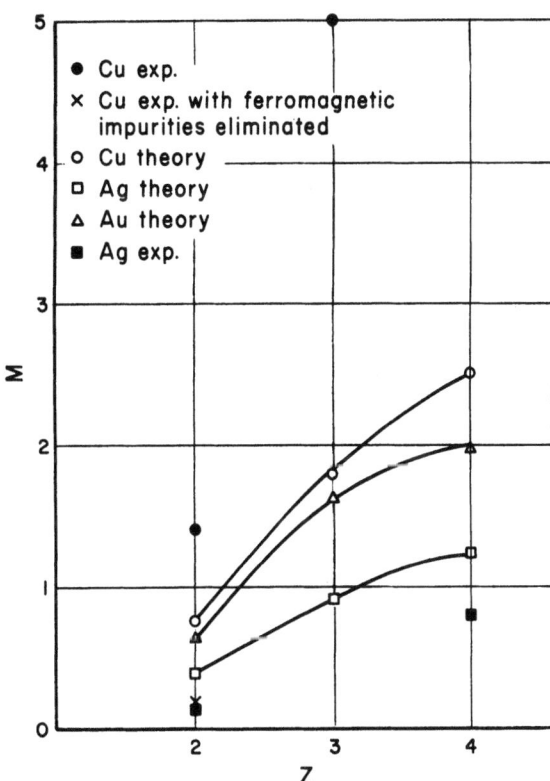

Fig. 1. Slope of electronic specific-heat coefficient vs. impurity-concentration curve. Experimental points are based on the most dilute alloys measured.

large; the theory should properly apply to the alkali metals, where a_0 is much smaller. It may be argued that the effects of truncation as well as correlation and the neglect of other Fourier coefficients of the potential are included by the choice of pseudo-potential, which differs considerably, e.g., from the value $-0.92\,\text{eV}$ calculated by Segall[7] for copper. Nonetheless, it is encouraging that the theory does produce estimates in reasonable agreement with experiment and accounts for the qualitative behavior of the results. It is possible that following, e.g., an approach presented by Cohen[9] for separating structural terms from the free energy of simple metals, the virtual crystal model may be given theoretical support for the study of bulk equilibrium properties.

References
1. J. Rayne, *Phys. Rev.* **108**, 22, 1957; **110**, 606, 1958.
2. B. A. Green, Jr. and H. V. Culbert, *Phys. Rev.* **137**, A1168, 1965.
3. H. Montgomery and G. P. Pells, Conference on the Electronic Structure of Alloys, University of Sheffield, England, 1963.
4. H. Jones, *Phys. Rev.* **134**, A958, 1964.
5. M. L. Glasser, *Phys. Rev.* **134**, A1296, 1964.
6. A. Houghton, *J. Phys. Chem. Solids* **20**, 289, 1961.
7. J. A. Ziman, *Advan. Phys.* **10**, 1, 1961.
8. B. W. Veal and J. Rayne, *Phys. Rev.* **130**, 2156, 1963.
9. M. H. Cohen (to be published).

SPECIFIC HEAT OF A DILUTE PALLADIUM–COBALT ALLOY AND OF PURE PALLADIUM

B. M. Boerstoel, F. J. du Chatenier, and G. J. van den Berg

Kamerlingh Onnes Laboratorium
Leiden, The Netherlands

The specific heat of a dilute palladium–cobalt alloy has been investigated between 0.1° and 30°K. Below 1°K, Miedema and B. M. Boerstoel have measured the specific heat by means of a method using adiabatic demagnetization, which method has been described by du Chatenier and Miedema in a paper presented at this conference. Apart from a few improvements, the experimental procedure of the measurements above 1°K is the same as already published before.[1,2]

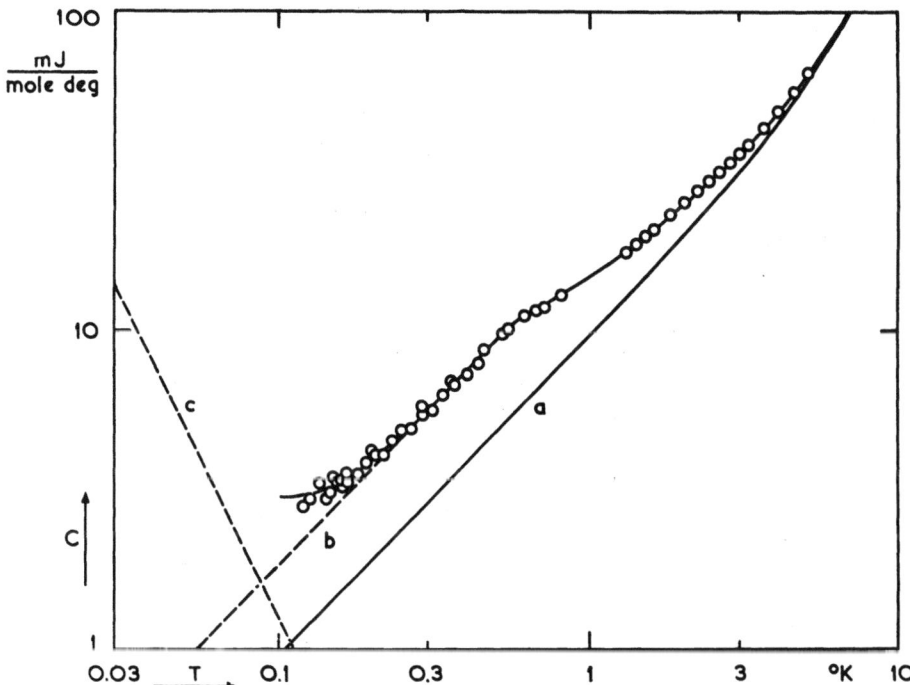

Fig. 1. Specific heat *vs.* temperature of a dilute palladium–cobalt alloy and of pure palladium: (a) pure palladium; (b) extrapolated linear term; (c) $1/T^2$ tail of the hfs contribution.

Below about 12°K, the specific heat of this alloy shows a large anomaly. The curve for C vs. T has a maximum at approximately 1.9°K. The specific heat below 10°K is presented on a logarithmic scale in Fig. 1.

In the temperature region between 0.1° and 0.5°K, the specific heat can be described by $C = 18.4T + 0.012_5 T^{-2}$ mJ/mole-deg. As can be seen below, the electronic specific heat of pure palladium can be represented by $C_e = 9.57T$ mJ/mole-deg. Consequently, there is a magnetic contribution to the specific heat of palladium–cobalt which is proportional to the temperature. Such contributions have already been found for copper–manganese, gold–manganese, and copper–chromium alloys.[3]

The existence of these contributions indicate the presence of a partition function $p(H)$ for the internal fields. Marshall[4] has given theoretical considerations about the internal fields in dilute alloys. According to these considerations, we must expect the specific heat to be proportional to the temperature at sufficiently low temperatures, provided $p(0)$ has a nonvanishing value. Apparently, this is also applicable to the dilute palladium–cobalt alloy. If $p(0)$ is inversely proportional to the concentration c, this specific heat contribution will be independent of c. So far, nothing can be said about this dependence; measurements on alloys with a higher cobalt concentration are in progress.

Below 0.25°K, an extra $1/T^2$ contribution to the specific heat has been found, which is ascribed to hfs effects of the cobalt. From this contribution the splitting parameter, $\theta_A = AS/k$, of cobalt can be calculated, assuming that the spin value of the cobalt ion is equal to $\frac{3}{2}$. The value for $\theta_A = 0.029 \pm 0.002°K$ is comparable in magnitude with the values for cobalt salts given by Bleaney and Ingram[5] and is larger than the value for cobalt metal of Heer and Erickson.[6]

The accuracy in a preliminary determination of the concentration ($c = 0.13$ at.%, approximately) by means of X-ray fluorescence is not better than 10%. With this value for the concentration, a spin value of 1.5_4 for the cobalt ion follows from entropy calculations, assuming the orbital angular momentum to be quenched. From the shape of the specific heat curve, it is not possible to derive a Curie temperature for this palladium–cobalt alloy. Magnetization experiments of Bozorth et al.[7] established the Curie temperature of a palladium–cobalt alloy containing 0.1 at.% Co as 7°K.

As the specific heat data of palladium given by several investigators[8–11] do not agree, the specific heat of a pure palladium sample of about 50 g has also been measured between 1.3° and 30°K. The purity of the palladium used was 99.999%, principal impurities being iron and silicon, each amounting to 0.0002%.

Values for the constants of the electronic and the lattice specific heat have been deduced from a plot of C/T vs. T^2 below 5°K. These are given in Table I, together with the data of other investigators.

Table I

	γ (mJ/mole-deg^2)	$\theta_D(0)$ (°K)	Purity (%)
This work	9.57 ± 0.07	267 ± 8	99.999
Rayne[11]	9.87 ± 0.11	299 ± 12	99.999
Rayne[11]	9.64 ± 0.08	297 ± 9	99.98
Hoare and Yates[10]	9.31 ± 0.05	274 ± 3	~99.99
Clusius and Schachinger[8]	13.0	275	?
Pickard and Simon[9]	13.0 ± 0.5	275	?

As can be seen from the table, there is reasonable agreement with the results of Hoare and Yates and with the γ-values of Rayne. Rayne's $\theta_D(0)$-values differ from our value. It is interesting to note that a plot of $\theta_D(T)$ vs. T between 1.3° and 30°K shows only a very shallow minimum at 10°K [$\theta_D(10) = 263°K$].

Our results on palladium–cobalt confirm Rayne's suggestion,[11] that magnetic impurities in palladium may give rise to considerable changes in the specific heat at low temperatures. Presumably the large deviations of the γ-values of Clusius and Schachinger[8] and Pickard and Simon[9] from the present value are due to such impurities.

Further results and more details about the experimental procedure will be published in *Physica*.

Acknowledgments

This investigation is a part of the research program of the Stichting voor Fundamenteel Onderzoek der Materie (F.O.M.) and was made possible by financial support from the Nederlandse Organisatie voor Zuiver Wetenschappelijk Onderzoek (Z.W.O.) and the Nederlandse Centrale Organisatie voor Toegepast Natuurwetenschappelijk Onderzoek (T.N.O.).

References

1. J. de Nobel and F. J. du Chatenier, *Physica* **25**, 969, 1959; *Commun. Kamerlingh Onnes Lab.*, Leiden, No. 317c.
2. J. de Nobel and F. J. du Chatenier, *Physica* **29**, 1231, 1963; *Commun. Kamerlingh Onnes Lab.*, Leiden, No. 336c.
3. G. J. van den Berg and J. de Nobel, *J. Phys. Radium* **23**, 665, 1962.
4. W. Marshall, *Phys. Rev.* **118**, 1519, 1960.
5. B. Bleaney and D. J. E. Ingram, *Proc. Roy. Soc. (London) Ser. A.* **208**, 143, 1951.
6. C. V. Heer and R. A. Erickson, *Phys. Rev.* **108**, 896, 1957.
7. R. M. Bozorth, P. A. Wolff, D. D. Davis, V. B. Compton, and J. H. Wernick, *Phys. Rev.* **122**, 1157, 1961.
8. K. Clusius and L. Schachinger, *Z. Naturforsch.* **2a**. 90, 1947.
9. G. L. Pickard and F. Simon, *Proc. Phys. Soc. (London) Ser. A.* **61**, 1, 1948.
10. F. E. Hoare and B. Yates, *Proc. Roy. Soc. (London) Ser. A.* **240**, 42, 1957.
11. J. A. Rayne, *Phys. Rev.* **107**, 669, 1957.

HEAT CAPACITY OF ORDERED AND DISORDERED COPPER–PLATINUM BELOW 4.2°K

John Rayne*
Carnegie Institute of Technology
Pittsburgh, Pennsylvania

and

Barton Roessler
Westinghouse Research Laboratory
Pittsburgh, Pennsylvania

The order–disorder transformation in the copper–platinum system, which occurs in the alloy of 50 at.%, is unusual since the number of unlike nearest neighbors in the ordered condition is the same as it is, on the average, in the random disordered condition.[1] Consequently, an explanation of the tendency toward ordering, which is based on a decreased interaction energy for unlike nearest neighbors (quasi-chemical theory), is not applicable.

The effect of superlattice formation on the Brillouin zone structure of an alloy has been discussed by both Muto[2] and Slater.[3] When a superlattice forms in an alloy, the X-ray structure factor for certain planes is no longer equal to zero as it is in the disordered condition. These extra superlattice reflections give rise to a reduction in the size of the basic Brillouin zone which can lower the electronic energy of the alloy by interacting with the electrons near the Fermi surface. Since all of the electrons in the alloy partake of this interaction and since its existence does not require that the number of unlike nearest neighbors change, it is an attractive explanation for the tendency toward ordering in copper–platinum and has been elaborated upon in further detail by Nicholas.[4] Although it is difficult to predict in detail what effects the Brillouin zone–Fermi surface interaction due to superlattice formation should have on the electronic structure, it is possible that the density of states curve could be changed. An earlier investigation[5] of the low-temperature heat capacity of Cu_3Au showed no difference in the electronic specific heat coefficient between the ordered and disordered conditions. It was felt, in view of the unusual nature of the transformation in copper–platinum, that a similar investigation in copper–platinum would be of interest.

Copper of 99.999% purity (Asarco A-58) and platinum sponge of 99.999% (Johnson-Mathey 1010) in the appropriate amounts were induction-melted in a high-purity graphite crucible under a protective atmosphere of helium. The resulting ingot was homogenized in the crucible by holding between 1250° and 1350°C for 5 hr. Chemical analysis showed no significant composition difference between the top and bottom of the ingot; the mean concentration of copper was 24.51 wt.%, the stoichiometric composition being 24.57 wt.%. A spectroscopic analysis showed

---
* Supported in part by the National Science Foundation.

iron to be present at less than 4 ppm, and manganese and cobalt both less than 1 ppm. To produce long-range order, the specimen, sealed in a Vycor tube under vacuum, was held for approximately 24 hr at a series of temperatures starting at 850°C and decreasing approximately 50°C every 24 hr down to 490°C. It was held for 4 days at 490°C, then cooled to 200°C in steps over the next two weeks, and finally furnace-cooled to room temperature. An X-ray back-reflection photograph using copper radiation and a nickel filter showed prominent superlattice lines and indicated the specimen to be well-ordered.[6] The specific heat of the ordered specimen was measured in the manner previously described.[7] The same specimen, sealed under vacuum in a Vycor tube, was then disordered by holding above the critical temperature for long-range order (T_c = 815°C) for 2.5 hr and then quenching from 887°C into a NaCl brine at -4°C. An X-ray back-reflection photograph of the quenched specimen showed that the superlattice reflections had disappeared. After the specific-heat measurements on the disordered specimen were completed, a second sample was taken for spectroscopic analysis as a check that the specimen had not become contaminated by any of the treatments to which it had been subjected, i.e., the ordering heat treatment, quenching, cleaning, etc. This analysis showed that the ferromagnetic impurities, iron, cobalt, and manganese, if present, were less than 4 ppm and most likely less than 1 ppm.

The results, presented in Fig. 1 and plotted in the usual form, viz., C/T vs. T^2, give two straight lines. Thus, the low-temperature heat capacity is of the form

$$C = \gamma T + A\left(\frac{T}{\theta}\right)^3 \tag{1}$$

the first and second terms representing the electronic and the lattice heat capacity, respectively. The intercept, γ, and the Debye temperature, θ, for the ordered and disordered states, shown in Fig. 1, have been determined from a least-squares fit of a straight line to the data points. There is clearly a large decrease in the γ-value for the ordered alloy compared to that of the disordered alloy. This behavior is to

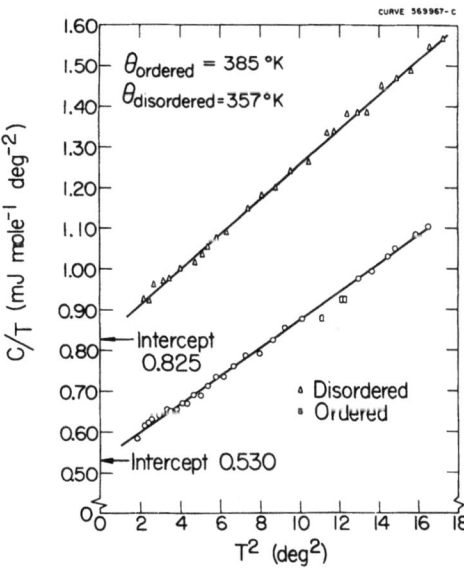

Fig. 1. Molar heat capacity divided by temperature, C/T, as a function of the temperature squared, T^2, for disordered and ordered copper–platinum.

be contrasted with that observed for Cu_3Au, where the γ-value remains unchanged on ordering.[8] The observed increase in the Debye temperature on ordering is also consistent with that observed in Cu_3Au.

It is of interest, in understanding the electronic structure of copper–platinum, to compare the relative change in the γ-value which occurs when platinum is added to copper to that which occurs when palladium is added to silver[9] and when platinum is added to gold.[10] In order to make this comparison, the γ-values should be adjusted for the difference in lattice parameter between the pure solvent and the alloy. Accordingly, if we assume the rigid band model[11]

$$\gamma^* = \gamma_{obs}\left(\frac{a}{a_0}\right)^2 \quad (2)$$

where γ^* is the γ-value which would be observed if the alloy had the same lattice parameter as the pure solvent, γ_{obs} is the value of γ measured for the alloy, and a and a_0 are the lattice parameters of the alloy and the pure solvent, respectively. This comparison is made in Table I, which shows that the corrected electronic specific-heat coefficient, γ^*, has increased relative to the value for the pure solvent, γ_0, much more in silver–palladium and in gold–platinum than in disordered copper–platinum. It is, of course, true that the comparison is complicated by the possible differences in short-range order in the two alloys. It shows, however, that a very clear difference exists, whereas it would seem reasonable to expect these two alloys to be quite similar in their electronic structures since both palladium and platinum are known to have 0.6 holes in their d-bands.[12] In particular, it is known that considerable short-range order exists in quenched copper–platinum.[6] Since ordering decreases the γ-value, the smaller relative increase in γ on alloying, noted in Table I, may be due to this effect.

The rhombohedral ordered copper–platinum structure can be thought of as a stacking of close-packed planes alternately composed of either pure platinum or pure copper.[13] It has been suggested[14] that with this atomic arrangement an ordering of the spins of the platinum atoms could give rise to a magnetic transition. Although it is not known with any degree of certainty how such a transition might affect the electronic specific heat, it should be pointed out that in Ni_3Mn, in which the ordered state is ferromagnetic, the electronic specific heat decreases on going from the disordered to the ordered condition.[15] The decrease in γ observed in ordered copper–platinum may, then, be a similar effect due to magnetic ordering, either ferromagnetic or antiferromagnetic. There appears to be no data available to resolve this question.

Table I. The Electronic Specific Heat Coefficients in Disordered Copper–Platinum, in Silver–Palladium, in Gold–Platinum, and in Pure Copper, Silver, and Gold*

	γ_{obs}	γ^*	γ_0	Difference
Copper–platinum disordered	0.825	0.75	0.69	0.06
Silver–palladium	1.69	1.80	0.64	1.16
Gold–platinum	2.95	2.84	0.87	1.97

* Units are expressed in mJ/mole-deg^2.

It is, of course, also possible that the Brillouin zone–Fermi surface interaction is responsible for the decrease in γ which occurs on ordering. With our present understanding of the effects of ordering on the electronic structure of alloys, it is not possible to predict either the magnitude or sign of the expected change in γ.

Acknowledgments

The authors wish to thank Mr. G. Beck for assistance in obtaining the X-ray patterns and Mr. G. E. Martin for help with specimen preparation. The encouragement of Dr. W. A. Tiller was also greatly appreciated. Partial support of this research by the Air Force Office of Scientific Research under contract number AF 49(638)-1029 is gratefully acknowledged. One of us (J.R.) would like to acknowledge his partial support by the National Science Foundation.

References

1. B. E. Warren, *Frontiers in Chem.* **5**, 101, 1948.
2. T. Muto, *Sci. Rept. Inst. Phys. Chem. Res., Tokyo*, **34**, 377, 1938.
3. J. C. Slater, *Phys. Rev.* **84**, 179, 1951.
4. J. F. Nicholas, *Proc. Phys. Soc. (London)* A **66**, 201, 1953.
5. J. A. Rayne, *Phys. Rev.* **108**, 649, 1957.
6. C. B. Walker, *J. Appl. Phys.* **23**, 118, 1952.
7. J. A. Rayne, *Phys. Rev.* **107**, 669, 1957.
8. J. A. Rayne, *Phys. Rev.* **108**, 649, 1957.
9. F. E. Hoare and B. Yates, *Proc. Roy. Soc. (London) Ser. A* **240**, 42, 1957.
10. D. W. Budworth, F. E. Hoare, and J. Preston, *Proc. Roy. Soc. (London) Ser. A* **257**, 250, 1962.
11. N. F. Mott and H. Jones, *The Theory of the Properties of Metals and Alloys*, Dover Publications, Inc., New York (1958), p. 168.
12. N. F. Mott and H. Jones, *The Theory of the Properties of Metals and Alloys*, Dover Publications, Inc., New York (1958), p. 195.
13. L. Guttmann, *Solid State Physics, Vol. 3*, F. Seitz and D. Turnbull (eds.), Academic Press, Inc., New York (1956).
14. P. A. Flinn, private communication.
15. J. E. Goldman, *Rev. Mod. Phys.* **25**, 108, 1953.

11
SOLIDS AND METALS AT LOW TEMPERATURES

11.1. Rare Gas Solids, Solid Hydrogen, Solid Methane

THERMODYNAMICS OF THE INERT GAS SOLIDS USING THREE-PARAMETER INTERATOMIC POTENTIALS

J. W. Leech and J. A. Reissland

Queen Mary College, University of London
London, England

Introduction

Lattice dynamical calculations are simplified by the assumption that two-body forces act between the various atoms of the solid in question. Here, we report on calculations of this kind based on a modified form of Mie–Lennard-Jones type potential designed to avoid some of the anomalies[1] resulting from the use of the more familiar type of expression.

The Three-Parameter Potential

In our investigations we use a three-term inverse power potential:

$$\phi = \frac{A}{r^m} + \frac{B}{r^n} + \frac{C}{r^6} \tag{1}$$

We have improved our previous method[2] (in which we chose a theoretical value for C) by using recent experimental measurements of the shear modulus of argon by Jones and Sparkes.[3] These values, together with experimental values for the sublimation energy and lattice spacing at absolute zero, enable us to determine the parameters A, B, and C appearing in (1). For convenience, we replace A and B by ε, the well depth, and σ, the equilibrium spacing. We have investigated a range of values for m and n but concentrate here on $m = 12$ and $n = 9$, since this combination gives the most acceptable results. Our calculated values for ε and σ are shown in Table I, where they are compared with other values calculated (a) assuming a Mie–Lennard-Jones 12–6 potential and using only sublimation energy and lattice spacing data and (b) assuming a theoretical value for C due to Margenau.[7]

Table I. Calculated Values of Potential Parameters for Argon*

	$-\varepsilon$	σ
(a) 12–6 potential	3.926	3.818
(b) 12–9–6 potential using C_{MARG}.	4.419	3.776
(c) 12–9–6 potential using experimental shear modulus†	4.428	3.776

* ε in units of 10^{-22} cal, σ in Å, C in 10^{-18} cal-Å^6.
† (c) yields a value $C = -1.311$ which compares with $C_{MARG} = -1.33$.

Thermodynamics

We have used the above potentials to calculate thermal expansion, specific heat, and elastic constants. The evaluation of the thermal expansion is described in Leech, Peachey, and Reissland.[4] The specific heat and Debye theta are obtained as in Horton and Schiff,[5] and the elastic constants as in Horton and Leech.[1] Results are presented in Figs. 1 and 2 and in Table II and are compared with those based on the use of a two-parameter (12–6) potential. In the case of the elastic constants, calculations have been made assuming only a range of possible values of the parameter C. The predictions of the new method could, however, be obtained from these by interpolation.

Comments

The effect of the added term in the potential expression (1) is to produce a deeper potential well. The necessity for this in the gas phase has been noted previously by Munn[6] in work on virial coefficients. Figures 1 and 2 show that the

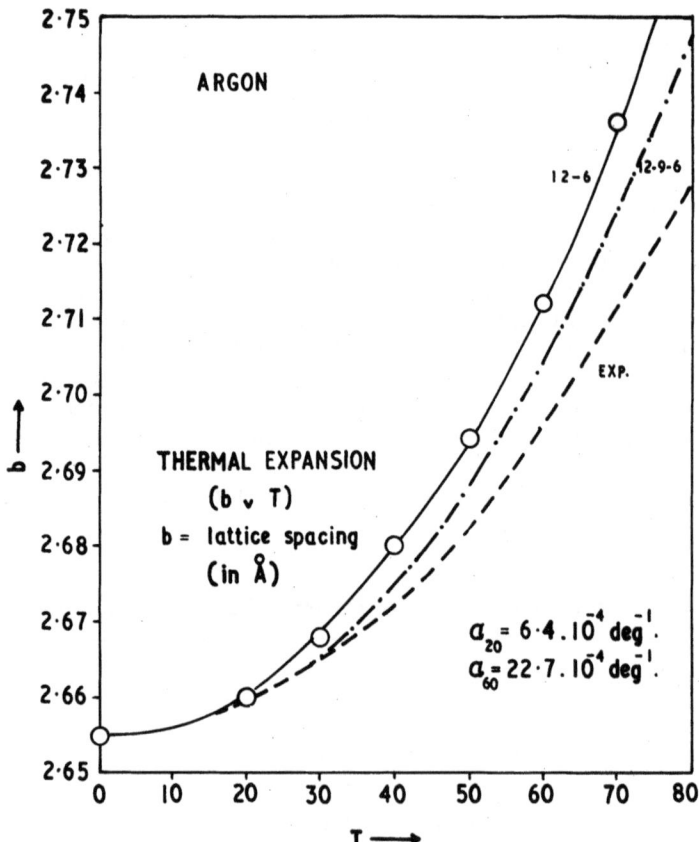

Fig. 1.

Table II. Calculated Values for Elastic Constants of Argon at Absolute Zero*

m–n–6	$-C$, 10^{-19} cal Å^6	$V_l(110)$, 10^{13} Å sec^{-1}	$V_t(100)$, 10^{13} Å sec^{-1}	$V_t(110)$, 10^{13} Å sec^{-1}	C_{11}, 10^{-22} cal Å^{-3}	C_{44}, 10^{-22} cal Å^{-3}	C_{12}, 10^{-22} cal Å^{-3}	G, 10^{-22} cal Å^{-3}	θ_0, °K
12–6 (NNO)	0	1.420	1.019	0.740	8.534	4.389	3.900	3.396	87.74
12–6 (AN)	0	1.370	1.062	0.656	7.942	4.768	4.299	3.242	84.85
12–9–6 (AN)	1.11	1.550	1.183	0.757	10.169	5.916	5.317	4.138	96.20
12–9–6 (AN)	1.33	1.521	1.163	0.741	9.793	5.723	5.144	3.987	94.40
12–9–6 (AN)	1.66	1.485	1.139	0.721	9.335	5.487	4.934	3.804	92.16

*All 12–9–6 results given in this table were obtained by assuming the quoted values for the parameter C.

modified potential gives improved agreement between calculated and observed values of specific heat and thermal expansion. Agreement with the shear modulus is built-in to the calculation and agreement is also found with predicted theoretical values for the coefficient of the van der Waals term of the potential expression. Anharmonic effects have still to be taken into account, but the present conclusion is that harmonic lattice dynamical calculations on argon are more realistic when based on a modified Mie–Lennard-Jones potential of type (1).

Work has also been carried out on neon, krypton, and xenon with broadly similar results, though neon appears to be too highly anharmonic for any form of harmonic calculation to be really successful. In the case of substances other than

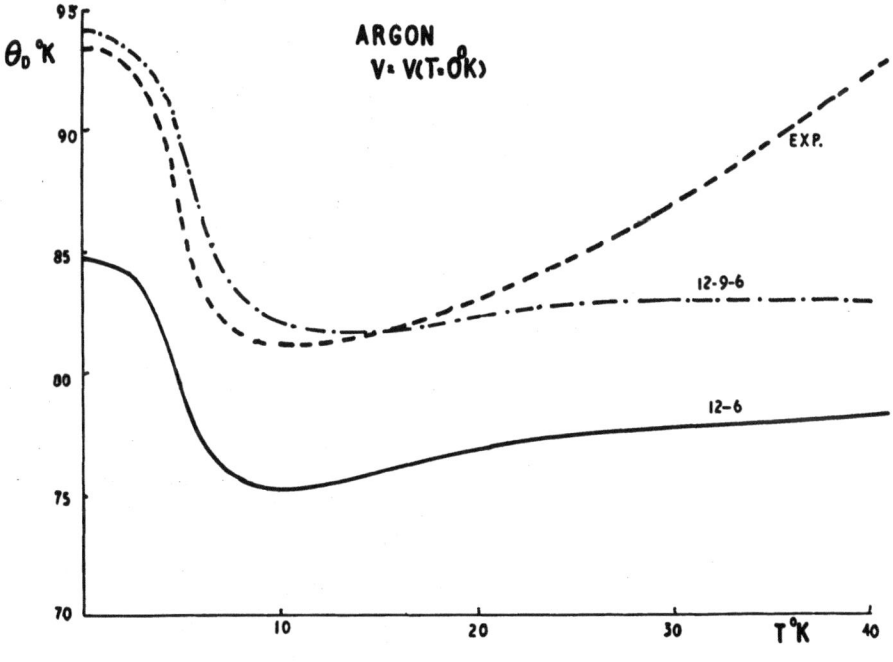

Fig. 2.

argon, it is necessary to assume values for the parameter C since there are, as yet, no experimental values for the shear modulus at absolute zero.

All the present calculations include effects due to all neighbors. Our finding is that harmonic calculations of a wide range of thermodynamic properties of argon, krypton, and xenon yield realistic values when based on a two-body potential of the form (1).

References

1. G. K. Horton and J. W. Leech, *Proc. Phys. Soc. (London)* **82**, 816, 1963.
2. J. W. Leech and J. A. Reissland, *Proceedings of the Eighth Conference on Low Temperature Physics*, Butterworths, London (1963).
3. G. O. Jones and A. R. Sparkes, *Phil. Mag.* (in press).
4. J. W. Leech, C. J. Peachey, and J. A. Reissland, *Phys. Letters* **10**, 69, 1964.
5. G. K. Horton and H. Schiff, *Proc. Roy. Soc. (London) Ser. A* **250**, 248, 1959.
6. R. J. Munn, *Proc. Phys. Soc. (London)* **40**, 1439, 1964.
7. H. Margenau, *Rev. Mod. Phys.* **11**, 1, 1939.

THE CRYSTAL STRUCTURES OF ARGON AND ITS ALLOYS*

C. S. Barrett

*Institute for the Study of Metals, University of Chicago
Chicago, Illinois*

and

Lothar Meyer

*Institute for the Study of Metals and Department of Chemistry
University of Chicago, Chicago, Illinois*

Solid argon is stable in the face-centered cubic structure at all temperatures, but it is found that it can also crystallize in a metastable form with hexagonal close-packed structure.[1] Plastic deformation introduces many stacking faults, particularly in the hcp phase, and eventually reduces samples to faulted fcc structure. Small amounts of nitrogen, oxygen, or CO stabilize the hcp phase at temperatures near the melting point.

Published phase diagrams for the argon–nitrogen and argon–oxygen systems require appreciable modification. The phases and phase transformations have been determined by X-ray diffraction. The technical details are published elsewhere.[2] The gases used were of research purity. The samples were mixed in a small high-pressure bomb; the inlet tube extended to the bottom, was closed at the end, and was provided with many small holes that forced the gas into the bomb in radial jets for good mixing.

Figure 1 shows the argon–nitrogen phase diagram. The solid phase in equilibrium with the liquid between 1 and 100% nitrogen is close-packed hexagonal in spite of the fact that pure argon is face-centered cubic. The argon-rich solid solutions with 1 to 55% nitrogen undergo a strain-induced martensitic transformation from hcp (stable at high temperatures) to fcc at low temperatures. The highest temperatures at which this transition becomes possible are those at which the combination of thermal expansion plus solid solution expansion gives a lattice constant for the fcc phase of $a_0 = 5.486$ Å, which corresponds to that obtained by extrapolating the curve of a_0 for pure argon to a point 5° above the melting point. The c/a value for the hcp argon-rich phase is 1.633 within experimental uncertainty at all compositions and temperatures.

The α-nitrogen solid solution forms spontaneously and isothermally from the β-nitrogen solid solution at nitrogen contents above 77.5% nitrogen. The transformation appears to be that known in metallic systems as a "massive" type, and involves only short-range diffusion in the phase interface. The curve of distance

---

* This work was supported in part by U.S. Army Research Office Grant ARO(D)-31-124-G152, and National Science Foundation Grant GP-2650 and by facilities provided in part by the Advanced Research Projects Agency.

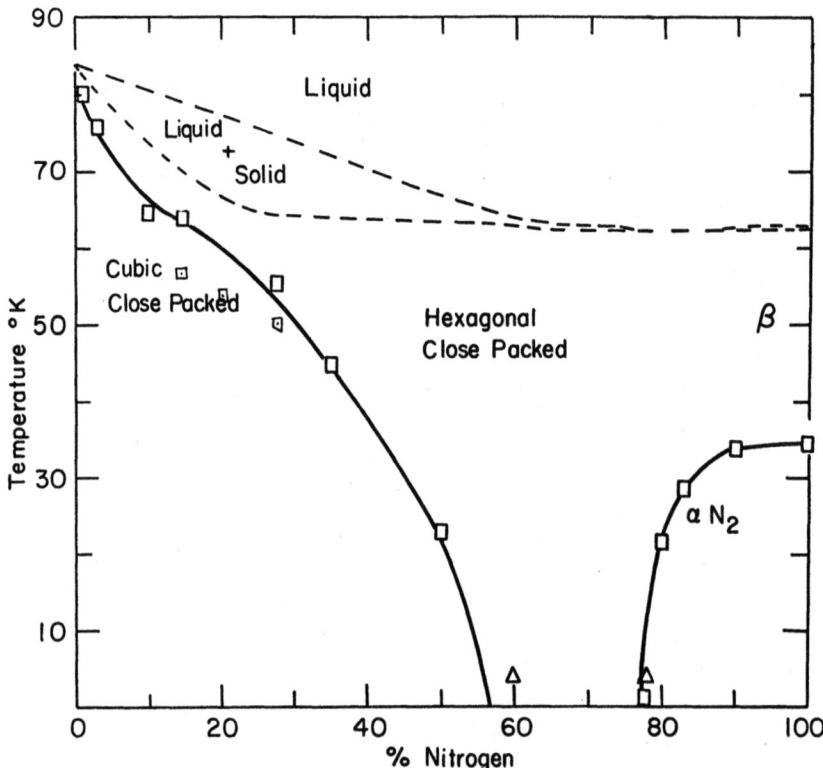

Fig. 1. Argon–nitrogen phase diagram. Open squares indicate upper limit for strain-induced martensitic transformation. Open triangles indicate no transformation detected. Filled squares indicate spontaneous "massive" transformation temperatures. Liquidus and solidus curves are from Long and Di Paolo.[5]

between nearest molecular centers in the β-phase is not an extension of the curve for the argon-rich phase. The α-phase, which is cubic, is not continuous with the cubic argon-rich solid solution; samples with 55–75% nitrogen do not transform from the hexagonal form even when cold worked at 4.2°K.

Both the argon-rich and the nitrogen-rich ends of the argon–nitrogen system reveal features of fundamental importance. On the argon-rich side of the phase diagram, these results have a direct bearing on the question of the relative stability of the fcc and hcp structures, a question which has been treated theoretically in great detail. Calculations of the static lattice energy by summing, pairwise, the nearest-neighbor interactions based on a Mie–Lennard-Jones potential have invariably led to the result that for the noble gas elements, the hcp structure has a lower free energy than the fcc, contrary to experimental determinations (except for helium). These calculations have indicated that the difference in free energy is very small. Likewise, considerations of lattice vibrations and anharmonicity of the vibrations have not changed the sign of the computed difference in free energy between these two phases.

Qualitative success in computing the relative free energy of these two structures was recently achieved by Jansen,[3] who took three-body interactions into account.

In all computations of lattice energies the difference between cubic and hexagonal close-packing appears to be small, usually even small compared to thermal energies near the melting point, kT_m. It might be expected from such calculations, then, that stacking faults would be very common in the noble gas elements when they solidify, the cubic sequence of close-packed atom layers (ABCABC...) and the hexagonal sequence (ABAB...) being almost equally probable and intimately mixed. Our observations are in distinct contradiction to such an expectation. We have observed remarkably sharp X-ray reflections from metastable hexagonal argon, even when the cubic phase was present in the sample simultaneously. Crystals of argon-rich solid solutions of nitrogen in argon also yielded sharp reflections in the as-frozen condition, both at high and at low temperatures.

Stacking faults are infrequent in the as-frozen samples, but are very numerous in the deformed samples, especially at temperatures and compositions near the line of equal free energy of the two phases, where it is possible to destroy the crystalline diffraction pattern by cold work, leaving only an amorphous substance.

Thermally induced reversion to the high-temperature phase occurs at a slower rate than the recovery of sharpness in the diffraction peaks. Therefore, the energy barrier to phase change appears to be higher than the barrier preventing the annealing out of stacking faults. The latter process requires, of course, merely the collapse or rearrangement of existing stacking faults, whereas the former presumably requires the formation and movement of half-dislocations on alternate close-packed planes.

The β–α phase transition in solid nitrogen is thought to be related to the interaction of quadrupole moments.[4] Calculations of the quadrupole interaction have been of a type corresponding to the theory of paramagnetism for dipoles. We think, however, that the evidence favors an analog to the Néel point of antiferromagnetism of dipoles, a coherent ordering of the quadrupole moments over more than one unit cell, and the appearance of a strong internal field corresponding to the Weiss field in magnetism. The change in crystal structure on heating seems to be a consequence of the loss of quadrupole ordering. The volume change is small. If size effects were of primary importance in the α–β transition, one would expect the α-nitrogen structure to be stabilized by adding argon because the molar volume of α-nitrogen is smaller than that of β-nitrogen, and the argon atom is smaller than the nitrogen molecule. On this basis, the α–β transition curve should have an upward trend as argon is dissolved in nitrogen, yet Fig. 1 shows the opposite: In fact, the stability range of the α-phase is terminated by dissolving about 25% argon in the nitrogen, i.e., when about 1 argon atom per unit cell is present. Argon cannot participate in the quadrupole ordering and destroys its coherence.

Figure 2 is a preliminary version of the argon–oxygen diagram. Again, we find some quite striking features. Argon is cubic; γ-oxygen, the high-temperature phase of oxygen, is also cubic. In spite of this fact, only 1% oxygen transforms the lattice to hcp which is stable near the solidus line up to 65%. The argon-rich side shows again a martensitic transformation to fcc at lower temperatures. However, between 20 and 60% oxygen, only the hexagonal phase is stable above hydrogen temperatures. The change in nearest-neighbor distance by alloying oxygen to the argon is very small, so that the phase change cannot be explained as a consequence of expanding lattice constants.

The oxygen-rich side is somewhat complex since oxygen has two low-temperature phases. Our investigations are not quite complete. We have established with certainty that the rhombohedral β-phase of oxygen shows a closed field as a function

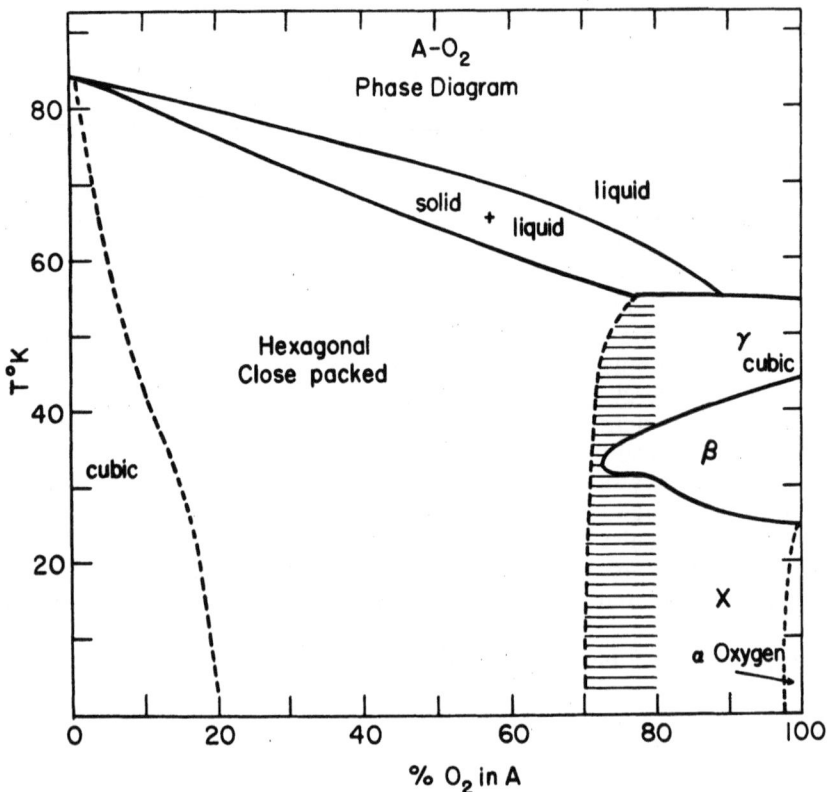

Fig. 2. Preliminary argon–oxygen phase diagram.

of argon content. The orthorhombic α-phase of oxygen disappears by adding about 5% argon. The borderlines between the oxygen structures and the hcp as well as the two-phase regions are still somewhat uncertain. A new phase, marked × in Fig. 2, seems to be sandwiched between the hcp and the α-oxygen field.

References

1. L. Meyer, C. S. Barrett, and P. Haasen, *J. Chem. Phys.* **40**, 2744, 1964.
2. C. S. Barrett and L. Meyer, *J. Chem. Phys.* **41**, 1078, 1964.
3. L. Jansen, *Phys. Letters* **4**, 91, 1963.
4. O. Nagai and T. Nakamura, *Progr. Theoret. Phys. (Kyoto)* **24**, 432, 1960.
5. H. M. Long and F. S. Di Paolo, *Chem. Eng. Progr.* **59**, 30, 1963.

SPECTROSCOPIC OBSERVATION OF THE ONE-PHONON SPECTRUM OF SOLID ARGON

G. O. Jones and J. M. Woodfine

*Queen Mary College
London, England*

Estimates by Davies and Stimson[1] of the far-infrared absorption in solid argon resulting from two-phonon induced dipoles suggested that there might be measurable absorption in 1 cm of the solid. We have found no measurable frequency-dependent absorption in solid argon between 70 and 500 μ over path lengths up to 25 cm. However, when fractions of the order 1% of krypton or xenon were introduced, there was absorption over a broad band, centered at about 180 μ. The absorption is roughly proportional to the amount of impurity and stronger with the heavier atom (xenon) as impurity. The location of the absorption was the same for both krypton and xenon in argon. The results suggest that we are observing the one-phonon spectrum of solid argon directly as a result of dipoles induced by the impurity atoms. Their random distribution destroys the zero-wavelength selection rule which determines the one-phonon absorption in polar crystals. The form of the spectrum may be compared with the calculated spectrum of Grindlay and Howard,[2] based on a 12–6 potential. Broadly, the experimental spectrum has the appearance of a "smoothed" version of the calculated spectrum, but with a "tail" which extends well beyond the cutoff frequency. This may be due to anharmonicity or other influence of the high temperature of the experiments, which have so far been carried out successfully only near the triple-point. Doniach[3] has made an estimate of the magnitude of the impurity-induced dipole on a simple shell model, in fair agreement with that which may be estimated from the present results.

References

1. R. O. Davies and R. F. Stimson, Conference on Lattice Dynamics, Copenhagen, 1963.
2. J. Grindlay and R. Howard, Conference on Lattice Dynamics, Copenhagen, 1963.
3. S. Doniach, private communication.

VARIATION WITH TEMPERATURE OF THE VELOCITY OF TRANSVERSE ELASTIC WAVES IN SOLID ARGON

G. O. Jones and A. R. Sparkes

*Queen Mary College
London, England*

The variation with temperature of the velocity of transverse elastic waves, v_t, in solid argon has been determined over the range 18°–62°K by measuring the resonant frequencies of torsional oscillations of a polycrystalline rod held only at one end. A small extrapolation leads to an estimate of 917 m/sec for v at 0°K—considerably higher than the estimate (810 m/sec) obtained by extrapolation by Barker and Dobbs[1] from results in the neighborhood of 70°K. In terms of the calculations of Horton and Leech,[2] based on the interatomic potential

$$\theta(r) = \frac{A}{r^m} - \frac{B}{r^6}$$

the best fit is obtained with $m = 13$ and with nearest-neighbor interactions only considered. Leech and Reissland[3] have recently employed our results in fitting to a 12–9–6 potential, leading to a value of specific heat in good agreement with experiment.

References

1. J. R. Barker and E. R. Dobbs, *Phil. Mag.* **46**, 1069, 1955.
2. G. K. Horton and J. W. Leech, *Proc. Phys. Soc. (London)* **82**, 816, 1963.
3. J. W. Leech and J. A. Reissland, private communication.

VARIATION WITH TEMPERATURE OF THE REFRACTIVE INDEX AND LORENTZ–LORENZ FUNCTION OF SOLID ARGON

A. J. Eatwell and G. O. Jones

Queen Mary College
London, England

The refractive index n of solid argon has been measured over the range 20°–83.8°K by the spectrometric method of minimum deviation. In terms of the Lorentz–Lorenz function

$$\frac{n^2 - 1}{n^2 + 2} \cdot \frac{1}{\rho}$$

(which on the Lorentz local field model is a measure of the atomic polarizability), the results indicate an atomic polarizability, α decreasing with increasing density ρ, reversing the conclusion of Jones and Smith[1] based on measurements made only near the triple-point (83.8°K). At the highest densities the variation in α is very roughly as $\rho^{-\frac{1}{3}}$. The results may be discussed in terms of a shell model of the type employed by Dick and Overhauser.[2] Qualitatively, and classically, we may say that increasing the density, or decreasing the shell radius, increases the frequency, ω_0, of oscillations of the shell relative to the core, and hence reduces polarizability at optical and lower frequencies. Doniach and Huggins[3] have recently estimated the change in ω_0 due to exciton–phonon coupling, with results in good agreement with $1/r^{12}$ law for the short-range forces. Dispersion data obtained in the present investigation lead to estimates of ω_0 corresponding to wavelengths in the neighborhood of 700 Å, to be compared with about 1000 Å, the location of the main ultraviolet band (Baldini[4]). We can also estimate the static dielectric constant accurately by extrapolation, leading to values 1.63_5 at 50°K and 1.59_5 at the triple-point. The latter value is in good agreement with the value 1.599 obtained directly by Amey and Cole.[5]

References

1. G. O. Jones and B. L. Smith, *Phil. Mag.* **5**, 355, 1960.
2. B. G. Dick and A. W. Overhauser, *Phys. Rev.* **112**, 90, 1958.
3. S. Doniach and R. Huggins, private communication.
4. G. Baldini, *Phys. Rev.* **128**, 1562, 1962.
5. R. L. Amey and R. H. Cole, *J. Chem. Phys.* **40**, 146, 1964.

THE SPECIFIC HEAT OF SOLID NEON*

C.-H. Fagerstroem and A. C. Hollis Hallett

Department of Physics, University of Toronto
Toronto, Ontario, Canada

Measurements were made on spectroscopically pure, naturally occurring neon in the temperature range 1.7° to 28°K. The gas was condensed in a metal calorimeter of volume $\sim 16 \text{ cm}^3$; the tube connecting the calorimeter with the gas supply was fitted with a valve so placed in the cryostat that the dead space above the condensed gas was as small as possible. A manganin heater was wound around the body of the calorimeter and a carbon thermometer, embedded in a copper sleeve with araldite, was clamped inside a re-entrant copper block which formed part of the bottom of the calorimeter. A mechanical heat switch permitted cooling of the calorimeter.

The carbon thermometer was calibrated against helium[1] and hydrogen[2] vapor pressure thermometers, and interpolation between 4° and 14°K was made using the Clement and Quinnell[3] formula. Extrapolation above 20°K was made, also using the triple point of oxygen (54.363°K) as an additional fixed point.[4] The accuracy of the extrapolation above 20°K may be judged by the fact that the triple point of neon was found to be at 24.547°K compared with 24.544°K found by Grilly,[5] and the β–γ transition in solid oxygen to be at 43.78°K compared with 43.80_0°K.[4]

The measurements actually yield C_s, the specific heat at the saturation vapor pressure; the correction required to deduce C_p was found to be negligible and has been ignored. Above 20°K, corrections arising from the vaporization of the solid become important, reaching a value 10% of C_p at the triple point. The correction has been calculated using measurements of the vapor pressure[5] and the virial coefficients.[6] The data above 7°K, with these corrections applied, are shown in Fig. 1; in the same figure, the data of Clusius et al.[7] are also indicated by the dashed curve only where they do not agree with the present data. The increasing scatter in the results with increasing temperature arises mainly from the fact that the thermal conductivity of neon decreases with increasing temperature, resulting in a lag in the response of the thermometer to the input of heat. Also, just below the triple point where contributions to the specific heat due to thermal generation of vacancies in the lattice can be expected, smaller temperature intervals were used to achieve better resolution of the curve of C_p versus T; this further increases the scatter in the results at these temperatures.

Corrections to obtain C_v from the measured C_p could not be obtained with any degree of certainty because the compressibility χ of solid neon is known only at 4°K,[8] although the expansion coefficient α has been measured between 8°K and the triple point T_t.[9] It has been assumed that a law of corresponding states[10] holds for

---

* Research supported by the National Research Council of Canada, the Ontario Research Foundation, and the University of Toronto.

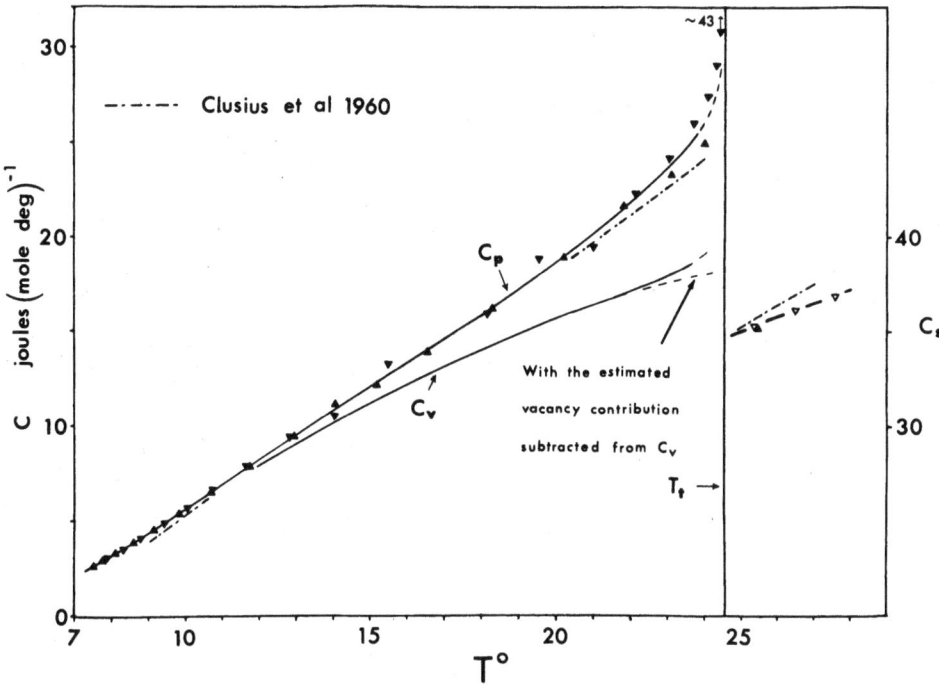

Fig. 1. The variation of the specific heat of solid neon with temperature between 7° and 28°K. The scale for C_s on the right refers to the measurements in the liquid region above T_t.

the compressibility and that $\chi_s(T)/\chi_s(0°K)$ varies with T/T_t in the same manner as it is known to vary for argon.[10] Although the accuracy of this procedure is doubtful, particularly above 15°K, it at least takes some account of the probable variation of χ_s with temperature and will have to suffice until more extensive measurements of χ_s have been made. The variation with temperature of C_v, deduced with this correction, is also shown in Fig. 1.

Below about 7°K the difference between C_p and C_v is negligible, and the results are more conveniently shown, as in Fig. 2, as the variation of the effective Debye temperature θ with temperature. It is clear from this figure that the curve has a minimum at about 9.5°K, and rises as the temperature falls. Extrapolation of the curve to 0°K to obtain θ_0 gives a value of $75 \pm 0.5°K$; this is indicated by the shaded region in Fig. 2 and also by the extrapolation of a curve of C_v/T^3 versus T^2.

The heat of sublimation at 0°K, L_0, has been determined in the usual manner by an integration on the gas and solid phases between 0°K and T_t, taking for the latent heat of sublimation at T_t the value 2107 J/mole obtained from the slope of the vapor pressure curve at T_t and the molar volume of the vapor deduced from the virial coefficient data.[6] L_0 is found to be 1855 J/mole, with an estimated error of less than 2%.

Just below T_t, effects are undoubtedly present due to the thermal generation of vacancies. Because the triple point of neon is as low as $\theta_0/3$, the lattice C_v rises rapidly with temperature in this region, and anharmonic effects are also undoubtedly

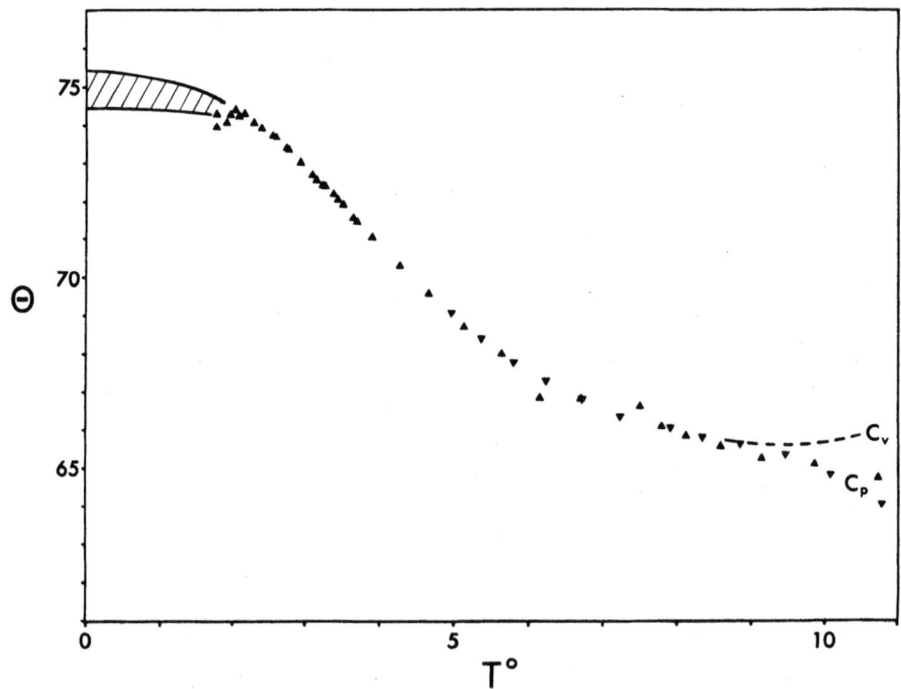

Fig. 2. The variation of the effective Debye temperature θ with temperature below 11°K. The shaded region represents the range of possible extrapolations of the curve to 0°K.

present to an unknown extent. Consequently, little more than a guess at the contribution due to the formation of vacancies can be made using a graphic method[10]; the enthalpy of formation is estimated to be 2000 J/mole, with a possible error of 10%.

When such contributions due to vacancy formation are subtracted from the values of C_v, a curve of θ versus T can be drawn for the whole temperature range. When the values of $\theta(T)$ are corrected[10] for the significant variation of molar volume of neon with temperature,[9] the shape of the curve of $\theta(T)/\theta_0$ versus T/θ_0 is found to be more similar to that observed for argon than that for krypton.[10]

References

1. F. G. Brickwedde, H. van Dijk, M. Durieux, J. R. Clement, and J. K. Logan, *J. Res. Natl. Bur. Std. A* **64**, 1, 1960.
2. H. W. Woolley, R. B. Scott, and F. G. Brickwedde, *J. Res. Natl. Bur. Std.* **41**, 379, 1948.
3. J. R. Clement and E. H. Quinnell, *Rev. Sci. Instr.* **23**, 213, 1952.
4. H. J. Hoge, *J. Res. Natl. Bur. Std.* **44**, 321, 1950.
5. E. R. Grilly, *Cryogenics* **2**, 226, 1962.
6. W. H. Keesom and J. A. van Lammaren, *Physica* **1**, 1161, 1934.
7. K. Clusius, P. Flubacher, U. Piesbergen, K. Schleich, and A. Sperandio, *Z. Naturforsch. A* **15**, 1, 1960.
8. J. W. Stewart, *J. Phys. Chem. Solids* **1**, 146, 1956.
9. L. H. Boltz and F. A. Mauer, *Advances in X-Ray Analysis*, Vol. 6, Plenum Press, New York (1963), p. 242.
10. R. H. Beaumont, H. Chihara, and J. A. Morrison, *Proc. Phys. Soc. (London)* **78**, 1462, 1961.

NUCLEAR MAGNETIC RESONANCE IN SOLID HYDROGEN AND DEUTERIUM UNDER HIGH PRESSURES

Samuel A. Dickson and Horst Meyer[†]

Department of Physics, Duke University
Durham, North Carolina

In this paper we present nuclear magnetic resonance measurements in solid hydrogen and deuterium between 1.5° and 4.2°K and under approximately hydrostatic pressures between 0 and 5500 atm. This work is an extension of earlier measurements by McCormick and Fairbank.[1,2]

The interest in such measurements is due to the great compressibility of both solids, which makes easy the observation of a change of several properties strongly dependent on density. For instance, the transition temperature T_λ from the state of free rotation to that of cooperative orientation of $\mathbf{J}$,[3] as well as the ortho–para conversion, is expected to change with density. The second moment M_2 of the nuclear resonance line above the transition temperature T_λ is caused primarily by intermolecular magnetic dipole–dipole interaction[4] and hence is expected to be proportional to the square of the density ρ. The preliminary results of McCormick[2] essentially verified these predictions for solid hydrogen.

We have measured these three quantities using the same pressure-generating equipment that McCormick used. The pressure was applied to the solidified gas as a uniaxial stress. Since solid hydrogen and deuterium are found to have a low shear strength[5] and thus exhibit plastic flow, the uniaxial stress resulted in the material being subjected to a hydrostatic pressure. The sample was compressed into an RF coil designed so that the solid can flow outside as well as inside this coil. The nuclear resonance signal was detected by a low-level Robinson oscillator followed by a phase-sensitive detector. The frequency used was about 14 Mcps for hydrogen and 6 Mcps for deuterium and was measured by a Hewlett–Packard electronic counter. Care was taken to avoid saturation and resulting distortions of the line.[6] The observations were made with $H_{RF} \simeq 6 \cdot 10^{-3}$ G.

Results

Ortho-Para Conversion in Hydrogen. The ortho concentration c was taken to be proportional to the area under the NMR line and was measured as a function of time at a constant density. It was normalized to $c_0 = 0.75$ at the beginning of the experiment ($t = 0$), because hydrogen at room temperature ortho concentration was always used for the experiment. The concentration could always be expressed as a function of time as $1/c = 1/c_0 + kt$, which is the integrated form of the rate equation

[†] Alfred P. Sloan Fellow.

first proposed by Cremer and Polanyi.[7] The conversion constant k was found to be roughly proportional to (ρ/ρ_0),[2,4] where ρ_0 is the density at zero pressure. This is in satisfactory agreement with recent data of Ahlers,[8] who found a dependence $(\rho/\rho_0)^{\frac{4}{3}}$. However, our experimental conversion constants were higher at all densities than those found by Ahlers. The conversion rates so determined were used to compute c in the rest of the experiments, where c enters as a parameter of the quantities being measured. The concentrations in these experiments varied between 0.55 and 0.75.

Line-Width Measurements Above T_λ for Hydrogen. So far, the line-width has been the parameter studied by experimenters,[4,6,9] but it is the second moment M_2 that has to be compared with theoretical predictions.[10] We have therefore made a study of the relation between these two quantities and found that for selected samples of lines at different densities and ortho concentrations one has at 4.2°K:

$$\Delta H_{ms} = (2.95 \pm 0.15)\frac{M_2^{\frac{1}{2}}h}{g\beta}[G] \qquad (1)$$

where ΔH_{ms} is the distance (in gauss) between the points of maximum slope, β is the nuclear Bohr magneton, and g is the corresponding Landé nuclear factor. For a Gaussian line, the numerical factor is 2.0. The hydrogen line is appreciably sharper than the Gaussian one. The empirical factor in equation (1) changes with temperature, and a systematic investigation of M_2 as a function of T is presently being made.

Harris[11] has calculated the effect of zero point vibrations and the rotational movement on the second moment and has shown it to be smaller than the present uncertainty due to lattice parameters. Measurements of the line-width as a function of density and concentration C showed that at 4.2°K:

$$\Delta H_{ms} = (7.1 \pm 0.3)\left(\frac{\rho}{\rho_0}\right)C^{\frac{1}{2}}[G] \qquad (2)$$

This is of the same form as expected from the theory of second moments, where the numerical factor is calculated to be approximately 6.5 for the hcp lattice.[12] (Kogan et al.[12] found $a = 3.78$ Å, $c = 6.16$ Å for hydrogen, and $a = 3.54$ Å, $c = 5.91$ Å for deuterium under zero pressure); the form is also that which would be expected from use of equation (1). Hence, the agreement is almost within the combined errors.

The Transition Temperature T_λ for Hydrogen. For a given ortho concentration and density, T_λ was determined by extrapolation as the temperature where the line structure due to intramolecular dipole–dipole interaction had vanished. It was possible to represent our thirty experimental points by the empirical relation:

$$T_\lambda(c, \rho) = (-1.23 + 3.78c)\left(\frac{\rho}{\rho_0}\right)^{2.0} \qquad (3)$$

Such a dependence on density is approximately that expected if the interaction potential is due to intermolecular electric quadrupole–quadrupole interaction,[13] in which case one would have $T_\lambda \propto (\rho/\rho_0)^{\frac{5}{3}}$. There is good agreement between our results, the point obtained by McCormick (if c is assumed to be 0.68), and the data from specific-heat measurements.[14] (It is known, however, that the temperature of the peak maximum is higher than that determined from the disappearance of the NMR structure.) This is shown in Fig. 1, where the ratio $T_\lambda(c, \rho)/T_\lambda(c, \rho_0)$ is plotted against the relative density.

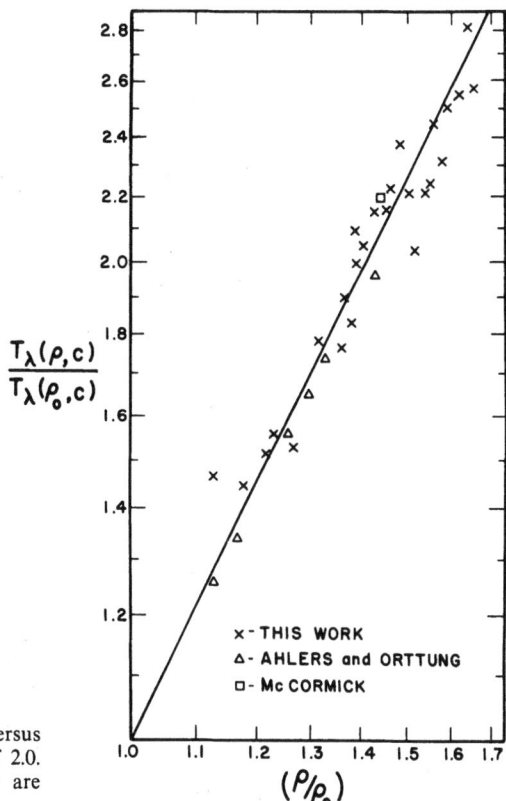

Fig. 1. Plot of the ratio $T_\lambda(\rho, c)/T_\lambda(\rho_0, c)$ versus relative density. The solid line has a slope of 2.0. The points from specific heat measurements are deduced from Fig. 8 of reference 14.

Deuterium. We have conducted a study of the line-width and the second moment as a function of density at various temperatures. As in solid hydrogen above T_λ, the line-width but not the moment M_2 was found to be temperature independent. The line-shape, however, is much closer to the Gaussian one than is that of hydrogen. It was found for several densities that at 4.2°K:

$$\Delta H_{ms} = (2.1 \pm 0.1) M_2^{\frac{1}{2}} \frac{h}{g\beta} [G] \qquad (4)$$

The line-width as a function of density is plotted in Fig. 2. Again, its linear dependence on ρ is in agreement with that expected from the theory of second moments and the data seem to extrapolate to zero line-width for zero density. This would mean that there would be no density-independent broadening, such as intramolecular broadening. There is an appreciable discrepancy between the experimental and theoretical second moment. This has already been found by Sugawara and coworkers[15] at normal density and has so far not been explained.

In addition to a study of the line-width, a search was made for a possible transition and splitting of the paradeuterium signal. No such phenomenon was observed over the range of temperatures and pressures covered. Recently, specific-heat measurements on deuterium have shown the existence of a λ transition for high para concentrations.[16] Hence, it might be possible to observe T_λ on enriched

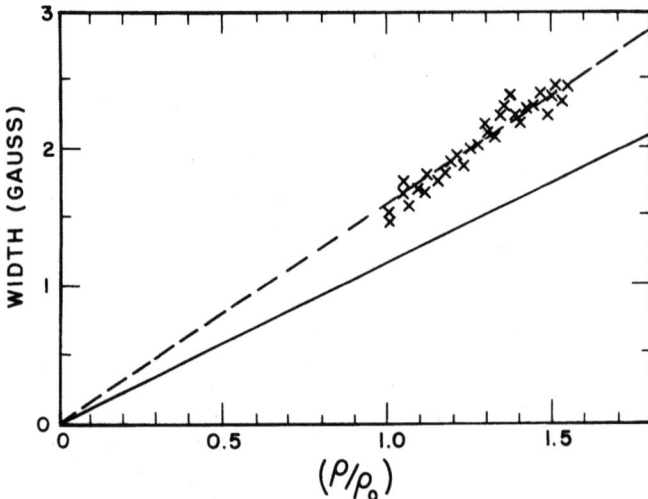

Fig. 2. Width of the deuterium line as a function of relative density at 4.2°K. The solid line is calculated from theory of second moments, assuming an hcp lattice.[12]

samples by means of NMR. We have built an apparatus to study this transition for both hydrogen and deuterium between 0.4° and 2°K. Also, we intend to study the line-widths of deuterium diluted with parahydrogen (this is effectively for $\rho/\rho_0 < 1$) to attempt to find an explanation for the above-mentioned discrepancy between theory and experiment.

Acknowledgments

We are indebted to Dr. P. N. Dheer, Mr. J. Pullman, and E. Morgan for help in several experiments. We are indebted to A. B. Harris for several stimulating discussions and to Dr. W. D. McCormick for advice on the high-pressure system. We are indebted to Dr. Ahlers for attracting our attention to the recent paper by Kogan et al. This work has been supported by a grant of the Army Research Office (Durham) and by a contract with the Office of Naval Research. A more complete technical report of this work has already been issued.

References

1. W. D. McCormick and W. M. Fairbank, *Bull. Am. Phys. Soc.* **3**, 166, 1958.
2. W. D. McCormick, thesis, Duke University (1959).
3. F. London, *Phys. Rev.* **102**, 168, 1956.
4. J. Hatton and B. V. Rollin, *Proc. Soc. (London) Ser. A* **199**, 222, 1949.
5. J. W. Stewart, *Phys. Rev.* **97**, 578, 1955.
6. T. Sugawara, *Sci. Rept. Res. Inst. Tohoku Uni. Ser. A* **8**, 95, 1956.
7. E. Cremer and M. Polanyi, *Z. Physik. Chem. B* **21**, 459, 1933.
8. G. Ahlers, *J. Chem. Phys.* **40**, 3123, 1964.
9. G. W. Smith and C. F. Squire, *Phys. Rev.* **111**, 188, 1958.
10. J. H. Van Vleck, *Phys. Rev.* **74**, 1168, 1948.
11. A. B. Harris (to be published).
12. V. S. Kogan, A. S. Bulatov, and L. F. Yakimenko, *Zh. Eksperim. i Teor. Fiz.* **46**, 148, 1964. (*Soviet Phys. JETP (English Transl.)* **19**, 107, 1964.)
13. T. Nakamura, *Progr. Theoret. Phys.* (Kyoto) **14**, 135, 1955.
14. G. Ahlers and W. H. Orttung, *Phys. Rev. A* **133**, 1642, 1964.
15. T. Sugawara, Y. Masuda, T. Kanda, and E. Kanda, *Sci. Rept. Res. Inst. Tohoku Uni. Ser. A* **7**, 67, 1955.
16. G. Grenier and D. White, *J. Chem. Phys.* **40**, 3015, 1964.

A SEARCH FOR ISOTOPIC PHASE SEPARATION IN HYDROGEN – DEUTERIUM MIXTURES*

E. M. de Castro, D. Husa, J. R. Gaines,† and J. G. Daunt

The Ohio State University
Columbus, Ohio

This work was originally started in the hope that information obtained from nuclear magnetic resonance experiments would provide additional data on the type of transition known as isotopic phase separation. In particular, it seemed probable that NMR experiments could readily distinguish the difference between a transition from the randomly mixed state (the high temperature equilibrium configuration) to the isotopically ordered state involving rearrangement of the isotopes into interpenetrating sublattices and a transition involving only a clumping together into small domains. The system hydrogen–deuterium was chosen for investigation, since Kogan, Lazarev, and Bulatova[1] had reported observing such a phase separation at the relatively high temperature of 16.4°K. At such high temperatures, the diffusion rates are large enough[2] so that domains of macroscopic size can form in a relatively short time. Since the line-width is dependent on the concentration of orthohydrogen molecules, one would expect to observe a transition from a simple line into a composite line with both a very broad and a very narrow component in the event a phase separation occurred. At high temperatures $T > 12°K$, pulsed NMR techniques were employed, since the true line-width was swamped by magnetic field inhomogeneities. No such transitions were observable from the T_1 and T_2 data. At temperatures below 12°K, the solid absorption line becomes very broad and steady-state (broad line) techniques are applicable.

The proton line-width at 4.2°K as a function of the concentration of normal hydrogen is shown in Fig. 1. The observed width of the line is very largely dependent on the magnitude of the radio frequency field used to induce the transitions between the spin states (hydrogen). This point was emphasized by Sugawara,[3] who found that the observed proton line-width in normal hydrogen actually decreased with increasing RF power as opposed to the usual saturation broadening of a liquid resonance line as predicted by the Bloch equations. Redfield[4] observed similar behavior in metallic solids and developed a theory of saturation to account for his results.

More recently, Provotorov[5] formulated a theory to describe NMR saturation in crystals. His expression for the absorption component of the complex susceptibility (χ'') is:

$$\chi''(\Delta) = k \frac{g(\Delta)}{1 + \pi\gamma^2 H_1^2 T_1 g(\Delta)[1 + (\hbar^2\Delta^2 T_1')/(H_0^2 T_1)]} \quad (1)$$

---

* Work supported in part by a grant from the National Science Foundation and a contract with the Office of Naval Research.
† Alfred P. Sloan Fellow.

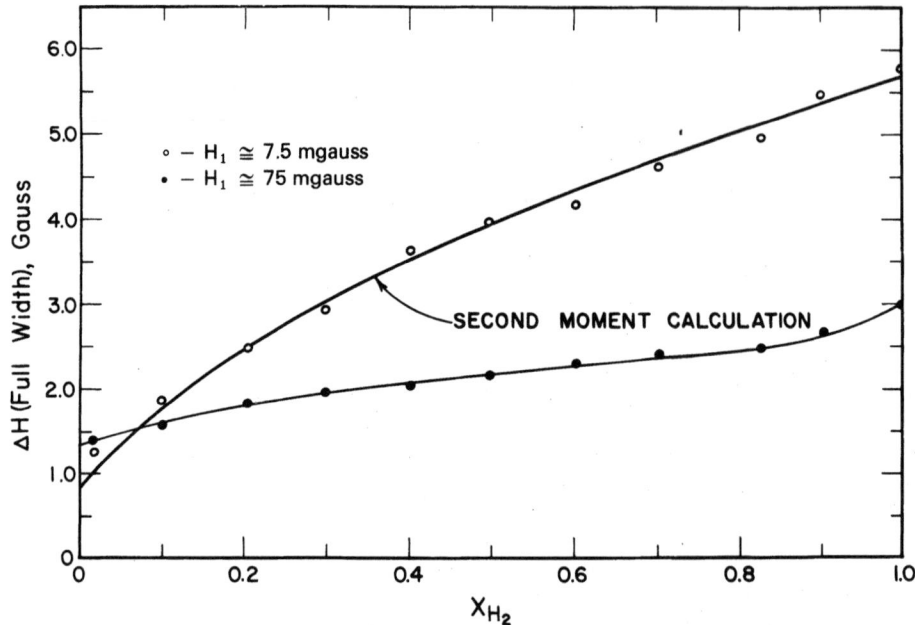

Fig. 1.

where $g(\Delta)$ is the absorption line-shape function. The predictions of this equation are: (A) For negligible saturation, χ'' is a faithful replica of $g(\Delta)$. (B) For appreciable saturation, χ'' should approach a Lorentzian-type curve

$$\chi'' \to \frac{k}{\pi \gamma^2 H_1^2 T_1}(1 + \Delta^2 \tau^2)^{-1} \tag{2}$$

with a line-width governed by the time constant τ. τ is essentially a "spin-lattice" relaxation time characteristic of the return of the local field distribution to equilibrium after the absorption of a quantum from the RF field and the accompanying change in the dipole–dipole energy of $\hbar\Delta$. Since τ is independent of H_1, the line-width as a function of H_1 should possess a nonzero lower limit. (C) $(\chi'')_{max}$ should be proportional to $(H_1^2)^{-1}$. (D) χ'' becomes more narrow with increasing RF power.

The results obtained at 4.2°K with solid mixtures of hydrogen–deuterium are in good agreement with the above predictions even though there is the additional complication of rotation of the ortho molecule to contend with. The uppermost curve of Fig. 1 represents the line-width data taken at the minimum value of RF power (in the "no saturation" region), while the lower curve represents the line-width data obtained with maximum RF power. H_1^2 changes by a factor of 100 between the two curves. The upper curve fits the expression

$$(\Delta\omega)_\chi = (\Delta\omega)_{pure}\{\chi + k(1 - \chi)\}^{\frac{1}{2}} \tag{3}$$

rather well over the entire concentration range. This is an interesting development in that the expression (3) assumes that the lattice sum

$$\sum_{j<k} \frac{1}{r_{jk}^6}(1 - 3\cos^2\Theta_{jk})$$

does not change appreciably as a function of concentration.

The effect of rotation of the orthohydrogen molecule on the observed resonance behavior is rather pronounced. Gutowsky and Pake[6] have carried out extensive investigations on the shape and temperature dependence of the proton line in hindered rotators. Their results for the line-width are very similar to ours in that the line-width (as measured from the integral of $d\chi''/d\omega$) is narrow at high temperatures, becomes broader as the rotation is perturbed, and eventually reaches a constant value at lower temperatures. The increasing width of the line is accompanied by the appearance of a composite structure on the derivative, characteristic of the transition from a state of free rotation to a state of hindered rotation.

If one starts at 4.2°K with a simple line and reduces the temperatures quickly to 1.1°K, the line becomes composite in a short period of time (the time necessary to sweep through the line is sufficient), which is more in keeping with the idea of a hindering of the rotation than with the idea of phase separation. Once the line has become composite, the shape remains essentially constant in time. Figure 2 shows the shape of the derivative curve at 1.1°K for various starting concentrations. It should be noted that the line is quite composite and that the magnitude of the broad hump relative to the magnitude of the central line is quite dependent on the starting concentration. It is the shapes of these curves that are so indicative of an isotopic phase separation, since they represent derivatives of curves that could easily be interpreted as a folding of two distinct intensity distributions, one characteristic of a narrow resonance line originating from domains rich in hydrogen. These curves, however, bear striking resemblance to the derivative curves obtained by Sugawara and coworkers[7] in pure hydrogen with varying orthohydrogen concentrations. Thus, if one attempts to interpret the hydrogen–deuterium composite lines as being indicative of phase separation, one must concede that the separation also occurs between orthohydrogen and parahydrogen in pure hydrogen.

To check on this possibility, the following experiment was performed: Pure hydrogen was allowed to remain in the sample chamber until the orthohydrogen concentration was approximately 55% (the ortho–para conversion rate being between 1 and 2% per hour). This sample was then frozen at 12°K and lowered to 4.2° and 1.1°K in slow steps over several hours. To obtain line-shapes such as those observed by Sugawara and coworkers,[7] the concentrations of orthohydrogen in the conjectured two phases must be nearly pure orthohydrogen in one phase (to give a line of width > 6 G) and nearly pure parahydrogen in the other (to give a line narrower than 1 G). Thus, in domains containing nearly pure orthohydrogen, one expects to observe the cooperative transition characteristic of a hindering of rotation (known as the "λ-splitting") at 1.1°K. The amplifier gain was increased and an exhaustive search made in the wings of the line for other components characteristic of this transition. No such fragments were seen, even though the signal-to-noise ratio was extremely high.

Heat capacity and warming rate measurements were made on a sample consisting of 20% parahydrogen and 80% normal deuterium. This sample was chosen because of the very pronounced composite line obtained below 4°K in a mixture of 20% normal hydrogen and 80% normal deuterium. If a pronounced anomaly

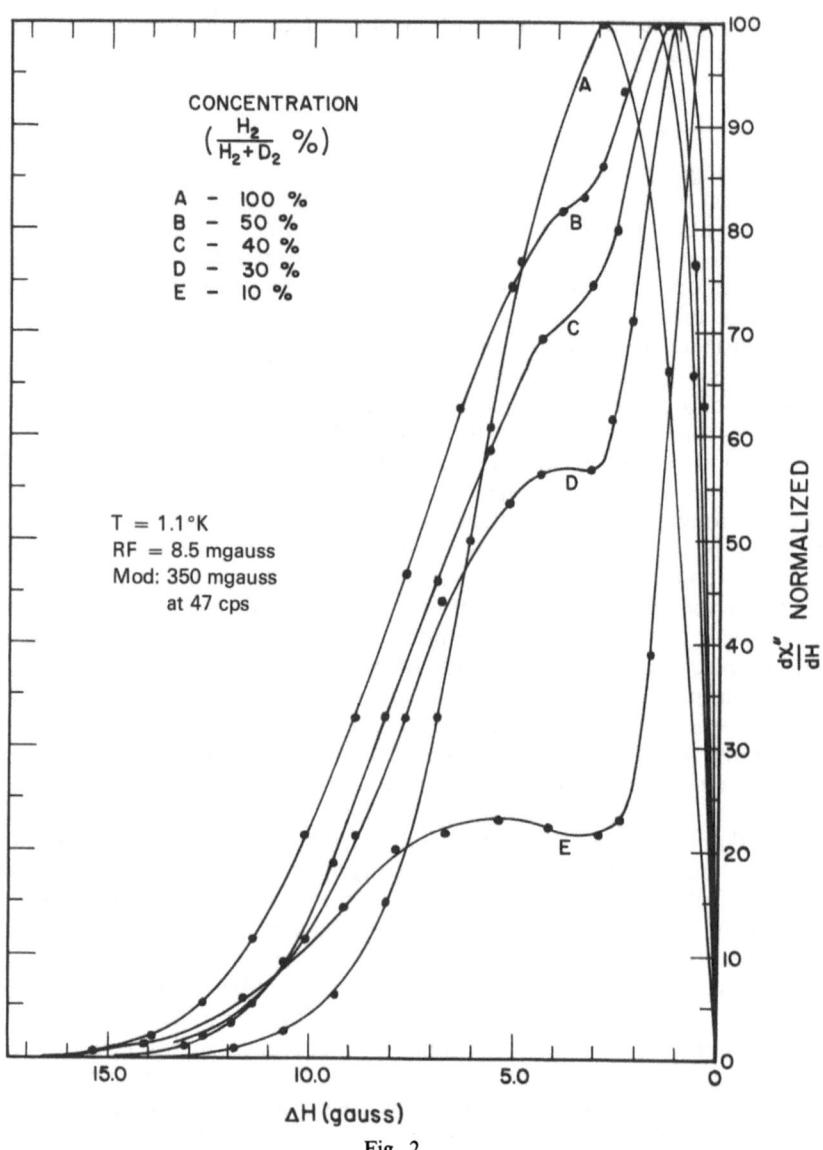

Fig. 2.

$(\Delta C/R \cong 1)^8$ occurred in the heat capacity below 4°K, one could conclude that a phase separation had indeed taken place, since no rotating hydrogen molecules were present. The warm-up rate was smooth, indicating that the heat capacity was a continuous function of the temperature, but the heat capacity measurements did reveal a slight bump between 2.5° and 3.5°K. This bump is due to the anomalous heat capacity of the rotating deuterium molecules in normal deuterium, as shown by the measurements of Grenier and White.[9]

In conclusion, then, the NMR data in hydrogen–deuterium mixtures can be adequately accounted for by extrapolation of the properties of the pure components.

In particular, the composite resonance lines are characteristic of a hindered rotator and no concrete evidence for isotopic phase separation has been found. It should be noted that although phase separation may be energetically favored below 4°K the diffusion rates are so small that the time required to produce observable domains can be extremely long.

References

1. V. S. Kogan, B. G. Lazarev, and R. S. Bulatova, *Zh. Eksperim i Teor. Fiz.* **34**, 238, 1958. (*Soviet Phys. JETP* (*English Transl.*) **7**, 165, 1958.)
2. M. Bloom, *Physica* **23**, 767, 1957.
3. T. Sugawara, *Sci. Rept. Res. Inst. Tohoku Uni. Ser. A* **8**, 95, 1956.
4. A. Redfield, *Phys. Rev.* **98**, 1787, 1955.
5. B. N. Provotorov, *Soviet Phys. JETP* (*English Transl.*) **14**, 1126, 1962.
6. H. S. Gutowsky and G. E. Pake, *J. Chem. Phys.* **18**, 162, 1950.
7. T. Sugawara, Y. Masuda, T. Kanda, and E. Kanda, *Sci. Rept. Res. Inst. Tohoku Uni. Ser. A* **7**, 67, 1955.
8. D. O. Edwards, A. S. McWilliams, and J. G. Daunt, *Phys. Rev. Letters* **9**, 195, 1962.
9. G. Grenier and D. White, *J. Chem. Phys.* **40**, 3015, 1964.

LIQUID–SOLID PHASE EQUILIBRIA IN THE HYDROGEN–DEUTERIUM SYSTEM[*]

David White and J. R. Gaines[†]

*Departments of Chemistry and Physics, The Ohio State University
Columbus, Ohio*

The experimental investigation of excess thermodynamic properties[1] of liquid solutions of the hydrogen isotopes shows that the deviations from Raoult's law are positive. The magnitudes of the deviations for hydrogen–deuterium mixtures are in accord with the theory of Prigogine, Bingen, and Bellemans,[2] which predicts a continuous range of liquid and solid solutions down to approximately 2°K. However, the recent experiments of Kogan, Lazarev, and Bulatova[3] clearly show a phase separation in the hydrogen–deuterium system at approximately 16.4°K. The phase diagram shows a peritectic at this temperature corresponding to the coexistence of a liquid solution in equilibrium with two solid solutions of different composition. The results of recent nuclear magnetic resonance experiments[4] in hydrogen–deuterium solid mixtures are not in agreement with those of Kogan and collaborators.[3] In order to clarify the phase behavior of mixtures of hydrogen and deuterium in the liquid and solid state, we have re-examined this system calorimetrically.

The calorimeter employed in these experiments has already been described in detail.[5] Two types of measurements were made:

(1) Heat capacity measurements of several solid mixtures of parahydrogen and normal deuterium from approximately 8°K to their melting points (in order to determine whether or not a phase separation occurs in the solid).

(2) Heating curves of several mixtures of parahydrogen and normal deuterium (i.e., temperature–time measurements for a constant rate of energy input into the calorimeter) from the solid to the liquid state (in order to establish the phase boundaries in the region where the liquid and solid coexist).

The heat capacities of the solid mixtures are shown in Fig. 1. The two solid lines represent the heat capacities of pure normal deuterium and pure parahydrogen. It can be seen from Fig. 1 that, to a reasonable approximation, the heat capacities of the mixtures can be represented as a linear combination of the heat capacities of the pure components. This type of behavior can be ascribed either to a continuous range of nearly ideal solid solutions or to a mechanical mixture of the pure components. We believe the former represents the actual physical situation.

The heating curves for several mixtures are shown in Fig. 2. The points A correspond to the discontinuities determined by numerical analysis of the data. The low temperature discontinuity should correspond to the melting point of the mixture, the high temperature one to the freezing point. The phase diagram constructed from points A is shown in Fig. 3 by the broken lines. It should be possible

---

[*] This work was supported in part by the United States Atomic Energy Commission, Division of Research, and in part by the National Science Foundation.
[†] Alfred P. Sloan Fellow.

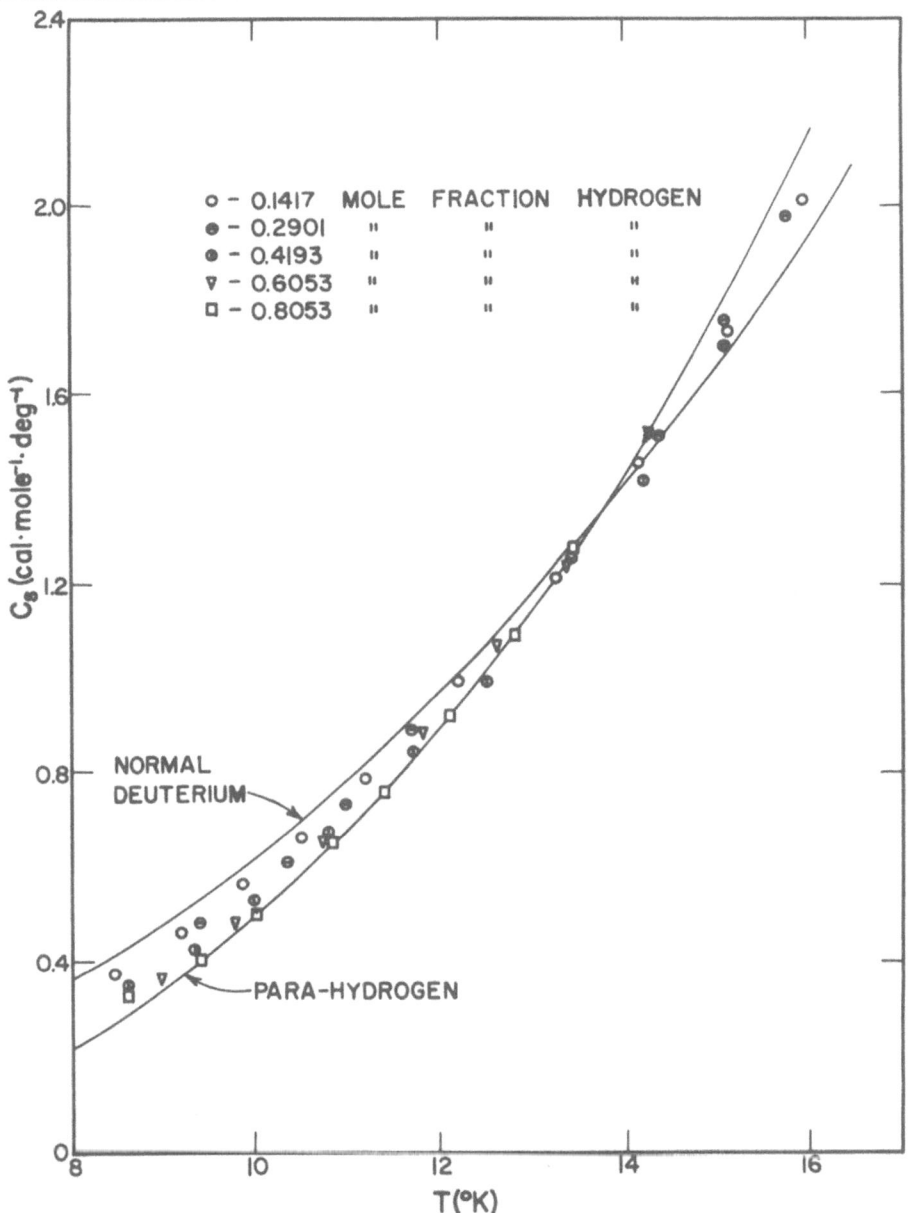

Fig. 1. Heat capacity curves for pure normal deuterium and pure parahydrogen.

on the basis of this phase diagram, the measured heat capacities of the solid mixtures, the heat capacities of the liquid mixtures, and the heats of fusion* to reconstruct the heating curves shown in Fig. 2. The calculated heating curve in the two-phase region of the 80.53% parahydrogen mixture is shown by the broken line (Fig. 2). The agreement with experiment is poor. By an iterative procedure in which the phase diagram

* The heat capacities of the liquid mixtures and the heats of fusion were estimated from the thermodynamic properties of the pure components, assuming ideal solution behavior.

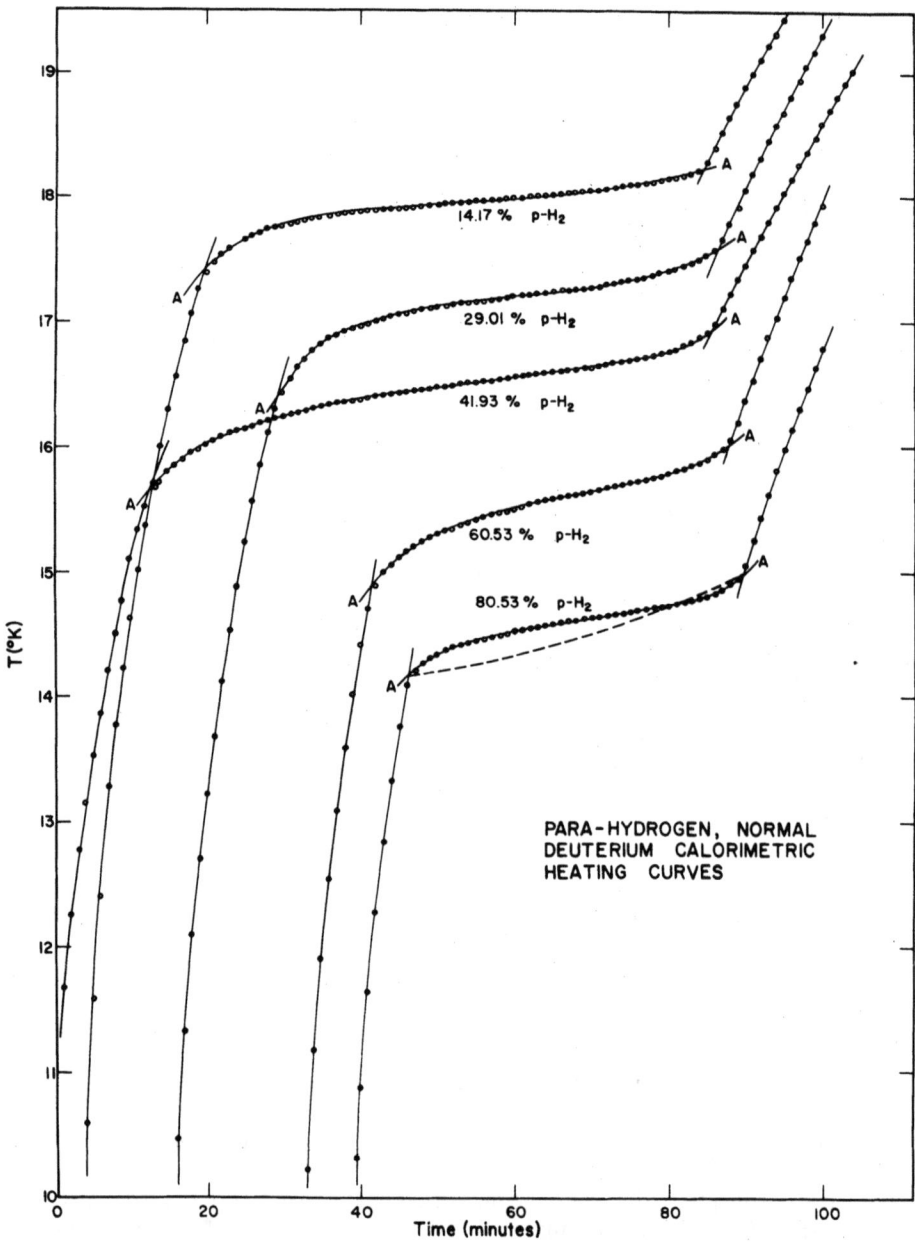

Fig. 2. Calorimetric heating curves for normal deuterium and parahydrogen.

(Fig. 3) is successively altered, it is possible to get good agreement between the experimental and calculated heating curves over nearly the entire two-phase region. The only serious disagreement occurs in the vicinity of the melting points. The new phase diagram based on these calculations is shown by the solid lines (open circles) in Fig. 3. It represents the phase diagram, consistent with the experimental heating curves, and shows no evidence of the peritectic suggested by Kogan et al.[3]

The reason for the large discrepancy between the apparent and true melting points can be ascribed to phase segregation, which occurs just before the initial complete solidification of the mixture prior to heating experiments. The solidification in these experiments occurred over a period of 2 hr. In the presence of an appreciable liquid phase, there is good equilibration between the liquid and solid. However, just prior to complete solidification, the composition of the remaining liquid is rich in hydrogen content with respect to the solid phase. This liquid composition is shown in Fig. 3 for the 60.53% parahydrogen mixture as point A on the corrected phase diagram. On freezing this liquid, equilibration with the remainder of the solid can occur only by solid diffusion. Since the process is very slow, a phase segregation occurs. It can be seen from Fig. 3 that, upon reheating such a mechanical mixture, the first solid to melt is the phase-segregated material, rich in hydrogen. Its melting point is given by point B (Fig. 3), which is in excellent agreement with the "apparent" freezing points (lower broken curve, Fig. 3) observed experimentally. The existence of such a phenomenon has been verified by direct

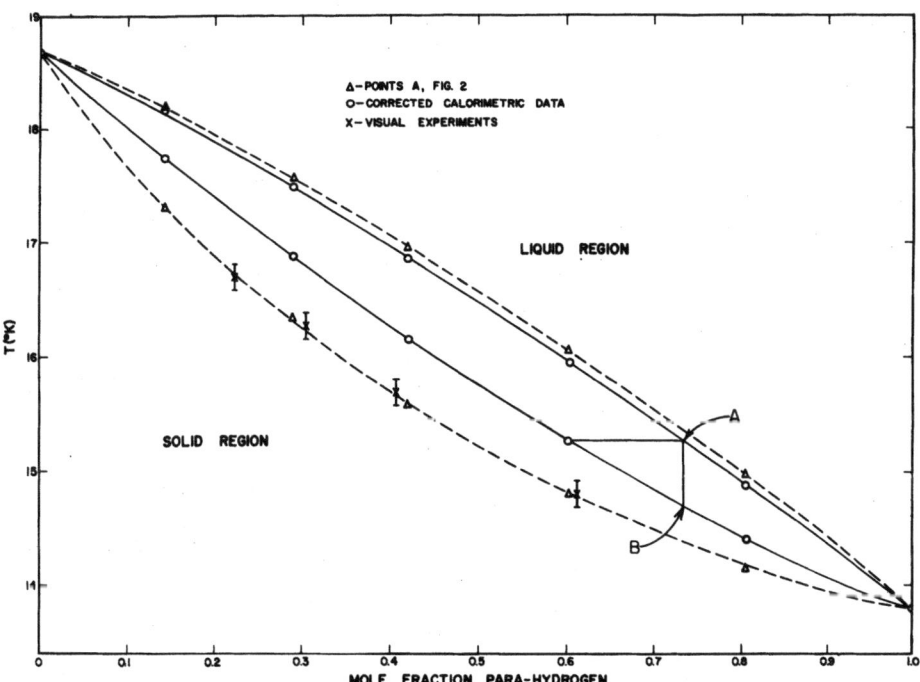

Fig. 3. Phase diagram (temperature, °K, versus mole fraction) of parahydrogen.

visual observation of the temperature at which the first liquid appears upon heating of the solid mixture. The apparent freezing points determined in these visual experiments are shown by the crosses in Fig. 3.

There is also a small discrepancy between the apparent and true freezing points. This is probably due to superheating of the calorimeter, which is especially pronounced ($\approx 0.05°K$) in the two-phase region. The corrected freezing points are given by the upper solid line (Fig. 3).

The experimental phase boundaries (Fig. 3) can be used to obtain an upper limit for the temperature at which phase separation in either the liquid or solid state can be expected to occur. The phase boundaries, when combined with known thermodynamic properties of the pure components in the solid and liquid state, permit the calculation of the ratio of the activity coefficients of a given component in the liquid and solid states at any temperature where two-phase equilibrium exists. Since the compositions of the two phases in equilibrium differ, the excess Gibbs free energy function (G^E) cannot be evaluated without specifying the form of the excess function. Assuming regular solution behavior for both liquid and solid solutions:[6]

$$G_l^E = \omega_l \chi_l (1 - \chi_l)$$
$$G_s^E = \omega_s \chi_s (1 - \chi_s) \qquad (1)$$

where χ is the mole fraction of one of the components and ω_l, ω_s are constants for the liquid and solid, respectively. Between 15.5° and 17.0°K, the above constants (evaluated from the phase data) are found to be

$$\omega_l = 24.8 \pm 3$$
$$\omega_s = 26.8 \pm 3$$

The regular solution model for the upper limit of the temperature T_c, at which phase separation occurs, gives

$$T_c \leqslant \frac{\omega}{2R} \qquad (2)$$

which in the case of liquid and solid solutions is 6.3° and 6.8°K, respectively. Even though phase separation in solid solutions of hydrogen and deuterium may be thermodynamically favored at this temperature, the transition would be extremely difficult to detect due to the extremely small value of the translational self-diffusion coefficient[7] (assuming thermally activated processes). The values of ω determined from the phase equilibria data are subject to considerable uncertainty. This parameter is very sensitive to the exact location of the phase boundaries, and furthermore there is no real justification for the relation in equation (1) used for the excess Gibbs function. The value of ω_l given above yields (for an equimolar mixture of hydrogen and deuterium) $G^E = 6.2$ cal/mole at an average temperature of 16.5°K. This is nearly three times larger than the experimental value at 20°K,[1] which also indicates that the values of T_c given above can only be considered as upper limits.

References
1. For review see D. White and C. M. Knobler, *Ann. Rev. Phys. Chem.* **14**, 251, 1963.
2. I. Prigogine, R. Bingen, and A. Bellemans, *Physica* **20**, 633, 1954.

3. V. S. Kogan, B. G. Lazarev, and R. F. Bulatova, *Zh. Experim. i Teor. Fiz.* **34,** 238, 1958: (*Soviet Phys. JETP* (*English Transl.*) **7,** 165, 1958): R. F. Bulatova, V. S. Kogan, and B. G. Lazarev, *Zh. Eksperim. i Teor. Fiz.* **34**, 1492, 1959 (*Soviet Phys. JETP* (*English Transl.*).
4. J. R. Gaines, E. M. de Castro, and J. G. Daunt, *Phys. Letters* **8**, 167, 1964.
5. G. Grenier and D. White, *J. Chem. Phys.* **40**, 3015, 1964.
6. E. A. Guggenheim, *Thermodynamics*, North Holland Publishing Co., Amsterdam (1950), p. 210.
7. M. Bloom, *Physica* **23**, 767, 1957.

PHASE SEPARATION IN SOLID HYDROGEN–DEUTERIUM AND ^{3}He–^{4}He MIXTURES

R. A. Coldwell-Horsfall*

University of Sheffield
Sheffield, England

Introduction

Over the last few years there has been a considerable increase in experimental work on the inert gas solids, in particular on solid hydrogen and helium, for which anharmonic effects are expected to be important, owing to their large zero-point energy. Therefore, it seems worthwhile to consider once again the question of phase separation in isotopically disordered systems, taking into account anharmonic effects.

A general expression has been obtained for the lowest order contribution to the free energy of an isotopically disordered lattice from anharmonic terms in the expansion of the potential energy of the lattice in powers of the nuclear displacements. The configuration average of this expression has been taken, retaining terms up to second order in the parameter μ, which represents the deviation of the isotopic mass from that of the mean mass lattice.

In this paper, we shall outline the steps involved in obtaining the general expression for the configuration average of the anharmonic contribution to the free energy, and we shall give the results obtained for a linear chain of atoms with nearest-neighbor interactions only, at the absolute zero of temperature. We use the parameters appropriate to hydrogen and helium to obtain a value for the difference between the ground-state energy of the disordered system and that of the pure isotopes in the case of hydrogen–deuterium and ^{3}He–^{4}He solid mixtures.

Theory

The equation of motion for a lattice with isotopic defects can be expressed formally as

$$[\omega^2 \mathbf{M} - \boldsymbol{\phi}]\mathbf{u} \equiv [\mathbf{L} - \delta\mathbf{L}]\mathbf{u} = 0$$

where $\mathbf{L}$ is the operator for a perfect lattice of mass $\bar{M}$ in the harmonic approximation and $\delta\mathbf{L}$ represents the difference between the mass M_l on the lth lattice site and the mass $\bar{M}$.

Now, elements of the inverse matrix $[\omega^2 \mathbf{M} - \boldsymbol{\phi}]^{-1} \equiv \mathbf{U}(\omega^2)$ say, appear in the expression for the anharmonic contribution to the free energy which is obtained by perturbation theory using the disordered lattice in the harmonic approximation as the unperturbed system, and $\mathbf{U}$ is the solution of

$$\mathbf{U} = \mathbf{G} + \mathbf{G}\delta\mathbf{L}\mathbf{U}$$

where

$$\mathbf{G} = \mathbf{L}^{-1}$$

---

* Now R. A. MacDonald. Present address: National Bureau of Standards, Washington, D.C.

therefore, an iterative solution for **U** enables us to express the anharmonic contribution to the free energy in terms of the eigenvectors and eigenvalues of the perfect harmonic lattice. In order to proceed further analytically, we have to make some assumption about the isotopic disorder which allows us to use this iterative solution for **U** to as low an order as possible.

We consider a randomly disordered two-component lattice with mean mass $\overline{M}$ defined by

$$\overline{M} = pM_1 + (1-p)M_2$$

where p is the probability that the site l is occupied by the mass M_1. We introduce the quantities μ_i, defined by

$$\mu_i = 1 - \frac{M_i}{\overline{M}}$$

which are related to $\delta \mathbf{L}$ by

$$\delta \mathbf{L} = \mu_i \omega^2 \overline{M} \mathbf{I}$$

and have average values defined by

$$\overline{\mu^n} = p\mu_1^n + (1-p)\mu_2^n$$

By definition of the mean mass, $\overline{\mu}$ is zero; therefore, when the configuration average of the free energy is taken, the lowest-order effect of disorder is given by terms in $\overline{\mu^2}$. The iterative solution for **U** is therefore taken to second order in $\delta \mathbf{L}$.

We now consider a linear chain of atoms with nearest-neighbor interactions only, at the absolute zero of temperature. We assume that the interatomic potential may be well represented by a Mie–Lennard-Jones (6–12) potential

$$\phi(r) = 4(r^{-12} - r^{-6})$$

where r, the interatomic spacing, is expressed in units of σ, the spacing for which $\phi(r)$ is zero, and the energy is expressed in units of ε, the depth of the potential. We use the de Boer parameter[1]

$$\Lambda = \frac{h}{\sigma\sqrt{M\varepsilon}}$$

to express the configuration average of the ground-state energy per particle as follows:

$$\langle E_0 \rangle = \phi(r) + \frac{\Lambda}{\pi 2} \sqrt{\phi''(r)}(1 + 0.3927\overline{\mu^2}) + \frac{\Lambda^2}{8\pi^4} \frac{\phi^{IV}(r)}{\phi''(r)}(1 + 0.8983\overline{\mu^2})$$

$$- \frac{\Lambda^2}{64\pi^5}\left[\frac{\phi'''(r)}{\phi''(r)}\right]^2 (4.4455 + 3.0164\overline{\mu^2})$$

The first two terms in this equation come from the harmonic contribution to the free energy, $\langle E_h \rangle$; the last two terms are the lowest order anharmonic contribution, $\langle E_A \rangle$, arising from quartic and cubic terms, respectively, in the expansion of the potential energy of the lattice. In the systems we chose to consider, namely a 1:1 mixture of hydrogen and deuterium for which $\overline{\mu^2} = \frac{1}{9}$ and a 4:3 mixture of ^{3}He–^{4}He for which $\overline{\mu^2} = \frac{1}{48}$, the values of Λ corresponding to the mean mass $\overline{M}$ are $\Lambda_{HD} = 1.42$ and $\Lambda_{He} = 1.08$.

We wish to compare $\langle E_0 \rangle$ to the ground-state energy of the separated isotopes. In order to do this, the isotopic mass M_i must be expressed in terms of $\bar{M}$ and μ_i. The ground-state energy per particle of the pure isotope, E_0^i, is then given by

$$E_0^i = \phi(r) + \frac{\Lambda}{\pi^2}\sqrt{\phi''(r)}(1 + \tfrac{1}{2}\mu_i + \tfrac{3}{8}\mu_i^2)$$

$$+ \frac{\Lambda^2}{8\pi^4}(1 + \mu_i + \mu_i^2)\left\{\frac{\phi^{IV}(r)}{\phi''(r)} - \frac{4.4455}{8\pi}\left[\frac{\phi'''(r)}{\phi''(r)}\right]^2\right\}$$

We now have to determine the appropriate value of r for each pure isotope and mixture from the one-dimensional form of the equation of state,[2] namely

$$-P = \frac{1}{\sigma^2 r^2}\left[\phi'(r) + \frac{\Lambda_i}{2\pi^2}\frac{\phi'''(r)}{\sqrt{\phi''(r)}}\right]$$

In this equation we consider only the harmonic contribution to the ground-state energy; the anharmonic contribution $\langle E_A \rangle$ is evaluated at r_0, the equilibrium separation of the static lattice. This is consistent with our perturbation theoretic approach.[2] It is clear that the factor $\sqrt{\phi''(r)}$ restricts the range of values which r may assume. For the Mie–Lennard-Jones (6–12) potential, the restriction is $r < 1.24$. Since the molar volumes of solid hydrogen and helium at the lowest possible pressures correspond to a value of $r \sim 1.28$,[3–5] a large pressure is necessary to reduce the lattice spacing to a value $r < 1.24$. For two such pressures, we obtained values of r from the equations of state and calculated the corresponding values of $\langle E_h \rangle$ and E_h^i. The results are given in Table I. Experimental values of the spacing are also given where available.

The difference between the energy of the disordered system and that of the separated isotopes is given by

$$\Delta E = \langle E_0 \rangle - [pE_0^1 + (1-p)E_0^2]$$

and the results for each system at the two chosen pressures are given in Table II, together with this energy difference in the quasi-harmonic approximation, ΔE_h, for comparison.

The energy difference for the hydrogen–deuterium system agrees with that obtained by Prigogine et al.[6] on the basis of an additivity hypothesis for the pure isotopes, using measured quantities only (2.5 cal/mole). Prigogine et al.[6] show that the main contribution to the free energy of mixing arises from the changes in molar volume which occur on mixing the isotopes, this energy being at least an order of magnitude greater than the change in energy due to isotopic mass disorder at a given molar volume.*

* See note added in proof.

Table I

	r_1	$\langle E_h \rangle_1$	r_2	$\langle E_h \rangle_2$	r_1 (experimental)	r_2 (experimental)
Hydrogen–deuterium	1.17397	−0.28041	1.15610	−0.16059		
Hydrogen	1.19953	−0.35065	1.17249	−0.16460	1.20	1.19
Deuterium	1.15508	−0.28592	1.14042	−0.19040	1.16	1.15
^{3}He–^{4}He	1.18015	0.27711	1.14391	0.82690		
^{3}He	1.19086	0.19963	1.15022	0.82187		
^{4}He	1.16803	0.33306	1.13541	0.82090	1.16	1.14

Table II

	$\langle E_A(r_0)\rangle$	$\langle E_0\rangle_1$	$\langle E_0\rangle_2$	ΔE_1	ΔE_2	$(\Delta E_h)_1$
Hydrogen–deuterium	0.66710	0.38668	0.5065			
Hydrogen	0.87123	0.52058	0.70663	0.03478	0.01382	0.03787
Deuterium	0.46913	0.18321	0.27872	(2.54 cal/mole)		
^{3}He–^{4}He	2.55031	2.82742	3.37721			
^{3}He	2.85223	3.05185	3.67410	0.01792	0.00306	0.02030
^{4}He	2.15330	2.48636	2.97420	(0.36 cal/mole)*		

*See note added in proof.

Results

In Tables I and II, the subscripts 1 and 2 denote values corresponding to a pressure of 900 atm in the hydrogen–deuterium system, 800 atm in the ^{3}He–^{4}He system, and a pressure of 1100 atm in both systems, respectively. All values are given in the reduced units defined in the text. The experimental values of r for hydrogen and deuterium are taken from Stewart[4] ($T = 4°K$) and those for ^{4}He from Domb and Dugdale[5] ($T = 0°K$).

Concluding Remarks

We have obtained estimates of the change in energy on mixing the isotopes of hydrogen and helium, respectively, in as simple and consistent a way as possible. The application of a one-dimensional model to these solids is unrealistic but is not a serious limitation. However, the procedure which we have adopted, namely to treat the anharmonic terms in the expansion of the potential energy as a perturbation on the energy of the harmonic lattice, is indeed questionable when applied to hydrogen and helium for which the parameter Λ is of the order of unity. The zero-point energy and lowest-order anharmonic contribution to the ground state energy are of comparable size in both cases. In a system with $\Lambda \ll 1$, so that the perturbation approach is valid, the effect of isotopic disorder is negligible, e.g., in ^{20}Ne–^{22}Ne, $\Lambda = 0.6$ and $\overline{\mu^2} = \frac{1}{440}$. It is clear that a different starting point is necessary if perturbation theory is to be useful in treating these strongly anharmonic solids.

Note Added in Proof

The energy difference for the ^{3}He–^{4}He system yields a phase separation temperature of 0.37°K, in agreement with the results of Edwards et al.[7]

References

1. J. de Boer, *Physica* **14**, 139, 1948.
2. G. Leibfried and W. Ludwig, *Solid State Physics*, Vol. 12, Academic Press, Inc., New York (1961), p. 339.
3. G. Ahlers, Report UCRL 10757, University of California, Berkeley (1963).
4. J. W. Stewart, *J. Phys. Chem. Solids* **1**, 146, 1956.
5. C. Domb and J. S. Dugdale, *Progress in Low Temperature Physics*, Vol. II, North Holland Publishing Co., Amsterdam (1957).
6. I. Prigogine, R. Bingen, and J. Jeener, *Physica* **20**, 383, 1954.
7. D. O. Edwards, A. S. McWilliams, and J. G. Daunt, *Phys. Rev. Letters* **9**, 195, 1962.

PHASE DIAGRAM OF SOLID METHANE*

J. S. Rosenshein and W. M. Whitney†

Department of Physics, Massachusetts Institute of Technology
Cambridge, Massachusetts

Introduction

Two phase transitions of higher order have been detected in solid CH_4 by previous investigators. The first study of the phase boundaries was made by Trapeznikova and Miljutin,[1] who used a calorimetric technique that employed helium as the pressure-transmitting medium. Their measurements, which were carried out at temperatures in the range 18° to 30°K and at pressures up to 2000 kg/cm², yielded a phase diagram composed of two lines that can be represented by the equations: $T_1 = 4.84 \times 10^{-3}p + 20.4$, $T_2 = 8.33 \times 10^{-3}p + 10$, where T_1 and T_2 are the transition temperatures of what we shall refer to as the first and second transitions, respectively, and p is the pressure in kg/cm². Measurements by both Stevenson[2] and Stewart,[3] who used the piston-displacement technique of Bridgman,[4] indicated that the second transition line T_2 did not extend to zero pressure, but leveled out below 18°K at a nearly constant pressure; however, Stewart found this pressure to be 2000 kg/cm², a value 1000 kg/cm² greater than that obtained by Stevenson.

Calorimetric measurements by other workers[5,6] have shown that the deuterated methanes CD_4 and CH_3D both have two higher-order transitions at vapor pressure, but until the measurements of Colwell, Gill, and Morrison (CGM),[7] there was no evidence for the existence of a second transition in CH_4 at pressures below 1000 kg/cm². Recently, these authors (CGM) have measured the specific heat of CH_4 at its vapor pressure at temperatures below those reached by previous investigators. Their work reveals a broad anomaly in the specific heat that has its maximum value at approximately 8°K. This transition was thought to be the equivalent of the second transition observed in the deuterated methanes, but its breadth contrasted greatly with the sharpness of the corresponding transitions in those substances.

The measurements reported here were undertaken to investigate the behavior of the second transition line in solid CH_4, and to resolve the disparity between the results of the piston-displacement technique and those of the calorimetric measurements.

Experimental Procedure

In our experiment, the sample is cooled to the desired starting temperature, and then allowed to warm up with constant heat input. The temperature of the sample is measured as a function of time. A plot of the inverse warming rate against temperature yields a curve that roughly follows the specific heat of the sample and

---

* This work was supported in part by the Advanced Research Projects Agency and in part by the Research Laboratory of Electronics, M.I.T., under tri-service support.

† Now at the Jet Propulsion Laboratory, California Institute of Technology, Pasadena, California.

its container. The anomalous specific heat at a phase transition causes a departure of the warming-rate curve from its otherwise smooth behavior.

The variable-temperature cryostat and calorimeter employed in this experiment are described elsewhere.[8] The sample chamber is a hollow cylinder made from heat-treated beryllium–copper alloy (Berylco 25), with an inner space that can accommodate 0.7 mole of solid methane. The chamber is isolated from the rest of the cryostat by a vacuum space, except for a stainless steel capillary tube that connects the chamber to the pressure source. Methane is condensed into the chamber through this tube, and pressure is applied hydrostatically by using fluid helium as the pressure-transmitting medium. The sample chamber is heated with a manganin wire wound around its outer surface. The temperatures of the sample chamber and calorimeter jacket are measured with 0.1-W ohmite carbon resistors, calibrated against a gas thermometer built into the calorimeter. The accuracy of the temperature values quoted is considered to be $\pm 0.1°$ below 30°K, but we can measure temperature changes of less than 5 mdeg. As the sample warms up, we record the time interval Δt required for a given change in the resistance of the carbon thermometer, and derive the corresponding temperature change ΔT from the slope of the resistance vs. temperature curve.

The methane used in this experiment is C.P. grade CH_4, purchased from the Matheson Co. Further purification of the gas is achieved by repeatedly solidifying it and pumping off the residual vapors until the pressure over the solid at 77°K is close to that appropriate to the pure substance.

Results and Discussion

A typical warming curve is shown in Fig. 1, where we have plotted the inverse warming rate $C'_p = \Delta t/\Delta T$ against the temperature T from data taken at 562.5 kg/cm². Transitions at 14.0° and 23.35°K are clearly in evidence. The break in the

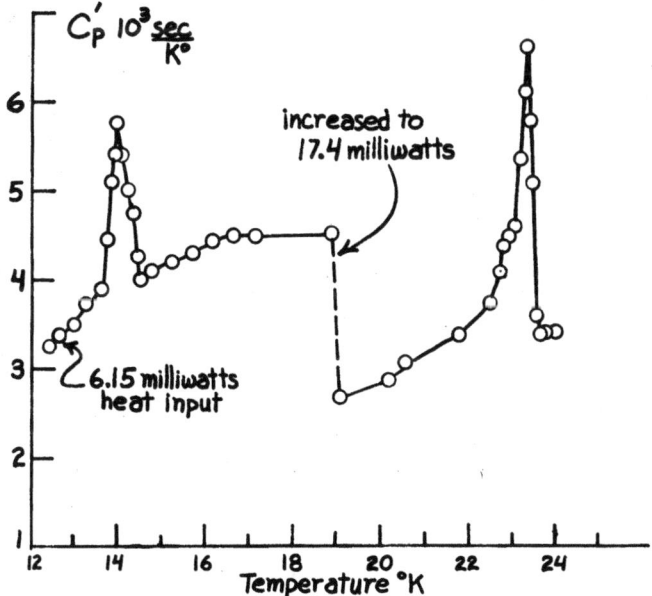

Fig. 1. Warming curve at 562.5 kg/cm², 0.7 mole CH_4.

warming curve, indicated by a dotted line, does not represent a transition, but a change in the warming rate produced by increasing the heat input.

The characteristics of the anomalies in C'_p observed at higher pressures are similar to those shown in Fig. 1. At lower pressures, the first transition remains sharp and well-defined, but the second broadens, and at 246 kg/cm² has a width of nearly 2°K. Furthermore, there is some evidence that the width of the second transition is larger for higher heating rates, although this relationship has not been established conclusively. Such a broadening would, however, be consistent with the findings of CGM, which indicate long thermal equilibration times associated with the second transition that they observed at vapor pressure.

At a pressure of 155 kg/cm², the second transition is not discernible in our experiments; any anomalies in the behavior of C'_p near the expected position of the phase change lie within the scatter of the individual data points. The lowest warming rate used near 10°K at this pressure was 2×10^{-4} °K/sec.

From the positions of the peaks of the transitions, taken from the warming curves, we construct the $p-T$ diagram shown in Fig. 2. The helium melting curve[9] is also shown, to indicate the lower limit in temperature for the application of hydrostatic pressure to the sample at a given pressure. The horizontal dashed lines show the path traced out in the $p-T$ plane during each warmup. The cross at 1160 kg/cm², where one of these lines intersects the helium melting curve, represents the point at which a warming arrest, associated with a first-order transition in the helium pressure-transmitting medium, was observed.

The first transition line can be represented by the equation $T_1 = 4.18 \times 10^{-3} p + 20.4$, whose slope is in good agreement with the value derived from the results

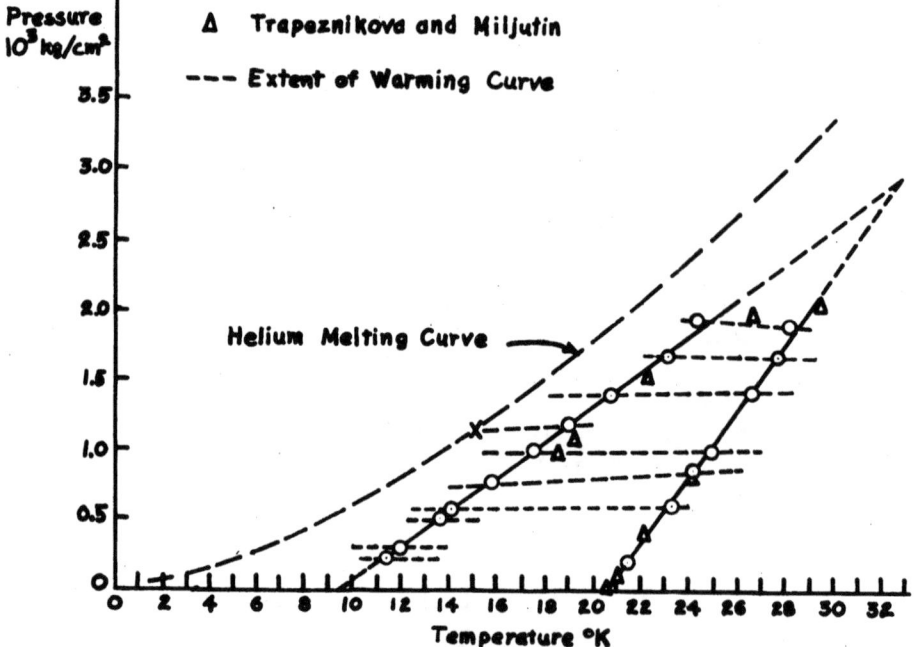

Fig. 2. Methane phase diagram.

of Trapeznikova and Miljutin. For the second transition, our data lie along the line $T_2 = 7.82 \times 10^{-3} p + 9.5$. This equation yields values of T_2 at a given pressure that lie somewhat lower than those derived from the points given by Trapeznikova and Miljutin. The line for T_2 intersects the temperature axis at 9.5°K. Approximately the same temperature is obtained by extrapolating the position of T_2, plotted as a function of CD_4 concentration in CH_4–CD_4 mixtures, to 100% CH_4.[5] Our intercept is somewhat higher than the position of the peak observed by CGM at vapor pressure, but since their peak has a half-width of 7°K, its position is somewhat indeterminate, and the discrepancy may not be significant. The two phase lines T_1 and T_2 meet in a triple point at 33°K and 2990 kg/cm^2, in agreement with the results of the three previous studies.[1-3]

Conclusions

We find that the phase line corresponding to the second transition observed in CH_4 at pressures above 1000 kg/cm^2 by Trapeznikova and Miljutin does not level out at lower pressures, but extends almost linearly downward to pressures below 280 kg/cm^2. When extrapolated, this line intersects the temperature axis at approximately the location of the specific-heat anomaly discovered by CGM. We have not found evidence in our experiment for the nearly horizontal phase line reported by Stevenson at 1000 kg/cm^2. (Our measurements do not extend above 2000 kg/cm^2.) The phase diagram of CH_4 now appears to be quite similar to that found by Stewart[3] for CD_4, which exhibits a triple point at 40°K and 3000 kg/cm^2. It therefore seems likely that the second transition in CH_4 corresponds to those observed calorimetrically at vapor pressure in the deuterated methanes.

References

1. O. N. Trapeznikova and G. A. Miljutin, *Nature* **144**, 632, 1939.
2. R. Stevenson, *J. Chem. Phys.* **27**, 656, 1957.
3. J. W. Stewart, *J. Phys. Chem. Solids* **12**, 122, 1959.
4. P. W. Bridgman, *The Physics of High Pressures*, G. Bell and Sons, Ltd., London (1949).
5. Bartholome, Drikos, and Eucken, *Z. Physik. Chem.* **B39**, 371, 1938.
6. K. Clusius and L. Popp, *Z. Physik. Chem.* **B46**, 63, 1940.
7. Colwell, Gill, and Morrison, *J. Chem. Phys.* **36**, 2223, 1962.
8. J. S. Rosenshein, unpublished Ph. D. Thesis, MIT (1963).
9. R. L. Mills and E. R. Grilly, *Phys. Rev.* **99**, 480, 1955.

PROTON MAGNETIC-RESONANCE ABSORPTION IN SOLID METHANE*

R. P. Wolf† and W. M. Whitney‡

*Department of Physics, Massachusetts Institute of Technology
Cambridge, Massachusetts*

From properties of the nuclear magnetic-resonance absorption lines in molecular crystals, conclusions can be drawn about the symmetry of the crystalline field at the sites of the resonant nuclei and about the motion of their molecules. We here report some results of a continuing study of the proton resonance in solid methane, CH_4, and three deuterated modifications CH_3D, CH_2D_2, and CHD_3. The principal objective of this research is to obtain information about the dynamical states of the molecules above and below the λ-points, with the expectation that such knowledge will contribute toward an elucidation of the mechanisms of the phase transitions in these solids.

Our measurements extend over the temperature region 2°–90°K. The proton resonance is excited at 18.24 Mcps and detected with a balanced Anderson bridge. All stated characteristics of the resonance line are derived from traces of the derivative of the absorption.

Each of our samples was prepared from commercially bottled gases, C.P., or reagent grade. No effort was made to remove oxygen or other impurities, although each sample was permanently sealed in a glass container to prevent further contamination. A long glass tube, extending down through the helium dewar and into a sample holder, is fastened to each container. Upon cooling, the gas condenses into the tip of this tube. The sample holder is insulated from the liquid-helium bath by a vacuum space. The temperature of the sample is regulated with a heater and measured with a carbon resistor that has been calibrated against a gas thermometer.

The ease with which a molecule can change its position or orientation in the lattice diminishes with decreasing temperature, and the change is often reflected in the behavior of such properties as the width and second moment of the absorption line. Values of the second moment S_2 for the four solids under study are plotted against temperature in Fig. 1. The second moment is approximately constant over a temperature range that extends from 25° to 60°K. Below this region it increases; above, it rapidly falls toward a lower limit set by the inhomogeneity of the magnetic field. The behavior of the line width ΔH is similar. In regions where S_2 and ΔH are changing, the line shape is also changing. At the lowest temperatures,

---

* Research supported by the Advanced Research Projects Agency, Contract SD-90. This work is based in part upon a Ph.D. thesis submitted by R. P. Wolf to the Department of Physics, M.I.T., September 1963.

† Present address: Harvey Mudd College, Claremont, California.

‡ Present address: Jet Propulsion Laboratory, California Institute of Technology, Pasadena, California.

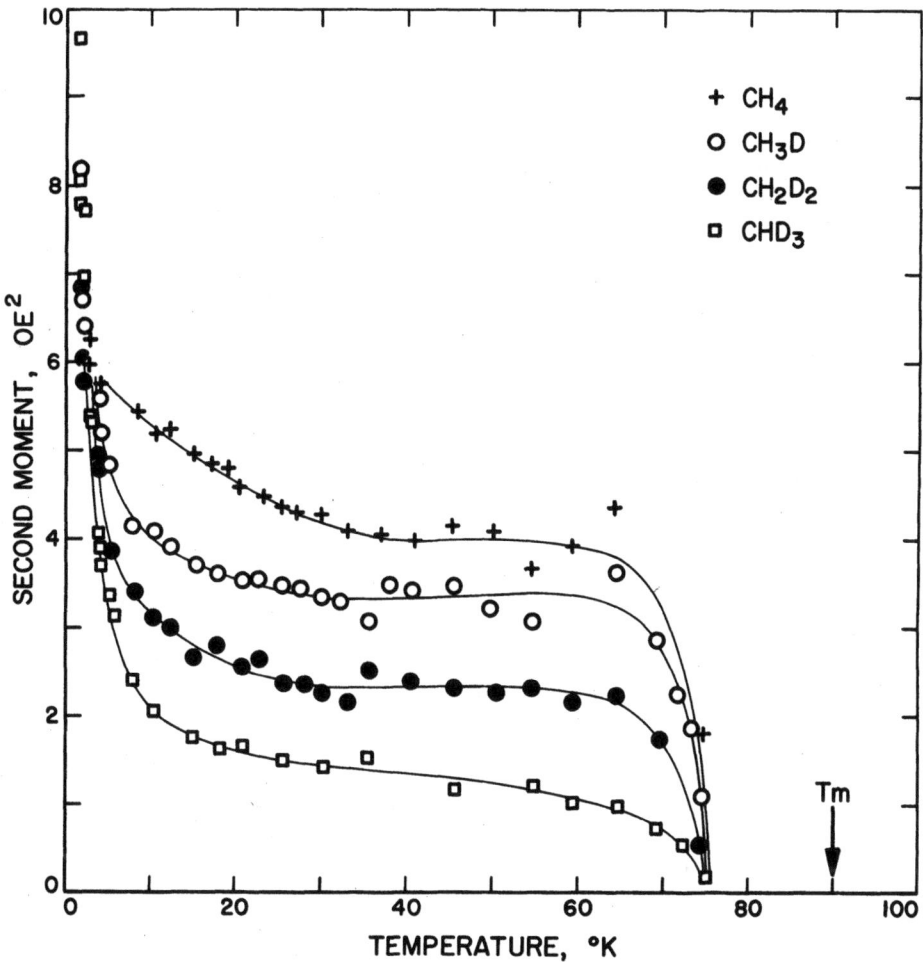

Fig. 1. Second moment vs. temperature. Data above 75°K are not shown.

the absorption curves are approximately Gaussian, but the shape differs somewhat from one sample to the next at a given temperature. Over the intermediate temperature region, the lines for all four samples are some 40% wider than would be expected for Gaussian lines with the same second moment. Above 65°K, the line shapes become predominantly Lorentzian in character until limited by the magnet inhomogeneity.

The vanishing of line width at temperatures above 65°K, a phenomenon first observed in CH_4 by Thomas, Alpert, and Torrey (TAT),[1] has been attributed by Waugh[2] to molecular self-diffusion in the solid. Our measurements lend qualitative support to this interpretation, in that the decrease in line width and second moment begins at roughly the same temperature for all four solids. Measurements of T_2 carried out by Bloom and Sandhu[3] from 57° to 110°K provide more direct confirmation, and show that the height of the energy barrier inhibiting diffusion is 3.2 kcal/mole for the four substances. Our results are consistent with this value;

however, as Andrew[4] points out, barrier heights cannot be reliably derived from values of the second moment in regions where the line shape is changing.

Molecular diffusion is well frozen out below 60°K. From this temperature down to 25°K, the second moment and line width increase only slightly with decreasing temperature. The small values of S_2 and ΔH suggest that the line is broadened entirely by intermolecular dipole–dipole interactions, the intramolecular contributions being averaged to zero by rapid molecular reorientation. Two types of evidence support this assertion: Below 30°K, we have obtained absorption lines from two samples of CH_4 diluted with krypton to 75 and 50% CH_4. At 30°K, the second moments for these two samples and for 100% CH_4 lie on a straight line that extrapolates to zero, within experimental error, for 100% Kr. This behavior is consistent with the absence of intramolecular broadening for nearly pure CH_4. At lower temperatures the three points still fall on a straight line, but a finite intercept at 100% Kr marks the presence of an intramolecular contribution that increases with decreasing temperature.

As a second test of the assumption of isotropic reorientation, we have calculated the intermolecular second moments for a number of structures, most of them ones that have been suggested for the lowest phase of CH_4, on the basis of the hypothesis that each molecule flips rapidly and randomly from one equivalent orientation to another. For the two possible cubic structures, $F\bar{4}3m$ and $P2_13$ (Nagamiya[5]), $S_2 = 4.4$ and $3.9\ Oe^2$, respectively. For most of the tetragonal structures proposed by Savitsky and Hornig,[6] S_2 lies in the range 3.8 to $4.0\ Oe^2$. These values agree satisfactorily with the average experimental second moment of $4.0\ Oe^2$ for CH_4 over the temperature region 30°–60°K. The accuracy of our experimental result is not good enough to enable us to select any one structure as being more suitable than another, although it seems likely that $F\bar{4}3m$ can be ruled out for this temperature region, and perhaps also C_{4v}^4 ($S_2 = 4.3\ Oe^2$). The experimental values of S_2 for CH_3D, CH_2D_2, and CHD_3 over the intermediate region are also consistent with isotropic reorientation. Bloom and Sandhu[3] have concluded from their pulse measurements that there are no intramolecular contributions to the second moments; however, their values of S_2 are somewhat larger than ours.

At temperatures for which the molecular thermal energy is small in comparison with the energy barriers, molecular reorientation will be frozen out, and the second moments should assume values several times those observed above 10°K. The values of S_2 shown in Fig. 1 are in fact rising at the lowest temperatures reached, but it is surprising that even at 1.8°K, the second moments of three of the solids investigated are still much smaller than would be predicted for a rigid lattice. The behavior of the fourth solid, CHD_3, is also surprising, but for a different reason. Below 2°K, the second moment for this substance is $8\ Oe^2$ and still increasing. This value exceeds the intramolecular contribution of $1.6\ Oe^2$, leaving a remainder of more than $6\ Oe^2$ to be attributed to the interaction between spins on neighboring molecules. Since the rigid lattice intermolecular second moment for CH_4 is of the same order of magnitude, it is clear that the structure of CHD_3 must be one that places several protons in close proximity. Structures with fewer than four nonequivalent molecules per unit cell can probably be ruled out.

We observe no abrupt change in either the line width or the second moment in the vicinity of the upper or lower phase transitions for any of the four solids studied. In this respect our measurements confirm the findings of TAT.

It is a matter of speculation whether or not the meta, ortho, and para forms of the CH_4 molecule, i.e., the isomeric nuclear spin species corresponding to total nuclear spin, $I = 2$, 1, and 0, respectively, can preserve their identity in the solid. It is known that the ortho form of hydrogen ($I = 1$) slowly converts to the para species ($I = 0$) in solid hydrogen. Looking for such conversion in CH_4, we measured the intensity of the CH_4 resonance as a function of time at 4.2°K. We found that the intensity increases by 30% over a period of 4 hr, with a time constant of approximately 90 min. A change of this magnitude can be accounted for by assuming that substantial numbers of ortho and para molecules are converting to meta methane, the form stable at very low temperatures.

We have also observed increases in the line width and second moment with time in CH_4 at 4.2°K, but not in the three deuterated samples. This effect, which has a characteristic time constant of approximately $\frac{1}{2}$ hr in C.P. CH_4, may well be a manifestation of the sluggish approach to equilibrium that Colwell, Gill, and Morrison[7] noted in their specific heat measurements in CH_4 below its lower transition, although the changes that they refer to apparently occupy many hours rather than 30 min. In our study, the equilibration time was twice as long for reagent grade as for C.P. CH_4, indicating a dependence upon impurities, possibly paramagnetic oxygen.

Acknowledgments

We have profited from conversations with Prof. J. S. Waugh and Prof. Myer Bloom. We are especially indebted to Prof. Waugh for the opportunity to carry out these measurements in his laboratory, and to Dr. James H. Loehlin, who assembled the NMR spectrometer.

References

1. Thomas, Alpert, and Torrey, *J. Chem. Phys.* **18**, 1511, 1950.
2. J. S. Waugh, *J. Chem. Phys.* **26**, 966, 1957.
3. M. Bloom and H. S. Sandhu, *Can. J. Phys.* **40**, 292, 1961.
4. E. R. Andrew, *J. Phys. Chem. Solids* **18**, 9, 1961.
5. T. Nagamiya, *Progr. Theoret. Phys. (Kyoto)* **6**, 702, 1951.
6. G. B. Savitsky and D. F. Hornig, *J. Chem. Phys.* **36**, 2634, 1962.
7. Colwell, Gill, and Morrison, *J. Chem. Phys.* **36**, 2223, 1962; **39**, 635, 1963.

ULTRASONIC PROPAGATION IN SOLID METHANE

A. A. Thiele* and W. M. Whitney†

Department of Physics, Massachusetts Institute of Technology
Cambridge, Massachusetts

and

C. E. Chase

National Magnet Laboratory,‡ Massachusetts Institute of Technology
Cambridge, Massachusetts

We present here a preliminary account of recent measurements of the velocity and attenuation of sound in solid methane, CH_4. The work was motivated in part by studies of the proton magnetic resonance in CH_4 by Thomas, Alpert, and Torrey.[1] Using pulse methods, they measured the spin-lattice relaxation time T_1, and found that it exhibits an anomalous peak at 20.4°K, the temperature of the (upper) phase transition. Their results over the temperature region above and below the peak have been analyzed by Tomita,[2] who derived a relationship between T_1 and the average time τ_c spent by a molecule in one of its equivalent orientations in the lattice before it flips to another. Near T_λ, $\tau_c \approx 10^{-7}$ sec, corresponding to a frequency spectrum that extends over a range of several megacycles. We were led to a study of the sound velocity and attenuation at frequencies of this order of magnitude to look for evidence, direct or indirect, of coupling between the rotational motions of the molecules and the vibrations of the lattice. In this paper, we show that anomalies in both the velocity and absorption do in fact occur near the λ-point, and that the relaxation time characterizing the absorption is of the same order of magnitude as the molecular flipping time τ_c. We have not yet been successful, however, in establishing a relationship between our results and those of the NMR investigations.

The portion of the apparatus where the solid methane samples are grown and in which the sound measurements are made is enclosed within a length of precision-bore glass tubing. The ultrasonic pulses are propagated between two X-cut quartz crystals, resonant at approximately 4 Mcps, which are embedded in the sample. Sliding smoothly within the glass tube is a movable support, the holder for the upper quartz crystal, whose vertical position can be adjusted from the top of the apparatus. The lower crystal is attached to the end of a copper rod that is soldered to, and extends through, a glass-to-metal seal at the bottom of the tubing. During condensation of the sample, the lower end of the rod is immersed in liquid nitrogen.

---

* Supported in part by the Advanced Research Projects Agency under Contract SD-90, and in part by the Research Laboratory of Electronics, M.I.T., supported jointly by U.S. Army Signal Corps, Office of Naval Research, and Air Force Office of Scientific Research under Contract DA36-039-sc-78108.
† Present address: Jet Propulsion Laboratory, C.I.T., Pasadena, California.
‡ Supported by the U.S. Air Force Office of Scientific Research.

The design of the crystal holders is such that the two crystals are held parallel to less than 0.1 sound wavelength.

Methane samples are grown from the vapor phase at a controlled pressure of 4–5 cm of mercury. Condensation begins on the lower quartz crystal and proceeds upward and outward. By adjusting the power supplied to a heater wound around the outside of the glass tube, we can prevent the solid methane from coming into contact with its inner wall and can control the rate of growth of the sample. With this apparatus, we have been able to grow clear crystals up to 1 cm in length in a few hours. It is not known whether or not we obtain a single crystal by this procedure; however, when the growing conditions are altered, as for instance by varying the deposition pressure for a short time, a surface discontinuity is produced that makes it easy to distinguish that part of the crystal grown before from that grown after the change. Such imperfections gradually disappear with aging of the crystal. Each of the two smaller samples with which our experimental results were obtained was annealed at liquid-nitrogen temperatures for at least 24 hr before being cooled to 4.2°K with liquid helium. Contraction of the sample during this stage occasionally destroyed contact with the upper quartz crystal, making it necessary to start the entire process over again. Once satisfactory contact was obtained, measurements were made as the sample warmed up from liquid-helium temperatures, typical warming rates being 5–10°K/hr.

The temperature variation of the velocity was measured by comparing the phase of the pulse at the receiving crystal with that of a CW reference signal, using electronic gear previously employed.[3,4] Changes of 0.1% of the total velocity could be easily resolved. The attenuation measurements, which are considerably less precise, were made simply by observing the height of the received signal on the oscilloscope screen.

The best results were obtained with a sample 3.9 mm long, for which we made measurements both at 3.8 and 11.6 Mcps. Data obtained at 3.8 Mcps with a second crystal 7.8 mm long are also presented, but poor contact between the upper quartz crystal and the solid methane made reliable attenuation measurements impossible at 12 Mcps for this sample. The crystal spacing was measured with a cathetometer.

The velocity is shown in Fig. 1 as a function of temperature. Since we measure directly the delay of the ultrasonic signal relative to some fixed but arbitrary reference phase, the velocity must be known at one temperature in order to convert delay measurements to velocity values. Measurements of the path length and of the total transit time, which are difficult to make because of the short path, lead to the value $u = 2900$ m/sec $\pm\ 10\%$ at 4.2°K. For convenience, we have assumed that $u_0 = 2900$ m/sec exactly, and also that u is independent of frequency at the lower temperatures.

Despite differences from one run to the next, the data at both frequencies exhibit similar behavior. With increasing temperature, the velocity decreases slowly at first, then drops quite rapidly just below T_λ. Above the λ-point u is still decreasing, but much less rapidly than below. The total change over the temperature range 0°–77°K is approximately 20%. The behavior close to T_λ is shown in the inset for one run at 3.8 Mcps and 3.9 mm. It is apparent that the velocity is continuous across the transition, but that its slope is not.

We believe that the difference between the 4 and 12 Mcps results from T_λ up to 30°K is significant, and indicates dispersion in the velocity amounting to as much as 1% of u_0. Within experimental error, there is no net dispersion over the range 0°–20°K. The systematic differences among the various runs below the

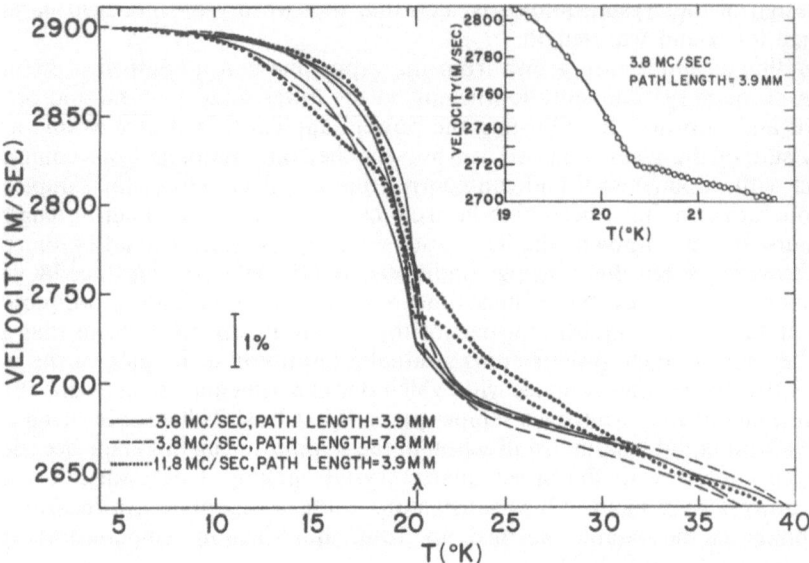

Fig. 1. Velocity of sound in solid CH_4 as a function of temperature. Experimental points have been omitted for clarity; spacing and scatter of data are shown for a typical run in the inset.

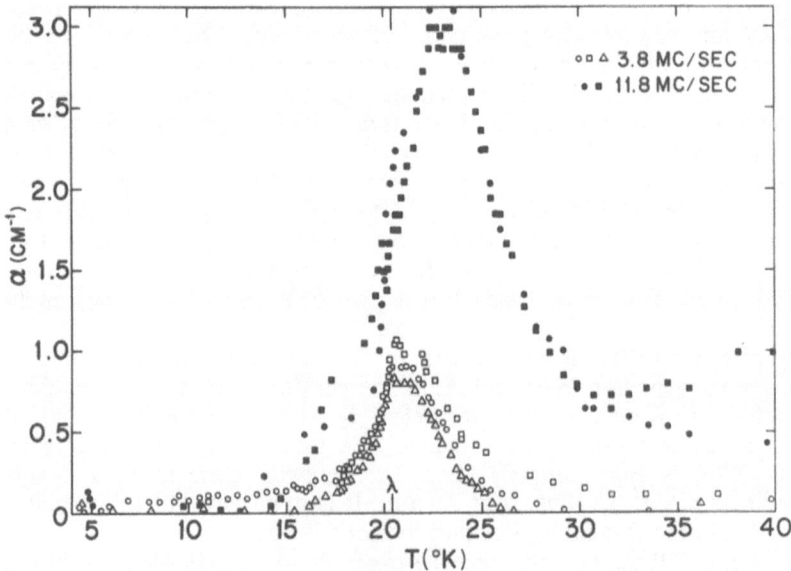

Fig. 2. Attenuation of sound in solid CH_4 as a function of temperature.

λ-point are not well understood, but appear to be connected with the presence of a long equilibrium time in this region. This effect, if real, may be a further manifestation of the sluggishness with which thermal equilibrium is achieved in solid CH_4 near its lower transition.[5]

The temperature variation of the attenuation at 4 and 12 Mcps is shown in Fig. 2. In order to obtain the values plotted, we have arbitrarily set $\alpha = 0$ at the lowest temperatures reached. The data above the λ-point are consistent with a single relaxation process whose characteristic time τ increases sharply with decreasing temperature, and whose relaxation strength is approximately constant. At T_λ, $\tau \sim 5 \times 10^{-8}$ sec, a value that compares well with Tomita's value $\tau_c \sim 10^{-7}$ sec at the same temperature. The agreement may, however, be coincidental. A plot of $\ln \tau$ against $1/T$ yields a straight line from which we estimate a characteristic temperature of $\sim 220°K$. From the variation of τ_c over the same temperature region, on the other hand, Tomita derived a value of $34°K$. It therefore appears unlikely that the interactions that induce relaxation of the lattice modes can be directly identified with those that bring about thermal equilibrium between the spin system and the lattice.

Acknowledgment

We express our appreciation to Professor Uno Ingard for his interest in this work and for helpful discussions.

References

1. Thomas, Alpert, and Torrey, *J. Chem. Phys.* **18**, 1511, 1950.
2. K. Tomita, *Phys. Rev.* **89**, 429, 1953.
3. C. E. Chase, *Phys. Fluids* **1**, 193, 1958.
4. W. M. Whitney and C. E. Chase, *Phys. Rev. Letters* **9**, 243, 1962.
5. Colwell, Gill, and Morrison, *J. Chem. Phys.* **36**, 2223, 1962.

11.2. Specific Heat, Phonon Scattering, Ionic Solids

SURFACE CONTRIBUTION TO THE SPECIFIC HEAT OF CRYSTALS AT LOW TEMPERATURES*

A. A. Maradudin and M. Ashkin

Westinghouse Research Laboratories
Pittsburgh, Pennsylvania

Introduction

A lattice dynamical calculation of the low-temperature specific heat of a crystal possessing free surfaces is presented. A pair of adjacent [100] free surfaces is created in a nearest and next-nearest neighbor central force model of a simple cubic crystal by setting all atomic interactions which link the two planes equal to zero. The negatives of these interactions are treated as a perturbation on the Hamiltonian of the uncut crystal. The change in the specific heat of the crystal resulting from the introduction of the pair of free surfaces is expressed as a contour integral over the Green's function of the perturbed crystal. In the limit of low temperatures, only the long wavelength components of the Fourier transform of the Green's function are required, and these can be evaluated analytically if the relation on the atomic force constants which corresponds to elastic isotropy is imposed. The result for the surface contribution to the specific heat C_s has the form $C_s = bST^2$, where S is the surface area, T is the absolute temperature, and b is a numerical constant. The value for b obtained in the present calculation differs somewhat from the value obtained earlier by Dupuis, Mazo, and Onsager, who studied a semi-infinite isotropic elastic continuum.

Theory

Existing calculations of the contribution to the specific heat of a finite crystal arising from the presence of free boundaries have been based on the theory of the vibrations of isotropic elastic continua.[1,2] In this work, we report the results of a lattice dynamical calculation of the specific heat of a finite crystal in which the free surfaces are treated as defects in a perfect, periodic crystal whose displacements satisfy the cyclic boundary condition in a macrocrystal which contains ($L^3 = N$) atoms.

In the harmonic approximation, the specific heat of an imperfect Bravais crystal can be written in a form well suited for its evaluation at low temperatures as

$$C_v(T) = k \sum_{n=1}^{\infty} n \sum_s (\beta\hbar\omega_s)^2 e^{-n\beta\hbar\omega_s} \qquad (1)$$

---

* This research was supported by the Advanced Research Projects Agency, Director for Materials Sciences, and was technically monitored by the Air Force Office of Scientific Research under Contract AF 49(638)–1245.

where ω_s is the frequency of the sth normal mode, and the other symbols have their usual meanings. The Green's function for the crystal is defined by[3]

$$G_{\alpha\beta}(ll';\omega^2) = \frac{1}{\sqrt{M_l M_{l'}}} \sum_s \frac{B_\alpha^{(s)}(l) B_\beta^{(s)}(l')}{\omega^2 - \omega_s^2} \qquad (2)$$

where M_l is the mass of the lth atom, and $B_\alpha^{(s)}(l)$ is the α-Cartesian component of the eigenvector of the dynamical matrix corresponding to the frequency ω_s. Using the orthonormality of the $\{B_\alpha^{(s)}(l)\}$,

$$\sum_{l\alpha} B_\alpha^{(s)}(l) B_\alpha^{(s')}(l) = \delta_{ss'} \qquad (3)$$

we can introduce the function $F(\omega^2)$ which is defined by

$$F(\omega^2) = \sum_s \frac{1}{\omega^2 - \omega_s^2} = \sum_{l\alpha} M_l G_{\alpha\alpha}(ll;\omega^2) \qquad (4)$$

This function has a simple pole with residue $(2\omega_s)^{-1}$ at each of the normal mode frequencies of the crystal. It follows, therefore, that we can rewrite the sum over s in equation (1) as a contour integral over the function $F(\omega^2)$

$$C_v(T) = \frac{2\beta^2 \hbar^2 k}{2\pi i} \sum_{n=1}^{\infty} n \int_c dz z^3 e^{-n\beta\hbar z} F(z^2) \qquad (5)$$

where c is a closed, counterclockwise contour which encloses the poles of $F(z^2)$. If we denote by $F_0(\omega^2)$ the function analogous to $F(\omega^2)$ for the perfect crystal, in terms of $\Omega(-\omega^2) = F(\omega^2) - F_0(\omega^2)$, the change in the specific heat due to the imperfections in the crystal is

$$\Delta C_v(T) = \frac{2\beta^2 \hbar^2 k}{\pi} \sum_{n=1}^{\infty} n \int_0^{\infty} dy y^3 \sin n\beta\hbar y \Omega(y^2) \qquad (6)$$

In writing this expression, we have used the fact that $\Omega(-\omega^2)$ is an even function of ω which goes to zero like ω^{-4} as $|\omega| \to \infty$ to choose for the contour C an infinite semicircle in the right half plane together with the imaginary axis. Only the integration down the imaginary axis gives a nonvanishing contribution.

From the theory of the asymptotic properties of Fourier integrals,[4] it is known that, if the function $\Omega(y^2)$ has as its only singular behavior the form

$$\Omega(y^2) \sim A \ln |y| + o(\ln |y|) \qquad (7)$$

in the limit as $|y|$ tends to zero, then the asymptotic behavior of the integral in equation (6) in the limit as $\beta \to \infty$ is

$$\int_0^{\infty} dy y^3 \sin n\beta\hbar y \Omega(y^2) \sim \frac{3\pi A}{n^4 \beta^4 \hbar^4} + o(\beta^{-4}) \qquad (8)$$

The specific heat, therefore, has as the leading term in its low-temperature expansion

$$\Delta C_v(T) \sim 6Ak\zeta(3)\left(\frac{kT}{\hbar}\right)^2 \qquad (9)$$

That the function $\Omega(y^2)$ has its only singularity for real y as y tends to zero, and that it is of the form given by equation (7) for two-dimensional lattices, has been shown by J. Mahanty et al.[5]

An expression for the function $\Omega(y^2)$ can be obtained from the results of recent work of A. A. Maradudin[6] for the particular case of a simple cubic crystal with nearest and next-nearest neighbor central force interactions in which a pair of adjacent [100] free surfaces is created by setting equal to zero all interactions which cross the plane $z = \frac{1}{2}a_0$, where a_0 is the lattice parameter. The result is

$$\Omega(y^2) = -\frac{1}{NM}\sum_{\mathbf{k}j}\frac{1}{[y^2 + \omega_j^2(\mathbf{k})]^2}\sum_{ll'}u_l(\mathbf{k}j)[\mathbf{I} - \mathbf{m}(k_x, k_y; -y^2)]_{ll'}^{-1}u_{l'}(\mathbf{k}j) \tag{10}$$

with

$$m_{ll'}(k_x, k_y; -y^2) = \frac{1}{NM}\sum_{\mathbf{k}zj}u_l(\mathbf{k}j)\frac{1}{y^2 + \omega_j^2(\mathbf{k})}u_{l'}(\mathbf{k}j) \tag{11}$$

and

$$u_l(\mathbf{k}j) = 2L\frac{\{\phi''[x(l)]\}^{\frac{1}{2}}}{x(l)}[\mathbf{x}(l) \cdot \mathbf{e}(\mathbf{k}j)]\sin\tfrac{1}{2}\mathbf{k} \cdot \mathbf{x}(l) \tag{12}$$

In these expressions, M is the atomic mass, which is the same for all atoms, $\omega_j(\mathbf{k})$ is the frequency of the normal mode of the unperturbed (uncut) crystal described by the wave vector $\mathbf{k}$ and branch index j, $\mathbf{e}(\mathbf{k}j)$ is the associated unit polarization vector, $\mathbf{x}(l)$ is a translation vector of the unperturbed crystal, and $\phi(r)$ is the interatomic potential. The vector $\mathbf{x}(l)$ assumes only the values $a_0(00\bar{1})$, $a_0(\bar{1}0\bar{1})$, $a_0(0\bar{1}\bar{1})$, $a_0(10\bar{1})$, $a_0(01\bar{1})$, which we label 1, 2, 3, 4, 5, respectively. The sum over $\mathbf{k}$ in equation (10) is carried out over the first Brillouin zone of the unperturbed crystal.

Equation (6) together with equations (10)–(12) could be used in a purely numerical calculation by the specific heat at any temperature. However, an analytic expression for $\Delta C_v(T)$ is more interesting to have, and it is possible to obtain the small y limiting behavior of $\Omega(y^2)$ analytically, provided that we make that choice for the force constants $\phi''(a_0)$ and $\phi''(\sqrt{2}a_0)$ which corresponds to elastic isotropy of the lattice in the long-wavelength limit, viz., $\phi''(a_0) = \phi''(\sqrt{2}a_0)$. The small $|y|$ behavior of $\Omega(y^2)$ is determined by the small $\mathbf{k}$ terms in the sum on the right side of equation (10). In this limit, we have

$$\omega_1^2(\mathbf{k}) = C_l^2 k^2 \tag{13a}$$

$$\omega_{2,3}^2(\mathbf{k}) = C_t^2 k^2 \tag{13b}$$

$$\mathbf{e}(\mathbf{k}l) = (k_x/k, k_y/k, k_z/k) \tag{13c}$$

where $C_l = [3\phi''(a_0)/a_0\rho]$ is the speed of sound for longitudinal waves and $C_t = C_l/(3)^{\frac{1}{2}}$ is the speed of sound for transverse waves. $\rho = M/a_0^3$ is the mass density. Explicit expressions for the eigenvectors of the transverse modes are not required because they can be eliminated by the use of the closure condition satisfied by the eigenvectors.

For its evaluation, it was convenient to rewrite $\Omega(y^2)$ as

$$\Omega(y^2) = -\frac{S}{4\pi^3}\frac{C_t^2}{a_0}\sum_{ll'}\frac{1}{ll'}\int dk_x dk_y I_{ll'}(k_x, k_y; y^2)D_{ll'}(k_x, k_y; y^2) \tag{14}$$

$$I_{ll'}(k_x, K_y; y^2) = \sum_j \int dk_z \frac{[\mathbf{l}\cdot\mathbf{e}(\mathbf{k}j)][\mathbf{l}'\cdot\mathbf{e}(\mathbf{k}j)]}{(y^2 + c_j^2 k^2)^2}\sin\frac{a_0}{2}\mathbf{k}\cdot\mathbf{l}\sin\frac{a_0}{2}\mathbf{k}\cdot\mathbf{l}' \tag{15}$$

$$D_{ll'}(k_x, k_y; y^2) = [\mathbf{I} - \mathbf{m}(k_x, k_y; -y^2)]_{ll'}^{-1} \tag{16}$$

In equation (14), S is the surface area ($S = 2L^2 a_o^2$) and we have introduced the dimensionless vector $\mathbf{l}$ by the relation $\mathbf{x}(l) = a_o \mathbf{l}$.

The integrals $I_{ll'}(k_x, k_y; y^2)$ can be evaluated analytically, as can the matrix elements $m_{ll'}(k_x, k_y; -y^2)$. The inversion of the matrix $\mathbf{I} - \mathbf{m}$ can be accomplished by first block-diagonalizing it into a 3×3 and a 2×2 matrix by a similarity transformation. To lowest order in $a_o[O(a_o^{-1})]$ the inverse matrix $\mathbf{D}(k_x, k_y; y^2)$ is obtained from the inverse of the 2×2 matrix thus obtained, and this inverse can be obtained analytically. The integrals over k_x and k_y in equation (14) are most easily evaluated in polar coordinates, because the integrand is a function of k_x and k_y only, through the combination $r^2 = (k_x^2 + k_y^2)$, and because only the small $|y|$ behavior of $\Omega(y^2)$ is desired. The expression for $\Omega(y^2)$ obtained in this way is

$$\Omega(y^2) = \frac{S}{4\pi} \frac{1}{C_t^2} \frac{11}{12} \ln |y| + o(\ln |y|) \tag{17}$$

Combination of equation (17) with equations (7) and (9) yields

$$\Delta C_v(T) = 3\pi \frac{k^3}{\hbar^2} \zeta(3) \frac{11}{6 C_t^2} S T^2 \tag{18}$$

This is the change in the specific heat of a cyclic crystal when free surfaces of area S are introduced into it. This result differs from that obtained by Dupuis, Mazo, and Onsager[2] in that in their expression the factor of 11 on the right side of equation (18) is replaced by 20. Their calculation is sufficiently different from ours that we have not been able to determine the origin of this difference.*

For a typical value of C_t, 3×10^5 cm/sec, and a temperature of 0.1°K, the surface contribution to the specific heat equals the bulk contribution when the ratio of crystal volume to surface area is of the order of 10^{-5} cm.

References

1. R. Stratton, *Phil. Mag.* **44**, 519, 1953.
2. M. Dupuis, R. Mazo, and L. Onsager, *J. Chem. Phys.* **33**, 1452, 1960.
3. A. A. Maradudin, in: *Phonons and Phonon Interactions*, T. A. Bak (ed.), W. A. Benjamin, Inc., New York (1964).
4. M. J. Lighthill, *Fourier Analysis and Generalized Functions*, Cambridge University Press, Cambridge (1958).
5. J. Mahanty, A. A. Maradudin, and G. H. Weiss, *Progr. Theoret. Phys.* (Kyoto) **24**, 648, 1960.
6. A. A. Maradudin, Westinghouse Research Laboratories Scientific Paper No. 63-129-103-P9. This Green's function is simply related to the double time Green's function calculated by A. A. Maradudin and A. A. Melngailis, *Phys. Rev.* **133A**, 1188, 1964.
7. A. A. Maradudin, E. W. Montroll, and G. H. Weiss, *Theory of Lattice Dynamics in the Harmonic Approximation*, Academic Press, Inc., New York (1963), pp. 214-215.

---

* A more recent work by Stratton, *J. Chem. Phys.* **37**, 2972, 1962, in which errors in his earlier paper[1] are corrected, shows that his result for the surface specific heat now agrees with that of Mazo, Dupuis, and Onsager. However, the origin of the disagreement between our results and theirs has not yet been found.

PHONON SPECTROSCOPY DOWN TO 0.3°K*

W. D. Seward

*Laboratory of Atomic and Solid State Physics, Cornell University
Ithaca, New York*

Introduction

At the present time, it is impossible to investigate lattice vibrations in the frequency range above 10^{10} cps using ultrasonic techniques. This range is, however, accessible in the thermal conductivity experiment. One can further look upon the thermal conductivity as broad-band spectroscopy in which the source spectrum is swept in frequency by varying the temperature at which the conductivity is measured. That this interpretation is possible can be seen from the expression for the conductivity in the Debye and relaxation-time approximations as given by Klemens,[1]

$$K(T) = \tfrac{1}{3} \int_0^{\omega_D} v_s^2 \tau(T, \omega) \frac{\partial C(T, \omega)}{\partial \omega} d\omega \tag{1}$$

Here, v_s is the sound velocity, C is the specific heat, and $\tau(T, \omega)$ is the relaxation time of phonons with the frequency ω. We can think of $\partial C(T, \omega)/\partial \omega$ as the source spectrum, $\tau(T, \omega)$ as a reciprocal cross section for scattering of phonons of frequency ω, and $K(T)$ as an integrated transmission. $\partial C(T, \omega)/\partial \omega$ has a maximum at about

$$\omega_m = \frac{4kT}{\hbar} \tag{2}$$

hence, it can be varied over a wide range of phonon frequencies by varying T. We shall call ω_m the dominant phonon frequency. By comparing the thermal conductivity as a function of T of a host crystal before and after a scattering center is added, one can deduce the frequency dependence of the scattering mechanism.

This method of looking at the thermal conductivity was particularly rewarding for Pohl in his measurements on KCl containing NO_2^- molecules.[2] He found a dip in the conductivity at 7°K which indicated that the NO_2^- scattered phonons with a Lorentzian resonance cross section. The resonant frequency needed to fit the data was equal to the dominant phonon frequency at 7°K, that is, a frequency corresponding to 30°K ($\omega = 3.75 \times 10^{12}$ sec^{-1}). Further experiments on NO_2^- in KBr showed no such dip, but there was reason to believe that it might exist below 1°K. Largely for this reason, a ^{3}He cryostat was built to operate in the temperature range 0.3–3°K providing adequate overlap with existing ^{4}He cryostats. In this temperature range, the dominant phonon frequency can be varied between $v = 2.6 \times 10^{10}$ and $v = 2.6 \times 10^{11}$ cps.

---

* Work supported by the U.S. Atomic Energy Commission and the Advanced Research Projects Agency.

Apparatus

The ^{3}He refrigerator is of conventional design.[3] The primary temperature standards are as follows: from 4.2°K to 1.2°K, the vapor pressure of ^{4}He; from 1.6°K to 0.6°K, the vapor pressure of ^{3}He; and below 0.8°K, a paramagnetic salt is used. A recent addition to the cryostat is a Honeywell germanium resistance thermometer which has worked well between 0.3–4.2°K. The thermal conductivity measurement is made in the steady state and is an extension of that which has been described elsewhere.[4,5] Speer $\frac{1}{2}$-W, 470-Ω, carbon resistors were used to measure the temperature gradient along the crystal. The thermal conductivity can be measured from 3°K down to about 0.3°K in one day using a single condensation of the 2-liter (STP) ^{3}He supply.

Measurements on Undoped Crystals

In pure crystals at temperatures well below the thermal conductivity maximum, the only important phonon scattering mechanism is the boundary. Assuming that this scattering is diffuse and that the crystal is infinite in length, Casimir,[6] treating the phonons as a Knudsen gas, showed that the conductivity should be essentially $\frac{1}{3}Cv_s l$. Here, C is the specific heat, l is the crystal diameter, and v_s is an appropriate average sound velocity. This equation has been extended by Berman, Simon, and Ziman[7] to include multiple reflections and effects due to finite length.

Preliminary measurements with the ^{3}He cryostat were made on undoped crystals to verify the Casimir relation and to serve as a check on the apparatus. Some of these data for 0.3–2°K, plus the high temperature data, are shown in Fig. 1. The curves for KCl and KI represent the highest purity at present attained by our crystal-growing facility,* whereas the LiF is Harshaw-grown. All crystals were sandblasted to ensure diffuse scattering at the boundary. KCl, LiF, and KBr (omitted for clarity) have a conductivity below 1°K which is proportional to T^3, hence, to the specific heat, as predicted by the Casimir theory. The measured constants of proportionality for these crystals range from perfect agreement to 40% lower than that calculated. The agreement is quite good considering the errors involved in taking v_s as the Debye sound velocity and those due to ignoring the finite length of the sample. The measured conductivity of KI agrees with the calculated value at 1°K; however, the conductivity is not proportional to T^3 but rather $T^{2.7}$. It appears that this is due to residual impurity scattering in the KI. It does not seem likely that KI would show specular reflection at these temperatures if LiF with the same surface preparation does not.

The fact that the measured conductivity is T^3 in three of the four undoped crystals we take as proof that the thermal conductivity can be measured and, in fact, is a meaningful quantity for alkali halides in the ^{3}He temperature range. It is interesting to note that the conductivity of the LiF crystal is $T^{3.00 \pm 0.03}$ over $3\frac{1}{2}$ orders of magnitude and that the dominant phonon wavelength at 0.3°K is about 2000 Å. It is clear then that specular reflection is not significant in this crystal for wavelengths as great as 2000 Å.

Measurements on Crystals Doped with Molecular Impurities

In contrast to frequency-independent boundary scattering, we now consider the extremely frequency-dependent scattering of phonons due to molecules.

---

* These and the doped crystals were grown by J. Francis and D. Bower in the Crystal-Growing Facility of the Materials Science Center at Cornell University.

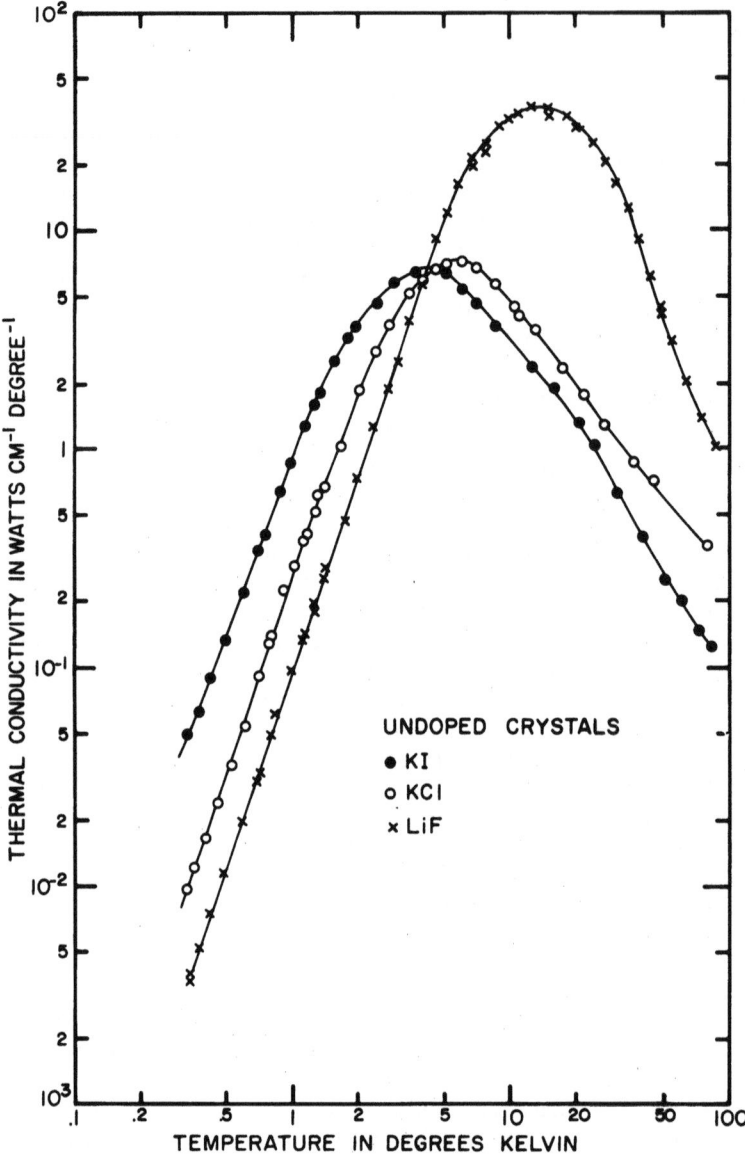

Fig. 1. Thermal conductivity of undoped LiF, KCl, and KI. These crystals have the following cross-sectional areas: LiF, 0.286 cm^2; KCl, 0.262 cm^2; and KI, 0.270 cm^2.

Figure 2 shows the thermal conductivity for undoped KCl and for KCl containing three concentrations of CN$^-$ molecules. The conductivity of the doped crystals is drastically reduced and exhibits distinct dips at about 5°K and 0.5°K. KBr crystals containing CN$^-$ also showed very strong phonon scattering with dips in the same places, although they were not so well resolved as in KCl. The concentration of CN$^-$ was determined from infrared optical absorption measurements calibrated by a chemical analysis of one crystal.

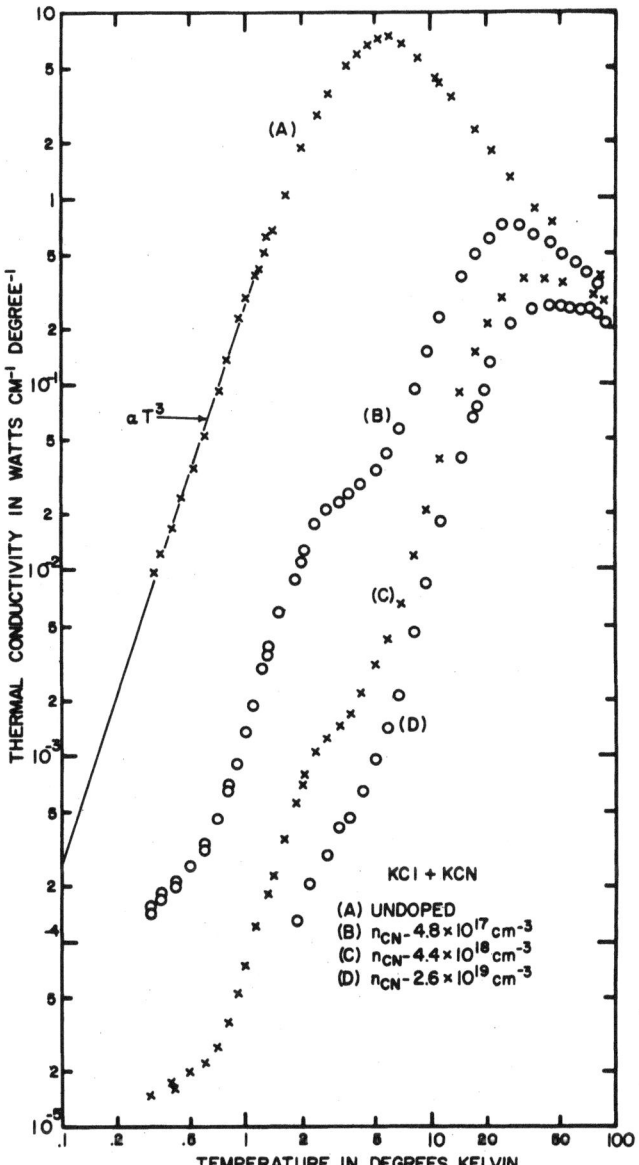

Fig. 2. Thermal conductivity of KCl doped with CN⁻ molecules. (Note added in proof: The above concentrations should be multiplied by a factor of 1.9.)

In an attempt to explain the thermal conductivity data, V. Narayanamurti[8] has investigated the temperature dependence of the infrared spectra of these crystals. At room temperature, he finds in the vibrational spectra a band due to the CN⁻ which is characteristic of a freely-rotating molecule in that it shows P and R maxima with the missing central line. Below 10°K, however, a strong central line, the Q branch, develops with sidebands characteristic of a librating molecule. Measurements at intermediate temperatures indicate that free rotation begins at about 20–30°K in both KCl and KBr. Independent calculations of this barrier height for rotation using measured librational splittings also give an energy corresponding to 20–30°K.

A possible model for the phonon scattering is one in which a phonon of energy about equal to the barrier height can classically excite the molecule over the potential barrier, resulting in a resonance-type absorption. The effect on the thermal conductivity would be greatest at about 5–7°K, since at these temperatures the dominant phonons have energies of about 20–30°K [see equation (2)]. The dip in the conductivity at 5°K is then in agreement with our model.

There remains the dip at 0.5°K. We suggest that it is due to a resonance absorption by the split levels of the ground state of the molecule.[9] This splitting, somewhat analogous to the ammonia inversion, arises because there is a finite probability that the CN^- can tunnel between equivalent equilibrium orientations. Rough calculations of this splitting are reasonably close to the observed energy of the dip. The splitting has not been resolved in the infrared spectra, although it should be within the instrumental resolution. This may be the result of a strong interaction with the phonons.

A second molecule that has been investigated is the NO_2^- molecule in the host lattices KCl, KBr, and KI. The thermal conductivity data above 1°K for NO_2^- in KCl and KBr have already been discussed.[2,10] We have extended these measurements to 0.3°K and also measured KI containing NO_2^-. The previously predicted dip in KBr was in fact found at 0.8°K, but KI had no structure down to 0.3°K. Our model, based on the infrared spectra, agrees completely with these results.

We assume that the cavity in which the NO_2^- is sitting is a halogen-ion lattice site. The variation of the barrier height with host crystal then seems plausible, since one would expect the barrier to decrease as the cavity is made larger, hence, less confining. In addition, a molecule in a large cavity should have weaker coupling to the lattice. This is observed in the infrared spectra of NO_2^- in KI as a lack of broadening at 2°K. It is also reflected in the thermal conductivity. The conductivity of KCl containing NO_2^- is reduced much more than that of KI with the same concentration of NO_2^-.

Acknowledgments

I am grateful to Prof. R. O. Pohl and V. Narayanamurti for their guidance and great help in explaining these data. I also thank the other members of the thermal-conductivity group at Cornell for their help with some of the measurements. The financial support of the Advanced Research Projects Agency and the Atomic Energy Commission is gratefully acknowledged.

References

1. P. G. Klemens, in: *Solid State Physics*, Vol. 7, F. Seitz and D. Turnbull (eds.), Academic Press, Inc., New York (1958), p. 14.
2. R. O. Pohl, *Phys. Rev. Letters* **8**, 481, 1962.
3. H. Weinstock, unpublished Ph. D. Thesis, Cornell University (1962).
4. G. A. Slack, *Phys. Rev.* **105**, 832 1957.
5. M. V. Klein, unpublished Ph. D. Thesis, Cornell University (1961).
6. H. B. G. Casimir, *Physica* **5**, 495, 1938.
7. R. Berman, F. Simon, and J. Ziman, *Proc. Roy. Soc. (London) Ser. A* **220**, 171, 1953.
8. V. Narayanamurti, *Phys. Rev. Letters* **13**, 693, 1964.
9. See M. V. Klein,[5] pp. 142–144; M. Wagner, *Phys. Rev.* **133**, A750, 1964 (see p. 758).
10. R. O. Pohl, *Z. Physik* **176**, 358, 1963 (in English).

THE EFFECT OF SPIN-PHONON INTERACTIONS ON THE THERMAL CONDUCTIVITY OF CHROMIUM DOPED MgO

L. J. Challis and D. J. Williams

Department of Physics, University of Nottingham
Nottingham, England

Recent measurements at low temperatures on dielectrics and semiconductors containing paramagnetic ions have shown that the thermal conductivity in zero field can be very much less than that expected assuming only mass-defect scattering.[1-4] The conductivity has also been shown to be field-dependent[5,7-11] and to decrease to a minimum value as the spins become tuned to the dominant phonon frequency.[5,6,9,10]

In the present work, experiments have been made on two specimens, A and B, of magnesium oxide containing about 0.3% chromium. The specimens were both mounted with their axes ([001] direction) horizontal. The field could be rotated in a horizontal plane, and the orientation of the specimens was such that the plane of rotation of the field was the $(1\bar{1}0)$ crystallographic plane of A($\pm 5°$) and the (010) plane of B($\pm 2°$).

In Fig. 1, the thermal conductivity of specimen A is shown for certain values of magnetic field applied transverse to the axis ($\mathbf{H} \parallel [110]$). The conductivity values of B are similar, although somewhat greater, and the zero field values of this specimen are shown in the figure by a broken line. In zero field, and below about 2.3°K, the conductivity increases linearly with temperature, and, at 2°K, the phonon mean free paths in specimens A and B are 9×10^{-3} and 12×10^{-3} cm, respectively, compared with the specimen diameters of about 4×10^{-1} cm.

In Fig. 2, the fractional increase in thermal resistance $(W_H - W_0)/W_0$ is given at a few temperatures for specimens A and B as a function of transverse field, i.e., for A, $\mathbf{H} \parallel [110]$ and for B, $\mathbf{H} \parallel [100]$. The resistance of A at 1.23°K reaches its maximum value in a field of 7.5 kOe, and that of B at 1.28°K in 5.5 kOe. The fields available were insufficient for any maximum to be observed in the measurements made at higher temperatures. If it is assumed that the maximum resistance is reached when a pair of spin levels is separated by an amount equal in energy to that of the dominant phonons, then, if this separation is $g\beta H$, the maximum should occur when $g\beta H = CkT$. At low temperatures, the constant C was found to have a value of about 2.5 in both cobalt lanthanum nitrate[9] and iron-doped MgO.[10] This value gives effective g-values for the scattering ion in the present work of 6.2, $\mathbf{H} \parallel [110]$ and 8.8, $\mathbf{H} \parallel [100]$, so that, if we assume the crystal field to have axial symmetry, $g_\parallel$ ([100]) is approximately equal to 8.8 and $g_\perp$ approximates 0. The temperature dependence of the conductivity in zero field is considerably less than T^3 at all temperatures below 4°K. There is also a change in the temperature dependence at 3°K. This may indicate the existence of a resonant scattering process which has its greatest effect just below this temperature. This corresponds to a zero-

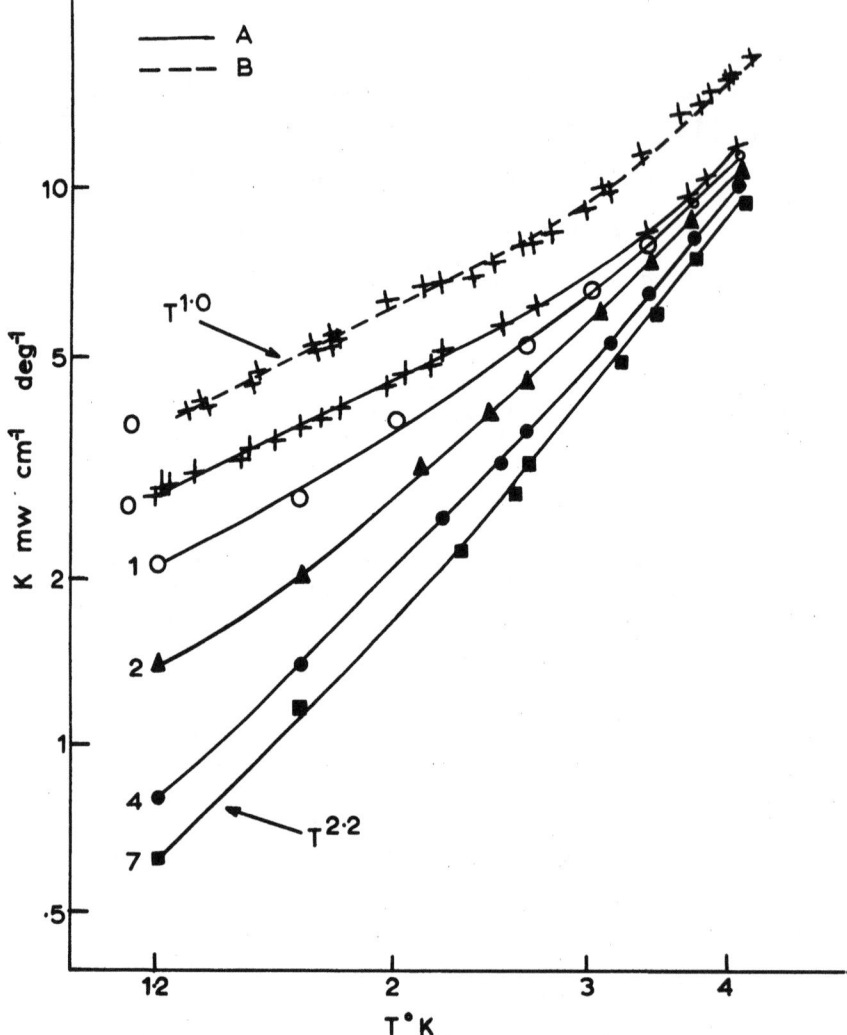

Fig. 1. The thermal conductivity of specimens A and B as a function of temperature. The transverse field in each case is given in kOe by the figure to the left of the curve.

field splitting of approximately 5 cm^{-1}. The maximum in the resistivity obtained in magnetic fields at 1.25°K corresponds to a splitting of approximately 2 cm^{-1}. It does not seem possible for this to correspond to the same pair of levels but with their splitting reduced by field, as this picture would lead to maxima in the resistivity at higher temperatures in even lower fields. A possible explanation is that, whereas in zero field, the scattering is between the two levels separated by 5 cm^{-1}, in a nonzero field, the degeneracy of the lower level is removed and the predominant scattering is between these spin levels with effective g-values given above. At present, no other measurements have been made to determine effective g-values in any other directions, although the anisotropy is apparent in the variation of the conductivity with

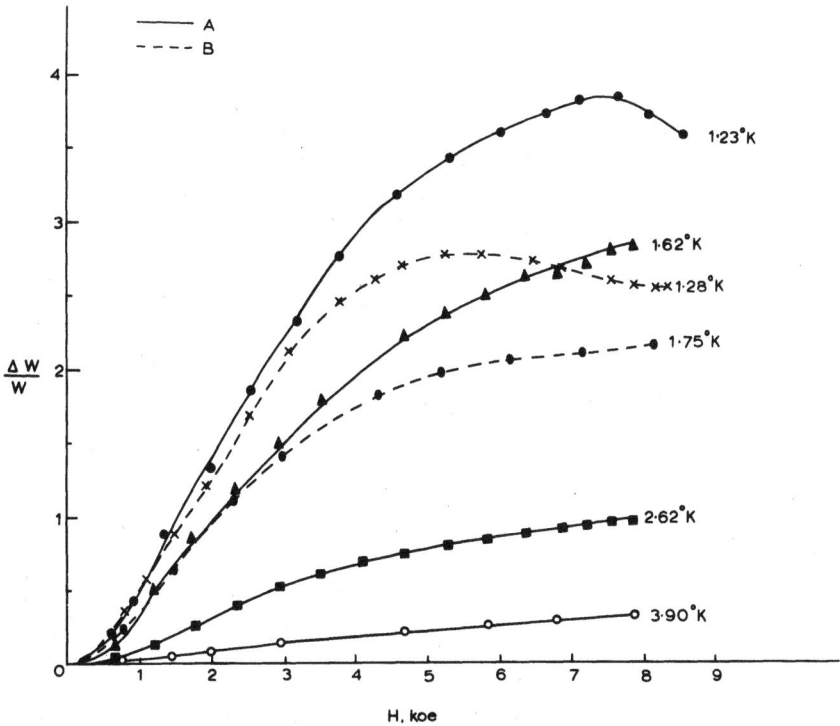

Fig. 2. The fractional increase in thermal resistance $[\Delta W/W_0 = (W_H - W_0)/W_0]$ as a function of transverse field.

field direction. For example, at 1.75°K and in a field of 8 kOe, the conductivity of B drops from a value of 1.755 mW-cm^{-1}/deg when the field is in the [100] direction to a minimum value of 1.505 mW-cm^{-1}/deg close to the [201] direction. It then rises to a maximum of 1.555 mW-cm^{-1}/deg at an angle close to the [101] direction. Since this latter direction corresponds to the transverse field direction of A, a comparison may be made between the values of $(W_H - W_0)/W_0$ for the two specimens under similar conditions ($H = 8$ kOe, $T = 1.75$°K, $H/\!/[101]$ or [110]). For B, this is a measured value; for A, it is obtained by slight extrapolation of values taken from the curves of Fig. 1. The values agree to within about 7% which suggests that $(W_H - W_0)/W_0$ and so the conductivity is not strongly dependent on the angle between the field and the heat current.

We next consider the identity of the ion responsible for the scattering. The largest amounts of paramagnetic material other than chromium (0.3%) in three specimens of similar color (dark green) to A and B taken from the same crystal are (in ppm): 100, iron; 10, manganese; <2, nickel; <2, copper; <20, vanadium; and <5, cobalt. The present results may be compared with those of Morton and Rosenberg[10] on an MgO specimen containing 2.7% iron (specimen D1). In zero field and at 1.5°K, the conductivities of A and B are only about $\frac{1}{10}$ of that of D1 which contains about 270 times more iron. This suggests strongly that the phonon scattering in A and B is not predominantly by isolated iron ions. This conclusion is supported by the further differences shown in Table I and also by the fact that the

Table I. Comparison of MgO Specimens

Specimen	A(Cr 0.3)	B(Cr 0.3)	D1(Fe 2.7)
Temperature dependence ($T = 1.5°K$, $H = 0$)	$T^{1.0}$	$T^{1.0}$	$T^{3.6}$
$((W_H - W_0)/W_0)_{max}$ ($T \sim 1.25°K$)	3.85*	2.77†	0.2‡
Magnetic field for maximum resistance, kOe ($T \sim 1.25°K$)	7.5*	5.5†	13

* $H_\perp/\!/[110]$ $H_\perp : H_\perp \dot{Q}$
† $H_\perp/\!/[100]$ $H_{11} : H/\!/Q$
‡ H_{11}, axis not stated.

thermal conductivities at these temperatures of MgO specimens containing similar amounts of iron to A and B, but negligible amounts of chromium, are about 50 times larger.[10]*

Additional support is provided by the work of Marshall, Rampton, and Rowell.[12] They found that, while they were barely able to detect a pulse of microwave phonons (9.5 kcps) after it had passed through specimen A, a pulse was detected without difficulty after passing through a very light green specimen from the same crystal. This, from an analysis of a similar sample, is believed to contain about 0.01% chromium and very similar quantities of all other paramagnetic impurities to A and B. While this evidence is uncertain to the extent of the possible variations in the bonding between the MgO and the transducer, it suggests that it is again the additional chromium that is responsible for the microwave phonon scattering.

We conclude, therefore, that the phonon scattering in A and B is largely by a chromium ion which would appear to be even more strongly coupled to the lattice than Fe^{2+}. The chromium was introduced into the MgO as Cr^{3+} (Cr_2O_3) and in equilibrium is thought to exist largely in this form. The ESR spectrum of chromium-doped MgO has been studied in detail,[13-17] but so far no chromium ions other than Cr^{3+} have been detected in this way, although the Cr^{2+} ion has been detected by optical absorption.[18] The spin-lattice relaxation time of Cr^{3+} (3×10^{-4} sec in MgO containing 0.05% Cr)[19] would appear to be too long to account for the observed behavior. Moreover, the g-value of 1.98 is much smaller than that indicated by the present work. It has been suggested that Cr^{2+}, which, like Fe^{2+}, is a non-Kramers ion, should be strongly coupled to the lattice.[20] If Cr^{2+} or Fe^{2+} is the scattering ion, the anisotropy suggests the presence of neighboring ions or vacancies, or a Jahn–Teller effect.

We note that both a Jahn–Teller effect and a large zero field splitting occur in the energy level picture for Cr^{2+} in MgO proposed by K. W. H. Stevens (private communication), although in this picture it appeared that transitions in which $S_z \neq 0$ would be relatively weak, in disagreement with the present interpretation. Another possibility is that the scattering is by Cr^{4+}, which has a very short spin-lattice relaxation time in Al_2O_3.[21]

The Cr^{2+} content in lightly-doped MgO can be increased by X-irradiation.[14,15,17] Accordingly, the thermal conductivity of A was measured shortly after it had been irradiated for two hours (tube voltage = 38 kV, current = 12 mA), but no detectable change was found.

The present work is to be continued in higher fields.

* A. F. Cohen; the results are quoted in ref. 3.

Acknowledgments

We wish to thank many of our colleagues for useful discussions, and we are particularly grateful to Dr. P. M. Rowell who suggested this material and loaned specimens with which he had kindly been supplied by J. Burrow of the H. H. Wills Laboratory, Bristol University. We also wish to acknowledge help given in various ways by Dr. B. Henderson and D. I. Matkin of A.E.R.E., Harwell, and Dr. K. J. Standley and his colleagues of the Department of Physics, University of Nottingham, Nottingham, England.

References

1. H. M. Rosenberg and B. Sujak, *Phil. Mag.* **5**, 1299, 1960.
2. R. Orbach, *Phil. Mag.* **5**, 1303, 1960.
3. G. A. Slack, *Phys. Rev.* **126**, 427, 1962.
4. G. A. Slack, *Phys. Rev.* **133**, A253, 1964.
5. I. P. Morton and H. M. Rosenberg, *Phys. Rev. Letters* **8**, 200, 1962.
6. R. Orbach, *Phys. Rev. Letters* **8**, 393, 1962.
7. B. Dreyfus, A. Lacaze, and F. Zadworny, *Compt. Rend.* **254**, 3337, 1962.
8. B. Dreyfus and F. Zadworthy, *J. Phys. Radium* **23**, 490, 1960.
9. R. Berman, J. C. F. Brock, and D. J. Huntley, *Phys. Letters* **3**, 310, 1963.
10. I. P. Morton, unpublished Ph. D. Thesis, Oxford University (1964).
11. M. G. Holland, *Proceedings of the International Conference on Semiconductor Physics*, Academic Press, Paris, 1964, p. 713.
12. F. G. Marshall, V. W. Rampton, and P. M. Rowell, private communication.
13. W. Low, *Phys. Rev.* **105**, 801, 1957.
14. J. E. Wertz and P. Auzins, *Phys. Rev.* **106**, 484, 1957.
15. J. E. Wertz, P. Auzins, J. H. E. Griffiths, and J. W. Orton, *Discussions Faraday Soc.* **26**, 66, 1958.
16. J. H. E. Griffiths and J. W. Orton, *Proc. Phys. Soc. (London)*, Ser. A **73**, 948, 1959.
17. J. E. Wertz, J. W. Orton, and P. Auzins, *J. Appl. Phys. Suppl.* **33**, 322, 1962.
18. R. L. Hansler and W. G. Segelken, *J. Phys. Chem. Solids* **13**, 124, 1960.
19. J. G. Castle and D. W. Feldman, *Phys. Rev.* **121**, 1349, 1960.
20. S. A. Al'tshuler, B. I. Kochelaev, and A. M. Leushin, *Soviet Phys.—Usp. (English Transl.)* **4**, 880, 1962.
21. R. H. Hoskins and B. H. Soffer, *Phys. Rev.* **133**, 490, 1964.

OBSERVATIONS OF A PHONON BOTTLENECK IN $CuCs_2(SO_4)_2 \cdot 6H_2O$ AT TEMPERATURES NEAR 0.1°K

A. R. Miedema and K. W. Mess

Kamerlingh Onnes Laboratorium
Leiden, The Netherlands

Experiments are reported concerning the spin-lattice relaxation in $CuCs_2(SO_4) \cdot 6H_2O$ at temperatures near 0.1°K. It will be shown that the results on the spin-lattice heat contact strongly indicate the presence of phonon-bottleneck effects. Data on the energy transfer between the spin system and the lattice heat waves have been obtained, while the spin temperature (T_S) was either higher or lower than the average lattice temperature (T_L).

Figure 1 shows the apparatus used. The figure describes an experiment in which T_S is less than T_L. The apparatus consists of a cooling salt (A), a thermal switch (B), a Curie's law thermometer (C), a copper ring (D) in which heat can be generated by means of eddy currents, and the $CuCs_2(SO_4)_2 \cdot 6H_2O$ sample (E).

Generally, the experimental procedure is as follows. After adiabatic demagnetization of the chrome–alum cooling salt, the magnet is moved vertically and the sample is magnetized. The heat of magnetization is transferred to the cooling salt through the lead thermal switch, which is kept in the normal state by the stray field of the magnet. Upon demagnetizing the sample, its magnetic spin system attains a very low temperature, while the lead thermal switch becomes superconductive, so that a thermally isolated system of sample–heater–thermometer is obtained. The spin-lattice heat contact is studied by generating a constant heat flux in the copper ring, which results in a difference between the temperatures T_S and T_L, as measured by the thermometer (C) and as derived from the magnetization of the copper spins in the sample, respectively. For different runs the heat flux is varied from 1–10^3 erg/sec, H is varied from 0.6–5 kOe, while T_S and T_L lie between 0.05–0.5°K and T_S/T_L reaches values as low as 0.2.

Practically the same apparatus has been used for experiments with relatively high spin temperatures ($T_S > T_L$), the thermal switch being replaced by a copper wire. The spin temperature decreases continuously from above 1°K down to 0.1°K after application of the field. The heat flow has been derived from the observed rate of change of the magnetization of the spin system, while the lattice temperature is observed by means of the magnetic thermometer (C).

The results for the heat flow at a given value of the applied field and the temperature difference, obtained with the two methods ($T_S < T_L$ or $T_S > T_L$) have been found to agree within 20%.

In case of a severe phonon-bottleneck, the spin-lattice heat contact may be described by

$$\dot{Q} = \frac{1}{t_{ph}} \left(\frac{3\omega^2 \Delta\omega}{2\pi v_0^3} \right) \left[\frac{\hbar\omega}{\exp(\hbar\omega/kT_S) - 1} - \frac{\hbar\omega}{\exp(\hbar\omega/kT_L) - 1} \right] \tag{1}$$

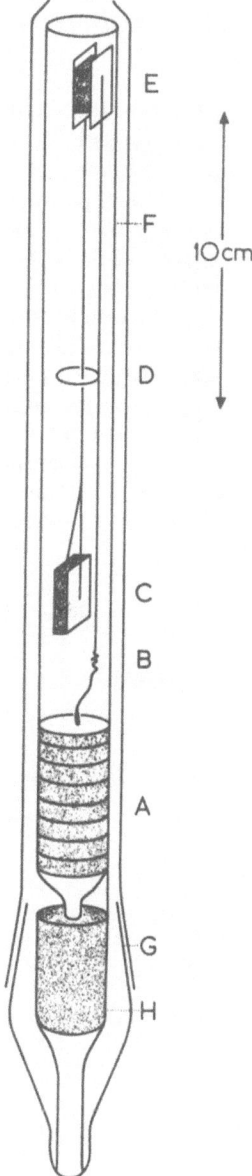

Fig. 1. Apparatus for studying spin-lattice relaxation at very low temperatures: (A) chromium-alum cooling salt, (B) lead thermal switch, (C) CeMg-nitrate thermometer, (D) heater, (E) sample, (F) glass thermal shield, (G) ground joint, and (H) guard fill.

Here, ω is the average angular frequency of the phonons on speaking terms with the spin system (phonons o.s.t.); $\hbar\omega - g\mu_B H$. The number of phonons o.s.t. per unit volume is given by $3\omega^2 \Delta\omega / 2\pi v_0^3$ in which $\Delta\omega$ is related to the width of the electron spin resonance line and v_0 is the velocity of sound. The term $\hbar\omega/[\exp(\hbar\omega/kT) - 1]$ corresponds to the average energy per oscillator and t_{ph} is the time for the phonons to exchange energy with the other, more numerous, oscillators.

Replacement of $\hbar\omega/[\exp(\hbar\omega/kT_S) - 1]$ by T'_S, $\hbar\omega/[\exp(\hbar\omega/kT_L) - 1]$ by T'_L, ω by $g\mu_B H/\hbar$, and $\Delta\omega$ by $g\mu_B \Delta H/\hbar$ yields

$$Q = At_{ph}^{-1} H^2 \Delta H (T'_S - T'_L) \qquad (2)$$

The constant A equals 8×10^{-3} erg/kOe³-cm³-°K. At relatively high temperatures, $T'_S = T_S$ and $T'_L = T_L$, so that for a given magnetic field the heat flow will be proportional to $T_S - T_L$.

The experimentally observed temperature dependence of the spin-lattice heat contact was in agreement with equation (2) for not too low values of T_S (i.e. $g\mu_B H/kT_S \leq 2$). Keeping the heat flow constant, both T_L and T_S continuously increase, so that $(T'_S - T'_L) \approx (T_S - T_L)$ remains constant.

A comparison of the values of $Q/(T'_S - T'_L)$ obtained in magnetic fields of 0.5–5 kOe is given by Fig. 2. Both $Q/(T'_S - T'_L)$ and H are on a logarithmic scale. Apart from an anomaly near $H = 4$ kOe, the observed values of $Q/(T'_S - T'_L)$ are roughly proportional to $H^{1.7}$ over the whole magnetic field range, and one may say that the curve tends to a H^2 dependence at the high field side.

The dashed line in Fig. 2 corresponds to a value for $At_{ph}^{-1}\Delta H$ of 120 erg-kOe⁻²-cm⁻³/°K-sec. If ΔH is taken for the width of the E.S.R. line ($\Delta H = 400$ Oe), we obtain $t_{ph} = 40$ μsec for a crystal of 1 cm thickness.

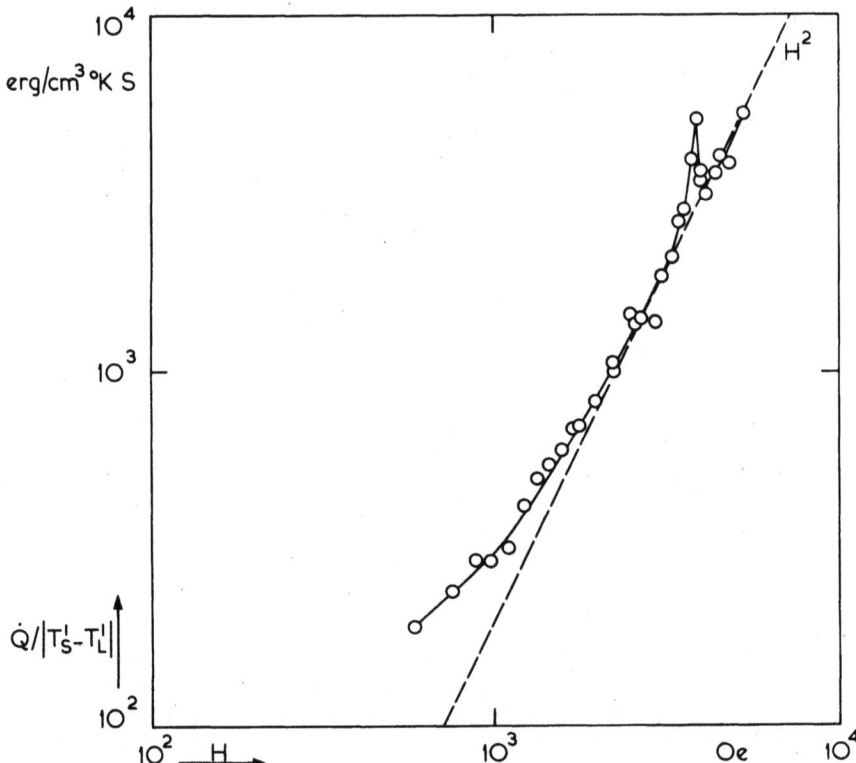

Fig. 2. The field dependence of the spin-lattice heat contact in $CuCs_2(SO_4)_2 \cdot 6H_2O$. Both $Q/(T'_S - T'_L)$ and H are plotted on a logarithmic scale. The dashed line corresponds to a H^2 dependence.

The agreement between our results on $CuCs_2(SO_4)_2 \cdot 6H_2O$ and equation (1) strongly indicates that phonon-bottleneck effects determine the heat flow. The phonons on speaking terms, which are in good thermal contact with the spins, exchange energy with the other lattice oscillators in a characteristic time t_{ph} of about 40 μsec. This time is found to be practically field- and temperature-independent for fields of 1–5 kOe and temperatures of 0.1–1°K.

Changing the total area of the metal in contact with the crystal by a factor of six, the same value for t_{ph} was found. Furthermore, we have found that t_{ph} is independent of the concentration of the copper ions, the same value of 40 μsec being obtained for a 1:10 zinc-diluted crystal. Neither treating the crystal surface with varnish nor growing a diamagnetic shell of ZnCs–tutton salt around the CuCs–tutton salt crystal appreciably affected the value of t_{ph}.

This may suggest that the energy transfer between lattice oscillators does not occur essentially at the crystal surface, as was suggested by Jeffries et al.,[1] for CeMg nitrate. In E.S.R. experiments ($H = 3$ kOe) at temperatures down to 0.3°K, they observed phonon-bottleneck effects, corresponding to a value for t_{ph} of 0.2 μsec for a crystal of 1 mm thickness. Since this time roughly equals the time needed for phonons to traverse the crystal field thickness, this indicates the phonon–phonon relaxation taking place at the crystal surface.

We have studied CeMg nitrate at lower temperatures and in lower magnetic fields. Our results quantitatively agree with those of Jeffries et al. in that $t_{ph} \approx 3$ μsec for a crystal of 1 cm thickness.

At very low temperatures, a strong dependence of the line width on temperature must be expected, since, in a completely magnetized spin system, the line broadening disappears. A reduction of the line width produces, according to equation (2), a reduction of the spin-lattice energy transfer, i.e., $Q/(T'_S - T'_L)$ is considerably lowered. We could describe the low-temperature data for all fields assuming that ΔH depends only on the value of the Boltzmann exponent; thus

$$\Delta H = (\Delta H)_T = f\left(\frac{g\mu_B H}{kT_S}\right)$$

The values of this function for different values of $\Delta E/kT_S$ are given in Table I. One may see that the line width is practically independent of temperature for $\Delta E/kT_S$ values up to 2.

Table I. The Apparent Line Width of the E.S.R. Line as a Function of Field and Temperature in the Case of a Highly Magnetized Spin System

$g\mu_B H/kT_S$	ΔH	$g\mu_B H/kT_S$	ΔH
1.0	1.00	3.0	0.58
1.5	1.00	3.5	0.49
2.0	0.88	4.0	0.39
2.5	0.80	4.5	0.30
2.7	0.70		

The line width is found to be a function of H/T_S only and has been normalized to unity for $g\mu_B H/kT_s \ll 1$.

A sudden reduction of the spin-lattice relaxation time at a sharply defined field (as shown in Fig. 2, $H = 3.8\,\text{kOe}$) has been found earlier in $Co(NH_4)_2(SO_4)_2 \cdot 6H_2O$ diluted with zinc by Van den Broek et al.[2] In their experiments, the field was found to be dependent on the orientation with respect to the crystal axes, so that $g\mu_B H$, the magnetic splitting, was constant. The energy splittings obtained by Van den Broek et al. were $\Delta E = 0.43\,k$ and $0.71\,k$ for two different cobalt concentrations, which are comparable with the present value for $CuCs_2(SO_4)_2 \cdot 6H_2O$ of $0.54\,k$. A possible reason for this anomaly in the spin-lattice heat contact may be a peak in the phonon spectrum, superposed to the normal Debye spectrum.

References
1. P. L. Scott and C. D. Jeffries, *Phys. Rev.* **127**, 32, 1962; R. H. Ruby, H. Benoit, and C. D. Jeffries, *Phys. Rev.* **127**, 51, 1962.
2. J. Van den Broek, L. C. Van der Marel, and C. J. Gorter, *Physica* **25**, 371, 1959.

THE EFFECT OF ELECTRON–PHONON SCATTERING ON THE THERMAL CONDUCTIVITY OF InSb

L. J. Challis, J. D. N. Cheeke, and D. J. Williams

Department of Physics, University of Nottingham
Nottingham, England

The thermal conductivity of both *n*- and *p*-type germanium has been shown to be strongly dependent on the excess carrier concentration n_{ex} at temperatures below that of the conductivity maximum.[1-4] Moreover, measurements made on *p*-type specimens of silicon,[5] indium antimonide,[6,7] and gallium arsenide[8] and on specimens of both *n*- and *p*-type gallium antimonide[9] all containing more than 10^{16} carriers/cm^3 show roughly similar behavior, e.g., a temperature dependence faster than T^3. It appears that the effect of the carriers is to scatter phonons, and the heat current due to the carriers has been shown to be negligible. The only exceptions to this behavior in the semiconductors examined so far are specimens of *n*-type InSb (1.4×10^{17} cm^{-3})[7] and *n*-type GaAs (5.0×10^{17} cm^{-3})[8] whose thermal conductivities both in magnitude and temperature dependence are close to those expected, assuming only boundary scattering.

As yet, little information exists on the variation of conductivity with n_{ex} for semiconductors other than germanium. In the present work, the thermal conductivity of six *p*-type specimens (germanium-doped) covering the range 6.9×10^{14}–4.5×10^{17} carriers/cm^3 and five *n*-type specimens (tellurium-doped) covering the range 3.8×10^{14}–4.0×10^{18} carriers/cm^3 have been measured in the temperature region 1.2–4.2°K. The specimens were single crystals and uncompensated $\{(N_A + N_D)/(N_A - N_D) \sim 1\}$ and their surfaces were sandblasted with 44–79 mesh powder. The carrier concentrations of the specimens are given in Table I. The

Table I. Thermal Resistances of InSb Specimens at 1.5°K

Specimen	Type	Number of carriers, cm^{-3} (at 77°K)	Thermal resistivities at 1.5°K (w^{-1} cm deg)		
			W_{meas}	W_b	W_d
1	p	6.9×10^{14}	2.29	1.23	1.06
2	p	3.3×10^{15}	3.47	0.86	2.61
3	p	5.2×10^{15}	6.62	1.25	5.37
4	p	2.0×10^{16}	21.2	0.98	20.2
5	p	1.3×10^{17}	29.8	1.27	28.5
6	p	4.5×10^{17}	27.6	1.47	26.1
7	n	3.8×10^{14}	2.41	1.46	0.95
8	n	5.0×10^{15}	1.54	1.29	0.25
9	n	5.7×10^{16}	1.07	1.06	0.01
10	n	1.4×10^{17}	2.23	1.17	1.06
11	n	4.0×10^{18}	2.57	1.04	1.53

temperature differences were measured using carbon resistors in a modified Wheatstone bridge. The calculation of the temperature differences using a digital computer has been described elsewhere.[10]

The thermal conductivities of the p-type specimens and one n-type specimen (No. 9) are shown in Fig. 1. It is observed that whereas above approx. 2°K the conductivity falls steadily with n_{ex}, the curves corresponding to the higher concentration specimens cross over below this temperature. The temperature dependence of the conductivity $(d \log K/d \log T)$ increases with n_{ex} to a maximum at approx. 10^{16} cm^{-3}, and then begins to fall again. The rise in slope could be the consequence of the gradual predominance of scattering by a process which by itself would result in a conductivity varying as approx. T^5, but the decrease above approx. 10^{16} cm^{-3} suggests a change in the nature of this process. The n-type specimen No. 9 ($n_{ex} = 5.2 \times 10^{16}$ cm^{-3}) had the highest conductivity measured, and the lowest conductivity of the n-type specimens was that of No. 11 which had a value about 90% of that of specimen No. 1 (p-type). The remaining n-type specimens had conductivities between those of No. 9 and No. 11, but for the sake of clarity the results are not included in Fig. 1. The thermal conductivity of all the specimens was also measured in transverse magnetic fields up to 8 kOe, but no detectable changes were observed.

Since the linear dimensions of the specimens differ somewhat, the effects of boundary scattering must be separated out before quantitative comparison may be made. In this preliminary analysis, this has been done by assuming that the thermal resistivities are additive. The resistivities arising from boundary scattering were calculated from the Casimir formula[11] and corrections made for the finite lengths of the specimens.[12,13] The resistivities calculated in this way are greater than the measured values for all of the n-type specimens and the p-type specimen No. 1. It is necessary to assume that 50% of the phonons are specularly reflected ($f = 0.5$) in order to bring the calculated resistivity of specimen No. 9 equal to the measured value. Since the surfaces of the specimens were similarly treated, we have used this value of f throughout. The measured resistivities W_{meas}, boundary resistivities W_b, and the differences W_d, are given in Table I at 1.5°K. For the n-type specimens, the values of W_d are quite small, and the apparent minimum in W_d is probably not significant. In n–InSb, the ionization energy is extremely small, perhaps zero, so that the electrons are all excited to the conduction band and are degenerate, as in a metal. The value of the resistivity observed, approx. $\leq 2W^{-1}$ cm deg, is very much less than that of a typical metal. For example, the resistivity of the phonon heat flow in lead at 1.5°K is approx. $340 \, W^{-1}$ cm-deg.[14] This difference is accountable, since the resistivity has been shown[15] to be proportional to m^{*2} and in n–InSb, m^* is approx. $0.013 \, m$. A similar argument explains the conductivity observed for bismuth.

In Fig. 2, W_d is plotted against n_{ex} for the p-type specimens. The curves at 1.3 and 1.5°K show maxima in the resistance between 10^{16} and 10^{17} cm^{-3}, but no maximum appears at 2°K, although the curve is becoming flatter. The existence of such maxima at the lower temperatures is, of course, apparent from the crossing of the curves below 2°K. Similar curves are given for n- and p-type germanium.[1–4] At 2°K, a maximum is observed in p-type, but not in n-type, germanium (either arsenic- or antimony-doped) for which the conductivity curves of the higher concentration specimens remain parallel at least down to 1.4°K, which is the limit of the experimental data.

A conductivity varying more rapidly than T^3 indicates that the scattering

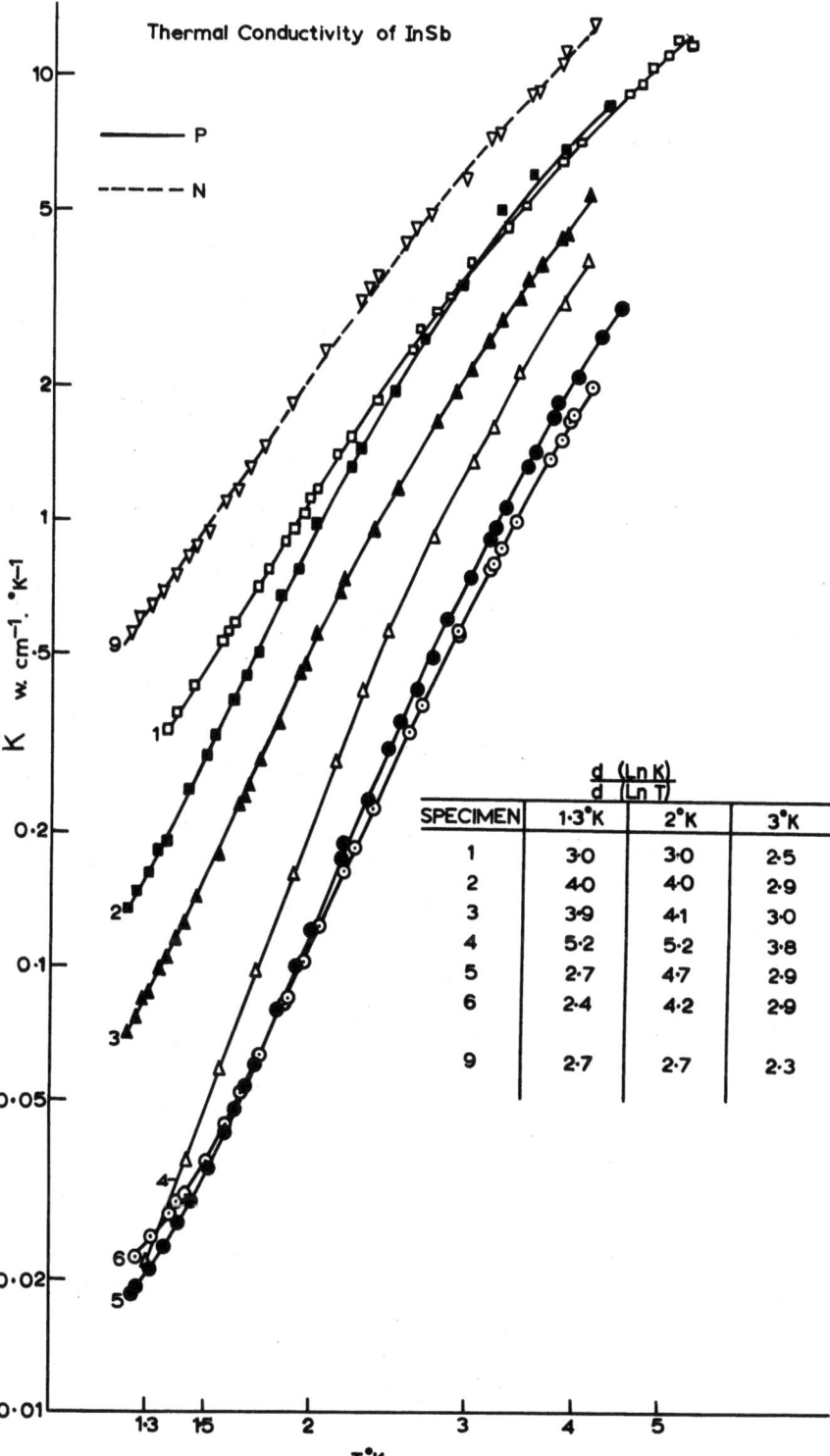

Fig. 1. The thermal conductivity of InSb. Solid line for p-type (1–6); broken line for n-type (9).

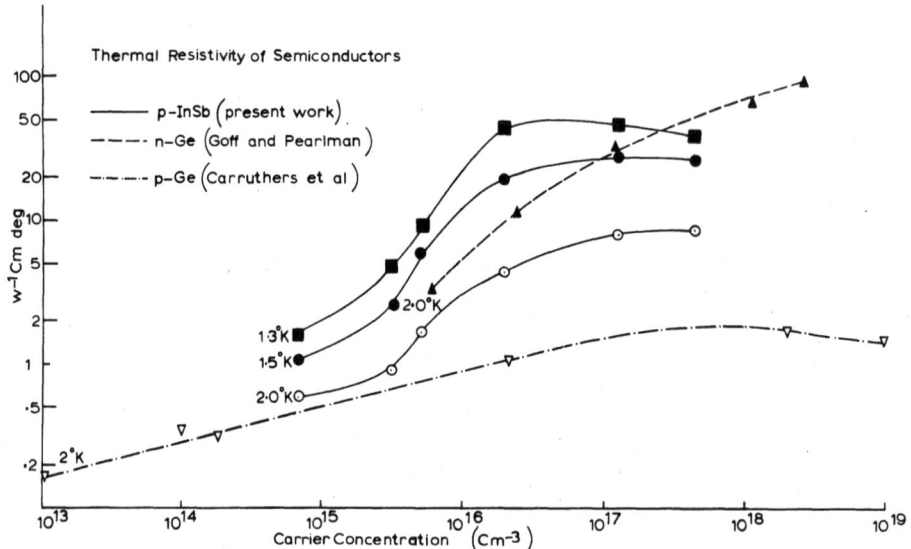

Fig. 2. The thermal resistivity (W_d) as a function of carrier concentration. Solid line for p-InSb (present work), broken line for n-germanium,[1] and broken dotted line for p-germanium.[2]

increases as the frequency is reduced, as is the case above the resonant frequency in a resonant scattering process. A change in slope to one varying less rapidly than T^3 indicates that the resonant frequency has been passed. Such a change is observed in the more highly doped specimens of p-InSb at about 1.5°K. No change occurs in the less highly doped specimens down to 1.3°K, which again suggests a different scattering process and one with a lower resonant frequency.

At sufficiently low concentrations, the impurity atoms should act as independent scattering centers. In n-germanium, the ground state of a bound electron is split by an amount 4Δ by the valley-orbit interaction. Keyes[16] noted that since this splitting was approx. 6.6°K for antimony in germanium and 48°K for arsenic in germanium, phonon scattering could result from transitions between these levels. He also observed that at helium temperatures, the dominant phonon wavelength λ was comparable with the mean diameter ω_0 of the electron orbit, so that there would be a large scattering cross section from these transitions. This phenomenon is, therefore, not unlike that of magnetoacoustic resonance and the same resonance condition $\lambda_0/2 \sim 2r_0$ or $\omega_0 \sim c/r_0$ applies. Keyes' calculation was made assuming $\hbar\omega \ll 4\Delta$, but the theory has been extended by Griffin and Carruthers[17] to the region of the true resonance, $\hbar\omega \sim 4\Delta$, and quite good agreement has been obtained with the experimental results for n-germanium. The theory predicts that the change in slope of the conductivity due to the geometrical resonance should occur at about 1.2°K, which is just below the limit of the experimental data. When the dominant phonon energy is approx. 4Δ, a further dip in the conductivity should occur. A dip at 7°K was indeed observed by Goff,[3] and by taking the dominant phonon energy to be $6kT$, Griffin and Carruthers obtained good agreement with the value for 4Δ of 48°K quoted earlier. Similar dips have recently been observed by Holland[8] in manganese-doped GaAs at about 8°K.

It appears possible that this mechanism could account for the conductivity of moderately doped p-InSb and other semiconductors with similar conductivities.

As yet, however, no change in slope which could correspond to geometrical resonance has been observed in moderately doped semiconductors within the temperature range of measurement, although the resonance condition would suggest that such a change should have been observed for p-germanium and probably also for p-InSb. However, if the true resonance frequency lies fairly close to, but below, the geometrical resonance frequency, the decrease in scattering that occurs when the frequency falls below geometrical resonance could be offset by the increase at true resonance. This would displace the change in slope of the conductivity to a lower temperature than that corresponding to geometrical resonance alone. Griffin and Carruthers used this argument to account for the absence of any detectable change in slope of the conductivity of p-germanium down to 0.3°K, and it could apply equally to p-InSb. The argument requires the ground state splitting in both cases to be less than 10^{-3} eV.

An alternative explanation of the scattering process has been put forward by Ziman.[15,1] (The model proposed by Pyle[18] does not seem able to account for the present results or those on n-germanium.[7]) In Ziman's theory, it is assumed that the scattering is by carriers in an impurity band. The model is not applicable, therefore, to the lower concentration specimens, but would seem more appropriate for the higher concentration specimens than a model which assumes independent scattering centers. At the lower temperatures, these specimens are degenerate so that phonon scattering, which is limited to electrons within kT of the Fermi energy, increases linearly with temperature. Eventually the distribution becomes nondegenerate. The principal limitation to scattering is now that the momentum of a scattering electron must be at least half that of the phonon. The mean phonon momentum increases linearly with temperature and, as the temperature increases, the number of electrons that can scatter drops rapidly. Hence, as the temperature rises, the scattering rate increases to a maximum and then falls. The maximum (and, hence, the change in slope of the conductivity) occurs at a temperature which increases with T_F, as was found in p-germanium. Unfortunately, the curves given by Ziman do not extend to concentrations for which the model would seem applicable in p-InSb. However, a very rough estimate suggests that the changes in slope of the conductivity of specimens No. 5 and No. 6 should occur at approx. 0.8°K and 1.3°K, whereas in fact these both occur at approx. 1.4°K. If the model were still valid for specimen No. 4, the change would occur at approx. 0.3°K.

The model suggests that at a given temperature the thermal resistance should rise as n_{ex} is increased. Furthermore, that, if the temperature is low enough so that the distribution is degenerate, i.e., the region below the slope change is reached, a limiting value of resistance should be attained. It is only possible for the curves to cross, and so it is only possible for there to be a maximum in the resistance if one of the other parameters in the model changes with n_{ex}. We note[19] that the impurity band in p-InSb overlaps the valence band at $n_{ex} \sim 2 \times 10^{17}$ cm^{-3}, and it seems likely that one result of this would be a decrease in the average effective mass. For a degenerate distribution, the resistance varies with m^{*2}, so that the observed fall in thermal resistance could be the result of this band overlap. This could also explain the maximum observed in p-germanium and the absence of a maximum in n-germanium for which no change of slope was observed down to approx. 1.3°K.

Acknowledgments

We are very grateful to Dr. J. B. Mullin, O. Jones, and Dr. J. B. Harness of R.R.E. Malvern for supplying the specimens used in this investigation.

References

1. J. A. Carruthers, T. H. Geballe, H. M. Rosenberg, and J. M. Ziman, *Proc. Roy. Soc. (London) Ser. A* **238**, 502, 1957.
2. J. F. Goff and N. Pearlman, *Proceedings of the Seventh International Conference on Low Temperature Physics*, University of Toronto Press, Toronto, Canada (1961), p. 284.
3. J. F. Goff, unpublished Ph.D. Thesis, Purdue University (1962).
4. J. A. Carruthers, J. F. Cochran, and K. Mendelssohn, *Cryogenics* **2**, 160, 1962.
5. J. C. Thompson and B. A. Younglove, *J. Phys. Chem. Solids* **20**, 146, 1961.
6. T. H. Geballe, *J. Appl. Phys.* **30**, 1153, 1959.
7. L. J. Challis, J. D. N. Cheeke, and J. B. Harness, *Phil. Mag.* **7**, 1941, 1962.
8. M. G. Holland, *Proceedings of the International Conference on the Physics of Semiconductors*, Academic Press, Paris, 1964, p. 713.
9. M. G. Holland, *Phys. Rev.* **134**, A471, 1964.
10. L. J. Challis, *Cryogenics* **2**, 23, 1961.
11. H. B. G. Casimir, *Physica* **5**, 495, 1938.
12. R. Berman, F. E. Simon, and J. M. Ziman, *Proc. Roy. Soc. (London) Ser. A* **220**, 171, 1953.
13. R. Berman, E. L. Foster, and J. M. Ziman, *Proc. Roy. Soc. (London) Ser. A* **231**, 130, 1955.
14. H. Montgomery, *Proc. Roy. Soc. (London) Ser. A* **244**, 85, 1958.
15. J. M. Ziman, *Phil. Mag.* **1**, 191, 1956; **2**, 292, 1957.
16. R. W. Keyes, *Phys. Rev.* **122**, 1171, 1961.
17. A. Griffin and P. Carruthers, *Phys. Rev.* **131**, 1976, 1963.
18. I. C. Pyle, *Phil. Mag.* **6**, 609, 1961.
19. R. F. Broom and A. C. Rose-Innes, *Proc. Phys. Soc. (London) Ser. B* **69**, 1269, 1956.

GRÜNEISEN γ VS. TEMPERATURE FROM ELASTIC COEFFICIENTS

K. Brugger

*Bell Telephone Laboratories, Inc.
Murray Hill, New Jersey*

Anharmonic effects in solids are usually expressed in terms of the Grüneisen constant γ, a measure of the volume dependence of the phonon frequencies. In the anisotropic Debye model,[1] where the lattice vibrations are replaced by the standing wave modes of a dispersionless continuum, γ becomes a function of elastic parameters and of temperature only:

$$\gamma(T) = \frac{\sum_p \int \gamma_p(T, \theta, \varphi) D_p(T, \theta, \varphi) \, d\Omega}{\sum_p \int D_p(T, \theta, \varphi) \, d\Omega} \qquad (1)$$

The index p and the angles θ and φ specify the polarization and the direction of propagation of the wave modes. The γ_p's can be calculated from second and third order elastic coefficients.[2] The functions D are given by

$$D_p(T, \theta, \varphi) = \left[\frac{T}{\Theta_p(\theta, \varphi)}\right]^3 \int_0^{\Theta_p(\theta, \varphi)/T} x^4 e^x (e^x - 1)^{-2} \, dx \qquad (2)$$

and the characteristic temperatures Θ are related to the sound speeds S for each mode by

$$\Theta_p(\theta, \varphi) = \frac{\hbar}{k} \left(\frac{6\pi^2}{V_0}\right)^{\frac{1}{3}} S_p(\theta, \varphi) \qquad (3)$$

where V_0 is the volume of the primitive unit cell. In the T^3-region for specific heats

$$\gamma(T \ll \Theta) = \frac{\sum_p \int \gamma_p S_p^{-3} \, d\Omega}{\sum_p \int S_p^{-3} \, d\Omega} \qquad (4)$$

and at high temperatures, where all D functions approach a common limit,

$$\gamma(T \gg \Theta) = \frac{1}{3N} \int \gamma_p \, d\Omega \qquad (5)$$

where N is the number of primitive unit cells in the crystal. Sheard[3] evaluated these limits for several substances and found fair agreement with experimental values.

Table 1. γ Values for Selected Modes in Cubic Crystals

Mode*	Number of modes	$-\gamma$†
αL	3	$\dfrac{1}{6c_{11}}[3c_{11} + 2c_{12} + C_1]$
$\alpha T, \gamma T''$	12	$\dfrac{1}{6c_{44}}[c_{11} + 2c_{12} + 2c_{44} + C_2]$
βL	4	$\dfrac{1}{6(c_{11} + 2c_{12} + 4c_{44})}[5c_{11} + 10c_{12} + 8c_{44} + 4(C_1 + C_2 - C_3) - C_1]$
βT	8	$\dfrac{1}{6(c_{11} - c_{12} + c_{44})}[5c_{11} + 4c_{12} + 2c_{44} + C_2 + 2C_3]$
γL	6	$\dfrac{1}{3(c_{11} + c_{12} + 2c_{44})}[2c_{11} + 3c_{12} + 2c_{44} + C_1 + C_2 - C_3]$
$\gamma T'$	6	$\dfrac{1}{3(c_{11} - c_{12})}[2c_{11} + c_{12} + C_3]$

* α—cube edges, β—body diagonals, γ—face diagonals, L—longitudinal, T—transverse, T'—along face diagonal, and T"—along cube edge.
† $C_1 = c_{111} + 2c_{112}$; $C_2 = c_{144} + 2c_{155}$; $C_3 = \frac{1}{2}(c_{111} - c_{123})$.

What information on the γ vs. T curves can we obtain from the Debye model in the intermediate range where the largest variations occur?* To avoid a complicated computer program for the evaluation of equation (1), two crude approximations are introduced: (1) Only room-temperature values of the elastic coefficients are used; (2) the D functions, shaped like specific heat curves, are replaced by step functions

$$d_p(T) = 0 \quad \text{for} \quad T < 0.248\,\Theta_p$$
$$d_p(T) = D(\infty) \quad \text{for} \quad T > 0.248\,\Theta_p \tag{6}$$

* After completion of this work, we became aware of a related treatment by J. G. Collins, *Phil. Mag.* **8**, 323, 1963.

Table II. Values and Characteristic Temperatures

Mode	Germanium		Silicon		KCl	
	γ	Θ	γ	Θ	γ	Θ
αL	1.36	412	1.11	725	2.35	342
$\alpha T, \gamma T''$	0.61	296	0.32	503	−0.48	134
βL	1.30	465	1.06	805	1.26	276
βT	0.36	254	0.12	510	2.11	196
γL	1.28	452	1.09	786	1.63	294
$\gamma T'$	0.17	230	−0.05	402	2.59	220

in which the step occurs where $D(T) = \tfrac{1}{2}D(\infty)$; and (3) instead of integrating over the solid angle, one sums over a number of discrete directions assigning the same weight to each. $\gamma(T)$ can then be written as

$$\gamma(T) = \frac{1}{N'} \sideset{}{'}\sum_{p,n} \gamma_p(n) \qquad (7)$$

where n is the index for the directions. The prime over the sum implies the prescription that terms for which $d_p(T) = 0$ are omitted. N' is the number of terms in the sum.

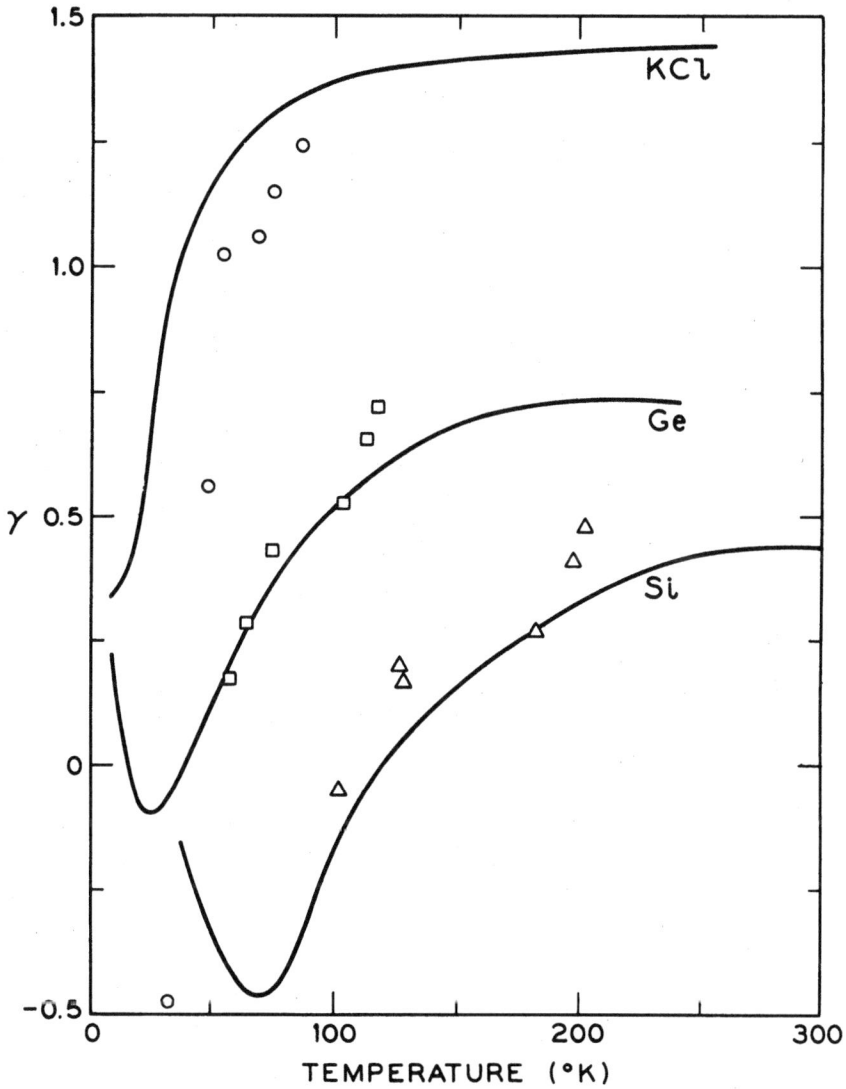

Fig. 1. Grüneisen γ vs. temperature for germanium, silicon, and KCl. Dots calculated from elastic data. Experimental curves after Gibbons[6] and White.[7]

We evaluate equation (7) for germanium, silicon and KCl. These are among the few substances for which values of the second and third order elastic coefficients and experimental γ curves are available. All three are cubic. The cube edges, the face and body diagonals, represent a suitable set of directions along which the $\gamma_p(n)$ are evaluated (Table I). The numerical γ values and the characteristic temperatures for these modes are listed in Table II. They are calculated from the latest sound speed measurements under pressure by McSkimin and Andreatch[4] for germanium and silicon, and from older measurements by Lazarus[5] for KCl. These data determine six points of the γ vs. T curves for each substance. They are shown in Fig. 1, which also contains the experimental curves. The calculation of the $\gamma_p(n)$ values involves differences of numbers of the same order of magnitude, amplifying experimental errors. The very low-lying point at 33°K for KCl is to be blamed on this. The agreement between calculated points and experimental curves is quite good, considering the crudeness of the approximations. In particular, the region of the main ascent is well-depicted. A more painstaking evaluation should yield very useful results, especially if also low-temperature elastic data are available.

References

1. J. M. Ziman, *Electrons and Phonons*, Clarendon Press (Oxford University Press, Fair Lawn, New Jersey) (1960).
2. K. Brugger, *Phys. Rev.* **137**, A1826, 1965.
3. F. W. Sheard, *Phil. Mag.* **3**, 1381, 1958.
4. H. J. McSkimin and P. Andreatch, *J. Appl. Phys.* **35**, 3312, 1964.
5. D. Lazarus, *Phys. Rev.* **76**, 545, 1949.
6. D. F. Gibbons, *Phys. Rev.* **112**, 136, 1958.
7. G. K. White, *J. Australian Inst. Metals* **8**, 134, 1963.

THE SCATTERING OF PHONONS BY SMALL-ANGLE GRAIN BOUNDARIES

K. A. McCarthy

Department of Physics, Tufts University
Medford, Massachusetts

A grain boundary formed by two crystallites whose misfit angle is less than 10° may be described[1] as an array of dislocations. In crystals such as sodium chloride, it is possible to introduce internal grain boundaries that consist almost entirely of edge dislocations. The theory[2] of thermal conductivity predicts that for alkali halide crystals, in a temperature region where only one scattering mechanism is present, the conductivity in crystals containing grain boundaries will vary as the cube of the temperature, whereas the conductivity in crystals containing random dislocations will vary as the square of the temperature. The latter dependence has been observed in lithium fluoride by Sproull, Moss, and Weinstock.[3]

The thermal conductivity of sodium chloride crystals containing as many as 200 grain boundaries/cm has been measured in a temperature region from 1.4–77°K. The grain boundaries were introduced into Optovac crystals by bending cleaved crystals (0.8 × 0.8 × 6.0 cm) in flexure about an [001] axis. Slight wetting of the crystal surfaces during deformation minimized the propagation of surface cracks. The plastically-deformed crystals were heated to a temperature of approximately 760°C for two hours, and were then slowly cooled for about fifty hours. The crystals were observed to polygonize, and then to recrystallize with many small-angle boundaries nearly parallel to each other. Figure 1 is a diagram which represents a recrystallized sodium chloride crystal, and which contains sketches of the grain boundaries and shows the position of the polygonized regions in the upper left-hand part of the sketch. The grain boundaries lie in {110} planes. The slip direction in sodium chloride is [110], and thus the dislocations forming the grain boundaries are edge dislocations. The orientation of the grain boundaries relative to the crystalline sample is determined from etch pits on the surface, one of which is sketched in the upper left-hand corner of Fig. 1.

The experimental data of the thermal conductivity K versus temperature for the deformed sodium chloride crystal in a temperature region where the grain boundaries would be the only scattering mechanism, give the relationship, $K = 2.29 \times 10^{-3} T^3$ W/cm-deg. This temperature dependence is consistent with that of scattering by grain boundaries, rather than that by dislocations.

Klemens has treated the case of scattering by randomly oriented boundaries, and has derived the relationship[4]

$$K_G = \frac{144\pi k^4 T^3}{\varepsilon_G \gamma^2 \theta^2 v^2 h^3} \int_0^\infty \frac{x^4 e^x \, dx}{(e^x - 1)^2}$$

where ε_G is the number of grain boundaries per unit length, γ is Grüneisen's constant, and θ is the misfit angle. The misfit angle of the boundaries in the deformed crystal

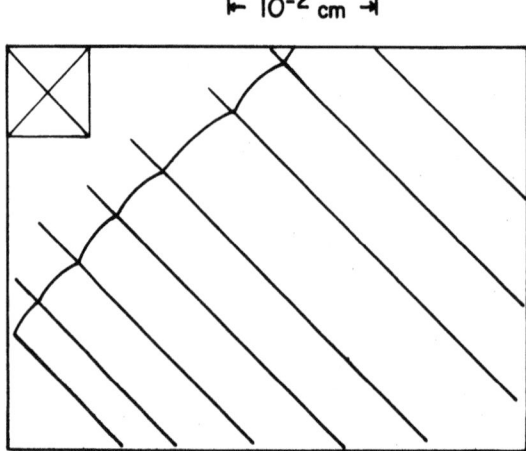

Fig. 1. A diagram which represents the etched surface of a sodium chloride crystal deformed in flexure and annealed at 760°C. The resulting grain boundaries are sketched, and the segments of the polygonized region are shown in the upper left-hand region.

is approximately 0.3°, and the number of boundaries/cm is approximately 250. These results are quite inconsistent with Klemens' calculations, since approximately 10^8 boundaries/cm would be needed to satisfy the theory of scattering by random boundaries.

Casimir[5] considered that, in scattering by external boundaries, the effective value of the mean free path was approximately that of the smallest dimension of the crystal and was independent of the phonon frequency. The conductivity K_B due to the boundary is expressed as

$$K_B = \frac{4\pi k^4 L T^3}{v^3 h^3} \int_0^\infty \frac{x^4 e^x \, dx}{(e^x - 1)^2}$$

In the deformed crystal, if an analysis similar to that of Casimir is applied to the experimental results, the mean free path equals 8.2×10^{-3} cm, and the conductivity is independent of the phonon frequency. The spacing between the grain boundaries equals the reciprocal of 250/cm or 4×10^{-3} cm. With the order-of-magnitude agreement, it is reasonable to assume that the internal grain boundaries scatter phonons in a manner similar to external grain boundaries in a single crystal. Closer agreement might be obtained if more were known of the surface of the grain boundary. Berman, Foster, and Ziman[6] have shown in sapphire single crystals that the mean free path for a polished external boundary is greater than for a sand-blasted external boundary; in the polished surface, the surface imperfections are small when compared with the phonon wavelength. In the recrystallized material, one would expect few imperfections in the boundary, and, thus, almost entirely specular reflection of the phonons.

Acknowledgments

The author wishes to thank W. J. Burke and R. A. Kashnow for their helpful discussions. This work was supported in part by the National Science Foundation.

References
1. W. Shockley and W. T. Read, *Phys. Rev.* **75**, 692, 1949.
2. P. G. Klemens, *Proc. Phys. Soc. (London) Ser. A* **68**, 1113, 1955.
3. R. L. Sproull, M. Moss, and H. Weinstock, *J. Appl. Phys.* **30**, 334, 1959.
4. See G. A. Slack, *Phys. Rev.* **105**, 837, 1957.
5. H. B. G. Casimir, *Physica* **5**, 495, 1938.
6. R. Berman, E. L. Foster, and J. M. Ziman, *Proc. Roy. Soc. (London) Ser. A* **231**, 130, 1955.

PHOTODIELECTRIC EFFECTS IN THALLOUS HALIDES AT LOW TEMPERATURES*

I. Lefkowitz † and A. D. Yoffe
Cavendish Laboratory
Cambridge, England

and

R. P. Lowndes and D. H. Martin
Physics Department, Queen Mary College
London, England

Introduction

Dielectric measurements on thallous halide single crystals have shown striking photoconductive and persistence effects at helium temperatures. The increases in the dielectric loss, expressed as the change in the AC conductance, are over six orders of magnitude, and the increases in the dielectric constant are up to factors of fifty. Measurements over a wide range of frequencies show a relaxation mechanism to be present. Optical measurements on thin thallous halide crystals at 4°K show the existence of exciton states whose dissociation temperatures lie close to the temperatures of the onset of the dielectric increases. The results are discussed in terms of polarizable centers and the dissociation and migration of excitrons.

The photodielectric effect is the increase in the dielectric constant of a material when it is excited by suitable radiation. Two principal theories have been put forward to explain the effect. One[1,2] suggests that the dielectric changes are due to photoconductivity and that the presence of free carriers in certain parts of the dielectric effectively reduces the dimensions of the material and, hence, causes the increase in the measured dielectric constant. The other theory[3] suggests that the dielectric-constant increases are caused by the presence in the excited material of a large number of polarizable centers. Measurements by Kronenberg and Accardo[4] indicate that both theories are correct and that some materials show photodielectric effects which are associated with photoconductivity, while other materials show a photodielectric effect which suggests the trapped-electron picture. Data on the photodielectric effect have previously been confined to phosphors of the ZnS:Cu type and powders of the ZnO:CdS type. Except for work on ZnO,[5] no large photodielectric effects have been reported in single crystals.[6] We have found large photodielectric effects in both single crystals and powders of the thallous halides, the changes occurring at much lower temperatures than all previous work and which appear to be due to different mechanisms at different temperatures.

Results and Discussion

The results given here are typical results obtained with TlBr single crystals and powder samples from four different sources. Results have been obtained for

---

* Supported in part by the United States Department of the Army through its European Research Office.

† Present address: Pitman-Dunn Institute, Frankford Arsenal, Philadelphia, Pennsylvania.

thallous chloride as well, but are not as complete as those for thallous bromide. The source used for excitation in all the measurements was a mercury discharge lamp used in conjunction with suitable filters. The dielectric measurements are accurate to 2% at low frequencies (approx. 100 kcps), but decreases in accuracy to about 5% were noted at higher frequencies.

For all unexcited samples, no dielectric anomaly was found from room temperature down to 1.5°K, the value of the low-frequency dielectric constant at 4°, 18°, and 295°K being 33.1, 33.0, and 29.6, respectively. The conductance of the specimens in the dark was measured to be less than 10^{-11} ohm^{-1}. For excited specimens, the dielectric constant and loss increased rapidly at low temperatures reaching peak values around 15°K. The dielectric constant showed increases up to fifty times the dark value, while the dielectric loss, expressed as the equivalent AC conductance, changed from less than 10^{-11} to 10^{-5} ohm^{-1}. Figure 1 shows a

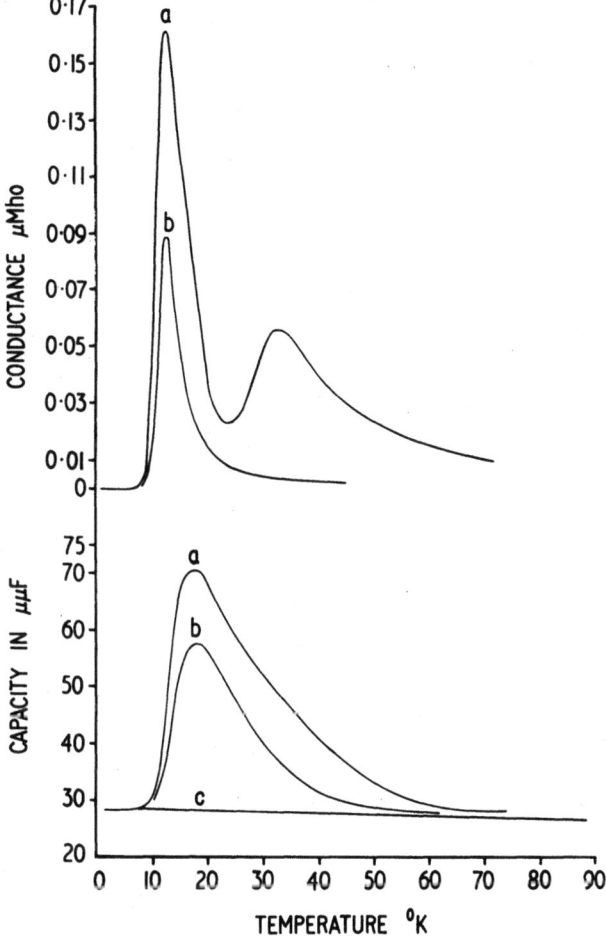

Fig. 1. Variation of the capacity and conductance of a TlBr crystal under different conditions. Curve (a) shows the changes for a continuously-illuminated crystal; curve (b) is for a crystal illuminated at 4°K, and then warmed in the dark; curve (c) is for a nonilluminated crystal.

typical variation in the capacity and conductance with temperature for an excited TlBr single crystal under different conditions. Curve (a) is for a continuously excited specimen and shows a peak in the dielectric constant at 18°K and peaks in the conductance at 15° and 36°K; a further peak not shown in the diagram is also found for the conductance at 120°K. Curve (b) is for a specimen excited at 4°K and then warmed at a steady rate in the dark and shows the peaks in the dielectric constant and dielectric loss at 18° and 15°K, respectively, but no further peaks at higher temperatures in the dielectric loss. Measurements on powder samples of TlBr show the dielectric constant and loss peaks at 18° and 15°K, but not other dielectric loss peaks for either continuously-excited or current-glow measurements.

Measurements made as a function of applied field frequency in the range 100 cps to 5 Mcps at fixed temperatures show the capacity changes to increase with decreasing frequency and the conductance to increase with increasing frequency.

When both single-crystal and powder samples are excited at a fixed temperature to an equilibrium state, the dielectric changes saturate for quite moderate intensities of illumination. After cessation of excitation, the dielectric changes drop to about 80% of their excited value, and then persist for considerable periods of time for measurements at 4°K; measurements at liquid-air temperatures show the dielectric changes to persist for times of the order of seconds only.

The persistance of the dielectric changes at low temperatures and their appearance in the current-glow measurements appear to indicate that they are due to trapped electrons. The electrons produced by the excitation have moved through the conduction band and have then been trapped in some metastable state just below the lowest energy state of the conduction band. The trap depth associated with the dielectric changes at 15°K has been determined to be 0.03 eV. Such a small binding energy for the trapped electrons means that they would be displaced by large amounts from their mean positions in the presence of an applied field, the displacement leading to a polarization of the electron trap. This polarization would lead to increases in the dielectric constant and dielectric loss, if the displacement motion was out of phase with the applied field variation. One would expect the system to relax as the applied field frequency approaches the natural vibrational frequency of the electron trap and the preliminary measurements indicate a relaxation frequency of $10^{5\pm1}$ cps.

The dielectric changes at higher temperatures have shown no persistence and would appear to be due to photoconductivity. It is interesting to note that recent low-temperature optical measurements[7] on thin single crystals of TlBr show a series of delocalized exciton states with thermal binding energies corresponding to dissociation temperatures of 130° and 40°K, which might be correlated with our conductance peaks at 120° and 36°K. To explain the dielectric constant increases, however, would require an exciton diffusion of several millimeters, which is unlikely in our crystals, although migrations of this order have been found in other crystals.[8,9]

Conclusion

The thallous halides have shown photodielectric effects at different temperatures which appear to be due to different mechanisms. Further investigations are being carried out and will be reported later.

References

1. H. Kallman, B. Kramer, and A. Perlmutter, *Phys. Rev.* **89**, 700, 1953.
2. H. Kallman, B. Kramer, and F. Mark, *Phys. Rev.* **99**, 1328, 1955.
3. G. F. J. Garlick and A. F. Gibson, *Proc. Roy. Soc. (London) Ser. A* **188**, 485, 1947.
4. S. Kronenberg and C. A. Accardo, *Phys. Rev.* **101**, 989, 1956.
5. J. Rous, *J. Phys. Radium* **15**, 176, 1954.
6. G. F. J. Garlick, *J. Appl. Phys. Suppl.* **4**, 585, 1955.
7. I. Lefkowitz, R. P. Lowndes, and A. D. Yoffe, (article in preparation).
8. M. Balkanski and R. D. Waldron, *Phys. Rev.* **112**, 123, 1958.
9. G. Diemer, G. J. van Gurp, and W. Hoogenstraaten, *Philips Res. Rept.* **13**, 458, 1958.

THE STATIC DIELECTRIC CONSTANT OF ALKALI HALIDES AT LOW TEMPERATURES*

M. C. Robinson† and A. C. Hollis Hallett

*Department of Physics, University of Toronto
Toronto, Ontario, Canada*

The static dielectric constant, ε_s, of single crystals of NaCl, KCl, and KBr have been measured between 4.2° and 300°K. The specimens were in the form of disks, 25 mm in diameter and 5 mm thick, on the faces of which thin, metallic films were evaporated to form a parallel-plate capacitor. The film on one face was scored to produce a guard-ring, and the temperature and capacitance of the system were measured as the system was allowed to warm slowly from 4.2°K. The capacitance was measured by means of a transformer bridge[1] operated at 1592 cps, the sensitivity of which was such that relative changes of 1×10^{-6} in ε_s could be detected. The absolute error in ε_s was estimated to be 0.5%; the relative error was approx. 0.03%.

The variation of ε_s of NaCl [111] (i.e., with the electric field along the [111] direction) with temperature is shown in Fig. 1. The shape of the curve is typical of that found for all the crystals examined; the temperature at which the minimum occurred, T_m, was different for each crystal, as shown in Table I. The room-temperature values of ε_s agree with those of Haussühl[2] to within 1% but vary from the older, but usually accepted, values[3] by as much as 5%. The values of $(1/\varepsilon_s)(\partial \varepsilon_s/\partial T)_p$ at 285°K agree with those of Bosman and Havinga[4] to within 2%.

The difference between the static and optical dielectric constant, $(\varepsilon_s - \varepsilon_\infty)$, is given by the relation[5]

$$\varepsilon_s - \varepsilon_\infty = \left(\frac{\varepsilon_\infty + 2}{3}\right)^2 \frac{4\pi N(e^*)^2}{\bar{M}\omega_0^2} + G \tag{1}$$

where ω_0 is the frequency of the long-wave lattice vibrations in the transverse optical mode, N the number of ion pairs with reduced mass $\bar{M}$ per unit volume, and e^* is the effective ionic charge. G is a correction term due to the anharmonicity of the lattice vibrations; $G/(\varepsilon_s - \varepsilon_\infty)$ varies from about 2% for NaCl to 7% for KCl at room temperature.[6] Making reasonable estimates of ε_∞ and G at 80°K it is found that the observed decrease in ε_s from 300° to 80°K is mainly due to the increase[7] in ω_0; e^*/e (e is the electron charge) appears to decrease by about 2% from a value approx. 0.8 at room temperature, as suggested by theory.[8]

However, at lower temperatures, the results are in complete disagreement with existing theory. If ε_s is considered to be a function of T and of volume $V(T, p)$, then

$$\left(\frac{\partial \varepsilon_s}{\partial T}\right)_p = \left(\frac{\partial \varepsilon_s}{\partial V}\right)_T \left(\frac{\partial V}{\partial T}\right)_p + \left(\frac{\partial \varepsilon_s}{\partial T}\right)_V = \frac{\alpha}{\beta}\left(\frac{\partial \varepsilon_s}{\partial p}\right)_T + \left(\frac{\partial \varepsilon_s}{\partial T}\right)_V \tag{2}$$

---

* Research supported by the National Research Council of Canada, the Ontario Research Foundation, and the University of Toronto.

† Present address: Department of Physics, McGill University, Montreal, Quebec, Canada.

The Static Dielectric Constant of Alkali Halides at Low Temperatures

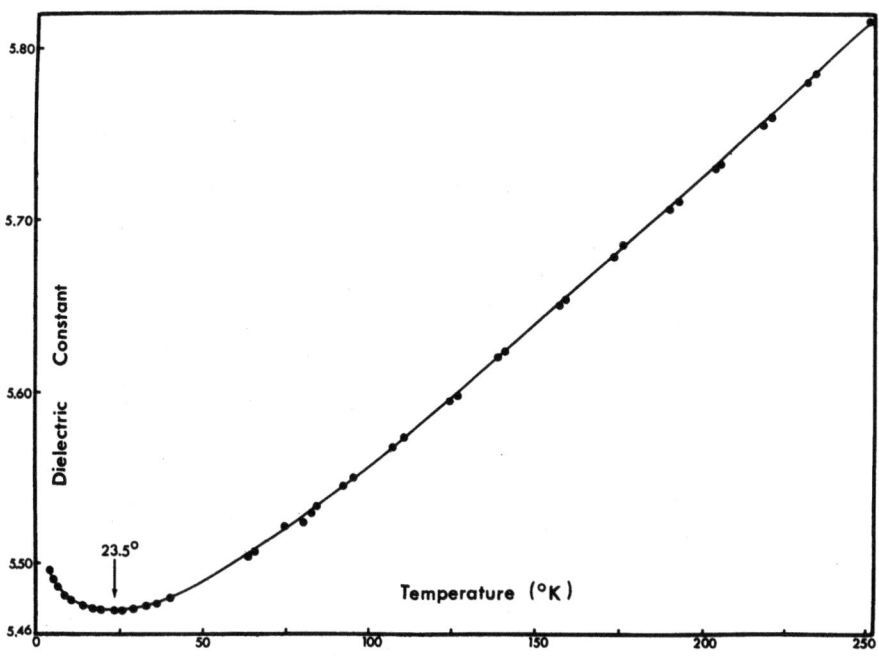

Fig. 1. The variation with temperature of the dielectric constant of NaCl with the electric field parallel to the [111] direction in the crystal.

where β is the compressibility and α the coefficient of volume expansion. At room temperature[4,6] $(\alpha/\beta)(\partial \varepsilon_s/\partial p)_T$ is greater than $\partial \varepsilon_s/\partial T)_V$; however, at 4.2°K, α is smaller by a factor[9] of about 4000, while measurements show that $(\partial \varepsilon_s/\partial p)_T$ is essentially unaltered in magnitude. Therefore, since for NaCl and KCl $(\partial \varepsilon_s/\partial T)_p$ is larger at 4.2°K, it follows that $(\partial \varepsilon_s/\partial T)_V = (\partial \varepsilon_s/\partial T)_p$ to a very good approximation at low temperatures. Furthermore, Szigeti[6] has shown that $(\partial \varepsilon_s/\partial T)_V = (\partial G/\partial T)_V = (\partial G/\partial T)_p$. An analysis of the expression[6] for G clearly shows that, at 4.2°K, $(\partial G/\partial T)_p$ is negligible and cannot possibly account for the relatively large values of $(\partial \varepsilon_s/\partial T)_p$

Table I

Representative values of the static dielectric constant ε_s, the temperature derivative $(1/\varepsilon_s)(\partial \varepsilon_s/\partial T)_p$, and the temperatures T_m at which the minimum is observed.

Crystal	T_m, °K	ε_s (300°K)	$\varepsilon_s(T)/\varepsilon_s$ (300°K)			$10^4(1/\varepsilon_s)(\partial \varepsilon_s/\partial T)_p$ (°K)$^{-1}$		
			4.2°K	T_m °K	80°K	4°K	80°K	285°K
NaCl [100]	25.4 ± 0.2	5.9171	0.9310	0.9250	0.9343	−20	2.6*	3.18*
NaCl [110]	25.9 ± 0.2	5.8881	0.9340	0.9273	0.9363	−22		
NaCl [111]	23.5 ± 0.2	5.9090	0.9301	0.9263	0.9352	−13		
KCl [100]	19.2 ± 0.2	4.8108	0.9345	0.9321	0.9419	−9	2.4	2.90
KBr [100]	7.8 ± 1.0	4.8808	0.92745	0.92744	0.9400	—	2.5	3.13

* Mean.

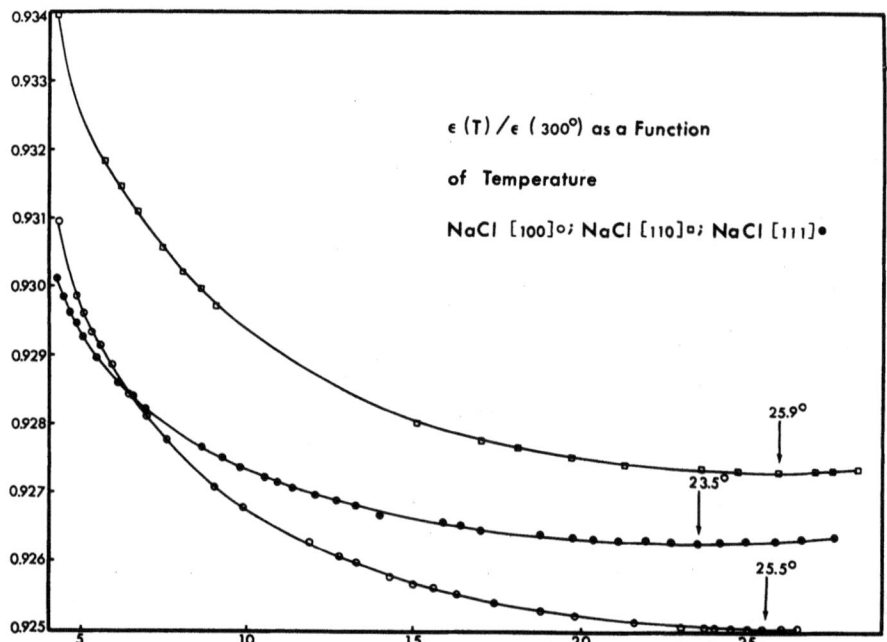

Fig. 2. The variation with temperature at low temperatures of the dielectric constant of NaCl with the electric field parallel to different crystal directions. The positions of the minima in the curves are indicated by arrows.

or cause it to be negative. The results obtained between 4.2° and 10°K fit quite well an empirical equation of the form $\varepsilon_s = K_1 + K_2/T$, where K_1 and K_2 are constants. This form suggests the presence of impurities, such as OH, with a permanent dipole moment which is free to rotate in the lattice. However, the infrared absorption spectrum of the crystals failed to indicate even one-twentieth of the amount of OH necessary to explain the magnitude of K_2 required to fit the data.

The detailed variation of ε_s with temperature between 4.2°–30°K is shown in Fig. 2 for each of the three NaCl crystals. It is clear from this figure (and from the values in Table I) that the differences between the three curves are much greater than experimental error. The apparent anisotropy might be due to noncentral forces between the ions of the crystal, but the results of further experiments with other alkali halides are necessary before definite conclusions can be drawn.

References

1. M. C. McGregor, J. F. Hersch, R. D. Cutkosky, F. K. Harris, and F. R. Kotter, *Trans. Inst. Radio Engr.* **1**(7), 253, 1958.
2. S. Haussühl, *Z. Naturforschung* **12A**, 445, 1957.
3. M. Born and K. Huang, *Dynamical Theory of Crystal Lattices*, Oxford University Press, Oxford, England (1956).
4. A. J. Bosman and E. E. Havinga, *Phys. Rev.* **129**, 1593, 1963.
5. B. Szigeti, *Trans. Faraday Soc.* **45**, 155, 1949; *Proc. Roy. Soc. (London), Ser A* **252**, 217, 1959.
6. B. Szigeti, *Proc. Roy. Soc. (London) Ser. A* **261**, 274, 1961.
7. M. Hass, *Phys. Rev.* **119**, 633, 1960; A. D. B. Woods, B. N. Brockhouse, R. N. Cowley, and W. Cochran, *Phys. Rev.* **131**, 1025, 1963.
8. B. G. Dick, Jr. and A. W. Overhauser, *Phys. Rev.* **112**, 90, 1958.
9. P. P. M. Meincke and G. M. Graham, *Proceedings of the Eighth International Conference on Low-Temperature Physics*, Butterworths, London (1963), p. 401.

THE THERMAL EXPANSION AND RELATED PROPERTIES OF RUBIDIUM BROMIDE AT LOW TEMPERATURES

B. W. James* and B. Yates*

*Department of Physics, Bradford Institute of Technology
Bradford, United Kingdom*

The linear thermal expansion coefficient α of rubidium bromide has been measured in the range 20°–273°K as part of an investigation of the thermal expansion of alkali halides with a rock-salt structure, consisting of ions of comparable mass. The single-crystal specimen, obtained from L. Light & Co. Ltd., with an impurity content of the order of 50 ppm, consisted of a hollow cylinder, 2.7 cm OD, 1.8 cm ID, and 1 cm high. This was investigated using an improved version of the interferometric technique described by Yates and Panter,[1] employing photographic recording, and yielding results with an average scatter of 0.45% in α, which are believed to be accurate in an absolute sense to within 2% at all temperatures.

The experimental results are shown graphically in Fig. 1, and the smoothed values are given in Table I, along with values of the Grüneisen parameter defined by

$$\gamma = 3\alpha V / C_p \chi_s$$

where V is the gram molecular volume (taking $\rho_{293} = 3.35_3$ g/cm³), C_p is the gram molecular heat[2] and χ_s is the adiabatic compressibility. The accuracy of these values of γ is estimated to decrease from 9% at 20°K to 4% at 270°K. In the absence of experimental compressibility data below room temperature, values of χ_s required for the calculation of γ were obtained from a linear interpolation between the experimental bulk modulus of Reinitz[3] at 295°K and the value at 0°K calculated from the theoretical results of Krishnan and Roy.[4] After using the data to calculate C_v from C_p, using the thermodynamic relationship

$$C_v = C_p \left[1 - \frac{1}{1 + (\chi_s C_p / 9V\alpha^2 T)} \right]$$

values of θ_D corresponding to values of C_v in the approximate range 10–30°K were evaluated, and a graph of θ_D^2 versus $1/T^2$ was extrapolated to $1/T^2 = 0$ to derive $\theta_\infty = 140 \pm 2$°K. $\theta_0 = 147 \pm 7$°K was obtained by using the expansion data from the present investigation and the theoretical values of c_{11}, c_{12} and c_{44} at 0°K in applying the tables of de Launay.[5] Following the procedure described by Barron, Leadbetter, and Morrison,[6] a parameter $T_r = 0.2\theta_\infty$ was chosen in plotting γ against $t = [1 + (T/T_r)^2]^{-1}$ and extrapolating to $t = 0$ to give γ_ω. Although the variations of C_v and α with temperature displayed no obvious peculiarities, the graph of γ against t turned upwards, with t diminishing at the higher temperature

---
* Present address: The Department of Pure and Applied Physics, Royal College of Advanced Technology, Salford, Lancashire, United Kingdom.

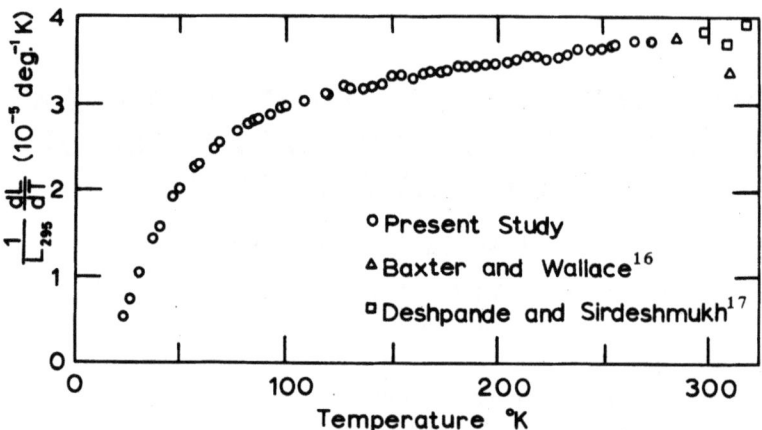

Fig. 1. Linear thermal expansion coefficient of RbBr.

end. This behavior may not be unique, since the less precise results[1] for KI ($\theta_D \doteq 130°K$) suggest a similar trend. Analysis of the results for RbBr between 30–180°K gives a value for γ_∞ of 1.40, while the results between 230–270°K give a value 1.53. This lies within the range 1.50–1.76 predicted by Blackman[7] on the basis of the point-charge ion model, in which equality of masses of the ions was assumed. Extrapolation of a graph of γ against T to $T = 0$ yields a value of γ_0 somewhere between 0.56 and 0.60. This is well below the values predicted by the point-charge ion model, and, in the absence of experimental data on the pressure variation of the elastic moduli, a comparison with the predictions of the anisotropic continuum model of Sheard[8] cannot be made.

The moment $\overline{v^2}$ of the frequency distribution was calculated from

$$\theta_\infty = (h/k)(5\overline{v^2}/3)^{1/2}$$

The coefficients of the Thirring expansion

$$\theta_D^2 = \theta_\infty^2 [1 - A(\theta_\infty/T)^2 + B(\theta_\infty/T)^4 - \ldots]$$

Table I. Linear Thermal Expansion Coefficient and Grüneisen Parameter for RbBr (Smoothed Values)

T (°K)	α (deg$^{-1}\cdot 10^6$)	γ	T (°K)	α (deg$^{-1}\cdot 10^6$)	γ
10	(0.70)	(0.61)	140	32.33	1.35
20	3.48	0.71	160	33.23	1.36
30	9.48	0.95	180	33.99	1.37
40	15.72	1.10	200	34.72	1.39
50	20.23	1.18	220	35.42	1.40
60	23.15	1.22	240	36.10	1.42
80	27.20	1.29	260	36.70	1.44
100	29.67	1.32	270	36.98	1.45
120	31.23	1.34			

were obtained from the intercept and slope of a graph of

$$[1 - (\theta_D/\theta_\infty)^2]/(\theta_\infty/T)^2$$

against $(\theta_\infty/T)^2$, the range $2.5 < (\theta_\infty/T) < 6$ being chosen when drawing the tangent, the range in which Barron, Berg, and Morrison[9] conclude that the anharmonicity of the vibrations is still small. These workers describe the procedure by which the other even moments $\overline{v^4}$ and $\overline{v^6}$ were then calculated from A, B, and θ_∞. The moments $\overline{v^{-1}}$ and $\overline{v^{-2}}$ were obtained from the expression given by Hwang[10]

$$\int_0^\infty (C_v/T^n)\,dT = 3Nk \cdot \Gamma(n+1) \cdot \zeta(n) \cdot (h/k)^{1-n} \overline{v^{1-n}}$$

where $\Gamma(n+1)$ is the gamma function, $\zeta(n)$ is the Riemann zeta function, and $(1 < n < 4)$. The evaluation of the integral between 0–10.5°K involved some estimation, in the absence of experimental specific-heat data in this region. After observing that the temperatures of the maxima in graphs of C_v/T^3 against T for KI, NaI, KBr, and KCl[11] were closely similar to the temperatures of the minima T_M in the corresponding graphs of θ_D against T, the temperature of this maximum in the case of RbBr was obtained by interpolation in a graph of T_M against θ_0 for these salts. The corresponding value of θ_D was then estimated by interpolation in a graph of θ_D, at T_M, against θ_0 for these salts, from which C_v/T^3 was calculated. The experimental evidence suggested a somewhat higher value, and a compromise was made in the drawing of the curve in the region of the maximum in the plot of C_v/T^3 against T for the integration. C_v/T^3 at $T = 0°K$ was taken as $929/\theta_0^3$ cal/mole/deg.[4]

The corresponding frequencies $v_D(n)$ were then calculated from

$$v_D(n) = \{[(n+3)\overline{v^n}]/3\}^{1/n}$$

where $n \neq 0$ and $v_D(n)$ is the maximum frequency of the Debye distribution with the same nth moment as the actual crystal. $v_D(0)$ was obtained from the geometric mean frequency v_g of the spectrum, which was calculated from the slope of a graph of T against $\exp[(S/3Nk) - 1]$ in the region of θ_∞, using an expression derived by Salter.[12] Meanwhile,

$$v_D(-3) = k\theta_0/h$$

The general features of the variation of $v_D(n)$ with n for RbBr, values of which are given in Table II, are qualitatively similar to those of other alkali halides,[9,13] the curve bending sharply below $n \doteq -0.5$, while the more gradual variation of $v_D(n)$

Table II. Frequencies of the Debye Distribution with the Same nth Moment as the Actual Crystal

n	$v_D(n)$ (sec$^{-1} \cdot 10^{12}$)	n	$v_D(n)$ (sec$^{-1} \cdot 10^{-12}$)
−3	3.06 ± 0.15	2	2.92 ± 0.03
−2	2.76 ± 0.06	4	3.20 ± 0.10
−1	2.68 ± 0.03	6	3.43 ± 0.17
0	2.66 ± 0.03		

with n above $n \doteq 0.5$ allows an interpolated value of $v_D(1)$ to be derived for the calculation of the zero-point energy E_z;

$$E_z = 3N h \overline{v^1}/2 = 587 \pm 6 \text{ cal/mole}$$

which is 6% lower than the value calculated from the approximation

$$E_z = 9Nk\theta_\infty/8$$

of Domb and Salter.[14]

Barron, Leadbetter, Morrison, and Salter[15] have indicated how the equivalent temperature $\theta^M(T)$ for the Debye–Waller effect may be calculated from thermodynamic data. The relationships

$$\theta_0^M = [hv_D(-1)]/k$$

$$\theta_\infty^M = [hv_D(-2)]/k$$

were substituted in the low- and high-temperature expansions given by these workers, to get the initial curvature near 0°K and the variation above about $0.3\theta_D$, respectively. The correction to the high-temperature results so calculated, arising from thermal expansion, yielded values of $\theta^M(T, V)$ from

$$\theta^M(T, V)/\theta^M(T, V_0) = (V_0/V)^{\gamma(-2)}$$

where

$$\gamma(-2) = \frac{\int_0^\infty \gamma C_v T^{-3} \, dT}{\int_0^\infty C_v T^{-3} \, dT}$$

These results, which are of use in the analysis of complex scattering processes, are given in Fig. 2. The estimated error in $\theta^M(T, V)$ at 0°K is $\pm 1\%$, increasing to approximately $\pm 2\%$ at 270°K.

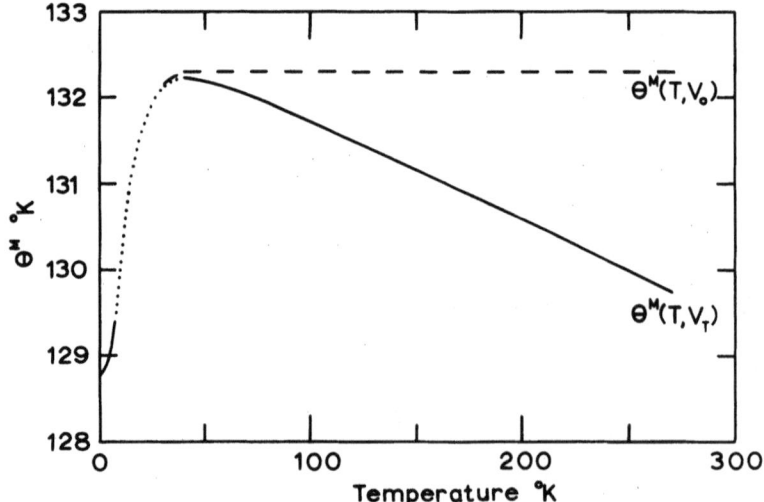

Fig. 2. The equivalent temperature for the Debye–Waller effect as a function of temperature. The solid curves refer to constant pressure, the dashed curve refers to constant volume (V at 0°K), and the dotted curve is interpolated.

Acknowledgments

We are indebted to Dr. R. K. Angus for valuable discussions, and to Drs. T. H. K. Barron, A. J. Leadbetter, and J. A. Morrison for access to their unpublished work.

References

1. B. Yates and C. H. Panter, *Proc. Phys. Soc. (London)* **80**, 373, 1962.
2. K. Clusius, J. Goldmann, and A. Perlick, *Z. Naturforschung* **4a**, 424, 1949.
3. K. Reinitz, *Phys. Rev.* **123**, 1615, 1961.
4. K. S. Krishnan and S. K. Roy, *Proc. Roy. Soc. (London) Ser. A* **210**, 481, 1952.
5. J. de Launay, *J. Chem. Phys.* **30**, 91, 1959.
6. T. H. K. Barron, A. J. Leadbetter, and J. A. Morrison, *Proc. Roy. Soc. (London) Ser. A* **279**, 62, 1964.
7. M. Blackman, *Proc. Phys. Soc. (London) Ser. B* **70**, 827, 1957.
8. F. W. Sheard, *Phil. Mag.* **3**, 1381, 1958.
9. T. H. K. Barron, W. T. Berg, and J. A. Morrison, *Proc. Roy. Soc. (London) Ser. A* **242**, 478, 1957.
10. J. L. Hwang, *J. Chem. Phys.* **22**, 154, 1954.
11. W. T. Berg and J. A. Morrison, *Proc. Roy. Soc. (London) Ser. A* **242**, 467, 1957.
12. L. Salter, *Proc. Roy. Soc. (London) Ser. A* **233**, 418, 1955.
13. T. H. K. Barron and J. A. Morrison, *Proc. Roy. Soc. (London) Ser. A* **256**, 427, 1960.
14. C. Domb and L. Salter, *Phil. Mag.* **43**, 1083, 1952.
15. T. H. K. Barron, A. J. Leadbetter, J. A. Morrison, and L. S. Salter, *Inelastic Scattering of Neutrons in Solids and Liquids*, Vol. 1, International Atomic Energy Agency, Vienna, (1963), p. 49.
16. G. P. Baxter and C. C. Wallace, *J. Am. Chem. Soc.* **38**, 259, 1916.
17. V. T. Deshpande and D. B. Sirdeshmukh, *Acta Cryst.* **14**, 353, 1961.

FAR-INFRARED IMPURITY MODES IN POTASSIUM IODIDE*

A. J. Sievers

*Laboratory of Atomic and Solid State Physics, Cornell University
Ithaca, New York*

Introduction

The spectrum of acoustic and optic lattice modes in potassium iodide is divided into two frequency regions.[1] In the perfect crystal, normal mode frequencies are not allowed in the intervening gap. Introducing impurities into the perfect lattice alters this description.[2] For different combinations of impurity mass and force constants, localized impurity modes are expected in the gap between the optic and acoustic branches, above the optic branch or in both regions. The amplitude associated with the localized impurity mode rapidly decreases with increasing distance from the defect.

An impurity mode above the optic branches was first observed for the hydride ion impurity in alkali halide crystals by Schaefer.[3] We have confined our investigation to the region below the transverse optic branch. A number of impurities have been studied in potassium iodide. Impurity-induced absorption has been observed both at frequencies located in the forbidden gap and also at frequencies in the acoustic spectrum.

We have used a grating monochromator with 12-in. optics to scan from 100 to 10 cm^{-1} (3×10^{12} to 3×10^{11} cps). At room temperature, absorption from the low-frequency wing of the T.O. mode in KI (T.O. = 101 cm^{-1} at 300°K) extends throughout much of the far-infrared region. This intrinsic absorption occurs via a second order process in which one photon creates one phonon and destroys another.[4] The absorption process depends on the number of thermally-excited phonons in the crystal and is "quenched" at low temperatures. With 1-cm thick potassium iodide crystals cooled to 2°K, transmission measurements have been readily obtained up to 85% of the T.O. mode frequency (T.O. = 108 cm^{-1} at 4.2°K).[5]

Halogen Ion Impurities

The impurities—hydride, fluoride, chloride, and bromide—have been introduced into KI by different methods.

The hydride ion or U-center impurity was formed by first additively coloring KI by the Van Doorn technique[6] and then heating the crystal in a hydrogen atmosphere. The final crystals were colorless. Measurement of the ultraviolet U-band in thin platelets gave some indication of the U-center concentration (about 10^{18} per cm^3). The far-infrared absorption spectrum measured with 1-cm thick crystals is shown in Fig. 1. A band composed of three broad absorptions is centered around 62 cm^{-1}. Photochemical conversion of U- to F-centers by ultraviolet irradiation is found to decrease the absorption by a factor of 2. The original

*Supported in part by the Advanced Research Projects Agency of the U.S. Atomic Energy Commission.

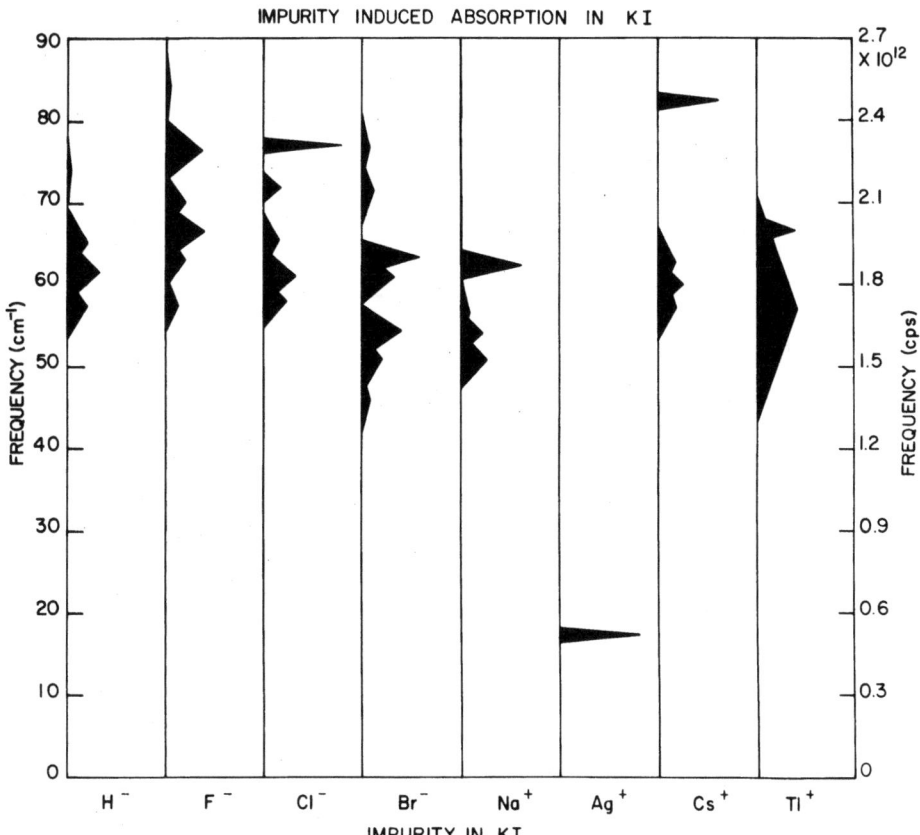

Fig. 1. Impurity-induced absorption in KI at 2°K. By comparing the doped crystal with the pure crystal, the intrinsic lattice absorption has been divided out. The absorption frequencies and relative line-widths are to be examined. The figure is somewhat pictorial in that the abscissa in each column is only approximately the induced absorption coefficient. For the sake of clarity, the strength of the sharp lines has been decreased with respect to the broad bands. The frequency region studied was 90–10 cm^{-1}. The resolution of the monochromator was usually set at $v/\Delta v = 36$.

far-infrared absorption strength is restored by heating the crystals. Both U-centers and OH$^-$-centers behave in this characteristic manner.[6,7] Although the presence of U-centers in the crystals has been definitely established, the concentration of OH$^-$ has not yet been determined. Thus, only a tentative assignment of the far-infrared absorption to the U-center is possible at this time.

One concentration of fluoride and one of bromide (both less than 0.5 mol. %) have been grown in air* by the Kyropoulos technique. A complex absorption spectrum has been found in the far-infrared for each system. To date, neither spectrum has been studied in detail, but our preliminary measurements are given in Fig. 1.

* Approximately 10^{17} OH$^-$ impurities per cm^3 are contained in our air-grown KI crystals. About five times this concentration was introduced into a KI crystal, and no far-infrared absorption was observed. We have concluded that the hydroxide impurities do not contribute to the observed absorption spectra of the air-grown crystals shown in Fig. 1.

Somewhat more detailed experiments have been carried out with KI:Cl⁻. A number of impurity concentrations have been grown by the Kyropoulos technique in an argon atmosphere.* The far-infrared absorption spectrum shown in Fig. 1 consists of a narrow line at 77 ± 1 cm^{-1} and a band composed of three broad lines centered at 61 cm^{-1}. From higher resolution measurements ($\nu/\Delta\nu \approx 80$), the absorption at 77 cm^{-1} has a width at one-half maximum absorption equal to 1% of the absorption frequency. The line intensity is found to vary linearly with

* These high-purity crystals were grown by the Crystal Growing Laboratory of the Material Science Center at Cornell University.

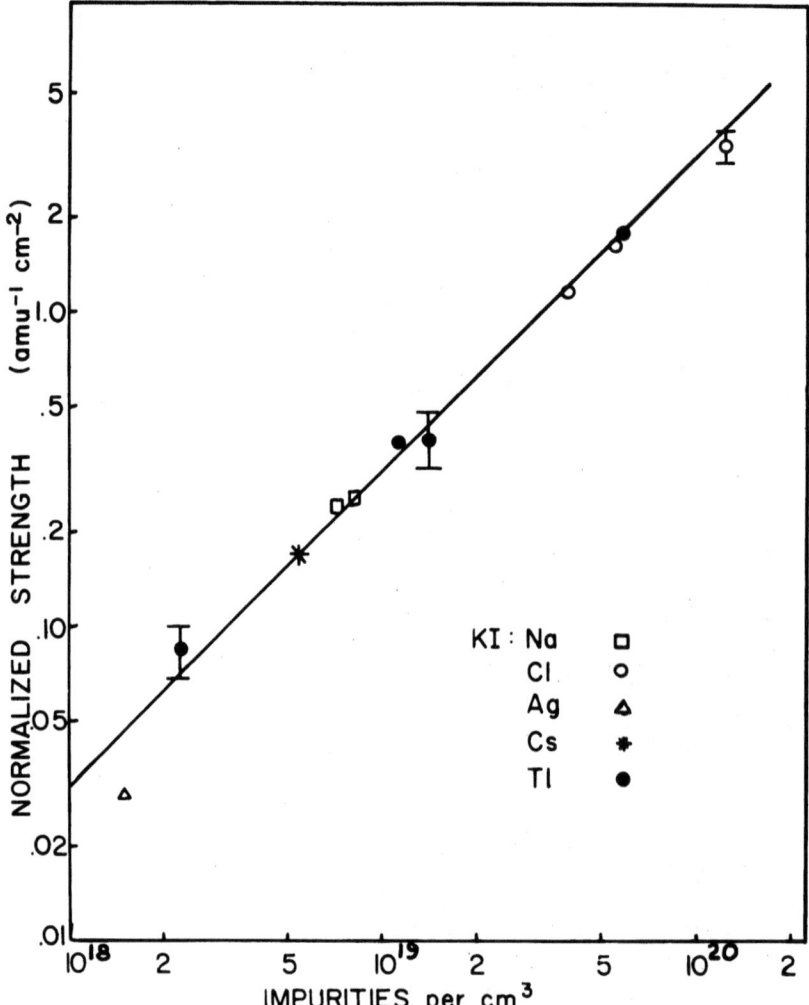

Fig. 2. Normalized absorption strength in doped KI versus impurity concentration. The absorption strength is the total area under an impurity-induced absorption curve. The normalization consists of dividing each strength by the atomic weight (in amu) of the impurity ion. The solid line has slope 1 and depicts a strength which varies linearly with impurity concentration. The estimated errors are typically about 20% of the given strengths.

impurity concentration; and also the total intensity, line plus band, shown in Fig. 2, varies linearly with impurity concentration.

Metal-Ion Impurities

Two sodium concentrations in KI have similar far-infrared absorption spectra: a narrow absorption line at 63.5 ± 1 cm^{-1}, and a broad band centered at 53 cm^{-1}, (see Fig. 1). For the narrow line, the full-line width at one-half maximum absorption is about 4% of the absorption frequency. Both of these crystals were grown in air* by the Kyropolous technique.

The far-infrared absorption spectrum of silver-doped potassium iodide† is somewhat different than found for other impurities. As shown in Fig. 1, only one low-frequency absorption has been observed. The line occurs at 17.5 ± 0.3 cm^{-1} with a width at half-maximum which is about 5% of the absorption frequency. This absorption frequency corresponds to one-sixth the Debye temperature and is well within the acoustic continuum.

The highest-frequency impurity mode in the forbidden gap has been observed in cesium-doped KI.* As shown in Fig. 1, a sharp line is observed at 83.5 ± 1 cm^{-1} and a broad band is centered at 60 cm^{-1}. The full width of the sharp line at one-half maximum absorption is 1% of the absorption frequency.

Four concentrations of KI:Tl$^+$ have been studied. Three crystals were grown at Cornell† and the fourth was obtained from Harshaw Chemical Company. The same spectrum was observed for all crystals. A sharp line is located at 64.5 ± 1 cm^{-1} and a broad band is centered at 55 cm^{-1}. The total strength varies linearly with thallium concentration, as shown in Fig. 2.

Discussion

The phonon-dispersion curves for KI have been calculated by Karo and Hardy.[1] The maximum frequency of the transverse acoustic branch occurs between 30 and 40 cm^{-1}. The cutoff for the longitudinal acoustic branch is between 55 and 65 cm^{-1}. For small impurity concentrations, most phonon modes should not be strongly perturbed.[2] From Fig. 1, the impurity-induced absorption occurs in three frequency regions: (1) The absorption associated with silver impurities occurs below the T.A. branch with no absorption observed at higher frequencies. (2) For all other impurities, absorption occurs above the T.A. branch, but within the L.A. branch. (3) For many of these same systems, absorption is observed in the frequency gap between the acoustic and optic branches. From Fig. 2, the sum of the strengths of both absorptions is the same order of magnitude as expected for an electric dipole transition. Also, the line and band intensities vary linearly with impurity concentration; hence, both absorptions probably stem from the same impurity center. Finally, the total, impurity-induced, integrated absorption appears to vary linearly with impurity mass.

An identification of the sharp lines observed in the gap with localized modes is compatible with our experimental results. The low-frequency absorption observed for silver impurities probably arises from an approximate localized mode in the acoustic continuum.[8,9] The large number of absorption lines observed in the L.A.-frequency region is not yet understood. From our low temperature measurements, the acoustic modes, as expected, are not infrared-active for pure crystals. Possibly, an impurity mode which rises out of the acoustic band and becomes infrared-active

---
* See footnote on p. 1171.
† See footnote on p. 1172.

leaves behind complementary, infrared-active, acoustic modes. Both modes then would contribute to the far-infrared absorption. We hope to resolve some of these questions by measuring far-infrared transmission in doped-alkali halides as a function of strain, temperature, and hydrostatic pressure.

Acknowledgments

We thank J. A. Krumhansl and R. O. Pohl for helpful discussions, and also C. D. Lytle for reducing much of the data.

References

1. A. M. Karo and J. R. Hardy, *Phys. Rev.* **129**, 2024, 1963.
2. A. A. Maradudin, E. W. Montroll, and G. H. Weiss, *Solid State Physics*, F. Seitz and D. Turnbull (eds.), Academic Press, Inc., New York (1963).
3. G. Schaefer, *J. Phys. Chem. Solids* **12**, 233, 1960.
4. E. Burstein, *Phonons and Phonon Interactions*, T. A. Bak (ed), W. A. Benjamin, Inc., New York (1964) p. 276.
5. G. O. Jones, D. H. Martin, P. A. Mawer, and C. H. Perry, *Proc. Roy. Soc.* (*London*) *Ser. A* **261**, 10, 1961.
6. J. H. Schulman and W. D. Compton, *Color Centers in Solids*, The Macmillan Company, New York (1962), p. 49.
7. J. Rolfe, *Phys. Rev. Letters* **1**, 56, 1958.
8. R. Brout and W. Visscher, *Phys. Rev. Letters* **9**, 54, 1962; W. M. Visscher, *Phys. Rev.* **129**, 28, 1963.
9. P. G. Dawber and R. J. Elliott, *Proc. Phys. Soc.* (*London*) *Ser. A* **81**, 453, 1963.

LINE WIDTH TRANSITIONS IN THE DEUTERON MAGNETIC RESONANCE OF POLYCRYSTALLINE ND₄Cl AND ND₄Br

V. Hovi and P. Pyykkö

Wihuri Physical Laboratory, University of Turku
Turku, Finland

The nuclear magnetic resonance in ND₄Cl and ND₄Br has been investigated previously by Chiba.[1] His paper gives for these salts in powder form the room-temperature line widths of the deuteron magnetic resonance. Furthermore, the deuteron-quadrupole coupling constant in the [ND₄]⁺ ion has been briefly discussed by Bersohn.[2] He predicts this constant to be greater than 200 kcps. Chiba's results indicate rapid reorientation of the [ND₄]⁺ ions at room temperature. Therefore, line width transitions can be expected to occur at lower temperatures. The object of the present investigation was to determine these transition temperatures and, in the same connection, to find the principal origin of the line broadening above the transition temperatures.

The line broadening caused by magnetic dipole–dipole interactions was calculated from the well-known Van Vleck second moment formula.[3] By assuming all the intraionic interactions to be averaged to zero, we have

$$\overline{(\Delta H^2)} = \frac{3}{5} g_D^2 \beta^2 I_D(I_D + 1) \sum_{k \atop \text{inter}} r_{jk}^{-6} + \frac{4}{15} g_X^2 \beta^2 I_X(I_X + 1) \sum_{k'} r_{jk'}^{-6} \qquad (1)$$

where β is the nuclear magneton, and g_D, g_X, I_D, and I_X are the g factors and the spin quantum numbers of the deuteron and of the halogen nuclei, respectively. The first term in equation (1) represents the contribution caused by the interionic deuteron–deuteron interactions, and the second one that by the deuteron–halogen interactions. The third contribution, caused by the interionic deuteron–nitrogen interactions, is found to be less than one percent of the total second moment, and can be neglected. The chlorine and bromine nuclei, occurring in nature, have two isotopes with slightly different g factors. Thus, the factors g_{Cl}^2 and g_{Br}^2 were approximated by $p_{Cl^{35}} g_{Cl^{35}}^2 + p_{Cl^{37}} g_{Cl^{37}}^2$ and by $p_{Br^{79}} g_{Br^{79}}^2 + p_{Br^{81}} g_{Br^{81}}^2$, respectively. The quantities p represent, of course, the probabilities of finding the corresponding nuclei. By using the lattice parameter values of ND₄Cl and ND₄Br,[4,5] accepting for the nitrogen–deuterium distance the value 1.03Å,[6] and assuming the line shapes to be Gaussian, the line widths, defined as distances between the inflection points, were calculated from equation (1). In ND₄Cl, all the [ND₄]⁺ ions were assumed to have the same orientation.[7] In the case of ND₄Br, below the II ⇌ III transition temperature, the slight tetragonal distortion from the CsCl structure was neglected. Furthermore, the orientations of the [ND₄]⁺ ions in ND₄Br were assumed to be a-b-a-b along one cubic axis and a-a-a-a along the other two.[7] The calculated line width values are given in Table I.

Table I. Line Widths (Between Maximum Slopes) of the Deuteron Magnetic Resonance Above the Line Width Transition Temperature in Polycrystalline ND_4Cl and ND_4Br

Compound	Observed line width kcps	Calculated line width kcps
ND_4Cl	0.87 ± 0.05*	0.87‡
	0.96 ± 0.05†	0.89¶
	0.74 ± 0.04*	1.11‖**
ND_4Br		0.69‖††

* Sample delivered by Carl Roth AG(Germany).
† Sample prepared in this laboratory by repeated exchange with heavy water.
‡ At 291°K.
¶ At 195°K.
‖ Both at 200 and 243°K.
** Deuteron–bromine dipole–dipole interactions included.
†† Deuteron–bromine dipole–dipole interactions excluded.

The experimental results were obtained by means of a Varian electromagnet (type V-4012-3B) and a Robinson marginal oscillator[8] of this laboratory. The frequency of this oscillator was stabilized by shunting the limiter output with a quartz crystal. The frequency used was 5.24 Mcps. Magnetic slow and fast sweeps were used. The magnetic field was directly calibrated on the recording chart by a calibrator which automatically followed the proton magnetic resonance in the same field.[9] The temperature measurements were carried out by means of two thermistors.

Two ND_4Cl and one ND_4Br samples were used for our observations. One of the ND_4Cl samples and the ND_4Br sample were delivered by Carl Roth AG (Germany). The isotopic purity of deuterium in these samples was guaranteed to be at least 98%. The other ND_4Cl sample was prepared in this laboratory by repeated exchange of NH_4Cl with heavy water. The calculated isotopic purity was then 98.9%.

The experimental results are shown in Fig. 1 and Table I. The line width seems to be constant from room temperature down to 200°K for ND_4Cl and down to 172°K for ND_4Br. These temperatures are here called line width transition temperatures. Their maximum error was estimated to be ±10°K. Below the transition temperature, the line was found to broaden with decreasing temperature, and, at about 175°K for ND_4Cl and 150°K for ND_4Br, the lines were no longer observable. Above the line width transition temperature, the lines had a closely Gaussian shape. In addition, the transition II ⇌ III both in ND_4Cl and in ND_4Br had no observable effect on the lines. The line widths and their estimated errors above the line width transition temperatures are given in Table I. The line broadening is illustrated by two experimental derivative curves of ND_4Br in Fig. 2.

Because of the large, quadrupole, coupling constant of the deuteron in the $[ND_4]^+$ ion, the line broadening below the line width transition temperature obviously is caused by electric intraionic quadrupole interactions. These are no longer averaged out as the reorientations of the $[ND_4]^+$ ions become less probable.

Above the line width transition temperature, the calculated line widths for ND_4Cl agree well with the experimental data. It can be concluded that the line broadening in ND_4Cl is caused by the interionic, magnetic, deuteron–deuteron and deuteron–halogen, dipolar interactions. For ND_4Br, however, the agreement

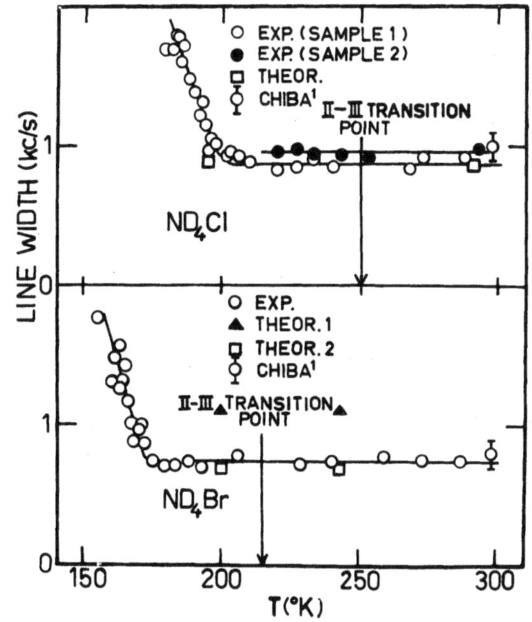

Fig. 1. The line width of the deuteron magnetic resonance, defined as the distance between the inflection points, in polycrstalline ND$_4$Cl and ND$_4$Br as a function of temperature. Each experimental point corresponds to the average of 3–17 independent observations. In the case of ND$_4$Cl, sample 1 was delivered by Carl Roth AG, while sample 2 was prepared in this laboratory by repeated exchange with heavy water. The theoretical values are taken from Table I.

between theory and experiment is not as good. The spin-lattice relaxation time T_1 of the Cl^{35} nuclei in NH$_4$Cl is 1 msec at 280°K, whereas the same quantity of the Br^{81} nuclei in NH$_4$Br is 0.059 msec at 285°K.[10] The T_1 values of the halogen nuclei in the heavy ammonium halides probably are not very much different. Because of this, it is presumable that, due to the small value of T_1 of the bromine nuclei, the deuteron–bromine dipolar interactions are averaged out. If we take into account only the interionic deuteron–deuteron interactions, we obtain for the line width of ND$_4$Br a value which is in reasonable agreement with the observed results. As the line width seems to be constant above the line width transition temperature,

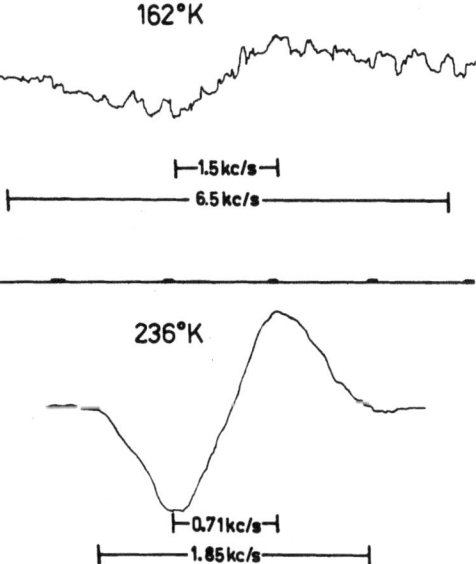

Fig. 2. The observed derivative curves of the deuteron magnetic resonance in ND$_4$Br at temperatures of 162 and 236°K. The scales corresponding to the two curves are different. The distance between the calibration marks is 2.35 G.

no line width contributions arising from the spin-lattice relaxation of the deuterons have been considered in the present investigation.

The higher line width transition temperatures of ND_4Cl and ND_4Br, as compared to those of the ordinary ammonium halides,[11,12] are attributed to the increase of the momentum of inertia of the ammonium ion.

Acknowledgments

Warm thanks are due to Miss Ulla Järvinen for her help during the measurements.

One of us (P.P.) is indebted to the University of Turku, Turku, Finland, for a scholarship awarded for 1964.

References

1. T. Chiba, *J. Chem. Phys.* **36**, 1122, 1962.
2. R. Bersohn, *J. Chem. Phys.* **32**, 85, 1960.
3. J. H. Van Vleck, *Phys. Rev.* **74**, 1168, 1948.
4. L. Vegard and S. Hillesund, *Avhandl. Norske Videnskaps-Akad. Oslo, I. Mat.-Naturv. Kl.* No. 8, 1942.
5. A. Smits and D. T. Tollenaar, *Z. physik. Chem.* **B52**, 222, 1942.
6. H. A. Levy and S. W. Peterson, *Phys. Rev.* **86**, 766, 1952.
7. J. Itoh, R. Kusaka, and Y. Saito, *J. Phys. Soc. Japan* **17**, 463, 1962.
8. F. N. H. Robinson, *J. Sci. Instr.* **36**, 481, 1959.
9. L. Niemelä, *J. Sci. Instr.* **41**, 646, 1964.
10. J. Itoh and Y. Yamagata, *J. Phys. Soc. Japan* **17**, 481, 1962.
11. A. H. Cooke and L. E. Drain, *Proc. Phys. Soc. (London) Ser. A* **65**, 894, 1952.
12. H. S. Gutowsky, G. E. Pake, and R. Bersohn, *J. Chem. Phys.* **22**, 643, 1954.

X-RAY INVESTIGATION OF THE MODIFICATIONS II AND III OF NH$_4$Br AT TEMPERATURES BETWEEN 22° AND −125°C

V. Hovi, K. Heiskanen, and M. Varteva

Wihuri Physical Laboratory, University of Turku
Turku, Finland

The observed modifications of NH$_4$Br can be summarized as follows:[1-3]

Mod. I (NaCl structure) $\xleftrightarrow{137.2°C}$ Mod. II (CsCl structure)

Mod. II (CsCl structure) $\xleftrightarrow{-38.1°C}$ Mod. III (tetragonal, nearly CsCl structure)

Mod. III (tetragonal, nearly CsCl structure) $\xleftrightarrow{\text{approx. }-180°C}$ Mod. IV (CsCl structure)

In the present investigation attention was paid to the transition from modification II to III. This transition belongs, in spite of a small change in structure, to the higher-order transitions.

The object of the present work was to investigate, by means of X-ray diffraction, the structures and lattice parameters of NH$_4$Br at temperatures between room temperature and −125°C. From the lattice parameters, the molar volume (and its change at the transition) and the cubic thermal-expansion coefficients of the different modifications were calculated as functions of temperature.

For the observations, a low temperature Debye–Scherrer camera was constructed. Figure 1 shows the vertical cross section of this camera. In the camera, the main parts are the standard Unicam cassette (H) (diameter, 19 cm) and the brass body (Q). In order to keep the temperature of the camera constant, the water jacket (O) was built inside the brass body. The specimen holder (L) was fastened to the body with bearings and rotated by the motor (P). The brass cone (N) guided the flowing cold gas between the body and the cassette.

The cassette was fitted with some auxiliary devices. In the collimator tube (B), two circular, lead plates (C) were placed. The two holes in these plates divided the X-ray beam into two parallel beams which hit the bipartite preparation (E). The undiffracted beams were collected in the beam trap (G). The circular iron plate (D), which had a hole in the middle for the preparation and the flowing gas, prevented the diffracted beams from mixing into each other. By using this arrangement, it was possible to take simultaneous X-ray photographs of the upper part (NH$_4$Br) and of the lower part (NaCl) of the preparation in the capillary.

The copper–constantan thermocouple (F), close to the top of the specimen, and the sodium chloride in the lower part of the latter, were used for determining the temperatures of the sample.

The dry nitrogen gas, which flowed through a spiral of copper tube, was cooled by liquid air in a Dewar flask. Then, the cooled gas was guided by an insulated,

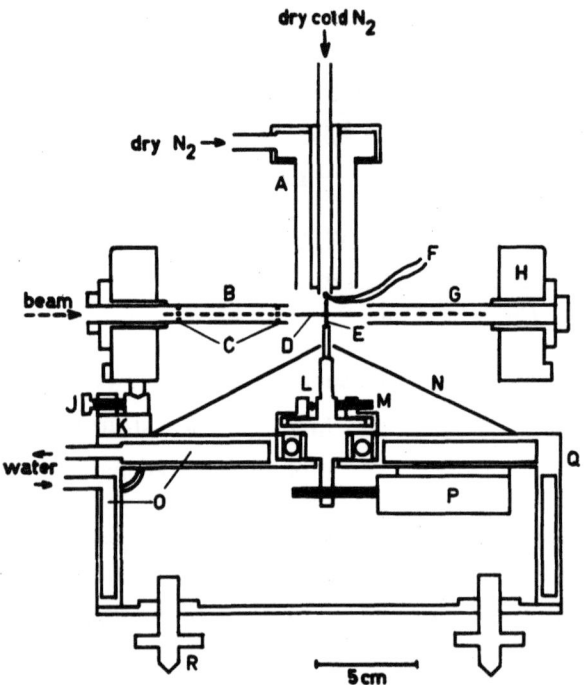

Fig. 1. Vertical cross section of the low temperature Debye–Scherrer camera. (A)—nozzle for the flowing dry nitrogen gas, (B)—collimator of the X-ray beam, (C)—circular lead plates, (D)—circular iron plate, (E)—bipartite specimen, (F)—copper-constantan thermocouple, (G)—direct beam trap, (H)—cassette, (J)—screws for centering the cassette, (K)—cassette holders, (L)—specimen holder, (M)—screws for centering the specimen, (N)—brass cone for guiding the flowing gas, (O)—water jacket, (P)—motor, (Q)—brass body, and (R)—screws for adjusting the camera.

German-silver tube and by a nozzle (A) to the specimen. In order to prevent the atmospheric moisture from condensing on the specimen capillary, dry nitrogen gas at room temperature was blown through the outer tube to the surroundings of the sample.

The NH_4Br and the NaCl used for this investigation were "analytical reagent" and "pro analysi" grades and were produced by the British Drug Houses, Ltd., London and E. Merck, AG, Darmstadt, Germany, respectively.

In order to analyze the diffraction lines of the different modifications of NH_4Br, predictions for these lines were calculated. For this purpose, the experimental results of Weigle and Saini[4] at 18 and −71.5°C were used.

Table 1. Values of the Lattice Parameter of NH_4Br at Room Temperature

Temperature, °C	Lattice parameter, Å	Authors
18	4.051	Weigle and Saini[4]
20	4.059	Donnay and Nowacki[7]
Room temperature	4.055	Sagel[8]
20	4.0587 ± 0.0002	Deshpande and Sirdesmukh[9]
22	4.0574 ± 0.0005	Present work

Table II. Values of the Lattice Parameters Measured Previously for Modification III of NH_4Br

Temperature, °C	Lattice parameter, $a_4 = a_1\sqrt{2}$ Å	Lattice parameter a_3 Å	Authors
−100	6.007	4.035	Ketelaar[10]
−71.5	5.713	4.055	Weigle and Saini[4]
−145	5.697	4.046	Weigle and Saini[4]

The photographs were taken by using Cu $K_{\alpha\beta}$ radiation from a Siemens Kristalloflex IV diffraction unit from this laboratory. The influence of the film shrinkage on the final results was taken into account in the usual way. The positions of the diffraction lines ($\theta > 50°$) were measured with an accuracy of 0.01 mm. The lattice parameters were determined from the X-ray photographs by normal methods.

The temperature of the specimen was measured by a copper–constantan thermocouple calibrated by means of a Muller bridge. The temperature was also checked by using the lattice parameter values of NaCl.[5] For the temperatures, the means of the values obtained from these two methods were taken.

One should point out that Riaño and Amoros[6] have determined experimentally the lattice parameter of NaCl at temperatures below 0°C. However, their results differed somewhat from those obtained in the present investigation for the same quantity. Thus, it was not possible for us to use the results of Riaño and Amoros for the indication of the sample temperatures.

At 22°C, we obtained for the lattice parameter of NH_4Br the value 4.0574 ± 0.0005 Å. This and previous results are given in Table 1. For the parameters of the tetragonal cell, $a_4 = a_1\sqrt{2}$ and a_3, only the values given in Table II have been presented in earlier papers. The results obtained in the present investigation for the modifications II and III are given in Tables III and IV. The estimated error in these lattice parameter values is ±0.001 Å. Figure 2 illustrates lattice parameters of

Table III. Values of the Lattice Parameters in the Present Investigation for Modification II of NH_4Br

Temperature, °C	Lattice parameters, $a_1 = a_2 = a_3$ Å
22	4.057
2	4.054
−4	4.052
−18	4.051
−25	4.050
−34	4.048
−37	4.047

Table IV. Values of the Lattice Parameters Measured in the Present Investigation for Modification III of NH$_4$Br

Temperature, °C	Lattice parameters, $a_1 = a_2$ Å	Lattice parameter, a_3 Å	Lattice parameter, $a_4 = a_1\sqrt{2}$ Å
−42	4.048	4.060	5.725
−43	4.048	4.062	5.725
−51	4.047	4.062	5.723
−57	4.046	4.062	5.722
−72	4.045	4.061	5.720
−83	4.043	4.061	5.718
−90	4.042	4.061	5.716
−108	4.039	4.057	5.712
−125	4.037	4.053	5.709

NH$_4$Br as functions of temperature. From the measured lattice parameters, we found the following expressions for the molar volume V (in cm^3/mole) as a function of temperature t:

$$V = 40.113 + 4.82 \cdot 10^{-3} t \text{ (Mod. II)}$$

$$V = 40.042 - 2.34 \cdot 10^{-3} t - 3.64 \cdot 10^{-5} t^2 \text{ (Mod. III)}$$

It is interesting to note that the molar volume of NH$_4$Br increased by 0.40% with decreasing temperature from −37 to −42°C. Also, Smits et al.[11] found a similar phenomenon in their dilatometric investigation. For a temperature interval of 1.5°C, which includes the transition temperature, one can calculate from their observations an increase of 0.1 cm^3/mole in the molar volume with decreasing temperature. This is about 70% of the increase found in the present investigation.

From the measured lattice parameters, we found the following values for the cubic thermal-expansion coefficients, expressed in (°C)$^{-1}$: $1.2 \cdot 10^{-4}$ (Mod. II) and $-6 \cdot 10^{-5} - 1.8 \cdot 10^{-6} t$ (Mod. III). A more detailed account of these results has been published elsewhere.[12]

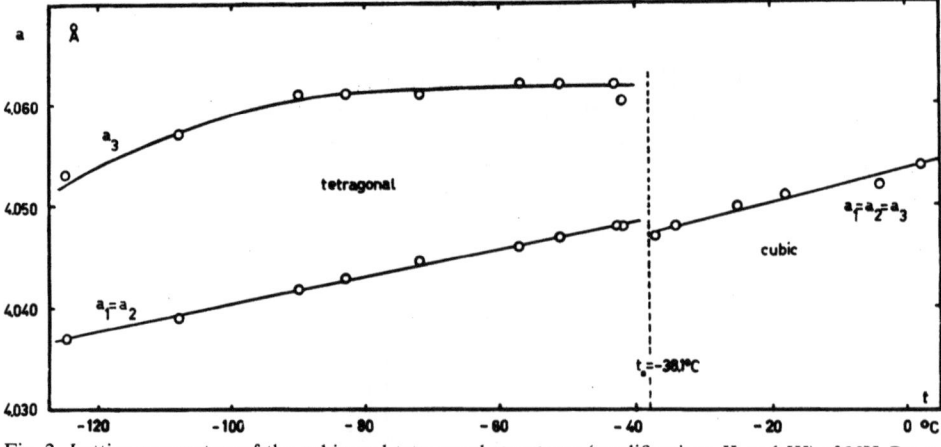

Fig. 2. Lattice parameters of the cubic and tetragonal structures (modifications II and III) of NH$_4$Br as functions of temperature.

References

1. J. Pöyhönen, *Ann. Acad. Sci. Fennicae Ser. A VI*, No. 58, 1960, Finland.
2. H. P. Klug and W. W. Johnson, *J. Am. Chem. Soc.* **59**, 2061, 1937.
3. C. C. Stephenson and H. E. Adams, *J. Chem. Phys.* **20**, 1658, 1952.
4. J. Weigle and H. Saini, *Helv. Phys. Acta* **9**, 515, 1936.
5. A. Eucken and W. Dannöhl, *Z. Elektrochem.* **40**, 814, 1934.
6. E. Riaño and J. L. Amoros, *Bol. Real Soc. Espan. Hist. Nat.* **58**, 181–198, 1960.
7. J. D. H. Donnay and W. Nowacki, *Geol. Soc. Am. Mem.*, No. 60, 1954; ref. in *Acta Cryst.* **14**, 353, 1961.
8. K. Sagel, *Tabellen zur Röntgenstrukturanalyse, Vol. VII*, Springer-Verlag, Berlin (1958), pp. 48, 62.
9. V. T. Deshpande and D. B. Sirdesmukh, *Acta Cryst.* **14**, 353, 1961.
10. J. A. A. Ketelaar, *Nature* **134**, 250, 1934.
11. A. Smits, J. A. A. Ketelaar, and G. J. Muller, *Z. Physik. Chem.* **A175**, 359, 1936.
12. V. Hovi, K. Heiskanen, and M. Varteva, *Ann. Acad. Sci. Fennicae. Ser. A VI*, No. 144, 1964, Finland.

X-RAY INVESTIGATION OF THE TRANSITION I → II OF NH$_4$I AT TEMPERATURES BETWEEN 22° AND −163°C

V. Hovi and M. Varteva

Wihuri Physical Laboratory, University of Turku
Turku, Finland

The observed modifications of NH$_4$I can be summarized as follows:[1-4]

Mod. I (NaCl structure)   approx. −16°C   Mod. II (CsCl structure)

Mod. II (CsCl structure)   approx. −42°C   Mod. III (tetragonal, nearly CsCl structure)

In the present investigation, attention was paid to the transition between modifications I and II. This transition is known to belong to the first-order transitions.

From the available literature, we have not found any systematic investigation dealing with the lattice parameters of NH$_4$I. However, some data have been reported. The previous values are given in Table I.

Only a few observations of the kinetics of the transition I → II have been reported. Smits and Muller[5] found, by using a dilatometric method, that it was necessary to keep the NH$_4$I sample some months at −78°C to transform completely modification I into modification II. By carrying out specific-heat measurements at temperatures between 14–300°K, Stephenson et al.[3] observed that one cannot cool modification I below −73°C without changing it into a lower modification.

Table I. Values of the Lattice Parameters Measured for the Modifications I, II, and III of NH$_4$I

Modification	Temperature, °C	Lattice parameter, Å	Authors
I	Room temperature	7.198*	Bartlett and Langmuir[10]
	Room temperature	7.244 ± 0.003	Havighurst et al.[11]
	Room temperature	7.259	Sagel[12]
	+22	7.2591 ± 0.0007	Present work
II	Below −17	4.37	Simon and v. Simson[13]
		4.379	Sagel[12]
	−16	4.335	Present work
III	−65	4.317†	Smits and Tollenaar[14]
	−80	4.315†	Smits and Tollenaar[14]

* The measured value 5.090 Å is multiplied by $\sqrt{2}$.
† Tetragonal structure.

The object of the present work was to investigate, by means of X-ray diffraction, lattice parameters of NH_4I at temperatures between 22 and $-39°C$, to determine the change of the molar volume at the transition I → II, and to obtain values for the cubic thermal-expansion coefficients of the modifications I and II. Furthermore, our intention was to observe, by using a low-temperature X-ray diffractometer, the velocity of the transition I → II of NH_4I as a function of supercooling and of particle size.

The method used for the measurements of the lattice parameter is explained in an earlier paper.[6]

The NH_4I and NaCl used for this investigation were grade "pro analysi" produced by the J. T. Baker Chemical Co., USA, and the E. Merck AG, Darmstadt, Germany, respectively. At 22°C, we found for the lattice parameter of NH_4I the value 7.2591 ± 0.0007 Å. This and previous values are given in Table I. The results obtained in the present investigation for the modifications I and II are shown in Table II. The estimated error in the lattice parameter values is ± 0.001 Å.

On the basis of the measured lattice parameters, we calculated values for the molar volume. According to the previous investigations, the values of the transition temperature at the transition I → II should be between $-12.5°$ and $-17.6°C$ (cp. Arell and Alare[7]). Assuming the transition temperature to be between $-14°$ and $-18°C$, the molar volume of NH_4I seems to increase by $16.96 \pm 0.02\%$ with increasing temperature. This is in good agreement with the results obtained for the corresponding transition of NH_4Cl by means of dilatometric (19.31%)[8] and X-ray methods (19.43%),[9] and for that of NH_4Br by means of the dilatometric method (18.85%).[8]

From the measured lattice parameters, we found the values $0.92 \cdot 10^{-4}$ (Mod. I) and $1.81 \cdot 10^{-4}$ (Mod. II), in $°C^{-1}$, for the cubic thermal-expansion coefficients.

For the kinetic measurements, a low temperature diffractometer, modified from that of Peisl and Waidelich,[15] was constructed. When the Geiger-tube moved, by means of two switches in the angle interval 9.5–11°, reflection (111) of modification I repeatedly appeared in the diffraction pattern. By decreasing the temperature of the specimen below the transition temperature, the transition started, and besides reflection (111), reflection (100) of modification II appeared (Fig. 1). With increasing

Table II. Values of the Lattice Parameters Measured in the Present Investigation for the Modifications I and II of NH_4I

Modification	Temperature, °C	Lattice parameter, Å
I	+22	7.259
	+9	7.256
	−1	7.254
	−5	7.253
	−13	7.251
II	−3	4.338
	−10	4.337
	−16	4.335
	−21	4.334
	−26	4.332
	−29	4.332
	−39	4.329

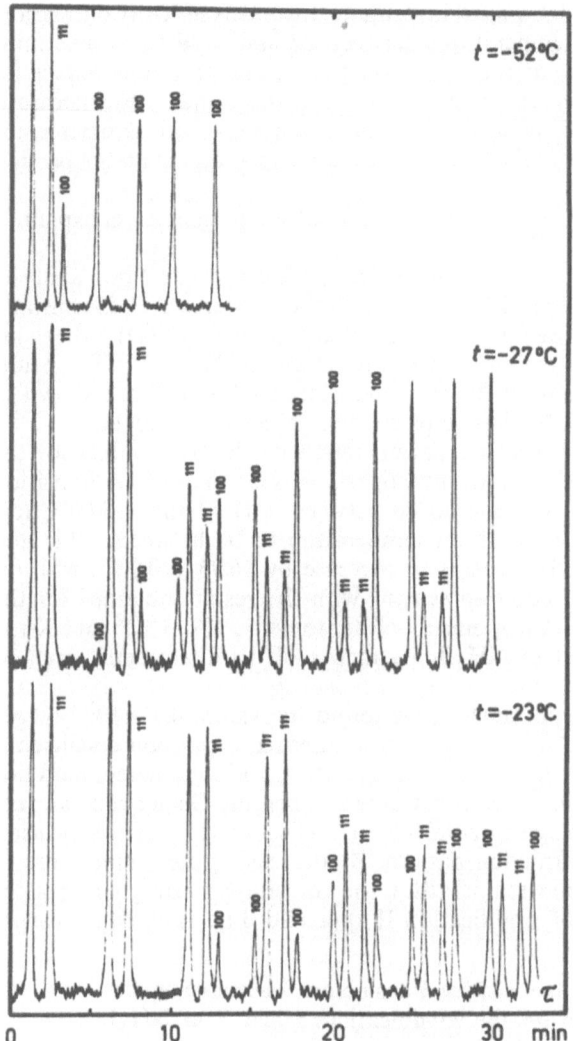

Fig. 1. Diffraction patterns representing the transition I → II of NH$_4$I at different temperatures. Each curve corresponds to the same particle size (0.100–0.300 mm).

time, reflection (111) disappeared, while the intensity of reflection (100) approached its maximum. With the assumption that intensity of the X-ray reflection of a modification in the specimen is proportional to the quantity of that same modification, we can determine from the intensities the percentage composition of the specimen as a function of time.

The samples were of the same quality as before. From the powdered salt, three specimens, having particle sizes of 0.100–0.300, 0.050–0.075, and 0.010–0.033 mm, were separated by sifting. The fourth specimen was prepared from the last one by changing its temperature twenty times over the transition. The average particle size of the heat-treated specimen was 0.001–0.004 mm, which was determined by the electron microscope from this laboratory.

In the present work, we select for the so-called zero-instant the mean of the instant when the temperature of the sample is passing $-16°C$ and of that when the

Table III. Empirical Functions of the Half-Transition Time for Different Particle Sizes

Particle size, mm	Half-transition time, min
0.100–0.300	$\left(\dfrac{58.97}{\Delta T}\right)^{1.522}$
0.050–0.075	$\left(\dfrac{66.88}{\Delta T}\right)^{1.804}$
0.010–0.033	$\left(\dfrac{96.42}{\Delta T}\right)^{2.727}$
0.001–0.004	$\left(\dfrac{117.6}{\Delta T}\right)^{4.016}$

sample has reached the measuring temperature. The time between the zero-instant and the instant when the sample contains 50% of both modifications is here called the half-transition time and is denoted by $\tau_{1/2}$.

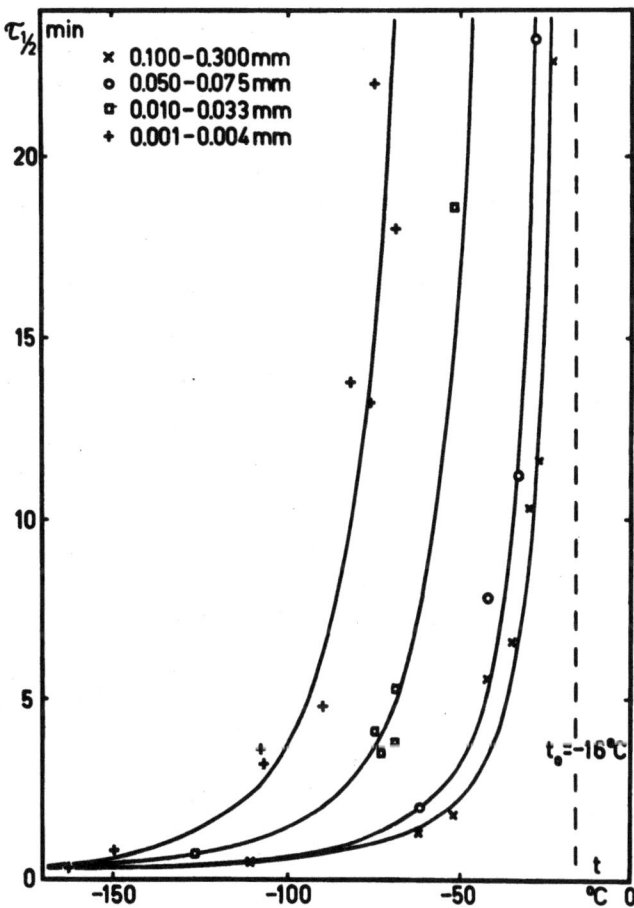

Fig. 2. The half-transition time $\tau_{1/2}$ (50% of both modifications I and II) of NH_4I as a function of temperature and particle size.

We select for the half-transition time the simple expression $\tau_{1/2} = (a/\Delta T)^b$, where a and b are adjustable parameters, and ΔT is the absolute value of the difference between the measuring and transition temperatures. The quantity ΔT indicates the supercooling of the sample. Of course, when ΔT approaches zero, $\tau_{1/2}$ becomes infinite. If we accept the value $-16°C$ for the transition temperature, we obtain, by smoothing the experimental data by means of the selected expression for $\tau_{1/2}$, the results that appear in Table III for different particle sizes.

Figure 2 shows the smoothed curves of $\tau_{1/2}$ from which the following conclusions were drawn:

1. Below $-100°C$, the influence of the temperature on the transition velocity is rather small. However, the transition always requires a certain minimum time independent of the particle size.
2. At a few ten-degree intervals below the transition temperature, $\tau_{1/2}$ seems to be strongly dependent on temperature.
3. At constant temperature, $\tau_{1/2}$ is considerably greater for smaller particle sizes.
4. When $\tau_{1/2}$ is constant, the supercooling is greater for a sample containing smaller grains than for a coarse-grained sample. A more detailed account of these results will be published in *Phys. Kondens. Materie* **3**, 305–310, 1965.

References

1. H. P. Klug and W. W. Johnson, *J. Am. Chem. Soc.* **59**, 2061, 1937.
2. F. Simon, C. v. Simson, and M. Ruhemann, *Z. Physik. Chem.* **129**, 321, 1927.
3. C. C. Stephenson, L. A. Landers, and A. G. Cole, *J. Chem. Phys.* **20**, 1044, 1952.
4. R. W. G. Wyckoff, *Crystal Structures, Vol. I,* Interscience Publishers, Inc., New York (1948).
5. A. Smits and G. J. Muller, *Z. Physik. Chem.* **B36**, 140, 1937.
6. V. Hovi, K. Heiskanen, and M. Varteva, *Ann. Acad. Sci. Fennicae Ser. A VI*, No. 144, 1964.
7. A. Arell and O. Alare, *Phys. Kondens. Materie* **2**, 423, 1964.
8. J. Pöyhönen, *Ann. Acad. Sci. Fennicae Ser. A VI*, No. 58, 1960.
9. K. Mansikka and J. Pöyhönen, *Ann. Acad. Sci. Fennicae Ser. A VI*, No. 118, 1962.
10. G. Bartlett and I. Langmuir, *J. Am. Chem. Soc.* **43**, 84, 1921.
11. R. J. Havighurst, E. Mack, Jr., and F. C. Blake, *J. Am. Chem. Soc.* **46**, 2368, 1924.
12. K. Sagel, *Tabellen zur Röntgenstrukturanalyse, Vol. VIII,* Springer-Verlag, Berlin (1958), pp. 48, 62.
13. F. Simon and C. v. Simson, *Naturwissenschaften* **14**, 880, 1926.
14. A. Smits and D. Tollenaar, *Z. Physik. Chem.* **B52**, 22, 1942.
15. H. Peisl and W. Waidelich, *Z. Angew. Phys.* **11**, 474, 1959.

HEAT PULSES IN ALKALI HALIDES AT LOW TEMPERATURES

R. J. von Gutfeld and A. H. Nethercot, Jr.

IBM Watson Research Center
Yorktown Heights, New York

Further studies of the propagation of heat pulses, similar to those previously reported for quartz and sapphire,[1] have been made on several alkali halides. These include NaCl, KCl, KBr, and KI. The objective was to determine the nature of the heat-pulse propagation in high-purity crystals at temperatures near that at which the thermal conductivity is a maximum. It is at such temperatures that observation of "second sound" in insulating crystalline media,[2] if it occurs, is most likely to be observed. Since superconducting film thermal detectors are used, this temperature must be compatible with the operating temperature of the detector used (approx. 8.0°K). This condition is approximately met by several of the alkali halides. At lower temperatures, energy transport should be by direct rectilinear propagation at the acoustic velocity if the defect and boundary scattering rates are sufficiently slow.

The crystals used came from several sources, some obtained from the Harshaw Chemical Company and some grown and purified at this laboratory from reagent-grade material. As in our previous work, a thin film of constantan (approx. 500 Å thick) served as the heating element through which a pulse of current (up to 1 A and about 0.1 μ sec long) was passed. On the opposite face of the crystal, a thin film bolometer of a suitable superconducting alloy and biased by a small DC current served as the thermal detector. The output signal was amplified and displayed on a CRO. Indium–tin was used as the detector at 3.8°K and lead–bismuth was used at 8.0°K. The thermal time constant was measured to be less than 0.02 μ sec for the 3.8° detector on NaCl crystals, but was probably considerably longer for the 8.0° detector (it had previously been measured to be approx. 0.3 μ sec on quartz crystals).

The heat pulses previously observed[1] were quite sharp for sapphire, while there was a broad background present in addition to the sharp pulses for quartz. These sharp pulses in both cases arrived at times determined by the acoustic energy velocities, and, hence, were caused by direct, uninterrupted, phonon propagation. For the quartz, the scattering was verified to be predominantly a small-angle scattering, such as that expected for large scale defects. The somewhat broader pulses observed in quartz at 8.0°K were probably due to the slower response speed of the lead–bismuth detector, rather than to a thermal broadening of the heat pulses due to phonon–phonon interactions. In the present experiments, the heat pulses observed for the best NaCl and KCl crystals were qualitatively similar to those observed for quartz, since they showed some sharp structure superposed on a broad scattered background. The heat pulses observed for the KBr and KI crystals, on the other hand, showed only the broadened, scattered type of response, presumably because of some type of defect scattering.

The type of heat pulses observed in NaCl depended strongly on the sample preparation. The effects were very similar to those observed by Klein[3] in measurements of the thermal conductivity. He had observed that the thermal conductivity was approximately two orders of magnitude larger for chlorine-treated crystals grown from reagent-grade material than for single crystals obtained from Harshaw. This was presumably due to the effects of treating NaCl with chlorine in reducing the OH radical impurity (or the oxygen band). Figure 1 shows the corresponding differences observed in the received heat pulses for the two types of crystals. It will be noted that the onset of the heat pulse in the Harshaw crystal occurs some 50–70 μ sec after the end of the excitation and that no sharp structure is visible. It is apparent that the diffusion equation describes the heat flow well. In contrast, the chlorine-treated crystal exhibited sharp heat pulses arriving at times corresponding to the velocities for the various polarizations of the acoustic vibrations (Fig. 2). Similar to the quartz, only a rather small fraction of the total energy flow is apparently by uninterrupted, rectilinear, phonon propagation, the remainder of the phonons being scattered by some defect mechanism. From data on the thermal conductivity K and specific heat C of NaCl at low temperatures, the value of the phonon mean free path l for assumed isotropic scattering can be calculated from the formula $K = \frac{1}{3}Cvl$ to be of the order of 0.5 cm. Thus, since the crystal length was only 0.6 cm, one might have expected more of the energy to arrive within the sharp pulse and less to be scattered than was observed. However, if the scattering is mainly forward scattering of the small-angle type (as was found to be the case for quartz), this discrepancy can easily be reconciled since the calculated 0.5-cm mean free path then corresponds to a much shorter small-angle mean free path. For technical reasons, confirmatory measurements of the angular dependence of the scattering have not been made as they were for quartz.

The velocities of the pulses are in better agreement with the wave (energy) velocities, rather than with the phase velocities, as had previously[1] also been found for quartz. These wave velocities were estimated by calculating the phase velocities from the elastic constants[4] extrapolated to 0°K. The wave velocities were then roughly estimated by a graphical method[5] from the phase velocities for the special azimuthal angles $\phi = 0°$ and 45°, for which these phase velocities are stationary with respect to ϕ.

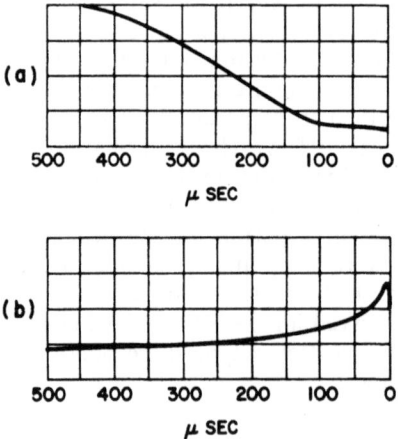

Fig. 1. NaCl at 3.8°K. (a) Harshaw crystal, 1.00 cm length. (b) Chlorine-purified crystal, 0.59-cm length.

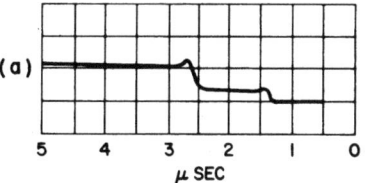
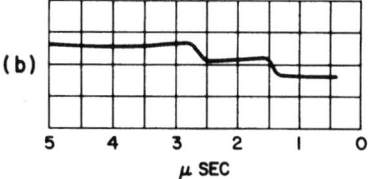
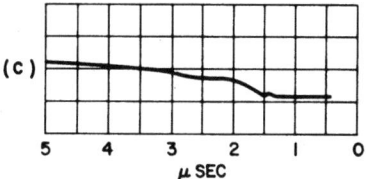

Fig. 2. Purified crystals of NaCl and KCl. (a) NaCl, 0.59-cm length at 3.8°K. (b) NaCl, 0.59-cm length at 8.0°K. (c) KCl, zone-refined, 0.49 cm at 3.8°K.

The pulse shape was also examined at 8.0°K. Although the pulses were indeed broader (Fig. 2b), this could have been caused by the slower detector. In any event, the discrete pulses arriving at the acoustic velocity were still present, and, thus, a considerable fraction of the phonons even at this elevated temperature do not have their flow effectively interrupted. Certainly, for these phonons the scattering rate does not increase with temperature as $\omega^4 \approx (kT/\hbar)^4$, as predicted for Rayleigh scattering, or as ωT^4, as predicted by Landau–Rumer for non-collinear phonon–phonon interactions. Thus, a significant number of phonons must collide only with large scale defects (since such scattering is relatively independent of frequency and hence of temperature) or suffer only collinear or near collinear phonon–phonon normal process collisions[6-8] (since these would not appreciably change the velocity of the energy flow). In any case, the phonons within the sharp pulses do not show any shifts in velocity at all comparable to those predicted for second sound. The scattered background phonons, of which there is a large majority, also do not show any particular changes as the temperature is raised from 3.8 to 8.0°K. Thus, the conclusion is either that the temperature (or the crystal purity or the crystal size) is not such that normal process collisions dominate over all sources of phonon momentum loss, or that the normal process collisions are not at all isotropic, but instead may be predominantly collinear or near collinear (as has recently been proposed).[6-8]

The results for zone-refined KCl were rather similar to those for NaCl, with the background scattering being somewhat greater (Fig. 2c) relative to the sharp pulses. Only traces of the sharp pulses were present in Harshaw KCl. Thus, the defect scattering appears to be somewhat larger for the best KCl used than for the best NaCl, even though the chemical purity of the zone-refined KCl is undoubtedly much greater. Calculations of the energy velocities from the known elastic constants[9] again gave better agreement with the observed heat-pulse velocities than did the phase velocities.

Crystals of KI obtained from Harshaw and of KBr treated with bromine gas were also investigated at 3.8°K. For both of these crystals, the scattering was so great that the thermal flow was principally a diffusive one, with no discernible sharp structure present. The onset of the pulses occurred at close to 1 μ sec for both crystals. These times are approximately those expected for travel at the acoustic velocity, and, thus, the phonon mean free path is not too much less than the sample dimensions for at least the low-frequency phonons. The mean free path is certainly not nearly as short as that found for the Harshaw NaCl, where the onset of the pulse was delayed by approx. 50 μ sec. It could also be mentioned that the evaporation of good detectors and heaters on these materials was much more difficult than on KCl and NaCl.

It is planned to continue this work, particularly in the directions of producing a faster detector near 8°K so that smaller broadenings can be reliably measured and of producing higher temperature detectors so that NaCl and KCl can be investigated at temperatures well above their thermal-conductivity maxima.

Acknowledgments

We would like to thank W. K. Schug for help with the evaporations, Dr. W. Bron for supplying some of the crystals, and Dr. W. E. Donath for help with the computer program for computing the phase velocities.

References

1. R. J. von Gutfeld and A. H. Nethercot, Jr. *Phys. Rev. Letters* **12**, 641, 1964.
2. J. C. Ward and J. Wilks, *Phil. Mag.* **42**, 314, 1951.
3. M. V. Klein, *Phys. Rev.* **122**, 1393, 1961.
4. W. C. Overton, Jr., and R. T. Swim, *Phys. Rev.* **84**, 758, 1951.
5. M. J. P. Musgrave, *Rept. Progr. Phys.* **22**, 74, 1959.
6. S. Simons, *Proc. Phys. Soc. (London) Ser. A* **82**, 401, 1963.
7. H. J. Maris, *Phil. Mag.* **9**, 901, 1964.
8. R. Nava, R. Azrt, I. Ciccarello, and K. Dransfeld, *Phys. Rev.* **134**, 581, 1964.
9. M. H. Norwood and C. V. Briscoe, *Phys. Rev.* **112**, 45, 1958.

11.3. Metals at Low Temperatures

TEMPERATURE DEPENDENCE OF THE ELECTRON MEAN FREE PATH IN TIN AT LIQUID-HELIUM TEMPERATURES

V. F. Gantmakher* and Yu. V. Sharvin

Institute for Physical Problems of the Academy of Sciences of the USSR
Moscow, USSR

Introduction

To obtain detailed information on the electron mean free path in anisotropic metals, the elaboration of a method for investigating separate groups of electrons is needed. We made an attempt to use a previously described method[1] for investigating the temperature dependence of the mean free path of electrons in tin.

First of all, it is necessary to give an exact definition of the quantity which is measured in experiments of this kind. If the constant magnetic field is inclined at a small angle to the surface of a single-crystal plate of thickness d, it is possible to observe a series of peaks in the curve of the dependence of the RF impedance upon the magnetic field. A peak occurs when the electrons in the vicinity of an elliptic limiting point on the Fermi surface make an integer number n of turns moving from one surface of the specimen to the other. The intensity of the effect depends on the number of electrons that have been accelerated initially while moving in the first skin-layer parallel to the surface of the sample and have gotten into the skin-layer on the opposite side, moving also along the surface.

The selection of electrons according to the direction of their velocity is, therefore, made twice, at the beginning and at the end of the trajectory. Due to the scattering processes, part of the effective electrons is removed (the replacement owing to the scattering of other electrons is negligible[2]). Because of the smallness of the skin-layer δ, scattering by rather small angles is sufficient. The magnitude of this angle depends on δ/d, n, and the angle θ between the magnetic field and the surface. The approximate evaluation gives $\theta \sim 10^{-3}$, assuming $\delta \sim 10^{-4}$ cm (frequency $\sim 3 \cdot 10^6$ cps). The interaction with the phonon at helium temperatures scatters the electron by the angle of the order of ρ_{ph}/R, where ρ_{ph} is the phonon momentum and R is the radius of curvature of the Fermi surface. In our case, $\rho_{ph}/R \sim 10^{-2}$. According to this estimation, a single interaction with a phonon is sufficient to make an electron ineffective. This is more so for other scattering processes where the electrons are deflected by large angles. It was shown by Gantmakher and Kaner[1] that the amplitude A of the effect depends on the lengths of electron trajectories—$s \sim d/\varphi$ as $e^{-s/l}$. It can be assumed that l, measured according to this formula, corresponds to the path length between two elementary scattering events.

* Present address: Institute for Physics of Solids, Academy of Sciences of the USSR, Moscow, USSR.

Experimental Results

The amplitude A has been measured as a function of temperature for the elliptic point I (by the notation of Gantmakher and Kaner[1]) situated at the Fermi surface near the direction [100]. This point is located presumably at the side-surface of the tube joining two corrugated planes in the fourth zone. The function $(\varphi/d)\ln A$ equals $1/l(T) + C$ where C is an arbitrary additional constant. The limiting value of this function at $T \to 0$ can be defined by extrapolating with sufficient accuracy. The temperature dependence of the quantity $1/l(T) - 1/l(0)$ thus found is shown in Fig. 1 (logarithmic scale). The straight line drawn through the points corresponds to the law $T^{3.3}$ which is near to the Bloch's cubic law for electron–phonon scattering.

Because of the insignificance of electron–electron collisions in the temperature range under consideration, these experimental results can be regarded as a confirmation that in our experiments the path length between two elementary collisions was indeed measured, and the scale of the ordinate in Fig. 1 represents $1/l_{ph}$, i.e., the probability of an electron–phonon interaction per unit of length. A small deviation from the cubic law is due presumably to the insufficient smallness of angle θ. Consequently, the effectiveness of a single electron–phonon collision falls with decreasing temperature. In the case of DC conductivity of bulk specimens, this leads to the well-known law $\rho \sim T^5$ which is valid for tin at helium temperatures.[3] Our data can be compared to the evaluation of the electron–phonon path length l_T obtained from the heat conductivity data in the free-electron approximation. According to the data of Zavaritsky,[4] $l_T \sim T^{-3}$ for tin, and l_T is 10–20 times smaller than our value. This may indicate that, at the investigated point of the Fermi surface, the phonon scattering is considerably smaller than the average value over the surface.

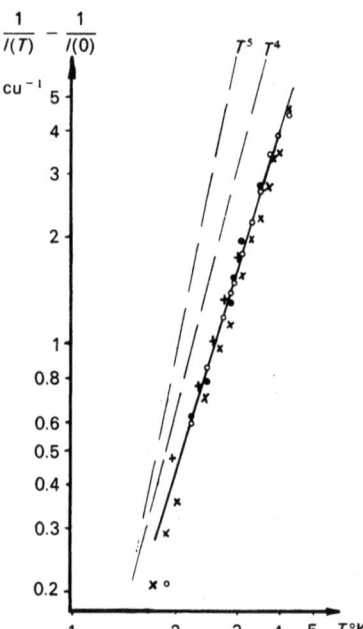

Fig. 1. Surface of specimen is parallel to (010). RF electric field $\mathbf{E}\|[100]$. Constant magnetic field $\mathbf{H}\|(001)$. $\times, \varphi = 4°2'$; $+, \varphi = 2°45'$; $\bigcirc, \varphi = 3°10'$; $\bullet, \varphi = 4°15'$.

References

1. V. F. Gantmakher, and E. A. Kaner, *Zh. Experim. i Teor. Fiz.* **45**, 1430, 1963. (*Soviet Phys. JETP (English Transl.)* **14**, 988, 1963.)
2. M. Ya. Azbel and E. A. Kaner, *Zh. Experim. i Teor. Fiz.* **32**, 895, 1957. (*Soviet Phys. JETP (English Transl.)* **5**, 730, 1957.)
3. V. B. Zernov and Yu V. Sharvin, *Zh. Experim. i Teor. Fiz.* **36**, 1039, 1959. (*Soviet Phys. JETP (English Transl.)* **9**, 737, 1959.)
4. N. V. Zavaritsky, *Zh. Experim. i Teor. Fiz.* **39**, 1571, 1960. (*Soviet Phys. JETP (English Transl.)* **12**, 1093, 1960.)

HEAT CONDUCTIVITY OF PURE METALS BELOW 1°K

G. Davey, K. Mendelssohn, and J. K. N. Sharma

The Clarendon Laboratory
Oxford, England

Two years ago a liquid helium-3 cryostat was set up for the accurate measurement of thermal conductivities below 1°K. In order to check the equipment, it was decided to measure the heat conductivity of copper since this was expected to yield a straight line passing through the origin. The results obtained showed definite deviations from this behavior, and it was, therefore, thought that the apparatus was faulty. Accordingly, a great number of modifications were introduced, such as changing and reversing the thermometers, varying the method of measuring the ^{3}He vapor pressure, etc. None of these altered the results obtained, and it was further found that similar irregularities occurred in other samples under investigation.

There exist few other data with which our results can be compared, and, as far as we know, none that can claim, in the temperature region concerned, the accuracy of 1% at which our measurements are aimed. In fact, plotting our own data with an accuracy of 5% instead of 1%, the irregularities observed in our measurements effectively disappeared. We thus came to the conclusion that our measured heat conductivities, though anomalous, were not in disagreement with any existing data,[1-4] including our own earlier data. It therefore seems quite possible that the deviations of the heat conductivity of pure metals from strict proportionality to the absolute temperature may be a real effect.

Our first measurements[5] were carried out on copper, gold, iron, and titanium, thereby covering a considerable variety of metals investigated. With the exception of the gold sample, which followed the expected linear dependency, all these metals showed anomalies in the thermal conductivity in the neighborhood of 0.7°K.

This work has now been extended to silver, palladium, platinum, and nickel the results of which are shown in Fig. 1. All these metals again exhibit deviations from the expected behavior and again at roughly the same temperature. In the earlier work on copper, we had noted that the effect was enhanced by greater purity of the sample, and we therefore again measured gold, which had originally followed a straight line. The sample used was of higher purity than that in the first experiments.

The new results, which are also given in Fig. 1, show not only the higher absolute values due to greater purity, but also very pronounced irregularities. It therefore appears, that the effect, if real, is a quite general phenomenon, since we have so far not found a metal which, when sufficiently pure, is free from these irregularities.

At first sight, it must seem disturbing that the irregularities should occur at roughly the same temperature in all metals investigated, and this again could indicate some fault in the method. However, closer analysis shows that this

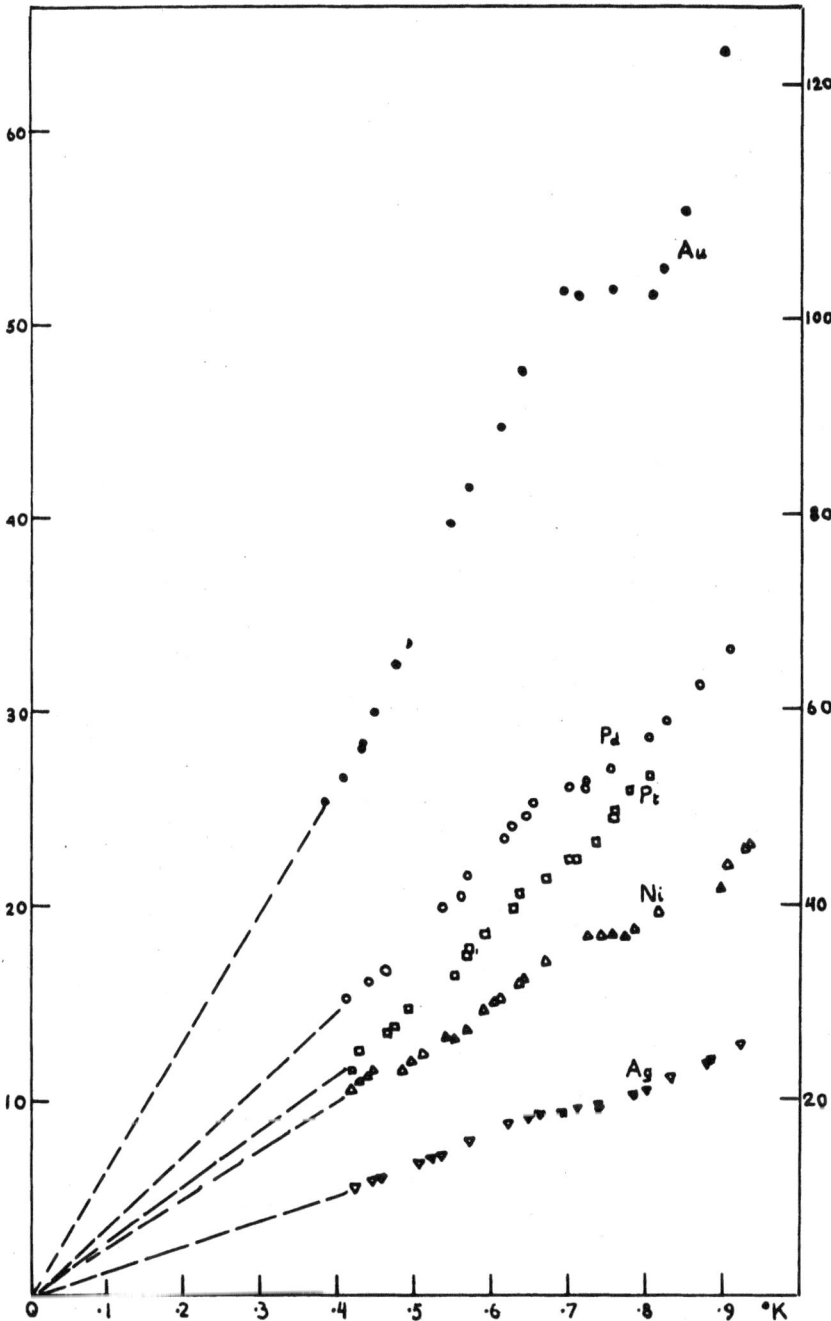

Fig. 1. Heat conductivity of silver, nickel, platinum, and palladium (left ordinate) and of gold (right ordinate) below 1°K. The ordinates are given in the unit watt × 10^{-2}.

coincidence is only apparent because of the relatively short temperature range covered. In fact, our first results already showed that the lowest points of each curve did not extrapolate linearly to absolute zero, suggesting that the anomalous behavior was still present at the lowest temperatures of our range.

This fact is further emphasized by our latest measurements which, particularly in the cases of palladium and nickel, make it likely that the irregularity near 0.8°K is followed by another in the neighborhood of 0.5°K. In fact, it now appears that instead of a marked deviation at a certain temperature, the thermal conductivities gently oscillate with varying temperature around a strict linear dependency.

In addition to the results given in Fig. 1, a wire of pure aluminum was measured in both the normal and the superconductive states. The results, which are given in Fig. 2, show a pronounced double oscillation with temperature in the normal state which makes it clear that the effect is not merely an irregularity centered at about 0.7°K. Moreover, extrapolation from the lowest measured value to absolute zero results in a slope which is steeper than the general trend of the data, leading us to suspect further irregularities below 0.4°K. It is interesting to note that a similar behavior, though far less pronounced, can be seen in the superconductive state.

It should be mentioned that the measured points given in all our results were, in each case, obtained in a number of successive runs, and, in one case, the specimen was removed from the apparatus and remeasured after it was again assembled. Completely consistent results were obtained under all these circumstances. An additional check was made on the equipment by measuring a sapphire single crystal which had been used in Dr. R. Berman's experiments.[6] The data obtained by us on this specimen between 0.4 and 1°K when plotted against T^3 give, to an accuracy of 1%, a straight line passing through the origin. In addition, the absolute value determined by us agrees completely with that obtained at higher temperatures in Dr. Berman's apparatus.

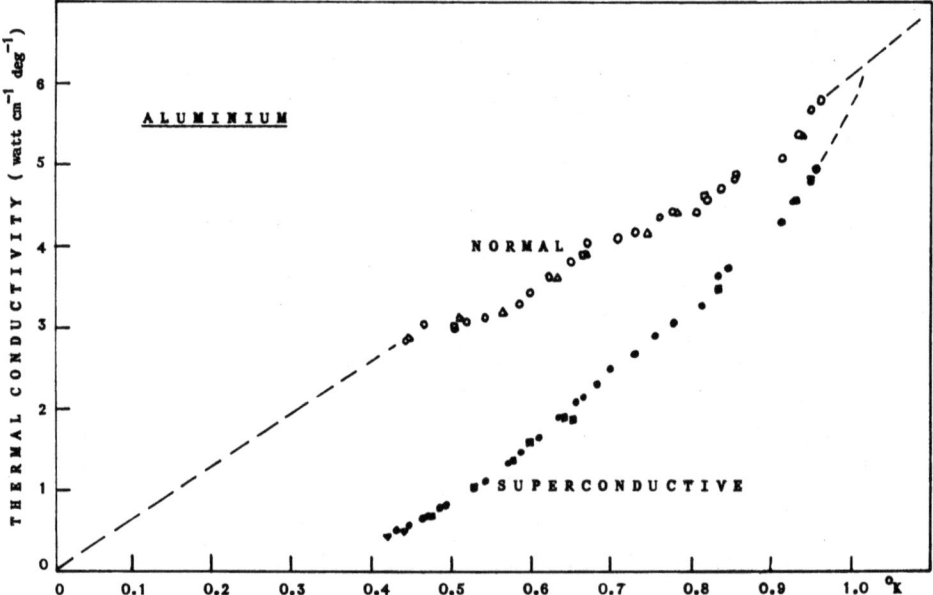

Fig. 2. Heat conductivity of aluminum below 1°K in the normal and superconductive states.

Experiments are under way to measure in the same temperature range the electrical resistivity of the specimens used in the heat-conductivity work. The first results have been obtained by Dr. I. Yoshida on the specimen of iron, and these give a value which within 3% accuracy is independent of temperature. Applying the Wiedemann–Franz law to the heat-conductivity data, it appears that an anomaly in the electrical resistivity of iron, corresponding to that in the thermal conduction, should have been easily noticed.

References

1. J. Nicol and T. P. Tseng, *Phys. Rev.* **92**, 1062, 1953.
2. N. Phillips, *Phys. Rev.* **100**, 1719, 1955.
3. A. Connolly, thesis, Oxford (1960).
4. A. Dupre, A. van Itterbeek, and L. Michiels, *Phys. Letters* **8**, 99, 1964.
5. G. Davey and K. Mendelssohn, *Phys. Letters* **7**, 183, 1963.
6. R. Berman, J. C. F. Brook, and D. J. Huntley, private communication.

FAILURE OF MATTHIESSEN'S RULE IN PLUTONIUM

E. King and J. A. Lee

Atomic Energy Establishment
Harwell, England

and

K. Mendelssohn and D. A. Wigley

The Clarendon Laboratory
Oxford, England

In the last few years we have investigated the resistivities at low temperatures of metals of the actinide series.[1] It has been found that, beginning with uranium, the resistivity becomes increasingly anomalous with rising atomic number. In plutonium this anomaly is very pronounced and essentially similar in all three crystal phases which we have been able to investigate. These are α-plutonium, which is stable below 112°C and has monoclinic structure, β-plutonium, which is body-centered monoclinic and stable between 112–184°C and δ-plutonium, a face-centered cubic modification which is stable between 320–450°C. The β-samples[2] used were quenched into liquid nitrogen at which temperature they retain their stability, and the δ-samples were stabilized at room temperature and below by alloying with a few percent of aluminum. In view of the considerable self-heat (2–3 mW/g) and the extreme toxicity of plutonium, special cryogenic techniques have to be applied, which have been described elsewhere.[3]

All three modifications have at room temperature a negative temperature coefficient of the resistivity and a resistivity maximum at low temperatures. This is at about 100°K for the α-phase, at 25°K for the β-phase, and between 150–200°K for the δ-phase, when extrapolated to zero aluminum content. Below the maximum, a sharp drop in the resistivity occurs and that of β-plutonium shows, even at 1.5°K, a considerable temperature variation of 3 μΩ cm/deg.

In addition to these observations, it has been found that if samples of plutonium are held at low temperatures, the resistivity will increase with time, due to the self-damage sustained by their own alpha-activity.[4] We have studied this effect in all three phases of the metal in experiments at liquid-helium temperatures lasting well over 8000 hr. The rate of this increase in resistivity depends on the crystal modification and on the isotopic content, but we have been able to show that, for all three phases, it decreases with accumulated damage, tending ultimately to saturation.[5] As a typical example, the resistivity of a sample of α-plutonium is shown in Fig. 1 as function of temperature. The full curve gives the resistivity as observed on first cooling and shows clearly the resistivity maximum, as well as the rapid fall of resistance with decreasing temperature. This particular specimen was held at 4.5°K for nearly 5000 hr, and during this time the original resistivity increased threefold, the increase having practically reached saturation. It was then warmed up fairly

Failure of Matthiessen's Rule in Plutonium

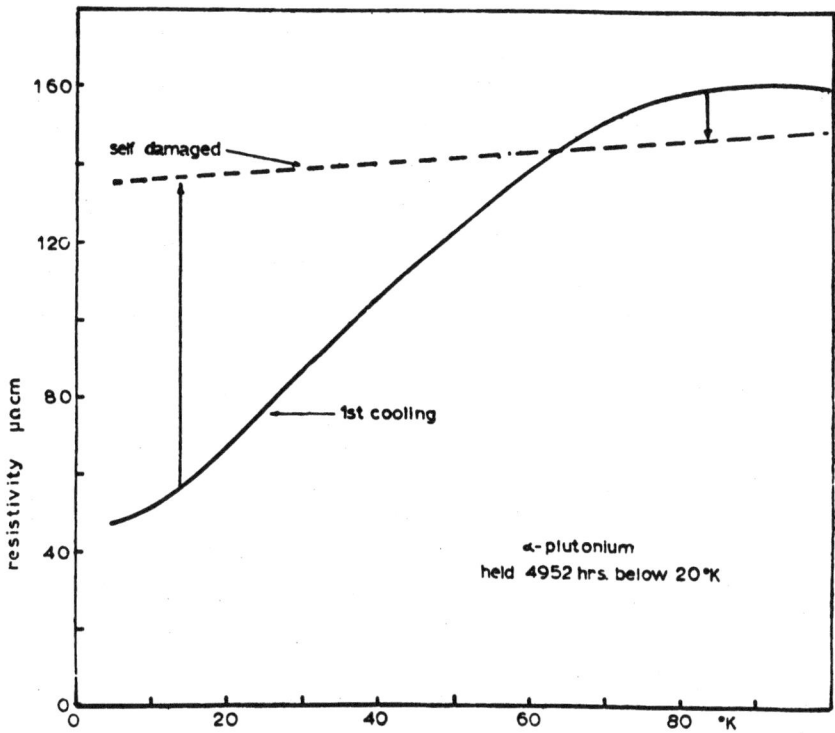

Fig. 1. Resistivity of α-plutonium. Full curve resistivity on first cooling. Dashed curve: resistivity after 5000 hr below 20°K.

rapidly, so as to avoid annealing effects, and its resistivity was recorded during this process. As can be seen from the dashed curve, the resistivity has now lost its strong temperature dependence, increasing only slightly between helium temperatures and those of liquid air. Values at higher temperature could not be obtained because annealing began to proceed at a very rapid rate.

A most remarkable feature of these results is that, from about 50°K upwards, the resistivity of the heavily damaged sample is lower than that in the undamaged state. This means that the additional resistance due to the accumulation of damage is not additive[6] as would be expected on the basis of Matthiessen's rule. In order to investigate this behavior in some detail, another sample of α-plutonium was held at helium temperatures for a little over 3000 hr and periodically heated up to 35°K. Separate experiments had shown that up to this temperature any annealing effects are negligibly small and will not influence the results. At each warming-up of the sample, its resistivity was carefully measured step by step over the range between 4.5–35°K, and the results are shown in Fig. 2. The deviations from Matthiessen's rule are best shown by the dashed curves which give the differences between the accumulated resistivity and the original one which had been obtained on first cooling. These curves, which, if the rule were obeyed, should be straight lines parallel to the abscissa, are all strongly curved.

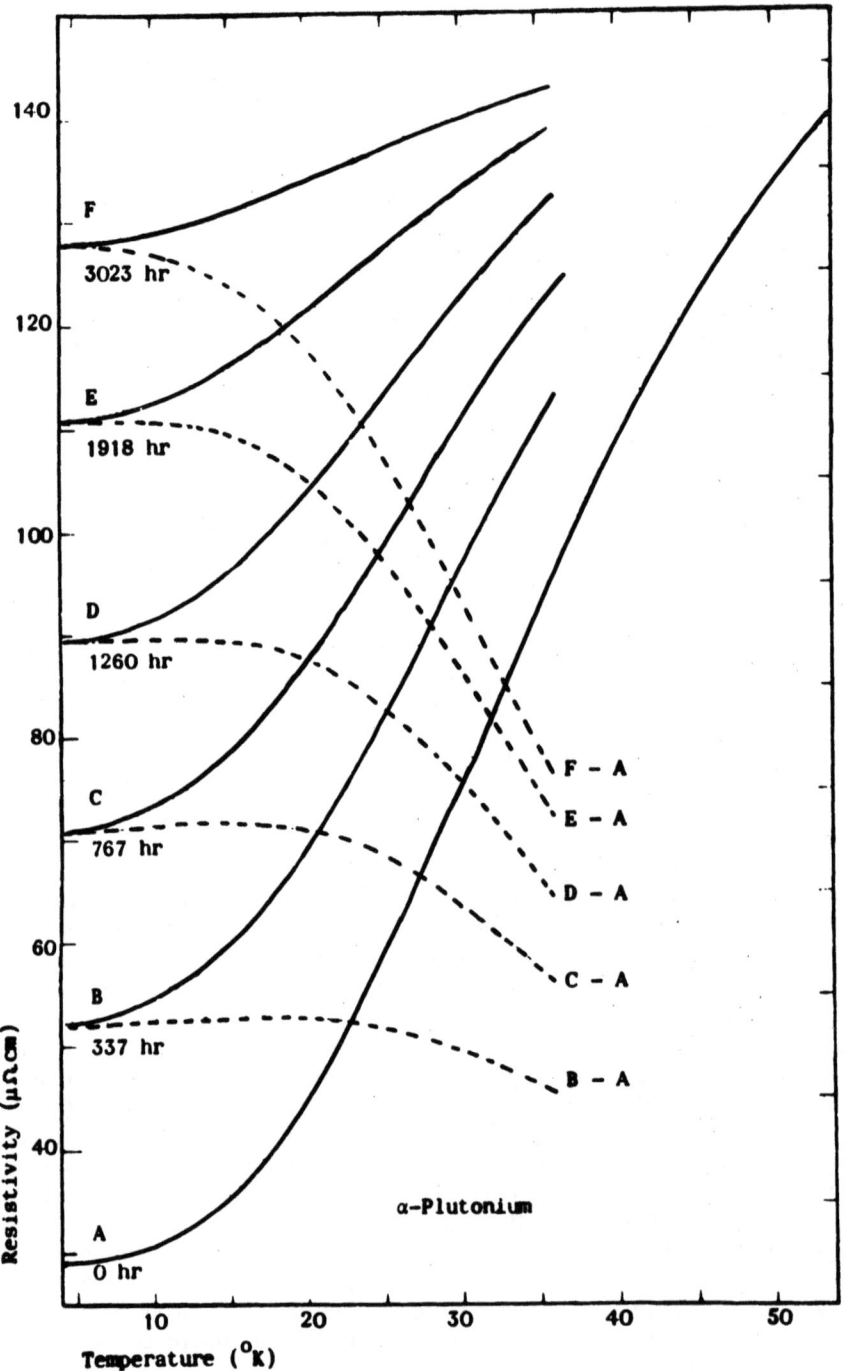

Fig. 2. Dependence on temperature resistivity of α-plutonium on first cooling (A) and after various holding times at 4.5°K (B–F).

Similar results, of even more pronounced deviations from Matthiessen's rule, were obtained on β-plutonium.[7] The results on δ-plutonium show the same behavior although, owing to its generally lesser temperature dependence, the effect is not as large as in the α- and β-phases.

All these results suggest that the increase in resistivity caused by radioactive self-damage is not simply due to the scattering of electrons on the accumulated point defects. In any such mechanism, we should expect Matthiessen's rule to be followed, at least for the smaller damage concentrations. On the contrary, we must assume that the damage primarily affects the mechanism which is responsible for the anomalously high temperature dependence of the resistivity in the undamaged state. At present, there exists no satisfactory explanation for the anomalous resistivity of plutonium, but it should be noted that the anomaly is present in all three phases, indicating that it is not connected with one particular crystal structure. This fact may make explanations based on a complicated band model less convincing. Our own view is that the resistance anomaly may be due to scattering produced in some form of magnetic ordering, although it should be noted that the absence of strong anomalies in the magnetic susceptibility seems to rule out a simple form of antiferromagnetism. On the other hand, magnetic ordering appears to occur in some metals, such as chromium and α-manganese, without showing itself in the susceptibility. More work is clearly required before it can be decided whether the co-operative effect operating in plutonium is a magnetic one.

We conclude then from our results that the strong deviations from Matthiessen's rule show that the observed effect of radiation damage is due to the annihilation of the co-operative phenomenon which is responsible for the anomalous resistivity of plutonium. It is, therefore, of interest to study the effect of self-damage on the resistivity of other metals. We have, therefore, investigated a specimen of neptunium and of uranium-233. In both substances, we were able to establish an increase of resistivity while they were being held at helium temperatures. However, the alpha-activity of neptunium is a hundred times smaller than that of α-plutonium, and the resistivity increase was 300 times smaller. The self-heat of uranium-233 is one-tenth that of α-plutonium, and here the resistivity was found 80 times smaller. With these small effects on the resistivity, no checks can be made of Matthiessen's rule within experimental times of the order of 10,000 hr or less.

We hope to extend our work shortly to metallic protactinium, which has an activity of the same order as plutonium and may be free from anomalous effects in the resistivity. Such experiments should be able to clarify the results obtained on plutonium. Another approach to the problem which is being pursued by us is the study of the effect of radiation damage on substances showing similarity to plutonium. The nearest approach to similar behavior seems to be shown by α-manganese. Should these experiments yield a positive result, the effect of low temperature radiation damage on the resistivity may become a useful method for the detection of magnetic ordering.

References

1. J. A. Lee, G. T. Meaden, and K. Mendelssohn, *Cryogenics* **1**, 52, 1960; G. T. Meaden, *Proc. Roy. Soc. (London) Ser. A* **276**, 553, 1963.
2. E. King and J. A. Lee, *Cryogenics* **3**, 177, 1963.
3. G. T. Meaden and J. A. Lee, *Cryogenics* **1**, 33, 1960.
4. J. A. Lee, K. Mendelssohn, and D. A. Wigley, *Cryogenics* **2**, 183, 1962; C. E. Olsen and R. O. Elliott, *J. Phys. Chem. Solids* **23**, 1225, 1962.

5. J. A. Lee, K. Mendelssohn, and D. A. Wigley, *Phys. Letters* **1**, 325, 1962; E. King, J. A. Lee, K. Mendelssohn, and D. A. Wigley, *Acta. Met.* **12**, 111, 1964.
6. J. A. Lee, K. Mendelssohn, and D. A. Wigley, *J. Phys. Soc. Japan* **18**, Suppl. III, 313, 1963; *Cryogenics* **3**, 46, 1963.
7. E. King, *Cryogenics* **4**, 108, 1964.

PRESSURE DEPENDENCE OF ELECTRICAL CONDUCTIVITY AT LOW TEMPERATURES*

W. S. Goree and T. A. Scott

*Physics Department, University of Florida
Gainesville, Florida*

The pressure dependence of electrical resistivity has been widely studied experimentally to very high pressures in the vicinity of room temperature, but relatively little has been done at low temperatures. From a theoretical viewpoint, the low temperature region is the most interesting because there the temperature dependence of the pressure coefficient of the lattice resistivity $1d\rho_l/\rho_l dP$ is greatest, and also because the impurity resistivity coefficient is significant and measurable.

Important previous investigations include those of Hatton[1,2] to 5000 atm at 4.2°K using piston–cylinder apparatus with solid hydrogen as the pressure transmitting medium, and those of Dugdale and Gugan[3,4] to 3000 atm using solid helium frozen under pressure at constant volume.

We have repeated some of this earlier work and extended it using both experimental techniques. In one case, piston–cylinder apparatus driven by a 50-ton ram was employed with solid helium as the pressure transmitting medium. With this equipment, gold, silver, tungsten, and molybdenum at 4.2°K and tantalum at 4.5°K were studied to 8000 atm. Solid helium was used rather than hydrogen because of its superiority in transmitting uniform pressure. However, the soft metals are still seriously deformed by pressure gradients in the solid, even when the pressure is applied slowly. This results in a marked hysteresis in the resistance values for increasing and decreasing pressure, as noted by Hatton.[1] The hysteresis decreases fairly rapidly with repeated pressure cycles—for example, in one experiment on silver, the pressure coefficients obtained during the first three cycles were 6×10^{-6} atm^{-1}, 3×10^{-6} atm^{-1}, and 1.9×10^{-6} atm^{-1}, respectively. An asymptotic curve is approached which may be taken as the true result, but the interpretation is clearly clouded by this circumstance. The asymptotic initial slopes, $1d\rho/\rho dP$, obtained in this way at 4.2°K are shown in Table I. All the samples showed a positive pressure coefficient, characteristic of impurities or deformation. It should be remarked, too, that the curves for gold and silver show a markedly increasing slope at the higher pressures, which is probably erroneous and responsible for our quoting only the low pressure coefficients.

A comparison was made between solid hydrogen and helium in separate experiments on silver, and it was found that the initial hysteresis was much worse with solid hydrogen, but the asymptotic values of $1d\rho/\rho dP$ approached after many cycles appear to be approximately the same.

For the hard metals, hysteresis, not attributable directly to friction of the piston seal, is slight, even on the first cycle, and consequently the piston–cylinder method is satisfactory. For both tantalum and molybdenum a peculiar effect noted

* Supported by the National Science Foundation.

Table I

Sample	Diameter, in.	Purity, %	$\dfrac{R_{4\cdot 2}}{R_{77\cdot 3}}$	$\dfrac{1}{\rho_{4\cdot 2}}\dfrac{d\rho_{4\cdot 2}}{dP} \times 10^{-6}\,\text{atm}^{-1}$			$-\dfrac{d\ln\theta R}{d\ln V}$	$\dfrac{d\ln A}{d\ln V}$
				Piston–Cylinder		Gas		
				This work	Hatton			
Au	0.005	99.999	0.0253	10.3	9.3	9.04		
Ag	0.025	99.95	0.0740	3.1	4.3			
Ag	0.005	99.999	0.0308		4.3	2.16	1.8	−1.3
Ta	0.085	99.999	0.0314	2.3	−0.14			
Mo	0.085	99.999	0.0047	0.57				
W	0.085	99.999	0.452	1.3				
In	0.020	99.999	0.001		−8.2		2.4	−0.4

earlier by Hatton[2] for tantalum was observed; the resistance curve measured for decreasing pressure lies below that for increasing pressure, but joins on again at zero pressure. It is important to note, however, that our measurements on tantalum disagree with Hatton's as to the sign of the initial pressure dependence.

The problems associated with piston–cylinder apparatus have motivated us to discard it for a more tedious, but much more satisfactory, technique. Substantial and apparently hydrostatic pressure may be achieved with solid helium by first applying the pressure at a temperature such that the helium is fluid and then freezing under constant-pressure conditions. We have used a large helium gas pressure facility capable of pressures to 14,000 atm to study silver, gold, and indium to 6000 atm pressure at temperatures between 4.2–77.3°K. Because helium is so compressible and the zero-point energy large, it is possible to obtain accurate, reproducible, and almost isobaric temperature dependences as the pressurized solid is cooled.

Pressure influences electrical resistance through a variety of mechanisms, so that unfortunately it is nearly impossible from resistivity data alone to arrive at a detailed interpretation which is unambiguous. However, some progress may be made in terms of Bloch–Gruneisen theory, and also the qualitative comparison of different materials is valuable. Following Dugdale,[4] we shall suppose that the lattice resistivity at constant volume may be written as

$$\rho_l = \frac{A}{T} F\!\left(\frac{T}{\theta_R}\right)$$

where A is a "catch-all" factor depending on the details of the electron–phonon interaction and the nature of the Fermi surface, and θ_R is the characteristic temperature. From this equation, one readily deduces that

$$\frac{d\ln\rho_l}{d\ln V} = \frac{d\ln A}{d\ln V} - \frac{d\ln\theta_R}{d\ln V}\!\left(1 + \frac{d\ln\rho_l}{d\ln T}\right)$$

The quantities $d\ln\rho_l/d\ln V$ and $d\ln\rho_l/d\ln T$ are obtained from the experimental pressure and temperature data, respectively, after correcting the temperature data to constant volume. Thus, one may deduce the volume dependence of A and θ_R.

On the basis of this theory one expects the pressure dependence of the lattice resistivity to be negative, and furthermore, from the form of $F(T/\theta_R)$, to be approximately three times greater (negatively) at low temperature than at high. On the other hand, for reasons not well understood, the impurity resistivity usually has a positive pressure coefficient. This results in a crossing of the resistance isobars at a low temperature where the two coefficients become equal. For our gold and silver samples, this crossing occurs at about 14°K. For indium, which was reported by the supplier to have the same purity as the gold and silver samples, we were unable to measure accurately the residual resistance at 4.2°K, and it would appear that either the sample was purer than the others, or that impurities in indium do not affect the resistance so significantly as in gold and silver. In this regard, it should be noted that Hatton[2] also observed a negative pressure coefficient at 4.2°K.

In Fig. 1, we show $1 d\rho/\rho dP$ for gold, silver, and indium obtained with the gas system at the fixed normal boiling points of helium, hydrogen, and nitrogen and the triple points of hydrogen and nitrogen. At 14 and 20.4°K, measurements were made only with fluid helium up to the fusion pressure at those temperatures, while at all other temperatures the measurements were made to 6000 atm. The resistance was found to be a linear function of pressure in all cases. At 4.2°K, the impurity resistivity coefficient of gold is much larger than that of our silver sample, which in turn is larger than the value reported by Dugdale and Gugan[3] for copper, indicating perhaps a natural progression. The coefficient of the indium sample remains negative down to at least 14°K. Relevant to this, one should note in Table I the negative value obtained by Hatton at 4.2°K with his piston–cylinder apparatus.

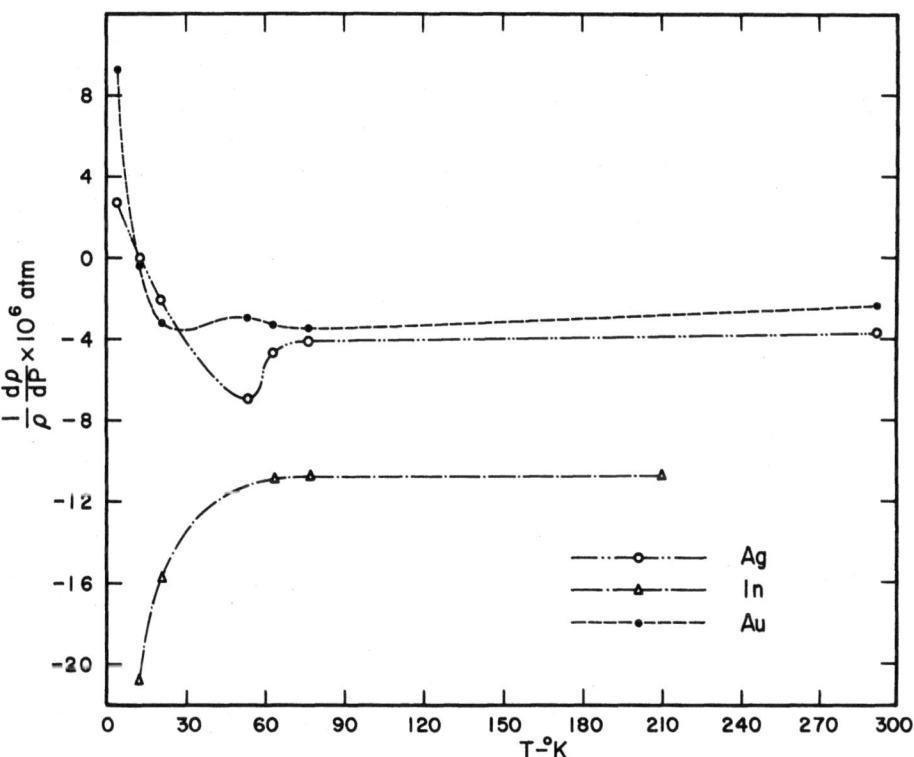

Fig. 1. The measured pressure coefficient of total resistivity of gold, silver, and indium as a function of temperature.

In Fig. 2, we show our present data for $d \ln \rho_l/d \ln V$ versus $(1 + d \ln \rho_l/d \ln T)$. In the case of silver, a straight line can be drawn through the points to as low as 20°K, as shown, but the 14°K-point falls far off this line. However, this apparently good straight-line fit is probably misleading because one does not expect the theory invoked to be very good below about $\theta_R/4$. The slope obtained gives a Gruneisen constant of 3.39 which is too large, and in disagreement with Dugdale's[4] value of 2.4 obtained from Bridgman's pressure data. Most likely, only the 77° and 63°K measurements should be considered in selecting the straight line; these give a lower slope and intercept, as quoted in Table I, which are in fair agreement

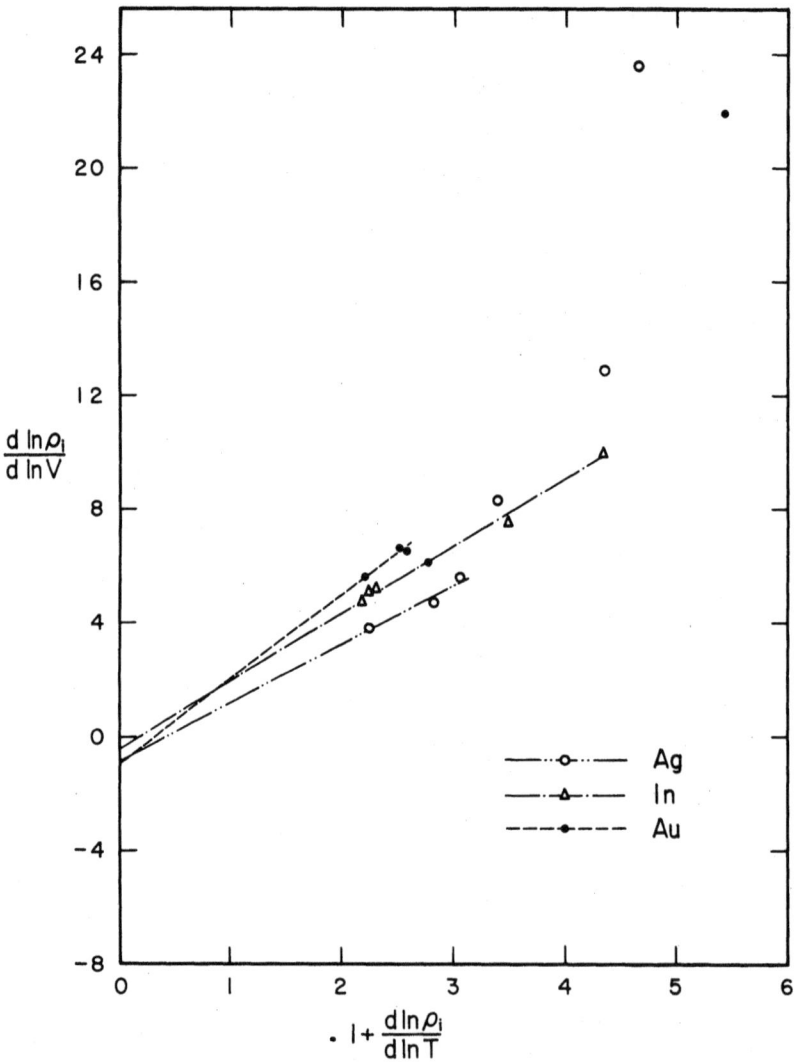

Fig. 2. The logarithmic volume coefficient of the lattice resistivity at constant density of gold, silver, and indium shown as a function of the logarithmic temperature coefficient.

with Dugdale. In the case of indium, all the points fit a straight line, yet it is known from thermal expansion measurements[5] on indium that the Gruneisen constant begins to vary below about 50°K. The case of gold shows much greater variations at the temperatures measured, and we have not attempted to draw any quantitative conclusions. Further measurements and analysis are required and are in progress.

Table I contains a summary of the most pertinent data. An interesting feature is the comparison of the pressure coefficients at 4.2°K for gold and silver obtained with the gas systems and that obtained with piston–cylinder apparatus by Hatton and by us.

References

1. J. Hatton, *Phys. Rev.* **100**, 1784, 1955.
2. J. Hatton, *Phys. Rev.* **103**, 1167, 1956.
3. J. S. Dugdale and D. Gugan, *Proc. Roy. Soc. (London) Ser. A* **241**, 397, 1957.
4. J. S. Dugdale, *Science* **134**, 77, 1961.
5. N. Madaiah and G. M. Graham, *Can. J. Phys.* **42**, 221, 1964.

PHASE TRANSFORMATION IN SODIUM AND ITS EFFECT ON THE ELECTRONIC SPECIFIC HEAT

D. L. Bhattacharya and Edward A. Stern

University of Maryland
College Park, Maryland

Sodium undergoes a martensitic phase transformation on cooling (from bcc to faulted hcp structure) at around 35°K, as was first found by Barrett.[1] The amount of transformation estimated from the area of the X-ray diffraction peaks was 7% for a coarse-grained sodium sample. Later, measurements of other properties of sodium[2] showed that, for normal sodium samples used in calorimetry or other experiments, the amount of spontaneous transformation on cooling is around 50%. There is one observation[3] which showed that a cast sample of sodium did not transform at all on cooling to liquid-helium temperatures for the first time.

It was pointed out by Stern[4] that the results for the specific heat of sodium, as determined calorimetrically until 1960, show that all experimental results agree as to the Debye temperature of about 158°K, but there is considerable variation in γ, the electronic heat coefficient. He analyzed the first four experimental results given in Table I, and came to the conclusion that the different experimental values for γ are due to the presence of a mixture of the two phases of sodium. These phases contribute separately to the total specific heat and their relative amounts are variable from sample to sample because of differences in thermal history. Martin[5] subsequently undertook an extensive series of measurements of the electronic specific heat of sodium in the temperature range 0.4–1.5°K. He used samples with different thermal histories and concluded (under the assumption that approximately 50% transformation takes place in the first cooling, and the transformation is inhibited on warming up the sample only to liquid-nitrogen temperature, so that only 25% transforms on second cooling) there is no significant variation in γ on thermal cycling. Hence, γ is unaffected by the phase transformation. It should be mentioned that Martin reports also that his first four measurements gave a statistically significant higher value of γ from the average of the final runs with cast and block specimens. He did not attribute this to the phase transformation.

In view of these observations, it was felt that a more controlled determination of γ is necessary. It is known that Young's modulus of sodium undergoes an abrupt change when a phase transformation occurs,[6] and so can serve as an indicator for the transformation. We have, therefore, constructed a cryostat in which Young's modulus (resonant frequency for longitudinal vibrations) of the specimen can be measured with an electrostatic drive system. The heat capacity of the same sample can be measured with the same cryostat at liquid-helium temperatures using the cooling-curve method.[7]

The transformation behavior of sodium bars was followed in the range 70–20°K by measuring the resonant frequency. The strongest resonance was chosen for the

Table I. Comparison of the Electronic Heat Coefficient γ and the Effective Thermal Mass Ratio m^*/m_0 and Debye Temperature θ_D of Sodium Obtained by Different Investigators

Temperature range	Investigator	γ mJ/°K^2-mole	m^*/m_0	θ_D(°K)	Sample preparation
1.6–4.2°K	Parkinson and Quarrington[a]	1.80 ± 0.18	1.64 ± 0.16	158	block
1.6–4.2°K	Roberts[b]	1.37 ± .0.04	1.25 ± 0.03	158	cast
0.4–2°K	Gaumer and Heer[c]	1.32	1.21	158	cast
0.15–1°K	Lien and Phillips[d]	1.45	1.33	156	cast
0.4–1°K	Martin[e]	1.40 ± 0.02	1.28 ± 0.02	157 ± 2	cast
		1.36 ± 0.03	1.25 ± 0.03	155.5 ± 3	block
4.2°K	Grimes and Kip[f]	–	1.24 ± 0.02*	–	single crystal with (111) plane surface

* Cyclotron mass ratio.
[a] D. H. Parkinson and J. E. Quarrington, *Proc. Phys. Soc. (London) Ser. A* **68**, 762, 1955.
[b] L. M. Roberts, *Proc. Physi. Soc. (London) Ser. B* **70**, 744, 1957.
[c] R. E. Gaumer and C. V. Heer, *Phys. Rev.* **118**, 955, 1960.
[d] W. H. Lien and N. E. Phillips, *Phys. Rev.* **118**, 958, 1960.
[e] D. L. Martin, *Phys. Rev.* **124**, 438, 1961.
[f] C. C. Grimes and A. F. Kip, *Phys. Rev.* **132**, 1991, 1963.

purpose. The specimen was mounted in a nylon frame and supported at the midpoint with nylon screws. A heater–germanium thermometer capsule was attached to the center by drilling a tapered hole in the specimen. The specimen was surrounded by an evacuated can and cooled slowly toward the liquid-helium bath temperature mainly through a thick copper wire cast with the sample. Cooling rate was adjusted with constantan wires soldered at the other end of the copper wire.

A capacitance bridge working at a frequency of 47.6 Mcps was used to detect the capacitance change between the drive electrode and the top surface of the specimen, the electrostatic drive being obtained from an oscillator–amplifier system suitably decoupled from the bridge. The resonance was observed on a cathode ray oscilloscope and the frequency monitored by a frequency counter and recorder.

In order to check our e.s. drive system and the bridge assembly, we measured Young's modulus of copper under identical conditions as in the sodium experiment. The copper used was 99.999% pure and in as-cast (as-obtained) condition. The resonant frequency of copper increased slowly with temperature and became constant below about 30°K. The accuracy of frequency determination is better than 0.05%.

The sodium used was of two purities—DuPont reactor grade (resistance ratio, 1100) and high-purity sodium† (resistance ratio, approx. 7000). Cylindrical bars of about 3-in. length and $\frac{3}{8}$-in. diameter were made by melting under oil in a stainless steel tube provided with a mild steel plug.

In our experiments with sodium, the material was prepared by casting the sodium under oil, cooling the melt in the furnace, cooling in air, quenching in liquid nitrogen, and storing at room temperature. It was found that in all cases the grain

† Kindly supplied by Dr. C. E. Taylor, Lawrence Radiation Laboratory, Livermore, California.

sizes of sodium were rather large (0.5–3 mm) and there was no appreciable grain growth in about an 8 hr time interval.

The experimental results to be described were therefore made on coarse-grained sodium samples, which would be expected to have fairly pronounced preferred orientation with respect to the axis of the cylindrical specimens. As such, it is not easy to compare the results from different samples, but the results obtained with a single sample show distinct features which we now summarize. The range of temperatures is 70–20°K.

1. The resonant frequency of a given sample increases as temperature decreases.
2. At around 35°K, the resonant frequency f_0 for a single mode of vibration usually drops from 1 to 10%, showing the onset of transformation. Damping increases considerably.

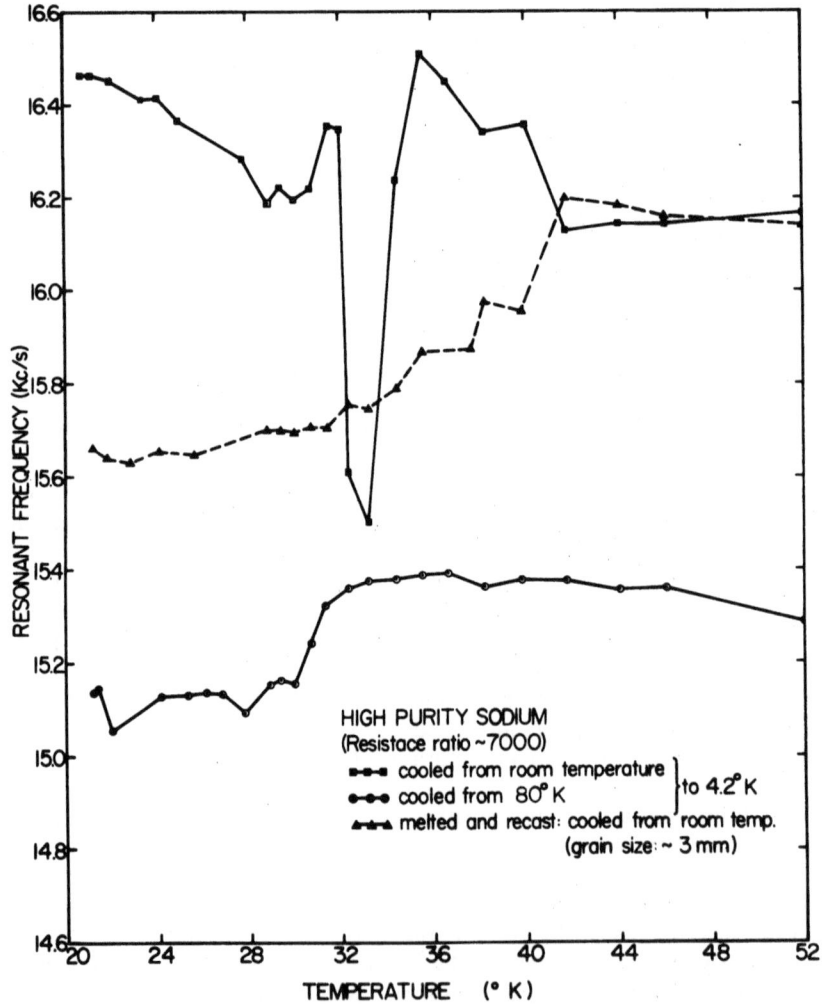

Fig. 1. Resonant frequency of vibration of high-purity sodium bar (approx. 3 in. long) as a function of temperature.

3. After 35°K, the resonant frequency increases again, but sometimes shows a decrease as the temperature is decreased.

On cooling to 1.5°K and warming up again, we find the following features:
1. The resonant frequency at 70°K is always smaller than the original cast samples. This happens both when the sample is warmed up to room temperatures and allowed to anneal for one week or so, or when the sample is kept surrounded by liquid nitrogen before cooling down again.
2. On thermal cycling, the fractional drop in frequency is not the same as in the first cooling, and the transformation does not occur at exactly the same temperature.
3. The drop in frequency at the transformation temperature after storing at liquid-nitrogen temperature can be smaller or greater than that of the first cooling.

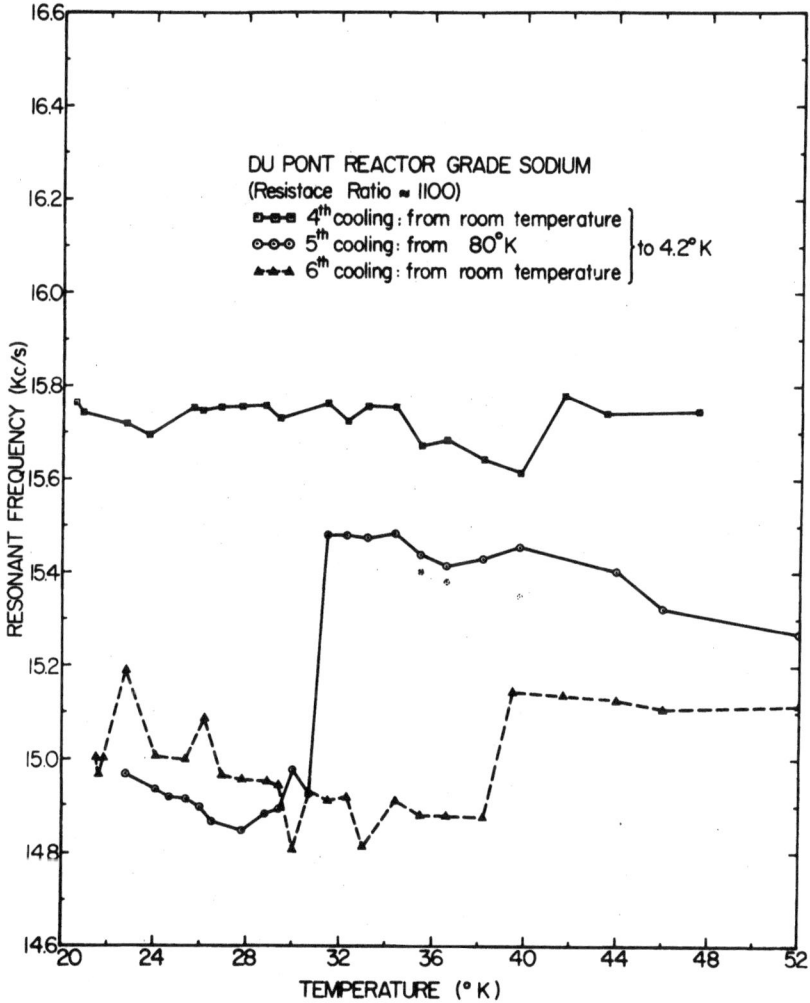

Fig. 2. Resonant frequency of vibration of DuPont reactor grade sodium bar (approx. 3 in. long) as a function of temperature.

These features are shown in Fig. 1 for a sample of highly-pure sodium and in Fig. 2 for an impure sample.

Our interpretation of these observations is as follows:

1. The drop in the resonant frequency at 70°K on thermal cycling is due to the alteration of the texture of the grains, Young's modulus being practically unaffected by internal stresses. The specimen recrystallizes on thermal cycling. The grain size of a thermally-cycled sample seems to become coarser than that of the cast specimen, and there is appreciable grain growth on storing at room temperature.
2. If the specimen has a pronounced texture, it will show a variable amount of transformation on cooling below the transformation temperature. This is observed as a difference in the magnitude of the frequency decrease at the transformation temperature. This also explains why the decrease in frequency is different in samples held at nitrogen temperatures.
3. The shift in the transformation temperature can be attributed to grain-size difference.
4. The fractional transformation in the specimen cannot be estimated from the drop in frequency. The grains in the sample were so large that any orientation change in the grains during transformation would change Young's modulus in an erratic fashion which depends on the initial orientations. This would obscure any effect due to a change in the average Young's modulus.
5. The average Young's modulus for the low temperature phase is smaller than that for the high temperature phase, while their Debye temperatures are in the opposite order. This is not inconsistent, because the Young's modulus depends mainly on the larger elastic constants, while the Debye temperature gives greater weight to the smaller elastic constants.

The electronic specific heats of various samples of sodium have been determined at liquid-helium temperatures in the same cryostat. The γ-values are [mJ/mole-(deg)^2 units] 1.38 ± 0.10 ($\theta_D = 155.5 \pm 2°K$) and 1.40 ± 0.10 ($\theta_D = 154 \pm 2°K$), respectively, for the first-cast and thermally-cycled samples. There is no significant difference between these samples, in agreement with Martin.[5] However, no definite conclusions can be drawn about the specific heat of each phase since the measurement of Young's modulus failed to monitor the amount of transformation.

Acknowledgment

Our thanks are given to Mr. J. J. Sabo for helping with the experiments.

References

1. C. S. Barrett, *Acta Cryst.* **9**, 671, 1956.
2. J. S. Dugdale, Geneva Conference on Electronic Properties of Metals at Low Temperatures (1958).
3. K. D. Alexopoulos and C. H. Shaw, *Ohio State University Technical Report* 809-3, 1961.
4. E. A. Stern, *Phys. Rev.* **121**, 397, 1961.
5. D. L. Martin, *Phys. Rev.* **124**, 438, 1961.
6. J. S. Dugdale, Geneva Conference on Electronic Properties of Metals at Low Temperatures (1958); L. Verdini, *Proceedings of the Third International Congress on Acoustics*, L. Cremer (ed.), Elsevier Publishing Company, Amsterdam.
7. J. K. Logan, J. R. Clement, and H. R. Jeffers, *Phys. Rev.* **105**, 1435, 1957.

EFFECT OF SPIN-ORBIT COUPLING ON THE HYPERFINE CONTACT CONTRIBUTION TO THE KNIGHT SHIFT

J. C. Appel

General Atomic Division of General Dynamics Corporation
San Diego, California

Introduction

In earlier nuclear magnetic resonance (NMR) measurements on mercury[1] and tin,[2] it was found that the Knight shift in the superconducting (SC) state does not vanish as $T \to 0$. If one assumes that the Knight shift is caused merely by the hyperfine (hf) contact interaction, and thus is proportional to the electron spin susceptibility, the experimental results are incompatible with the BCS theory, the ground-state wave function of which consists of quasibound electron pairs in singlet spin states. Several distinct explanations have been offered to resolve the discrepancy[3]; in particular, Anderson[4] and Ferrel[5] have shown that spin-reversing scattering, arising from spin-orbit coupling (SOC) effects in Coulomb scattering events, can account for a finite hf contact contribution to the Knight shift in the SC ground state. Consequently, NMR experiments were undertaken on two SC metals with small atomic numbers, vanadium and aluminum, in which spin-reversing scattering caused by atomic imperfections is expected to be small. For vanadium, Noer and Knight[6] observed no change in the Knight shift below T_c (transition temperature). This result is attributed to the dominant role played by the orbital part of the hf interaction. For aluminum, however, there is no unfilled d band and thus the orbital interaction[7] will be small compared with the hf contact interaction. Furthermore, a third contribution to the Knight shift, arising from exchange interactions between spin-polarized conduction electrons and core electrons and from hf contact interactions between spin-polarized core electrons and nuclear magnetic moments, may or may not be small[8]; in any case, however, it does not contribute to the Knight shift in the BCS ground state. An NMR experiment on aluminum then provides a relevant test on the significance of spin-reversing scattering. If the lifetime of an electron in a "spin-up" or a "spin-down" state is $\tau \gg \hbar/kT_c$, the Knight shift should vanish as $T \to 0$. For aluminum, this condition requires a lifetime $\tau \gg 10^{-11}$ sec[4], which is not a particularly large time, so that a nearly vanishing Knight shift may reasonably be expected as $T \to 0$.[9] Careful experimental observations by Hammond[10] on aluminum films of 200 Å thickness show, however, that the Knight shift decreases by only 25% of its value at T_c, as $T \to 0$. In view of this result, the question was raised by Ferrel[11] whether SOC effects arising from the crystal potential produce a significant high-frequency contribution to the spin susceptibility. Let us recall that SOC forces, considered as a small perturbation, cause admixture of wave functions from bands with proper symmetry into the conduction band wave functions. A resulting two-component Bloch spinor $\varphi_{m\rho}(\mathbf{k}, \mathbf{r})$ contains a component from each proper band which is proportional to

$\lambda(\mathbf{k})/\Delta E(\mathbf{k})$ where λ = SOC energy, ΔE = band gap, m = band index, and ρ = spin index.[12] Usually, $\Delta E \gg kT_c$; but we shall find that SOC forces can be relevant only if $\lambda/\Delta E$ is of the order of 1.

General Theory

The detailed theoretical discussion of the effect of spin-orbit coupling on the hf contact contribution to the Knight shift is in preparation. In this note, we confine ourselves to a summary of the mathematical procedure and of our chief results. We proceed by:
1. Calculating the relativistic eigenfunctions of conduction electrons belonging to a simple band, in a static magnetic field $H = H_z$.
2. Computing, in the representation (1), the diagonal elements of the hf contact interaction between the conduction electron system and the dipole field of a nucleus with magnetic moment $\mathbf{\mu}_{nuc}$, considered as fixed in z direction.
3. Writing down the perturbation expansion of the free energy F in terms of procedure (1).

That part of the free energy which is linear in H, F_1, gives us the Knight shift, $\Delta H/H = -F_1/\mu_{nuc}H$. An alternative procedure of determining F_1 is Wilson's density matrix method,[13] which allows one to work with unperturbed Bloch states.

As for procedure (1), one is concerned with the solution of the Pauli equation describing the relativistic motion of a single conduction electron in an effective periodic potential $V(\mathbf{r})$ and in a homogeneous magnetic field H:

$$\left\{\frac{\mathbf{P}^2}{2m} + \frac{\hbar e}{4m^2c^2}(\boldsymbol{\mathscr{E}} \times \mathbf{P})\cdot\boldsymbol{\sigma} + \frac{\hbar e}{4im^2c^2}\boldsymbol{\mathscr{E}}\cdot\mathbf{P} - \frac{e\hbar}{2mc}\boldsymbol{\sigma}\cdot\mathbf{H}\right.$$
$$\left. + \left(1 + \frac{\mathbf{P}^2}{4mc^2}\right)\left[\frac{\hbar}{i}\frac{\partial}{\partial t} + V(\mathbf{r})\right]\right\}\Psi = 0 \quad (1)$$

Here e is the electron change, $e\boldsymbol{\mathscr{E}} = \nabla V(\mathbf{r})$, and $\mathbf{P} = \mathbf{p} - (e/c)\mathbf{A}$ is the canonical momentum. Equation (1) is gauge invariant. To find its solution for the stationary case, $\mathscr{H}_0\Psi = (\hbar/i)\partial\Psi/\partial t = E\Psi$, we find it convenient to extend Wannier's formalism[14] for Bloch electrons in a magnetic field to the relativistic Hamiltonian $\mathscr{H}_0$, which incorporates spin-orbit coupling, the s shift correction, and the mass–velocity correction. One then arrives at a system of two simultaneous equations for the two components of a Bloch-type spinor:

$$\varphi_{m\rho}(\mathbf{r}, \mathbf{k}) = e^{i\mathbf{k}\cdot\mathbf{r}}[u_{m\rho}(\mathbf{r}, \mathbf{k})\alpha + v_{m\rho}(\mathbf{r}, \mathbf{k})\beta] \quad (2)$$

and for the Fourier coefficients of the energy band function:

$$W_{m\rho\rho'}(\mathbf{k}) = \sum_{\mathbf{R}} \omega_{m\rho\rho'}(\mathbf{R}) e^{i\mathbf{k}\cdot\mathbf{R}} \quad (3)$$

where $\mathbf{R}$ is a lattice vector.

The functions $u_{m\rho}$, $v_{m\rho}$, and $\omega_{m\rho\rho'}$ can be found as power series expansions in H. The Hamiltonian $\mathscr{H}_0$ of the simultaneous equations couples only the two spinor components $u_{m\rho}$, $v_{m\rho}$ of the same band index m but of the same and of different spin index ρ. With the help of the field-dependent spinors $\varphi_{m\rho}(\mathbf{k}, \mathbf{r})$, an eigenfunction of equation (1) can be written in the form

$$\Psi_{m\rho}(\mathbf{r}) = \sum_{\rho'} \int_{\Omega^*} \mathscr{f}_{m\rho\rho'}(\mathbf{k})\varphi_{m\rho'}\left(\mathbf{r}, \mathbf{k} - \tfrac{1}{2}\frac{e}{\hbar c}\mathbf{H} \times \mathbf{r}\right) d\mathbf{k} \qquad (4)$$

(gauge: $\mathbf{A} = \tfrac{1}{2}\mathbf{H} \times \mathbf{r}$)

where the amplitude function $\mathscr{f}_{m\rho\rho'}$ is found as eigenfunction of the effective Hamiltonian $W_{m\rho\rho'}[\mathbf{K} = \mathbf{k} - \tfrac{1}{2}(ie/\hbar c)\mathbf{H} \times \partial/\partial \mathbf{k}]$, and Ω^* is the volume of the fundamental Brillouin zone. In order to arrive at tractable eigenfunctions $\Psi_{m\rho}(\mathbf{r})$, it is assumed that the effective Hamiltonian $W_{m\rho\rho'}$ is composed of an orbital part which is parabolic in $\mathbf{K}$ and a spin part which is found as first order term of the expansion of $W_{m\rho\rho'}(\mathbf{k})$ in terms of H. This part defines an effective g factor, $g_m(\mathbf{k})$, which depends on the direction cosines of $\mathbf{H}$. In this parabolic approximation, the validity of which will be discussed elsewhere, $\mathscr{f}_{m\rho\rho'}$ is the momentum-representation eigenfunction of a free electron in a magnetic field. The effect of the lattice and of dynamic electron–electron interactions is accounted for only roughly by a proper choice of the effective mass m^*. The wave functions $\Psi_{m\rho}(\mathbf{r})$ of the conduction electrons, found in the parabolic approximation, are used for the computation of the diagonal elements of the hf contact interaction which for the ith electron is given by $\mathscr{H}_{hf} = -(8\pi/3)$ $\boldsymbol{\mu}_{nuc} \cdot \boldsymbol{\mu}_e \delta(\mathbf{r}_i)$, where $\boldsymbol{\mu}_e = (g_0/2)\beta_e \boldsymbol{\sigma}$ with $\beta_e = -|e|\hbar/2mc$. Finally, the free energy of N conduction electrons interacting with a fixed nuclear moment is given by

$$F = F_0 + F_1 + \cdots \qquad (5)$$

where F_0 is the free energy of conduction electrons and nuclear moment in the field H, and where:

$$F_1 = \mathrm{tr}\{\rho_N(1, 2, \ldots, N)[\mathscr{H}_{nf}(1) + \mathscr{H}_{nf}(2) + \cdots \mathscr{H}_{nf}(N)]\} \qquad (6)$$

In this expression, ρ_N is the N-particle-density matrix operator. Since $\mathscr{H}_{hf}$ is a one-electron operator, and since electron–electron interactions are ignored, we can express F_1 in terms of the one-electron distribution function for Fermi-Dirac statistics. The final result can then be written in the form[15]:

$$\frac{\Delta H}{H} = K = \frac{16\pi}{3}\frac{\beta_e^2 g g_0}{4}\rho(\zeta)\sum_\rho |c_{\uparrow\rho}|^2 [|u_{m\rho;0}(k_\zeta)|^2]$$

$$+ \frac{16\pi}{3}\beta_e \sum_\mathbf{k}\sum_\rho |c_{\uparrow\rho}|^2 \left\{\sum_{m' \neq m}\sum_{\rho' \neq \rho} A_z(m, \rho, k; m', \rho', k)\right.$$

$$\left. \times [u^*_{m\rho;0}u_{m'\rho';0} - v^*_{m\rho;0}v_{m'\rho';0}] + \mathrm{compl.\ conj.}\right\} \qquad (7)$$

Here $\rho(\zeta)$ is the density of states at the Fermi surface, $g_0 = 2$, and $u_{m\rho;0} = u_{m\rho;0}(0, \mathbf{k})$. The spinor components u and v and the coefficients A_z are defined by the power series expansion of the field-dependent Bloch functions:

$$\varphi_{m\rho}(\mathbf{r}, \mathbf{k}) = \varphi_{m\rho;0}(\mathbf{r}, \mathbf{k}) + H^\alpha \varphi_{m\rho;\alpha}(\mathbf{r}, \mathbf{k}) + \cdots \qquad (8)$$

where

$$\varphi_{m\rho;\alpha}(\mathbf{r}, \mathbf{k}) = \sum_{m',\rho',k'} A_\alpha(m, \rho, k; m', \rho', k')\varphi_{m\rho';0}(\mathbf{r}') \qquad (9)$$

The coefficients $c_{\rho\rho'}$ are defined by the series expansion of the amplitude function $f_{m\rho\rho'}$ in equation (4):

$$f_{m\rho\rho'} = c_{\rho\rho'}[f_{m,0}(\mathbf{k}) + H^\alpha f_{m,\alpha}(\mathbf{k}) + \cdots] \tag{10}$$

where $f_{m,0}(\mathbf{k})$ is an eigenfunction of $W_{m\rho\rho}(\mathbf{K}; H = 0)$. If spin-orbit coupling is ignored, $g = g_0$, $c_{\rho\rho'} = 0$ for $\rho \neq \rho'$, and $v_{m\rho'0} = 0$, so that the first part of K_c becomes equal to the original expression of Townes, Herring, and Knight.[16] The second part vanishes since $A_z(\rho \neq \rho') = 0$ for this case. For the special case where spin-orbit coupling is *weak*, g depends in first order and $|c_{\rho\rho'}|^2$ and $|u_{m\rho;0}|^2$ in second order on a smallness parameter $\lambda_\zeta/\Delta E_\zeta$ where λ_ζ is an average spin-orbit coupling energy for Bloch electrons at the Fermi surface and where ΔE_ζ is an average energy gap between conduction band states at the Fermi surface and excited states to which the orbital angular momentum connects. The second part of K_c is proportional to $\lambda(E)/[\Delta E(E)]^2$; it depends on the coefficients A_z, which are defined as coefficients in the series expansion (9) for the first-order correction to the zero-field Bloch spinor $\varphi_{m\rho;0}$. If one assumes that the second spin-orbit contribution is positive, the total spin-orbit contribution to the Knight shift is given by

$$K^{so}\left[\delta g(\zeta) + \int_0^\zeta \frac{\rho(E)}{\rho(\zeta)} \frac{\lambda(E)}{[\Delta E(E)]^2} dE\right] K$$

It is this remaining part of K_c which remains unaffected when a metal becomes superconducting, and $T \to 0$. For aluminum the ratio $\lambda/\Delta E$ is of the order 0.01 and so is the ratio K^{so}/K.

Spin-Reversing Scattering

Whereas for the heavy metals tin and mercury there can be a significant high-frequency contribution to the Knight shift from SOC effects discussed in the preceding section, this contribution accounts for only a few percent of the Knight shift in aluminum. Therefore, in this case, one suspects another dominant source for the discrepancy. As for spin-orbit coupling effects, in addition to the high-frequency contribution to the Knight shift arising from the lattice, there is a broadening of the low-frequency contribution (corresponding to the Pauli paramagnetism of free electrons) due to orbital scattering of electrons at nonmagnetic imperfections.[4,5] If the broadening of the low-frequency contribution (which is a δ-function for free electrons) becomes comparable to the energy gap $\sim 3.5\, kT_c$, the Knight shift closely remains finite for the SC ground state. Anderson[4] has discussed this broadening in terms of scattered states, which are linear combinations of Bloch states. For the absorptive part of the spin susceptibility he finds the expression

$$\chi''(\omega) \propto \text{fct}(\omega\tau) = \frac{1}{1 + \omega^2\tau^2}$$

The parameter τ is determined by the "spread" of the scattered states in the spin variable, not in the momentum variable; τ can be considered as the average life time of a Bloch electron in a pure spin state. Anderson's final result for the Knight shift ratio in the normal and SC states depends on $\tau/\hbar k T_c$.

Let us roughly estimate τ. To that end we shall also take into account the effect of SOC effects arising from the lattice. Then, with orbital scattering events, there occur two matrix elements between Bloch spinor components which are of interest. The first element connects the large spinor components of $\varphi_{m\rho}(\mathbf{r}, \mathbf{k})$ and $\varphi_{m\rho}(\mathbf{r}, \mathbf{k}')$ via the SOC strength of the imperfection. The second connects a large and a small component with the help of the screened Coulomb field of an imperfection. Both are of comparable magnitude, and it is readily seen that $\tau \propto (\Delta E/\lambda)^2$. In particular, if one assumes that $(\lambda/\Delta E)^2$ is the probability for a spin-reversing scattering event, then for a thin film with thickness d, the lifetime $\tau \propto (d/v_\zeta)/(\Delta E/\lambda)^2$, where v_ζ is the velocity of an electron at the Fermi surface. In the special case of aluminum, the lifetime τ determined in this fashion is larger by an order of magnitude than the value derived from Hammond's experimental data, with the help of Anderson's formula.

To summarize the special application of our consideration to nontransition metal superconductors: Whereas, for tin and mercury, spin-orbit coupling effects arising from the periodic effective potential and from orbital scattering by nonmagnetic atomic imperfections can reasonably account for the finite Knight shift in tin and mercury as $T \to 0$, neither of the spin-orbit coupling effects is sufficiently strong to explain a finite Knight shift, of the magnitude observed by Hammond, in superconducting aluminum. It is conceivable that spin-reversing scattering caused by *paramagnetic* imperfections accounts for the experimental observations on aluminum.

Acknowledgments

It is a pleasure to thank Prof. W. Kohn for many stimulating discussions and for several constructive comments on the subject of this paper. In addition, I have enjoyed many beneficial and stimulating discussions with my colleague Dr. R. H. Hammond on NMR experiments in superconductors. I am indebted to Dr. Y. Yafet for correcting an error in equation (7).

References

1. E. Reif, *Phys. Rev.* **106**, 208, 1957.
2. G. M. Androes and W. D. Knight, *Phys. Rev.* **121**, 779, 1961.
3. J. M. Blatt, *Theory of Superconductivity*, Academic Press, Inc., New York (1964).
4. P. W. Anderson, *Phys. Rev. Letters* **3**, 325, 1959.
5. R. A. Ferrel, *Phys. Rev. Letters* **3**, 262, 1959.
6. J. Noer and W. D. Knight, *Rev. Mod. Phys.* **36**, 177, 1964.
7. In conjunction with NMR experiments on platinum, a detailed theoretical discussion of the orbital hf field, including SOC effects, has been presented by A. M. Clogston, V. Jaccarino, and Y. Yafet, *Phys. Rev.* **134**, A650, 1964.
8. G. D. Gaspari, W. M. Shyu, and T. P. Das, (*Phys. Rev.* **134**, A852, 1964) find that, even in Li, the core polarization shifts attributed to the s-part and to the p-part of the conduction electron wave function are significant; both parts have, however, opposite signs and, in fact, cancel one another.
9. Provided the hf contact interaction plays the dominant role.
10. R. H. Hammond and G. M. Kelly, *Rev. Mod. Phys.* **36**, 185, 1964.
11. R. A. Ferrel, University of Maryland, Technical Report 329 (unpublished).
12. R. J. Elliott, *Phys. Rev.* **96**, 266, 1954; Y. Yafet, in: *Solid State Physics*, Vol. 14, F. Seitz and D. Turnbull (eds.), Academic Press, Inc., New York (1963).
13. A. H. Wilson, *Proc. Cambridge Phil. Soc.* **72**, 1147, 1953.
14. G. H. Wannier, *Rev. Mod. Phys.* **34**, 645, 1962; G. H. Wannier and D. R. Fredkin, *Phys. Rev.* **125**, 1910, 1962; see also W. Kohn, *Phys Rev.* **115**, 1560, 1959; L. M. Roth, *J. Phys. Chem. Solids* **23**, 433, 1962; E. I. Blount, *Phys. Rev.* **126**, 1636, 1962.
15. For the derivation of this formula, see J. Appel, *Phys. Rev.* (to be published).
16. C. H. Townes, C. Herring, and W. D. Knight, *Phys. Rev.* **77**, 852, 1950.

12
LIQUID AND GASEOUS HYDROGEN

SUMMARY OF RECENT DETERMINATIONS OF THE ULTRASONIC VELOCITY IN FLUID PARAHYDROGEN

B. A. Younglove

Cryogenics Division, National Bureau of Standards
Boulder, Colorado

The ultrasonic velocity of compressed fluid hydrogen has been measured between 15° and 100°K and in the density range 0.015 to 0.085 g/cm³. Also, the ultrasonic velocity in normal and parahydrogen of the saturated liquid has been measured between 14.5° and 32.2°K.

The measured values can be compared to the value calculated from other thermodynamic quantities:

$$C^2 = V/\beta_s$$

where C is the velocity, V the specific volume, and β_s the adiabatic compressibility. Since the compressibility increases drastically in the region of the critical point the velocity of sound is usually computed from:

$$C(V, T) = V\left[\left(\frac{\partial P}{\partial T}\right)_v^2 T/C_v - \left(\frac{\partial P}{\partial V}\right)_T\right]^{\frac{1}{2}} \quad (2)$$

where P is the pressure, T the temperature, and C_v the specific heat at constant volume. Even with this formulation it is quite unusual that enough data exist for a liquid to permit a reliable calculation. However, extensive measurements on the thermodynamic properties of parahydrogen do exist.

Measurements were made with 10 Mc quartz crystals, using pulses as in the method of Greenspan and Tschiegg.[1] A few values on the saturated liquid were made with the standing wave method, as described by Borgnis,[2] at 7 Mc and for a different path length. Agreement was within 0.02%. Also, in an apparatus test, measurements on water between 25° and 33°C were found to be within 0.02% of Greenspan's data.

Some data for liquid parahydrogen of Van Itterbeek, Van Dael, and Cops[3] were compared at 20.5°K and were approximately 0.3% high.

Velocity calculations have been made at this laboratory by Weber (Fig. 1), based on previously reported PVT[4] and C_v[5] data. The calculated value was compared to the measured value at the temperature and density of each measured point, allowing extensive comparison. Agreement is within 0.5% with larger deviations near the phase boundaries, the largest being near the triple point and critical point. The measured values were generally 0.2 to 0.3% higher than the calculated values. The reason for such consistent deviation is not known, although

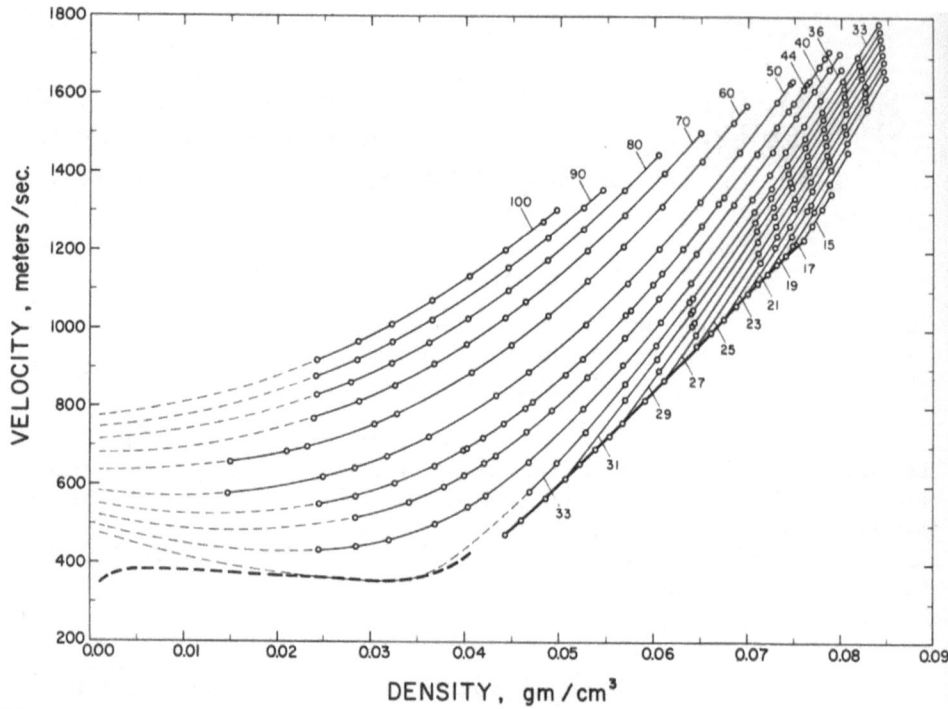

Fig. 1. Velocity of sound in liquid parahydrogen as a function of density for several isotherms. The numbers by the curves are the temperatures of the isotherms. The circles are data points. The light dashed lines are calculations based on other thermodynamic data. The heavy lines are values for the phase boundaries. The critical point is at 33°K and 0.031 g/cm^3, the boiling point at 20°K and 0.071 g/cm^3, and the triple point at 14°K and 0.077 g/cm^3.

it is in the correct direction for dispersion or diffraction[6] effects. Dispersion effects have been found in gaseous hydrogen,[7,8] but are considered negligible for the liquid. Errors introduced by diffraction effects are estimated to be less than 0.003%.

The ultrasonic velocity in saturated liquid normal hydrogen and in parahydrogen was slightly different when the two were compared at the same density

Table I. Coefficients for Representing Ultrasonic Velocity as a Polynomial in Density

Temperature, °K	A_0	A_1	A_2	A_3	A_4
15	-1.9517×10^3	$+4.1894 \times 10^4$			
19	-1.2084×10^3	$+2.60566 \times 10^4$	$+9.0743 \times 10^5$		
23	-1.1395×10^3	$+2.6856 \times 10^4$	$+7.8549 \times 10^4$		
27	-8.650×10^2	$+2.2243 \times 10^4$	$+1.0144 \times 10^5$		
31	-2.4010×10^3	$+1.26097 \times 10^5$	-2.28474×10^6	$+2.35010 \times 10^7$	-8.504×10^7
44	$+5.8576 \times 10^2$	-4.392×10^3	$+4.625 \times 10^2$	$+5.7882 \times 10^6$	-3.5252×10^7
60	$+6.1809 \times 10^2$	$+1.9048 \times 10^3$	-2.6070×10^4	$+4.5194 \times 10^6$	-2.4834×10^7
80	$+7.3826 \times 10^2$	-1.0335×10^3	$+1.8780 \times 10^5$	$+3.856 \times 10^5$	
100	$+7.6207 \times 10^2$	$+3.1274 \times 10^3$	$+1.1947 \times 10^5$	$+7.563 \times 10^5$	

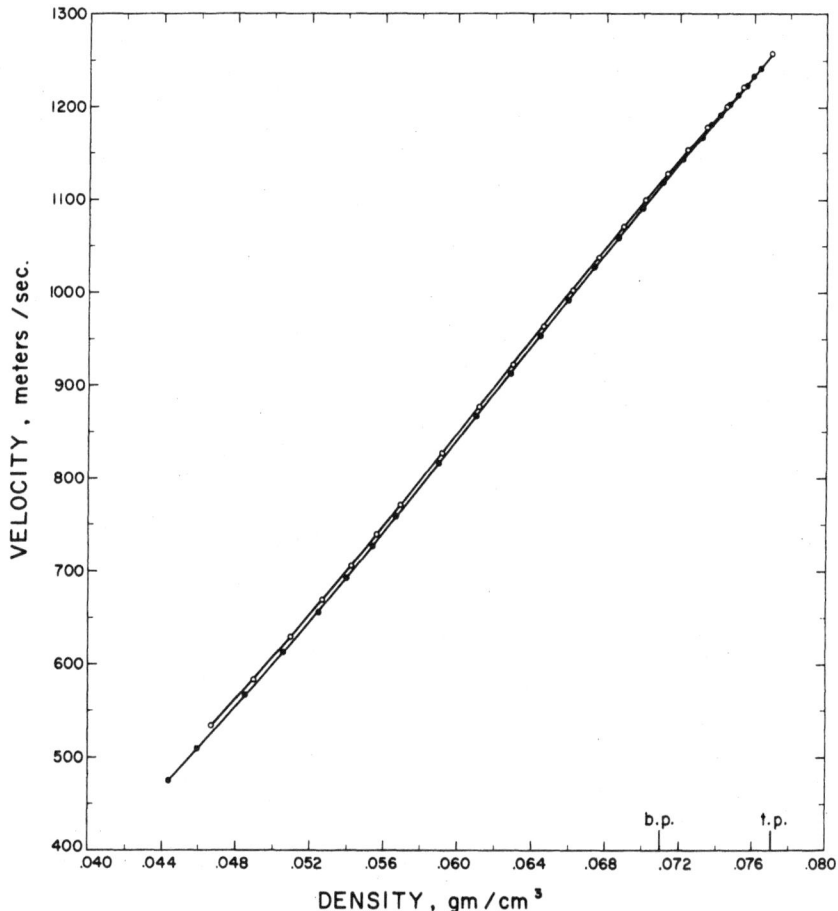

Fig. 2. Velocity of sound for liquid normal and parahydrogen on the saturated liquid phase boundary. The open circles are for normal and the closed ones for para. For reference, the critical point density is 0.031 g/cm³, the boiling point 0.071, and the triple point 0.077.

(Fig. 2). Liquid normal hydrogen has the greater velocity at the same density, and consequently a smaller compressibility. This difference between the two types of hydrogen is greater if they are compared at the same pressure; however, it still amounts to only about 0.7% at most.

The velocity along an isochore can be represented by a low-order polynomial in density, $\Sigma A_n(T)\rho^n$ (see Table I). Obvious failures result from attempts to use such correlating functions[9,10] as:

$$\frac{1}{C}\left(\frac{\partial C}{\partial P}\right)_T = A(T)\frac{1}{\rho}\left(\frac{\partial \rho}{\partial P}\right)_T$$

The most advanced basic calculation of velocity of sound of liquid parahydrogen is that by Reed and Henderson,[11] using energy levels calculated from the Lennard–Jones and Devonshire potential. However, their results are high by about 30%. The results described here will be published in longer form in the near future.

Acknowledgments

The author expresses appreciation to L. A. Weber for his extensive calculations on the velocity of sound.

References

1. M. Greenspan and C. E. Tschiegg, *J. Res. Natl. Bur. Ltd.* **59**, 249, 1957.
2. F. E. Borgnis, *J. Acoust. Soc. Am.* **24**, 19, 1952.
3. A. Van Itterbeek, W. Van Dael, and A. Cops, *Physica* **29**, 456, 1963.
4. R. D. Goodwin, D. E. Diller, H. M. Roder, and L. A. Weber, *J. Res. Natl. Bur. Std.*, *A* **67**, 177, 1963.
5. B. A. Younglove and D. E. Diller, *Cryogenics* **2**, 348, 1962.
6. H. J. McSkimin, *J. Acoust. Soc. Am.* **33**, 539, 1961.
7. A. Van Itterbeek and H. Zink, *Physica* **29**, 370, 1963.
8. H. D. Parbrook and W. Tempest, *J. Acoust. Soc. Am.* **30**, 985, 1958.
9. E. H. Carnevale and T. A. Litovitz, *J. Acoust. Soc. Am.* **27**, 547, 1955.
10. M. R. Rao, *Indian J. Phys.* **14**, 109, 1940.
11. R. D. Reed and D. Henderson, *Australian J. Chem.* **17**, 705, 1964.

MEASUREMENTS OF THE VISCOSITY OF PARAHYDROGEN

Dwain E. Diller

Cryogenics Division, National Bureau of Standards
Boulder, Colorado

The temperature and density dependence of the viscosity of parahydrogen has recently been measured between 14° and 100°K and at densities up to 2.8 times the critical density. We used the torsional crystal method developed by Welber[1] for measuring the viscosity of liquid helium. At frequencies near resonance, the logarithmic decrement Δ of the torsional crystal, defined as the ratio 1/2 energy dissipated/energy of vibration, can be related to the viscosity–density product of the fluid by:

$$(\eta\rho)_{\text{fluid}} = \left(\frac{M}{S}\right)^2 \frac{f_0}{\pi} (\Delta - \Delta_0)^2 \tag{1}$$

The symbols M, S, f_0, and Δ_0 refer to the mass, surface area, resonant frequency, and vacuum decrement, respectively. For our measurements, a cylinder of natural quartz 5.0 cm long and 0.3 cm in diameter was driven in the torsional mode at frequencies near 39.1 Kc. The logarithmic decrements were determined from measurements of the resistance at resonance and the bandwidth of the resonance curve. With the crystal surrounded by hydrogen, the bandwidths ranged from 0.15 to 3.2 cps ± 0.002 cps. Bandwidths in vacuum ranged from 0.005 cps at 14°K to 0.025 ± 0.001 cps at 100°K. Densities accurate to 0.1% were determined from measurements of temperature and pressure and the previously measured PVT relationship.[2] The overall precision of the viscosity measurements ranged from 0.2 to 2% and was better than 0.5% in most cases.

Figure 1 shows the parahydrogen viscosities measured at saturation and along 20 isotherms. Pressures up to 345 atmospheres were required to obtain the densities shown. Both the density and temperature dependence of the viscosity of parahydrogen are qualitatively similar to the behavior of nitrogen and argon reported by Zhdanova.[3,4] The isotherm plot shows that the density dependence at constant temperature is quite small at densities up to twice the critical but increases markedly thereafter. The density dependence at low densities is adequately described by Enskog's[5] rigid sphere theory. The high-density behavior is a topic to which considerable theoretical effort is still being devoted.[6]

The temperature dependence at constant density is also demonstrated in Fig. 1, although not so obviously as the density dependence. At densities up to about twice critical, the viscosity increases slowly with temperature. At densities greater than twice the critical, the viscosity decreases with temperature and the magnitude of the temperature dependence is much larger than at low densities. The change in the sign of the temperature dependence has been interpreted in terms of a change in the mechanism of momentum transfer.[7]

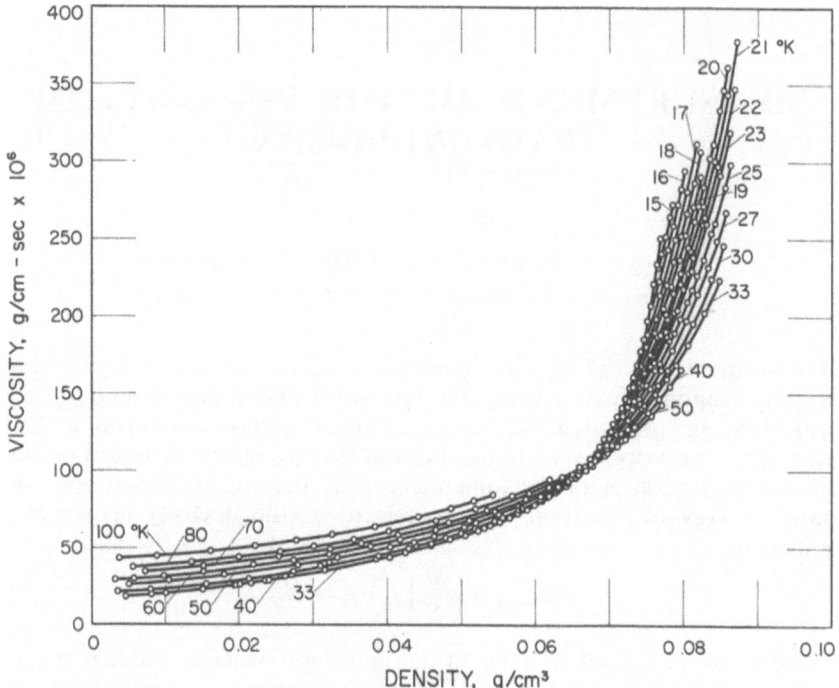

Fig. 1. The viscosity of parahydrogen vs. density.

The measurements in this report have been correlated with the assumption that the temperature and density dependence can be represented by:

$$\eta(T, \rho) = \eta_0(T) + \eta_\phi(T, \rho) \qquad (2)$$

where the first term is the viscosity in the low density limit and the second term has been called the residual or excess viscosity. Figure 2 shows the temperature dependence of the excess viscosity as isochoric plots of log (excess viscosity) vs. $1/T$. At low densities, the temperature dependence is very small and has been neglected in previous correlations.[8,9] The isochores of $\log \eta_\phi(T, \rho)$ vs. $1/T$ are linear over a wide range of temperatures and densities and can be represented by:

$$\log \eta_\phi(T, \rho) = \log A(\rho) + B(\rho)/T \qquad (3)$$

Log $A(\rho)$ and $B(\rho)$ have been determined from the intercepts and slopes of the isochores and further represented by:

$$\log A(\rho) = 5.7694 + \log \rho + 65.0\rho^{\frac{1}{2}} - 2.625 \times 10^5 \exp\left[-2.655\left(\frac{1}{\rho^{\frac{1}{2}}}\right)\right] \qquad (4)$$

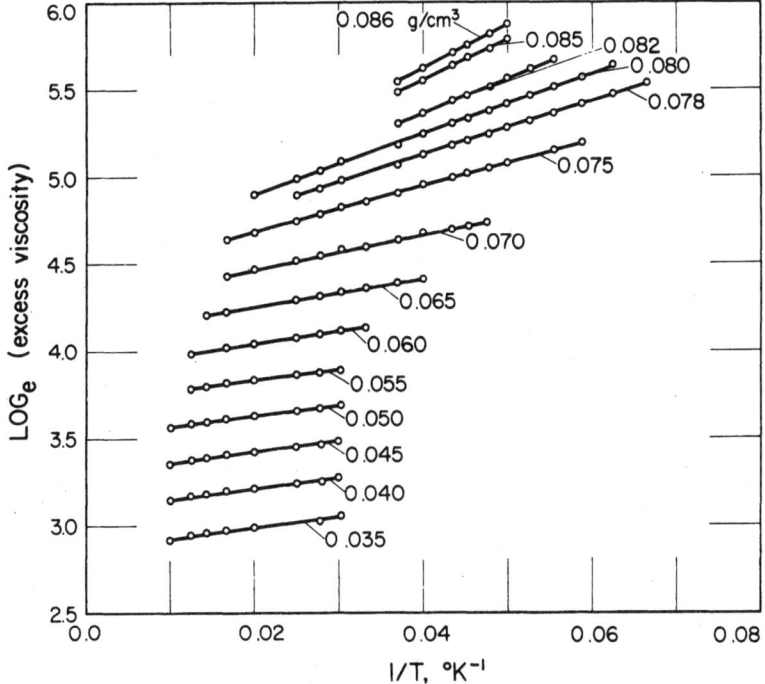

Fig. 2. Isochores of log (excess viscosity) vs. $1/T$.

where the units of $A(\rho)$ are g/cm-sec $\times 10^6$, and:

$$B(\rho) = T_0\left\{10.0 + 7.2\left[\left(\frac{\rho}{0.070}\right)^6 - \left(\frac{\rho}{0.070}\right)^{\frac{1}{2}}\right] - 17.63 \exp\left[-58.75\left(\frac{\rho}{0.070}\right)^3\right]\right\} \text{in } °K \quad (5)$$

where $T_0 = 1.0$.

The measurements in this report have been compared with viscosities computed from equations (2–5). The standard deviation is 0.7%. A comparison has also been made between viscosities computed from equations (2–5) and measurements along the 25°C isotherm reported by Michels.[10] The extrapolated viscosities agree with measured values to better than 3%.

References

1. B. Welber, *Phys. Rev.* **119**, 1816, 1960.
2. R. D. Goodwin, D. E. Diller, H M. Roder, and L. A. Weber, *J. Res. Natl. Bur. Std., A* **67**(2), 173, 1963.
3. N. F. Zhdanova, *Soviet Phys. JETP* (English Transl.) **4**, 749, 1957.
4. N. F. Zhdanova, *Soviet Phys. JETP* (English Transl.) **4**, 19, 1957.
5. D. Enskog, *Kgl. Svenska Vetenskapsakad. Handl.* **63**(4), 1922.
6. B. A. Lowry, S. A. Rice, and P. Gray, *J. Chem. Phys.* **40**(12), 3673, 1964.
7. H. S. Green, *The Theory of Molecular Fluids*, North Holland Publishing Co., Amsterdam (1952).
8. L. I. Stiel and G. Thodos, *Ind. Eng. Chem. Fundamentals* **2**(3), 233, 1963.
9. J. D. Rogers, K. Zeigler, and P. McWilliams, *J. Chem. Eng. Data* **7**(2), 179, April 1962.
10. A. Michels, A. Schipper, and R. H. Rintoul, *Physica* **19**, 1011–1028, 1953.

THE DIELECTRIC POLARIZABILITY OF FLUID PARAHYDROGEN

John W. Stewart*

Cryogenic Engineering Laboratory, National Bureau of Standards
Boulder, Colorado

Introduction

In the case of nonpolar substances such as hydrogen, the macroscopic polarizability (induced dipole moment per unit mass per unit electric field) P is related to the dielectric constant ε and the density ρ through the Clausius–Mossotti equation:

$$P = \frac{\varepsilon - 1}{\varepsilon + 2} \cdot \frac{1}{\rho} \tag{1}$$

Elementary theory predicts that the righthand side of equation (1), often called the Clausius–Mossotti function, should be a constant independent of density. Detailed verification of this requires accurate measurements of both ε and ρ over as wide a range of conditions as possible. Sufficiently accurate measurements of ρ have previously been available for *normal* hydrogen, but a really meaningful investigation of the constancy of P for fluid parahydrogen over a wide range of density in both the gaseous and liquid states has been made possible only recently by the high-precision density measurements in this laboratory by Goodwin and his co-workers.[1] Their values of ρ are estimated to have an absolute precision of better than 0.1%. They are internally consistent to 0.02%. For corresponding precision in the calculated values of P, the dielectric constant must be determined to considerably higher accuracy.

Experimental Technique

In the present investigation, the dielectric constant of fluid parahydrogen was measured by the capacitance ratio method over the density, pressure, and temperature ranges 0.0024 to 0.0796 g/cm^3, 2.6 to 326 atm, and 24° to 100°K, respectively. The measuring capacitor (nominal vacuum capacitance C_0, 60 pF) was wholly contained within a copper pressure bomb capable of withstanding 5000 psi internal pressure. A General Radio Type 1615-A capacitance bridge was employed to measure the capacitance to 1 or 2 parts in 6×10^5 at 1 and 10 kc. Shielded leads were used so that no correction for lead capacitance was necessary. Only ratios of capacitances were needed in the final results, since $\varepsilon = C/C_0$.

The pressure and temperature of the hydrogen within the cell were determined at the same time as the capacitance. A calibrated platinum resistance thermometer was mounted at atmospheric pressure in a well on the outside of the pressure chamber. When the temperature was changed, thermal equilibrium with the samples was attained in a short time through the copper walls. A thermocouple monitored

---

* National Academy of Sciences—National Research Council Post Doctoral Research Associate, 1962–63. Permanent address: Department of Physics, University of Virginia, Charlottesville, Virginia.

the temperature gradient along the length of the pressure bomb. A sensitive deadweight piston gauge was used to measure the pressure. The special electronically controlled regulating system designed and built by Goodwin[2] for the density measurements maintained the temperature of the system constant to $\pm 10^{-3}$ °K for as long as desired.

The presence of parahydrogen was verified by continuous sampling during filling of the stream of gas downstream from the catalyst in a differential thermal conductivity comparison instrument,[3] and by occasional vapor pressure measurements of the sample itself.

A total of 205 (P, T) points on twenty-two isotherms was investigated. The density corresponding to each measured pressure and temperature was determined through a computer program in accordance with the earlier measurements.[1] Small corrections were made for variations in density caused by the observed thermal gradients along the length of the pressure bomb and for the change of volume of the copper in the capacitor resulting from the hydrostatic pressure.

Results and Discussion

The resultant values of the Clausius–Mossotti function were calculated from (1). They do not appear to depend explicitly upon temperature. The results were fitted by least squares to a quadratic function of density:

$$\frac{1}{P} = A + B\rho + C\rho^2 \tag{2}$$

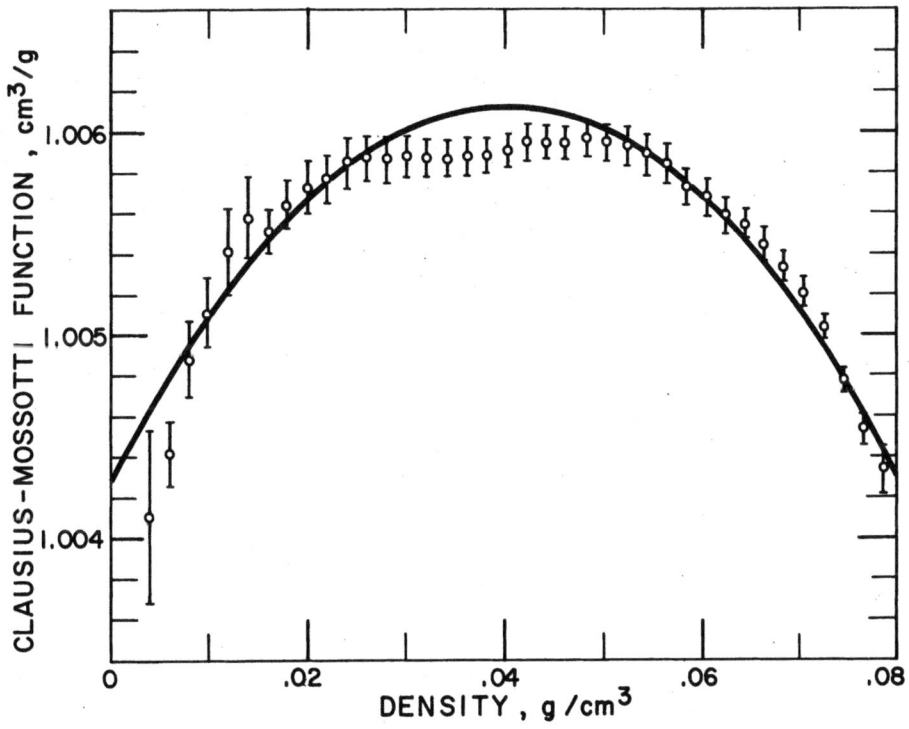

Fig. 1

The constants A, B, and C have the values: $A = (0.99575 \pm 0.00132)$ g/cm^3, $B = (-0.09069 \pm 0.02463)$, and $C = (1.1227 \pm 0.2895)$ g/cm^3. The maximum value of P occurs at $\rho = 0.040$ g/cm^3, and is 0.2% above the value at $\rho = 0$.

This fitted curve for the Clausius–Mossotti function is shown in Fig. 1. The points on this plot are interpolated and averaged values of the actual experimental observations. The deviations from the means at each density are shown as error bars.

The observed departure from constancy of P is believed to be a real effect. The reproducibility of the data for the same density at different temperatures and with separate fillings of the sample holder was of the order of 0.05%, except for the points close to the critical density (0.031 g/cm^3). It is impossible to achieve high precision PVT measurements near the critical point by these techniques because $(\partial \rho / \partial T)_P$ becomes infinite. The corrections to the calculated density for thermal gradient were much larger in this vicinity.

The initial rise of the Clausius–Mossotti function followed by a decrease is qualitatively similar to the behavior reported for other nonpolar gases,[4] including argon[5,6] and carbon dioxide.[7,8] Hamann[9] comments that the maximum of the Clausius–Mossotti function for these two and for ethylene all fall in the vicinity of the critical density. Here it is seen that parahydrogen, with a critical density only 1/15th that of carbon dioxide, shows the same behavior.

In view of the simplifying assumptions made in its derivation, it is quite remarkable that the Clausius–Mossotti equation holds as well as it does. For real substances, one can make qualitative estimates of the effect of deviations from these assumptions. For example, fluctuations of the induced molecular dipole moments should increase in proportion to density, and lead to the initial increase of the Clausius–Mossotti function with ρ. This might also be expected to lead to an inverse temperature effect, which, however, was not observed in the case of parahydrogen. At higher densities, short-range interactions between the molecules would assume increasing importance, leading to a subsequent decrease in the polarizability. The magnitude of the effect in the case of parahydrogen is too small to make quantitative correlation with existing theories of polarizability feasible.

A more detailed description of this work has been published elsewhere.[10]

References

1. R. D. Goodwin, D. E. Diller, H. M. Roder, and L. A. Weber, *J. Res. Natl. Bur. Std.*, A **67**, 173, 1963.
2. R. D. Goodwin, *J. Res. Natl. Bur. Std.*, C **65**, 231, 1961.
3. J. R. Purcell and R. N. Keeler, *Rev. Sci. Instr.* **31**, 304, 1960.
4. L. Jansen, *Phys. Rev.* **112**, 434, 1958.
5. D. R. Johnston, G. J. Oudemans, and R. H. Cole, *J. Chem. Phys.* **33**, 1310, 1960.
6. A. Michels, C. A. ten Seldam, and S. D. J. Overdijk, *Physica* **17**, 781, 1951.
7. D. R. Johnston and R. H. Cole, *J. Chem. Phys.* **36**, 318, 1962.
8. A. Michels and L. Kleerekoper, *Physica* **6**, 586, 1939.
9. S. D. Hamann, *Physico-Chemical Effects of Pressure*, Butterworths, London (1957), p. 99.
10. J. W. Stewart, *J. Chem. Phys.* **40**, 3297, 1964.

THE TEMPERATURE DEPENDENCE OF THE RELAXATION TIME FOR ROTATIONAL TRANSITIONS IN H_2, HD, AND D_2

H. F. P. Knaap, C. G. Sluijter, and J. J. M. Beenakker

Kamerlingh Onnes Laboratorium
Leiden, The Netherlands

The relaxation time for rotational–translational energy transfer for the hydrogen isotopes has been determined by sound-absorption measurements. For theoretical reasons, molecules with only two occupied rotational states are most suited for this, since in that case only one relaxation time is present. Such a two-level system can be found in hydrogenic molecules at low temperatures, because the very small moments of inertia of these molecules give rise to a large spacing of rotational-energy levels. For parahydrogen and orthodeuterium, only the levels $J = 0$ and $J = 2$ are then occupied. In this case, it is possible to determine uniquely the relaxation time for the $0 \to 2$ transition from the sound-absorption curve.

For a one-relaxation-time process, the excess absorption α' due to the relaxation phenomenon is, following Herzfeld and Litovitz,[1] given by

$$\alpha' \lambda (V_0/V)^2 = \pi \frac{(C_p - C_v)C'}{C_v(C_p - C')} \frac{\omega \tau'}{1 + \omega^2 \tau'^2} \tag{1}$$

with λ the wave length, V the sound velocity, V_0 the sound velocity at zero frequency, C_p and C_v the specific heat at constant pressure and volume, C' the rotational specific heat, $\omega = 2\pi f$, f is the frequency of the sound wave, and $\tau' = (C_p - C'/C_p)\tau$. τ is the true relaxation time that determines the rate at which energy is exchanged between the internal and external degrees of freedom.

By measuring at different frequencies one would obtain the curve given by equation (1). In equation (1), ω occurs always in the combination $\omega \tau'$. Under the assumption that transitions between rotational and translational energy states are caused only by collisions, the relaxation times τ will be inversely proportional to the collision rate. Hence, τ is proportional to $1/p$ and therefore the absorption can also be measured as a function f/p, where f/p differs from $\omega \tau'$ only by a constant independent of pressure and frequency. Experimentally, it is much easier to measure the absorption curve by varying the pressure than by varying the frequency.

The sound absorption is determined by measuring the decay in amplitude of a sound wave as a function of distance from the sound transmitter.[2] This is done by moving a detector at uniform speed through the sound field away from the transmitter. The level indicated by the detector is registered by means of a logarithmic level recorder. In the case of an exponential decay of the sound pressure as function of distance from the sound source, the curve registered on the recorder will be a straight line. From the slope of this line the sound-absorption coefficient can be determined when microphone driving speed and recorder paper speed are known.

By subtracting from the experimentally determined absorption α_{exp} the calculated absorption α_{class} due to friction and heat losses we find the absorption α' due to the relaxation phenomenon:

$$\alpha_{exp} - \alpha_{class} = \alpha'$$

The calculated absorption α_{class} is calculated from

$$\alpha_{class} = \frac{2\pi^2}{\gamma p V_0}\left(\tfrac{4}{3}\eta + \frac{\gamma - 1}{c_p}\lambda\right)f^2$$

where η is the shear viscosity, λ the coefficient of heat conductivity, $\gamma = C_p/C_v$, and c_p the specific heat per unit mass at constant pressure.

The quantity $\alpha'\lambda(V_0/V)^2$ is calculated and plotted on a double logarithmic scale vs. f/p (see Fig. 1). The theoretical curve has a universal shape, irrespective of the factor $(C_p - C_v)C'/(C_p - C')C_v$ and the value of τ'. By plotting the experimental points, one can determine easily whether one or more relaxation times are involved. This is done by moving a transparency with the theoretical single relaxation time curve over the measured points. If different relaxation times are present, the measured points will be on a broader curve. As can be seen from Fig. 1, the experimental points lie very well on the theoretical curve that is shifted so as to give the best fit. One can determine the values of $(f/p)_{max}$ and of $(C_p - C_v)C'/(C_p - C')C_v$. This latter quantity agrees within 2% with the value calculated from specific heat data. One can calculate from $(f/p)_{max}$ the relaxation time τ. In this way, measurements for parahydrogen at 77°, 90°, and 170°K and for orthodeuterium

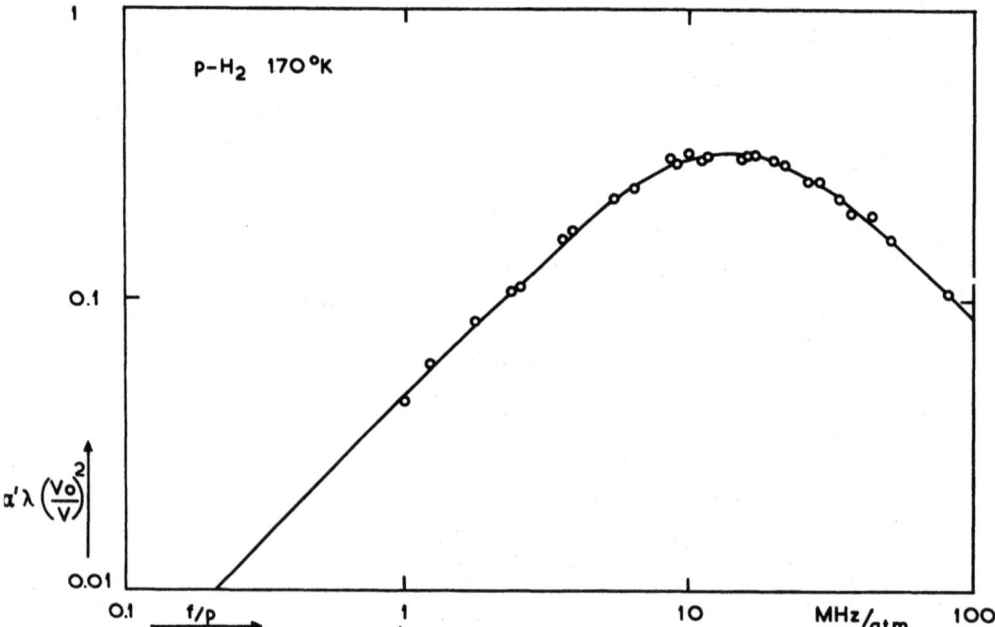

Fig. 1. Excess sound absorption for parahydrogen at 170°K as a function of f/p. ○, experimental points. ——, theoretical curve for a one relaxation-time process.

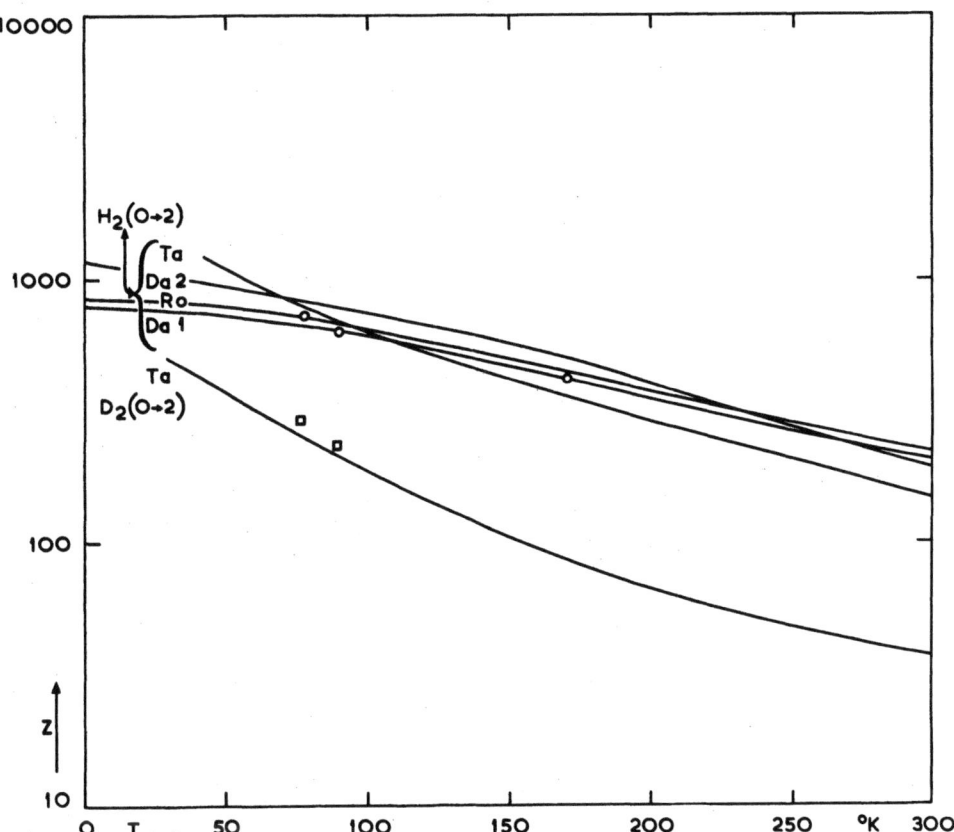

Fig. 2. The collision number Z as a function of temperature for parahydrogen and orthodeuterium. O, experimental points. ———, theoretical curves calculated from the collision cross sections as derived by Takayanagi (Ta),[5] Davison (Da1 and Da2),[6] and Roberts (Ro).[7]

at 77° and 90°K have been analyzed.[2,3] Theoretically, the relaxation time is obtained from $\tau^{-1} = f_{02} + f_{20}$, in which f_{02} and f_{20} are the transition probabilities for the $0 \to 2$ and $2 \to 0$ transition.[4] The quantity f_{02} is obtained by integration of the inelastic scattering cross section $Q_{02}(k)$ over the Maxwellian distribution of wave numbers k. The scattering cross section $Q_{02}(k)$ has been calculated by several authors (Takayanagi,[5] Davison,[6] and Roberts[7]) for various angle-dependent intermolecular potentials. The scattering cross section $Q_{02}(k)$ turns out to be proportional to the square of β, the parameter that determines the strength of the angle-dependent part of the intermolecular potential. In Fig. 2 a comparison is made between theory and experiment. We have plotted the collision number that is defined as $Z = \tau/\tau_c$, where τ_c is the average time between collisions. Z can be regarded as the average number of collisions necessary for reaching equilibrium between the rotational and translational degrees of freedom. By choosing an appropriate value of β, the experimental and theoretical results are found to be in agreement over a fairly large temperature range (see Fig. 1). This value of β appears to be $\beta = 0.11$ for both hydrogen and deuterium.[4]

For HD one has to expect a much greater nonsphericity due to the asymmetric mass distribution. This leads to much lower values of the relaxation time and Z.

This has been experimentally verified. Measurements are in progress over a wide temperature range from 20° to 300°K. Further research on mixtures has been begun.

References
1. K. F. Herzfeld and A. L. Litovitz, *Absorption and Dispersion of Ultrasonic Waves*, Academic Press, Inc., New York (1959).
2. C. G. Sluijter, H. F. P. Knaap, and J. J. M. Beenakker, *Commun. Kamerlingh Onnes Lab.*, Leiden, No. 3376; *Physica* **30**, 745, 1964.
3. C. G. Sluijter and R. M. Jonkman, *Physica* **30**, 1964 (to be published).
4. C. G. Sluijter, H. F. P. Knaap, and J. J. M. Beenakker, *Physica* **30**, 1670, 1964.
5. K. Takayanagi, *Proc. Phys. Soc. (London) Ser. A* **70**, 348, 1957; *Sci. Rept. Saitama Univ. Ser. A* **3**, 87, 1959.
6. W. D. Davison, *Discussions Faraday Soc.* **33**, 71, 1962.
7. C. S. Roberts, *Phys. Rev.* **131**, 209, 1963.

THE DIFFERENCE IN POLARIZABILITY BETWEEN ORTHOHYDROGEN AND PARAHYDROGEN IN THEIR GROUND STATES

H. F. P. Knaap, L. J. F. Hermans, and J. J. M. Beenakker

Kamerlingh Onnes Laboratorium
Leiden, The Netherlands

Normal hydrogen consists of two kinds of molecules: orthohydrogen (odd rotational states) and parahydrogen (even rotational states). Since conversion is in general very slow, orthohydrogen and parahydrogen can be treated as different gases. It is possible to prepare highly pure samples of both species.

At low temperature only the lowest energy states are occupied, i.e., $J = 1$ for orthohydrogen and $J = 0$ for parahydrogen. It has been shown that the rotation of the orthomolecule gives rise to a slightly larger internuclear distance through centrifugal stretching. Since the polarizability depends directly on the internuclear distance, one expects a difference in polarizability between orthohydrogen and parahydrogen. This difference was theoretically estimated to be of the order of 0.2% (see Bell[1] and Babloyantz and Bellemans[2]).

Knaap and Beenakker[3] used this difference in polarizability to show that there is a difference between the interaction constants of orthohydrogen and parahydrogen. The main influence arises through the London–van der Waals attraction term, which is roughly proportional to the square of the polarizability. These authors showed that by using a corresponding-states treatment based on the differences in parameters, one can account for a large part of the difference in the macroscopic properties of orthohydrogen and parahydrogen at low temperatures. The second virial coefficient, the viscosity, and the vapor pressure were treated.

It was thought necessary to provide direct experimental confirmation for the difference in polarizability. We used a Cole capacitance bridge[4] suitable for work at low temperatures. Two cells, A and B, which contain a condenser of $160\,pF$ are cooled to 20°K. At 20°K the ortho- and paramolecules are in the states $J = 1$ and $J = 0$, respectively. The cells are both filled with parahydrogen. Then one of the cells is pumped and normal hydrogen ($\tfrac{3}{4}$ orthohydrogen and $\tfrac{1}{4}$ parahydrogen) is admitted, and the change in capacitance caused by the change in dielectric constant ε of the gas is measured on a voltmeter. This change in capacitance is related to a change in the polarizability α by Clausius–Mossotti's relation

$$\frac{\varepsilon - 1}{\varepsilon + 2} = \tfrac{4}{3}\pi n \alpha \tag{1}$$

with n the particle density. Since $\varepsilon - 1$ is of the order of 4×10^{-3} for hydrogen (at 20°K and 1 atm), one has to be able to detect differences in dielectric constant or capacitance of 1 in 10^7 in order to measure a relative difference in α of 2×10^{-3} with a reasonable accuracy.

The results for the system parahydrogen–normal hydrogen are shown in Fig. 1. The deflection on a galvanometer is plotted as a function of the pressure of the gas in the cells. The systems parahydrogen–normal hydrogen and parahydrogen– 50% parahydrogen + 50% normal hydrogen have been investigated. The effect appears to be linear in the concentration of paramolecules. One would expect a straight line from equation (1), the slope of which would be proportional to the difference in α. However, at the rather high densities which we use, the density is not proportional to the pressure, since the contribution of the second virial coefficient B to the equation of state is important. For the deflection we now get the proportionality

$$u \sim p\left[\Delta\alpha - \left(\frac{\Delta B}{B} + \frac{\Delta\alpha}{\alpha}\right)\alpha B d\right] \qquad (2)$$

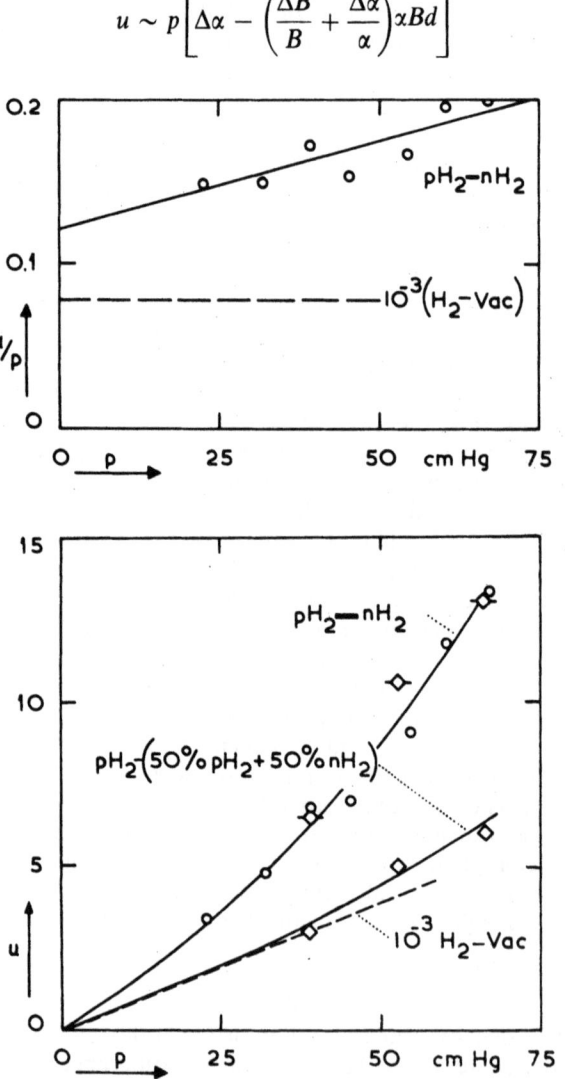

Fig. 1. Deflection as function of pressure. ○ pH_2–nH_2. ◇ pH_2–(50% pH_2 + 50% nH_2). ◇ calculated from ◇. ------- line indicating 1/1000 of the deflection for H_2–vacuum.

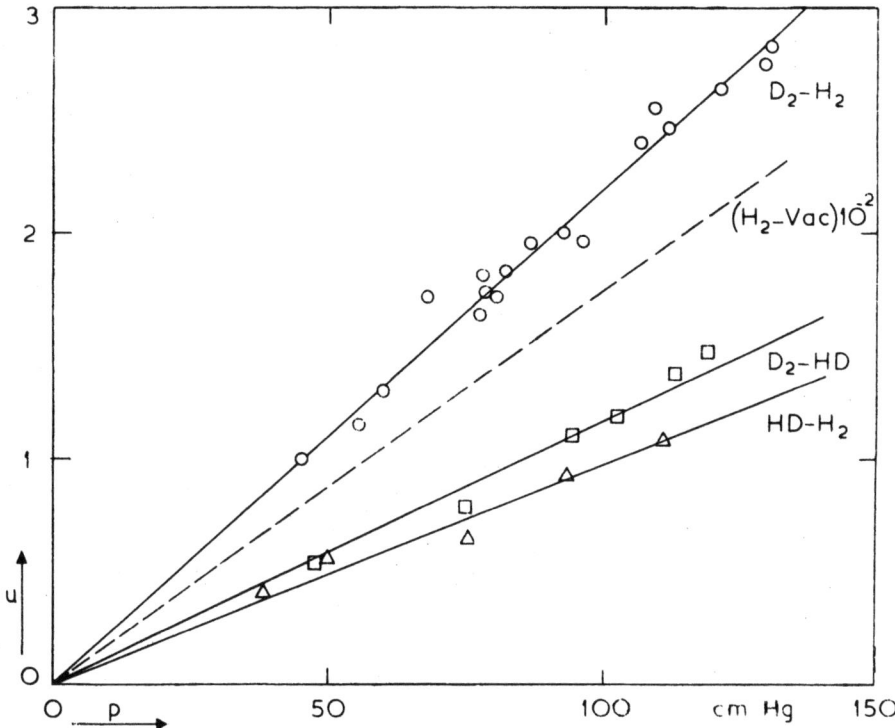

Fig. 2. Deflection as function of pressure. ○ D_2–H_2. □ D_2–HD. △ HD–H_2. ------ line representing 1/100 of the deflection for H_2–vacuum.

where d is the density and $\Delta\alpha$ and ΔB are the differences in polarizability and second virial coefficient. If one plots the quantity u/p as a function of density or pressure, one expects a straight line [see equation (2)]. The u/p value at $p = 0$ is proportional to $\Delta\alpha$, and the slope of the line is proportional to $\Delta B/B + \Delta\alpha/\alpha$. A calibration is provided by measuring the deflection when one cell is filled with hydrogen while the other is evacuated. A thousandth of this value is indicated in Fig. 1 by the dotted line.

The results for $\Delta\alpha/\alpha$ and $\Delta B/B$ are shown in Table I. Good agreement is obtained between theory and experiment for $\Delta\alpha/\alpha$. Furthermore, the value of $\Delta B/B$ compares favorably with the results of direct determination.[5]

Table I. Results for the System Parahydrogen–Normal Hydrogen at 20.4°K

	This research	Beenakker, Varekamp, and Knaap[5]	Babloyantz and Bellemans[2]
$\dfrac{\alpha_{nH_2} - \alpha_{pH_2}}{\alpha} \times 100$	0.15		0.15
$\dfrac{B_{nH_2} - B_{pH_2}}{B} \times 100$	1.0	1.0	

Table II. Polarizability of the Hydrogen Isotopes

	This research	Bell[1]	Ishiguro et al.[6]
$\dfrac{\alpha_{H_2} - \alpha_{D_2}}{\alpha} \times 100$	1.26	1.3	1.86
$\dfrac{\alpha_{HD} - \alpha_{D_2}}{\alpha} \times 100$	0.67		1.02
$\dfrac{\alpha_{H_2} - \alpha_{HD}}{\alpha} \times 100$	0.56		0.84

We also measured the difference in $\Delta\alpha/\alpha$ for the systems D_2–H_2, D_2–HD, and HD–H_2 at 77°K (see Fig. 2). In Table II, the results are compared with theoretical estimates[6] and the result of an indirect determination of $\Delta\alpha/\alpha$.[1]

References

1. R. P. Bell, *Trans. Faraday Soc.* **38**, 422, 1942.
2. A. Babloyantz and A. Bellemans, *Mol. Phys.* **3**, 313, 1960.
3. H. F. P. Knaap and J. J. M. Beenakker, *Commun. Kamerlingh Onnes Lab.*, Leiden, Suppl. No. 119b; *Physica* **27**, 523, 1961.
4. D. R. Johnston, G. J. Oudemans, and R. H. Cole, *J. Chem. Phys.* **33**, 1310, 1960.
5. J. J. M. Beenakker, F. H. Varekamp, and H. F. P. Knaap, *Commun. Kamerlingh Onnes Lab.*, Leiden No. 319a; *Physica* **26**, 43, 1960.
6. E. Ishiguro, T. Arai, M. Kotani, and Mizushima, *Proc. Phys. Soc. (London) Ser. A* **65**, 178, 1952.

13
TECHNIQUES

DIELECTRIC DISSIPATION MEASUREMENT BELOW 7.2°K*

W. H. Hartwig and D. Grissom

University of Texas
Austin, Texas

Introduction

The dielectric constant of a material in a sinusoidal electric field is expressed as:

$$K = K_1 - jK_2 \qquad (1)$$

The fraction of stored energy that is dissipated is given as $\tan \delta = K_2/K_1$. The equilibrium condition is

$$K(\omega) = K_\infty + \int_0^\infty \alpha(t) \exp(j\omega t)\, dt \qquad (2)$$

where $\alpha(t)$ is a characteristic decay function. $K_1(\omega)$ approaches K_∞ as $\omega \to \infty$ and K_s as $\omega \to 0$.

Dielectric dissipation is the irreversible process of converting electric field energy into lattice phonons. Many mechanisms may contribute to the total observed loss. However, they can be grouped into three categories: (1) relaxation between two or more stable equilibrium positions across potential barriers, (2) resonance about a single equilibrium position, and (3) drift conduction with an absence of equilibrium positions.

Dissipation Relations

Relaxation across an energy barrier H takes place with an average period $\tau = \beta \exp(H/kT)$. The approach to equilibrium is exponential according to:

$$\alpha(t) \propto \exp(-t/\tau) \qquad (3)$$

The classical Debye equations, as set up by Fröhlich[1] for a single relaxation time, show that the peak value of $\tan \delta$ is independent of temperature and occurs at lower frequency as temperature is decreased.

Resonance is characterized by the relation

$$\alpha(t) = \gamma \exp(-t/\tau) \cos(\omega_0 t + \Psi) \qquad (4)$$

where τ is a damping time and ω_0 is the resonant frequency. In the simple crystalline lattice, ω_0 is typically in the infrared. In more complex structures, ω_0 might be expected to have lower frequency components. Before resonance can be observed, however, the dominant relaxation and conduction mechanisms will have to be

---

* Work supported by the National Science Foundation.

suppressed. The Debye solution for resonance absorption, as predicted by (4), shows the peak value of tan δ increases slowly as T is reduced.

Conduction losses are determined by the concentration of free carriers. Since free carrier density is related to temperature by the Boltzmann factor, this mechanism will be effectively frozen out at cryogenic temperatures.

Superconductive Resonant Circuits

The surface resistance R_s of an ideal superconductor, derived from the London equations,[2] is given by

$$R_s/R_n = \left[\frac{\omega\tau\Phi}{1 + \omega^2\tau^2\Phi^2} \{(1 + \omega^2\tau^2\Phi^2)^{\frac{1}{2}} - 1\} \right]^{\frac{1}{2}} \tag{5}$$

where R_n is normal surface resistance, ω is frequency, τ is relaxation time of normal electrons, $\Phi(t)$ is density ratio of normal and superelectrons. The Q's of superconducting resonant circuits are taken to be related to surface resistance by $Q_n/Q_s = R_s/R_n$.

Four loss mechanisms contribute to a finite loaded Q: surface resistance, radiation, coupling, and dielectric dissipation. These are related by:

$$Q_L^{-1} = Q_S^{-1} + Q_R^{-1} + Q_C^{-1} + Q_D^{-1} \tag{6}$$

Radiation losses are eliminated by enclosing the circuit in a superconducting shield can. Hartwig[3] has shown that the coupling losses can be reduced to insignificance without a serious drop in sensitivity. The dominant loss occurs in the dielectric material placed in regions of high electric field, such as between the plates of the capacitor. It is readily shown that

$$\tan \delta = Q_D^{-1} \tag{7}$$

if the entire electric field is in the dielectric material. The loaded and unloaded Q's are then related simply as:

$$Q_L^{-1} = Q_u^{-1} + Q_D^{-1} \tag{8}$$

Measurement of Q

Very high values of Q are attainable with superconducting resonant circuits in the 1 to 1000 Mcps range. Numerous configurations have been constructed using electroplated lead as the superconductor. Since it is desirable to limit the volume of the experiment, lumped LC circuits are used for the lowest frequencies; helical resonators ideally bridge the range from 30 to 600 Mcps, and re-entrant cavities are suitable above this. The Q is measured by observing the envelope decay transient after the resonator input signal is gated off. Q is calculated from

$$Q = \pi f_0 \Delta t \tag{9}$$

where f_0 is the resonant frequency in cps and Δt is the exponential time constant. Frequency can be measured to within a few cps with a counter because of the narrow bandwidth and excellent frequency stability of these resonators. The decay time can be determined to $\pm 5\%$ error with a CRO having a calibrated sweep.

All configurations made are capable of tuning over a wide range without serious degeneration of Q. As a consequence, a tunable helical resonator or cavity can be used at its fundamental frequency, at the odd overtones (approximately harmonics), and can be tuned above these frequencies with suitable tuning slugs using the Meissner effect. This provides a basic capability to make almost continuous measurements of Q, with temperature and frequency as independent variables.

Measurement of K_1 and tan δ

Sensitivity. We define a figure of merit for measurement of tan δ using resonant decay techniques:

$$S_f = Q_u \tan \delta = (Q_u/Q_L) - 1 \tag{10}$$

where the subscript f refers to a dielectric which fills the entire electric field volume. For a partially filled capacitor,

$$S_p = \frac{S_f C_D}{C_D + C} \tag{11}$$

where $C_D + C$ is the total circuit capacity.

The helical resonantor shown in Fig. 1 is a quarter-wave transmission cavity terminated in the dielectric sample. Since the field geometry is unchanged by the presence of the sample, the cavity perturbation theory is valid, according to Spencer, LeCraw, and Ault,[4] even though the frequency change is not small. This leads to

$$\tan \delta = \frac{f_D(K_1 - 1)}{2Q_D \Delta f} \tag{12}$$

where f_D is the resonant frequency of the loaded resonator and Δf is the frequency perturbation. Furthermore:

$$S_p = \frac{S_f 2 \Delta f}{f_D(K_1 - 1)} \tag{13}$$

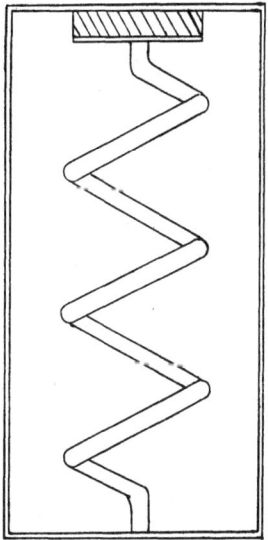

Fig. 1. Helical resonator configuration.

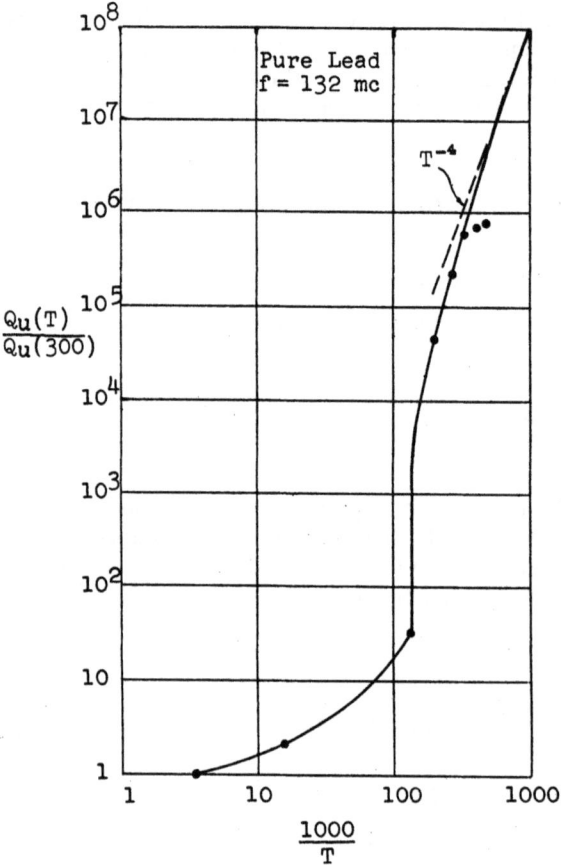

Fig. 2. Sensitivity relations for measurement of dielectric dissipation.

K_1 can be calculated directly from Δf if the dielectric completely fills the resonator and can be calculated by substitution otherwise. The improvement in sensitivity as temperature is reduced is seen from equation (14) and Fig. 2:

$$\frac{S(T)}{S(T_0)} = \left[\frac{Q_u(T)}{Q_u(T_0)}\right]\frac{\tan \delta(T)}{\tan \delta(T_0)} \tag{14}$$

Typical values of Q_u at 300°K range from 10^2 to 10^4, depending upon geometric configuration. The limit of accurate measurement is represented by $S \approx 1$, where the dielectric loss is equal to all other losses. This means that a loss tangent of 10^{-2} to 10^{-4} can be readily measured at room temperature if sample size and shape pose no problems. Q_u will increase as temperature is lowered, but the improvement for normal metals is less than two orders of magnitude due to dominance of defect scattering. Lead combines several characteristics which suit it to this application. Since it has the ability to anneal itself, thermal scattering is the dominant mechanism to well below the superconducting transition. Hartwig and Victor[5] have shown that impurity scattering will become important below 4.2°K for 0.9999 purity. The theoretical curve for pure lead at 132 Mcps illustrates the significant increase in sensitivity below the transition temperature. Sensitivity is also a function

Table I. Summary of Dielectric Properties Measurements

Material	f_u, Mcps	$Q_u \times 10^{-6}$	f_D, Mcps	$Q_L \times 10^{-6}$	K_1	tan δ	S
Empty	132.3	10.8	132.3				
Silica	132.3	9.8	98.3	0.0247	3.64	1.5×10^{-4}	392
Teflon	132.3	10.8	118.6	2.54	1.89	1.2×10^{-6}	3.1
Helium	162.1943 ± 0.0001	21.5 ± 0.5	158.3310 ± 0.0001	21.4 ± 0.5	1.04940	$< 10^{-9}$	< 0.25

of frequency, since $R_s \propto \omega^{\frac{3}{2}}$ according to equation (5). The experimental points indicate a departure from the London surface resistance asymptote, which is proportional to T^{-4}.

Results

Table I summarizes the measurements made on three materials: fused silica and Teflon disks, 1.5 by 0.125 in., placed as shown in Fig. 1; and liquid helium, which completely filled an unterminated resonator.

Discussion

The single observation of K_1 and tan δ (max) for helium is an excellent example of the value of this technique. The purpose of the test was to determine if dissipation of liquid helium would affect tan δ measurements on immersed dielectric samples. The new upper limit of 10^{-9} for tan δ was obtained at the threshold of sensitivity set by geometric configuration and frequency.

The best measurements of K_1 have been made by Chase, Maxwell, and Millett,[6] who reported a value of 1.0492 ± 0.0006. Their data confirmed a previous experiment by Grebenkemper and Hagen,[7] who also reported an upper limit on tan δ of 5×10^{-6} at 9.1 kMcps. Our measurement makes possible the correction of the uncertainty in the fifth significant figure and the addition of a sixth. The precision of the value reported was not determined, but the frequencies listed in Table I are correct to within 100 cps.

Acknowledgments

We wish to thank Professor J. C. Davis, C. R. Haden, W. E. Sayle, J. L. Stone, B. G. Streetman, and J. M. Victor for their valuable consultation and assistance in all phases of this work.

References

1. H. Fröhlich, *Theory of Dielectrics*, Oxford University Press, London (1949), Chapt. III.
2. F. London, *Superfluids*, Vol. I, Dover Publications, Inc., New York (1961), p. 29.
3. W. H. Hartwig, *Electronics*, February 22, 1963, pp. 43–47.
4. E. G. Spencer, R. C. LeCraw, and F. Ault, *J. Appl. Phys.* **28**(1), 130, 1957.
5. W. H. Hartwig and J. M. Victor, *Bull. Am. Phys. Soc.* **8**(8), 613, 1963.
6. E. Chase, E. Maxwell, and W. E. Millett, *Physica* **27**, 1129–1145, 1961.
7. C. J. Grebenkemper and J. P. Hagen, *Phys. Rev.* **80**, 89, 1950.

USE OF LOW-TEMPERATURE TECHNIQUES TO MEASURE GRAVITATIONAL FORCES ON CHARGED PARTICLES*

F. C. Witteborn, L. V. Knight, and W. M. Fairbank

Physics Department, Stanford University
Stanford, California

Direct measurements of the force of gravity on charged elementary particles have never been made because of the difficulty of isolating the particles from electric and magnetic fields. Furthermore, no direct measurements have been made of the gravitational force on antimatter. Use of superconducting persistent magnets, cryogenic pumping, and a number of other techniques has enabled us to reduce all vertical forces on a free electron to values comparable to that of gravity, even though the electric force from one electron 5 m away is equal to the gravitational force from the whole earth. An experiment is under way to measure the force of gravity on a free electron and eventually on a free positron. Electrons have already been observed to take 80 msec to travel 15 cm, which indicates that their average kinetic energy is 10^{-11} eV.

A schematic of the apparatus for electrons is shown in Fig. 1. A 1 msec pulse consisting of about 10^9 electrons is emitted from a tunnel cathode† at the bottom of a cylindrical copper drift tube. The electrons are guided by a homogeneous persistent superconducting magnet along the axis of the tube to a fourteen-stage electron multiplier detector. The number of electrons arriving in each millisecond time interval after emission of the pulse from the cathode is recorded electronically. The process is repeated every 400 msec until sufficient data are obtained. If no other force than gravity were acting on the electrons, the maximum observable time of flight would be $t_{max} = \sqrt{(2h/g)} = 175$ msec, where h is the field free length of the drift tube (15 cm) and g is the local gravitational constant. Electrons leaving the cathode with less energy than $mgh = 0.8 \times 10^{-11}$ eV (where m is the electron mass) would not reach the detector. If the effect of gravity on the time of flight distribution is to be seen, potential energy changes from all other causes must be made less than mgh over a considerable length of the drift tube.

In a metallic tube whose diameter is perfectly constant, the vertical forces are negligible except near the ends. Image forces and external fields fall off exponentially with increasing distance from the ends. In a real tube, diameter variations and contact potential differences can produce vertical electric fields. To minimize these effects, the OFHC copper drift tube was ground mechanically, electropolished, and then annealed under hydrogen atmosphere.

---

* Supported in part by the National Aeronautics and Space Administration.
† We are indebted to J. W. Hall II, of G.E. Electronic Components Division, Owensboro, Kentucky, for providing these devices.

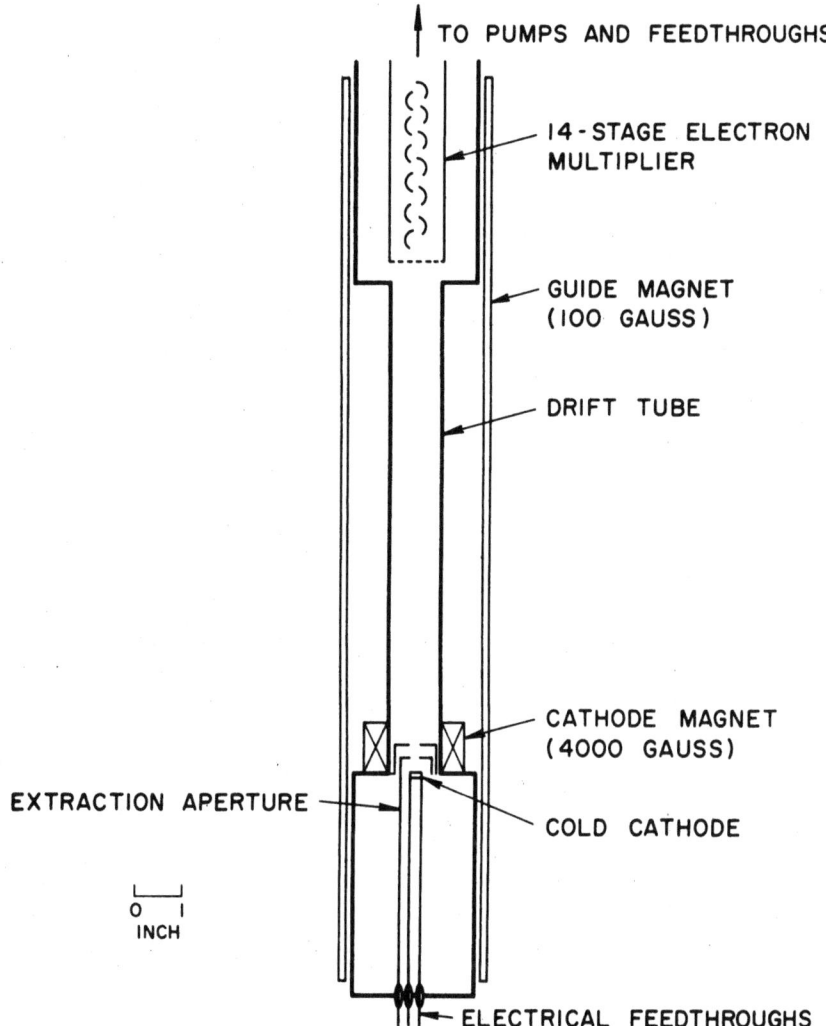

Fig. 1. Electron free-fall apparatus. Vacuum pumps, dewars, electronics, wires, and transfer tube are not shown.

Vertical forces may also arise from variations of the magnetic energy

$$E_{\text{mag}} = 2\mu_\beta B\left(n + \tfrac{1}{2} + \frac{s}{2}g_s\right)$$

where μ_β is the Bohr magneton, B is the magnetic field, g_s is the electron spin g-factor, n is a positive integer, and $s = \pm\tfrac{1}{2}$.[1] The magnetic energy of the ground state $(n = 0, s = -\tfrac{1}{2})$ is a factor of 10^3 smaller than the next higher state and of opposite sign. Emission of the electrons from inside a 4000 G field results in a gain by those not in the ground state of at least 5×10^{-5} eV as they enter the 100 G field in the drift tube, while those in the ground state lose kinetic energy. Thus, any electron going through the drift tube with less kinetic energy than 5×10^{-5} eV must be in the ground state and therefore will suffer minimum interaction with small magnetic field gradients.

In order to avoid interactions between slow electrons and background gases, the pressure was kept below 5×10^{-10} torr, as measured in the room-temperature part of the apparatus.

A thermoelectric voltage drop may arise from a temperature gradient along the drift tube. Thermopowers up to several microvolts per degree have been measured even for very pure copper samples.[2] This can be eliminated by plating the inside of the drift tube with a superconductor, but this has not yet appeared to be necessary.

An interesting aspect of the experiment concerns the effect of gravity on electrons in the metal walls of the drift tube. This may be just sufficient to set up an electric field that would cancel the gravitational force on free electrons. This field would have the opposite effect on positrons and a negligible effect on ions.

Several experimental runs have been made using the 1-in. diameter drift tube. Data from one run consisting of accumulated counts from 9000 pulses of electrons are shown in Fig. 2. After random noise has been subtracted out, the data fall very close to a CT^{-2} curve out to about 30 msec and slow electrons are apparently observed with flight times of up to 80 msec, corresponding to an energy of 10^{-11} eV. The cutoff at this point puts an upper limit to the various magnetic, electrostatic, and thermoelectric effects mentioned earlier. The CT^{-2} distribution is typical of most of the data taken when the apparatus was new. Later, a T^{-1} component appeared. Recently, progressively less delayed detector signals were seen which finally disappeared altogether. Since the drift tube must be occasionally exposed to air to replace the cathode, a nonuniform layer of adsorbed gas may have formed. This could cause contact potential differences of 0.1 eV or more.

Several tests were performed in attempts to check whether the delayed detector signals were actually caused by slow electrons. A grid in front of the detector stopped all signals when a small negative voltage was applied. The signals were not affected by positive voltages on the grid. When the potential of the cathode was raised to over $+2$ V, no signals reached the detector. When more than -2 V was applied, only fast particles reached the detector. This is evidence against delayed emission from the cathode. It also is evidence against the formation of slow negative ions in the drift tube, although not against their formation on the surface of the emitting film. Increasing the magnetic field gradient lowered the maximum flight times approximately as would be expected for slow electrons in the ground state. An attempt was made to produce a uniform axial electric field by running a current through the walls of the drift tube. Unfortunately, when this method was finally tried, we were no longer getting the CT^{-2} distribution. It did enable us to tentatively link the T^{-1} distribution with light negative ions, but the drift tube resistance at 4.2°K has not yet been measured well enough to permit an accurate mass estimate. A cutoff at 175 msec in the T^{-1} distribution has been observed in several instances. This is where gravity should cut off the distribution of any ion regardless of mass.

It will now be shown that the CT^{-2} curve is just what would be expected for slow electrons. The energy spread of electrons emitted from the cathode is about 2 eV, with no sharp spikes in the distribution. The cathode is biased relative to the drift tube so that the slow electrons come from near the middle of that distribution. Over any small energy range (such as that represented by times of flight greater than 1 msec), the number of electrons with energy between E and $E + dE$ is $n(E)\, dE = K\, dE = Kmv\, dv$, where K is constant. If no forces act on the electron,

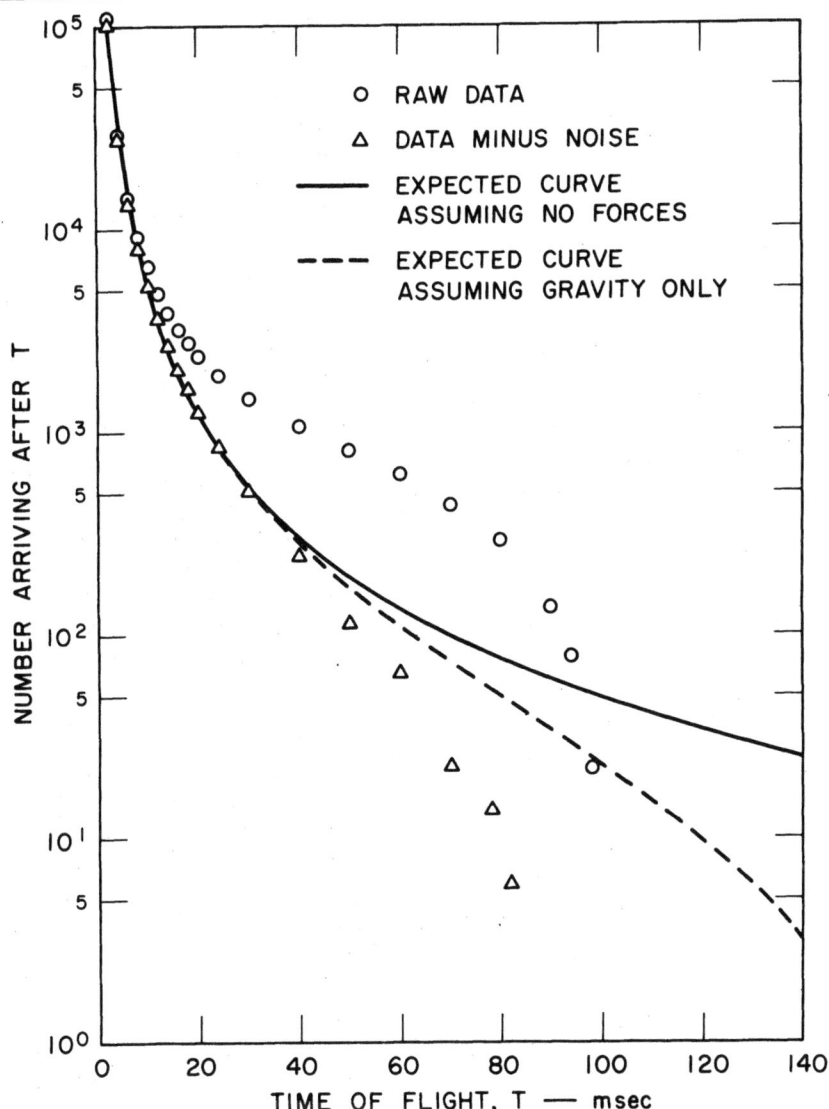

Fig. 2. Experimental and theoretical flight-time distributions for slow electrons.

$v = h/t$, so $n(E)\, dE = (-Kmh^2\, dt/t^3)$. The number of electrons reaching the detector after time T is thus

$$N(T) = \int_0^{E(T)} n(E)\, dE = -Kmh^2 \int_\infty^T \frac{dt}{t^3} = \frac{Kmh^2}{2T^2} = CT^{-2}$$

In summary, we believe we have observed electrons with energies as low as 10^{-11} eV and have actually measured the force of gravity on negative ions. This apparatus had been intended to test only the feasibility of such an experiment on an electron. The results are encouraging enough that a larger apparatus is being built which should be capable of detecting forces on the electron of 0.1 times that of gravity.

The production of low-energy electrons in the magnetic ground state also presents an ideal system for measuring the anomalous magnetic moment of free electrons. The measurement of the anomaly in the electron g-factor or g-2 requires only a comparatively simple modification of the apparatus.

Two separated oscillating fields are placed between the source and the detector. This is the Ramsey method of excitation used in atomic and molecular beam experiments.[3] A magnetic mirror (i.e., an increase in magnitude of the magnetic field) that will repel any slow electron not in the ground state is created between the second oscillating field and the detector. The electrons reach the excitation region in the ground state. If a transition occurs before an electron reaches the mirror, it will not be detected and a resonance is indicated. Cyclotron resonance occurs at $\omega_c = 2\mu_\beta B/\hbar$ and a spin transition at $\omega_s = g_s\mu_\beta B/\hbar$. Since the fields are produced by persistent current magnets, B remains constant and $g_s - 2 = 2(\omega_s - \omega_c)/\omega_c = 2(f_s - f_c)/f_c$. Thus, g-2 is completely determined by the measurement of these two frequencies.

With Ramsey excitation, the accuracy is largely dependent on three factors: (1) the oscillator stability, (2) the time the particle spends between the separated fields, and (3) the homogeneity of the average field experienced by the electrons as they pass between the two oscillating fields. The oscillator and magnet types available, plus the data obtained on slow electrons, indicate a possible accuracy of one part in 10^8 or better for g-2.

References

1. L. D. Landau and E. M. Lifshitz, *Quantum Mechanics: Non-Relativistic Theory*, Addison-Wesley Publishing Co., Inc., Reading, Mass. (1958).
2. A. V. Gold, D. K. C. MacDonald, W. B. Pearson, and I. M. Templeton, *Phil. Mag.* **5**, 765, 1960.
3. N. F. Ramsey, *Molecular Beams*, Oxford University Press, Fair Lawn, New Jersey (1956).

A REALIZATION OF A LONDON–CLARKE–MENDOZA TYPE REFRIGERATOR[1]

P. Das, R. de Bruyn Ouboter, and K. W. Taconis

Kamerlingh Onnes Laboratorium
Leiden, The Netherlands

The refrigerator uses the heat of mixing at constant osmotic pressure between ^{3}He and ^{4}He, and has attained a temperature of 0.22°K in preliminary runs. The eight (probably) essential parts of the cycle are the condenser (C), expansion valve (V), heat exchanger (I), mixing chamber (M), superfluid duct (D), helium mixture funnel (F), evaporator (E), and circulation pump (see Fig. 1).

In the condenser (C) the ^{3}He gas is cooled from room temperature down to the temperature of the outer helium bath, 1.3°K, and is liquefied. It enters the heat exchanger (I) after passing the expansion valve (V) and is further cooled successively by thermal contact with the evaporator (E) and by counterflow with the helium mixture funnel (F); subsequently, it arrives in the mixing chamber (M). Here, the ^{3}He joins the upper layer of high ^{3}He concentration, which is in equilibrium with the lower (phase-separated) layer of low ^{3}He concentration. At the interface between the two layers, the mixing takes place; the lower layer can feed itself with pure superfluid ^{4}He through a special superfluid duct (D) connected to the evaporator (E). The mixture produced streams to the evaporator through the helium mixture funnel (F), which is part of the heat exchanger (I). In the funnel much extra cold is produced by expansion of the diluted ^{3}He on its way to the evaporator (E) under the influence of the temperature rise. The evaporator is kept at about 0.8°K by pumping off the ^{3}He vapor from it by means of a diffusion pump at room temperature. The output of the rotary pump which backs the diffusion pump is connected to the condenser so that the cycle is closed. The thermodynamics of the refrigeration cycle described is fully discussed by K. W. Taconis and R. de Bruyn Ouboter in Chapter II of *Progress in Low Temperature Physics*, volume IV (1964), edited by C. J. Gorter.

In Fig. 2 the refrigeration cycle is presented in the T–X diagram operating between 0.1° and 0.8°K. All quantities shown in this diagram refer to the case in which 1 mole of solute (^{3}He) is taken through the whole cycle. V/X is the volume of solution containing 1 mole of solute (^{3}He).

We will discuss the enthalpy balance of the whole cycle. The numbers between brackets refer to the position in question in Fig. 2.

Evaporator to Condenser [1] → [2]

$T = 0.8$°K. In the evaporator [1] ^{3}He is evaporated ($C_V/C_L \gg 1$), and in the condenser [2] the pure ^{3}He vapor is condensed. The condensation energy is greater than the evaporation energy, and the system loses an amount of energy

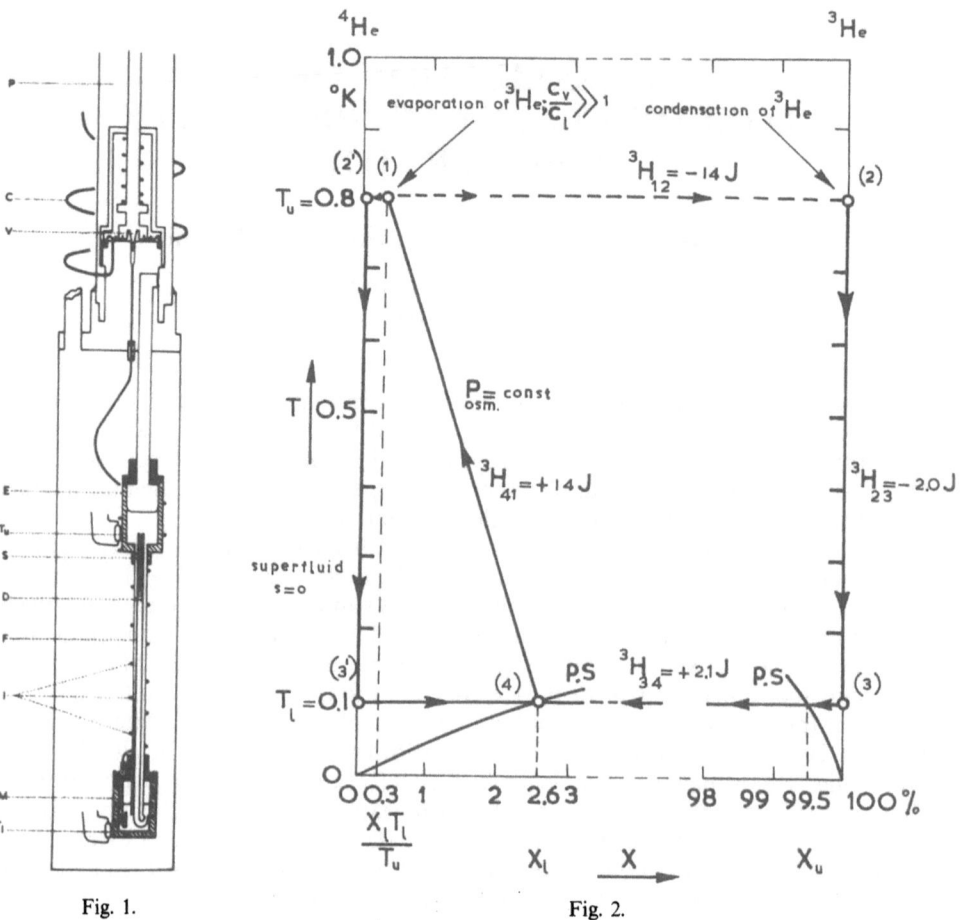

Fig. 1.

Fig. 2.

equal to

$$(-H_3^0 - \int_0^{T=0.8} C_3^0\,dT) - (-NE_{03} - \tfrac{3}{2}RT) \approx -7\,J$$

In addition, the work done in the evaporator [1] is $-\int P_{\text{osm}}\,dV = -RT_U \approx -7\,J$ to keep the mixture in [1] at the fixed concentration of 0.3%. Thus, the total loss of energy is $-14\,J$.

Heat Exchanger [2] → [3]

The liquid ^{3}He has to be cooled from 0.8° to 0.1°K by an amount of energy equal to:

$$-\int_{0.1}^{0.8} C_3^0\,dT = -2.0\,J$$

This is done by a heat exchanger, using the amount of cold produced during [4] → [1].

Mixing Chamber [3] → [4]

$T = 0.1°K$. The ^{3}He solution is diluted by adding superfluid ^{4}He $(S = 0)$ through a superleak $([2'] → [3'] → [4])$; cooling can result from the external work done during the expansion of the solute "gas" ^{3}He $(\approx P_{osm} \times V/X_1 \approx RT_L)$ and from the heat of transition from the concentrated to the dilute phase $(\approx H^E/X_1 \approx \frac{3}{2}RT_1)$. In total, an amount of heat of $\approx \frac{5}{2}RT_1 \approx +2.1\ J$ can be absorbed at 0.1°K.

Heat Exchanger Backflow [4] → [1]

The steady-state condition requires that the osmotic pressure $P_{osm} = RTX$ is constant along the capillary leading from [4] to [1], since one may neglect the fountain pressure at low enough temperatures. In [4] we have a 2.6% mixture at 0.1°K, so at [1] at 0.8°K we have a mixture with a concentration of $2.6 \times 0.1/0.8 \approx 0.3\%$. During this process an amount of heat

$$\int_{0.1}^{0.8} C_{P_{osm}}\, dT = \int_{0.1}^{0.8} \tfrac{5}{2}R\, dT \approx +14\ J$$

can be absorbed when 1 mole of ^{3}He is transported.

We like to remark that it may be possible that the cooling capacity is somewhat disappointing. This may be due to the fact that at absolute zero the effective potential well of an ^{3}He atom in an ^{4}He II surrounding $(10\% → NE_{03} = -23.6\ J/mole)$ is somewhat deeper than the effective potential of an ^{3}He atom in an ^{3}He surrounding $(H_{03} = -21.5\ J/mole)$. Hence, the heat of mixing for a dilute mixture is not simply $\frac{3}{2}RX$ but is given by the relation:[2]

$$H^E = X[NE_{03} + \tfrac{3}{2}RT - H^0_{03} - \int_0^T c^0_3\, dT]$$

There is no restriction with respect to the sign for the heat of mixing at absolute zero. If H^E/X becomes negative, we will find a finite stratification concentration at absolute zero. Certainly, the heat of mixing is very small in the limit $T → 0°K$. The thermodynamic analysis of the cooling cycle mentioned above was based on the assumption that at absolute zero the heat of mixing was zero and that at finite temperature for a dilute mixture $H^E/X \approx \frac{3}{2}RT$, or if the osmotic pressure was kept constant $(H^E/X) \approx \frac{5}{2}RT$.

References

1. H. London, G. R. Clarke, and F. Mendoza, *Phys. Rev.* **128**, 1992, 1962.
2. R. de Bruyn Ouboter, K. W. Taconis, C. le Pair, and J. J. M. Beenakker, *Physica* **26**, 853, 1960 (especially equation 14 and Fig. 17).

MIX
Papier aus verantwortungsvollen Quellen
Paper from responsible sources
FSC® C105338

If you have any concerns about our products,
you can contact us on
ProductSafety@springernature.com

In case Publisher is established outside the EU,
the EU authorized representative is:
Springer Nature Customer Service Center GmbH
Europaplatz 3, 69115 Heidelberg, Germany

Printed by Libri Plureos GmbH
in Hamburg, Germany